Polymer Nanocomposites

Polymer Nanocomposites

Processing, Characterization, and Applications

Joseph H. Koo, Sc.D.

Second Edition

New York Chicago San Francisco
Athens London Madrid
Mexico City Milan New Delhi
Singapore Sydney Toronto

Library of Congress Control Number: 2019937653

McGraw-Hill Education books are available at special quantity discounts to use as premiums and sales promotions or for use in corporate training programs. To contact a representative, please visit the Contact Us page at www.mhprofessional.com.

Polymer Nanocomposites: Processing, Characterization, and Applications, Second Edition

1 2 3 4 5 6 QVS 23 22 21 20 19

ISBN 978-1-260-13231-1
MHID 1-260-13231-5

This book is printed on acid-free paper.

Sponsoring Editor
Summers Scholl

Editorial Supervisor
Stephen M. Smith

Production Supervisor
Lynn M. Messina

Acquisitions Coordinator
Elizabeth M. Houde

Project Manager
Poonam Bisht, MPS Limited

Copy Editor
Mohammad Taiyab Khan, MPS Limited

Proofreader
A. Nayyer Shamsi, MPS Limited

Art Director, Cover
Jeff Weeks

Composition
MPS Limited

This book is dedicated to my wife, Penelope,
and my daughter, Hilary, whose
moral support and encouragement
made possible the completion.

About the Author

Joseph H. Koo, Sc.D., is Senior Research Scientist and Director of the Polymer Nanocomposites Technology Laboratory at The University of Texas at Austin. He is also Founder, Vice President, and CTO of KAI, LLC, in Austin. Dr. Koo is a Fellow of the Society for the Advancement of Material and Process Engineering (SAMPE), an advisory committee member of the National Space and Missile Materials Symposium (NSMMS), an Associate Fellow of the American Institute of Aeronautics and Astronautics (AIAA), and Editor-in-Chief of *Flame Retardancy and Thermal Stability of Materials.*

Contents

Part 3 Opportunities and Trends for Polymer Nanocomposites

Preface

Nanotechnology is the study and control of matter at dimensions of roughly 1–100 nanometers, where unique phenomena enable novel applications. Nanotechnology is also the design, fabrication, and application of nanostructures. The excitement surrounding nanoscale science and technology gives us unique opportunities to develop revolutionary materials. Nanoscale science and technology is an emerging field that encompasses nearly every discipline of science and engineering. Nanophase and nanostructured materials, a new branch of materials science research, are attracting a great deal of attention because of their potential applications in areas such as polymer nanocomposites, electronics, optics, catalysis, ceramics, magnetic data storage, and others.

The research in nanotechnology is evolving and expanding very rapidly, which makes it an impossible task for a volume of this size to cover all aspects of the field. This second edition is focused primarily on polymer nanocomposites, based on my own research as well as the research experiences of others. I have outlined a logical progression of what polymer nanocomposites are, how they are made, what their assets and deficiencies are, what they are designed for, and to what extent they are commercially exploited. Each specialized chapter is part of a sequence of developing polymer nanocomposites. I have been using the first edition, published by McGraw-Hill in 2006, since the spring of 2007 to teach a semester course offered once a year at The University of Texas at Austin. As a result of my class, I developed this expanded volume and presented each chapter in a different manner than in my first book. One of the primary goals of this text is to aid the readers who wish to engage in research and development of polymer nanocomposites in their professions.

This text consists of 12 chapters (divided into three parts). Part 1, *Nanomaterials, Processing, and Characterization*, contains five chapters. Chapter 1 provides an introduction to nanotechnology. Chapter 2 consists of an overview of the most commonly used nanomaterials and their origins, manufacturing, properties, and current and potential applications. Chapters 3 to 5 cover selection of resin matrix and nanomaterials and processing and characterizing of multifunctional polymer nanocomposites.

Part 2, *Multifunctional Properties of Polymer Nanocomposites*, consists of six chapters (Chaps. 6 to 11) that discuss the important properties and current applications of polymer nanocomposites. Polymer nanocomposites can be thermoplastic-based, thermoplastic elastomer–based, or thermoset-based. Mechanical, thermal, flammability, ablation, electrical, and other enhanced properties, such as tribological, permeability, and oil field application, of polymer nanocomposites are discussed in these chapters.

In the final chapter (Chap. 12), which is the only chapter in Part 3, *Opportunities and Trends for Polymer Nanocomposites*, a brief overview of the opportunities, trends, and challenges for nanomaterials and polymer nanocomposites is given.

Throughout the book, the theme is developed that polymer nanocomposites are a whole family of polymeric materials whose properties are capable of being tailored to meet specific applications. This volume serves as a general introduction to researchers just entering the field and to scholars from other subfields seeking information. It is well-suited as a textbook for a one-semester course tailored to upper-level undergraduate and entry-level graduate students, or for a professional short course.

During my preparation for and writing of this book, a great deal of help in terms of information and images was provided by a number of people and their organizations, whom I would like to thank, in no particular order: Dr. Richard A. Vaia of Air Force Research Laboratory/WPAFB; Mr. Max L. Lake, Mr. Patrick Lake, and Dr. Carla Leer Lake of Applied Sciences; Dr. Alexander Dole of Arkema; Mr. Joseph Ventura of Bayer MaterialScience; Mr. Thomas Pitstick of Carbon Nanotechnologies; Dr. Nate Hansen of Conductive Composites; Mr. Alan Hedgepeth and Dr. Stephan Sprenger of Evonik AG; Dr. Joseph D. Lichtenhan of Hybrid Plastics; Dr. Sanjay Mazumdar of Lucintel; Professor Brian Wardle of MIT; Professors Giovanni Camino and Alberto Fina of Polytechnic of Turin, Alessandria, Italy; Dr. Doug Hunter of Southern Clay Products; Mr. Kiyoshi Shimazaki and Ms. Chiharu Kanayama of Thinky; and Mr. Tim Ferland of ANF Technology.

Many thanks are due to my friends Dr. Louis A. Pilato of Pilato Consulting/KAI, LLC, and Dr. Maurizio Natali of University of Perugia/KAI, LLC, for collaborating with me on numerous research projects, a tutorial class, and an online course that I offered at The University of Texas at Austin's Center for Lifelong Engineering Education. These activities are used as examples throughout this book. I would also like to thank Dr. Charles Y-C Lee and Dr. Jocelyn Harrison of Air Force Office of Scientific Research/NL, and Dr. Shawn Phillips, Dr. Rusty Blanski, and Dr. Steven A. Svejda of Air Force Research Laboratory/Propulsion Directorate, Edwards AFB, California, for supporting our numerous research programs. I would also like to acknowledge research programs sponsored by DOD (AFOSR, AFRL, AMRDEC, DTRA, MDA, NAVAIR, NAVSEA, NSWC, and ONR), DOE, DOT, EPA, FAA, NASA, and private companies for the past 20+ years. I am thankful to all my colleagues, especially to students who took my Polymer Nanocomposites class for the past 13 years and provided excellent inputs for this volume and to students in my research group who contribute to our research programs. Special thanks go to Mr. Rishabh Shah (my current research student) for obtaining permissions from different sources to use their data throughout this book. Many thanks go to Ms. Summers Scholl of McGraw-Hill for providing the opportunity to publish this second edition and particularly her staff for tolerance and assistance.

Joseph H. Koo, Sc.D.
Department of Mechanical Engineering
Texas Materials Institute
Center for Nano and Molecular Sciences
The University of Texas at Austin
Austin, Texas, USA
E-mail: jkoo@mail.utexas.edu
Website: http://www.me.utexas.edu/~koo

Polymer Nanocomposites

PART 1

Nanomaterials, Processing, and Characterization

CHAPTER 1

Introduction to Nanotechnology

1.1 Definition of Nanotechnology

The Subcommittee on Nanoscale Science, Engineering, and Technology (NSET), Committee on Technology (CoT), National Science and Technology Council (NSTC), defines nanotechnology as [1]:

Nanotechnology is the understanding and control of matter at dimensions between approximately 1 and 100 nanometers, where unique phenomena enable novel applications. Encompassing nanoscale science, engineering, and technology, nanotechnology involves imaging, measuring, modeling, and manipulating matter at this length scale.

A nanometer is one-billionth of a meter. A sheet of paper is about 100,000 nanometers thick; a single gold atom is about a third of a nanometer in diameter. Dimensions between approximately 1 and 100 nanometers are known as the nanoscale. Unusual physical, chemical, and biological properties can emerge in materials at the nanoscale. These properties may differ in important ways from the properties of bulk materials and single atoms or molecules.

Mauro Ferrari, Professor of Molecular Medicine at the University of Texas Health and Science Center, of Experimental Therapeutics at the M.D. Anderson Cancer Center, and of Bioengineering at Rice University, provides the following definition [2]:

At the nanoscale there is no difference between chemistry and physics, engineering, mathematics, biology or any subset thereof. An operational definition of nanotechnology involves three ingredients: (1) nanoscale sizes in the device or its crucial components, (2) the man-made nature, and (3) having properties that only arise because of the nanoscopic dimensions.

Professor Ferrari points out how the line between disciplines becomes blurred at such a small scale. This is why nanotechnology is characterized in such broad terms—it bridges the competencies of all the sciences.

Professor Robert Langer, one of 13 Institute Professors at the Massachusetts Institute of Technology, defines nanotechnology in simpler terms [2]:

Nanotechnology is concerned with work at the atomic, molecular, and supramolecular levels in order to understand and create materials, devices and systems with fundamentally new properties and functions because of their small structure.

The common thread among these definitions is the mention of the molecular scale, or nanoscale. The aim of this chapter is to develop intuition and explore the importance of this scale.

The extent of U.S. government activity alone can be gauged from the fact that 20 federal agencies and departments participated in the National Nanotechnology Initiative (NNI website: www.nano.gov) in fiscal year (FY) 2019. The NNI, established in FY 2001, is a U.S. government research and development (R&D) initiative involving 20 federal departments, independent agencies, and commissions working together toward the shared and challenging vision of "a future in which the ability to understand and control matter at the nanoscale leads to a revolution in technology and industry that benefits society" [1, 3]. The NNI is managed within the framework of the NSTC. The NSET Subcommittee of the NSTC CoT coordinates planning, budgeting, program implementation, and review of the progress for the initiative. The National Nanotechnology Coordination Office (NNCO) acts as the primary point of contact for information on the NNI. The NNI website provides public outreach on behalf of the NNI and promotes access to and early application of the technologies, innovations, and expertise derived from NNI activities.

The four NNI goals are as follows [1]:

- Advance a world-class nanotechnology research and development program
- Foster the transfer of new technologies into products for commercial and public benefit
- Develop and sustain educational resources, a skilled workforce, and a dynamic infrastructure and toolset to advance nanotechnology
- Support responsible development of nanotechnology

Federal organizations with the largest investments are the following:

- National Science Foundation (NSF): fundamental research and education across all disciplines of science and engineering
- National Institutes of Health (NIH): nanotechnology-based biomedical research at the interaction of life and physical sciences
- Department of Energy (DOE): fundamental and applied research providing a basis for new and improved energy technologies
- Department of Defense (DOD): science and engineering research advancing defense and dual-use capabilities
- National Institute of Standards and Technology (NIST): fundamental research and development of measurement and fabrication tools, analytical methodologies, metrology, and standards for nanotechnology

Other agencies and agency components investing in mission-related nanotechnology research are the Department of Homeland Security, Food and Drug Administration, Environmental Protection Agency, National Aeronautics and Space Administration, National Institute for Occupational Safety and Health, Consumer Product Safety Commission, Department of Transportation (including the Federal Highway

Administration), and U.S. Department of Agriculture (including the National Institute of Food and Agriculture, the Forest Service, and the Agricultural Research Service).

The 2019 federal budget provides more than $1.4 billion for the NNI, affirming the Administration's continuing commitment to a robust U.S. nanotechnology effort. Nearly half (43%) of this budget is allocated for applied R&D and support for the Nanotechnology Signature Initiatives (NSIs), reflecting an increased emphasis within the NNI on commercialization and technology transfer. The cumulative NNI investment since FY 2001, including the 2019 request, now totals more than $27 billion. There has been nearly a decade of major investment and progress in nanoscience and nanotechnology and it has been summarized in a *Special Report to the President and Congress on the Fourth Assessment of the National Nanotechnology Initiative* [3, 4] and *The National Nanotechnology Initiative Supplement to the President's 2019 Budget* [1]. More details on the NNI budget, program plans, and progress toward NNI goals are provided in Chap. 12.

1.2 Brief History of Nanotechnology

Japanese scientist Norio Taniguchi first coined the term *nanotechnology* in 1974 in reference to small-scale semiconductor interactions, but he was unaware of the earlier uses of the technology and the concept [5]. One example of the such earlier use of nanotechnology is medieval artists making stained glass between AD 500 and 1450. Little did they know that the metallic additives to molten glass used for coloration acted as nanoparticles, scattering light in different ways to produce different colors [5].

The idea of nanotechnology was first introduced to the public in 1959 by Professor Richard Feynman in his esteemed lecture series entitled "There's Plenty of Room at the Bottom" in 1959 [4–8]. In this talk, Feynman describes a future where scientists control matter with atomic precision [6]. He anticipates, in seemingly prophetic fashion, the creation of microscopes that could resolve individual molecules, densely packed computer chips, and miniature robots that could aid in surgery and drug delivery [5, 6].

Part of Feynman's dream became realized in 1981 with the development of the scanning tunneling microscope (STM) capable of observations with 0.1-nm resolution [7]. The STM enabled scientists to directly observe manipulations at the molecular scale and thereby served a key fundamental tool for nanotechnology research.

Advances in microscopy have led to a number of other high-resolution visualization and characterization techniques, including transmission electron microscopy (TEM), scanning electron microscopy (SEM), atomic force microscopy (AFM), wide-angle x-ray diffraction (WAXD), and small-angle x-ray scattering (SAXS) [10]. These instruments will be described in Chap. 5.

Feynman's vision of nanoscale microchips came true in 2001 when IBM constructed the first logic gate using carbon nanotubes [7]. This sort of miniaturization in the realm of computing enabled by nanotechnology is what has given birth to the smartphones, tablets, and ultrabooks we enjoy today. Other discoveries and applications anticipated by Feynman include the use of nanoscale devices in medicine. Readers should refer to the NNI website www.nano.gov under "Nanotechnology 101/Nanotechnology Timeline," which traces the development of nanotechnology from first concept to the latest development.

1.3 What Is the Significance of *Nanoscale Materials*?

Nanoscale (nanoscopic scale) materials, which can be either stand-alone solids or subcomponents in other materials, are less than 100 nm in one or more dimensions. To put this dimension in perspective, a nanometer (nm) is one-billionth of a meter, and one-millionth of a millimeter, about four times the diameter of an atom, is the corresponding length scale of interest. Truly, grasping the scale of the nanometer is difficult for our macro-oriented brains, but real-world comparisons can help give us a sense. For example, the blink of an eye is to a year as a nanometer is to a meter stick [6]. Figures 1.1 [7] and 1.2 show some interesting examples. Figure 1.1 shows some examples: a single-walled carbon nanotube is about 1 nm in diameter; DNA is about 2.5 nm in diameter; a nanoparticle is about 4 nm in diameter; a bacterium is about 2.5 μm in length; large raindrop is about 2.5 mm in diameter; a single strand of hair is about 100 μm in diameter; an ant is about 4 mm long; a large raindrop is about 2.5 mm in diameter; a house is 10 m wide; and the Indianapolis Motor Speedway is 4 km per lap. Additional interesting materials for comparison are a human red blood cell (10,000 nm), a cell of *E. coli* bacteria (1,000 nm), a viral cell (100 nm), a polymer coil (40 nm), quantum rods (Q-rods) (30 nm in length) with an aspect ratio of 10:1, and a quantum dot (QD) (7 nm in diameter), all shown in Fig. 1.2.

To illustrate the inherent value of manipulating matter at such a scale, Daniel Ratner, a bioengineering professor at the University of Washington, proposes a helpful thought experiment [4]. Suppose we have a 3 × 3 × 3 foot cube of pure gold. If we were to bisect this cube in each dimension, we would have eight smaller cubes. These new cubes would exhibit the same intrinsic properties as the original—each would still be heavy, shiny, and yellow with the same chemical and structural properties. If we were to continue bisecting until we had cubes sized on the order of microns (10^{-6} of a meter), the inherent bulk properties of the material would still remain constant. This is not specific to gold; the same holds true for steel, plastic, ice, or any pure solid. However, if we were to reach the nanoscale, quantum effects would begin to dominate and the gold's properties, including its color, melting point, and intermolecular chemistry, would change. These quantum effects had been "averaged out of existence" in the bulk material [8]. At this nanoscale, the force of gravity gives way to van der Waals forces, surface tension, and other quantum forces [9].

1.4 Why Is This Nanoscale So Special and Unique?

Nanomaterials are also known as *nanoscale structures, nanoscale materials, nanophase, or nanoparticles* in the literature. To distinguish nanomaterials from bulk, it is crucial to demonstrate the unique properties of nanomaterials and their prospective impacts in science and technology. Nanomaterials can be classified as one-dimensional nanoscale structures (platelet-type), two-dimensional nanoscale structures (fiber-type), and three-dimensional nanoscale structures (particulate-type). These three most important nanomaterials are illustrated schematically in Fig. 1.3. The above nanomaterials are described in details in Chap. 2. Nanoscale structures have very high surface-area-to-volume and aspect ratios, making them ideal for use in polymer nanocomposites (PNCs).

Wave-like properties of electrons inside matter and atomic interactions are influenced by materials' variations on the nanoscale length. By creating nanoscale structures it is possible to control the fundamental properties of materials, such as their melting temperature, magnetic properties, charge capacity, and even their color, without

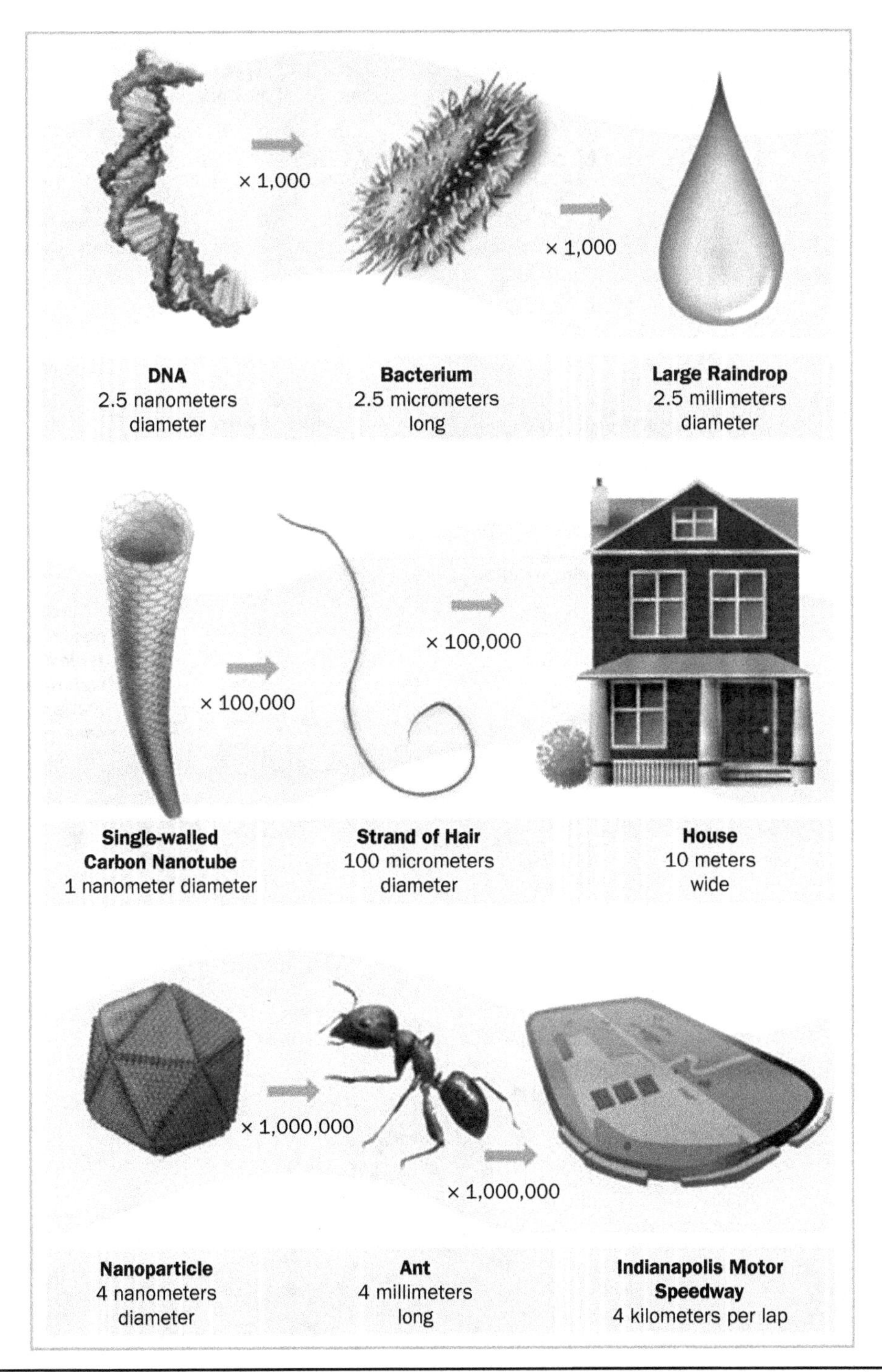

FIGURE 1.1 Visual examples of the size and scale of nanotechnology [7]. (*From www.nano.gov.*)

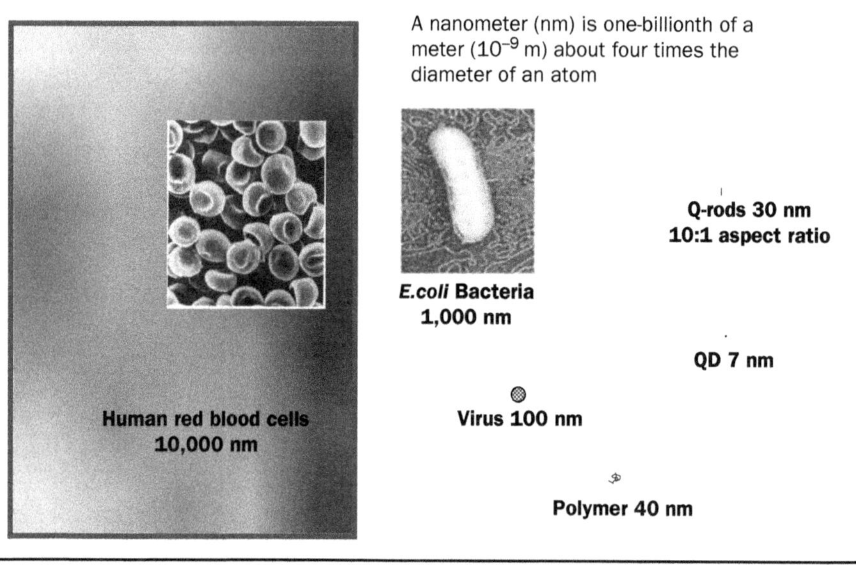

FIGURE 1.2 What *nano* really means. (*Courtesy of R. Vaia.*)

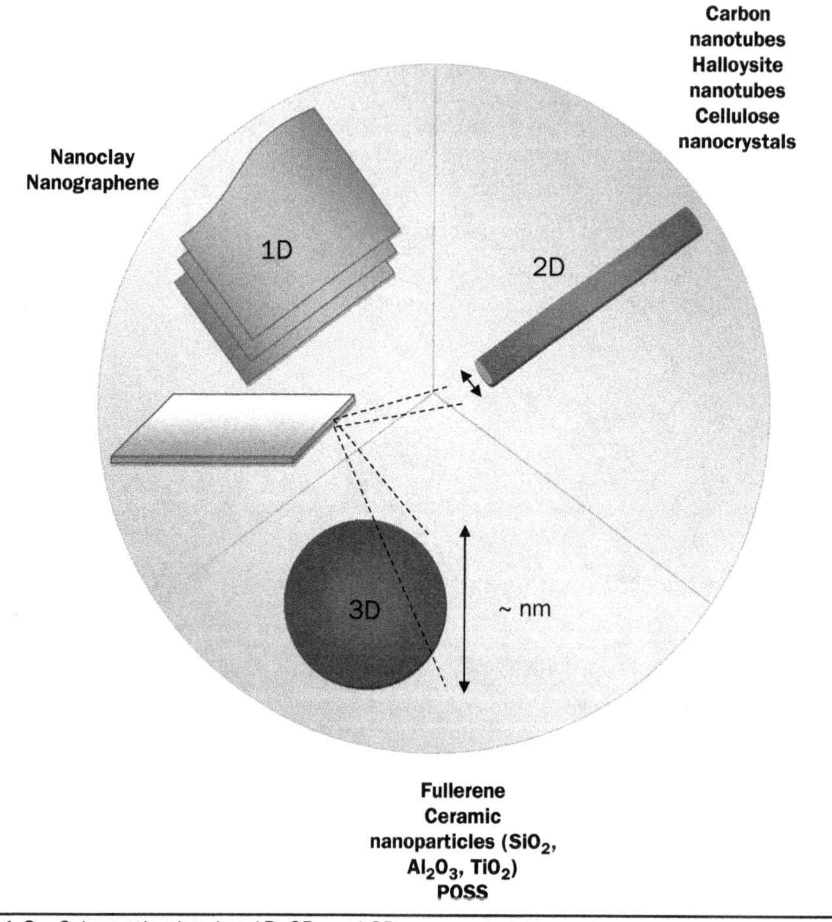

FIGURE 1.3 Schematic showing 1D, 2D, and 3D nanoscaled dimensions.

changing the chemical composition of the nanoscale structures. This will lead to new, high-performance polymers and nanotechnologies that were impossible previously.

For the past five decades, we have been working with macrocomposites, such as filled polymers or fiber-reinforced polymer matrix composites (PMCs), where the length scale of the polymer fillers or the fiber diameters is in micrometers (Fig. 1.4). The reinforcement length scale is in micrometers, and the interface of fillers is close to the bulk polymer matrix. During the past two decades, we discovered nanocomposites, where the length scale of the reinforcement (nanoparticles) is in nanometer scale (Fig. 1.4). These nanocomposites have *ultra-large interfacial area per volume,* and the distance between the polymer and filler components is extremely short. Polymer coils and chains are 40 nm in diameter, and the nanomaterials are on the same order of magnitude as the polymer. As a result, molecular interaction between the polymer coils/chains and the nanomaterials will give PNCs very unusual material properties that conventional polymers do not possess.

Before discussing the properties of nanomaterials, it may be advantageous to describe an example demonstrating the elementary consequences of the small size of nanoparticles [10]. The first and most important consequence of a small particle size is its huge surface area, and in order to obtain an impression of the importance of this geometric variable, the surface-area-to-volume ratio should be addressed. Assuming spherical particle, the surface area a of one particle with diameter D is $a = \pi D^2$, and the corresponding volume v is $v = \pi D^3/6$. Therefore, the surface-area-to-volume ratio is

$$R = a/v = 6/D \tag{1.1}$$

This ratio is inversely proportional to the particle size and as a consequence the surface area increases with decreasing particle size. The same is valid for the surface per mol A, a quantity that is extremely important in thermodynamic considerations.

$$A = na = M/(\rho\pi D^3/6)\, \pi D^2 = 6M/\rho D \tag{1.2}$$

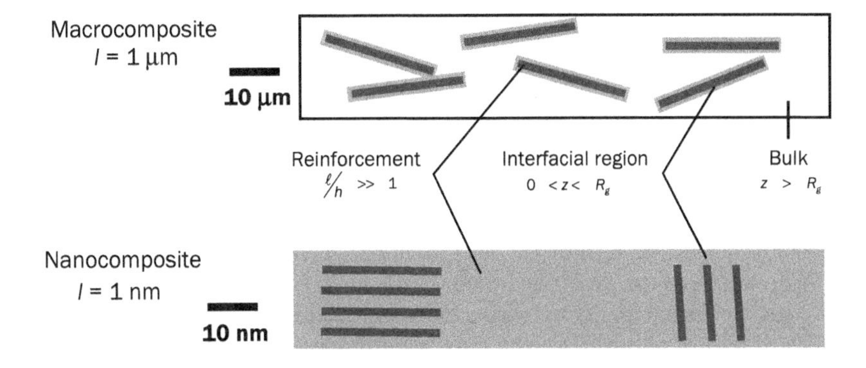

Characteristic

Ultra-large interfacial area per volume
High fraction interfacial (interphase) material
Short distances between components

Nanopolymer Nanoinorganic

FIGURE 1.4 Nanoscale materials uniqueness. (*Courtesy of R. Vaia.*)

In Eq. (1.2), n is the number of particles per mol, M is the molecular weight, and ρ is the density of the material. Similar to the surface-area-to-volume ratio, the area per mol increases inversely in proportion to the particle diameter. Hence, huge values of surface area are achieved for particles that are only a few nanometers in diameter.

The unique properties and improved performance of nanomaterials are determined by their size, surface structures, and interparticle interactions. The role played by particle size is comparable to the role of the particle's chemical composition, adding another parameter for designing and controlling particle behavior. To fully understand the impacts of nanomaterials in nanoscale science and technology, one needs to learn why nanomaterials are so special!

The excitement surrounding nanoscale science and technology provides unique opportunities to us to develop revolutionary materials. Nanoscale science and technology is a relative young field that encompasses nearly every discipline of science and engineering. Nanoscale structures are a new branch of materials research attracting a great deal of attention because of its potential applications in chemical catalysis, computing, imaging, material synthesis, medicine, printing, and many other fields.

1.5 How Polymer Nanocomposites Work

The advent of academic, government, and industrial nanotechnology research has spurred significant growth opportunities for nanocomposites for the consumer, defense, aerospace, and health industries [11–15]. This is evident from the number of journal papers reporting on nanotechnology scientific research; and the number of patents issued by the U.S. Patent and Trademark Office (PTO) grew significantly in the past 30 years. Statistical data collected by Orbit Database during 2018 show a total of 19,773 nanotechnology patents were filed at the U.S. PTO, of which 3,413 were issued by universities. Among these patents, 1,432 were granted patents and 1,981 were published patent applications, altogether comprising 17% of the total nanotechnology patents of the U.S. PTO [13]. This is a good indication of how widespread the nanotechnology research activities have been across several disciplines. This book mainly focuses on a branch of nanotechnology research involving only the polymers, specifically on PNCs.

The major objective of PNC research is to explore the different methods to achieve significant enhancement of properties over those of the traditional matrix polymers using only a small fraction of nanomaterials. The nanomaterials are defined to have at least one dimension less than 50 nm [16]. These nanomaterials are expected to create new physical and chemical properties beyond what traditional filled polymeric materials possess. It is well-known from decades-long PNC research that some nanomaterials improve only one type of property, such as mechanical, thermal, flammability, or electrical, while a selected few nanomaterials can improve multifunctional properties at the same time.

The uniqueness of these novel PNCs will enable the circumvention of classic polymer performance by accessing new properties and exploiting synergies between PNCs that only occur when the nanoscale morphology and the fundamental physics associated with a property coincide. They represent a radical alternative to traditional filled polymers. In contrast to traditional polymer systems where reinforcement is on the order of microns, PNCs are exemplified by discrete constituents on the order of a few nanometers. PNCs consist of improved mechanical, thermal, ablation, flammability, electrical, optical, permeability, charge dissipation, chemical resistance, magnetic, catalytic, and other properties (as discussed in detail in later chapters of this book). PNCs can differ completely from those of its component materials, thereby opening up unprecedented technologies. A common example of nanomaterials in currently available PNCs are carbon nanotubes, which can help to enhance electrical and thermal conductivities, increase stiffness, and enhance crack deflection, and toughness. Other common nanomaterials used in PNCs are TiO_2 and ZnO for optical properties, nitrides and carbides for hardness and wear resistance, nanoclays for flammability, and nanometals for color.

The challenges in the field of PNC research are threefold. First, the identification of the proper nanomaterial–polymer pairs is essential and critical. Otherwise, the desired properties cannot be realized. Second, the nanomaterials must be nano-dispersed in the matrix polymers. Otherwise, anticipated properties cannot be achieved. Third, not all true PNCs are cost-effective. The cost constraint largely governs the scope of current PNC research and commercialization in industrial laboratories. Some novel, innovative academic PNC research will never be commercialized, due to the prohibitive cost of nanomaterials, time-consuming processing and fabrication steps, and sometimes the lack of proper market. Of the three challenges presented in this paragraph, the first two challenges relate to academic research and the third challenge relates to industrial research with PNC preparation and commercialization.

In the physical form, these nanomaterials (nanoparticles) are usually available as agglomerates. In the case of platelet-type (see Chap. 2 for more detailed descriptions), a large number of individual platelets are held together as microscopic agglomerates by attractive forces originated from electrostatic, ionic, and van der Waals forces. The state of the agglomeration of nanomaterials is usually dictated by the manufacturing methods, and the particle agglomerates serve as the starting material in the preparation of PNCs. The agglomerates, when combined with the matrix polymer in solution or melt state, are subjected to several dispersion forces, including shear, ultrasonic, and centrifugal, to disperse individual particles to the nanoscale in the matrix polymer (see Chap. 4 for detailed discussions). The transformation from initial microscopic particle agglomerates to individual nanoparticles by dispersion forces has served challenges to engineers and scientists for the past two decades.

Uniform and individual dispersion of these nanoparticle agglomerates produces *ultra-large interfacial area per volume* between the nanoparticle and the matrix polymer. PNCs fundamentally differentiate from traditional filled polymers due to the immense internal interfacial area and the nanoscopic nature of the nanomaterials. As a result, the overall performance of PNCs cannot be understood by simple scaling rules that apply to traditional polymeric composites. New combinations of properties derived from the nanoscale structures of PNCs provide opportunities to circumvent traditional performance associated with conventional reinforced polymers, thus epitomizing the promise of PNCs.

Numerous examples can be found in the literature demonstrating substantial improvements in multifunctional properties of PNCs. Its value comes from providing value-added properties not present in the neat resin, without sacrificing the inherent processability and mechanical properties of the neat resin. The preparation of a blend or composite with multifunctionality requires a trade-off between desired performance, mechanical properties, cost, and processability. The term *nanocomposite* when used to describe properties comparisons is intended to relate to traditional unfilled and filled polymers and not fiber-reinforced PMCs. PNCs may provide matrix resins with *multifunctionality,* but they should not be considered in the near and intermediate term as a potential replacement for current state-of-the-art carbon fiber–reinforced PMCs.

1.6 Strengths and Weaknesses of Nanoparticles

This is a relatively young field that has huge growth potential. In the last 20–30 years, extensive research has been conducted in understanding how nanoparticles affect the chemical nature of compounds and their overall effect on a material's properties. The most dramatic effect noticed by nanoparticles is that when working on this level, the surface area of the nanoparticle is tremendous for its size. In a study conducted by the University of Rostock and the Fritz Haber Institute of the MPG in Germany [16], the surface area and diameter of gold catalyst nanoparticles were measured. Five samples of nanoparticles were synthesized, varying in shape: icosahedral, cuboctahedral, decahedral, truncated octahedron, and spherical. Given this small size and these various shapes, these nanoparticles have very large surface area. The surface area is inversely proportional to the diameter of the nanoparticle. In this study, the smallest particles averaging 1.6 nm in diameter had a surface area of 173 m^2/g. If a hefty spoonful—about 2.5 g—of this nanoparticle was flattened out, it could cover an entire basketball court. With a large surface area per volume, a nanoparticle is able to make numerous bonds with the surrounding molecules, far more than a particle on the micron scale. The ability to form so many bonds leads the nanoparticles to create an extremely solid molecular structure that, in turn, increases the mechanical properties of materials. Due to having a large surface-area-to-volume ratio, fewer nanoparticles can be incorporated in materials that are able to outperform the same materials made by conventional mean using the micron scale.

Although nanoparticles have many strengths, there are some drawbacks. Although the cost of creating nanoparticles has declined within the last few years, it is still high. Mass-producing nanoparticles is a difficult task. Taking the production of nanoparticles in a laboratory to high-level production for industrial use is a very challenging process. With this challenge come higher prices. Furthermore, even though nanoparticles can enhance materials, there is a limit. Because of the bonds the nanoparticles can create, when they are added to a liquid the viscosity of the liquid increases. If too high a weight percent of nanoparticles is added to a mixture, the viscosity could increase to a level at which the material is unable to be utilized. This is especially important when working with a thermoset resin, which needs fluidity to be able to penetrate the fibers that are designed to reinforce. Another problem for nanoparticles has to do with the fact that sometimes a nanoparticle with very promising mechanical properties is created, but those properties do not transfer to a material to which this nanoparticle is introduced.

1.7 Safety of Nanoparticles

Due to the size of nanoparticles, it is important to study the effect that these nanomaterials have on a human body. Nanoparticles combine the properties of solids with the ability to move—a property of molecules [17]. Until recently, most items synthetically created by humans did not pose the threat that nanoparticles do. These materials were too large to penetrate the human body in the manner that a nanoparticle can. Nanoparticles are small enough to diffuse into a biological system through a lipid-based barrier that is effective against larger particles. With these small particles, reactive surfaces can enter a biological system and interfere with existing chemical transformation pathways. They can disrupt the body chemistry at the cellular level. The people most at risk are those who handle and process nanoparticles.

There are currently safety measures in place at different laboratories and companies for the treatment of nanoparticles, but there is no national standard. With an increase in use of nanoparticles, there will be an increase in the number of people who come in contact with the raw particles. There are regulations in place for the handling and transportation of these materials, but these regulations govern bulk materials, not nanoparticles. The NNI is the government body in charge of monitoring nanotechnologies. It created the Environmental, Health, and Safety (EHS) Research Strategy in 2011 to establish awareness and safety protocols.

Nanotechnology EHS (nanoEHS) activities have become a hallmark of the NNI, with R&D, policy, and regulation in this area extensively coordinated among federal agencies. NanoEHS activities, including relevant topics incorporated into the NSIs, account for over 10% of the NNI budget allocation in 2016. In 2014, the NNI released the *Progress Review on the Coordinated Implementation of the National Nanotechnology Initiative 2011 Environmental, Health, and Safety Research Strategy* [18], which summarizes the extensive, ongoing collaborations and recent accomplishments in this area. A key component of the strategy is extensive communication and coordination with academic, industrial, and international nanoEHS communities. Examples of these efforts include the U.S. participation in the Innovative Nanoscience and Nanotechnology program [19]; the U.S.-EU Communities of Research [20]; collaboration with dedicated multidisciplinary academic centers, such as the NSF's Centers for Environmental Implications of Nanotechnology at the University of California, Los Angeles, and Duke University; and strong U.S. participation in the development of nanoEHS-related standards. With more research on the effects of nanoparticles on the health and the environment, the industry will most likely become more regulated in the future.

1.8 Overview of the Book

This book is aimed at the following readers:

- Advanced students and instructors in the fields of science and engineering.
- Professional scientists and engineers, who may be trained in more traditional disciplines but need to learn about this emerging technology.
- Policymakers and management looking for an understanding of the scientific challenges, prospective uses, and emerging markings for nanomaterials and PNCs.

The overall goal is to capture the multidisciplinary and multifunctional flavor of nanomaterials and PNCs while providing in-depth discussion of selected areas.

The book is divided into 12 chapters:

- Part 1, *Nanomaterials, Processing, and Characterization,* consists of Introduction to Nanotechnology (Chap. 1); An Overview of Nanomaterials (Chap. 2); Selecting Resin Matrix and Nanomaterials for Applications (Chap. 3); Processing of Multifunctional Polymer Nanocomposites (Chap. 4); and Structure and Property Characterization (Chap. 5).
- Part 2, *Multifunctional Properties of Polymer Nanocomposites,* consists of Mechanical Properties of Polymer Nanocomposites (Chap. 6); Thermal Properties of Polymer Nanocomposites (Chap. 7); Flammability Properties of Polymer Nanocomposites (Chap. 8); Ablation Properties of Polymer Nanocomposites (Chap. 9); Electrical Properties of Polymer Nanocomposites (Chap. 10); and Widespread Properties of Polymer Nanocomposites (Chap. 11).
- Part 3, *Opportunities and Trends for Polymer Nanocomposites,* consists of Opportunities, Trends, and Challenges for Nanomaterials and Polymer Nanocomposites (Chap. 12).

The most commonly used nanomaterials in the market are discussed, including (a) *one-nano-dimensional-scale materials,* such as montmorillonite (MMT) clays, nanographene platelets (NGPs), and layered double hydroxides (LDHs); (b) *two-nano-dimensional-scale materials,* such as carbon nanofibers (CNFs), carbon nanotubes (CNTs) [multiwalled carbon nanotubes (MWNTs), double-walled carbon nanotubes (DWNTs), and single-walled carbon nanotubes (SWNTs)], halloysite nanotubes (HNTs), and nickel nanostrands (NiNs); and (c) *three-nano-dimensional-scale materials,* such as polyhedral oligomeric silsesquioxane (POSS®), nano-silica (*n*-silica), nano-alumina (*n*-alumina), nano–titanium dioxide (*n*-TiO_2), nano–magnesium hydroxide [*n*-$Mg(OH)_2$], nano–silicon carbide (*n*-SiC), nano-silver (*n*-Ag), nano–zinc oxide (*n*-ZnO), and others, used to disperse into the families of polymers, such as thermoplastics, thermosets, and thermoplastic elastomers.

This book focuses on research activities in the areas of (a) developing *processes* to disperse nanomaterials uniformly in the different types of polymers; (b) using wide-angle x-ray diffraction (WAXD), transmission electron microscopy (TEM), scanning transmission electron microscopy (STEM), scanning electron microscopy (SEM), and optical microscopy techniques to characterize nanomaterials and PNC *structures;* (c) studying the *processing-structure-property* relationships of these types of novel PNCs; and (d) characterizing properties and evaluating the *performance* of these PNCs using established laboratory devices and techniques based on ASTM, UL, and other established industry standards. The ASTM and UL test protocols were adapted to characterize the physical, mechanical, thermal, flammability, electrical, optical, tribological, permeability, and other properties of the thermoplastic-, thermoset-, and elastomer-based nanocomposites. Selected laboratory devices, such as a microscale combustion calorimeter (MCC), a cone calorimeter (CC), a simulated solid rocket motor (SSRM), an oxyacetylene test bed (OTB), a subscale solid rocket motor, in situ ablation recession and thermal sensors, and char strength sensors, were used to evaluate the materials' performance under extreme environments.

In these studies, for solid thermosetting reactive prepolymers or thermoplastics with solid nanomaterials, four different processing methods were used to disperse nanomaterials in the polymers. These methods were (a) solution intercalation, (b) melt intercalation, (c) three-roll milling, and (d) centrifugal mixing. For liquid thermosetting reactive prepolymers or thermoplastics with solid nanomaterials, four different processing methods were used; they were (a) in situ polymerization, (b) emulsion polymerization, (c) high-shear mixing, and (d) ultrasonication. The WAXD, TEM, SEM, and STEM analyses were used to characterize the degree of dispersion of these nanomaterials in the host polymer. These imaging techniques allowed us to screen formulations and distinguish compositions that exhibited either favorable or unfavorable nanodispersed nanomaterial or polymer blends. Furthermore, these analytical techniques facilitated and provided guidelines in the scale-up of favorable compositions.

1.9 Summary

In today's world, nanotechnology is all around us but remains invisible to our eyes because of its operating scale. From its origins in the vivid imagination of Professor Richard Feynman, nanotechnology has evolved into a multidisciplinary field that is affecting positive change in many aspects of our lives. Thanks to smart nanoshells, our bodies are safer from cancer and other diseases. Nanoscale semiconductors have enabled the dramatic miniaturization of computers, allowing them to now fit in the palm of our hand. The energy used to power these devices has become more environmentally friendly with advances in solar and wind energy capture thanks to nanotechnology. Today's airplanes are lighter and use less fuel, and spacecraft are becoming capable of increasingly distant voyages.

The rapid growth in the field of nanotechnology is due to its great potential in improving a vast number of properties in materials with just a small concentration of these nanoparticles. Polymer is an important material for engineers to use them in many applications that can benefit from its flexibility, lightweight, and cost efficiency. Adding nanoparticles to the polymer matrix can greatly improve the bulk mechanical, thermal, flammability, ablation, and electrical properties, as reported in the literature. Still, although the benefits of nanotechnology have been praised, it should be noted that there are certain safety and health precautions that must be taken when handling these nanoparticles, such as when exposed to carbon. As a result, industry, government, and academic organizations should emphasize on the nanosafety by providing guidelines and guidance to protect the health of people and the environment.

Nanotechnology plays important roles in the transformation of science, technology, and society marked by the convergence across technical fields and domains of human activity [21]. It is one of the fastest-growing technologies. However, since it is a relatively new technology, nanotechnology science and engineering raise many important questions, especially at the intersection of technology and society [22]. Discussed throughout this book is the uniqueness of polymer nanostructured materials compared to other materials, such as micro-sized composites. This book also reviews studies that show different processes, ratios, and kinds of combinations of nanoparticles that perform at the highest levels of quality. With the discovery of this new technology, it is to our advantage to further our studies of it. Advancements in nanotechnology will create unique challenges and new opportunities in the future [23].

The possible impact of nanotechnology and nanomaterials on future products and services is tremendous. This emerging field has led to an enormous expansion in research and development related to nanomaterials, devices, and systems. It also has raised new questions as to how these products will influence the environment, human health, business, education, and other areas of societal impact. As with any emerging technology, a huge amount of disconnected information is emerging, often spread across multiple disciplines. The influence of these nanomaterials is truly multidisciplinary, and progress will depend on cross-communication between fundamental science and diverse applications. This work is an attempt to provide readers with a broader perspective by offering information on diverse nanomaterials and examples of how these materials can affect polymer properties. It will also aim to help envision future products and possible applications.

An extensive list of resources is included in this chapter for further reading; the majority of these publications are referenced throughout this book.

1.10 Study Questions

1.1 How difficult is it to obtain uniform dispersity of nanomaterial in a polymer?

1.2 What is the difference between polymer nanocomposites and polymer composites?

1.3 Instead of using polymer in a polymer nanocomposite, can a metal or ceramic be used to create nanocomposites?

1.4 What specific interactions are more apparent at the nanoscale which create the special properties?

1.5 What is the exfoliation process?

1.6 When nanoscale materials scale to nanostructured materials, do they retain the same properties or do they change because they are no longer at individual nano size?

1.7 Can a nanocomposite be classified as a composite?

1.8 For which applications polymer nanocomposites can be used?

1.9 What are typical ratios of nanomaterials to polymers in polymer nanocomposites?

1.11 References

1. *The National Nanotechnology Initiative Supplement to the President's 2019 Budget*, National Science and Technology Council (NSTC), Committee on Technology (CoT), Subcommittee on Nanoscale Science, Engineering, and Technology (NSET), Washington, DC, August 2018.
2. Nanotechnology (2006, January 1) Retrieved December 15, 2014, from http://www.nature.com/nnano/journal/v1/n1/full/nnano.2006.77.html.
3. *National Nanotechnology Initiative Strategic Plan*, National Science and Technology Council (NSTC), Committee on Technology (CoT), Subcommittee on Nanoscale Science, Engineering, and Technology (NSET), Washington, DC, February 2014; www.nano.gov/2014StrategicPlan, p. 5.
4. *Report to the President and Congress on the Fourth Assessment of the National Nanotechnology Initiative*, Executive Office of the President; President's Council of Advisors of Science and Technology (PCAST), April 2012.
5. Athavale A. (2011) Nanotechnology: Origin and history. *TechTalk@KPITCummins*, 4(3):5–8. Retrieved December 15, 2014, from http://www.kpit.com/downloads/tech-talk/tech-talk-july-september-2011.pdf#page=3.
6. Feynman, R. P. (February 1960) There's plenty of room at the bottom, Annual Meeting at American Physical Society at the California Institute of Technology, December 29, 1959, first published in *Caltech Engineering and Science*, 23(5):22–36.
7. Size of the nanoscale. (n.d.) Retrieved December 15, 2014, from http://www.nano.gov/nanotech-101/what/nano-size.
8. Ratner, M., and Ratner, D. (2003) *Nanotechnology: A gentle introduction to the next big idea.* Upper Saddle River, NJ: Prentice Hall, pp. 1–18.
9. What is nanotechnology? (n.d.) Retrieved December 15, 2014, from http://www.nano.gov/nanotech-101/what/definition.
10. Koo, J. H. (2006) *Polymer Nanocomposites—Processing, Characterization, and Applications.* New York, NY: McGraw-Hill.
11. Vollath, D. (2008) *Nanomaterials: An Introduction to Synthesis, Properties, and Applications.* Weinheim, Germany: Wiley-VCH Verlag GmbH, pp. 1–20.
12. Li, X., Chen, H., Dang, Y., Lin. Y., Larson, C. A., and Roco, M. C. (2008) A longitudinal analysis of nanotechnology literature: 1976–2004. *J. Nanopart. Res.*, 10(3):22.
13. StatNano, website: www.statnano.com.
14. Li, X., Hu, D., Dang, Y., Chen, H., Roco, M. C., Larson, C. A., and Chan, J. Nano (2009) Mapper: An Internet knowledge mapping system for nanotechnology development, *J. Nanopart. Res.*, 2009;11:529–552.

15. Jancar, J., Douglas, J. F., Starr, F. W., Kumar, S. K., Cassagnau, P., Lesser, A. J., and Sterstein, S. S. (2010) Current issues in research on structure-property relationships in polymer nanocomposites. *Polymer,* 51:3321–3343.
16. Rocco, M. C. (1998) Nanoparticle and nanotechnology research in the USA. *J. Aerosol. Sci.,* 29:749–760.
17. Janz, A., Kockritz, A., Yao, L., and Adsorption, A. M. (2010) Surface area: Fundamental calculations on the surface area determination of supported gold nanoparticles. *Langmuir,* 26(9):6783–6789.
18. *Progress Review on the Coordination Implementation of the National Nanotechnology Initiative 2011 Environmental, Health, and Safety Research Strategy.* National Science and Technology Council (NSTC), Committee on Technology (CoT), Subcommittee on Nanoscale Science, Engineering, and Technology (NSET), Washington, DC, June 2014; http://www.nano.gov/2014EHSProgressReview.
19. The SIINN (Safe Implementation of Innovative Nanoscience and Nanotechnology) ERA-NET is a coordination activity funded by the European Commission within the 7th Framework Programme, http://www.siinn.eu. It promotes the safe and rapid transfer of European research results in nanoscience and nanotechnology (N&N) into industry applications.
20. The U.S.-EU organization bridges nanoEHS research efforts, http://www.us-eu.org.
21. Stark, W. J. (2011) Nanoparticles in biological systems. *Angew. Chem. Int. Ed.,* 50:1242–1250.
22. Roco, M., and Bainbridge, W. (2013) The new world of discovery, invention, and innovation: Convergence of knowledge, technology, and society. *Springer Science+Business Media Dordrescht,* 15:1946.
23. Tahan, C., Leung, R., Zenner, G., Ellison, K., Crone, W., and Miller, C. (2006) Nanotechnology and society: A discussion-based undergraduate course. *Am. J. Phys.,* 74:443.

1.12 Further Reading

Abdi, F., Garg, M. eds. (2017) *Characterization of Nanocomposites: Technology and Industrial Applications.* Singapore: Pan Stanford Publishing.

Agarwal, A., Bakshi, S. R., and Lahiri, D. (2011) *Carbon Nanotubes—Reinforced Metal Matrix Composites.* Boca Raton, FL: CRC Press.

Ajayan, P. M., Schadler, L. S., and Braun, P. V. (2003) *Nanocomposite Science and Technology.* Weinheim, Germany: Wiley-VCH Verlag GmbH.

Al-Malaika, S., Golovoy, A., and Wilkie, C. A. (1999) *Chemistry and Technology of Polymer Additives.* Malden, MA: Blackwell Science.

Bauhofer, W., and Kovacs, J. Z. (2009) A review and analysis of electrical percolation in carbon nanotube polymer composites. *Compos. Sci. Technol.,* 69(10):1486–1498.

Bhat, B. N., ed. (2108) *Aerospace Materials and Applications.* Reston, VA: AIAA.

Beall, G. W., and Powell, C. E. (2011) *Fundamentals of Polymer-Clay Nanocomposites.* Cambridge, UK: Cambridge University Press.

Bhattacharya, S. N., Gupta, R. K., and Kamal, M. R. (2008) *Polymeric Nanocomposites—Theory and Practice.* Cincinnati, OH: Hansen.

Brandt, M. ed. (2017) *Laser Additive Manufacturing: Materials, Design, Technologies, and Applications.* Oxford, UK: Elsevier.

Callister, W. D., Jr. (2007) *Materials Science and Engineering—An Introduction.* 7th ed. New York, NY: Wiley.

Cao, G. (2004) *Nanostructures and Nanomaterials—Synthesis, Properties & Applications.* London: Imperial College Press.

Chou, T. W., Gao, L., Thostenson, E. T., Zhang, Z., and Byun, B.-Y. (2011) An assessment of the science and technology of carbon nanotube-based fibers and composites: A review. *Compos. Sci. Technol.,* 70(1):1–19.

Dai, L. ed. (2006) *Carbon Nanotechnology–Recent Developments in Chemistry, Physics, Materials Science and Device Applications.* Amsterdam, the Netherlands: Elsevier.

Di Ventra, M., Evoy, S., and Heflin, J. R., Jr. (2004) *Introduction to Nanoscale Science and Technology.* Norwell, MA: Kluwer Academic Publishers.

Duffa, G. (2013) *Ablative Thermal Protection System Modeling.* Reston, VA: AIAA.

Endo, M., Strano, M. S., and Ajayan, P. M. (2008) Potential applications of carbon nanotubes. *Top. Appl. Phys.,* 111:13–61.

Fecht, H.-J., and Werner, M. eds. (2004) *The Nano-Micro Interface—Bridging the Micro and Nano Worlds.* Weinheim, Germany: Wiley-VCH Verlag GmbH.

Friedrich, K., Fakirov, S., and Zhang, Z. eds. (2005) *Polymer Composites—From Nano- to Macro-Scale.* New York, NY: Springer.

Galimberti, M. ed. (2011) *Rubber-Clay Nanocomposites.* Hoboken, NJ: Wiley.

Geckeler, K. E., and Nishide, H. eds. (2010) *Advanced Nanomaterials.* Vols. 1 & 2. Weinheim, Germany: Wiley-VCH Verlag GmbH.

Geckeler, K. E., and Rosenberg, E. (2006) *Functional Nanomaterials.* Stevenson Ranch, CA: American Scientific Publishers.

Gerard, J.-F. ed. (2001) *Fillers and Filled Polymers.* Weinheim, Germany: Wiley-VCH Verlag GmbH.

Gibson, R. F. (2010) A review of recent research on mechanics of multifunctional composite materials and structures. *Compos. Struct.,* 92(12):2793–2810.

Gogotsi, Y. ed. (2006) *Nanotubes and Nanofibers.* Boca Raton, FL: CRC Press.

Green, M. J., Behabtu, N., Pasquali, M., and Adams, W. (2009) Nanotubes as polymer. *Polymer,* 50:4979–4997.

Guo, Z., and Tan, L. (2009) *Fundamentals and Applications of Nanomaterials.* Norwood, MA: Artech House.

Gupta, R. K., Kennel, E., and Kim, K.-J. eds. (2010) *Polymer Nanocomposites Handbook.* Boca Raton, FL: CRC Press.

Harper, C. A. ed. (2002) *Handbook of Plastics, Elastomers, & Composites.* 4th ed. New York, NY: McGraw-Hill.

He, G-Q., Yan, Q-L., Liu, P-J., and Gozin, M. eds. (2019) *Nanomaterials for Rocket Propulsion Systems.* Oxford, UK: Elsevier.

Hornyak, G. L., Tibbals, H. F., Dutta, J., and Moore, J. J. (2009) *Introduction to Nanoscience & Nanotechnology.* Boca Raton, FL: CRC Press.

Hull, T. R., and Kandola, B. K. eds. (2009) *Fire Retardancy of Polymers: New Strategies and Mechanisms.* Cambridge, UK: RSC Publishing.

Iijima, S. (1991) Helical microtubules of graphitic carbon. *Nature,* 354:56–58.

Karn, B., Mascinangioli, T., Zhang, W., Colvin, V., and Alivisatos, P. eds. (2005) *Nanotechnology and the Environment—Applications and Implications.* ACS Symposium Series 890. Washington, DC: American Chemical Society.

Koo, J. H. (2016) *Fundamentals, Processing, and Applications of Polymer Nanocomposites.* Cambridge, UK: Cambridge University Press.

Kostoff, R. N., Johnson, J. A., Murday, J. S., Lau, C. G. Y., and Tolles, W. M. (2006) The structure and infrastructure of the global nanotechnology literature. *J. Nanopart. Res.* 8(3–4):301–321.

Kostoff, R. N., Murday, J. S., Lau, C. G. Y., and Tolles, W. M. (2006) The seminal literature of nanotechnology research. *J. Nanopart. Res.* 8(2):193–213.

Krishnamoorti, R., and Vaia, R. A. eds. (2005) *Polymer Nanocomposites—Synthesis, Characterization, and Modeling.* ACS Symposium Series 804. Washington, DC: American Chemical Society.

Le Bras, M., Camino, G., Bourbigot, S., and Delobel, R. eds. (1998) *Fire Retardancy of Polymers—The Use of Intumescence.* Cambridge, UK: RSC Publishing.

Le Bras, M., Wilkie, C. A., Bourbigot, S., Duquesne, S., and Jama, C. eds. (2005) *Fire Retardancy of Polymers—New Applications of Mineral Fillers.* Cambridge, UK: RSC Publishing.

Liu, L., Ma, W., and Zhang, Z. (2011) Macroscopic carbon nanotubes assemblies: Preparation, properties, and potential applications. *Small,* 7(11):1504–1520.

Marijnissen, J., and Gradon, L. eds. (2010) *Nanoparticles in Medicine and Environment—Inhalation and Health Effects.* New York, NY: Springer.

McNally, T., and Potschke, P. (2011) *Polymer-Carbon Nanotube Composites: Preparation, Properties, and Applications.* Cambridge, UK: Woodhead Publishing.

Mittal, V. ed. (2010) *Optimization of Polymer Nanocomposites Properties.* Weinheim, Germany: Wiley-VCH Verlag GmbH.

Mensah, T., Wang, B., Winter, J., and Davis, V. (2018) *Nanotechnology Commerialization: Manufacturing Processes and Products.* New York, NY: Wiley.

Mittal, V. ed. (2010) *Polymer Nanotube Nanocomposites—Synthesis, Properties, and Application.* Salem, MA: Wiley & Scrivener Publishing.

Mittal, V. ed. (2011) *Thermally Stable and Flame Retardant Polymer Nanocomposites.* Cambridge, UK: Cambridge University Press.

Mittal, V. ed. (2012) *Characterization Techniques for Polymer Nanocomposites.* Weinheim, Germany: Wiley-VCH Verlag GmbH.

Miziolek, W., Karna, S. P., Mauro, J. M., and Vaia, R. A. eds. (2005) *Defense Applications of Nanomaterials.* ACS Symposium Series 891. Washington, DC: American Chemical Society.

Morgan, B., and Wilkie, C. A. (2007) *Flame Retardant Polymer Nanocomposites.* Hoboken, NJ: Wiley.

Mukhopadhyay, P., and Gupta, R. K. eds. (2013) *Graphite, Graphene and Their Polymer Nanocomposites.* Boca Raton, FL: CRC Press.

Mukhopadhyay, S. M. ed. (2012) *Nanoscale Multifunctional Materials.* Hoboken, NJ: Wiley.

Nelson, J. K. ed. (2010) *Dielectric Polymer Nanocomposites.* New York, NY: Springer.

Paulo Davim, J. ed. (2013) *Tribology of Nanocomposites.* New York, NY: Spring.

Pinnavaia, T. J., and Beall, G. W. eds. (2000) *Polymer-Clay Nanocomposites.* New York, NY: Wiley.

Priest, S. H. ed. (2012) *Nanotechnology and the Public—Risk Perception and Risk Communication.* Boca Raton, FL: CRC Press.

Pusch, R., and Yong, R. N. (2006) *Microstructure of Smectite Clays and Engineering Performance.* New York, NY: Taylor & Francis.

Roco, M. (2011) The long view of nanotechnology development: The NNI at 10 years. *J. Nanopart. Res.*, 13:427–445.

Schnorr, J. M., and Swager, T. M. (2011) Emerging applications of carbon nanotubes. *Chem. Mater.*, 23(3):646–657.

Schulz, M. J., Kelkar, A. D., and Sundaresan, M. J. eds. (2006) *Nanoengineering of Structural, Functional, and Smart Materials.* Boca Raton, FL: CRC Press.

Seal, S. ed. (2008) *Functional Nanostructures—Processing, Characterization, and Applications.* New York, NY: Springer.

Shatkin, J. A. (2008) *Nanotechnology: Health and Environmental Risks.* Boca Raton, FL: CRC Press.

Spitalskya, Z., Tasisb, D., Papagelis, K., and Galiotis, C. (2010) Carbon nanotube-polymer composites: Chemistry, processing, mechanical, and electrical properties. *Prog. Polym. Sci.*, 35:357–401.

StatNano, website: www.statnano.com.

Tjong, S. C. (2009) *Carbon Nanotube Reinforced Composites—Metal and Ceramic Matrices.* Weinheim, Germany: Wiley-VCH Verlag GmbH.

Tomanek, D., and Enbody, R. J. (2000) *Science and Application of Nanotubes.* New York, NY: Kluwer Academic/Plenum Publishers.

Vaia, R. A. (2002) Polymer nanocomposites open a new dimension for plastics and composites. *AMPTIAC Newsl.*, 6(1):17–24.

van Krevelen, D. W., and te Nigenhuis, K. (2012) *Properties of Polymers.* 4th ed. Amsterdam, the Netherlands: Elsevier.

Vilgis, T. A., Heinrich, G., and Klippel, M. (2009) *Reinforcement of Polymer Nano-Composites—Theory, Experiments and Applications.* Cambridge, UK: Cambridge University Press.

Vollath, D. (2008) *Nanomaterials: An Introduction to Synthesis, Properties, and Applications.* Weinheim, Germany: Wiley-VCH Verlag GmbH.

Wang, L. ed. (2000) *Characterization of Nanophase Materials.* Weinheim, Germany: Wiley-VCH Verlag GmbH.

Wang, Z. L., Liu, Y., and Zhang, Z. eds. (2003) *Handbook of Nanophase and Nanostructured Materials, Vol. 4: Materials Systems and Applications (II).* New York, NY: Kluwer Academic/Plenum Publishers.

Wardle, L. (2012) 3D Nano-engineering composites and 2D polymer nanocomposites: Processing and properties. In *ASC Series on Advances in Composite Materials, Nanocomposites,* Vol. 2, T.-W. Chou and C. T. Sun, eds. Lancaster, PA: DEStech Publishing.

Wardle, L., Koo, J. H., Odegard, G. M., and Seidel, G. D. (2018) Advanced nanoengineering materials. In *Aerospace Materials and Applications,* B. Bhatt, ed. Reston, VA: AIAA, pp. 275–304.

Yao, N., and Wang, Z. L. eds. (2005) *Handbook of Microscopy for Nanotechnology.* Boston, MA: Kluwer Academic Publishers.

Zhang, J. Z. (2009) *Optical Properties and Spectroscopy of Nanomaterials.* Singapore: World Scientific Publishing.

CHAPTER 2

An Overview of Nanomaterials

2.1 Introduction

The increasing availability of nanomaterials offers the promise for researchers to develop novel polymer composite materials with new and improved properties. Nanomaterials lie at the intersection of material science, physics, chemistry, biology, medicine, and for many of the most interesting applications. With the increasing development and availability of nanomaterials, new materials discover use in numerous applications, such as electronics, aerospace, military, pharmaceuticals, and medicine. One unique research area of particular interest is the development of polymer nanocomposites (PNCs). PNCs are created from a combination of inorganic nanomaterials and organic polymers; they offer properties that are representatives of both components. They are also known as *hybrid materials.* For traditional composite materials, they often exhibit localized heterogeneity that can lead to the loss of desired properties. When the micrometer-sized inorganic particles are incorporated into the polymer, the concept of true hybridization at the molecular level is not achieved. The nanomaterials make it possible to increase the interaction between the organic and inorganic phases by several orders of magnitude. In this chapter, numerous types of nanomaterials are introduced in terms of their origin, background, structure, surface treatment, manufacturing, fabrication, properties, features, potential applications, and suppliers.

Polymer nanocomposites are composites in which interphases dominate the composite properties because of the very small size of the nanomaterial component, at least one dimension of it is in nanometer scale. This polymer can be thermoplastic, thermoset, or elastomer. The polymer–nanomaterial interface is a key determinant of PNC properties. The interaction of nano-sized particles with a polymer molecule is truly a molecular-level interaction. The chemical surface reactivity and bulk mixing dynamics would influence the incorporation of nanomaterials into the host polymer. Since their invention more than two decades ago, PNCs have become a new and important class of materials that not only offer superior performance as compared to conventional composites, but also provide multifunctional properties that extend their scope of application to new areas [1–10]. PNCs provide major enhancements in numerous properties, such as mechanical, thermal (stability and conductivity), flammability, ablation, electrical, optical, tribological, permeability, and chemical resistance. These multifunctional properties of PNCs are discussed in detail in numerous references [1–10]

Table 2.1 Characteristics of Nanoparticles to Polymers

Improved properties	Disadvantages
• Mechanical properties (tensile strength, stiffness, toughness)	• Viscosity increase (limits processability)
• Gas barrier	• Dispersion difficulties
• Synergistic flame-retardant additive	• Optical issues
• Dimensional stability	• Sedimentation
• Thermal expansion	• Black color when different carbon-containing nanomaterials are used
• Thermal conductivity	
• Electrical conductivity	
• Ablation resistance	
• Chemical resistance	
• Reinforcement	
• Other multifunctional properties	

and in subsequent chapters of this book. The properties and performance of PNCs can be affected by some important factors, such as the following:

- Nanomaterials and their surface treatments
- Polymer matrix—that is, thermoplastic, thermoset, or elastomer
- Methods of preparation of PNCs
- Structure and morphology of PNCs

It is a very complex matter to fully understand why the enhancement of properties occurs in PNCs. To illustrate this point, several advantages and disadvantages of incorporating nanomaterials into the polymer matrix to create a new PNC material are summarized in Table 2.1.

2.2 Types of Nanomaterials

There are many different types of commercially available nanomaterials that can be incorporated into the polymer systems to form PNCs. Depending on the application, the researcher must determine the type of nanomaterial to provide the desired effect. A brief discussion of the most commonly used nanomaterials in the literature is included in this chapter. In general, these nanomaterials can be categorized based on their nanoscale dimensions, such as one-, two-, or three-dimensional nanoscale materials (Fig. 2.1), and are further discussed later in this chapter.

One-nanoscale-dimension (lamellar) nanomaterials:

- Montmorillonite (MMT) clays
- Nanographene platelets (NGPs)
- Layered double hydroxides (LDHs)

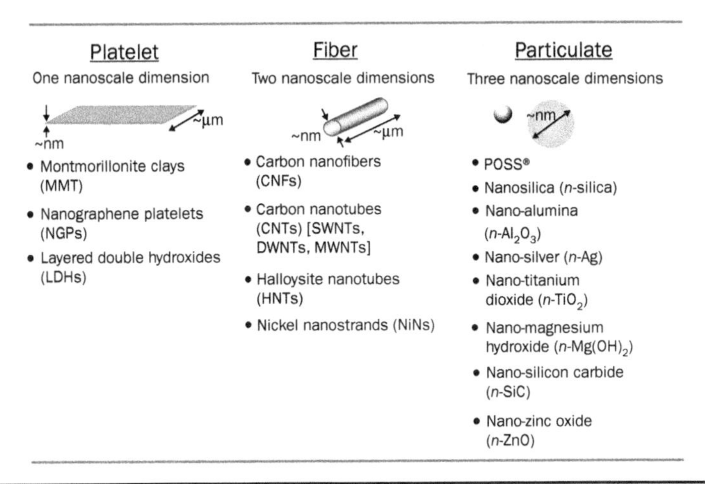

Figure 2.1 Overview of nanomaterials grouped in terms of their nanoscale dimensions.

Two-nanoscale-dimensions (fiber) nanomaterials:

- Carbon nanofibers (CNFs)
- Carbon nanotubes (CNTs) [single-walled (SWNTs), double-walled (DWNTs), and multiwalled (MWNTs)]
- Halloysite nanotubes (HNTs)
- Nickel nanostrands (NiNs)
- Aluminum oxide nanofibers (Nafen™)

Three-nanoscale-dimensions (particulates) nanomaterials:

- Polyhedral oligomeric silsesquioxanes (POSS®)
- Nanosilica (*n*-silica)
- Nano-alumina (*n*-alumina)
- Nano–titanium dioxide (*n*-TiO_2)
- Nano–magnesium hydroxide [*n*-$Mg(OH)_2$]
- Nano–silicon carbide (*n*-SiC)
- Nano-silver (*n*-Ag)
- Nano–zinc oxide (*n*-ZnO)
- Others (*n*-$BaSO_4$, *n*-$BaTiO_3$, *n*-$CaCO_3$, *n*-Fe_2O_3, *n*-Fe_3O_4, nano-diamond, and *n*-ZnS)

2.2.1 One Nanoscale Dimension in the Form of Lamellar

Montmorillonite (MMT) Clays

Origin Nanoclay is one of the most widely studied nanomaterials in a variety of polymer matrices for various applications [1–3, 9, 11–13]. Bentonite (natural clay) is most commonly formed by the in situ alteration of volcanic ash (Fig. 2.2). Another, less common method is the hydrothermal alteration of volcanic rocks. Bentonite contains

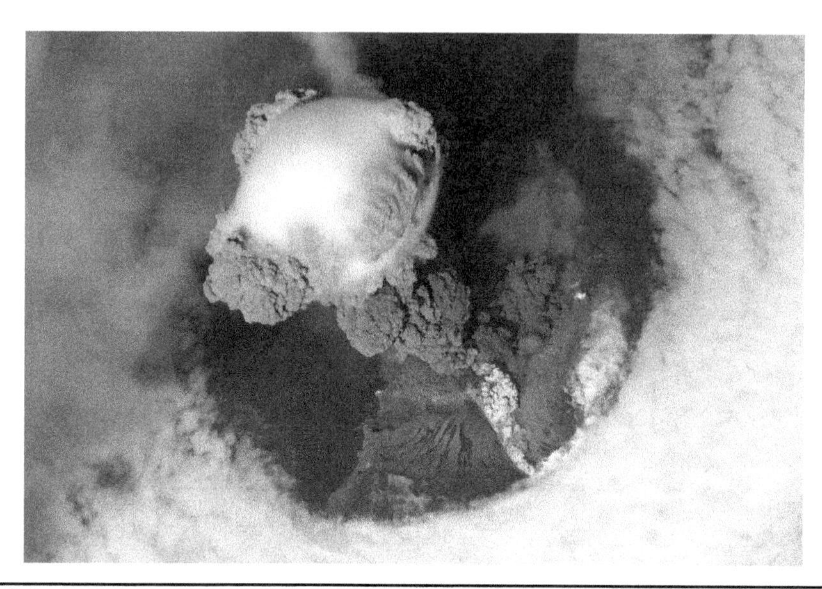

Figure 2.2 Bentonite comes from the deposition and alternation of volcanic ash from inland sea beds from 85 million years ago.

montmorillonite (MMT) but can also contain glass, mixed-layer clays, illite, kaolinite, quartz, zeolite, and carbonates. Clay soil has particle sizes of less than 2 μm. The *expanding clays* are phyllosilicates, smectite, and MMT, and the *nonexpanding clays* are talc, kaolin, and mica. The expanding clays are more versatile than the nonexpanding clays. The starting material of Southern Clay Products' (SCP is located in Gonzales, Texas, now known as BYK Additives & Instruments, a member of ALTANA, website: www.byk.com) commercial clay products comes from volcanic eruptions in the Pacific Ocean and the western United States during the Cretaceous period about 85–125 million years ago. During that time, magma—molten rock mixture underneath the Earth's crust—had forced its way to the surface and, under extreme pressure, exploded into the atmosphere in a series of global volcanic eruptions. Airborne ash from the eruption clouds subsequently fell on the Earth's surface, forming deposits characterized by high-volume bedding of well to moderately sorted ash. The parent ash was deposited under marine conditions—in other words, most of it landed in seas and oceans—and there were some minor accumulations in alkaline lakes [11]. With time, the ash beds in inland sea areas of the western United States became what is now the clay-producing areas of Montana, Wyoming, and the Dakotas. Opinions differ concerning the process and time of alternation of the ash to clay. It is estimated that the resulting deposits in Wyoming alone contain over 1 billion tons of available clay [11, 12].

Structure Silica is the dominant constituent of the MMT clays, with alumina being essential. The chemical structure of the MMT clay is illustrated in Fig. 2.3, showing its sheet structure consisting of layers containing the tetrahedral silicate layer and the octahedral alumina layer. The tetrahedral silicate layer consists of SiO_4 groups linked together to form a hexagonal network of the repeating units of composition Si_4O_{10}. The alumina layer consists of two sheets of closely packed oxygens or hydroxyls, between which octahedrally coordinated aluminum atoms are imbedded in such a position that they are equidistant from six oxygens or hydroxyls. The two tetrahedral layers sandwich the octahedral layer,

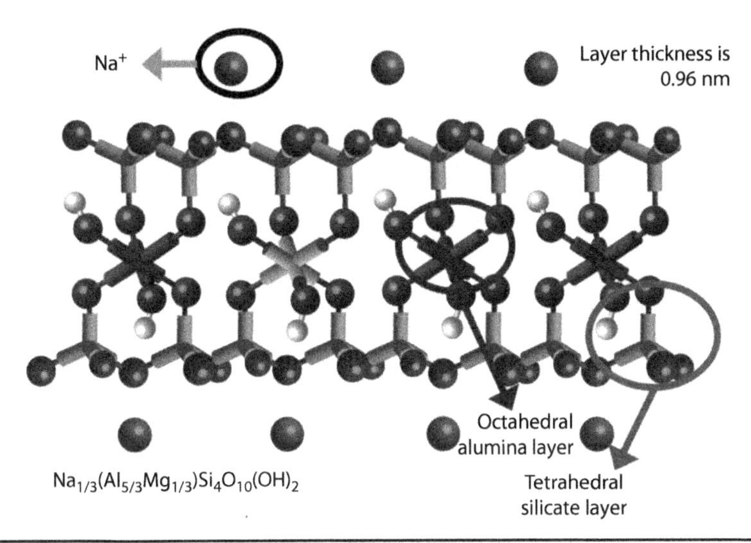

Figure 2.3 Chemical structures of montmorillonite nanoclays.

sharing their apex oxygens with the latter. These three layers form one clay sheet that has a thickness of 0.96 nm. The chemical formula of the MMT clay is $Na_{1/3}(Al_{5/3}Mg_{1/3})Si_4O_{10}(OH)_2$. In its natural state, Na^+ cation resides on the MMT clay surface.

Organic Surface Treatment Because layered silicates are hydrophilic materials, they must be made organophilic (hydrophobic) to become compatible with the host polymers that are hydrophobic polymers. Without organic treatment, layer silicates will only disperse, and phase separate in the presence of very polar polymers. Organic treatment is typically accomplished via ion exchange between inorganic alkali cations on the clay surface with the desired organic cation. The organic treatment, at the interface between inorganic silicate and organic polymer, is a vital part of the PNC, and therefore, it must be tailored to synthetic conditions. Synthetic methods for PNC preparation include solvent mixing, in situ polymerization, and melt compounding. One of the most industry-friendly methods of making PNC is the use of melt compounding. The polymer and organically treated clay are heated to the melting point of the polymer, and the two are mixed together using compounding equipment, such as an extruder or a mixing head. There are several types of clay surface treatments:

- Quaternary ammonium salts based on textile antistatic agents
- Alkyl imidazoles, which provide improved thermal stability
- Coupling and tethering agents—reactive diluents, can be used as functional amino compounds
- Cation types containing phosphorous ionic compounds

Achieving exfoliation of organo-montmorillonite clay in various polymer continuous phases is a function of the surface treatment of the MMT clays and the mixing efficiency of the dispersing apparatus or equipment. For hydrophilic nanoclay, such as BYK's Cloisite® Na^+, it is recommended to dissolve it either in water using high-shear mixing or to add Cloisite Na^+ directly into a water-soluble, aqueous polymer solution. Three types of mixing equipment are recommended to disperse hydrophobic

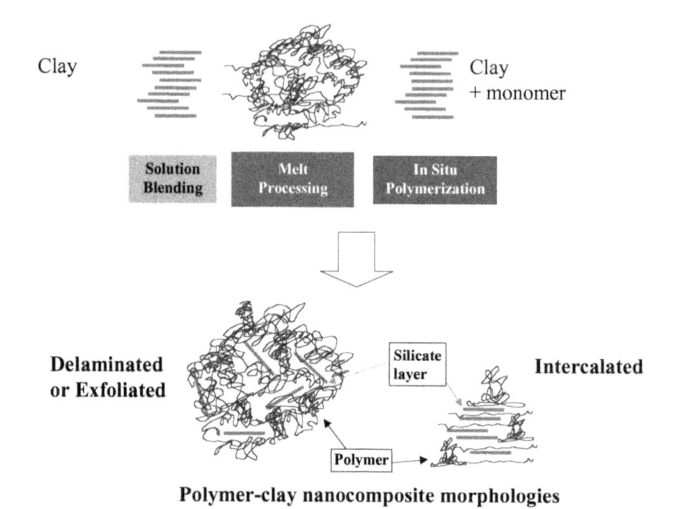

FIGURE 2.4 Schematic showing the preparation of polymer-clay nanocomposites. (*Courtesy of J. Gilman.*)

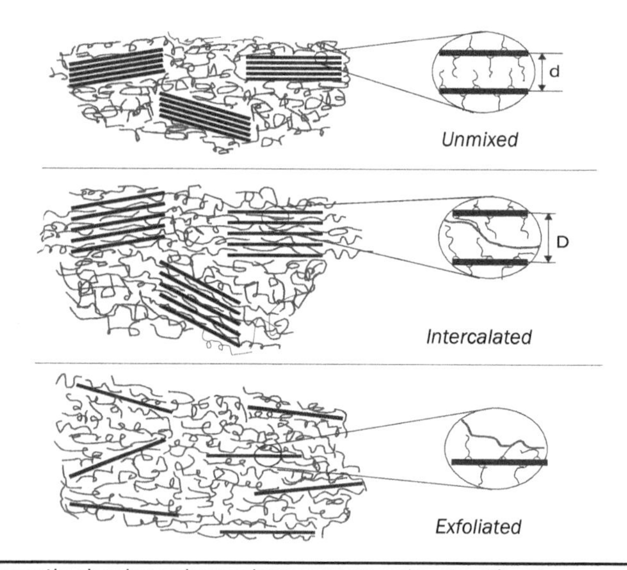

FIGURE 2.5 Schematic showing polymer-clay nanocomposite classifications.

nanoclays, such as BYK's Cloisite Na^+, Ca^{++}, 3, 5, 10A, 11, 15, 20, 30B, 93, and 116, into different resin systems. The recommended equipment are as follows: high-shear mixer or three-roll mill for liquid resins, Brabender mixer for viscous resins, or twin-screw extruder for solid resins.

The preparation of polymer-clay nanocomposites is clearly illustrated in Fig. 2.4. Depending on the physical state of the polymer, one can incorporate the clay into the polymer by solution blending, melt-blending, or in situ polymerization processes to form polymer-clay nanocomposites. Polymer-clay nanocomposites can be classified morphologically into (a) unmixed, (b) intercalated, and (c) exfoliated (delaminated) states, as shown in Fig. 2.5. The most desirable morphological state for the polymer-clay

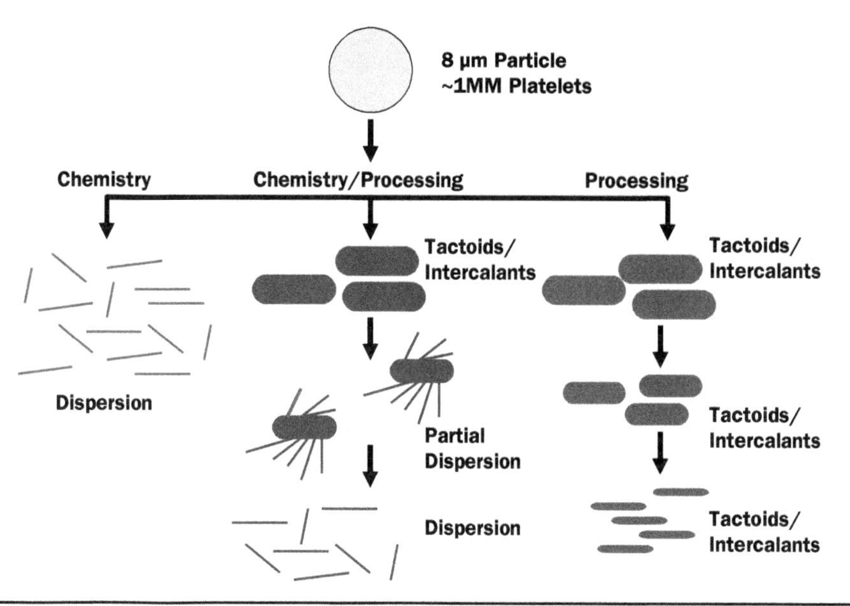

Figure 2.6 Different dispersion mechanisms of dispersing nanoclays. (*Courtesy of SCP.*)

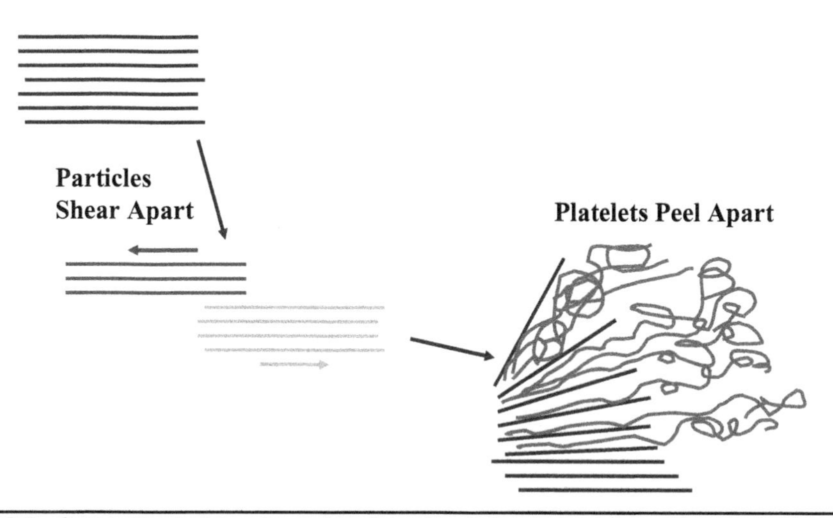

Figure 2.7 Schematic diagram showing particle shearing and platelets peeling apart. (*Courtesy of SCP.*)

nanocomposites is exfoliation, followed by intercalation. The processing challenge of nanoclay is to disperse the 8-µm particles into more than 1 million platelets using the proper processing technique and conditions. The dispersion mechanism can be (a) via a chemistry route, as in in situ polymerization; (b) via a processing route, as in extrusion; or (c) by combining the chemistry and the processing routes (the optimal way), as summarized in Fig. 2.6.

Figures 2.7 and 2.8 further illustrate the concept of particles shearing apart and platelets peeling apart schematically (Fig. 2.7) and in the form of transmission electron micrographs (Fig. 2.8). Transmission electron microscopy (TEM) is the most useful tool

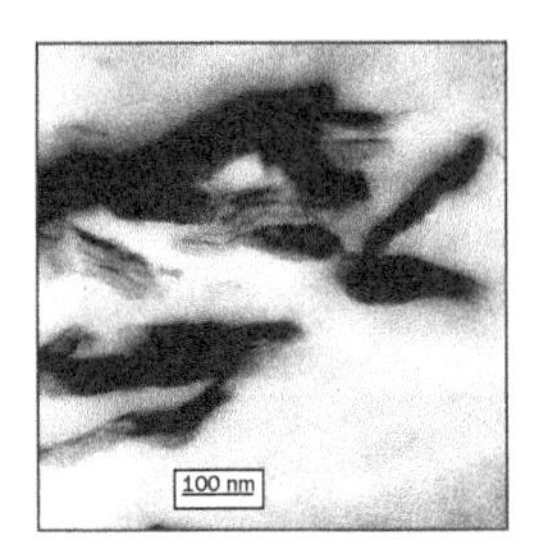

Particles Shear Apart

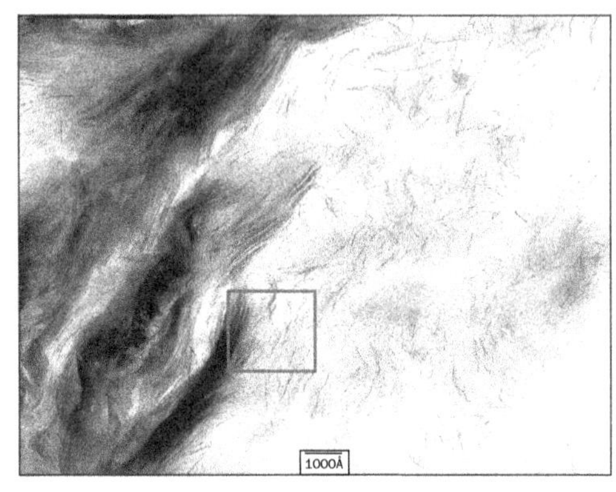

Platelets Peel Apart

Figure 2.8 TEM micrographs showing particle shearing and platelets peeling apart. (*Courtesy of SCP.*)

to determine the degree of dispersion of the nanomaterials in the polymer matrix. This will be discussed in more detail in Chap. 5.

Organoclay Suppliers There are several worldwide suppliers for organoclays. In the United States, the two major suppliers are BYK Additives & Instruments and Nanocor®. BYK produces Cloisite, Laponite®, and Nanofil® types of clay materials. Cloisite and Nanofil materials are based on naturally occurring MMT, while Laponite is synthesized clay. Product information can be obtained at the company's website www.nanoclay.com [12]. Nanocor, located in Chicago, IL, produces Nanomer® nanoclays, I.xxE (e.g., I.28E and I.30E), masterbatches/concentrates, and nanocomposites. Product information can be obtained at Nanocor's website www.nanocor.com [13]. Other organoclay suppliers are Kunimine Industries Co., Ltd. (Japan) (website: www.kunimine.co.jp); Laviosa Chimica Minerariain (Italy) (website: www.laviosa.it); and several Chinese companies.

Cloisite Product Lines BYK manufactures and markets Cloisite additives [11, 12]. These MMT organoclays are surface modified to allow complete dispersibility and miscibility with many different resin systems which they were designed to improve. Cloisite MMT clays, such as Cloisite Na^+, Ca^{++}, 3, 5, 10A, 11, 15, 20, 30B, 93, and 116, are referenced in this section. Several of these MMT clays are mentioned throughout this book.

Cloisite Na^+ is a natural bentonite. This polymer additive improves various polymer physical properties, such as reinforcement, heat deflection temperature (HDT), coefficient of linear thermal expansion (CLTE), and barrier properties. Typical properties, dry particle size, density, and organic modified data of Cloisite Na^+, Ca^{++}, 5, 10A, 11, 15, 20, 30B, 93, and 116 are shown in Tables 2.2 and 2.3.

Cloisite Ca^{++} is a natural bentonite. It is designed to be used as an additive for plastics and rubber to improve various physical properties, such as reinforcement, coefficient of thermal expansion (CTE), synergistic flame retardancy, and barrier.

Cloisite 3 has no data available at the BYK's website.

Table 2.2 Typical Physical Properties of Cloisite Nanoclay

MMT clay	Moisture (%)	Dry particle size, d_{50} (μm)	Color	Packed bulk density (g/L)	Density (g/cc)	X-ray results, d_{001} (nm)
Cloisite Na+	4–9	<25	Off-white	568	2.86	1.17
Cloisite Ca++	4–9	<10	Off-white	625	2.8	1.55
Cloisite 5	<3	<40	Off-white	480	1.77	3.27
Cloisite 10A	<3	<10	Off-white	265	1.9	1.9
Cloisite 11	<3	<40	Off-white	265	2.0	1.84
Cloisite 15	<3	<10	Off-white	165	1.66	3.63
Cloisite 20	<3	<10	Off-white	175	1.77	3.16
Cloisite 30B	<3	<10	Off-white	365	1.98	1.85
Cloisite 93	<3	<40	Off-white	440	1.88	2.79
Cloisite 116	8–13	<15	Off-white	340	2.8	1.25

Table 2.3 Cloisite MMT Clay with Organic Modifier

MMT clay	Organic modifier
Cloisite Na+	Natural bentonite
Cloisite Ca++	Natural bentonite
Cloisite 5	Bis(hydrogenated tallow alkyl)dimethyl salt
Cloisite 10A	Benzyl(hydrogenated tallow alkyl)dimethyl salt
Cloisite 11	Benzyl(hydrogenated tallow alkyl)dimethyl salt
Cloisite 15	Bis(hydrogenated tallow alkyl)dimethyl salt
Cloisite 20	Bis(hydrogenated tallow alkyl)dimethyl salt
Cloisite 30B	Alkyl quaternary ammonium salt
Cloisite 93	Trialkyl ammonium salt
Cloisite 116	Natural bentonite

Cloisite 5 is a bis(hydrogenated tallow alkyl)dimethyl salt with bentonite. It is an additive for plastics and rubber to improve physical properties, such as reinforcement, CTE, synergistic flame retardancy, and barrier.

Cloisite 10A is benzyl(hydrogenated tallow alkyl)dimethyl salt with bentonite. It is designed to be used as an additive for plastics and rubber to improve various physical properties, such as reinforcement, CTE, synergistic flame retardancy, and barrier.

Cloisite 11 is benzyl(hydrogenated tallow alkyl)dimethyl salt with bentonite. It is designed to be used as an additive for plastics and rubber to improve various physical properties, such as reinforcement, CTE, synergistic flame retardancy, and barrier.

Cloisite 15 is bis(hydrogenated tallow alkyl)dimethyl salt with bentonite. It is designed to be used as additive for plastics and rubber to improve various physical properties, such as reinforcement, CTE, synergistic flame retardancy, and barrier.

Cloisite 20 is bis(hydrogenated tallow alkyl)dimethyl salt with bentonite. It is designed to be used as an additive for plastics and rubber to improve various physical properties, such as reinforcement, CTE, synergistic flame retardancy, and barrier.

Cloisite 30B is alkyl quaternary ammonium salt bentonite. It is designed to be used as an additive for plastics and rubber to improve various physical properties, such as reinforcement, CTE, synergistic flame retardancy, and barrier.

Cloisite 93 is trialkyl ammonium bentonite salt with bentonite. It is designed to be used as an additive for plastics and rubber to improve various physical properties, such as reinforcement, CTE, synergistic flame retardancy, and barrier.

Cloisite 116 is a natural bentonite. It is designed to be used as an additive for plastics and rubber to improve various physical properties, such as reinforcement, CTE, synergistic flame retardancy, and barrier.

Nanocor Product Lines Three general categories of products are available from Nanocor or its distributors: nanomer nanoclays, masterbatches/concentrates, and nanocomposites.

Nanomer nanoclays are supplied as microfine powders under the "I" series designation, ready to be used for different resins, such as polyamides, epoxy, urethanes, ethylene propylene diene monomer (EPDM), and other engineering resins.

Masterbatches/concentrates in resin systems, such as polypropylene homopolymer, polypropylene copolymer, linear low-density polyethylene (LLDPE), LDPE, HDPE, and ethylene-methyl acrylate (EMA) are supplied in high loading pellet forms. PNC nylon concentrate is designed for letdown into nylon 6, 6.6, and amorphous nylon.

Nanocomposites are supplied in pellet form, ready for injection molding, blow molding, and film casting, by Nanocor's industrial partners. Imperm® supplied by ColoMatrix Corporation is an ultrahigh barrier nylon, providing protection for oxygen sensitive products and CO_2 retention for carbonated soft drinks, waters, beers, and flavored alcoholic beverages. Durethan® KU2-2601 available from LANXESS Deutschland is a nylon 6 nanocomposite for films and paper coating, designed for medium barrier applications requiring excellent clarity. Aegis™ NC is a nylon 6 nanocomposite available from Honeywell Polymers, designed for film and paper coating, and for engineering applications. Aegis OX is a nylon 6 nanocomposite available from Honeywell Polymers, combined nanocomposite barrier and oxygen scavenger in one system used for beer bottles. NanoTUFF™ is a nylon 6 nanocomposite available from Nylon Corporation of America for applications where high modulus and cold weather impact are required. NanoSEAL™ is formulated for fuel systems (tanks and hoses) to meet California Air Resources Board fuel emissions standards, available from Nylon Corporation of America.

Nanographene Platelets (NGPs)

Origin Graphene is the name given to a flat monolayer of carbon atoms tightly packed into a two-dimensional (2D) honeycomb lattice, and is a basic building block for graphitic materials for all other dimensionalities (Fig. 2.9), as described by Geim and Novoselov [14]. It can be wrapped up into 0D fullerenes, rolled into 1D nanotubes, or stacked into 3D graphite. Theoretically, graphene (or *2D graphite*) has been studied for 60 years, and is widely used for describing properties of various carbon-based materials [14].

A thin flake of this ordinary carbon, just one atom thick, lies behind the 2010 Nobel Prize in Physics awarded to Andre Geim and Konstantin Novoselov. Both of them originally studied and began their careers as physicists in Russia, now they are professors at the University of Manchester, the United Kingdom. They have shown that carbon in such a flat form has exceptional properties that originate from the remarkable world of quantum physics. Graphene is a form of carbon. As a material, it is completely new, not only the thinnest ever but also the strongest. As a conductor of electricity, it performs as well as copper. As a conductor of heat, it outperforms all other known materials. It is almost completely transparent, yet so dense that not even helium, the smallest gas atom, can pass through it. Geim and Novoselov extracted the graphene from a piece of graphite, such as found in ordinary pencils. Using regular adhesive tape, they managed to obtain a flake of carbon with a thickness of just one atom. This was done at a time when many believed it was impossible for such thin crystalline materials to be stable.

A nanographene platelet (NGP) is an emerging class of nanomaterials. Jang and Zhamu reported an excellent review of processing of NGPs and NGP nanocomposites [15] and their review forms the basis of this section. A NGP is a nanoscale platelet

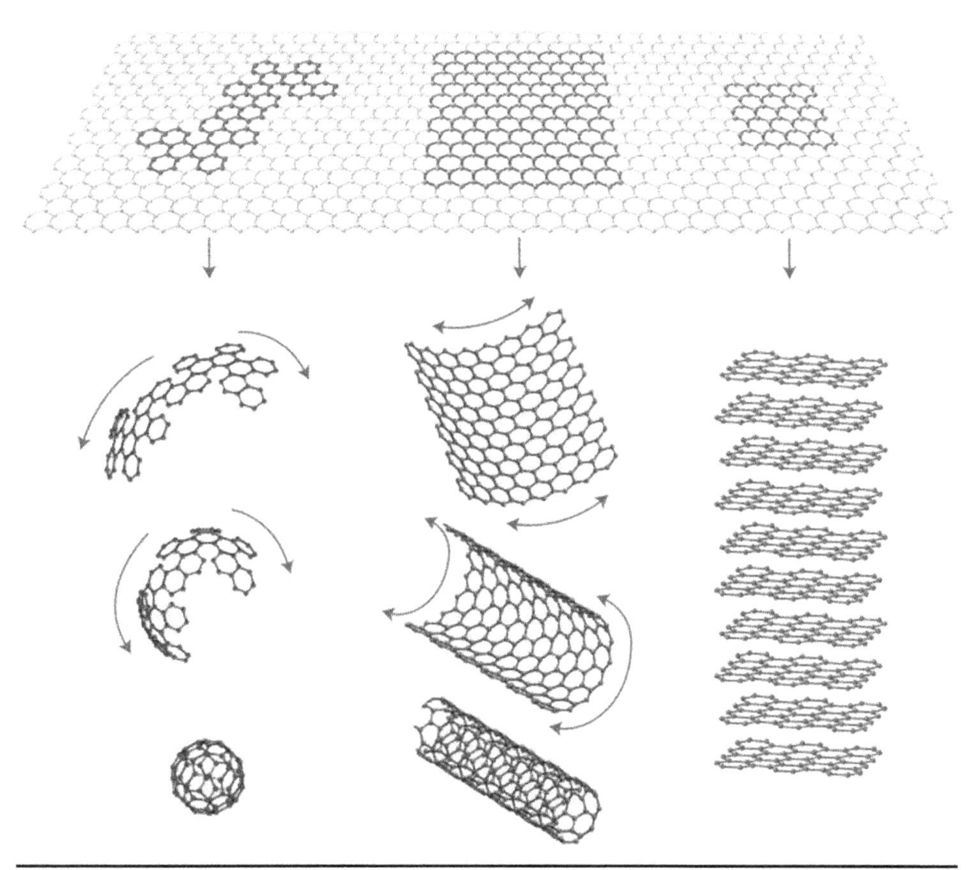

FIGURE 2.9 Mother of all graphitic forms. Graphene is a 2D building material for carbon materials of all other dimensionalities. It can be wrapped up into 0D buckyballs, rolled into 1D nanotubes, or stacked into 3D graphite [14].

composed of one or more layers of graphene plane with a platelet thickness from about 0.34 to 100 nm. NGPs are predicted to have a range of unusual physical, electrical, chemical, and mechanical properties [15]. Although practical electronic device application for graphene is not envisioned to occur within the next 5–10 years, its application as nanofiller in a composite material is imminent. The availability of processable graphene sheets in large quantities is essential to the success in exploiting composite and other applications.

For more than six decades, scientists have presumed that a single-layer graphene sheet (one atom thick) could not exist in its free state based on the reasoning that its planar structure would be thermodynamically unstable. Several groups worldwide have recently succeeded in obtaining isolated graphene sheets [16–24]. Several unique properties associated with these 2D crystals have been discovered [25–33]. In addition to single graphene sheets, double-layer or multiple-layer graphene sheets also exhibit unique and useful behavior. Single-layer and multiple-layer graphene sheets are also collectively referred to as nanographene platelets (NGPs).

Carbon nanotubes (CNTs) [this is discussed in the section "Carbon Nanotubes (CNTs)"] and NGPs exhibit some similar behaviors, but some vastly distinct properties. Selected physical and mechanical properties of NGPs, CNTs, and vapor-grown carbon fibers (VGCNFs) [also discussed in the section "Carbon Nanofibers (CNFs): Vapor-Grown Carbon Fibers (VGCFs)"] are presented in Table 2.4.

Table 2.4 Comparison of Physical Properties of CNTs, CNFs, and NGPs [15]

Properties	Single-walled carbon nanotubes	Carbon nanofibers	Nanographene platelets
Specific gravity	0.8 g/cc	1.8 (AG)–2.1 (HT) g/cc (AG = as grown; HT = heat treated) (graphitic)	1.8–2.2 g/cc
Elastic modulus	~1 TPA (axial direction)	0.4 (AG)-0.6 (HT) TPa	~1 TPa (in-plane)
Strength	50–500 GPa	2.7 (AG)-7.0 (HT) GPa	~100–400 GPa
Resistivity	5–50 μΩ cm	55 (HT)-1,000 (AG) μΩ cm	50 μΩ cm (in-plane)
Thermal conductivity	Up to 2,900 $Wm^{-1} K^{-1}$ (estimated)	20 (AG)-1,950 (HT) $Wm^{-1} K^{-1}$	5,300 $Wm^{-1} K^{-1}$ (in-plane), 6–30 $Wm^{-1} K^{-1}$ (c-axis)
Magnetic susceptibility	22 × 10^6 emu/g (radial); 0.5 × 10^6 emu/g (axial)	N/A	22 × 10^6 emu/g (perpendicular to plane); 0.5 × 10^6 emu/g (perpendicular to plane)
Thermal expansion	Negligible in the axial direction	~1 × $10^{-6} K^{-1}$ (HT, axial)	–1 × $10^{-6} K^{-1}$ (in-plane); 29 × $10^{-6} K^{-1}$ (c-axis)
Thermal stability	>700°C (in air); 2,800°C (in vacuum)	450–650°C (in air)	450–650°C (in air)
Specific surface area	Typically 10–200 m^2/g; up to 1,300 m^2/g	10–60 m^2/g	Typically 100–1,000 m^2/g; up to >2,600 m^2/g

Early and Recent Research on NGPs In the late 1980s, attempts to produce NGPs were made, although the significance of NGPs was not well recognized and they were known as *thin graphite flakes*. In 1988, Bunnell [34–36] developed entailed intercalating graphite with a strong acid to obtain a graphite intercalation compound (GIC), thermally exfoliating the GIC to obtain discrete layers by ultrasonization, mechanical shear forces, or freezing to layers into discrete flakes. Technically, the acid-treated graphite was actually graphite oxide (GO), rather than pristine graphite. Although most of the flakes presented are thicker than 100 nm [36], flakes as small as 20 nm were cited. Zaleski et al. [37] used air milling to further delaminate exfoliated graphite flakes to an average flake thickness of approximately 25 nm based on Jang and Zhamu's calculation [15]. Jang and Huang [18, 19] succeeded in isolating single-layer and multilayer graphene structures from partially graphitized polymeric carbon, which were obtained from a polymer or pitch precursor. Figure 2.10 shows polymeric carbon fibers obtained by intercalating and exfoliated carbonized polyacrylonitrile (PAN) fibers [15]. Graphene platelets extracted from these fibers using ball milling procedure are shown in Fig. 2.11 [15]. Horiuchi and colleagues [38–42] have done some significant work on the preparation of nanoscale graphite oxide (GO) platelets, which they termed as *carbon nanofilms*. According to Horiuchi et al. [38], single-layer graphene was detected.

In 2004, Novoselov et al. [16, 17] prepared single-sheet graphene by removing graphene from a graphite sample one sheet at a time using a "scotch-tape" method. Although this method is not amenable to large-scale production of NGPs, their work did spur globally increasing interest in nanographene materials, mostly motivated that graphene could be useful for developing novel electronic device [24–32]. Small-scale production of ultrathin graphene sheets on a substrate can be obtained by thermal decomposition–based epitaxial growth [29] and a laser desorption-ionization technique [43]. More promising techniques for mass production of NGPs are likely those involve chemical or thermal exfoliation of intercalated graphite or oxidized graphite.

Chen [44] exposed GO to 1,050°C for 15 seconds to obtain exfoliated graphite, which was then subjected to ultrasonic irradiation in a mixture of solution of water and

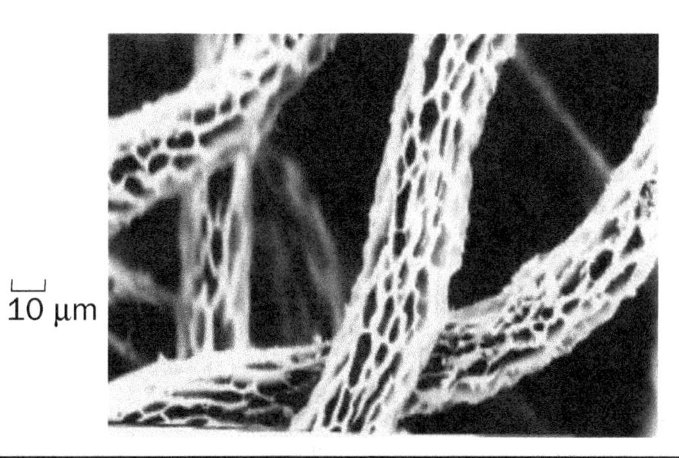

Figure 2.10 A SEM image of a partially exfoliated polymeric carbon fiber [15].

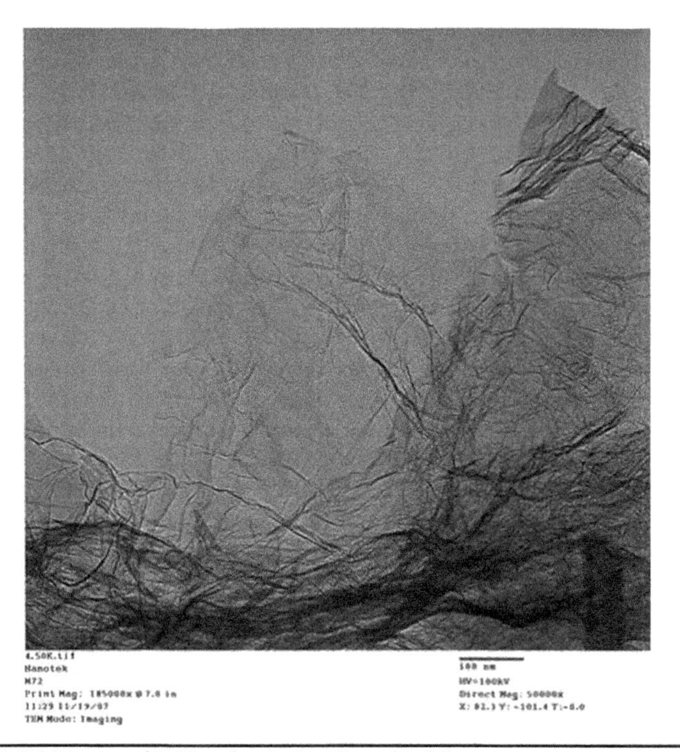

Figure 2.11 A TEM image of NGPs after ball milling of exfoliated graphite (which was obtained from sulfuric acid–nitric acid intercalation and thermal shock exposure at 1,050°C). Several graphene sheets collapsed to overlay one another after dispersion in acetone and then acetone removal during the TEM sample preparation procedure [15].

alcohol. Li et al. [24] followed a similar approach. Jang et al. [45] thermally exfoliated GIC to produce exfoliated graphite [15] and subjected exfoliated graphite to mechanical shear, such as ball milling, to obtain NGPs, which were mostly single- to five-layer structures [15]. The specific surface area of the samples prepared by this method was found to be in a range of 300–1,300 m^2/g, as measured by the BET method. Thermal exfoliation to produce nanostructured graphite was attempted by Petrik [46]. Drzal and Fukushima [47] conducted thermal exfoliation of intercalated graphite using microwaves as a heat source.

McAllister and colleagues [21–23] used thermal exfoliation of GO to obtain exfoliated graphite oxide platelets, which were found to contain a high production of single-layer graphene sheets, based on the BET method with nitrogen gas adsorption in the dry state and in an ethanol suspension with methylene blue dye as a probe. McAllister et al. [21] provided a good model to explain the exfoliation mechanisms and kinetics of GO.

Mack and colleagues [48, 49] developed a low-temperature process that involved intercalating graphite with potassium melt and contacting the resulting potassium-intercalated graphite with alcohol, producing violently exfoliated graphite containing many ultrathin NGPs. This process must be carefully conducted in a vacuum or an extremely dry glove box environment, since pure alkali metals, such as potassium and sodium, are extremely sensitive to moisture and pose an explosion danger. This process is not easily amenable to the mass production of nanoscale platelets. One major advantage of this process is that it can produce non-oxidized graphene sheets, since no acid/oxidizer intercalation or a high temperature is involved.

Geim and Novoselov [14] recently published an excellent progress article "The Rise of Graphene." Lu et al. [50] wrote a recent review "Graphene: Fundamentals and Functionalities." Mukhopadhyay and Gupta [51] edited a new book *Graphite, Graphene and Their Polymer Nanocomposite*, which is a great resource for this subject matter.

Potential Applications of NGPs and NGP Nanocomposites For scientific and engineering applications, anticipated features and benefits of NGP-based materials can be summarized as follows [15]:

- Nanographene exhibits many peculiar electronic, optical, magnetic, and chemical properties (others may not have yet been discovered) that enable many potential device applications [16, 17, 25–33].
- In addition to lower costs (compared to carbon nanotubes), another major advantage of graphene-based nanocomposites is their capability of forming a thin film, paper, or coating for electrostatic charge dissipation (ESD) and electromagnetic interference (EMI) applications. This occurs when the NGP loading exceeds the percolation threshold, so that NGPs form a network of electron transport path.
- Due to the ultrahigh thermal conductivity of NGPs (four times more thermally conductive, yet four times lower in density compared to copper), a nanocomposite thin film, paper, or coating can be used as a thermal management layer in a densely packed microelectronic device.
- High loading of NGPs (5–75 wt%) is incorporated into a polymer or carbon matrix, the resulting nanocomposite possesses an exceptionally high electrical conductivity for fuel cell bipolar plate applications [52, 53].
- NGP nanocomposites have a good combination of mechanical, stiffness, strength, micro-cracking resistance, electrical and thermal conductivities, and barrier performance at a minimal filler concentration. The mechanical properties of NGPs have been used in making golf balls [54] and micro-composite container for hydrogen storage [55].
- NGP composites (in the form of conductive paper/film/coating, structural adhesive, etc.) can be an integral part of lightning strike protection strategies for aircraft, telecommunication towers, and wind turbine blades.
- NGPs can be the component material for lithium ion battery electrodes. Self-assembled graphite oxide nanoplatelets and polyelectrolytes can be the cathode material that provides exceptionally high specific capacity [56].
- Ultrathin graphene films, being optically transparent and electrically conductive, are a potential alternative of the metal oxide winder electrodes for solid-state dye-sensitized solar cells [57].
- For supercapacitor electrode applications, NGP-based materials possess these desirable features: (a) the dimensions of platelets can be tailored to obtain NGPs with lower thickness about 0.34 nm and length (width) range of about 100 nm to 10 μm, yielding a specific area of 2,600 m^2/g; and (b) the surfaces of NGPs can be functionalized to achieve pseudo-capacitance–induced redox-like charge transfer.

Nanographene Platelet Suppliers There are numerous NGP suppliers worldwide. Several major suppliers are included in this section. Koo et al.'s study [58] describes the methodology used to characterize the as-received NGPs from several nanomaterial manufacturers. The microstructures of these NGPs were first examined using scanning electron microscopy at several different magnifications. The thermal stability of these NGPs was evaluated using thermogravimetric analysis with at least three different heating rates, such as 5, 10, 20, and 40°C/min in argon or nitrogen atmosphere. The residual mass and rate of mass loss of these NGPs were compared. The kinetics parameters, such as the activation energy and pre-exponential factor of these NGPs, were then calculated using an isoconversion technique. Eleven NGPs were characterized using the above methods. The similarities and differences of these NPGs were compared. Based on these facile techniques, one can rank the morphological and thermal characteristics of these NGPs. The methodology serves as a unique screening technique to select the appropriate NGP that can be chosen for whatever specific application is desired. It is summarized in Table 2.5 [59–68].

Layered Double Hydroxides (LDHs)

Origin The incorporation of polymers into layered double hydroxides (LDHs) to form polymer/LDH nanocomposites has been a subject of academic interest for more than 20 years according to Qiu and Qu's excellent review article on polymer/LDH nanocomposites [69]. These polymer/LDH nanocomposites often exhibit enhanced mechanical, thermal, optical, electrical, and flame retardancy properties. Among them, the thermal and flame retardancy properties are the most interesting and will be discussed in this section. Matusinovic and Wilkie [70] published another review using LDHs as a new class flame-retardant (FR) additive for polymers. These two review articles form the basis of the discussion of LDHs in this section. Layered double hydroxides are a different kind of layered crystalline filler for nanocomposites. LDHs combine the FR features of conventional metal hydroxide fillers (magnesium hydroxide and aluminum hydroxide) with those of layered silicate nanofillers (MMT). LDHs are considered to be a new emerging class of nanofillers for FR polymer nanocomposites. Generally, their chemical structure is represented by the formula $\left[M^{2+}{}_{1-x}M_x^{3+}\cdot(OH)_2\right]^{x+}\left[A^{m-}\right]_{x/m}^{x-}$ $2H_2O$, where M^{2+} is a divalent metal ion (such as Mg^{2+} or Zn^{2+}), M^{3+} is a trivalent metal ion (such as Al^{3+} or Cr^{3+}), and A^{m-} is an anion with valency m (such as COC_3^{2-}, Cl^-, or NO_3^-), and the value of x is equal to molar ratio of $M^{2+}/(M^{2+} + M^{3+})$ and is general in the range of 0.2 to 0.33 [69]. The typical structure of LDH materials is presented in Fig. 2.12 [69]. The structure of LDHs is based on that of magnesium hydroxide, brucite, in which Mg^{2+} ions are arranged in sheets, each magnesium ion being octahedrally surrounded by six hydroxide groups, whereas each hydroxide spans three magnesium ions. Isomorphous substitution for some fraction of the divalent ions of trivalent ions of comparable size (e.g., Al^{3+}, Fe^{3+}) forms mixed metal layers—that is, $[M_{1-x}^{2+}M_x^{3+}\cdot(OH)_2]^{x+}$—with a net positive charge. Electrically neutrality is maintained by the anion located in interlayer domains containing water molecules. These water molecules are connected to both the metal hydroxide layers and the interlay anions through extensive hydrogen bonding. The presence of anion and water molecules leads to enlargement of the basal spacing from 0.48 nm in brucite to about 0.78 nm in Mg-Al LDH. Because the interlayer ions are confined to the interlayer space by a relatively weak electrostatic force, they can be removed without destroying the layered structures of LDH. This anionic exchange

Table 2.5 Summary of Eleven Types of NGPs [58]

Nanographene platelets		Source	Remarks
NGP-1	Grafmax HC 11-1Q	Nacional de Grafite, Sao Paulo, Brazil (www.grafite.com)	Provided by Prof. Antonio F. Avila, Universidade Federal de Minas Gerais, who used this product for his hybrid nanocomposite studies [59, 60], which showed thermal stability improvement for his epoxy system.
NGP-2	TG679	Graftech International, Parma, OH (www.graftech.com)	Dr. S. G. Miller used a functionalized TG679 in her thesis studies in epoxy resin [61], which showed mixed results in mechanical properties.
NGP-3	Russian NGP	Dr. Howard K. Schmidt, Chemical and Biomolecular Engineering Department, Rice University, Houston, TX (left Rice University)	Dr. Schmidt obtained large quantities of NGP-3 from a Russian source; no technical information is available on this material [62].
NGP-4	XGnP™	XG Sciences, East Lansing, MI (www.xgsciences.com)	This is an earlier version of XGnP™ [63].
NGP-5	N006-010-P	Angstron Materials, Dayton, OH (www.angstronmaterials.com)	Thickness (*Z*) is about 10–20 nm, average *X* and Y is 14 μm, and specific surface area is 110 m^2/g [64].
NGP-6	0541DX	Skyspring Nanomaterials, Houston, TX (www.ssnano.com)	Thickness is 6–8 nm, average diameter is 15 μm, and specific surface area is 120–150 m^2/g [65].
NGP-7	Grade 2	Cheap Tubes, Brattleboro, VT (www.cheaptubesinc.com)	Thickness is <10 nm, average diameter is 15 μm, and specific surface area is 100 m^2/g [66].
NGP-8	Grade M	XG Sciences, East Lansing, MI (www.xgsciences.com)	Thickness is 6–8 nm, average diameter is 15 μm, and specific surface area is 120–150 m^2/g [63].
NGP-9	MEGD	Prof. Rod Ruoff, Department of Mechanical Engineering, The University of Texas at Austin, Austin, TX [recently moved to Ulsan National Institute of Science and Technology (UNIST), Ulsan, S. Korea]	Abundant technical information can be found on Prof. Ruoff's website at UNIST, Ulsan, S. Korea [67].
NGP-10	Grafmax HC 11	Nacional de Grafite, Sao Paulo, Brazil (www.grafite.com)	Nacional de Grafite provided this grade to us and has been supplying NGP to Prof. Avila's research [68]. D_{32} = 6.99 μm.
NGP-11	Grafmax HC 30	Nacional de Grafite, Sao Paulo, Brazil (www.grafite.com)	Nacional de Grafite provided this grade to us and has been supplying NGP to Prof. Avila's research [68]. D_{32} = 18.15 μm.

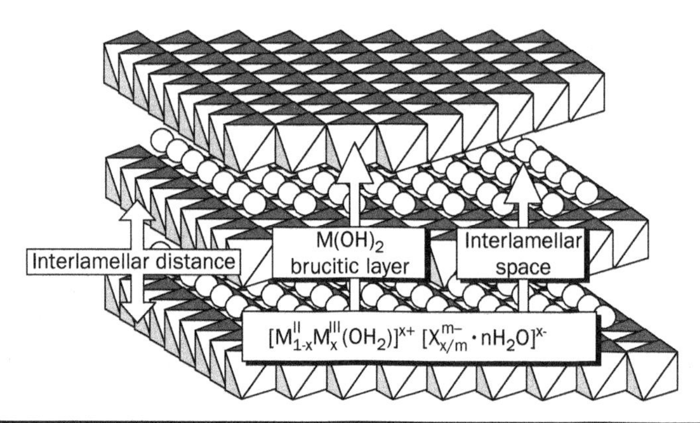

Figure 2.12 Layered crystal structure of layered double hydroxide [69]. (*Courtesy of ACS.*)

capacity is a key feature of LDHs, which make them unique as far as inorganic materials are concerned.

An interesting property of LDHs is that the identity of both the divalent and trivalent metal ions and their ratio, as well as the identity of the interlayer anion, are adjustable resulting in varying charge densities and interlayer spacing. This makes a great variety of possible LDHs. The metals include lithium, magnesium, zinc, copper, cobalt, calcium, iron, aluminum, and a host of others. A wide variety of anions may also be used, including both inorganic and organic anions. There is evidence that the identity of both the metals and the anions is important to obtain good dispersion. The ratios of M^{2+} to M^{3+} may range from 2 to a maximum of 5 or 6, but the most common stoichiometries are Mg_2Al and Mg_3Al.

Properties and Applications Due to their unique layered structure and highly tunable chemical composition based on different metal species and interlayer anions, LDHs have many interesting properties, such as unique anion-exchange ability, easy synthesis, high bond water content, *memory effect*, nontoxicity, and biocompatibility. Based on these properties, LDHs are considered very important layered crystals with potential applications in catalysis [71], controlled drugs release [72], gene therapy [73], improvement of heat stability and flame retardancy of PNCs [69, 70], controlled release or adsorption of pesticides [74], and preparation of novel hybrid materials for specific applications, such as visible luminescence [75], UV/photo stabilization [76], magnetic nanoparticle synthesis [77], or wastewater treatment [78].

High charge density, small intergallery distance, and strong hydrophilicity prevent good dispersion of LDH at the nanometer level within the polymer matrix. Monomer and polymer molecules cannot easily penetrate between the LDH layers, nor can the layers be easily and homogeneously dispersed in a hydrophobic polymer matrix. By modifying LDH with organic anions, the intergallery distance is increased and the layers become more organophillic, so monomer or polymer molecules can more easily penetrate between the layers, making LDHs suitable for use as nanofillers. Even after organic modification, the charge density is still high, and dispersion can be difficult. The identity of the metals and the anions controls dispersion. MMT has been studied much more than the LDHs and it is clear that the fire retardancy due to MMT is quite different than that from LDHs. For MMT, there is no reduction in the peak heat release rate (PHRR) unless the clay is well dispersed [70]. For an LDH-containing nanocomposite, there can be reduction even if the LDH is poorly dispersed. Also, the amount of MMT that must be used is around 3 wt%, while a much larger amount, about 10 wt%, is typically required for an LDH to lower the PHRR.

2.2.2 Two Nanoscale Dimensions in the Form of Fibers

Carbon Nanofibers (CNFs): Vapor-Grown Carbon Fibers (VGCFs)

Origin Carbon nanofibers (CNFs) are a form of VGCF which is a discontinuous graphitic filament produced in the gas phase from the pyrolysis of hydrocarbons [79–84]. With regard to properties of physical size, performance improvement, and product cost, CNFs complete a continuum bounded by carbon black, fullerenes, and single-wall to multiwalled carbon nanotubes on one end and continuous carbon fiber on the other [80], as illustrated in Fig. 2.13. Carbon nanofibers are able to combine many of the advantages of these other forms of carbon for reinforcement in commodity and high-performance engineering polymers. Carbon nanofibers have transport and mechanical properties that approach the theoretical values of single-crystal graphite, similar to the fullerenes, but they can be made in high volumes at low cost—ultimately lower than that of conventional carbon fibers. In equivalent production volumes, CNFs are projected to have a cost comparable to E-glass on a per-pound basis, yet they possess properties that far exceed those of glass and are equal to, or exceed, those of much more costly commercial carbon fiber. Maruyama and Alam published an excellent review of carbon nanotubes and nanofibers in composite materials [82].

Manufacturing Carbon nanofibers are manufactured by Applied Sciences Inc./ Pyrograf® Products (ASI) (Cedarville, Ohio; website: www.apsci.com), by pyrolytic decomposition of methane in the presence of iron-based catalyst particles at

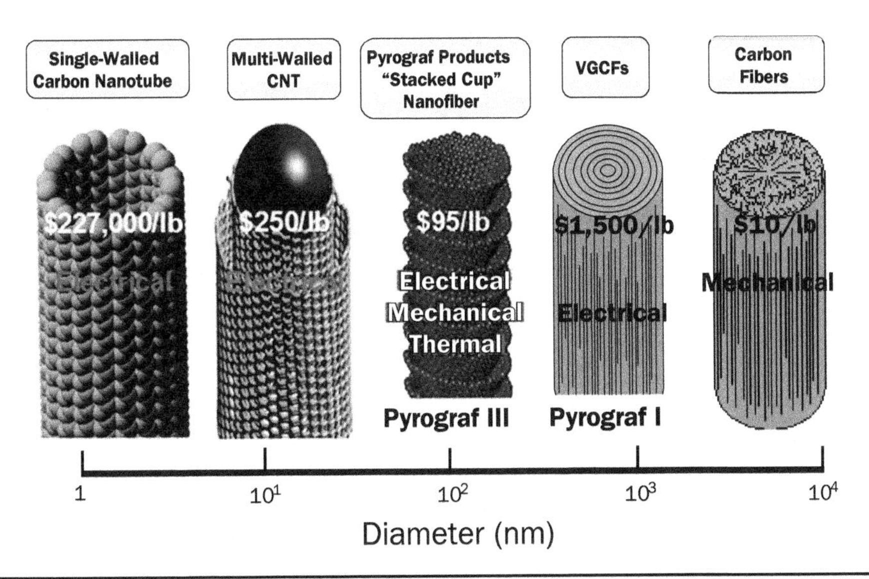

FIGURE 2.13 Sizes and costs from single-walled carbon nanotubes to conventional carbon fibers. (*Courtesy of ASI.*)

temperatures above 900°C. Typically, about 100 mg of powdered catalyst is placed in a ceramic boat, which is positioned in a quartz tube located in a horizontal tube furnace. The catalyst is reduced in a dilute hydrogen-helium stream at 600°C and quickly brought to the desired reaction temperature. Following this step, a mixture of hydrocarbon, hydrogen, and inert gas is introduced into the system, and the reaction is allowed to proceed for about 2 hours. This approach produces about 20 g of carbon fibers from the more active catalyst systems.

Properties Pyrograf-I, an early product is 10 μm in diameter, was the first generation of the Applied Sciences product. Pyrograf-III is a patented, very fine, highly graphitic CNF. Pyrograf-III is available in diameters ranging from 50 to 200 nm and lengths of 50–100 μm. Therefore, CNFs are much smaller than conventional continuous or milled carbon fibers (5–10 μm) but significantly larger than carbon nanotubes (1–10 nm). Compared to PAN and pitch-based carbon fiber, the morphology of CNFs is unique in that there are far fewer impurities in the filament, providing for graphitic and turbostatic graphite structures, and the graphene planes are more preferentially oriented around the axis of the fiber. Consequences of the circumferential orientation of high-purity graphene planes are a lack of cross-linking between the graphene layers and a relative lack of active sites on the fiber surface, making CNFs more resistant to oxidation and less reactive for bonding to matrix materials. Also, in contrast to carbon fiber derived from PAN or pitch precursors, CNFs are produced only in a discontinuous form, where the length of the fiber can be varied from about 100 μm to several centimeters, and the diameter is on the order of 100 nm. As a result, CNFs possess an aspect ratio of about 1,000.

Carbon nanofibers exhibit exceptional mechanical and transport properties, thus demonstrating their excellent potential as an abstract component for engineering materials. Table 2.6 lists the properties of VGCFs, both as grown and after a graphitizing heat treatment to 3,000°C. Note that, owing to the difficulty of direct measurements on the nanofibers, the values in Table 2.7 are measured on vapor-grown fibers that have

Table 2.6 Properties of CNF

Properties (units)	As grown	Heat treated
Tensile strength (GPa)	2.7	7.0
Tensile modulus (GPa)	400	600
Ultimate strain (%)	1.5	0.5
Density (g/cc)	1.8	2.1
Electrical resistivity (μΩ-cm)	1,000	55
Thermal conductivity (W/m-K)	20	1,950

Table 2.7 Thermoset Polyester/Pyrograf-III Composite Properties [81]

Fiber content (wt%)	Tensile strength (MPa)	Tensile modulus (GPa)	Electrical resistivity (ohm-cm)
17% PR-19	51.5	4.55	3.2
17% PR-19, OX	47.4	4.55	7.1
5% PR-19 in 10% ¼″ glass	44.1	11.52	5.0
5% PR-19, OX in 10% ¼″ glass	33.8	8.92	7.0

been thickened to several microns in diameter. Such fibers consist almost exclusively of chemical vapor deposition (CVD) carbon, which is less graphitic and more defective than the catalytically grown carbon core that constitutes the CNF. Thus, the properties listed in the table represent an estimate for the properties of CNF.

One of the goals for the broad utility of CNFs is to provide mechanical reinforcement comparable to that achieved with continuous tow carbon fiber at a price that approaches that of glass fiber reinforcement, and with low-cost composite fabrication methods, such as injection molding. Theoretical models [81] suggest that reinforcement by discontinuous fibers such as CNFs can closely approach that of continuous fibers, as long as the aspect ratio of the fibers is high, and the alignment is good. Work is ongoing to improve the mechanical benefits of CNFs through fiber surface modification to provide physical or chemical bonding to the matrix. Such modifications have resulted in strength and modulus improvements of four to six times the values of neat resin; however, these values are still a modest fraction of what may be anticipated from idealized fiber-matrix interphase and alignment of the fibers within the matrix. The more immediate opportunities for use in structural composites lie in the prospect for modifying the properties of the matrix material. For example, use of small volume loadings of CNFs in epoxy may allow for improvement of interlaminar shear strength of PAN or pitch-based composites. The CNF additives to fiberglass composites could provide benefits to a suite of properties, including thermal and electrical conductivity, coefficient of thermal expansion, and mechanical properties, as suggested by the data in Table 2.4. Figure 2.14 shows the TEM micrograph of an individual CNF with the hollow core in two distinct regions: catalytic and deposited.

Carbon nanofiber nanocomposites can offer multifunctional performance for several potential aerospace applications, including:

- EMI shielding, electrostatic painting, and antistatic
- Thermal conductivity of spacecraft, batteries, and electronics

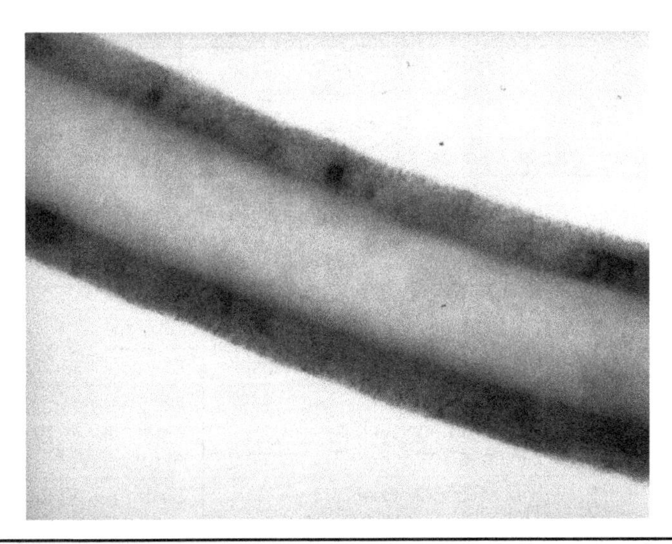

Figure 2.14 A TEM micrograph showing carbon nanofibers have hollow core and two distinct regions—catalytic and deposited. (*Courtesy of ASI.*)

- Improved mechanical properties in polymers (composite structure, injection-molded parts, tires)
- Reduced coefficient of thermal expansion (CTE) for spacecraft resin systems

Applied Sciences, Inc. has suggested that CNFs could be used for the following applications: (a) improving strength and modulus (Table 2.8); (b) lowering resistance (Table 2.9); (c) CTE control (Table 2.10); and (d) strain-compliant conductors (Table 2.11).

Table 2.8 CNF for Structural Applications for Better Strength and Modulus

Generic area	Military	Civilian
Synthetic rubber	Tank tracks, aircraft tires, ship structures	Automotive tires, aircraft tires, ship structures, sporting goods
Thermoplastic structures	Aircraft, satellite structures, ship structures, optical components	Auto body structures, sporting goods, optical components
Thermoset structures	Aircraft, satellite structures, ship structures, optical components	Auto body structures, sporting goods, optical components
Carbon/epoxy structures	Airframes, especially Z-direction enhancement	Sporting goods, automotives, aerospace structures

Table 2.9 CNF for Applications for Lower Resistance

Generic area	Military	Civilian
Static dissipation	Satellite charge control, fuel lines, hoses, tubing, ordnance	Fuel hoses and tubing for automotives, electronics assembly
Electrostatic paint spray	Aircraft, ground vehicles, ships, other structures	Auto body components, aircraft, other vehicles
Electromagnetic interference mitigation	Plane crash reduction, avionics, RF, secure facilities	High-speed computing, communications
Lightning strike mitigation	Aircraft (especially composite airframe), ships, radomes, ordnance bunkers	Aircraft, ground structures
Compliant contacts	Solar cells, electronics SEM	Electronics

Table 2.10 CNF for Applications for CTE Control

Generic area	Military	Civilian
Optics	Low-cost injection moldable mirrors, laser	Low-cost injection moldable optics
Structures	Satellite structures especially where stability in orbit is important, aircraft, ground vehicles, ships	Auto body components (match aluminum or steel) and for aerospace and ships
Electronics	MCT devices, electronic boards, high-power electronics	Electronic boards, computers, cooling components

Table 2.11 CNF for Applications for Strain Compliant Conductors

Generic area	Military	Civilian
Compliant contacts	Solar cell contacts for spacecraft, electronic boards	Electronic boards, solar cell contacts, high-temperature lead free solder alternative
Thermal control devices		Temperature actuators, temperature measurement devices

Glasgow et al. [83] demonstrated that the achievement of significant mechanical reinforcement in CNF composites requires high fiber loadings and is somewhat dependent on generating an appropriate interphase between the CNF and the matrix. Novel surface treatments under development have yielded good improvements in the tensile modulus and strength of CNF-reinforced polypropylene. Adding surface functional groups, particularly oxygen groups, has also demonstrated benefits for interphase development. Carboxyl and phenolic groups contributing to a total surface oxygen concentration in the range of 5–20 atom percent have been added to CNF used to fabricate epoxy polymer matrix composites, providing improved flexural strength and modulus. The effect of similarly functionalized CNF in bismaleimide (BMI) polymer matrix composites also shows promise. Data for propylene, epoxy, and BMI/CNF reinforced composites indicate that higher fiber volume loadings will find a role in structural composite markets as price and availability improve.

Recently, Tibbetts et al. [84] did a comprehensive review on the fabrication and properties of VGCNF/polymer composites. This review summarized the wide variety of composite properties achieved with VGCNF composites. When compounded with thermoplastics or thermosets, VGCNF can more than triple the polymer's tensile strength and modulus. Compressive strength is generally improved by an even large margin. Preliminary research provides some indication that a practical method may be found to improve the orientation of VGCNF to achieve even greater improvements. The most beneficial application of VGCNF is to add conductivity to polymer matrices. Resistivities of below 0.15 Ω cm may be achieved with a fiber loading of about 15 wt%, and a percolation threshold of below 1 wt% is possible. Furthermore, a tenfold improvement in thermal conductivity has been demonstrated in epoxy composites.

CNF suppliers Carbon nanofibers are manufactured by Applied Sciences Inc./Pyrograf Products (ASI), which is located in Cedarville, Ohio (website: www.apsci.com).

Carbon Nanotubes (CNTs)

Origin There are excellent reviews and publications on the synthesis and the physical properties of CNTs [85–89]. In this section, only a brief introduction is provided regarding the manufacturing, properties, and classification of carbon nanotubes. CNTs have attracted much attention in the last decade because of their unique potential uses for structural, electrical, and mechanical purposes [90]. Carbon nanotubes have high Young's modulus and tensile strength, and they can be metallic, semiconducting, or semimetallic, depending on the helicity and diameter [90]; some typical material properties of CNTs, CNFs, and NGPs are compared in Table 2.4.

Manufacturing Carbon nanotubes can be prepared by arc evaporation [91], laser ablation [92], pyrolysis [93], plasma enhanced chemical vapor deposition (PECVD) [94], and electrochemical methods [95, 96]. Carbon nanotubes were first synthesized in 1991 by Sumio Iijima on the carbon cathode by arc discharge [97]. However, the experimental discovery of single-walled carbon nanotubes occurred in 1993 [98, 99]. In 1996, a much more efficient synthesis route was developed, involving laser vaporization of graphite to prepare arrays on the order of single-walled carbon nanotubes [100]. The process offered major new opportunities for quantitative experimental studies of carbon nanotubes.

Properties Properties of carbon nanotubes have been studied extensively. Carbon nanotubes are excellent candidates for stiff and robust structures, because the carbon-carbon bond in the graphite is one of the strongest in nature. The TEM data revealed that carbon nanotubes are flexible and do not break upon bending [101]. The thermal conductivity of carbon nanotubes can be extremely high, and the thermal conductivity of individual carbon nanotubes was found to be much higher than that of graphite and bulk nanotubes [102]. The mechanical and transport properties of CNTs are compared to other conductive materials in Table 2.4. Overall, carbon nanotubes show a unique combination of stiffness, strength, and tenacity compared to other fibers which usually lack one or more of these properties (Table 2.4). Thermal and electrical conductivities are also very high and are comparable to other conductive materials (Table 2.4). Carbon nanotubes have a wide spectrum of potential applications. Examples include the use in catalysis, storage of hydrogen and other gases, biological cell electrodes, quantum resistors, nanoscale electronic and mechanical devices, electron field emission tips, scanning probe tips, flow sensors, and nanocomposites [103].

Classifications Carbon nanotubes have many structures, different in length types of helicity, and the number of walls or layers. Carbon nanotubes can be categorized by their structures as single-walled nanotubes (SWNTs), multiwalled nanotubes (MWNTs), and the newly established double-walled nanotubes (DWNTs), based on the number of walls present in the carbon nanotubes, as illustrated in Fig. 2.15. By definition, SWNTs are single-walled carbon nanotubes about 1 nm in diameter with micrometer-scale lengths;

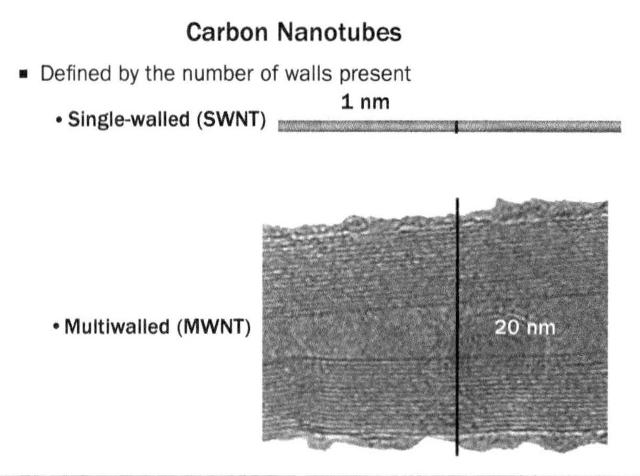

Figure 2.15 Definition of single- and multi-walled carbon nanotubes. (*Courtesy of CNI.*)

MWNTs are multiwalled carbon nanotubes with an inner diameter of about 2–10 nm, an outer diameter of 20–70 nm, and a length of about 50 μm; and DWNTs are double-walled carbon nanotubes with outer diameters of less than 2 nm and have lengths from several hundred nanometers to several microns.

Single-Walled Carbon Nanotubes Single-walled carbon nanotubes (SWNTs) are tubes of graphite that are normally capped at the ends. They have a single cylindrical wall. The structure of SWNTs can be visualized as a layer of graphite, a single atom thick, called graphene, which is rolled into a seamless cylinder. Most SWNTs have a diameter close to 1 nm; the tube length can be many thousands of times that. Single-walled carbon nanotubes are more pliable yet harder to make than multiwalled carbon nanotubes are. They can be twisted and bent into small circles or around sharp-angle bends without breaking.

Buckytubes are tubular fullerenes, polymers that are part of the fullerene family of carbon molecules discovered by Dr. Richard E. Smalley and colleagues in 1985. Buckytubes comprise single-walled carbon nanotubes and nested (endohedral or endotopic) SWNTs (i.e., one, two, or more tubular fullerenes nested inside another tubular fullerene, as depicted in Fig. 2.16.

Buckytubes are carbon nanotubes that are graphene (layers of graphite) rolled up into seamless tubes. Graphene consists in a hexagonal structure similar to chicken wire. Rolling up graphene into seamless tubes can be accomplished in various ways. For example, carbon-carbon bonds (like the wires in chicken wire) can be parallel or perpendicular to the tube axis, resulting in a tube where the hexagons circle the tube like a belt, but oriented differently. Alternatively, the carbon-carbon bond need not be either parallel or perpendicular, in which case the hexagons will spiral around the tube with a pitch, depending on how the tube is wrapped. Buckytubes have extraordinary electrical conductivity, thermal conductivity, and mechanical properties. They are the best electron field emitter. As polymers of pure carbon, they can be reacted and manipulated using the chemistry of carbon. This characteristic provides the opportunity to modify the structure and to optimize solubility and dispersion. Buckytubes are molecularly

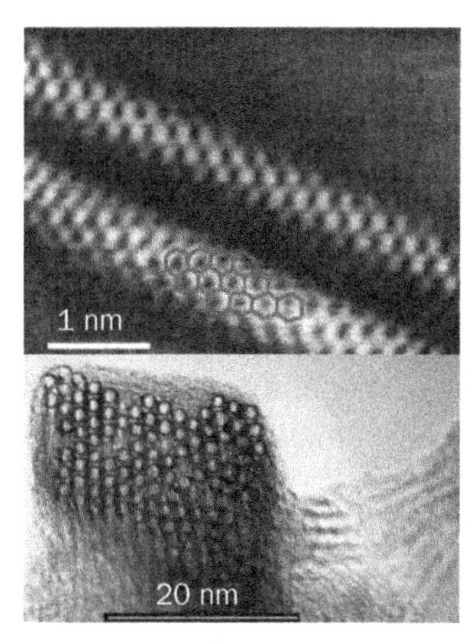

- SWNTs are perfect: each atom in its place

Ropes of SWNTs

- Caused by strong van der Waals forces between SWNT surfaces
- Enable self-assembly
- Make dispersion challenging

FIGURE 2.16 TEM micrographs showing SWNTs and ropes of SWNTs. (*Courtesy of CNI.*)

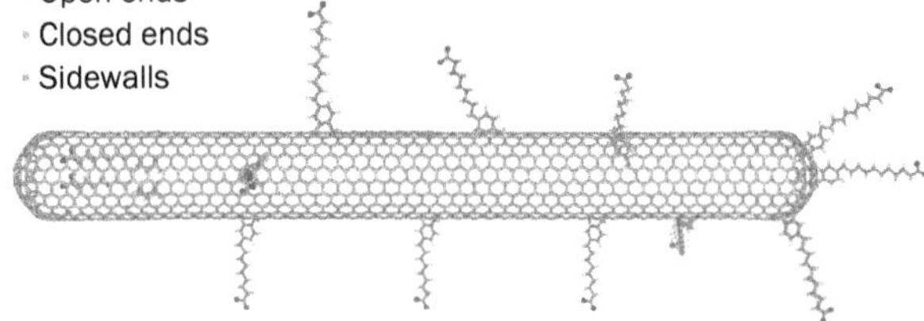

Figure 2.17 SWNTs can be customized and controlled using organic chemistry. (*Courtesy of CNI.*)

perfect, which means that they are free of property-degrading flaws in the nanotube structure. Their material properties can therefore approach closely the very high levels intrinsic to them. Buckytubes have extraordinary properties: their strength is about 100 times that of steel, their electrical conductivity is close to that of copper, their thermal conductivity is about three times their diameter, and they have tremendous accessible surface area. These extraordinary characteristics give buckytubes potential in numerous applications. In most applications, raw buckytubes will need to be customized. The customization can be precisely controlled using organic chemistry in open ends, closed ends, and sidewalls, as illustrated in Fig. 2.17.

Single-walled carbon nanotubes offer incredible opportunities in electrical properties, mechanical properties, thermal properties, and field emission, as follows:

- *Electrical properties:* Electrically conductive composites for electrostatic dissipation, shield, and conductive sealants; energy storage for super capacitors and fuel cells; electronic materials and devices for conductive inks and adhesives; electronic packaging; device and microcircuit components
- *Mechanical properties:* High-performance composites; coatings for wear-resistance and low friction; high-performance fibers; reinforced ceramic composites
- *Thermal properties:* Thermally conductive polymer composites; thermally conductive paints and coatings
- *Field emission:* Flat-panel displays; electron device cathodes and lighting

Multiwalled Carbon Nanotubes Multiwalled carbon nanotubes (MWNTs) can appear either in the form of a coaxial assembly of single-walled carbon nanotube similar to a coaxial cable or as a single sheet of graphite rolled into the shape of a scroll. MWNTs have an interior diameter of 2–10 nm, and exterior diameter of 20–75 nm, and a length of 50 μm [104]. They are produced by CVD synthesis of xylene-ferrocene at a temperature of 725°C with high purity greater than 95 percent. MWNTs are easier to produce in high volume than SWNTs. MWNTs are nanoscale carbon fibers with a high degree of graphitization. Technically they are neither fullerenes nor molecular. The structure of MWNTs is less well understood because of its greater complexity and variety. Regions of structural imperfection may diminish its desirable material properties. The challenge in producing SWNT on a large scale as compared to MWNTs is reflected in the

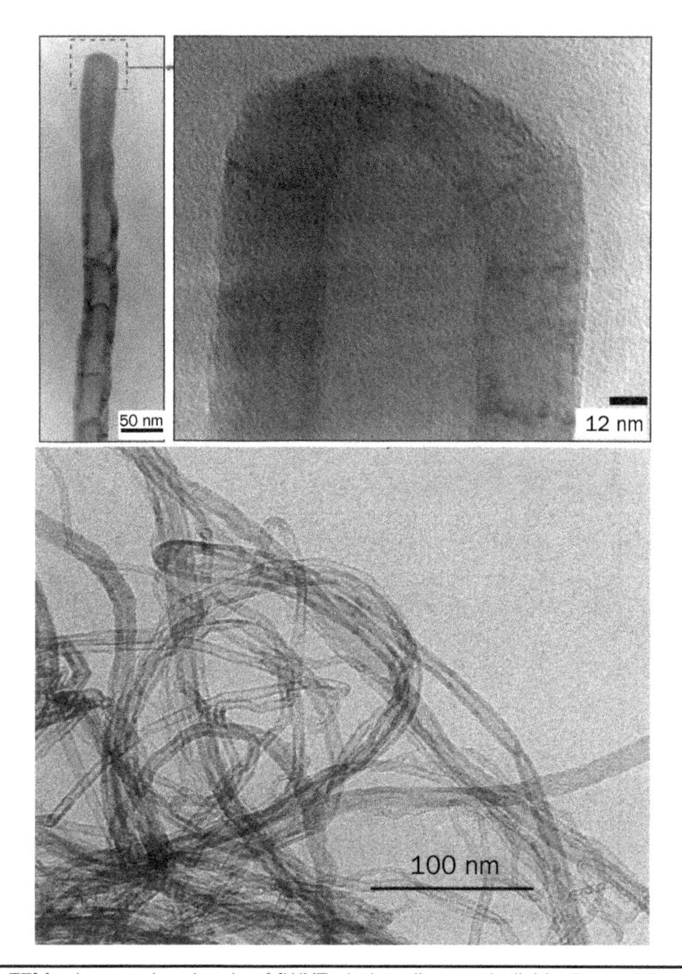

Figure 2.18 TEM micrographs showing MWNTs in bundles, as individual nanotube, and in high magnification. (*Courtesy of CNI.*)

price of SWNTs, which currently remains significantly higher than MWNTs. They are very good materials, with vastly superior properties over graphite and carbon black (Fig. 2.18) [104]. Single-walled carbon nanotubes have a performance of up to 10 times better, and are outstanding for every specific application.

Double-Walled Carbon Nanotubes Double-walled carbon nanotubes (DWNTs) are an important sub-segment of MWNTs. DWNTs combine similar morphology and other material properties of SWNTs, while significantly improve their resistance to chemicals. This property is especially important when functionality is required to add new properties to DWNTs. Since DWNTs are a synthetic blend of single-walled and MWNTs, they exhibit the electrical and thermal stability of MWNTs and the flexibility of the SWNTs. Since SWNTs are developed for specific applications, they are more susceptible to breakage if they have been functionalized. When structural imperfections are created, they can modify their overall mechanical and electrical properties. With DWNTs, only the outer wall is modified, thereby preserving their intrinsic properties. Research has shown that DWNTs have better thermal and chemical stability than SWNTs. As a result, a DWNT can be used for gas sensors and dielectrics, and to demanding applications, such as field emission displays, PNCs, and nanosensors.

Carbon Nanotubes Suppliers There are numerous carbon nanotubes suppliers worldwide and a few major suppliers are included in this section.

Nanocyl SA is a leading global manufacturer of specialty and industrial carbon nanotubes technologies, located in Sambreville, Belgium (website: www.nanocyl.com) [105]. It was founded as a spin-off from the Universities of Namur and Liege by Professor Nagy and Professor Pirard with the support of private investors in 2002. Nanocyl uses the "catalytic chemical vapor deposition (CCVD)" method for large-scale production of carbon nanotubes. Nanocyl focuses its technology and knowledge on the practical application of carbon nanotubes into new and existing materials, resulting in the improved performance of polymers, metals, composites, and biomaterials. Currently, Nanocyl produces SWNTs, MWNTs, and DWNTs, as well as masterbatches of PNCs. Nanocyl manufactures products, such as PLASTICYL™, EPOCYL™, AQUACYL™, THERMOCYL™, BIOTYL™, PREGCYL™, and STATICYL™ that facilitate the integration of Nanocyl™ NC7000 in its customers' applications. The company also offers other high-quality carbon nanotube research grades and technologies.

Nanocyl NC7000 series comprise thin MWNTs manufactured via CCVD process. They are available in powder form, and each nanotube has an average diameter of about 9.5 nm and average length of 1.5 μm, 90% carbon purity, and 10% metal oxide. This MWNT product has also been predispersed in other products, such as masterbatches (PLASTICYL), epoxy resins (EPOCYL), silicones (THERMOCYL, BIOTYL, and STATICYL), and aqueous dispersions (AQUACYL).

Bayer MaterialScience was marketing the nano-sized materials under the trade name Baytubes® since 2007 [106]. Bayer terminated Baytubes project in the middle of 2013 and seeks to divest its know-how in the area. Baytubes C 150P is currently supplied by Covestro. Bayer MaterialScience set up a second pilot production plant with an annual capacity of another 30 tons expanding the capacity to 60 tons in 2007, which made Bayer MaterialScience one of the top three carbon nanotubes producers worldwide. Baytubes differ from other natural or engineered nanotubes in many ways. They are of high purity (more than 95% standard grade; more than 99% can be achieved in high purity variations) and show no other residue than from the catalyst and catalyst carrier. In particular, TEM analysis indicates the absence of carbon soot in the product. They show a high bulk density consistent with the large agglomerate structure seen already with the naked eye, i.e., without using electron microscopy. The entangled nature of the Baytubes can only be observed with electron microscopy. The size of the agglomerates (more than 100 μm) in Baytubes facilitates the handling at the workplace by reducing the potential of exposure via inhalation.

Baytubes are extremely resilient. They might only have a quarter of the mass of steel, but they are five times stronger when subjected to mechanical loads. They also conduct electric current as effectively as copper does. The tiny tubes can be used to produce extremely robust plastics that are constructed in the same way as reinforced concrete is. Baytubes are already used in epoxy systems, thermoplastic systems, and coating systems. Final products are typically high-tech sport equipment, injection molded and extruded parts, as well as conductive coatings. An excellent example is the current use

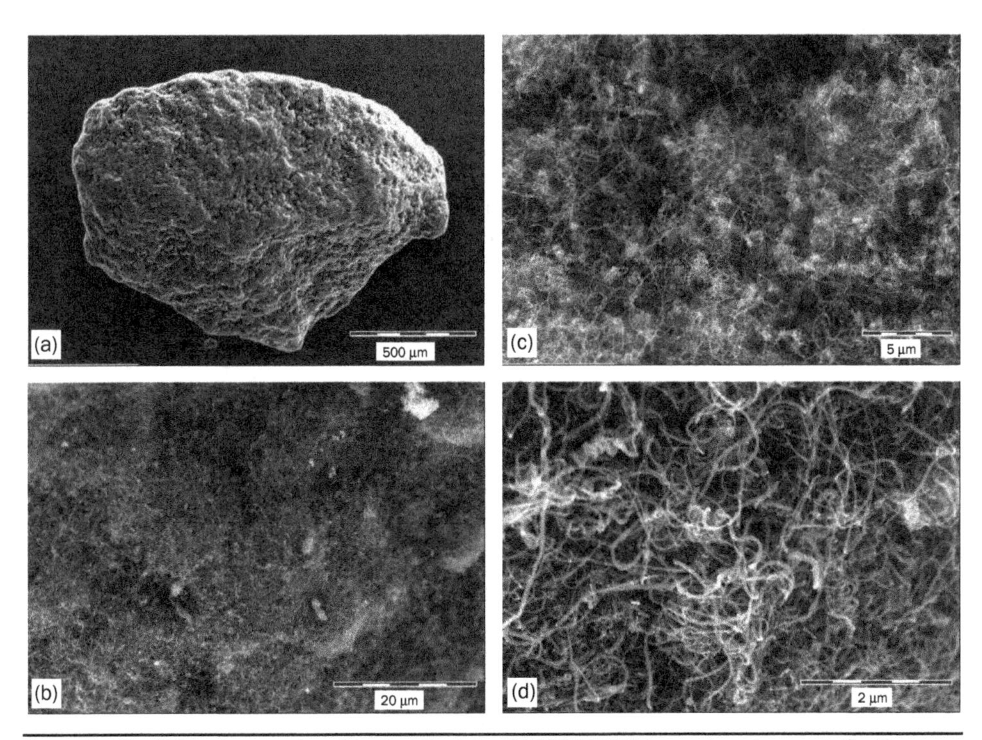

FIGURE 2.19 TEM micrographs of typical Baytube C150P MWNT in progressive magnifications from (a) to (d). (*Courtesy of Bayer MaterialScience.*)

of Baytubes in skis, ski poles, hockey sticks, baseball bats, and surfboards. Baytubes help improve the mechanical strength of some of those articles as well as reducing weight of some others. Baytubes can be used to make rotor blades for wind turbine plants lighter, longer, and thus more efficient. In the automotive industry, the use of Baytubes in plastics can avoid the need for expensive production steps. These examples will open the door to more demanding applications in energy and aircraft industry.

Baytubes are agglomerates of MWNTs with low outer diameter, narrow diameter distribution, and an ultrahigh aspect ratio (length-to-diameter ratio). Baytubes show excellent tensile strength and E-modulus, as well as exceptional thermal and electrical conductivity. Baytubes are produced in a high-yield catalytic process based on chemical vapor deposition. The process yields easy-to-handle agglomerates with high bulk density. The optimized process results in a high degree of purity (low concentration of residual catalyst and absence of free amorphous carbon). Baytubes C 150P has carbon purity greater than 95 wt%, an outer mean diameter of about 13 nm, an inner mean diameter of about 4 nm, and length is more than 1 µm. It has a bulk density of 130–150 kg/m^3. Figure 2.19a–d shows some representative SEM micrographs of Baytube C150P MWNT in progressive magnifications [106].

Arkema launched a research project in 2003 to study CNTs and their applications. In 2006, Arkema started up at its Lacq facility the first pilot laboratory capable of producing some 20 ton/year of CNTs, sampled under the trademark Graphistrength® (website: www.graphistrength.com) [107].

Arkema has developed in its laboratories and in collaboration with universities a continuous process for the synthesis of multiwalled carbon nanotubes for the large-scale manufacture of products with good quality consistency and perfect reproducibility. Arkema uses a continuous synthesis process based on CCVD in a fluidized bed-type reactor. Nanotubes grow on iron particles. This produces a black powder with characteristics that are perfectly controlled and stable over time.

Graphistrength C100 MWNT is a black powder with an apparent density of 50–150 kg/m^3, mean agglomerate size of 200–500 µm, C content greater than 90 wt%, mean number of walls of 5–15, an outer mean diameter of 10–15 nm, and length about 0.1–10 µm. Arkema particularly developed a range of innovative masterbatches that are easy to process within various thermoplastic, elastomer, and thermoset matrices. These high-tech masterbatches help optimize the application properties of end-products.

Arkema is currently testing on a pilot scale the use of agricultural ethanol as a source of carbon for the production of its Graphistrength MWNTs. Future production will be based on this renewable raw material.

Halloysite Nanotubes (HNTs)

Origin Halloysite nanotubes (HNTs), types of naturally occurring clay mineral with nanotubular structures, have attracted a lot of attention recently. Du et al. [108] did an excellent review on newly emerging applications of halloysite nanotubes. They stated that due to their unique characteristics, such as nanosized lumens, high L/D ratio, and low hydroxyl group density on the surface of HNTs, exciting applications have been developed for these cheap and abundantly deposited clays. HNTs have been widely used in controlled or sustained release, in nanoreactors or nanotemplates, and for the elimination of contaminants or pollutants. HNT-polymer nanocomposites exhibit improved properties, such as superior mechanical, flame retardancy, thermal stability, and reduced CTE properties. Liu et al. [109] provided an overview of the recent advances of HNT-polymer nanocomposite research, paying attention to interfacial interactions of the HNT-polymer nanocomposites. They summarized the characteristics of HNT-polymer nanocomposites and predicted the development of the potential applications in high-performance composites for aircraft and automobile industries, environmental protection, and biomaterials. Yuan et al. [110] reviewed the recent research advances on the properties and applications of HNTs, paying attention to the structures and morphology of HNTs and their changes, the formation of tubular structures, the physicochemical properties, and surface chemical modifications, and the HNT-based advanced materials, and related applications. Zhang et al. [111] reviewed applications and interfaces of HNT-polymer nanocomposites. They reviewed the recent development in functionalized HNTs with metal nanoparticles and the diverse applications of the resulting materials. The interfacial characteristics between HNTs and nanoparticles are emphasized by combing HNTs' advantages, such as nature, biocompatibility, and nontoxicity, which allow them to be expanded to applications, such as drug delivery system, multiphase nanocomposites in the near future. The above four reviews [108–111] form the basis of this section.

HNTs were first reported by Berthier in 1826 [112] as a dioctahedral 1:1 clay mineral of the kaolin group, and are found deposited in soils worldwide [108–119]. Many countries, such as China [116], France [117], Belgium [118], and New Zealand [119], have deposits of HNTs on their respective territory.

Based on the state of hydration, HNTs are generally classified into two groups: hydrated HNTs with a crystalline structure of 10 Å d_{001} spacing and dehydrated HNTs

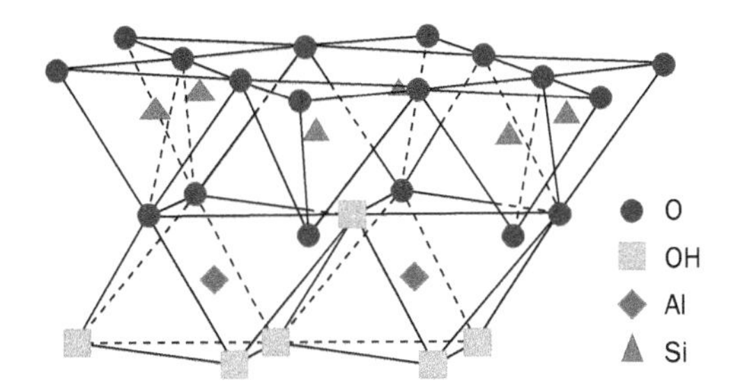

Figure 2.20 Crystalline structure of HNTs.

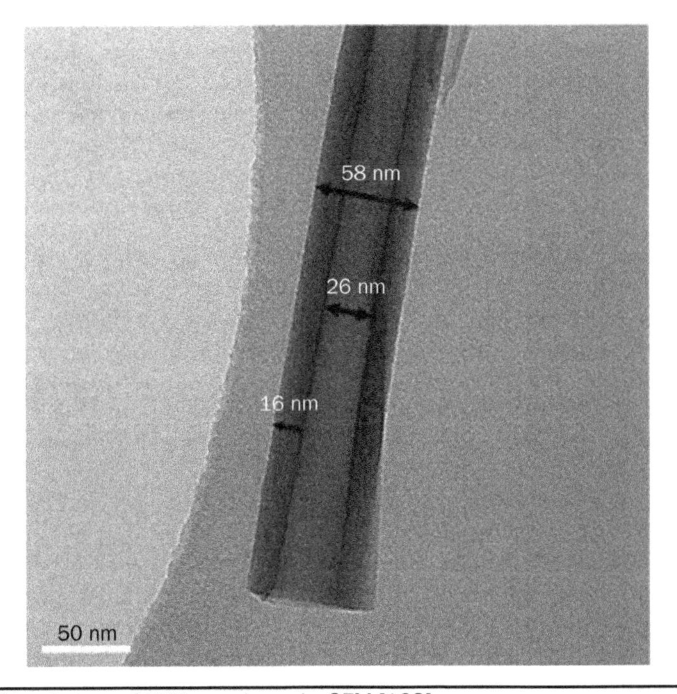

Figure 2.21 Morphology of HNTs as shown by SEM [123].

with 7 Å d_{001} spacing. Figure 2.20 shows a typical crystalline structure of HNTs. The presence or history of interlayer water in HNTs is one of the most important features to distinguish HNTs from kaolinite [114]. The d_{001} spacing of HNTs changes from 10 to 7 Å through an irreversible dehydration process. HNTs have a similar chemical composition to kaolinite, and the ideal chemical formula can be expressed as $Al_2Si_2O_5(OH)_4.nH_2O$, where n equals 2 and 0, representing the hydrated and dehydrated states, respectively [120, 121].

Due to the different crystallization conditions and geological occurrence, HNTs adopt different morphologies, such as tubular, spheroidal, and plate-like structures. The tubular structure is the most common and useful [114]. Typically, the inner and outer diameters of HNTs are 1–30 nm and 30–50 nm, respectively, and their lengths are 100–2,000 nm [122–124]. Some typical morphologies of HNTs in TEM are shown in Fig. 2.21 [123]. HNTs contain two types of hydroxyl groups, inner and outer hydroxyl groups, which are situated between layers on the surface of HNTs, respectively. Due to the multilayer structure, most of the hydroxyl groups are inner groups and only a few hydroxyl groups are located on the surface of the HNTs. The surface of HNTs is mainly composed of O–Si–O groups, and the siloxane surface can be confirmed from Fourier transform infrared spectra. As a result, when compared with other silicates, such as kaolinite and MMT, the density of surface hydroxyl groups of HNTs is smaller.

Properties HNTs have been widely used in controlled or sustained release, in nanoreactors or nanotemplates, and for the elimination of contaminants or pollutants [108]. Due to their characteristics such as nanoscale lumens, high aspect ratio, and low hydroxyl group density on the surface, HNTs have attracted attention of numerous researchers. Traditionally, due to the high aspect ratio and superior high-temperature resistance, HNTs were exploited to produce mainly high-quality ceramics, such as thin-walled porcelain or crucible products [125, 126]. Recently, scientists and engineers have discovered and developed a wide range of exciting new applications for this unique, cheap, and abundantly available naturally occurring clay with nanoscale lumens. Due to their unique characteristics, it is expected that HNTs would prove promising as reinforcing filler for polymer materials. Recently, an increasing number of studies have focused on the fabrication of HNT-polymer nanocomposites and their properties [108–111].

Fabrication of HNT-Polymer Nanocomposites For *thermoplastics*, melt-blending is an industrially friendly method to fabricate their composites with inorganics. Due to the ease of dispersion of HNTs, they can be dispersed relatively uniformly in thermoplastics by direct melt-blending, especially for polymers with high polarity, such as polyamides. Some HNT-polymer nanocomposites, such as polyamide 6 (PA6), PP, and poly(butylene terephthalate) have been produced successfully using melt-blending process without any surface modification of HNTs [108–111]. Compatibility between polyolefins and inorganics is challenging due to the great polarity discrepancy and the chemical inertness of polyolefins. To increase the compatibility between HNTs and PP, Du et al. proposed a two-step method of grafting PP chain onto the HNT's surface [108, 127]. These methods effectively increased the mechanical performance of HNT-PP nanocomposite, which may be used for other polymers. Other methods, such as hydrogen bonding and charge transfer mechanisms, have also been proposed for the fabrication of HNT-PP nanocomposites [108]. Due to their fewer surface hydroxyl groups, HNTs will expect to disperse better than other silicates, such as MMT and kaolinite. HNTs can be dispersed uniformly by direct melt-blending in polyamides with high polarity [108]. Du et al. used hydrogen bonding approach to disperse HNTs in PP. The morphology of the HNT-PP nanocomposites is shown in [127].

For *rubber/clay nanocomposites*, a good dispersion of the clay and strong interfacial interactions are two crucial factors in determining the performance of the nanocomposites [128]. Guo et al. utilized methacrylic acid (MMA) to improve the performance of HNT-SBR nanocomposites by direct blending [129]. Carboxylated butadiene-styrene rubber (xSBR) is a copolymer of styrene, butadiene, and a small amount of acrylic acid. Due to the presence of carboxyl groups, it is expected that HNTs will have good compatibility with xSBR. HNTs and xSBR have been used to prepare nanocomposites with strong interfacial interaction via hydrogen bonding [130]. To obtain good dispersion of HNTs in xSBR, a co-coagulation process of xSBR latex and HNT's aqueous solution was used [108]. Hybrids containing up to 65 wt% HNTs were fabricated in a SBR matrix with the in situ formation of zinc diborate by Guo et al. [131]. Apart from the mechanical strength and heat resistance comparable to general engineering plastics, the hybrids

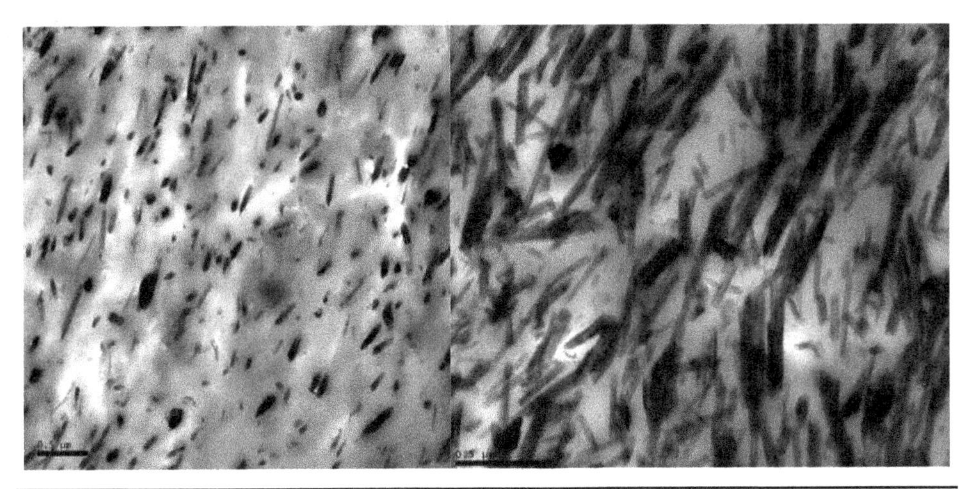

Figure 2.22 TEM images of PA6/HNT (100/10) nanocomposite (left) and SBR/HNT/MAA (100/40/2) nanocomposite (right) [108].

exhibit every good transparency, which is attributed to the excellent dispersion of nano-tubular clay, as shown in Fig. 2.22 [108].

For *thermosetting resins*, which are traditionally processed by casting or potting, it is rather difficult to achieve a uniform dispersion of HNTs. Liu et al. [124, 132] fabricated HNT-epoxy nanocomposites with good dispersion of HNTs via co-curing with cyanate ester resin and surface modification of HNTs.

Features Incorporation of HNTs into polymers will enhance the following properties:

- *Mechanical performance:* Mechanical properties, especially the modulus of PP, can be effectively improved using HNTs. The flexural properties, tensile strength, and impact strength all are noticeably improved with the addition of HNTs [108, 109].
- *Thermal stability and flame retardancy:* Apart from acting as reinforcements, HNTs have also been demonstrated as effective FR additives for PP as reported by Du et al. [133]. The thermal stability and flame-retardancy effects of HNTs on PP are believed to be for heat and mass transport and the presence of iron in the HNTs. This suggests that HNTs are promising as nonhalogen FR fillers.
- *Crystallization behavior:* Nanosized inorganic inclusions generally influence the crystallization process of polymer, such as PP nanocomposites [134, 135]. Ning et al. [136] and Du et al. [137] found that well-dispersed HNTs in PP matrix can serve as nucleation agents, resulting in an enhancement of overall crystallization rate. These results suggest that, similar to other nanosized inorganics, such as silica and MMT, HNTs serve as nucleation agents and facilitate the crystallization of nanocomposites [108, 109].
- *Reduced coefficient of thermal expansion:* Introduction of interfacial reactions in thermosetting systems may effectively reduce the thermal expansion of the cured resin. The coefficient of thermal expansion (CTE) of HNT-epoxy hybrids with low HNT concentration is substantially lower than that of the neat cured resin [131].

> It is believed that interfacial reactions take place during the curing of the hybrids between the aluminols and silanols and the cyanate ester. Consequently, the covalently linked interface formed is responsible for the improved performance and morphological characteristics of the hybrids [108–110].

In conclusion, due to the characteristics, such as nanosized lumens, high L/D ratio, and low hydroxyl group density on the surface, more exciting applications have been discovered for these unique, cheap, and abundantly deposited clays. So far, HNTs have been widely used in controlled or sustained release, in nanoreactors or nanotemplates, and for the elimination of contaminants or pollutants. Recently, research has focused on HNT-polymer nanocomposites. These investigations suggest that nanocomposites exhibit markedly improved properties, such as superior mechanical performance, higher thermal stability, better flame retardancy, and reduced CTE. HNTs possess promising prospects in the preparation of new structural and functional materials [108–110].

HNT Suppliers There are only a few HNT suppliers worldwide. In the United States, the main suppliers are NaturalNano (Rochester, New York; website: www.naturalnano.com) and Applied Materials Inc. (New York, New York; www.appliedminerals.com).

Nickel Nanostrands (NiNs)

Origin Nickel nanostrands were developed under Phase I of an Air Force Small Business Innovation Research (SBIR) program. Nickel nanostrands are strands of submicrometer-diameter nickel particles linked in chains and are micrometers to millimeters in length. Figure 2.23 shows the different magnifications of the nickel nanostrands [138]. They are analogous to carbon nanofibers or MWNTs but provide the additional properties of nickel, significantly expanding the variety of options available for developing future nanostructured technologies. They bring the additional electromagnetic, chemical, catalytical, and metallurgical properties associated with nickel to the nanostructured designer's toolbox. These new materials are

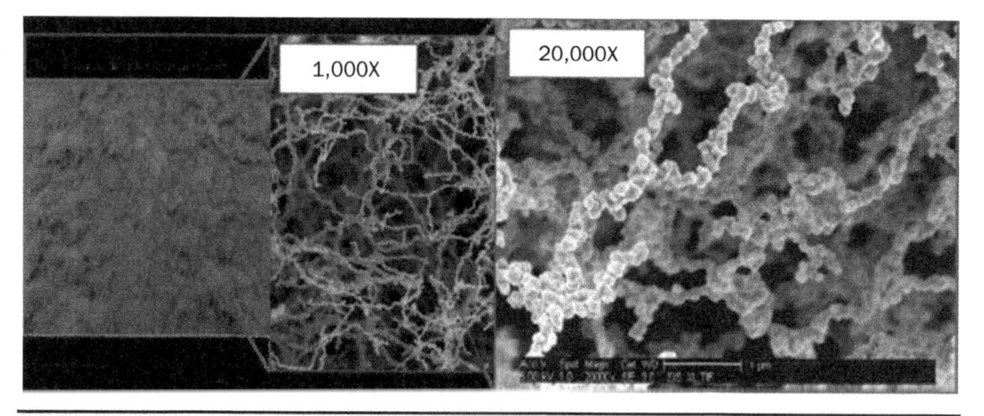

FIGURE 2.23 Images depicting nickel nanostrands in different magnifications. (*Courtesy of Conductive Composites.*)

manufactured by Conductive Composites (Heber City, Utah; website: www.conductivecomposites.com) [139]. Conductive Composites focuses on creating materials and technologies that enable electrical conductivity and electromagnetic capabilities in composites and polymers.

Properties These nickel nanostrands have already demonstrated their utility in creating conductive resins, paints, adhesives, and thermoplastics for a wide range of conductive polymer and conductive composite applications. Conductive Composites created paints with a sheet resistance less than 1 ohm per square, and adhesives and thermoplastics with conductivities of 40 S/cm and 150 S/cm, respectively.

Nickel nanostrands can also further enhance the conductivity of fiber- or particle-reinforced composites, thus providing a three-dimensional conductive lattice in otherwise insulating polymer resin matrix. A loading of only 2% by volume (2 vol%) in standard carbon prepreg will double the conductivity of a carbon fiber-reinforced composite. Nickel nanostrands were dispersed in different median, such as 0.5 vol% in water, 5.0 vol% in epoxy, and 10 vol% in urethane [139]. This is important for the Air Force and industry because infiltrated carbon composites have proven highly useful in lightning strike protection for aircraft and other structures. When nickel nanostrands are added to elastomers, the resulting composite exhibits remarkable changes in conductivity with respect to tensile or compressive strain. Since nanostrands are a magnetic material, they may be magnetically aligned, while the carrier is still in the liquid phase, yielding other unique applications, such as magnetically oriented inks or magnetically aligned conductive fibers. The unique microstructure and chemistry of nickel nanostrands could also lead to important advancements in filtering, catalysis, energy storage, and nanometallurgy. Though nickel nanostrands are similar to carbon nanofibers in terms of size and shape, their anticipated applications are quite different. The proprietary chemistry by which they are produced will no doubt result in a volume production pricing comparable to other nickel commodity powders.

Features and Applications Nickel nanostrands can be engineered to meet the diameter and length specifications in many fields of submicrometer and nanostructure designs. Diameters ranging from 50 nm up to 2 μm have been produced, with aspect ratio generally in the 50:1 to 500:1 range. Currently, Conductive Composites is in the process of producing nanostrands in the range of 10–30 nm in diameter. Further development could lead to advancements in nanotechnology directly benefiting the Air Force, aerospace community, and other industries. In general, nickel nanostrands have the following features:

- Highly branched, three-dimensional nanostructures that offer the ultimate balance of conductivity, electromagnetic, and mechanical properties
- Can be used as a high-performance additive or as a stand-alone material
- Proven performance advantages over carbon- and silver-based technologies
- Supplied in an as-manufactured continuous form or as a 3D powder
- Easily dispersed in fluid systems using low shear techniques
- Mixing and dispersion guide available

Present and potential applications and markets of nickel nanostrands are summarized as follows:

- Paints
- Thermoplastics
- Elastomers
- Adhesives
- Prepregs
- Energy storage
- Filtering
- Catalysis
- Thermosets
- Coatings

Currently, Conductive Composites offers three different grades of nickel nanostrands: nickel nanostrands, premium grade; nickel nanostrands; and fine nickel nanostrands. Table 2.12 summarizes some of their physical properties [138, 139].

Nickel Nanostrands Suppliers and Products. Conductive Composites is the sole manufacturer of nickel nanostrands and other technologies, which gives the company the ability to create a wide variety of highly electrically conductive fibers, nonwovens, plastics, resins, adhesives, polymer coatings, composites, and other systems. A summary of conductive material products is listed as follows:

- *Coated fiber (NiFiber):* Lightweight structural fibers achieve an ultimate combination of strength and conductivity.
- *Coated nonwoven sheets (NiShield):* Conductive lightweight layers with exceptional shielding and conductivity properties in a continuous roll format.
- *Nanostructured powders (Nanostrands):* Unique branched geometry and excellent percolation properties lead to highly conductive real-world properties.

All the materials on this list are available in commercial volume and the production lines of Conduct Composites run high-volume continuous-throughput manufacturing in ISO 9001 compliant facilities. The production processes of Conduct Composites are clean and "green" with no process exhaust effluents and only closed-loop fluid systems.

Aluminum Oxide Nanofibers

ANF Technology Ltd., a Surrey, U.K.-registered company with production facility in Tallinn, Estonia, and a representative office in California, produces aluminum oxide nanofibers branded as Nafen (website: www.nafen.eu) [140]. It consists of very long coaligned fibers 10–20 nm in diameter built of perfect alumina monocrystals in gamma/chi phase. Figure 2.24 shows several SEM and TEM micrographs of the Nafen aluminum

TABLE 2.12 Typical Properties of Nickel Nanostrands

Properties	Values
Bulk density	0.090–0.250 g/cc
Specific surface area (BET)	2–5 m^2/g
Purity	99.999%
Density	8.91 g/cc

FIGURE 2.24 SEM and TEM micrographs of the Nafen aluminum oxide nanofibers as grown. (*Courtesy of ANF Technology.*)

oxide nanofibers as grown [140]. Table 2.13 shows some typical physical and mechanical properties of Nafen [140].

Nafen is the only type of superior-grade nanofibers recently available on the market in industrial volumes at a reasonable price. ANF has teamed up with some of the world's leading technological research institutions and centers of excellence to uncover the potential benefits of Nafen. Testing has been carried out at Fraunhofer Institute of Polymeric Materials and Composites (PYCO, Germany), Moscow Institute of Steels and Alloys (MISiS), University of South Carolina, University of Wisconsin at Milwaukee, the University of Texas at Austin, University of Cambridge, and other industrial and research institutes. Initial results achieved provide strong evidence of significant performance enhancement with impregnating Nafen into major polymer systems, such as polycyanurate thermoset, epoxy, polyester acrylate, polyether, polyamide, and PVB for a variety of applications.

Table 2.13 Typical Physical and Mechanical Properties of NAFEN Aluminum Oxide Nanofibers

Properties	Values
Fiber diameter (nm)	10–20
Fiber length (mm)	1–150
Fiber tensile strength (GPa)	12
Fiber tensile modulus (GPa)	400
Specific surface area (BET) (m^2/g)	150
Phase	gamma/chi
Specific gravity (g/cm^3)	3.98
Bulk density (g/cm^3)	0.1–0.3

Courtesy of ANF Technology.

Several advantages are noted in comparing Nafen with other nanofibers:

- Inexpensive
- Easy to disperse
- Transparent, no color issues
- Low-loading levels below 5 wt%
- High aspect ratio more than 10
- Compatible with carbon and glass fibers
- Dielectric
- Flame-retardant
- Thermally conductive

Features and Applications When 0.1 wt% Nafen was incorporated in DER 330, a low-viscosity epoxy system, using ultrasonication technique and hardened with polyethylene polyamine (PEPA), significant improvements of fracture energy from 3 KJ/m^2 (cured neat epoxy) to 11 KJ/m^2 (epoxy + plastifier + 0.1 wt% Nafen) and flexural strength from 20 MPa (cured neat epoxy) to 50 MPa (epoxy + plastifier + 0.1 wt% Nafen) were achieved [140].

When 0.5–4 wt% Nafen was incorporated in DER 330, a low-viscosity epoxy system, using ultrasonication technique in ethanol and mechanically mixed, and hardened with diamino diphenyl sulfone (DDS), significant improvements of fracture energy from 6 MPa (cured neat epoxy) to 20 MPa (epoxy + 0.5 wt% Nafen) and shear strength from 38 KJ/m^2 (cured neat epoxy) to 60 KJ/m^2 (epoxy + 0.5 wt% Nafen) were also achieved [140].

When 1–5 wt% of Nafen was added to polycyanurate by direct mixing with a three-roll mill, improvements in Young's modulus (140 MPa at 0 wt% and 200 MPa at 5 wt% Nafen), tensile strength (200 N at 0 wt% and 350 N at 5 wt%), elongation at break (0.6 $MN/M^{3/2}$ at 0 wt% and 1 $MN/M^{3/2}$ at 5 wt% Nafen), and fracture toughness (7% at 0 wt% to 10% at 5 wt% Nafen) were observed [140].

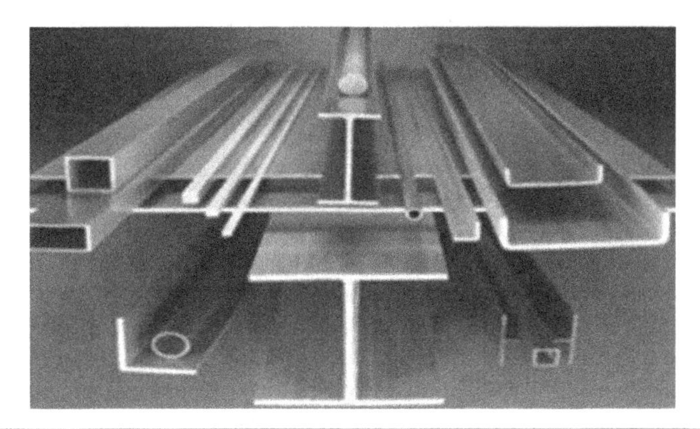

Figure 2.25 Polyether-Nafen coating can be used as anticorrosion metal coatings. (*Courtesy of ANF Technology.*)

Nafen was dispersed in ethanol using ultrasonication and mechanically mixed with PVB powder to produce a transparent film that possessed increased toughness and ultimate strength, but still remained transparent [140], which can be used for triplex safety glass [140].

When 0.5 wt% Nafen was predispersed in glycidyl methacrylate (GMA) at 10 wt% by mechanical mixing with the polyether. Hardness (6X) and Young's modulus were both increased with enhanced elasticity ensuring no cracking at bending points [140]. Polyether-Nafen coating can be used as anticorrosion coatings, topcoat materials, potting compounds, plywood patch, wire and cable coatings, and thermal break (Fig. 2.25) [140].

Aluminum Oxide Nanofibers and Products ANF is the sole manufacturer of Nafen. ANF also develops Nafen-based dispersions and compounds for polymeric products. ANF develops and supplies Nafen dispersions and compounds to potential customers. Currently, Nafen has shown promise in epoxy for fiber-reinforced composites, polycyanurate high-temperature applications, polyester acrylate UV-curable transparent coatings, polyether protective coatings, PVB transparent films, flame-retardant polyamide 11, and other systems. A summary of Nafen products is as follows:

- *Nafen aluminum oxide nanofibers:* Controlled diameter of 10 nm; narrow size distribution; polycrystal length up to 15 cm (block form); monocrystal length of 300 nm (mean average); uniaxial aligned or random oriented fiber blocks; industrial production volumes; chemically and thermally stable; 400^{+} GPa Young's modulus; well wetted and dispersed in water, alcohol, and other liquids; resistant to radiation; high specific surface area (155 m^2/g) for catalysis or functionalizing the surface
- *Nafen aluminum oxide nanofiber dispersions:* Incorporated specific loadings or masterbatches of Nafen (controlled diameter of 10 nm, monocrystal length of 300 nm) in polymers, water, solvents, and other liquids; industrial production volumes; easily processable; standard packaging; fiber surface treatments suitable for various formulation requirements

All the above materials are available in commercial volume from ANF Technology's production and process development facility in Tallinn, Estonia, as well as from ANF's distribution partners in the United States, Europe, and Asia.

2.2.3 Three Nanoscale Dimensions in the Form of Particulates

Polyhedral Oligomeric Silsesquioxanes (POSS)

Origin A departure from the use of one-dimensional and two-dimensional nanomaterials is the concept involving the creation of three-dimensional inorganic–organic structural materials. One approach to developing better materials is to create inorganic–organic composite materials in which inorganic building blocks are incorporated into organic polymers. In the early 1990s, a scientific team under the direction of Dr. Joseph Lichtenhan at Edwards Air Force Base in California invented the first new class of chemical feedstock to be developed in 50 years—polyhedral oligomeric silsesquioxanes or POSS. In 1998, Dr. Lichtenhan and his team spun this POSS technology out of the Air Force Research Laboratory and founded Hybrid Plastics, Inc., taking the first step in evolving from a small R&D facility to a full-scale manufacturing. In 2004, the company took its next major step and relocated from California to Hattiesburg, Mississippi (website: www.hybridplastics.com) [141].

Polyhedral oligomeric silsesquioxane (POSS) nanostructured materials, representing a merger between chemical and filler technologies, can be used as multifunctional polymer additives, acting simultaneously as molecular level reinforcements, processing aids, and flame-retardants. POSS nanoscopic chemicals provide unique opportunities to create revolutionary material combinations through a melding of the desirable properties of ceramics and polymers at the 1-nm length scale. These new combinations enable the circumvention of classic material performance trade-offs by exploiting the synergy and properties that only occur between materials at the nanoscale. They release no volatile organic compounds and therefore produce no odor or air pollution, while offering easy incorporation into existing manufacturing protocols. These elegant chemicals also push the technology into new applications, such as soluble yet biologically inert delivery vehicles for pharmaceuticals and medical applications, for example, biocompatible prosthetics.

Properties POSS technology is derived from a continually evolving class of compounds closely related to silicones through both composition and a shared system of nomenclature. POSS chemical technology has two unique features:

1. The chemical composition is a hybrid, intermediate ($RSiO_{1.5}$) between that of silica (SiO_2) and silicone (R_2SiO).
2. POSS molecules are physically large with respect to polymer dimensions and nearly equivalent in size to most polymer segments and coils.

POSS molecules can be thought of as the smallest particles of silica possible. However, unlike silica or modified clays, each POSS molecule contains covalently bonded reactive functionalities suitable for polymerization or grafting POSS monomers to polymer chains. Each POSS molecule contains nonreactive organic functionalities for solubility and compatibility of the POSS segments with the various polymer systems.

The chemical diversity of POSS technology is very broad and a large number of POSS monomers and polymers are currently available or undergoing development. POSS chemical technology is easy to use with monomers available in both liquid and solid form and they are soluble in most common solvents. Hence, POSS technology can be used in the same manner as common organics, in either monomer or polymeric (resin) form. POSS chemical feedstock can be added to nearly all polymer types (glassy, elastomeric, rubbery, semicrystalline, and crystalline) and compositions. Enhancements in the physical properties of polymers incorporating POSS segments result from POSS's ability to control the motions of the chains while still maintaining the processability and mechanical properties of the base resin. This is a direct result of POSS's nanoscopic size and its relationship to polymer dimensions.

The anatomy of a POSS nanostructured chemical is illustrated in Fig. 2.26 [141]. It is a precise, three-dimensional silicone-oxygen cage structure for molecular-level reinforcing of polymer segments and coils. Hybrid Plastics currently offers over a large variety of nanostructured chemicals [141]. The key purpose of POSS technology (Fig. 2.27) is to create hybrid materials that are tough, lightweight, and as easy to process as polymers, yet they have the characteristics of high-use temperature and oxidation resistance, like ceramics [141].

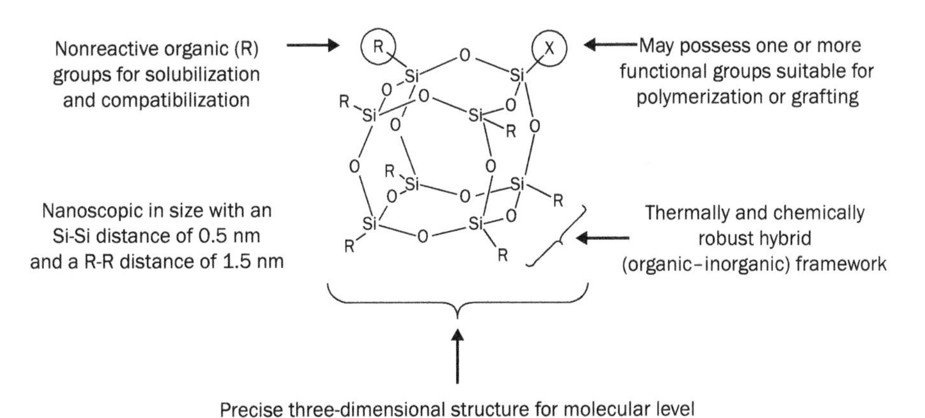

FIGURE 2.26 Anatomy of a POSS molecule. (*Courtesy of Hybrid Plastics.*)

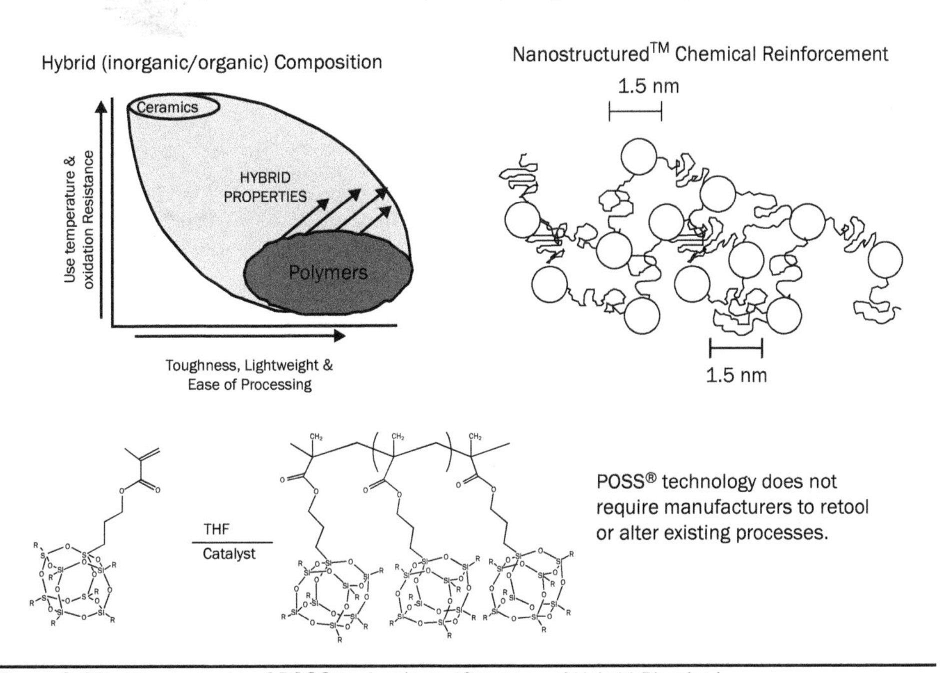

FIGURE 2.27 Key aspects of POSS technology. (*Courtesy of Hybrid Plastics.*)

Features Polyhedral oligomeric silsesquioxane materials are one type of hybrid inorganic–organic material of the form $(RSiO_{1/2})_n$ or R_nT_n, where organic substituents are attached to a silicon-oxygen cage [142]. The most common POSS cage is the T_8 (e.g., $Vinyl_8T_8$), although other cages with well-defined geometries include n = 6, 10, 12, 14, 16, and 18 [143]. By incorporating these Si-O cages into organic polymers, properties superior to the organic material alone are realized, offering exciting possibilities for the development of new materials [144–148]. Vinyl- and phenyl-based POSS materials have been used in composites that have shown excellent fire resistance performance [149].

Another example of POSS-molecular silica blends is shown in Fig. 2.28, where each of the four types of POSS has a different R group that was blended into 2 million molecular weight (MW) polystyrene. The morphology of the POSS-polystyrene nanocomposites was changed significantly, as shown in the TEM images in Fig. 2.28a–d. Figure 2.28a illustrates R = cyclopentyl (Cp_8T_8) in the POSS material when blended with the polystyrene, where a snowflake domain was formed. Fig. 2.28b shows R = cyclopentyl (CP_7T_8Styrenyl) in the POSS material when blended with the polystyrene, where a partial compatibility was formed. Figure 2.28c demonstrates R = styrenyl ($Styrenl_8T_8$) of the POSS material when blended with polystyrene, where a phase inversion occurred. Figure 2.28d illustrates R = phenethyl ($Phenethyl_8T_8$) in the POSS material when blended with 50 wt% loading of POSS into the polystyrene—a transparent material. This example shows that the POSS material can be customized to control the morphology and the properties of the resulting nanocomposite blends. A valuable reference on

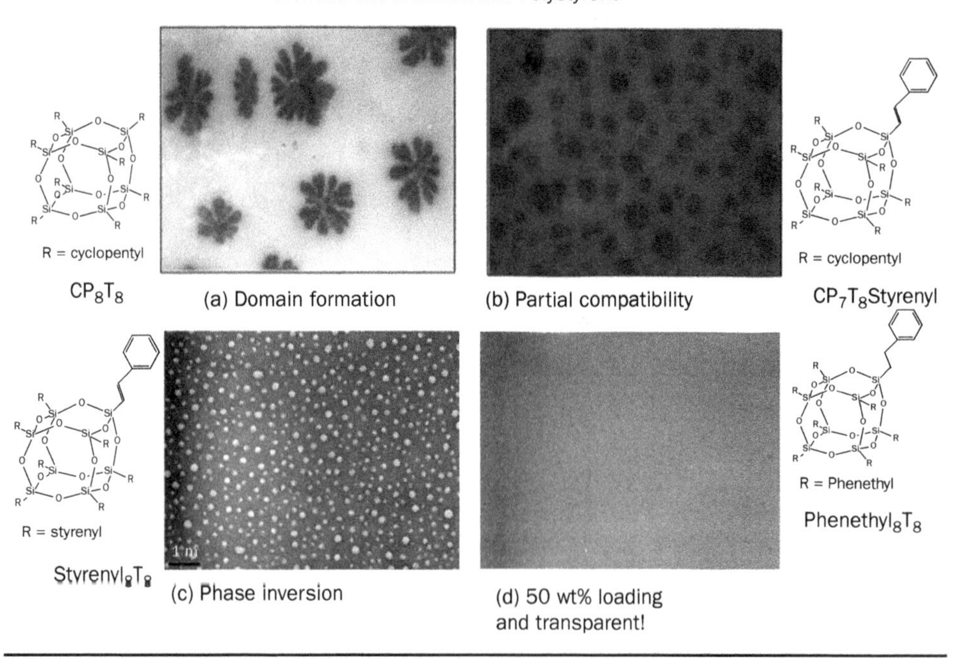

Figure 2.28 TEM micrographs showing the change of morphologies of polystyrene by changing the R group of the POSS materials. (*Courtesy of A. Lee.*)

the "applications of POSS" was recently published using POSS in plastics, composite materials, space materials, dental materials, tissue engineering, drug delivery, lithography, fuel cells, batteries, lubricants, liquid crystal, LED, sensor, photovoltaic, and biomedical devices [150].

POSS Suppliers and Products Hybrid Plastics, Inc. manufactures POSS® Nanostructured® chemicals, flow aids, dispersion aids, thermoset resins, thermoplastic masterbatch, and others. The full list of products is as follows:

- *R&D chemicals:* A variety of R&D quantities of POSS chemicals can be purchased using Hybrid Plastics' online catalog.
- *Bulk chemicals:* Large production quantities of a few of Hybrid Plastics' most popular POSS chemicals.
- *Thermosets:* Formulated thermosetting resins for coatings, adhesives, and composites.
- *Thermoplastics:* Masterbatch formulations of common thermoplastics for improved flow, reduced surface energy, better dispersion, and more.
- *Electronics:* POSS chemicals and formulated materials are manufactured for use in electronics applications.
- *POSS and nanopowders:* POSS chemicals with a metal in one corner for catalysis, and predispersed pigments/fillers.

Nanosilica (n-silica)

Origin Dispersed nanosilica (also known as silicon dioxide) ranges from 5 to 70 nm in size [151] and can be spherical or irregular in shape. Nanosilica can be produced by high-temperature hydrolysis [152, 153] or sol-gel (macrosurfaced nanosilica) [154] methods. Both methods are briefly discussed in this section.

In 1941, Degussa (now known as *Evonik*) patented a high-temperature hydrolysis process of metallic oxides to produce extremely fine-particle oxides. The method was converted into large-scale production in the 1950s and has become the process for the preparation of nanoparticles based on silicon dioxide, aluminum oxide, and titanium dioxide. Under TEM analyses, the primary particles of the three oxides show cubic forms with rounded off corners. All materials exhibit no internal surface. Silicon dioxide is produced under the trademark of AEROSIL®. The raw materials are chlorosilanes that are hydrolyzed in an oxygen-hydrogen flame. The resultant silicon dioxide occurs in aerosol and is subsequently separated from the gaseous phase. A special process is used to remove residual hydrochloric acid that is still absorbed on the material.

Properties AEROSIL is highly dispersed, amorphous, very pure silica produced by high-temperature hydrolysis of silicon tetrachloride in an oxyhydrogen gas flame [151–153]. It is a white, fluffy powder consisting of spherically shaped primary particles. AEROSIL OX 50 has the largest average primary particle size of 40 nm; AEROSIL 300 and 380 have the smallest size of 7 nm. The primary particles are spherical and free of pores. The primary particles in the flame interact to develop aggregates, which join together reversibly to form agglomerates. Figure 2.29 shows a TEM micrograph of AEROSIL 300 in which the primary particles, aggregates, and agglomerates can be clearly seen [153]. The average diameters of the primary particles are in the range of 7–40 nm, according to the AEROSIL grade. The specific surface areas range between

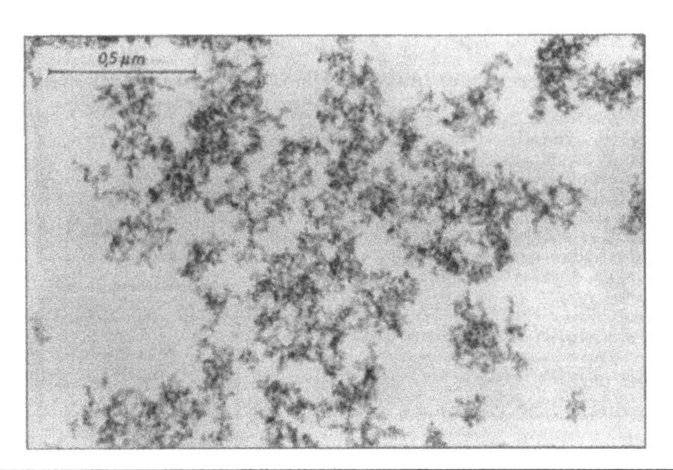

Figure 2.29 TEM micrographs of AEROSIL 300 showing particle in aggregates. (*Courtesy of Evonik.*)

50 m^2/g and 380 m^2/g. In contrast to precipitated silicas, AEROSIL does not have a clearly defined agglomerate size. Particle size distribution becomes wider as the average primary particle size increases, and the tendency to form agglomerates is reduced [153]. The extremely small particles naturally result in a large specific surface area, ranging from approximately 50 m^2/g (AEROSIL OX 50 has an average primary particle size of 40 nm) to 380 m^2/g (AEROSIL 380 has an average primary particle size of 7 nm). The size of the surface area is easily illustrated if one considers that approximately 20 g of AEROSIL 200 (average primary particle size is 12 nm) has the same surface area as a football field. AEROSIL consists entirely of amorphous silicon dioxide. It starts to sinter and turn into glass above 1,200°C. Crystallization only occurs after heat treatment. AEROSIL is nearly insoluble in water. It is also insoluble in acids. It does, however, dissolve in strong alkaline media to form silicates.

Applications Siloxane and silanol groups are situated on the surface of the AEROSIL particles. The latter is responsible for the hydrophilic behavior of the untreated AEROSIL. These silanol groups also determine the interaction of the AEROSIL with solids, liquids, and gases. One gram of AEROSIL 200 with a specific surface area of 200 m^2/g contains approximately 1 mmol of silanol groups. The surface of AEROSIL particles can be easily chemically modified by reacting the silanol groups with various silanes and siloxanes, resulting in hydrophobic AEROSIL particles. For example, AEROSIL R972 and AEROSIL R812 are made hydrophobic with dimethylsilyl or trimethylsilyl groups, which are stable against hydrolysis. Hydrophilic and hydrophobic grades of AEROSIL have proved successful for use in numerous applications, such as a reinforcing filler, a thickening and thixotropic agent, an anti-settling agent, and a free-flow aid. Hydrophobic types of AEROSIL are characterized by the following properties:

- Extremely low moisture absorption
- Ability to impart hydrophobicity to other systems
- Easier to disperse in organic media
- Effective rheology control in complex liquid

Table 2.14 Material Properties of AEROSIL R805 and R202 [153]

Properties	AEROSIL R805	AEROSIL R202
Specific surface area (BET) (m^2/g)	150 ± 25	100 ± 20
Carbon content (%)	4.5–6.5	3.5–5.0
Average primary particle size (nm)	12	14
Tapped density standard material (g/L)	50	50
Moisture 2 hours at 105°C	<0.5	<0.5
Ignition loss 2 hours at 1,000°C based on material dried for 2 hours at 105°C	5.0–7.0	4.0–6.0
pH value in 4% dispersion	3.5–5.5	4.0–6.0
Si02—content based on ignited material (%)	>99.8	>99.8

AEROSIL is a submicrometer amorphous silica that has been used successfully for decades as a thickening agent and as a thixotropic agent in the liquid system. It is also used to adjust the rheological properties of epoxy resins. It also serves as a rheological aid to provide effective, stable thickening and thixotropy of epoxy systems [151–153]. AEROSIL fumed silica for rheology control is widely used in silicone rubber, coatings, plastics, printing inks, adhesives, lubricants, creams, ointments, and toothpaste. Nelson and coworkers used this nanoparticle with polystyrene in their recent study demonstrating significant improvement in reduced flammability [155]. Properties of AEROSIL R805 and R202 are listed in Table 2.14 [153].

AEROSIL has proven track record in the following traditional applications:

- *In plastics and adhesives:* coating polymers, sealant polymers, film, gel coats and laminating resins, casting resins, cable compounds, adhesive, silicone rubber, fluorinated rubber.
- *In coatings:* solvent-based coatings (two-pack systems), solvent-based coatings (air drying systems), water-based coatings (clear coats and pigmented systems), UV-curing coatings, powder coatings; unsaturated polyester.
- *In printing inks:* letterpress printing, flexographic printing, offset printing, gravure printing.
- *As an external additive for toner:* negative toner (black); positive toner (black); color toner (negative).
- *In pharmaceuticals and cosmetics:* antiperspirants; aerosols, spray; coated tablets; creams, lotions; fragrances; powders; hair formulations, shampoos; lipsticks; nail polish; ointments, pastes; mascara, makeup; suppositories; tablets, capsules; toothpastes, dental powder.

The introduction of most nanoparticles into low-viscosity, reactive resin systems, such as epoxies, vinyl esters, cyanate esters, BMI, and phenolics, results in a rapid increase in viscosity as the nanoparticle content is increased to 5–7 wt%. This leads to limited use of the nanomodified reactive resin system in many low- to medium-viscosity demanding processing applications, such as vacuum-assisted resin transfer molding (VARTM),

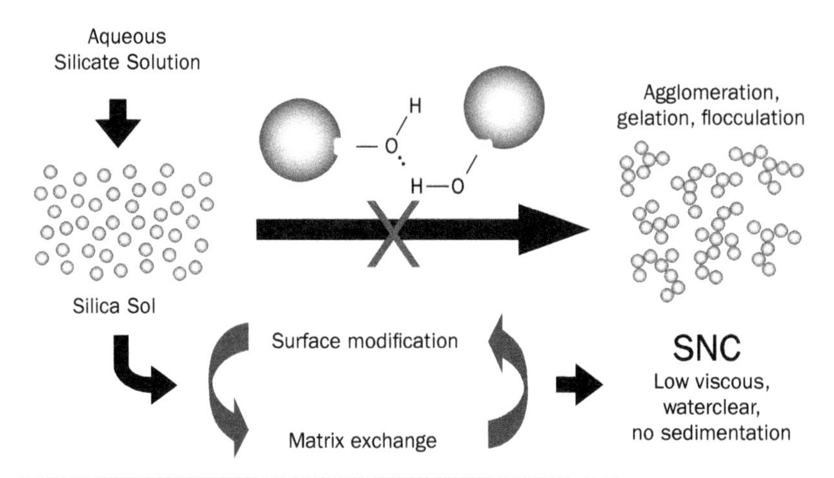

Figure 2.30 Hanse Chemie's route to produce nanosilicas. (*Courtesy of Hanse Chemie.*)

impregnation of fabric, filament winding, and other processing methods. Although there are instances that indicate 5–7 wt% nanoparticles are optimum, reduction in viscosity through the use of solvent and/or increased temperature allows reasonable dispersion of the nanoparticles into the resin matrix system. Questionable uniformity of nanoparticle dispersion using high shear mixing conditions followed by removal of solvent is an additional issue, as is "scale-up."

As mentioned earlier, there are processing problems (viscosity increase) associated with the introduction of 5–7 wt% nanoparticles into low-viscosity resin systems (epoxies, vinyl esters, etc.) A novel technique of preparing macrosurfaced silica appears to avoid the increased viscosity problem with the introduction of 5–7 wt% nanoparticles.

Macrosurfaced Nanosilica. The use of Hanse Chemie Technology [154, 156, 157] of macrosurface-treated spherical nanosilica will allow the introduction of ≥10 percent nanosilica with minimal increase in viscosity in the nanomodified reactive resin system and still provide multifunctionality to the resulting cured resin system. Macrosurfaced nanosilica is currently manufactured by Evonik Nanoresins GmbH (Geesthacht, Germany; website: www.nanoresins.ag). Evonik Nanoresins' proprietary synthesis of silica nanocomposites is shown in Fig. 2.30 [154]. A silicate solution is transformed into silica sol using the modified sol-gel–process method. The silicas were surface-modified to be compatible with selective polymer matrices to form transparent PNCs with low viscosity and no sedimentation. Evonik Nanoresins is able to create homogeneous dispersions of amorphous silicon dioxide (SiO_2) in standard, chemically unchanged organic monomers, prepolymers, and oligomers using this technique. The silica phase consists of discrete nanospheres (20 nm) with an extremely narrow size distribution [154]. Owing to agglomerate-free colloidal dispersion of the nanosilica particles, the nanocomposites are highly transparent, with low viscosity, and they do not show sedimentation even with SiO_2 loading up to 60%, as shown in a TEM micrograph in Fig. 2.31 [154]. This nanosilica can also be used as raw materials for coatings or varnishes and as additives for a wide variety of other applications.

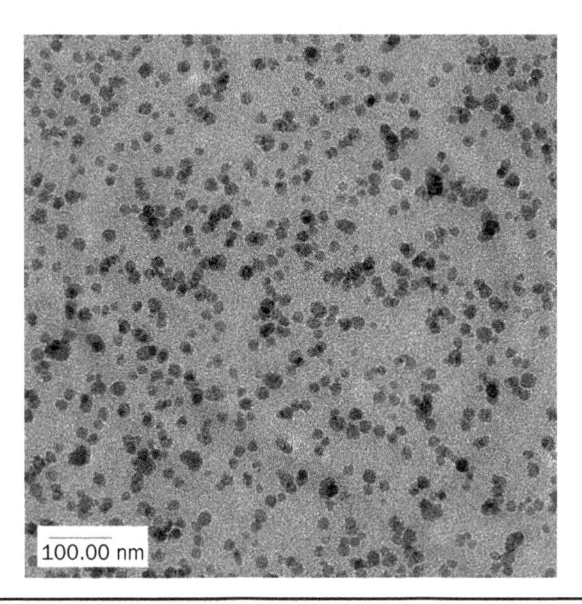

Figure 2.31 TEM micrographs showing Hanse Chemie's homogenous dispersion of nanosilica in polymer. (*Courtesy of Hanse Chemie.*)

It is the versatility that makes this sol-gel process so unique: the nanoparticles can be incorporated into a wide range of different polymer materials. Nanosilica particles thus produced are very well suited for further processing due to their low viscosity. They achieve specific improvement in the mechanical and thermal properties without the disadvantages usually associated with inorganic fillers. Thus, the toughness and hardness of a material can be increased without any loss in optical clarity.

Properties Evonik Nanoresins claims that these nanosilica particles can easily penetrate "closed meshed fabrics" and are therefore well suited for reinforcement of composites, especially VARTM. This characteristic would similarly apply to uniform impregnation of fabric in prepreg preparation. Furthermore, it is possible that with a further increase in nanosilica content to 15–20%, with a modest increase in viscosity and reasonable economics for the nanosilica, flame-resistant epoxy polymer matrix composites (PMCs) may emerge without jeopardizing the expected benefits of nano-modification, such as, among others, improved heat strength, higher T_g, and improved toughness. Evonik Nanoresins currently manufactures three nanocomposite products under the trade names of Nanocryl®, Nanopox®, and Nanopol®.

Nanocryl is created via low-viscosity dispersions of nanosilica in unsaturated (meth-) acrylates. It increases the surface hardness of coatings and their scratch and abrasion resistance without any loss in transparency and elasticity. It also improves other properties, such as gas permeability, cure shrinkage, and fracture toughness. Whenever acrylates are used, Nanocryl optimizes the qualities of coatings, adhesives, electronics, and high-performance. Figure 2.32 illustrates that there is no viscosity buildup of nanosilica in acrylate ester up to 50 wt%. Nanocryl may be used in all applications that

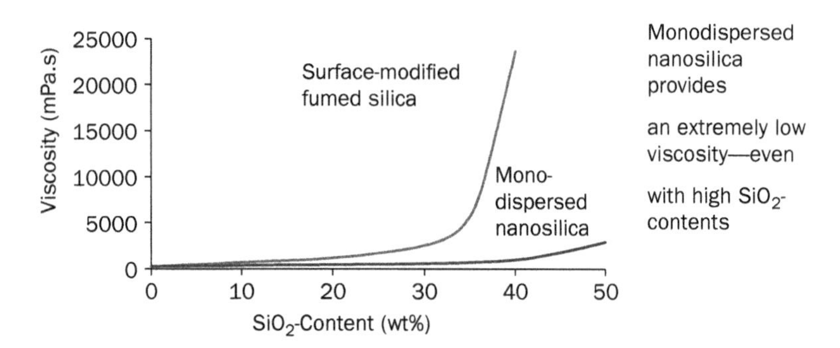

Figure 2.32 Viscosities of fumed silica and monodispersed nanosilica in acrylate ester. (*Courtesy of Hanse Chemie.*)

require radiation and peroxide curing [154]. The stability of the dispersions is affected by incompatible additives (gelation and flocculation).

Nanopox is made by dispersing nanosilicas in epoxy resins. Nanopox optimizes the quality of epoxy resins. With the nanosilicas only 20 nm in size, Nanopox penetrates even the tiniest structures, such as carbon fibers, thus increasing the material's fractural toughness and impact resistance. Combine Nanopox with standard epoxy resins and obtain the hardness and stiffness of much more expensive, highly cross-linked epoxies, along with toughness. It can be used for applications, such as coatings, structural adhesive, structural composites, casting resins (e.g., plotting, impregnation, encapsulation), electronics, and general purpose.

Nanopol is a product family of surface-modified nanosilicas dispersed in solvent carriers. Nanopol is a raw material that substantially improves the properties and application of solvent-based coating systems. Nanopol allows formulation of low-viscosity, sedimentation-free system even at high filler contents. The monodispersed nanosilicas provide optimum coating formulations which perform like silica-filled products while avoiding the disadvantages of conventional inorganic fillers. It offers outstanding scratch resistance and abrasion resistance without sacrificing other desired properties, such as transparency, gloss, and flexibility ofcoatings on wood, plastics, or metal substrates. Its major application is in coatings.

Applications A variety of applications can be considered with the diversity of Evonik Nanoresins nanosilica products ranging from epoxies, acrylates, polyurethanes, and silicone elastomers for coatings, fiber-reinforced composites, structural adhesives, to castings. The Evonik Nanoresins nanoparticles and reactive resin modifiers products provide the following general features:

Processing:

- Commercially available, stable dispersion for reactive systems, such as epoxies, acrylates, urethanes, and silicones
- Ease of processing, low viscosity, providing excellent wet-out of woven fabric with reduced heat of reaction and reduced cure shrinkage
- Cure rate unaffected
- Combine with typical fillers

Nanocomposite products:

- Enhanced scratch resistance, abrasion resistance without impairing transparency in UV-curing clear coatings
- Enhanced barrier properties to water vapor, gases, and solvents in UV-curing coatings
- Improved impact strength and elastic modulus of fiber-reinforced composites
- Increased toughness and tensile shear strength in epoxy-resin-based and other structural adhesives
- Reduced thermal yellowing, reduced thermal expansion, and inner tension
- Enhanced thermal conductivity or improved electrical insulation properties

Nano-Alumina (n-alumina)

Origin In 1941, Degussa (now known as *Evonik*) developed and patented a process where silicon tetrachloride was vaporized and hydrolyzed in an oxyhydrogen flame. AEROSIL nanosilicas were produced using this process. The basis of the AEROSIL process is the hydrolysis of the gaseous metallic chlorides under the influence of water that develops during the oxygen–hydrogen reaction, and at the temperature characteristics for such a reaction. The formation of the highly dispersed oxides takes place schematically, according to the following equations [152]:

Aluminum oxide C

$$4\,AlCl_3 + 6\,H_2 + 3\,O_2 \rightarrow 2\,Al_2O_3 + 12\,HCl$$

Nano–titanium dioxide P 25

$$TiCl_4 + 2\,H_2 + O_2 \rightarrow TiO_2 + 4\,HCl$$

Under TEM analyses, the primary particles of the three oxides show cubic forms, with the corners rounded off. Similar to AEROSIL particles, they have no internal surface.

Properties Aluminum oxide C has an average primary particle size of about 13 nm and a specific surface area of about 100 m^2/g. The aluminum oxides produced through precipitation of aluminum hydroxide from aluminate solution followed by calcinations consist of particles in the order of magnitude of 10 μm. High-surface aluminum oxide gels, in contrast to Aluminum oxide C, have a high proportion of internal surface. Figure 2.33 shows a TEM micrograph of Aluminum oxide C [152]. The specific surface is also a function of the specific weight, and for *heavy* oxides, these must be evaluated correspondingly higher.

The compacted apparent density of Aluminum oxide C is about 80 g/L. For use in powdered coatings or as a free-flow aid in powders, which tend to form lumps, a less compact material with a compacted apparent density of about 50 g/L is offered as Aluminum oxide CS. Through dispersion of Aluminum oxide C in water, followed by drying and grinding, a compacted apparent density of about 500 g/L can be attained. In combination with AEROSIL, Aluminum oxide C has proven to be effective in the thickening and the formation of thixotropizing of polar liquids. For the thickening and the formation of thixotropizing of water, AEROSIL COK 84, a mixture of 84%

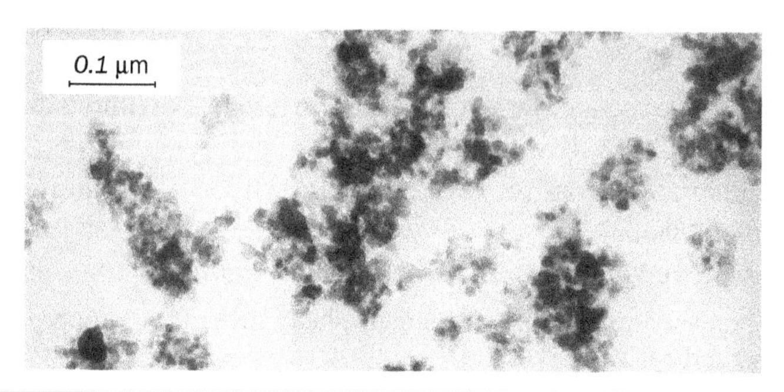

FIGURE 2.33 TEM micrographs of Aluminum oxide C. (*Courtesy of Evonik.*)

TABLE 2.15 Chemical Purity of DISPERAL and DISPAL Alumina Powders [159]

Impurity	PPM (Typical)
Na_2O	20
Fe_2O_3	100
SiO_2	120

AEROSIL 200 and 16% Aluminum oxide C, has been established as optimum for this application.

A partial listing of potential Aluminum oxide C applications includes production of aluminum nitride, inkjet papers, thermo-transfer printing, cable insulation, high-voltage insulators, transparent coatings, acrylate suspensions containing pigments, solder-resistant masks, photoresists for the production of ICs, coating of steel, heat insulation mixtures, powdered coating, and hair shampoo. Aluminum oxide nanoparticles have been used to form a nanocomposite for improved stereolithography [158].

Another Type of n-alumina Sasol pioneered processes utilizing alkoxide chemistry to convert primary aluminum metal into synthetic boehmite alumina of exceptional purity. DISPERAL® and DISPAL® are the trade names for the high purity, highly dispersible, boehmite alumina powders and sols/dispersions manufactured by Sasol in Brunsbuttel, Germany and in Lake Charles, Louisiana (website: www.sasolalumina.com). These alumina, which are nanosized in the dispersed phase, exhibit a unique combination of purity, consistency, and dispersibility that make them excellent materials for use in colloidal applications. These Sasol nano-alumina have traditionally been used in applications, such as sol-gel ceramics, catalysis, refractory materials, rheology control, and surface fractionizing. Other more recently developed uses include surface coating and paint detackification. Sasol's processes yield *n*-alumina with significantly lower levels of common impurities, such as iron, sodium, and silica (Table 2.15) [159]. Due to its extensive experience in its business, Sasol can tailor-made *n*-alumina (Fig. 2.34) [159]. Sasol is able to produce highly dispersible *n*-alumina with a wide range of physical properties, such as different crystalline size and shape: platelet, needlelike, and blocky type [159]. They can be in dispersed particle size ranging from 15 to 600 nm and resulting in translucent to opaque dispersions [160].

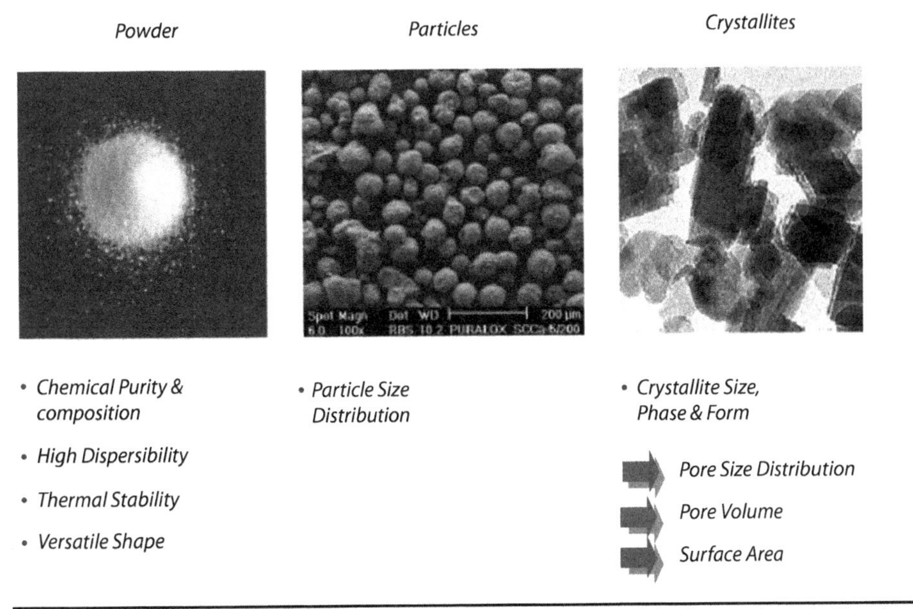

Figure 2.34 Sasol's tailor-made alumina in chemical purity, particle size, and crystallite size, phase, and form. (*Courtesy of Sasol.*)

Features The unique product characteristics of Sasol dispersible *n*-alumina can lead to many advantages in different applications. Some key features are as follows:

- *Synthetic n-alumina:* These high purity materials are produced under careful control to yield products with excellent quality control.
- *Powdered and predispersed n-alumina:* One can choose the either form that is most appropriate for his or her processing needs.
- *Highly dispersible powders:* Low-viscosity nanoparticle sols can be prepared at room temperature in 10–30 minutes.
- *Versatile n-alumina:* They can be employed under a variety of conditions, including low or high pH and low or high shear.
- *Dispersible n-alumina in organic media:* Examples include alcohols, dimethylformamide (DMF), and others.

N-alumina can be synthesized using sol-gel technique [159, 160], hydrothermal synthesis, precipitation, and combustion method [159, 160]. Like other nanoparticles, *n*-alumina aggregates to micrometer-sized particles and can be broken down by mechanical

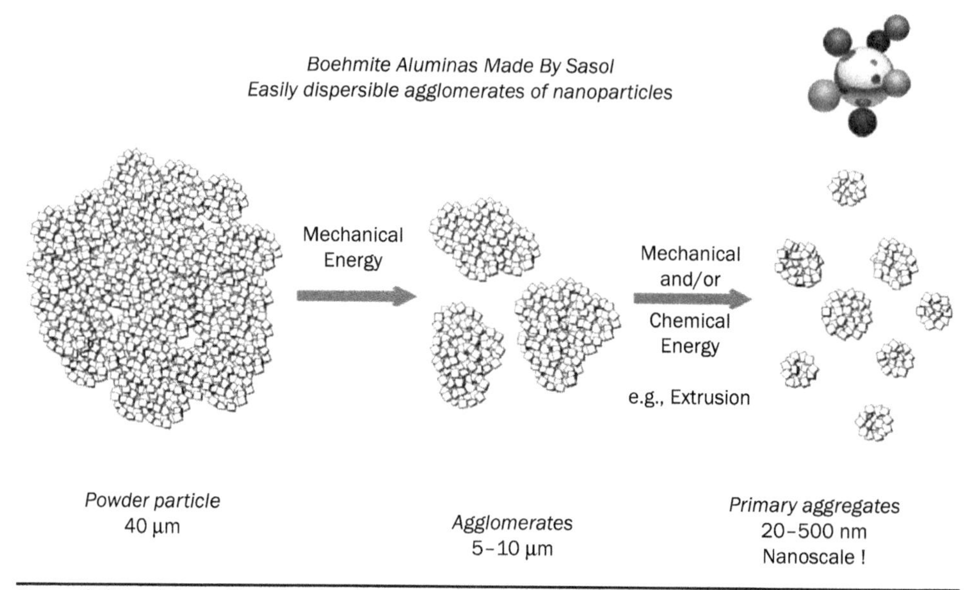

FIGURE 2.35 Schematics showing how *n*-alumina were dispersed by chemical and mechanical means. (*Courtesy of Sasol.*)

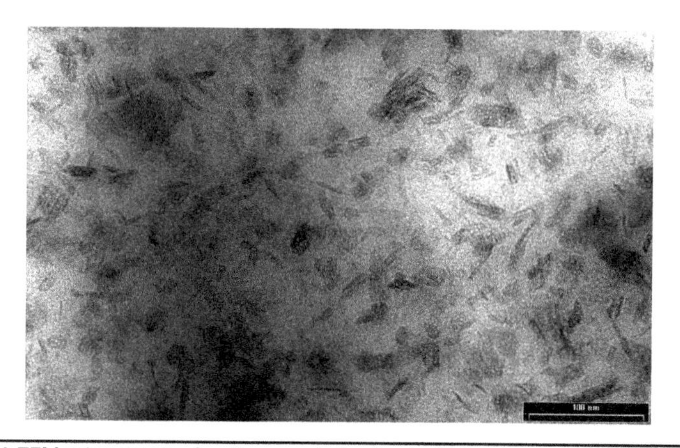

FIGURE 2.36 A TEM micrograph of a dispersed alumina in PA6 polymer. (*Courtesy of Sasol.*)

and chemical methods. Figure 2.35 shows a schematic representation of what occurs to the powder during dispersions using chemical attack and mechanical energy [160]. Figure 2.36 shows the TEM micrograph of a dispersible alumina at magnification of 180,000 [160]. *N*-alumina may improve hardness, thermal conductivity and stability, wear resistance, strength, and stiffness, among other characteristics.

Nano-Titanium Dioxide (n-TiO_2)

Origin *N*-titanium dioxide P 25 has an average primary particle size of about 21 nm and a specific surface of about 50 m^2/g [152]. It is therefore considerably more finely divided than the titanium dioxides that are produced on a major industrial scale according

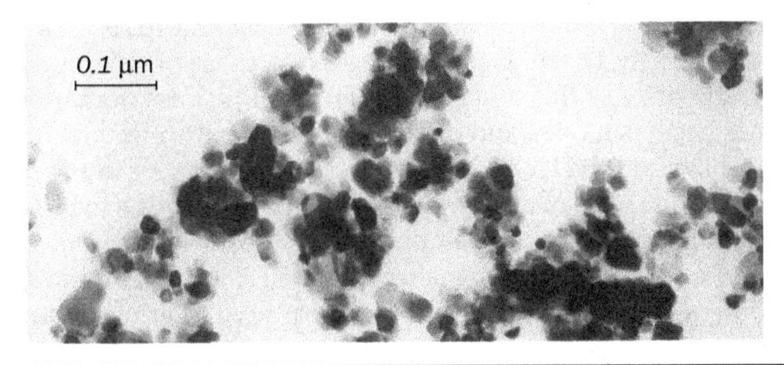

Figure 2.37 A TEM micrograph of Titanium dioxide P25. (*Courtesy of Evonik.*)

to the sulfate or chloride process for pigment applications, and that have a particle diameter of about 0.3 μm. Their surfaces consequently lie in the order of magnitude of 10 m^2/g. Figure 2.37 shows a TEM micrograph of n-TiO_2 P 25 [152].

Properties *N*-titanium dioxide P25 has been shown to be an excellent additive to improve the heat stability of a room temperature-vulcanizing (RTV) silicone adhesive/sealant. n-TiO_2 P25 can offer long-term heat stability at high temperatures. It also has the added benefit of flammability retardancy to RTV silicone using only a small amount of n-TiO_2 P25. It also can offer semi-reinforcing qualities, which larger-particle-size heat stabilizers cannot. For optimum properties, a minimum loading level of three parts is recommended.

Potential n-TiO_2 P25 applications consist of high-voltage insulation material, IC substrate boards, toners, florescent tubes, battery separators, and polystyrene.

Nano-Magnesium Hydroxide [n-Mg(OH)$_2$]

Origin Nano–magnesium hydroxide is used industrially as a non–hazardous alkali to neutralize acidic wastewaters. It also takes part in the block method of building artificial reefs. Solid $Mg(OH)_2$ also has smoke suppressing and flame-retardancy properties. This is due to the endothermic decomposition it undergoes at 332°C:

$$Mg(OH)_2 \rightarrow MgO + H_2O$$

The heat absorbed by the reaction acts as retardant by delaying ignition of the associated substance. The water released dilutes any combustible gases and inhibits oxygen from aiding the combustion. Common usage of $Mg(OH)_2$ as a FR additive includes plastics [161], roofing, and coatings. It is also used for folk remedy. *N*-magnesium hydroxide has platelet structure and it requires exfoliation; otherwise the mechanical properties of the resulting PNCs would reduce [161].

Global Market and Applications Nano–magnesium hydroxide has various applications ranging from flame-retardants, to health care, to functional fillers. Almost 70% of the n-$Mg(OH)_2$ is presently used as flame-retardant by cable and wire industries [162]. However, it also finds limited applications in other industries, such as in textile industry, because of its ease of dispersion, production of chemicals especially polymers, building material, and in health-care industry, because of its antibacterial properties. The emerging application areas where research efforts are underway are in weaponry, aerospace, defense, and specialty cables.

About 77% of the global consumption is in East Asia and North America. Despite its technical superiority, the high cost and less awareness of this product restricted its consumption. The price of this material is a factor of process of manufacturing along with the volumes besides its size and purity. The hub for the manufacturers of *n*-$Mg(OH)_2$ is located in China and the United States, accounting for 80% of the global capacity. There is large underutilization of existing production capacity due to less demand. The Indian market is still in progress of development due to lack of awareness.

The market for *n*-$Mg(OH)_2$ reached 646 metric ton in 2015 despite its high cost and lack of awareness. Global magnesium hydroxide market was valued at US$ 551.4 million in 2015 and is expected to register a moderate single-digit CAGR in terms of value during the forecast period, 2016–2026 [162].

Magnesium Hydroxide Nanoparticle Suppliers There are several U.S. suppliers of magnesium hydroxide nanopowder. Some of them include:

- *US Research Nanomaterials, Inc.*, located in Houston, Texas (website: www.us-nano.com). The $Mg(OH)_2$ nanoparticles produced by it are in the form of a white powder with 99% purity, average diameter of 10 nm, bulk density of about 0.23 g/cc, true density of 2.36 g/cc, and specific surface area of 90 m^2/g.
- *American Elements* is located in Los Angeles, California (website: www.americanelements.com). It produces four $Mg(OH)_2$ nanoparticles with 99%, 99.9%, 99.99%, and 99.999% purity. It is a white flaky powder, with a MW of 58.83, true density of 2.34 g/cc, bulk density of about 0.23 g/cc, melting point of 350°C, average diameter of 10 nm, and specific surface area of 90 m^2/g.
- *Nanostructured & Amorphous Materials, Inc.* is located in Houston, Texas (website: www.nanoamor.com). Its $Mg(OH)_2$ nanoparticles come with 99% purity. It is a white powder, with true density of 2.36 g/cc, bulk density of about 0.21 g/cc, average particle diameter of 15 nm, and specific surface area of 80 m^2/g.

There are also suppliers from China, such as Haicheng Xinnuoxier Mining Industry Company Ltd. that sells *n*–magnesium hydroxide particles in industrial quantities. Key players operating in the magnesium hydroxide market include Nedmag Industries Mining & Manufacturing B.V., Nabaltec AG, Huber Engineering Materials, Kyowa Chemical Industry Company Ltd., Konoshima Chemical Company Ltd., Tateho Chemical Industries Company Ltd., Baymag Inc., Akzo Nobel, Albemarle Corp., Atmatis GmbH, BASE SE, Chemtura Corp., China Minmetals Non-ferrous Metals Co. Ltd., Cytec Industries, Israel Chemicals, and others [162].

Nano-Silicon Carbide (n-SiC)

Origin Non-oxide ceramics, such as silicon carbide, boron carbide, and silicon nitride, are attractive structural materials for high-temperature applications because of their low density, very high hardness, and excellent thermal and chemical stability [163–166]. Despite these advantages they are susceptible to fast fracture during mechanical loading due to their inherently brittle nature. Silicon carbide nanoparticles possess high purity, narrow range particle size distribution, and larger specific surface area. It has chemical stability, high thermal conductivity, smaller thermal expansion coefficient, and better abrasion resistance. The micro hardness of SiC nanoparticles is 2,840 to about 3,320 kg/mm^2 and hardness is between corundum and diamond. Its mechanical strength is higher than corundum.

Applications Applications of silicon carbide nanoparticles are numerous. They can be used as follows:

- *As reinforcing phase for silver alloy:* By adding *n*-SiC particles as an effective reinforcing phase for silver alloy to improve material properties. With 1 wt% of *n*-SiC particles added to the silver alloy, tensile strength, thermal conductivity, and intensity, and abrasion resistance will greatly be improved.
- *To strengthen copper compound materials:* Copper matrix compound material of high intensity and high conductivity has been widely applied to electronic parts and equipment. The high intensity and high conductivity of copper alloy are seeming contradictory properties. Usually, mechanical properties are improved at the expense of thermal conductivity. The stable *n*-SiC dispersion-strengthened copper matrix material is an ideal solution to this problem. By adding a small amount of *n*-SiC to the matrix, it reinforced the copper alloy strength.
- *To enhance abrasion resistance in rubber tires and rubber products:* When *n*-SiC particles were added in small amount in rubber tires, the improvement in abrasion resistance of the tires by 15–30% was reported without alternating original rubber formula of degrading the performance of the tires. SiC nanopowder can also be applied to other rubber products with the requirements of abrasion resistance, heat dissipation, and heat resistant, such as rubber roller and fuser film sleeve for printers.
- *To improve performance of engineering polymers.* When *n*-SiC particles were treated by coupling agent and added about 10 wt% to engineering polymers, such as PI, PEEK, PTFE, improvements have been reported in abrasion, thermal conductivity, insulation, tensile strength, impact resistance, and high temperature resistance of these polymers.
- *To prepare ceramic matrix composites.* Ceramic matrix composite is prepared by compounding multiple nanoparticles with ceramic as matrix. Types of ceramic matrix include silicon nitride, silicon carbon, and others. These advanced ceramics are advantageous in high-temperature resistance, and are high intensity and rigid, light weight, and corrosion resistant, but quite brittle. Combining nanoparticles and ceramic matrix into a compound is an effective approach to improving the reliability of ceramics and producing nano-reinforced ceramic matrix composites with excellent tenacity. Ceramic matrix composites have been applied to tools, sliding components, engine parts, and energy device. It can be used as structural components where wear resistance, high-temperature resistance, corrosion resistance were sorted in machinery, chemical, energy aviation, space, and defense industries.
- *In radar and infrared stealth materials:* Nano-ceramic powder is a new type of ceramic infrared-absorbing agent, mainly including silicon carbide and silicon nitride nanopowders. They can absorb a larger range of wave band and their absorption strength is great. The compound of silicon carbon nanoparticles and nano-magnetic absorbing agent (i.e., magnetic metal powder) will have its absorption effect improved greatly.
- *In coating materials.* Incorporation of *n*-SiC to coating materials can improve the abrasion resistance, corrosion resistance, and oxidative stability properties. It has been shown that incorporation of nanoparticles of silicon carbide, zirconium carbide, titanium carbide, titanium nitride, and boron carbide into compound

coating of metal can provide superior abrasion resistance and self-lubrication. The achieved abrasion resistance is 100 times higher than steel bearing, friction coefficient being 0.06–0.1; in addition, *n*-SiC also provides the coating material with high-temperature stability and abrasion resistance. These coatings were applied to specific components of liquid rocket to extend the service life of these parts.

Nano-Silicon Carbide Suppliers There are several U.S. suppliers of nano–silicon carbide:

- *US Research Nanomaterials, Inc.* is located in Houston, Texas (website: www.us-nano.com). Its silicon carbide nanopowder (SiC, beta) is a grayish white powder with 99+% purity, manufactured by plasma CVD process with a cubic morphology, average diameter of less than 80 nm, bulk density of 0.05 g/cc, true density of 3.216 g/cc, specific surface area of 25–50 m^2/g, and zeta potential of –27.8 mV.
- *American Elements* is located in Los Angeles, California (website: www.americanelements.com). It has four SiC nanoparticles with 99%, 99.9%, 99.99%, and 99.999% purity. They are spherical high surface particles, with a MW of 40.1, true density of 3.22 g/cc, melting point of 2,730°C, diameter range of 10–150 nm, and specific surface area range of 10–75 m^2/g.
- *Nanostructured & Amorphous Materials, Inc.* is located in Houston, Texas (website: www.nanoamor.com). Their SiC beta nanoparticles comes with 97.5% purity, manufactured by plasma CVD process. It is a grayish-white powder with a spherical morphology, true density of 3.22 g/cc, bulk density of 0.068 g/cc, and particle diameter range of 45–55 nm.
- *NanoAmor* is located in Europe (website: www.nanoamor-europe.com). NanoAmor's SiC beta nanoparticles are 97+% purity with a melting point of 2,700°C. It is a black powder with a nearly spherical morphology, true density of 3.22 g/cc, bulk density of about 0.051 g/cc, an average particle diameter of 10 nm and a maximum particle diameter of 20 nm, and a specific surface area of 150–200 m^2/g.

There are also manufacturers/suppliers from China, such as Linyi Jinmeng Carborundum Co., Ltd. (website: www.lyjmsic.en.alibaba.com) and NaBond Technologies Co., Ltd. (www.nabond.com), that sell *n*–silicon carbide particles in industrial quantities.

Nano-Silver (n-Ag)

Origin Silver nanoparticles have unique optical, electrical, and thermal properties and are being incorporated into products that range from photovoltaic to biological, to chemical sensors. Examples include conductive inks, pastes, and fillers which utilize silver nanoparticles for their high electrical conductivity, stability, and low sintering temperatures. Additional applications include molecular diagnostics and photonic devices, which take advantage of the novel optical properties of these nanomaterials. An increasingly common application is the use of silver nanoparticles for antimicrobial coatings, and many textiles, keyboards, wound dressings, and biomedical devices now contain silver nanoparticles that continuously release a low level of silver ions to provide protection against bacteria. Understanding how the size, shape, surface, and aggregation state of the silver nanoparticles change after integration into a target application is critical for optimizing performance. Silver nanoparticle manufacturer offers precisely manufactured monodisperse silver nanoparticles that are free from agglomeration, making them ideal for research, development, and use in a variety of innovative applications. Each batch of nanoparticles is extensively characterized using TEM images, dynamic light scattering (for particle size analysis), Zeta potential measurements, and UV/Visible spectral analysis to ensure consistent materials in every batch [167].

Properties and Applications Silver nanoparticles (*n*-Ag) exhibit antimicrobial, conductive, and optical properties making them suitable for a wide variety of applications, such as coatings and ink, polymers and nanocomposites, printed electronics, medical devices, and molecular diagnostics [167].

Medical Applications Silver nanoparticles are widely incorporated into wound dressings, and are used as an antiseptic and disinfectant in medical applications and in consumer goods. They have a high surface area per unit mass and release a continuous level of silver ions into their environment. The silver ions are bioactive and have broad-spectrum antimicrobial properties against a wide range of bacteria. By controlling the size, shape, surface, and agglomeration state of the *n*-Ag particles, specific silver ion release profiles can be developed for a given application.

Conductive Composites Incorporation of *n*-Ag particles into polymers, composites, and adhesives increases the electrical and thermal conductivity of the material. Silver pastes and epoxies are widely utilized in the electronics industries. Silver nanoparticle–based inks are used to print flexible electronics and have the advantage that the melting point of the small *n*-Ag particles in the ink is reduced by hundreds of degrees compared to bulk silver. When sintered, these *n*-Ag particle–based inks have excellent conductivity.

Plasmonics Silver nanoparticles have unique optical properties because they support surface plasmons. At specific wavelengths of light, the surface plasmons are driven into resonance and strongly absorb or scatter incident light. This effect is so strong that it allows for individual nanoparticles as small as 20 nm in diameter to be imaged using a conventional dark field microscope. This strong coupling of metal nanostructures with light is the basis for the new field of plasmonics. Applications of plasmonic silver nanoparticles include biomedical labels, sensors, and detectors. It is also the basis for analysis techniques, such as surface enhanced Raman spectroscopy (SERS) and surface enhanced fluorescence spectroscopy.

Photovoltaics There is an increasing interest in utilizing the large scattering and absorption cross sections of plasmonic silver nanoparticles for solar applications. Since the *n*-Ag particles act as efficient optical antennas, very high efficiencies can be obtained when the nanoparticles are incorporated into collectors. Figure 2.38 shows that silver nanosphere optical properties can be tuned by precisely controlling the nanoparticle size [167]. In this figure, nanoComposix's NanoXact silver nanospheres with sizes ranging from 20 to 80 nm is shown.

Figure 2.38 Silver nanosphere optical properties can be tuned by precisely controlling the nanoparticle size. Here, NanoXact silver nanospheres with sizes ranging from 20 to 80 nm are shown. (*Courtesy of nanoComposix.*)

Nano-Silver Particle Suppliers There are several U.S. suppliers of *n*-silver particles:

- *nanoComposix, Inc.* is located in San Diego, California (website: www.nanocomposix.com). It specializes in silver nanoparticles and has several types of products: (1) precisely engineered silver nanospheres, (2) silver plates with tuned wavelength, (3) re-dispersible powders in customer's choice of solvent, (4) silica-coated spheres with increased versatility, (5) silver nanowires with high aspect ratio, and (6) custom nanoparticles with a particle size, material, or surface coating.
- *American Elements* is located in Los Angles, California (website: www.americanelements.com). It has five *n*-silver particles with 99%, 99.5%, 99.9%, 99.95%, and 99.99% purity. They are silver high surface particles with spherical morphology, cubic crystal phase, a MW of 107.87, true density of 10.49 g/cc, bulk density of 0.312 g/cc, melting point of 962°C, boiling point of 2,162°C, diameter range of 80–100 nm, and specific surface area of 5.37 m^2/g.
- *Nanogap Inc.* is located in Richmond, California (website: www.nanogap-usa.com). It has three types of silver products: (1) atomic quantum clusters (AQC) are stable clusters of between 2 and 150 atoms, with average particle size in the range of 0.5–2 nm, (2) NanoParticles are manufactured by different processes than those used to produce AQC, and are composed of larger particles, typically in the size range of 5–50 nm, and (3) NanoFibers are a high aspect ratio fibrous product supplied as a dispersion (containing a small amount of polymer to avoid aggregation).
- *NovaCentrix, LLC* is located in Austin, Texas (website: www.nanocentrix.com). NovaCentrix manufactures silver nanopowder for both the conductive ink market and for biological uses. It has an average particle size of 25 nm (using TEM image analysis) and with a specific surface area of 23 m^2/g. X-ray diffraction data for silver nanopowder show the material to be highly crystalline. Silver nanopowder is used in electrically conductive inks for injection and flexographic printing.

Nano–Zinc Oxide (n-ZnO)

Origin Nano–zinc oxide powders and dispersions are inorganic zinc oxide nanoparticles with antibacterial, antifungal, anticorrosive, catalytic, and UV-filtering properties [168, 169]. Particles are available in size range of 10–200 nm, shown by Fan and Lu in their review article of *n*-ZnO [169]. They are also available as a nanofluid. Nanofluids are generally defined as suspended nanoparticles in solution either using surfactant or surface charge technology.

Applications A variety of research is underway in the following areas:

- *Nano electronic and photonic materials:* MEMS, NEMS, and piezoelectrics
- *Bionanomaterials:* Biomarkers, biodiagnostics, and biosensors
- *Medical:* Sunscreen agent used in cosmetics, antibacterial, health protection, UV protection, and dental cements
- *High-tech:* High-temperature lubricant in gas turbine engines, PNCs, composites, FR additives, and textiles

- *Energy:* Electrodes for solar cells, fuel cell layers, and pigments for paints.
- *Environmental:* Environmental remediation, gas sensors, photocatalytic decontamination, attenuation of UV light, and demilitarization of chemical and biological warfare agents.

Nano-Zinc Oxide Particle Suppliers There are several U.S. suppliers of nano–zinc oxide:

- *US Research Nanomaterials, Inc.* is located in Houston, Texas (website: www.us-nano.com). Its ZnO nanopowder is a milky-white powder with 99+% purity in single crystal phase with a nearly spherical morphology, average particle diameter of 10–30 nm, true density of 5.61 g/cc, and specific surface area of 20–60 m^2/g.
- *American Elements* is located in Los Angles, California (website: www.americanelements.com). It has three zinc oxide nanopowders with 99.9% (powder), 99.9% (dispersion), and 99.99% purity. The *n*-ZnO are spherical, white powders, with a MW of 81.37, true density of 5.6 g/cc, melting point of 1,975°C, boiling point of 2,360°C, diameter range of 10–200 nm, and exact mass of 79.92 g/mol.
- *Sciventions* is located in Toronto, Canada (website: www.sciventions.com). Its ZnO nanoparticles are sold in suspensions in concentration of 1.5 mg/mL. It is a colorless, transparent suspension with UV absorption of 288 nm, pH of 7 to 8, and particle size of 1–10 nm (90%).

Other Nanoparticles

There are many other nanomaterials, such as barium sulfate ($BaSO_4$) [170–172], barium titanate ($BaTiO_3$) [173–175], calcium carbonate ($CaCO_3$) [176–178], Fe_3O_4 [179–181], Fe_2O_3 [182–184], nanodiamond [185–187], zinc sulfide (ZnS) [188–191], and others have been used to produce PNCs with enhanced material properties.

2.3 Summary

In this chapter we discussed several commonly used nanomaterials in one-, two-, and three-dimensional nanoscale materials needed to form PNCs. The origin, properties, features, and applications of these nanomaterials were presented, providing the readers with the fundamentals and properties of these nanomaterials and the knowledge to select the proper nanoparticles for their applications.

2.4 Study Questions

2.1 Give examples of other nanomaterials that are commonly used and are not discussed in this chapter and quote their references.

2.2 Why is organic treatment so important for montmorillonite nanoclays? Name and describe the dispersion mechanisms and equipment for nanoclay within the polymer. What are some advantages of BYK's Cloisite Na+ nanoclay?

2.3 For the nanoclay dispersion categories, is there a set standard for industries using this material? For example, when is an unmixed polymer-clay nanocomposite acceptable versus when an intercalated polymer-clay nanocomposite is acceptable?

2.4 What is the intended major goal of carbon nanofiber in the industry? Describe some of the multifunctional performance that carbon nanofibers can offer as well as the potential use for each of the factor listed.

2.5 Why have carbon nanotubes attracted so much attention for research? Why does carbon nanotube have such a stiff and robust structure? Why do materials like buckytubes and small-diameter carbon nanotubes possess such amazing properties?

2.6 Are there specific industries using specific carbon nanotubes (SWNTs, MWNTs, DWNTs) for their applications?

2.7 What are some special features of HNT-polymer nanocomposite properties enhancements?

2.8 What are some special features of nickel nanostrands?

2.9 What are the two unique structural factors that POSS possess and how do these factors enhance POSS? What is the key purpose of POSS technology and what are some of the property enhancements that POSS can offer?

2.10 How does nano–titanium dioxide compare to other flame-retardant and improved heat stability polymer nanocomposites?

2.11 How does nano-titanium influence sunscreen? For instance, does nano-titanium actually reflect UV rays? If so, does the amount of nano-titanium influence the SPF rating of different sunscreens?

2.12 What is the one important characteristic of nanosilica that stands out? Are there any problems that occur with the introduction of these nanoparticles into polymers?

2.5 References

1. Pinnavaia, T. J., and Beall, G. W., eds. (2000) *Polymer-Clay Nanocomposites*. New York: Wiley.
2. Krishnamoorti, R., and Vaia, R. A., eds. (2001) *Polymer Nanocomposites: Synthesis, Characterization, and Modeling*, ACS Symposium Series 804. Washington, DC: American Chemistry Society.
3. Koo, J. H. (2006) *Polymer Nanocomposites: Properties, Characterization, and Applications*. New York: McGraw-Hill.
4. Morgan, A. B., and Wilkie, C. A., eds. (2007) *Flame Retardant Polymer Nanocomposites*. Hoboken, NJ: Wiley.
5. Gupta, R. A., Kennel, E., and Kim, K. J., eds. (2010) *Polymer Nanocomposites Handbook*. Boca Raton, FL: CRC Press.
6. Mittal, V., ed. (2010) *Polymer Nanotube Nanocomposites: Synthesis, Properties, and Applications*. Hoboken, NJ: Wiley.
7. Mittal, V., ed. (2010) *Optimization of Polymer Nanocomposites Properties*. Weinheim, Germany: Wiley-VCH.
8. Mittal, V., ed. (2011) *Thermally Stable and Flame Retardant Polymer Nanocomposites*. Cambridge, UK: Cambridge University Press.
9. Beall, G. W., and Powell, C. E. (2011) *Fundamentals of Polymer-Clay Nanocomposites*. Cambridge, UK: Cambridge University Press.
10. Mittal, V., ed. (2012) *Characterization Techniques for Polymer Nanocomposites*. Weinheim, Germany: Wiley-VCH.

11. Briell, B. (2000) Nanoclay—Counting on Consistency, presented at *Nanocomposite 2000*. Southern Clay Products, Gonzales, TX.
12. Southern Clay Products, Gonzales, TX (website: www.nanoclay.com).
13. Nanocor, Chicago, IL (website: www.nanocor.com).
14. Geim, A. K., and Novoselov, K. S. (2007) The rise of graphene. *Nat. Mater.*, 6:183. doi: 10.1038/nmat1849.
15. Jang, B. Z., and Zhamu, A. (2000) Processing of nanographene platelets (NGPs) and NGP nanocomposites: A review. *J. Materials Science*, 43:5092–5101.
16. Novoselov, K. S., Geim, A. K., Morozov, S. V., et al. (2004) Electric field effect in atomically thin carbon film. *Science*, 306:666. doi: 10.1126/science.1102896.
17. Novoselov, K. S., Jiang, D., Schedin, F., et al. (2005) Two-dimensional atomic crystals. *Proc. Natl. Acad. Sci. U. S. A.*, 102:10451. doi: 10.1073/pnas.0502848102.
18. Jang, B. Z., and Huang, W. C. (2006) US Patent 7,071,258 (July 4, 2006).
19. Jang, B. Z. (2006) US Patent 11/442,903 (June 20, 2006); a divisional of 10/274,473 (October 21, 2002).
20. Schwalm, W., Schwalm, M., and Jang, B. Z. (2004) Local Density of States for Nanoscale Graphene Fragments. *American Physical Society*, Montreal, Canada.
21. McAllister, M. J., Li, J. L., Adamson, D. H., et al. (2007) Single sheet funcationalized graphene by oxidation and thermal expansion of graphite. *Chem. Mater.*, 19:4396. doi: 10.1021/cm0630800.
22. Li, J. L., Kudin, K. N., McAllister, M. J., et al. (2006) Oxygen-driven unzipping of graphitic materials. *Phys. Rev. Lett.*, 96:176101. doi: 10.1103/PhysRevLett.96.176101.
23. Schniepp, H. C., Li, J. L., McAllister, M. J., et al. (2006) Functionalized single graphene sheets derived from splitting graphite oxide. *J. Phys. Chem. B*, 110:8535. doi: 10.1021/jp060936f.
24. Li, X., Wang, X., Zhang, L., Lee, S., et al. (2008) Chemically derived, ultrasmooth graphene nanoribbon semiconductors. *Science*, 319:1229–1231. doi: 10.1126/science.1150878.
25. Novoselov, K. S., Geim, A. K., Morozov, S. V., et al. (2005) Two-dimensional gas of massless Dirac fermions in graphene. *Nature*, 438:197. doi: 10.1038/nature04233.
26. Zhang, Y., and Ando, T. (2002) Hall conductivity of two-dimensional graphite system. *Phys. Rev. Lett*, B65:245420.
27. Zhang, Y., Tan, Y. W., Stormer, H. L., et al. (2005) Experimantal observation of the quantum Hall Effect and Barry's phase in graphene. *Nature*, 438:201. doi: 10.1038/nature04235.
28. Zhang, Y., Small, J. P., Amori, M. E., et al. (2005) Electric field modulation of Galvanomagnetic properties of mesoscopic graphite. *Phys. Rev. Lett.*, 94:176803. doi: 10.1103/PhysRevLett.94.176803.
29. Berger, C., Song, Z., Li, T., et al. (2004) Ultrathin epotaxial graphite: 2D Electron gas properties and a route towards graphene-based nanoelectronics. *J. Phys. Chem. B*, 108:19912. doi: 10.1021/jp040650f.
30. Enoki, T., and Kobayashi, Y. (2005) Magnetic nanographite: an approach ot molecular magnetism. *J. Mater. Chem.*, 15:3999. doi: 10.1039/b500274p.
31. Heersche, H. B., Jarillo-Herrero, P., Oostinga, J. B., et al. (2007) Bipolar supercurrent in graphene. *Nat. Lett.*, 446:56. doi: 10.1038/nature05555.
32. Soon, Y. W., Cohen, M. L., and Louie, S. G. (2006) Half-metallic graphene nanoribbons. *Nat. Lett.*, 444:347. doi: 10.1038/nature05180.
33. Meyer, J. C., Geim, A. K., Katsnelson, M. I., et al. (2007) The structure of suspended grpahene sheets. *Nat. Lett.*, 446:60. doi: 10.1038/nature05545.

34. Bunnell, L. R., Sr. (1991) US Patent 987(4):175.
35. Bunnell, L. R., Sr. (1991) US Patent 019(5):446.
36. Bunnell, L. R., Sr. (1993) US Patent 186(5):919.
37. Zaleski, P. L., Derwin, D. J., Girkant, R. J., et al. (2001) US Patent 287(6):694.
38. Horiuchi, S., Gotou, T., Fujiwara, M., et al. (2004). Single graphene sheet detected in a carbon nanofilm. *Appl. Phys. Lett.*, 84:2403.
39. Horiuchi, S., Gotou, T., Fujiwara, M., et al. (2003) Carbon nanofilm with a new structure and property. *Jpn. J. Appl. Phys.*, 42(Part 2):L1073. doi: 10.1143/JJAP.42. L1073.
40. Hirata, M., and Horiuchi, S. (2003) US Patent 596(6):396.
41. Hirata, M., Gotou, T., and Ohba, M. (2005) Thin-film particles of graphite oxide. 2" Preliminary studies for internal micro fabrication of single particle and carbonnaceous electronic circuits. *Carbon*, 43:503. doi: 10.1016/j.carbon.2004.10.009.
42. Hirata, M., Gotou, T., Horiuchi, S., et al. (2004) Thin-film particles of graphite oxide. 1: High-yield synthesis and flexibility of the particles. *Carbon*, 42:2929.
43. Udy, J. D. (2006) US Patent Application No. 11/243,285 (October 4, 2005); Pub No. 2006/0269740 (November 30, 2006).
44. Chen, G. H. (2004) Preparation and characterization of graphite nanosheets from ultrasonic powder technique. *Carbon*, 42:753. doi: 10.1016/j.carbon.2003. 12.074.
45. Jang, B. Z., Wong, S. C., and Bai, Y. (2005) US Patent Appl. No. 10/858,814 (June 3, 2004); Pub. No. US 2005/0271574 (Pub. December 8, 2005).
46. Petrik, V. I. (2006) US Patent Appl. No. 11/007,614 (December 7, 2004); Pub. No. US 2006/0121279 (Pub. June 8, 2006).
47. Drzal, L. T., and Fukushima, H. (2006) US Patent Appl. No. 11/363,336 (February 27, 2006); 11/361,255 (February 24, 2006); 10/659,577 (September 10, 2003).
48. Mack, J. J., Viculis, L. M., Kaner, R. B., et al. (2005) US Patent 872(6):330.
49. Viculis, L. M., Mack, J. J., and Kaner, R. B. (2003) Intercalation and exfoliation routes to graphite nanoplatelets. *Science*, 299:1361. doi: 10.1126/science.1078842.
50. Lu, W., Soukiassian, P., and Boecki, J. (2012) Graphene: fundamentals and functionalities. *MRS Bulletin*, 37.
51. Mukhopadhyay, P., and Gupta, R. K., eds. (2013) *Graphite, Graphene and Their Polymer Nanocomposites.* Boca Raton, FL: CRC Press.
52. Jang, B. Z., Zhamu, A., and Song, L. (2006) US Patent Application No. 11/324,370 (January 4, 2006).
53. Song, L., Guo, J., Zhamu, A., et al. (2006) US Patent Application No. 11/328,880 (January 11, 2006).
54. Sullivan, M. J., and Ladd, D. A. (2006) US Patent 7,156,756 (January 2, 2007) and No.7,025,696 (April 11, 2006).
55. Jang, B. Z. (2007) US Patent 186(7):474.
56. Szabo, T., Szeri, A., and Dekany, I. (2005) Composite graphitic nanolayers prepared by self-assembly between finely dispersed graphite oxide and a cationic polymer. *Carbon*. 43(1), 87–94.
57. Wang, X., Zhi, L., and Mullen, K. (2008) Transparent, conductive graphene electrodes for dyesensitized solar cells. *Nano Lett.*, 8(1):323. doi: 10.1021/nl072838r.
58. Koo, J. H., Pinero, D., Hao, A., Lao, S. C., Johnson, B., Baek, M. G., Lee, J., et al. (2013) Methodology for assessment of the morphological and thermal characteristics of nanographene platelets, AIAA-2013-1584, presented at the *54th AIAA/ASME/ASCE/ AHS/ASC, SDM*, Boston, MA, April 8–11, 2013.

59. Ávila, A. F. Department of Mechanical Engineering, Universidade Federal de Minas Gerais, Belo Horizonte, Brazil (e-mail: aavila@netuno.lcc.ufmg.br).
60. Ávila, A. F. (2009) Composite laminates performance enhancement by nanoparticles dispersion: an investigation on hybrid nanocomposite. In *Composites Performance and Trends*, Columbus, F., ed. Hauppauge, NY: Nova Science Publishers.
61. Miller, S. G. (2008) Effects of nanoparticle and matrix interface on nanocomposite properties. Ph.D. dissertation, University of Akron, Akron, OH.
62. Schmidt, H. K. Chemical and Biomolecular Engineering Department, Rice University, Houston, TX.
63. XG Sciences, Inc. at East Lansing, MI (website: www.xgsciences.com).
64. Angstron Materials, LLC at Dayton, OH (website: www.angstronmaterials.com).
65. Skyspring Nanomaterials, Inc., Houston, TX (website: www.ssnano.com).
66. Cheap Tubes, Inc., Brattleboron, VT (website: www.cheaptubesinc.com).
67. Ruoff, R. Department of Mechanical Engineering, Ulsan National Institute of Science and Technology (UNIST), Ulsan, S. Korea.
68. Nacional de Grafite, Sao Paulo, Brazil (website: www.grafite.com).
69. Qiu, L., and Qu, B. (2011) Polymer/layered double hydroxide flame retardant nanocomposites. In *Thermally Stable and Flame Retardant Polymer Nanocomposites*, Mittal, V., ed. Cambridge, UK: Cambridge University Press, pp. 332–359.
70. Matusinovic, Z., and Wilkie, C. A. (2012) Fire retardancy and morphology of layered double hydroxide nanocomposites: a review. *J. Mater. Chem*, 22:18701–18704.
71. Choudary, B. M., Bharathi, B., Reddy, C. V., Kantam, M. L., and Raghavan, K. V. (2001) The first example of catalytic N-oxidation of tertiary amines by tungstate-exchanged Mg-Al layered double hydroxide in water: A green protocol. *Chemical Communications*, 1736–1737.
72. Choy, J. H., Kwak, S. Y., Jeong, Y. J., and Park, J. S. (2000) Inorganic layered double hydroxides as nonviral vectors. *Angewandte Chemie International Edition*, 39:4042–4045.
73. Desigaux, L., Ben Belkacem, M., Richard, P., Cellier, J., Leone, P., Cario, L., Leroux, F., et al. (2006) Self-assembly and characterization of layered double hydroxide/DNA hybrids. *Nano Letters*, 6:199–204.
74. Lakraimi, M., Legrouri, A., Barroug, A., de Roy, A., and Besse, J. P. (1999) Removal of pesticides from water by anionic clays. *J. Chim. Phys. – Chim. Biol.*, 96:470–478.
75. Yan, D., Lu, J., Wei, M., Ma, J., Evans, D. G., and Duan, X. (2009) A combined study based on experiment and molecular dynamics: Perylene tetracarboxylate intercalated in a layered double hydroxide matrix. *Physical Chemistry Chemical Physics*, 11:920–929.
76. Tian, Y., Wang, G., Li, F., and Evans, D. G. (2007) Synthesis and thermo-optical stability of methyl red-intercalated Ni-Fe layered double hydroxide material. *Materials Letters*, 61:1662–1666.
77. Lukashin, A. V., Vertegel, A. A., Eliseev, A. A., Nikiforov, M. P. Gornert, P., and Tretyakov, Y. D. (2003) Chemical design of magnetic nanocomposites based on layered double hydroxides. *Journal of Nanoparticle Research*, 5:455–464.
78. Mohan, D., and Pittman, C. U. (2007) Arsenic removal from water/wastewater using adsorbents – A critical review. *Journal of Hazardous Materials*, 142:1–53.
79. Tibbetts, G. G. (1984) Why Are Carbon Filaments Tubular? *J. Crystal Growth*, 66:632.
80. Lake, M. L., and Ting, J.-M. (1999) Vapor grown carbon fiber composites. In *Carbon Materials for Advanced Technologies*, Burchell, T. D., ed. Oxford, UK: Pergamon.

81. Tibbetts, G. G., Finegan, J. C., McHugh, J. J., Ting, J.-M., Glasgow, D. G., and Lake, M. L. (2000) Applications research on vapor-grown carbon fibers. In *Science and Application of Nanotubes,* Tomanek, E., and Enbody, R. J., eds. New York: Kluwer Academic/Plenum Publishers.
82. Maruyama, B., and Alam, K. (2002) Carbon nanotubes and nanofibers in composite materials. *SAMPE J.,* 38(3):59.
83. Glasgow, D. G., Tibbetts, G. G., Matuszewski, M. J., Walters, K. R., and Lake, M. L. (2004) Surface treatment of carbon nanofibers for improved composite mechanical properties. *Proc. SAMPE 2004 Int'l Symposium,* SAMPE, Covina, CA.
84. Tibbetts, G. G., Lake, M. L., Strong, K. L., and Rice, B. P. (2007) A review of the fabrication and properties of vapor-grown carbon nanofiber/polymer composites. *Comp. Sci. Tech.,* 67(7–8):1709–1718.
85. Terrones, M. (2003) Science and technology of the twenty-first century: Synthesis, properties, and applications of carbon nanotubes. *Ann. Rev. Mater. Res.,* 33:419.
86. Harris, P. J. F. (1999) *Carbon Nanotubes and Related Structures, New Materials for the Twenty-First Century.* Cambridge, UK: Cambridge University Press.
87. Tanaka, K., Yamabe, T., and Fukui, K. (1999) *The Science and Technology of Carbon Nanotubes.* Amsterdam, The Netherlands: Elsevier.
88. Saito, R., Dresselhaus, G., and Dresselhaus, M. S. (1998) *Physical Properties of Carbon Nanotubes.* London: Imperial College Press.
89. Dai, L., ed. (2006) *Carbon Nanotechnology.* Amsterdam, The Netherlands: Elsevier.
90. Dresselhaus, M. S., Dresselhaus, G., and Eklund, P. C. (1996) *Science of Fullerenes and Carbon Nanotubes.* San Diego, CA: Academic Press.
91. Ebbesen, T. W. (1994). Carbon Nanotubes. *Ann. Rev. Mater. Sci.,* 24:235.
92. Guo, T., Nikolaev, P., Thess, A., Colbert, D. T., and Smalley, R. E. (1995) Catalytic growth of of single-walled nanotubes by laser vaporization. *J. Phys. Chem.,* 55:10694.
93. Endo, M., Takeuchi, K., Igarashi, S., Kobori, K., Shiraishi, M., and Kroto, H. W. (1993) The production and structure of pyrolytic carbon nanotubes. *J Phys. Chem. Solids,* 54:1841.
94. Groning, O., Kuttel, O. M., Emmenegger, C., Groning, P., and Schlapbach, L. (2000) Field Emission Properties of Carbon Nanotubes. *J Vac Sci Technol,* B18:665.
95. Hsu, W. K., Hare, J. P., Terrones, M., Kroto, H. W., Walton, D. R. M., and Harris, P. J. F. (1995) Condensed-phase nanotubes. *Nature,* 377:687.
96. Hsu, W. K., Terrones, M., Hare, J. P., Terrones, H., Kroto, H. W., and Walton, D. R. W. (1996) Electrolytic formation of carbon nanostructures. *Chem. Phys. Lett.,* 262:161.
97. Iijima, S. (1991) *Nature,* 354:56.
98. Iijima, S., and Ichihashi, T. (1993) Single-Shell Carbon Nanotubes of 1-nm Diameter. *Nature,* 363:603.
99. Bethune, D. S., Kiang, C. H., de Vries, M. S., Gorman, G., Savoy, R., Vazquez, J., and Beyers, R. (1993) The discovery of single-wall carbon nanotubes at IBM. *Nature,* 363:605.
100. Thess, A., Lee, R., Nikolaev, P., Dai, H., Petit, P., Robert, J., Xu, C., et al. (1996) Crystalline ropes of metallic carbon nanotubes. *Science,* 273:483.
101. Ajayan, P. M., Stephan, O., Colliex, C., and Trauth, D. (1994) Aligned carbon nanotube arrays formed by cutting a polymer resin-nanotube composite. *Science,* 265:1212.
102. Kim, P., Shi, L., Majumdar, A., and McEuen, P. L. (2001) Thermal transport measurements of individual multiwalled nanotubes. *Phys. Rev. Lett.,* 87:215502.

103. Cao, G. (2004) *Nanostructures & Nanomaterials Synthesis, Properties & Applications.* London: Imperial College Press.
104. Smith, K. (2005, June) Carbon nanotechnologies, Inc., Houston, TX, personal communication.
105. Nanocyl, Sambreville, Belgium, Nanocyl™ SWNT, DWNT, MWNT (website: www.nanocyl.com).
106. Bayer MaterialScience, Leverkusen, Germany, Baytubes® MWNT (website: www.baytubes.com).
107. Arkema, Lacq, France, Graphistrength® MWNT (website: www.graphistength.com).
108. Du, M., Guo, B., and Jia, D. (2010) Newly emerging applications of halloysite nanotubes: a review. *Polym. Int.,* 59:574–582.
109. Liu, M., Jia, Z., Jia, D., and Zhou, C. (2014) Recent advances in research on halloysite nanotubes-polymer nanocomposite. *Progress in Polymer Science,* 39(8):1498–1525.
110. Yuan, P., Tan, D., and Annabi-Bergaya, F. (2015) Properties and applications of halloysite nanotubes: recent research advances and future prospects. *Applied Clay Science,* 112–113:75–93.
111. Zhang, Y., Tang, A., Yang, H., and Quyang, J. (2016) Applications and interfaces of halloysite nanocomposites. *Applied Clay Science,* 119:8–17.
112. Berthier, P. (1826) Analyse de l'halloysite. *Annales de Chimie et de Physique,* 32:332–325.
113. Prudencio, M. I., Braga, M. A. S., Paquet, H., Waerenborgh, J. C., Pereira, L. C. J., and Gouveia, M. A. (2002) Clay mineral as severages in weathered basalt profiles from central and southern Portugal: Climate. *Catena,* 49(1):77.
114. Joussein, E., Petit, S., Churchman, J., Theng, B., Righi, D., and Delvaux, B. (2005) Halloysite clay minerals-a review. *Clay Minerals,* 40:383.
115. Nakagaki, S., and Wypych, F. (2007) Nanofibrous and nanotubular supports for the immobilization of metalloporphyrins as oxidation catalysts. *J. Colloid. Interf. Sci.,* 315:142.
116. Wilson, I. R. (2004) Kaolin and halloysite deposits of China. *Clay Minerals,* 39:1.
117. Perruchot, A., Dupuis, C., Brouard, E., Nicaise, D., and Ertus, R. (1997) L'halloysite Kartstique: Comparasion des Gisements Types de Wallonie (Beligique) et du Perigord (France). *Clay Miner,* 32:271.
118. Kloprogge, J. T., and Frost, R. L. (1999) Raman microprobe spectroscopy of hydrated halloysite from Neogene Cryotokarst from Southern Belgium. *J. Raman Spectrosc.,* 30:1079.
119. Churchman, G. J., and Theng, B. K. G. (2002) Clay research in Australia and New Zealand. *Appl. Clay. Sci.,* 20:153.
120. Kautz, C. Q., and Ryan, P. C. (2003) The 10 Å and 7 Å halloysite transition in a tropical soil sequence. *Clay Miner,* 51:252.
121. Hillier, S., and Ryan, P. C. (2002) Identification of halloysite (7 Å) by ethylene glycol solvation: the 'MacEwan effect'. *Clay Miner,* 37:487.
122. Du, M. L., Guo, B. C., Cai, X. J., Jia, Z. X., Liu, M. X., and Jia, D. M. (2008) Morphology and properties of halloysite nanotubes reinforced polypropylene nanocomposites. *e-Polymers,* 130:1.
123. Ye, Y. P., Chen, H. B., Wu, J. S., and Ye, L. (2007) High strength epoxy nanocomposites with natural nanotubes. *Polymer,* 48:6426–6433.
124. Liu, M. X., Guo, B. C., Du, M. L., Cai, X. J., and Jia, D. M. (2007) Properties of halloysite nanotubes-epoxy resin hybrids and the interfacial reactions in the systems. *Nanotechnology,* 18:455703.

125. Imai, T., Naitoh, Y., Yamamoto, T., and Ohyanagi, M. (2006) Translucent nano mullite based ceramic fabricated by spark plasma. *Journal of the Ceramic Society of Japan*, 114:138–140.
126. (2007) *Am. Ceram. Soc. Bull.*, 86:A19 .
127. Du, M. L., Guo, B. C., Liu, M. X., and Jia, D. M. (2006) Preparation and characterization of polypropylene grafted halloysite and their compatibility effect of polypropylene/halloysite. *Polym. J.*, 38:1198.
128. Ma, J., Xiang, P., Mai, Y. W., and Zhang, L. Q. (2004) A novel approach to high performance elastomer by using clay. *Macromol. Rapid Commun.*, 25:1692.
129. Guo, B. C., Lei, Y. D., Chen, F., Liu, X. L., Du, M. L., and Jia, D. M. (2008) Styrene-butadiene rubber/halloysite nanotubes nanocomposites modified by methacrylic acid. *Appl. Surf. Sci.*, 255:2715.
130. Du, M. L., Guo, B. C., Lei, Y. D., Liu, M. X., and Jia, D. M. (2008) Carboxylated butadiene-styrene rubber/halloysite nanotube nanocomposites: interfacial interaction and performance. *Polymer*, 49:4871.
131. Guo, B. C., Chen, F., Lei, Y. D., Zhou, W. Y., and Jia, D. M. (2010) Tubular clay composites with high strength and transparency. *J. Macromol. Sci. B: Phys.*, 49:1.
132. Liu, M. X., Guo, B. C., Du, M. L., Lei, Y. D., and Jia, D. M. (2008) Natural inorganic nanotubes reinforced epoxy resin nanocomposites. *J. Polym. Res.*, 15:205.
133. Du, M. L., Guo, B. C., and Jia, D. M. (2006) Thermal stability and flame retardant effects of halloysite nanotubes on poly(propylene). *Eur. Polym. J.*, 42:1362.
134. Labour, T., Gauthier, C., Seguela, R., Vigier, G., Bomal, Y., and Orange, G. (2001) Influence of the beta crystalline phase of the mechanical properties characterization. *Polymer*, 42:7127.
135. Tordjeman, P., Robert, C., Marin, G., and Gerard, P. (2001) The effect of α, β crystalline structure on the mechanical properties of polypropylene. *Eur. Phys. J. E.*, 4:459.
136. Ning, N. Y., Yin, Q. J., Luo, F., Zhang, Q., Du, R., and Fu, Q. (2007) Crystallization behavior and mechanical properties of polypropylene/halloysite composites. *Polymer*, 48:7374.
137. Du, M. L., Guo, B. C., Wan, J. J., Zou, Q. L., and Jia, D. M. (2010) Effects of halloysite nanotubes on kinetics and activation energy of non-isotherm crystallization of polypropylene. *J. Polym. Res.*, 17:109.
138. "Nanostrand User Guide," Conductive Composites, Heber City, Utah, USA.
139. Conductive Composites, Huber City, Utah, USA (website: www.conductivecomposites.com).
140. ANF Technology Ltd, 44a the Green, Warlingham, Surrey CR6 9NA, UK (website: www.nafen.eu).
141. Hybrid Plastics, Inc., Hattiesburg, Mississippi, USA (website: www.hybridplastics.com).
142. Voronkov, M. G., and Vavrent'yev, V. I. (1982) Polyhedral oligosilsesquioxanes and their homo derivatives. *Top. Curr. Chem.*, 102:199–236.
143. Agaskar, P. A., and Klemperer, W. G. (1995) The higher hudros[herosiloxanes: synthesis and structures of $H_nSi_nO_{1.5n}$ (n=12, 14, 16, 18). *Inorg. Chim. Acta.*, 229:355–364.
144. Baney, R. H., Itoh, M., Sakakibara, A., and Suzuki, T. (1995) Silsesquioxanes. *Chem. Rev.*, 92:1409–1430.
145. Lichtenhan, J. D. (1995) Polyhedral oligomeric silsesquioxanes: Building blocks for silsesquioxane-based polymers and hybrid materials. *Comments Inorg. Chem.*, 17:115–130.

146. Lichtenhan, J. D. (1996) *Polymeric Materials Encyclopedia*, J. C. Salamore, ed., Boca Raton , FL: CRC Press, pp. 7769–7778.
147. Li, G. Z., Wang, L. C., Ni, H. L., and Pittman, C. U. (2001) Polyhedral oligomeric silsesquioxane (POSS) polymers and copolymers: A review. *Inorg Orgaomet Polym*, 11:123–154.
148. Phillips, S. H., Haddad, T. S., and Tomczak, S. J. (2004) Developments in nanoscience: polyhedral oligomeric silsesquioxane (POSS)-polymers. *Curr Opin Sol State Mater Sci*, 8:21–29.
149. Sorathia, U., and Perez, I. (2004) Improving fire performance characteristics of composite materials for naval applications. *Polymeric Materials: Science & Engineering*, 91:292–296.
150. Hartman-Thompson, C., ed. (2011) *Applications of Polyhedral Oligomeric Silsesquioxanes*, New York, NY: Springer.
151. Technical Bulletin AEROSIL® No. 27, Degussa AG, D-63403 Hanau-Wolfgang, Germany, 10/2001.
152. Technical Bulletin AEROSIL® No. 56, Degussa AG, D-63403 Hanau-Wolfgang, Germany, 10/1990.
153. Technical Bulletin AEROSIL® Fumed Silica, Degussa AG, D-63403 Hanau-Wolfgang, Germany, 9/2002.
154. Sprenger S., and Pyrlik, M. (2004, August) Nanoparticles in composites and adhesives: synergy with elastomers. *Proc. 11th International Conference on Composites/Nano Engineering*, Hilton Head Island, SC.
155. Yang, F., Yngard, R., and Nelson, G. L. (2005) Flammability of polymer-clay and polymer-silica nanocomposites. *J Fire Sciences*, 23:209–226.
156. U.S. Patent Application, 20040147029, dated July 29, 2004.
157. Cinquin, J., Bechtel, S., Schmidtke, K., and Meer, T. (2004, August) Polymer nanocomposites of aeronautic applications: from dream to reality?" *Proc. 11th International Conference on Composites/Nano Engineering*, Hilton Head Island, SC.
158. Pool, A. D., and Hahn, H. T. (2003) A nanocomposite for improved stereolithography. *Proc. 2003 SAMPE ISSE*, SAMPE, Covina, CA.
159. "Inorganic Specialty Chemicals-Alumina Nano-particles," Sasol NA, Houston, TX.
160. "Disperal®/Dispal®—High purity dispersible alumina," technical datasheet, Sasol NA, Germany.
161. Huang, H., Tian, M., Liu, L., Liang, W., and Zhang, L. (2006) Effect of particle size of flame retardancy of Mg $(OH)_2$-filled ethylene vinyl acetate copolymer composites. *Journal of Applied Polymer Science*, 100:4461–4469.
162. Transparency Market Research. (2018) *Magnesium Hydroxide Market—Global Industry Analysis, Size, Share, Growth, Trend and Forecast 2018-2026*. Transparency Market Research, Pune, India.
163. Yong, V., and Hahn, H. T. (2004) Kevlar/Vinyl ester composites with SiC nanoparticles. *Proc. 2004 SAMPE ISSE*, SAMPE, Covina, CA.
164. Sakka, Y., Bidinger, D. D., and Aksay, I. A. (1995) Processing of silicon carbide-mullite-alumina nanocomposites. *J. Am. Ceram. Soc*, 78(21):479–486.
165. Padhi, P., and Sachikanta, K. (2011) A novel route for development of bulk Al/SiC metal matrix nano composites. Department of Mechanical Engineering, Konark Institute of Science & Technology, Bhubaneswar, India & Central Tool Room of Training Center, Bhubaneswar, India.

166. Kassiba, A., Bouclé, J., Makowska-Janusik, M., and Errien, N. (2007) Some fundamental and applicative properties of [polymer/nano-SiC] hybrid nanocomposites. *Journal of Physics: Conference Series*, 79. Conf. Ser. 79012002. doi: 10.1088/1742-6596/79/1/012002.
167. Oldenburg, S. J. (2005) Silver nanoparticles: properties and applications, nanoComposix, San Diego, CA (website: nanocomposix.com).
168. Wang, Z. L. (2004) Zinc oxide nanostructures: growth, properties and applications. *J. Phys. Condens. Matter.*, 16:R829–R858. PII: S0953-8984(04)58969-5.
169. Fan, Z., and Lu, J. G. (2005) Zinc oxide nanostructures: synthesis and properties. *Journal of Nanoscience and Nanotechnology*, 5(10):1–13. doi: 10.1166/jnn.2005.182.
170. Ricker, A., Liu-Snyder, P., and Webster, T. J. (2008) The influence of nano MgO and $BaSO_4$ particle size additives on properties of PMMA bone cement. *International Journal of Nanomedicine*, 3(1):125–132.
171. Aninwene, G., Stout, D., Yang, Z., and Webster, T. J. (2013) Nano-$BaSO_4$: a novel antimicrobial additive to Pellethane. *International Journal of Nanomedicine*, 8: 1197–1205.
172. Aninwene, G., Stout, D. A., Yang, Z., and Webster, T. J. (2013) Nano $BaSO_4$: a novel means to create antimicrobial radiopaque thermoplastics. *Proc. 2013 AIChE Annual Meeting*, November 3–8, 2013, San Francisco, CA.
173. Chanmal, C. V., and Jog, J. P. (2008) Dielectric relaxations in PVDF/$BaTiO_3$ nanocomposites. *eXpress Polymer Letters*, 2(3):294–301.
174. Beltran, H., Maso, N., Cordoncillo, E., and West, A. R. (2007) Nanocomposite ceramics based on La-doped $BaTi_2O_3$ and $BaTiO_3$ with high temperature-independent permittivity and low dielectric loss. *Journal of Electroceramics*, 18(3–4):277–282.
175. Singh, K. C., and Jiten, C. (2013) Production of $BaTiO_2$, nanocrystalline powders by high energy milling and piezoelectric properties of corresponding ceramics. *Key Engineering Materials*, 547:133–138.
176. Chatterjee, A., and Mishra, S. (2013) Rheological, thermal and mechanical properties of nano-calcium carbonate ($CaCO_3$)/poly(methyl methacrylate) (PMMC) core-shell nanoparticles reinforced polypropylene (PP) composites. *Macromolecular Research*, 21(5):474–483.
177. Shelesh-Nezhad, K., Orang, H., and Motallebi, M. (2012) The effects of adding nano-calcium carbonate particles on the mechanical and shrinkage characteristics and molding process consistency of PP/nano-$CaCO_3$ nanocomposites. In *Polypropylene*, Dogan, F., ed. InTech, pp. 357–378. doi: 10.5772/35272. Available from: http://www.intechopen.com/books/polypropylene/the-effects-of-adding-nano-calcium-carbonate-particles-on-the-mechanical-and-shrinkage-character.
178. Sato, T., and Beaudoin, J. J. (2011) Effect of nano-$CaCO_3$ on hydration of cement containing supplementary cementitious materials. *Advances in Cement Research*, 23(1):33–43.
179. Hu, C., Mou, Z., Lu, G., Chen, N., Dong, Z., Hu, M., and Qu, L. (2013) 3D graphene-Fe_3O_4 nanocomposites with high-performance microwave absorption. *Physical Chemistry Chemical Physics*, 15:13038–13043.
180. Gu, H., Huang, Y., Zhang, X., Wang, Q., Zhu, J., Shao, L., Haldolaarachchige, N., et al. (2012) Magnetoresistive polyaniline-magnetite nanocomposites with negative dielectric properties. *Polymer*, 53:801–809.

181. Kalantari, K., Ahmad, M. B., Shemeli, K., and Khandanlou, R. K. (2013) Synthesis of talc/Fe_3O_4 magnetic nanocomposites using chemical co-precipitation method. *International Journal of Nanomedicine*, 8:1817–1823.
182. Mahapatra, A., Mishra, B. G., and Hota, G. (2013) Electrospun Fe_2O_3-Al_2O_3 nanocomposite fibers as efficient absorbent for removal of heavy metal ions from aqueous solution. *Journal of Hazard Materials*, 258–259:116–123.
183. Ortega, D., Garitaonandia, J. S., Barrera-Solano, C., Ramirez-del-Solar, M., Blanco, E., and Dominguez, M. (2006) γ-Fe_2O_3/SiO_2 nanocomposites for magneto-optical applications: nanostructural and magnetic properties. *Journal of Non-Crystalline Solids*, 352:2801–2810.
184. Menon, L., Patibandla, S., Bhargava Ram, K., Shkuratov, S. I., Aurongzeb, D., Holtz, M., Berg, J., et al. (2004) Ignition studies of Al/Fe_2O_3 energetic nanocomposites. *Applied Physics Letters*, 84(23):4735–4737.
185. Kidalov, S. V., Shakhov, F. M., and Vul, A. Y. (2007) Thermal conductivity of nanocomposites based on diamonds and nanodiamonds. *Diamond and Related Materials*, 16(12):2063–2066.
186. Mochalin, V. N., Shenderova, O., Ho, D., and Gogotsi, Y. (2011) The properties and applications of nanodiamonds. *Nature Nanotechnology*, 7:11–23, doi: 10.1038/NNAO.2011.209.
187. Neitzel, I. (2012) *Nanodiamond-Polymer Composites*. Ph.D. dissertation, Drexel University, Department of Materials Engineering, Philadelphia, PA.
188. Pugh-Thomas, D., Walsh, B. M., and Gupta, M. C. (2011) CdSe (ZnS) nanocomposite luminescent high temperature sensor. *Nanotechnology*, 22:185503, doi: 10.1088/0957-4484/22/18/185503.
189. Pan, S., and Liu, Z. (2012) ZnS-graphene nanocomposites: Synthesis, characterization and optical properties. *Journal of Solid Chemistry*, 191:51–56.
190. Ummartyotin, S., Bunnak, N., Juntaro, J., Sain, M., and Manuspiya, H. (2012) Hybrid organic-inorganic of ZnS embedded PVP nanocomposite film for photoluminescent application. *Computes Rendus Physique*, 13(9–10):994–1000.
191. Patil, B. N., and Acharya, S. A. (2013, May 12) Preparation of ZnS-graphene nanocomposites and its photocatalytic behavior for dye degradation. *Advanced Materials Letters*, doi: 10.5185/amlett.2013.fdm.16.

CHAPTER 3

Selecting Resin Matrix and Nanomaterials for Applications

3.1 Characteristics of Polymer Nanocomposites

There are two main challenges to developing polymer nanocomposites after the desired nanomaterials have been selected for the polymer of interest. First, the choice of nanomaterials requires an interfacial interaction and/or compatibility with the polymer matrix. Second, the proper processing technique should be selected to uniformly disperse and distribute the nanomaterials or nanomaterial aggregates within the polymer matrix. In Chap. 2, the different types of nanomaterials were briefly described. In this chapter, we will discuss a selection of resin matrix and nanomaterials for specific applications. A range of thermoplastic nanocomposites, thermoset nanocomposites, and elastomer nanocomposites is surveyed in this chapter. Selective polymer nanocomposites for different applications will be used to demonstrate the technical approach. Several of the functions of these materials are listed below:

- *Mechanical:* increased modulus, strength, increased toughness, elongation (in some cases); for detailed discussions refer to Chap. 6.
- *Thermal:* increased thermal resistance, enhanced thermal stability, higher glass transient temperature (T_g), increased heat deflection temperature (HDT), reduced coefficient of thermal expansion (CTE); for detailed discussions refer to Chap. 7.
- *Flammability:* low peak heat release rates, reduced CO, reduced smoke, passed UL 94 V-0; for detailed discussions refer to Chap. 8.
- *Ablation:* low erosion, enhanced charred strength, better insulation; for detailed discussions refer to Chap. 9.
- *Electrical:* improved thermal conductivity, lower resistivity (depends on the nanomaterials); for detailed discussions refer to Chap. 10.
- *Tribological:* reduced scratch resistance, better abrasion resistance; for detailed discussions refer to Chap. 11.

Selected nylon 6 nanocomposites, epoxy nanocomposites, and thermoplastic olefin (TPO) nanocomposites are presented in this chapter as typical examples for thermoplastic-, thermoset-, and elastomer-based nanocomposites. Physical, mechanical, and thermal properties as well as the performance of these polymer nanocomposite materials are discussed. In most cases, polymer nanocomposites exhibit multifunctionality.

3.2 Different Types of Polymer Nanocomposites

In general, polymers can be classified into three basic families of resin, namely *thermoplastics, thermosets,* and *elastomers*. Table 3.1 lists several characteristics of the thermosetting and thermoplastic resin systems.

3.2.1 Thermoplastic-Based Nanocomposites

Materials are often classified as metals, ceramics, or polymers. Polymers differ from the other materials in a variety of ways, but generally they exhibit lower densities and moduli. The lower densities of polymeric materials offer an advantage in applications where lighter weight is desired. The addition of thermally and/or electrically conducting fillers allows the polymer formulator the opportunity to develop materials from insulating to conducting type characteristics.

Thermoplastics are used in a vast array of products. In the automotive area, they are used for interior parts and in under-the-hood applications. The packaging applications area is a large area for thermoplastics, from carbonated beverage bottles to plastic wrap. Application requirements vary widely; fortuitously, plastic materials can be formulated to meet these different business opportunities. It remains for the designers to select from the array of thermoplastic materials that are available to meet the particular needs of their applications.

Table 3.1 Comparisons of Thermoplastic and Thermosetting Resin Characteristics

Thermoplastic resin	Thermosetting resin
• High MW solid	• Low MW liquid or solid
• Stable material	• Low to medium viscosity, requires cure
• Reprocessable, recyclable	• Cross-linked, nonprocessable
• Amorphous or crystalline	• Liquid or solid
• Linear or branched polymer	• Low MW oligomers
• Liquid solvent resistance	• Excellent environmental and solvent resistance
• Short process cycle	• Long process cycle
• Neat up to 30% filler	• Long or short fiber reinforced
• Injection/compression/extrusion	• Resin transfer molding (RTM)/filament winding (FW)/sheet molding compound (SMC)/prepreg/pultrusion
• Limited structural components	• Many structural components
• Neat resin + nanoparticles	• Neat or fiber reinforced + nanoparticles
• Commodity: high-performance areas for automotive, appliance housings, toys	• Commodity: advanced materials for construction, marine, aircraft, aerospace

Many thermoplastics have been nano-modified into polymer nanocomposites by incorporating nanomaterials into the polymers. Several of these polymer nanocomposites will be described in detail in Chaps. 6 through 11. A typical example of this class of thermoplastic-based nanocomposites is nylon 6 nanocomposites, discussed next.

Nylon 6 Nanocomposites

Mica-type silicates, such as montmorillonite, hectorite, and saponite, are attractive nanoclays [1–3] functioning as reinforcing fillers for polymers because of their high aspect ratio and unique intercalation and exfoliation characteristics. The incorporation of organoclays into polymer matrices has been known for many decades. In 1950, Carter et al. [4] developed organoclays that were surface-treated with organic bases to reinforce latex-based elastomers. In 1963, Nahin and Backlund [5] incorporated organoclay into a thermoplastic polyolefin matrix. Organoclay-modified composites were obtained with strong solvent resistance and high tensile strength by irradiation-induced cross-linking. In 1976, Fujiwara and Sakamoto [6] of Unichika patented the first organoclay hybrid nanocomposite. Ten years later, a Toyota research team disclosed improved methods to produce nylon 6–clay nanocomposites using in situ polymerization similar to the Unichika process [7–10]. They reported that these nylon 6–clay nanocomposites exhibited increased solvent resistance, decreased CTE, reduced permeability, and enhanced flame-retardant characteristics. A very comprehensive review of polymer/layered silicate nanocomposites was published by Ray and Okamoto of Toyota Technical Institute in 2003 [11]. These materials were commercialized in Japan by Ube and Unitika. Giannelis and colleagues [12–15] reported that nanocomposites can be obtained by direct polymer intercalation, where polymer chains diffuse into the space between the clay galleries. They also suggested that this approach can be combined with conventional polymer extrusion to decrease the time to form these hybrids by shearing clay platelets and leading to sample uniformity.

Dennis et al. [16] demonstrated that the degree of exfoliation and dispersion of layered silicate nanocomposites formed from nylon (polyamide) 6 by melt compounding. This polymer processing is affected by both the chemistry of the clay surface and the type of extruder and its screw design. Increasing the mean residence time in the extruder generally improves the exfoliation and dispersion. It appears that an optimum extent of back mixing is necessary as judged by the broadness of the residence time distribution and optimal shear intensity. However, excessive shear intensity or back mixing also causes poor exfoliation and dispersion. Shear intensity is required to start the dispersion process by shearing particles apart into tactoids or intercalants. Residence time in a low-shearing or mildly shearing environment is required to allow polymer to enter the clay galleries and peel the platelets apart (Chap. 2). The non-intermeshing twin-screw extruder used in the study yielded the best exfoliation and dispersion. Excellent exfoliation and dispersion can be achieved with both co-rotating and counter-rotating, intermeshing types of extruders when a fully optimized screw configuration is used. Figure 3.1 shows transmission electron microscopy (TEM) micrographs of selective nylon 6–clay nanocomposites using different extruder and screw configurations [16]. It is recognized that the extruder processing conditions are important variables that must be optimized to affect a high degree of exfoliation and dispersion.

Cho and Paul [17] prepared nylon 6–organoclay nanocomposites by melt compounding using a typical co-rotating twin-screw extruder and compared with nylon 6 composites containing glass fibers and untreated clay. TEM and WAXD indicate that

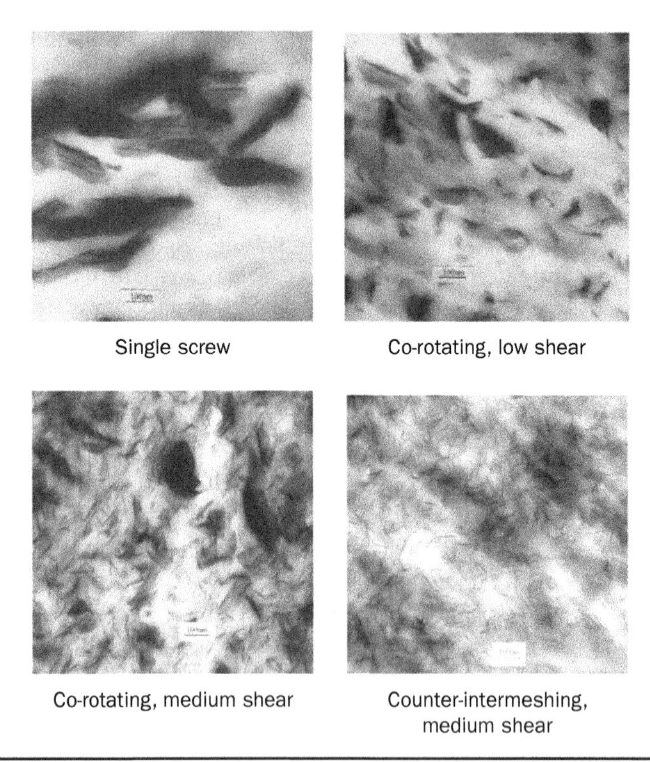

FIGURE 3.1 Transmission electron micrographs for elective nylon 6–Cloisite 15A nanocomposites using different extruders and screw configurations [16].

the organoclay was well dispersed into the nylon 6 matrix and the mechanical properties of these materials compare well with nylon 6–clay nanocomposites formed by in situ polymerization and melt processing. Melt processing of these nylon 6–clay nanocomposites in a single-screw extruder failed to give similar levels of dispersion or exfoliation.

A high degree of exfoliation by melt processing requires an adequate residence time in the extruder, and sufficient shear. Although highly exfoliated nylon 6–clay nanocomposites show continuous improvement in strength and modulus relative to the neat nylon 6 matrix as more organoclay is added, there is a loss of ductility beyond a certain organoclay loading. The type of organo-treatment and the speed of testing seem to have significant effects on ductility. Adding additional organoclay to nylon 6 significantly increases the ductile-to-brittle transition temperature. The mechanical properties of the organoclay nanocomposites showed greater values than glass fiber composites (Table 3.2) [17].

Fornes et al. [18] examined the effect of varying molecular weight of nylon 6 and nanoclay. Nylon 6 (low, or LMW; medium, or MMW; and high, or HMW) was prepared by a co-rotating twin-screw extruder. Wide-angle x-ray diffraction and TEM results collectively reveal a mixed structure for LMW-based nanocomposites having regions of intercalated and exfoliated clay platelets, whereas the MMW and HMW composites revealed highly exfoliated structures. Qualitative TEM observations were supported by quantitative analyses of high-magnification TEM images. The average number of platelets per stack was shown to decrease with increasing nylon 6 molecular weights.

TABLE 3.2 Mechanical Properties of Nylon 6 Composites [17]

Nylon 6 composites	Mineral content (%)	Izod impact strength (J/m)	Modulus (GPa)	Yield strength (MPa)	Elongation at break (%)*
Nylon 6	0	38 ± 4	2.66 ± 0.2	64.2 ± 0.8	200 ± 30
N6/glass fiber	5	53 ± 8	3.26 ± 0.1	72.6 ± 0.8	18 ± 1.3
N6/montmorillonite	5	40 ± 2	3.01 ± 0.1	75.4 ± 0.3	22 ± 6.0
N6/organoclay	3.16	38 ± 3	3.66 ± 0.1	83.4 ± 0.7	126 ± 25
N6/organoclay/ glass fiber	8	44 ± 3	4.82 ± 0.1	95.0 ± 0.9	8 ± 0.5

*Crosshead speed 0.51 cm/min.

The TEM particle density increased with increasing molecular weight, revealing clay platelet exfoliated for the nanocomposites in the order of HMW > MMW > LMW composites (Fig. 3.2) [18]. Tensile properties revealed superior performance for the higher molecular weight nylon 6 composites, with highest properties for HMW. Additionally, the HMW-based nanocomposites had the highest moduli, yield strengths, and elongation at break (Fig. 3.3) [18]. The notched Izod impact strength was relatively independent of clay concentration for the HMW and MMW composites, but gradually decreases for the LMW composites [18]. The melt viscosity, and consequently the shear stress, increased with increasing nylon 6 molecular weight. The higher shear stress is believed to have been the major contributor to exceptional exfoliation of clay platelets in the higher molecular weight nylon 6 matrix.

Figure 3.4 schematically illustrates the various roles of shear stress during the melt compounding of nanocomposites [18]. Initially, the stress should assist in rupturing large organoclay particles into dispersed stacks of silicate tactoids (Fig. 3.4a). As the extrudate travels in the extruder, transfer of the stress from the molten polymer to the silicate tactoids is proposed as shearing tactoids into smaller stacks of silicate platelets (Fig. 3.4b). Ultimately, individual platelets peel apart through a combination of shear and diffusion of polymer chains in the organoclay gallery (Fig. 3.4c).

The Paul Research Group [19] also studied the *structure–property* relationships for nanocomposites formed by melt processing a series of organically modified montmorillonite clays with high and low molecular weight nylon. The structure of the alkyl ammonium on the clay was systemically varied to determine how specific groups affect nylon 6 nanocomposite morphology and physical properties. As demonstrated by WAXD, galleries of the organoclays expand in a systematic manner to accommodate the molecular size and the amount of amine surfactant exchanged for the inorganic cation of the native montmorillonite. The density of the organic material in the galleries is in the range expected for organic liquids and solids. Both the modulus and the yield strength of the nanocomposites appear to decrease as the original organoclay *d*-spacing increases, in contrast to some earlier suggestions in the literature.

Three distinct surfactant structural effects that lead to a high amount of exfoliation, high stiffness, and increased yield strengths for nanocomposites based on high

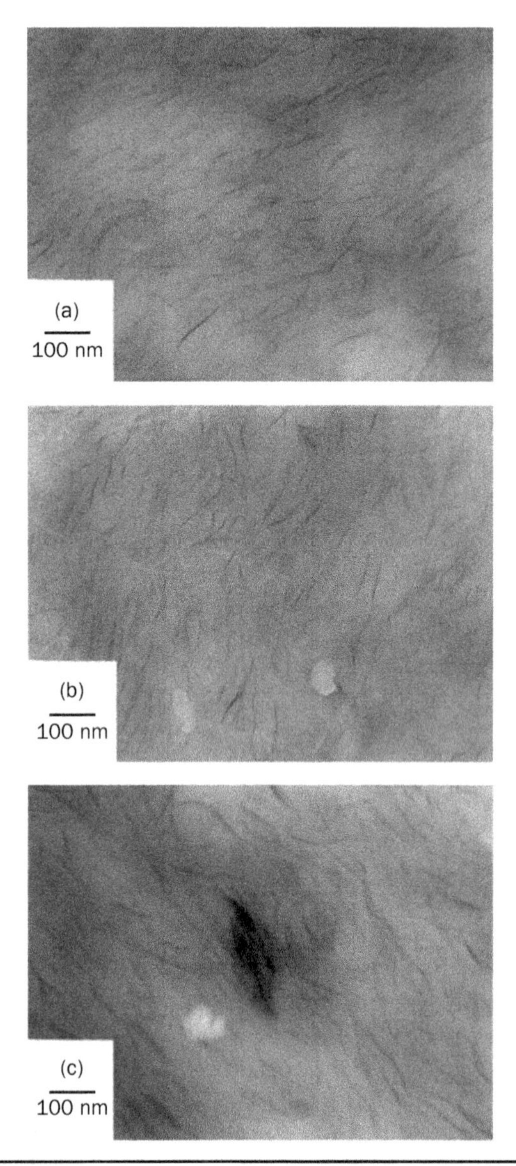

Figure 3.2 Transmission electron micrographs of melt-compounded nanocomposites containing about 3.0 wt% montmorillonite based on (a) HMW, (b) MMW, and (c) LMW nylon 6 [18].

molecular weight polyamide have been identified. They are (a) one long alkyl substitute group on the ammonium ion rather than two, (b) methyl groups rather than 2-hydroxy-ethyl groups on the amine, and (c) an equivalent amount of amine surfactant on the clay (to avoid an excess amount). Similar trends, but lower extends of exfoliation, were seen for a lower molecular weight grade of nylon 6. The authors proposed that these effects are due to the amount of exposed

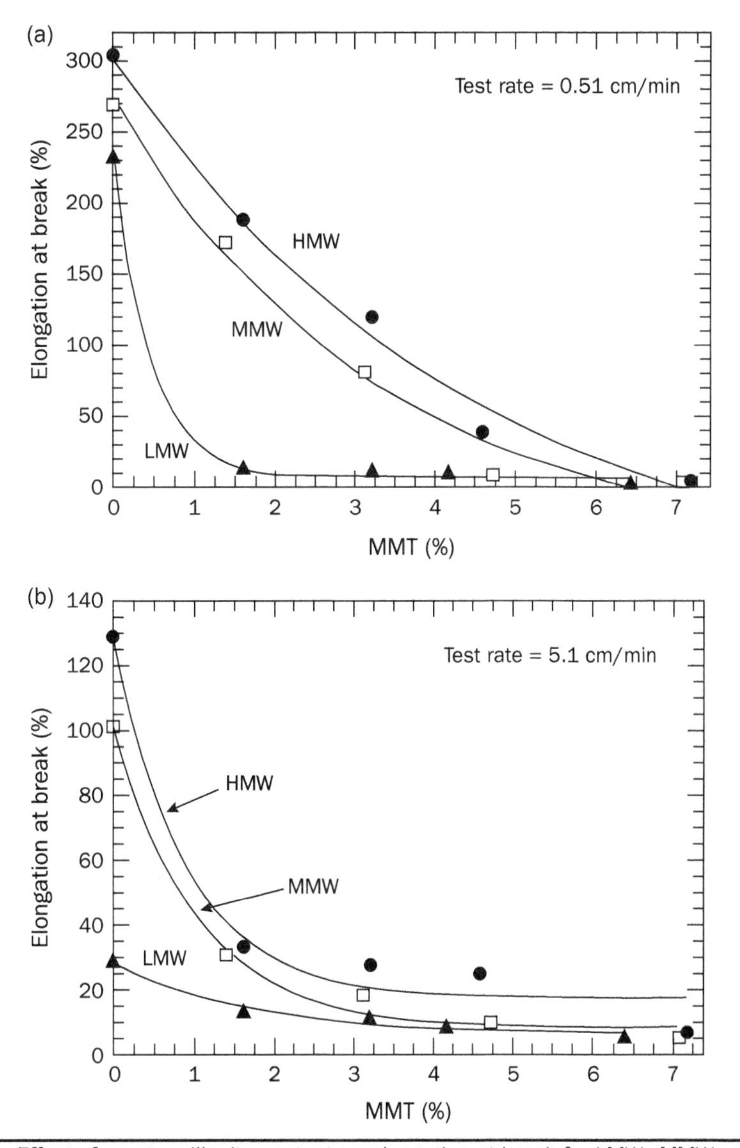

FIGURE 3.3 Effect of montmorillonite content on elongation at break for LMW-, MMW-, and HMW-based composites at a crosshead speed of (a) 0.51 cm/min and (b) 5.1 cm/min.

silicate surface. Alkyl ammonium ions, which cover a larger percentage of the silicate surface, shield desirable polar polyamide–polar clay interactions and ultimately lead to lesser platelet exfoliation. However, this finding does not imply that unmodified clay—that is, sodium montmorillonite—would be optimum. Organic surface modification of the clay is still required to overcome the cohesive forces between neighboring platelets, so that polymer intercalation and exfoliation can occur during melt processing. It is

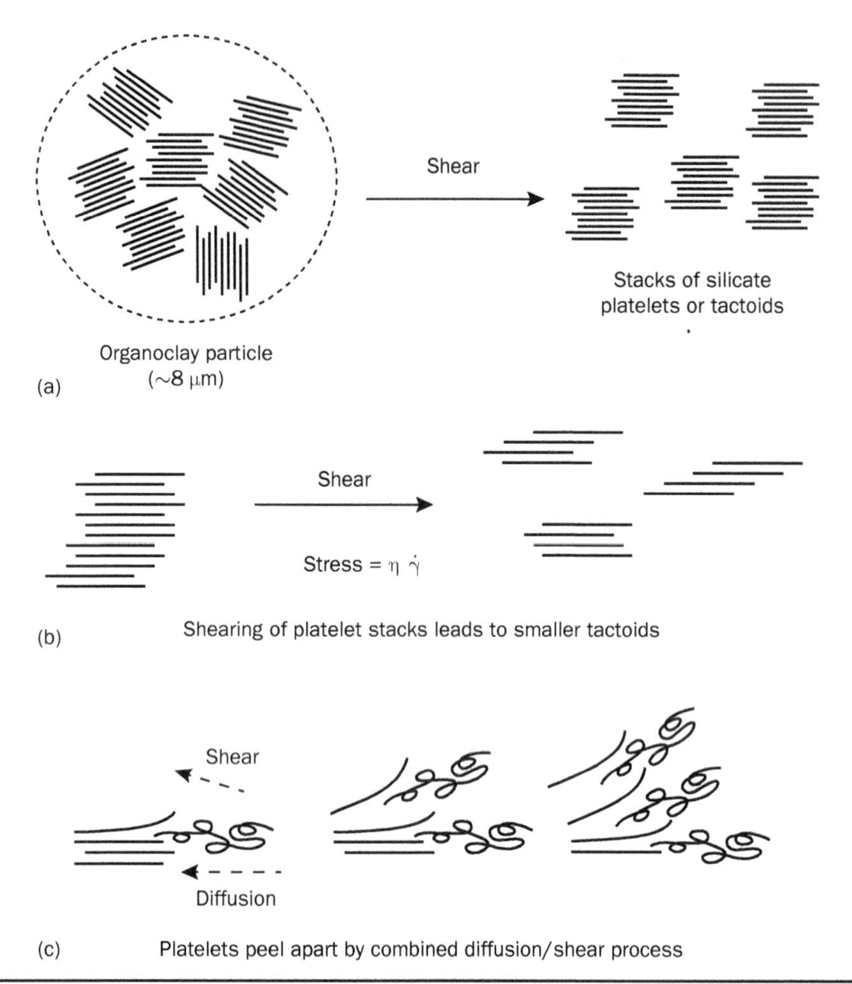

FIGURE 3.4 Stepwise mechanism of clay platelet exfoliation in the melt compounding of nanocomposites: (a) organoclay particle breakup, (b) clay tactoid breakup, and (c) platelet exfoliation [18].

possible that these observations may "only apply" to nylon 6 polymer. Thus, the nature of polymer–organoclay thermodynamic interactions may be different for other polymers and may have different *structure–property* effects. Continued studies in this area are encouraged.

The molecular weight of nylon 6 has a significant effect on nanocomposite morphology and mechanical properties. The higher molecular weight nylon 6 consistently leads to better exfoliation and greater reinforcement than lower molecular weight nylon 6. In summary, with proper selection of nylon 6, organoclay modifier, and processing conditions, one can produce high-performance nylon 6 nanocomposites by melt processing. These features contributed to the development of commercial nylon-type products by Toyota, Ube, Unitika, Honeywell, and others.

3.2.2 Thermoset-Based Nanocomposites

Thermosetting-type resins consist of solid, semisolid, or liquid organic reactive intermediate material that cures or cross-links into a high molecular weight product with no observable melting point. The basic characteristic of these intermediate thermosetting resins is that they will, upon exposure to elevated temperature from ambient to above 232°C (450°F), undergo an irreversible chemical reaction often referred to as *polymerization*, or *cure*. Each family member has its own set of individual chemical characteristics based on its molecular structure and its ability to either homopolymerize, copolymerize, or both. This transformation process separates the thermosets from the thermoplastics. The important beneficial factor lies in the inherent enhancement of the physical, thermal, chemical, and electrical properties of thermoset resins, because of that chemical cross-linking polymerization reaction, which contributes to their ability to maintain and retain these enhanced properties when exposed to severe environmental conditions.

An intermediate reactive thermosetting species is defined as a liquid, semisolid, or solid composition capable of curing of some defined temperature that can be ambient to several hundred degrees temperature and that cannot be reshaped by subsequent reheating. In general, these intermediate reactive thermosetting compositions contain two or more ingredients: a reactive resinous matrix with a curing agent that causes the intermediate material to polymerize (cure) at room temperature, or a low molecular weight resinous matrix and curing agent that, when subjected to elevated temperatures, will commence to polymerization and cure.

Several thermosetting resins have been nano-modified into polymer nanocomposites by introducing the nanomaterial into the thermosetting resin. These intermediate thermosetting nanocomposites can be impregnated into fiber reinforcements, such as glass, silica, quartz, carbon/graphite, aramid, poly(p-phenylene-2-6-benzobisoxazole) (PBO), polyethylene, boron, or ceramic and upon curing lead to laminates or nano-modified polymer matrix composites (PMCs). Several of these thermoset-based nanocomposites will be described in detail in Chaps. 6 through 11. An epoxy-based nanocomposite is a typical example for this class of material.

Epoxy Nanocomposites

Epoxy resins are widely used in commercial and military applications due to their high mechanical/adhesion characteristics, and solvent and chemical resistance combined with the versatility cure over a range of temperatures without the emission of volatile by-products. The properties of epoxy-based organic-inorganic composites can be fine-tuned by an appropriate choice of the structures of both epoxy prepolymer and hardener and of the type and amount of inorganic filler or fiber.

The use of inorganic nanoparticles can be particularly interesting to the common processing techniques used for conventional epoxy-based PMCs. Micrometer-sized inorganic particles are currently widely used for the reinforcement of epoxy matrices to lower shrinkage on curing, for thermal expansion coefficients, to improve thermal conductivity, and to meet mechanical requirements. The final properties of the PMCs are affected by several factors, such as intrinsic characteristics of each component, the contact, the shape and dimension of fillers, and the nature of the interface. A strong interface between matrix and filler is needed to achieve high performance. It requires that the load applied on the PMCs is mainly transferred to the fillers via the interface. To enhance these properties, the use of submicrometer particles can lead to a significant improvement of the mechanical properties of the PMCs. In the last decade, tremendous amount of research has been conducted in the preparation of submicrometer inorganic particles, leading to the possibility of preparing PMCs reinforced with nanofillers [20]. The enhanced modulus, decreased coefficient of thermal expansion, and potential rigid-phase toughening afford thermoset nanocomposite opportunities in PMCs. Epoxy resin reinforced with nanoparticles represents one of the most studied systems. Details of epoxy nanocomposite studies are included in this section.

Brown et al. [21] studied the role of various quaternary ammonium surfactants in the epoxy/diamine nanocomposite formation and processing techniques, which are necessary in ultimately fabricating PMCs with a nanocomposite matrix resin. The use of a hydroxyl-substituted quaternary ammonium modifier provides flexibility to combine both catalytic functionality (increases the intragallery reaction rate) and enhanced miscibility toward both components (resin and clay). Balancing these two factors is necessary to maintain rheological properties that are compatible with PMCs.

Initially, Brown et al. examined the effect of various processing steps on nanocomposite formation. The generalized processing procedure to fabricate the Epon 828 layered silicate nanocomposites is shown in [21]. For all the three organically modified layered silicates (OLS) of this study, intercalation and not exfoliation of Epon 828 was observed. This study shows autoclave processing (evacuated bag with 100 psi hydrostatic pressure) was useful in producing bubble-free uniform plaques, especially for systems containing greater than 10 wt% OLS. Figure 3.5 shows optical micrographs of the optically transparent plaques of autoclaved nanocomposites containing up to 20 wt% S30A in Epon 828/D400 resin, and turbid plaques containing up to 20 wt% B34 in Epon 828/D400 resin [21].

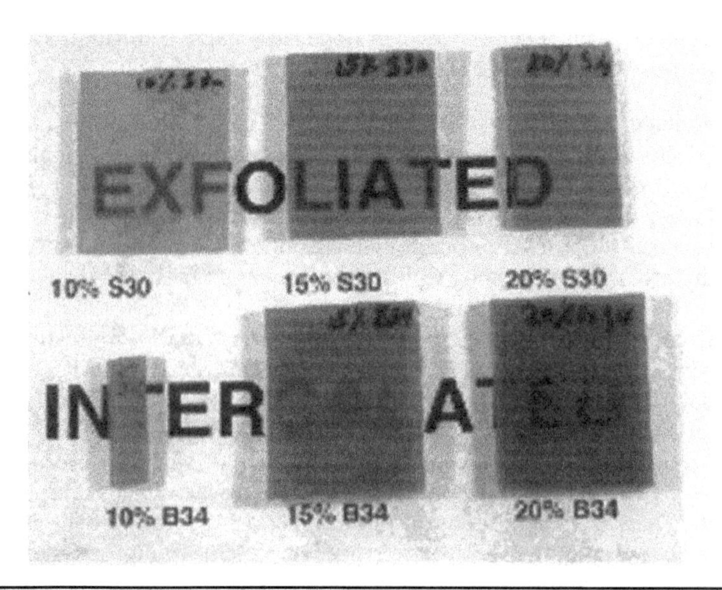

Figure 3.5 Optical image of 5-mm-thick plaques of Epon 828/D400 containing 10%, 15%, and 20% S30A and 10%, 15%, and 30% B34 [21].

Detailed TEM verified the intercalated structure of B34-containing resin (Fig. 3.6a,b) [21]. Individual layers, oriented perpendicular to the sample surface, appear as dark lines, with lateral size around 500 nm. Primary particles consisting of groups of parallel layers are larger with dimensions from 0.5 to 1.5 μm. Although the S30A nanocomposites were suggested as exfoliated by x-ray diffraction and the plaques were optically clear, a mixture of partially intercalated as well as exfoliated layers exists.

The temperature dependence of the dynamic shear moduli and tan (δ) from torsional bar measurements (0.1% strain 1 Hz) for unmodified Epon 828/D2000 were compared with nanocomposites containing 10 wt% OLS, which showed small differences at temperature range from 0°C to 50°C [21]. The temperature dependency of the dynamic shear moduli and tan (δ) at various concentrations of intercalated B34 and partially exfoliated S30A in Epon 828/D2000 were also compared [21]. Table 3.3 summarizes the glass transition temperature and dynamic shear moduli at various temperatures [21].

Incorporation of the layered silicate into the epoxy resulted in flammability [22] and ablation resistance [23]. When exposed to an open flame, unmodified Epon 828/D2000 burns, leaving oily and tacky residue. Resin containing intercalated B34 burns but produces a rigid graphitic char. Increased exfoliation of the layers, as in the S30A-containing resin, resulted in a self-extinguishing behavior upon removal of the flame. The resulting char exhibited a highly uniform microcellular microstructure consisting of 10- to 20-μm-diameter cells separated by 1- to 2-μm struts of silica-alumina-oxycarbide (Fig. 3.7) as determined by microprobe system [21].

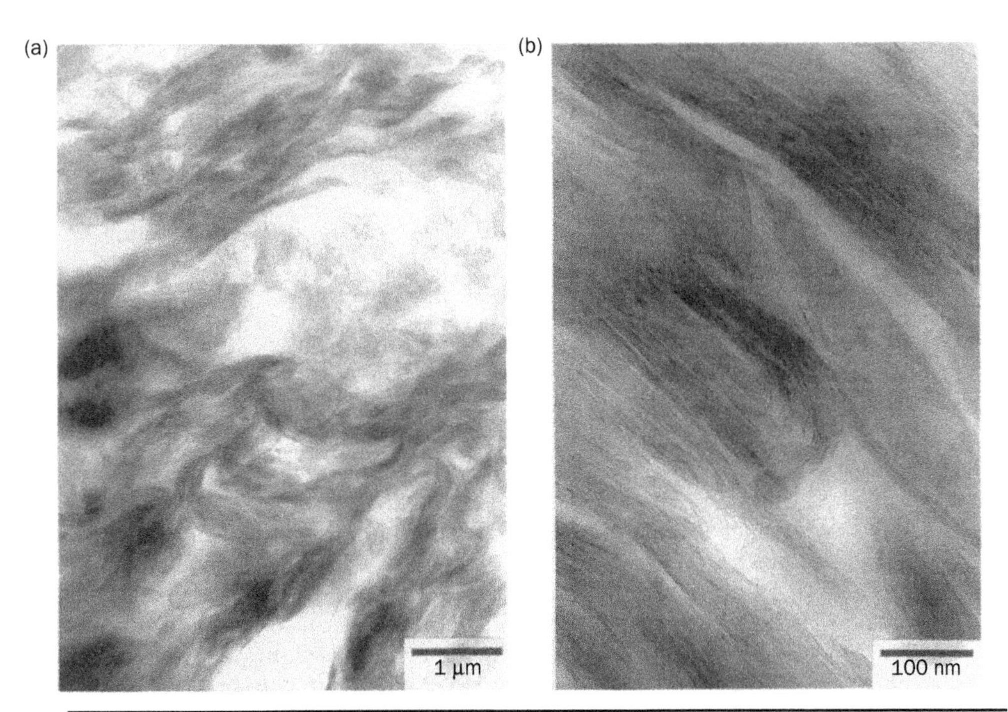

Figure 3.6 Bright-field TEM micrographs of 10% B34 in Epon 828 cured with D2000: (a) low magnification of large aggregate and (b) high magnification within aggregate [21].

Table 3.3 Dynamic Mechanical Summary for OLS/Epon 828 /D2000 [21]

	Wt% OLS			G (±25%)		
		T_g*	−140°C†	−60°C†	−25°C‡	0°C‡
828/D2000		−24	1.6	1.1	31	2.4§
B34	2.5	−18	1.7	1.3	35	1.2§
	5.0	−18	1.8	1.3	52	2.5§
	10	−18	1.9	1.4	80	3.1
S30A	2.5	−18	1.7	1.3	90	2.5§
	5.0	−22	1.8	1.3	50	1.0§
	10	−15	2.3	1.7	120	8.0
	15	−22	2.4	1.9	270	23
B24	10	−22	1.4	1.1	28	2.9§

*Plus or minus 3°C.
†In GPa.
‡In MPa.
§Lowest resolution for transducer about 1–2 MPa.

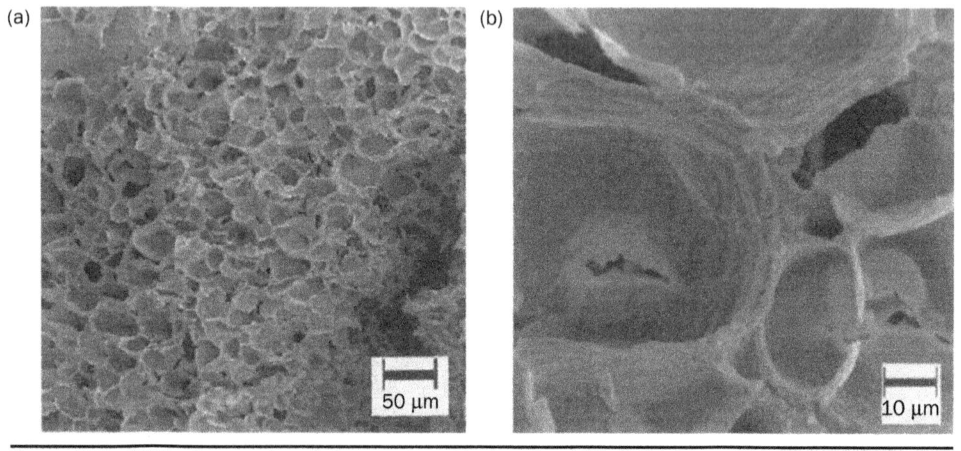

Figure 3.7 SEM micrographs of char produced by burning 10% S30A/Epon 828/D2000 in air [21].

This study suggests that surface modifiers should include a catalyst to enhance reactivity within the OLS crystallites. A critical balance must be maintained for use with PMCs. If miscibility between the OLS and thermoset mixture is great, unacceptably large viscosity increases and potential gelation of the system will occur through impingement of exfoliated sheets. If miscibility toward both components is insufficient, a non-stoichiometric mixture of reagents will be presented in the interlayer, and enhanced interlayer reactivity will not produce the desired network formation. Thus, optimal methodology for multicomponent thermosets would entail a choice of an OLS modifications that (a) partially compatibilizes the mixture, leading to a disordered intercalate containing both components and separation of the OLS aggregates into individual crystallites (slight viscosity increase), and (b) catalyzes interlayer reactivity to enable rapid layer separation before matrix gelation. Overall, this study further refines the criteria for organic modifier necessary to fabricate blends of nanoscopic inorganic phases in high-performance resins.

Most of the recent focus on epoxy systems has been on difunctional resins, predominantly using the diglycidyl ether of bisphenol-A (DGEBA). The more highly cross-linked systems are difficult to toughen and a smaller increase than those achieved for flexible and low glass transition epoxy systems may still prove to be beneficial in comparison to using higher concentrations of more commonly used fillers. Although DGEBA is widely used, most aerospace materials and other high-performance applications demand epoxies of higher functionality because of the required higher modulus and glass transition temperature.

Becker et al. [24] investigated the possibilities of improving the mechanical properties of multifunctionality epoxy resins through dispersion of octadecyl ammonium ion-modified layered silicates within the polymer matrix. Becker and colleagues used three different resins: (a) difunctional diglycidyl ether of bisphenol-A (DGEBA), (b) trifunctional triglycidyl *p*-amino phenol (TGAP), and (c) tetrafunctional tetraglycidyldiamino diphenylmethane (TGDDM) in their studies [24, 25]. All resins are cured with diethyltoluene diamine (DETDA). The morphology of the final, cured materials was probed by wide-angle x-ray scattering (WAXS), optical, and atomic force microscopy. The α- and β-relaxation temperatures of the cured systems were determined using dynamic mechanical thermal analysis. It was found that the presence of organoclay steadily decreased both transition temperatures with increasing filler concentration. The effect of different concentration of the alkyl ammonium–modified layered silicate on the toughness and stiffness of the different epoxy resins was analyzed. All resin systems have shown improvement in both toughness and stiffness of the materials through the incorporation of layered silicates, despite the fact that it is often found that these two properties cannot be simultaneously achieved.

The clay used in the studies of Becker and colleagues [24, 25] is a commercially available octadecyl ammonium ion–modified montmorillonite layered silicate (Nanomer I.30E) from Nanocor Inc. The three epoxy resins used are DGEBA, TGAP, and TGDDM. The hardener is a mixture of two DETDA. Structures of these materials are shown in Table 3.4 [24].

A wide range of epoxy nanocomposites with varying amounts of organoclay (0%, 2.5%, 5%, 7.5%, and 10%) by weight were prepared. WAXS traces of the clay concentrations show that the organoclay, with an initial *d*-spacing of 23 Å, is exfoliated in the DGEBA-based system (Fig. 3.8a) [24]. Only the 10% concentration shows a distinct peak correlating to a *d*-spacing of 48 Å. In contrast to the DGEBA-based system, resins of high functionality show distinctive peaks even at a lower organoclay concentration, indicating that these systems have a lower degree of exfoliated layered silicate. WAXS traces are shown in Fig. 3.8b for TGAP and in Fig. 3.8c for TGDDM [24].

Before investigating the microstructure of the nanocomposites, optical microscopy was applied to image the bulk surface. Optical micrographs for the DGEBA system containing 5% of organoclay show large particulate phases [24]. This finding indicates that even for DGEBA-based system, not all clay particles are fully dispersed in the polymer phase. Both optical microscopy and AFM images show that at least part of the silicate

Table 3.4 Epoxy Resins and Hardener as Used for Nanocomposite Synthesis [24]

Substance	Formula
DGEBA	O; CH_2—$CHCH_2O$—; CH_3; C; CH_3; —OCH_2CH—CH_2; O
TGAP	O; CH_2—$CHCH_2O$—; N; CH_2CH—CH_2; O; CH_2CH—CH_2; O
TGDDM	O; CH_2—CH—CH_2; CH_2—CH—CH_2; O; N; CH_2; N; CH_2—CH—CH_2; O; CH_2—CH—CH_2; O
DETDA	CH_3; NH_2; CH_3CH_2; CH_2CH_3; NH_2 CH_3; H_2N; NH_2; CH_3CH_2; CH_2CH_3

content remains in tactoids or stackings of layers, rather than forming a homogenous morphology through the whole material.

Dynamic mechanical thermal analysis was applied in two steps, from –100 to 50°C and from 50 to 300°C, to investigate the influence of the organoclay on the α- and β-transition. The value of glass transition temperature (T_g) decreased steadily with increasing organoclay concentrations. The reduction in T_g was found to be on the order of 15°C for TGAP- and DGEBA-based systems, and 20°C for the TGDDM-based system at an organoclay content of 10%. In summary, it appears that in all systems, incorporation of clay leads to a decrease in both glass transition and secondary relaxation temperatures, possibility due to a decrease in effective cross-link density.

Differential scanning calorimetry data of peak temperatures and percentage residual cure are summarized in Table 3.5 [24]. Differential scanning calorimetry traces for the DGEBA-based systems confirmed full cure, while the TGAP- and TGDDM-based nanocomposites show small peaks around 250°C, indicating further reaction. Although the amount of reaction is very small, the filled resin was more fully cured than the neat systems were.

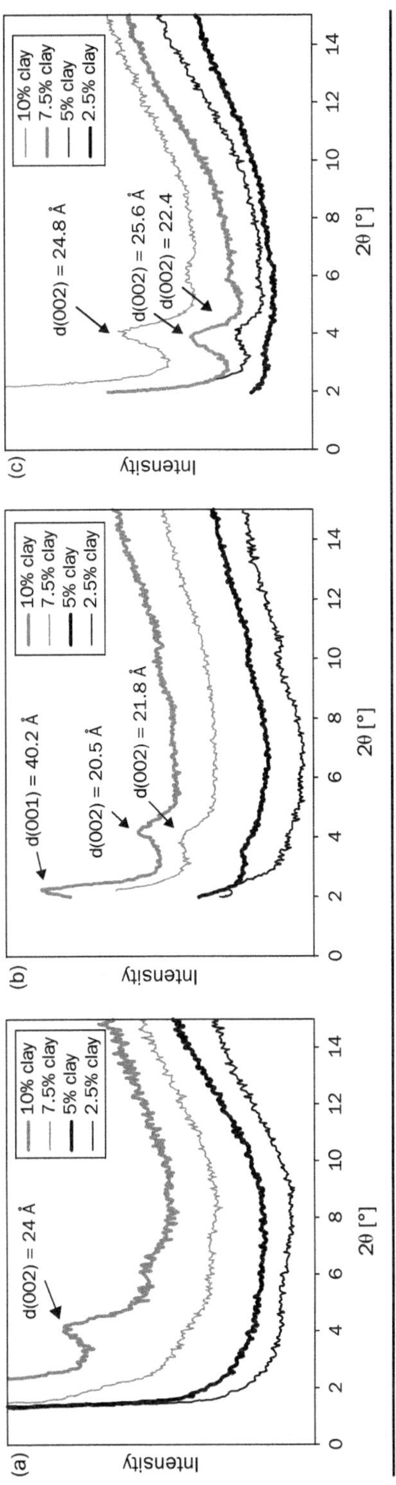

Figure 3.8 Wide-angle x-ray scattering of (a) DETDA-cured DGEBA, (b) DETDA-cured TGAP, and (c) DETDA-cured TGDDM nanocomposites containing 0–10% organoclay [24].

Table 3.5 DSC Results of Postcured Epoxy-Clay Nanocomposite Systems [24]

	TGAP		TGDDM	
Clay concentration (%)	**Peak (°C)**	**% Residual cure**	**Peak (°C)**	**% Residual cure**
0	266.4	5.1	253.7	4.3
2.5	270.2	2.9	237.4	0.7
5	251.0	3.0	248.5	3.7
7.7	253.7	2.8	233.0	1.1
10	255.7	1.6	229.5	0.6

Density results of the various nanocomposites after curing were determined by gas displacement pycnometry. Theoretical densities according to the rule of mixture were calculated based on an organoclay density of 1.68 g/cm^3. All densities are within a very small range of 1.19–1.23 g/cm^3. The densities are generally slightly below predicted values according to the rule of mixture. The lower density is possibly related to poorer packing of the polymer molecules near the silicate layers.

Modulus was determined using the three-point bend test. All resin systems show a monotonic increase in modulus with increasing organoclay concentrations. A normalized modulus was calculated by dividing the actual nanocomposite modulus value through the modulus of the unfilled system. The increase in modulus for all resin systems in a range of about 20 wt% for an organoclay concentration of 10 wt% is demonstrated [24]. The improved stiffness can be directly ascribed to the reinforcement of exfoliated, high-aspect-ratio platelets. In the work by Lan and Pinnavaia [26], it was found that a Jeffamine D2000-cured DGEBA nanocomposite increased the modulus by approximately 500% at an organoclay concentration of 10 wt%. The DGEBA system in this study shows only slightly greater relative improvements in modulus than the TGAP or TGDDM resin, respectively.

The fracture toughness, as quantified by stress intensity factor, K_{Ic} was determined using the compact tension test. Most toughening techniques show a loss in stiffness; however, in this study both toughness and stiffness were improved through the organoclay incorporation. The stress intensity factor indicates that DGEBA and TGDDM show an increase in toughness in a similar range, whereas improvement in the TGAP system is significantly lower.

Improvement in stiffness of the high-functionality epoxy resins is comparable with those achieved for the difunctional (DGEBA) resin system. Although it is often found that improvement in modulus compromises toughness of the material, both toughness and stiffness were improved through the incorporation of organoclay. The improved modulus and toughness make nanocomposite strategy an attractive alternative to the commonly used micrometer-sized fillers.

Becker et al. [25] extended the above research to study the relationship between cure temperature, morphological, and mechanical properties of the di-, tri-, and tetrafunctional, high-performance, epoxy layered-silicate nanocomposites. Wide angle x-ray scattering was used to monitor the different stages of organoclay exfoliation kinetics. It was found that some degree of conversion was required to obtain significant intercalation. The polymer nanocomposite morphology was probed using TEM, WAXD, and positron annihilation lifetime spectroscopy. Comparison of the nanocomposite formation based on three different epoxy resins with different structures and functionalities has shown that the structure and chemistry of the epoxy resin, its mobility, and reactivity are key factors controlling exfoliation, and therefore, the morphology of the cured nanocomposite. Two different effects have to be considered: improved dispersion and better exfoliation of the silicate platelets in the polymer matrix and changes in the nature of the reaction and thus the network formation. The difunctional DGEBA resin gave better exfoliation than the high-functionality resins. It was observed for all resins that higher cure temperatures led to improved intercalation and exfoliation. Tactoids were observed in all three resin systems. This is attributed to better catalysis of the intragallery reaction by organo-ions, which reside within the galleries. The structural mobility (diffusion) as indicated by the viscosity of the resin seemed less relevant.

High cure temperatures were found to improve clay exfoliation and simultaneously increased toughness and modulus in the case of DGEBA- and TGAP-based materials. The mechanical properties remained relatively unaffected by the cure temperature. It is assumed that the effect of cure temperature on the reaction chemistry and the effect on cross-link density also have a major impact on mechanical properties in addition to the changes in organoclay dispersion. More dramatic changes in the nanocomposite morphology may be required to significantly change the mechanical properties. Because the processing temperature window is limited by side reactions and thermal degradation, variation of cure temperature alone may not be sufficient to form a full-dispersion, *true* nanocomposite. Free volume properties did not vary significantly between resins or with curing temperature and generally followed the rule of mixtures, although there was a suggestion that the presence of clay leads to increased free volume. This finding was consistent with decreased T_g upon the addition of layered silicate and decreased cross-link density in interfacial regions of clay and epoxy matrix.

Wang et al. [27] developed a *slurry-compounding* process for the preparation of epoxy-clay nanocomposites. The microstructures of the nanocomposites were characterized by optical and transmission electron microscopy. It was found that clay was highly exfoliated and uniformly dispersed in the resulting nanocomposite. Characterization of mechanical and fracture behaviors revealed that Young's modulus increases monotonically with increasing clay concentration, while the fracture toughness shows a maximum at 2.5 wt% of clay. The micro-deformation and fracture mechanism were studied throughout TEM and SEM analyses of the microstructures of the arrested crack tips and damage zone. The initiation and development of the micro-cracks are the dominant micro-deformation and fracture mechanisms in these nanocomposites. The formation of a large number of micro-cracks and the increase in the fracture surface area due to crack deflection are the major toughening mechanisms.

Silica nanoparticles having different sizes were characterized by Bondioli et al. [28] using the sol-gel process. Epoxy-silica nanocomposites were prepared with nanosilicas (reaction of TEOS with methanol or ethanol) ranging from 1 to 5 wt%. Scanning electron microscopy analysis and tensile tests were carried out on these epoxy-silica nanocomposites. The results indicated the absence of particle aggregation and enhanced elastic modulus. Mechanical properties were also modeled using a finite element code (OOF [29]) able to construct a numerical model based on microstructural analyses of the material (Table 3.6) [28]. Table 3.6 summarizes the mechanical properties (elastic moduli experimentally determined and modeled) of nanocomposites [28]. A more reliable three-phase model was prepared by considering the presence of an interphase

Table 3.6 Experimental and Model Elastic Modulus of Epoxy-Silica Nanocomposites [27]

Composite	SiO_2 content (wt%)	*E* (experimental) (GPa)	Standard deviation (experimental) (GPa)	*E* (two-phase model) (GPa)
Unfilled epoxy	0	2.000	0.100	2.000
Epoxy/SiO_2 from ethanol	1	2.135	0.149	2.018
	3	2.272	0.181	2.062
	5	2.370	0.123	2.099
Epoxy/SiO_2 from methanol	1	2.428	0.065	2.019
	3	2.506	0.147	2.061
	5	2.537	0.202	2.104

layer surrounding the particles with the intermediate elastic properties between the epoxy and the inclusion, and a characteristic size proportional to the particle radius. The composite elastic moduli were calculated from the stress–strain curves. The results were recorded (Table 3.6), as a function of particle concentration. The elastic modulus increases with higher amount of reinforcement, but the values predicted by the computational model are considerably lower. It is worth noting that analytical equations, such as the Halpin–Tsai rule [30] or the Lewis–Nielsen equation [31], give estimates that are in good agreement with the numerical results, and thus differ from the experimental data. This means that analytical and numerical approaches commonly used for particulate composite are not suitable when a nanophase is involved.

A more refined model was used by Bondioli et al. by taking into account the interphase between the matrix and the particle as a third constituent phase. It was assumed that such an interphase had properties in between the polymeric matrix and the inorganic filler. It was hypothesized that the dimension of the interphase's extension was proportional to the particle size. It was set equal to half of the average diameter. Nanocomposites with nanosilicas were modeled, and the respective elastic moduli were evaluated with the same procedure previously described [28].

The numerical results are in good agreement with the experimental data, especially if the error bars are taken into consideration. These results demonstrate that such a model is a reliable tool for the prediction of elastic properties of epoxy-silica nanocomposites. Future development of this work should aim at the experimental characterization of the nanocomposite interphase to support the hypotheses of this research. The characteristic size and stiffness should be determined. This could be achieved by means of AFM techniques [32] or nanoindentation equipment [33].

3.2.3 Elastomer-Based Nanocomposites

Another important group of polymers is the elastic or rubberlike, known as *elastomers*. These elastomeric materials can be block copolymers or multiphase systems containing soft (low T_g) segments and hard segments (high T_g, possible crystalline). They are processable under thermoplastic conditions. A few elastomers have been nano-modified into elastomer nanocomposites, which will be described in detail in Chaps. 5 through 11. For completeness, some rubber materials (synthetic, isoprene, natural, etc.) that are nano-modified are included. A typical example of thermoplastic olefin nanocomposites is included in this section.

Thermoplastic Olefin (TPO) Nanocomposites

Most of the early work on the formation of nanocomposites focused on their preparation via in situ polymerization of caprolactam (Toyota group and others) in the presence of expanded silicate material. Recently, researchers have been investigating melt processing as the preferred method for the preparation of thermoplastic nanocomposites, such as nylon [15–18], polystyrene [11, 34], and polypropylene [35, 36]. These researchers have primarily been interested in the effect of processing conditions on the physical and mechanical property enhancement of the resulting polymer nanocomposites (PNCs).

Fasulo et al. [37] conducted extrusion trials designed to determine the role of processing conditions with regards to the dispersion of the clay-based filler systems in thermoplastic olefin (TPO) resin systems. The TPO resins were supplied by Basell Polyolefins, Inc. These resins were based on a blend of polypropylene homopolymer, impact-modified polypropylene, and ethylene-propylene-based elastomers. The organically modified MMT nanoclays were provided by Southern Clay Products, Inc. These fillers were refined from Wyoming bentonite clay and was modified using an ammonium-based surfactant.

Fasulo et al. examined over 300 varying nanocomposite formulations and evaluated their surface appearance. The surface evaluation consisted of visual ranking, controls, and microscopy. Surface imperfections of the injection-molded samples were examined visually and using scanning electron microscopy (SEM). Energy-dispersive x-ray spectroscopy (EDS) analysis was utilized to examine on the surfaces and cross-sections of these samples. Figure 3.9 shows representative surface with a numerical ranking [37].

Property evaluation included physical and mechanical properties. The physical properties all appeared to be relatively independent of the processing conditions. The only property that appeared to correlate with the amount of agglomerated material is the number of brittle failures from multiaxial instrumented impact, and even this result was minimal. Figure 3.10 shows two micrographs, the control formulation and a nanocomposite prepared using the same processing conditions [37]. Both materials exhibited with few imperfections. The imperfections visible on the control formulation can be attributed to either the coating process or poor elastomer dispersion.

Figure 3.11 shows micrographs of the painted surfaces from two of the worst-ranked processing conditions [37]. These imperfections are more numerous and larger than those observed on the control panel and can be attributed only to the surface imperfections observed on the substrate. For the low-high-low processing conditions, the imperfections are small but are visible in bright light. For the high-high-low processing conditions, the imperfections are large and visible in any light. These results provided a reasonable estimate of the size of surface imperfections required for the imperfection to be visible after the painting operation.

This chapter shows that certain processing conditions maximize the clay agglomeration whereas others will minimize it. A balance of processing parameters is required. The clay can agglomerate when the processing temperature is higher because there is opportunity to degrade the intercalant or surfactant that exists between the clay sheets prior to the wetting of the filler by the molten resin. Without the surfactant, the surface tension of the unmodified clay sheets leads to agglomeration. When the feed rate is

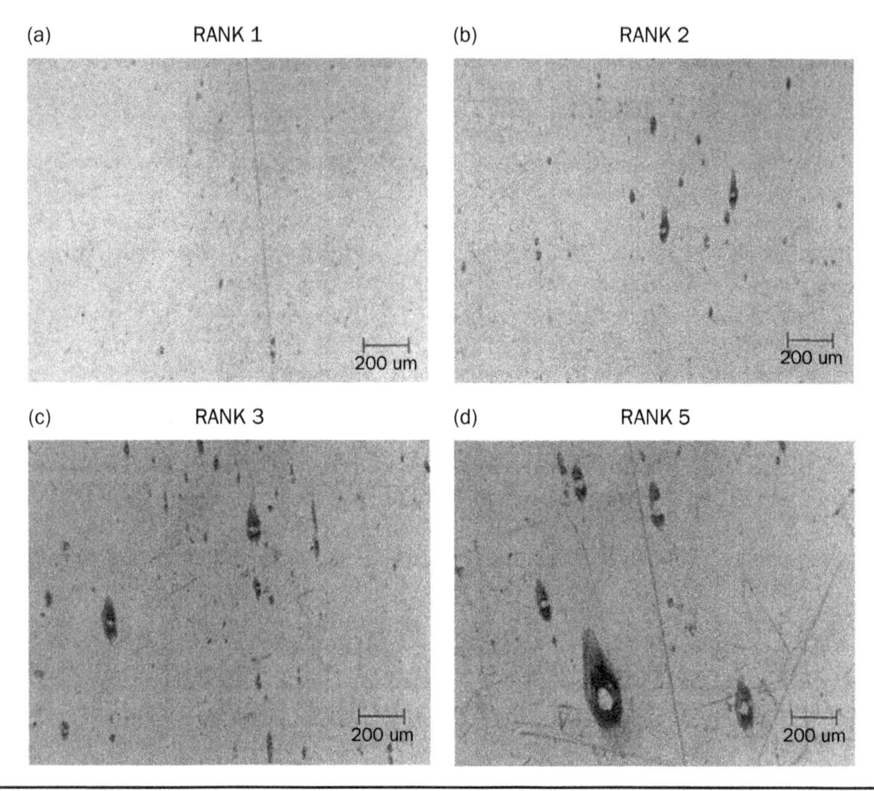

FIGURE 3.9 Micrographs showing representative surfaces with various ranks [37].

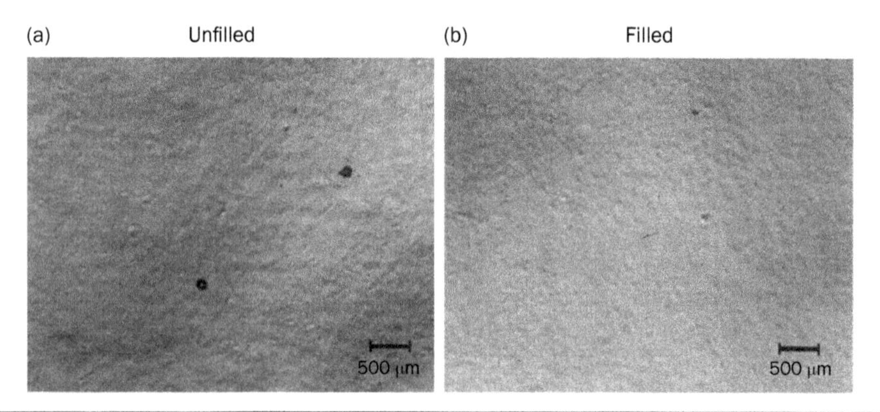

FIGURE 3.10 Micrographs of a coated unfilled formulation and a coated nanocomposite formulation [37].

high, there is great chance of forming a mass of clay that can then experience increased pressure as it is processed in the extruder and creates agglomeration. A low screw rotation speed imparts less energy to the clay sheets, leading to slower reduction in the breakdown in the height of the clay stacks, and ultimately reduces the exfoliation of the filler material.

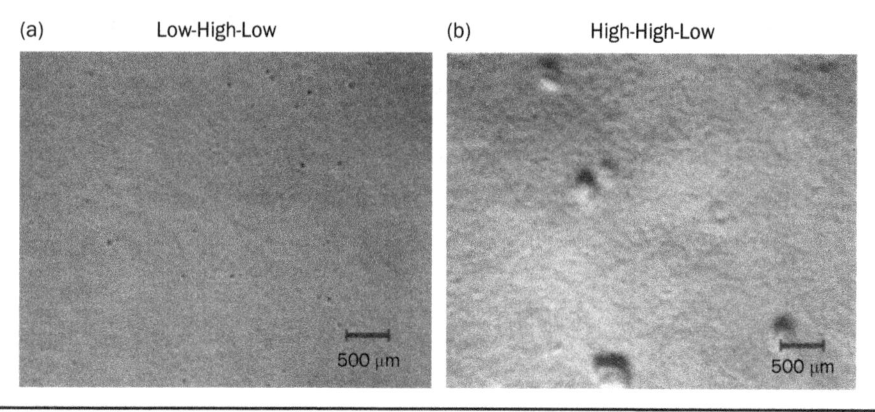

FIGURE 3.11 Micrographs of coated nanocomposite formulations processed under the worst conditions [37].

3.3 Summary

Several examples of thermoplastic-, thermoset-, and elastomer-based nanocomposites are presented in this chapter. These examples demonstrate that physical, mechanical, and other material properties will be enhanced with the selection of a nanomaterial compatible to the polymer. Suitable processing and curing conditions are essential to create polymer nanocomposites for a specific application. Characterization tools are useful to understand the behavior, properties, and performance of polymer nanocomposites.

Scientists and engineers can develop formulations for a range of polymer nanocomposites to fit their requirements and applications. Customer-designed polymer nanocomposites can be prepared to fit one's needs and fulfill one's requirements. The challenge is to select the baseline polymers with the proper nanomaterials to solve specific problems. Processing and morphological characterization are the keys to understanding the fundamentals of this new class of materials.

3.4 Study Questions

3.1 Why do polymer nanocomposites exhibit such multifunctionality? Provide examples using thermoplastic, thermoset, and elastomer nanocomposites.

3.2 Do different types of processing methods for polymer nanocomposites create different dispersions of nanoparticles within the polymer? In other words, is a specific processing method more reliable and create a better dispersion than other methods?

3.3 Is it possible to perform more than one nanocomposite processing method on a material?

3.4 Is it easier to exfoliate with lower wt% of nanoparticles in the composite rather than with higher wt%? Why (or why not)?

3.5 Why can't exfoliation or intercalation occur if MMT is modified with a surface agent, such as benzyl dimethyloctadecylammonium cation, or if too much water is added?

3.6 How do you classify whether a polymer nanocomposite is intercalated or exfoliated? In other words, what magnification visually expresses these classifications and to what extent can you decipher between the two? How reliable is determining between exfoliation and intercalation?

3.7 In which "under-the-hood" applications are thermoplastics used?

3.8 What are the applications of thermoset-based nanocomposites?

3.9 What are the differences between thermoplastic- and elastomer-based nanocomposites?

3.10 What are the applications of elastomer-based nanocomposites?

3.11 Do amorphous polymers usually have better impact resistance than more crystalline polymers?

3.12 Is there a significant relationship between electrical conductivity and polymer crystallinity?

3.13 Is it harder to disperse nanoparticles/fillers into branched thermoplastics?

3.14 Are properties of elastomers highly dependent on whether they are thermoplastic or thermosets?

3.15 How does the incorporation of the nanoparticle alter the thermosetting resin? Does it change the curing at all?

3.5 References

1. Pinnavaia, T. J. (1983) Intercalated clay catalysts. *Science,* 220:365–371.
2. Mehrotra, V., and Giannelis, E. P. (1990) Conducting molecular multilayers: intercalation of conjugated polymers in layered media. *Mater. Res. Soc. Symp. Proc.,* 171:39–44.
3. Giannelis, E. P. (1992) A new strategy for synthesizing polymer-ceramic nanocomposites. *J. Minerals,* 44: 28–30.
4. Carter, L. W., Hendricks, J. G., and Bolley, D. S. (1950) United States Patent No. 2531396 (assigned to National Lead Co.).
5. Nahin, P. G., and Backlund, P. S. (1963) United States Patent No. 3084117 (assigned to Union Oil Co.).
6. Fujiwara, S., and Sakamoto, T. (1976) Japanese Kokai Patent Application No. 109998 (assigned to Unichika K.K., Japan).
7. Fukushima, Y., and Inagaki, S. (1987) Synthesis of an intercalated compound of montmorillonite and 6-polyamide. *J. Inclusion Phenomena,* 5:473–482.
8. Okada, A., Fukushima, Y., Kawasumi, M., Inagaki, S., Usuki, A., Sugiyama, S., Kuraunch, T., et al. (1988) United States Patent No. 4739007 (assigned to Toyota Motor Co., Japan).
9. Kawasumi, M., Kohzaki, M., Kojima, Y., Okada, A., and Kamigaito, O. (1989) United States Patent No. 4810734 (assigned to Toyota Motor Co., Japan).
10. Usuki, A., Kojima, Y., Kawasumi, M., Okada, A., Fukusima, Y., Kurauch, T., Kamigaito, O. (1993) Synthesis of nylon 6-clay hybrid. *J. Mater. Res.,* 8:1179–1184.
11. Ray, S. S., and Okamoto, M. (2003) Polymer/layered silicate nanocomposites: a review from preparation to processing. *Prog. Polym. Sci.,* 28:1539–1641.

12. Vaia, R. A., Ishii, H., and Giannelis, E. P. (1993) Synthesis and properties of two-dimensional nanostructures by direct intercalation of polymer melts in layered silicates. *Chem. Mater.*, 5:1694–1696.
13. Vaia, R. A., Jandt, K. D., Kramer, E. J., and Giannelis, E. P. (1995) Kinetics of polymer melt intercalation. *Macromolecules*, 28:8080–8085.
14. Vaia, R. A., and Giannelis, E. P. (1997) Lattice model of polymer melt intercalation in organically modified layered silicates. *Macromolecules*, 30:7990–7999.
15. Vaia, R. A., and Giannelis, E. P. (1997) Polymer melt intercalation in organically-modified layered silicates: model predictions and experiment. *Macromolecules*, 30:8000–8009.
16. Dennis, H. R., Hunter, D. L., Chang, D., Kim, S., White, J. L., Cho, J. W., and Paul, D. R. (2001) Effect of melting processing conditions on the extent of exfoliation in organoclay-based nanocomposites. *Polymer*, 42:9513–9522.
17. Cho, J. W., and Paul, D. R. (2001) Nylon 6 nanocomposites by melt compounding. *Polymer*, 42:1083–1094.
18. Fornes, T. D., Yoon, P. J., Keskula, H., and Paul, D. R. (2001) Nylon 6 nanocomposites: the effect of matrix molecular weight. *Polymer*, 42:9929–9940.
19. Fornes, T. D., Yoon, P. J., Hunter, D. L., Keskula, H., and Paul, D. R. (2002) Effect of organoclay structure on nylon 6 nanocomposite morphology and properties. *Polymer*, 43:5915–5933.
20. Baraton, M. I., ed. (2002) Synthesis. In *Fictionalization and Surface Treatment of Nanoparticles*. Los Angeles, CA: American Science Publishers.
21. Brown, J. M., Curliss, D., and Vaia, R. A. (2000) Thermoset-layered silicate nanocomposites. Quaternary ammonium montmorillonite with primary diamine cured epoxies. *Chem. Mater.*, 12:3376–3384.
22. Gilman, J. W. (1999) Flammability and thermal stability studies of polymer layered-silicate (clay) nanocomposites. *Appl. Clay Sci.*, 15:31–59.
23. Vaia, R. A., Price, G., Ruth, P. N., Nguyen, H. T., and Lichtenhan, J. (1999) Polymer/layered silicate nanocomposites as high performance ablative materials. *Appl. Clay Sci.*, 15:67–92.
24. Becker, O., Varley, R., and Simon, G. (2002) Morphology, thermal relaxations and mechanical properties of layered silicate nanocomposites based upon high-functionality epoxy resins. *Polymer*, 43:4365–4373.
25. Becker, O., Cheng, Y-B., Varley, R. J., and Simon, G. P. (2003) Layered silicate nanocomposites based on various high-functionality epoxy resins: the influence of cure temperature on morphology, mechanical properties, and free volume. *Macromolecules*, 36:1616–1625.
26. Lan, T., and Pinnavaia, J. T. (1994) Clay-reinforced epoxy nanocomposites. *Chem. Mater.*, 6:2216–2219.
27. Wang, K., Chen, L., Wu, J., Toh, M. L., He, C., and Yee, A. F. (2005) Epoxy nanocomposites with highly exfoliated clay: mechanical properties and fracture mechanisms. *Macromolecules*, 38:788–800.
28. Bondioli, F., Cannillo, V., Fabbri, E., and Messori, M. (2005) Epoxy-silica nanocomposites: preparation, experimental characterization, and modeling. *J. Appl. Polym. Sci.*, 97:2382–2386.
29. Carter, W. C., Langer, S. A., and Fuller, E. R. (1998) *OOF Manual: Version 1.0*. http://www.ctcms.nist.gov/oof.

30. Barbero, E. J. (1998) *Introduction to Composite Materials Design (Solution Manual).* New York, NY: Taylor & Francis.
31. Lewis, T. B., and Nielsen, L. E. (1970) Dynamic mechanical properties of particulate-filled composites. *J. Appl. Polym. Sci.*, 14:1449–1471.
32. Gao, S.-L., and Mader, E. (2002) Characterisation of interphase nanoscale property variations in glass fibre reinforced polypropylene and epoxy resin composites. *Composites Part A: Applied Science and Manufacturing*, 33:559–576.
33. Wang, H., Bai, Y., Liu, S., Wu, J., and Wong, C. P. (2002) Combined effects of silica filler and its interface in epoxy resin. *Acta Materialia*, 50:4369–4377.
34. Hoffmann, B., Dietrich, C., Thomann, R., Friedrich, C., and Mulhaupt, R. (2002) *Macromol. Rapid Commun.*, 21:57.
35. Kawasumi, M., Hasegawa, N., Kato, M., Usuki, A., and Okada, A. (1997) Preparation and mechanical properties of polypropylene-clay hybrids. *Macromolecules*, 30:6333–6338.
36. Reichert, P., Nitz, H., Klinke, S., Brahdsch, R., Thomann, R., and Mulhaupt, R. (2000) Poly(propylene)/organoclay nanocomposite formation: influence of compatibilizer functionality and organoclay modification. *Macromol. Mater. Eng.*, 275:8–17.
37. Fasulo, P. D., Rodgers, W. R., Ottaviani, R. A., and Hunter, D. L. (2004) Extrusion processing of TPO nanocomposites. *Polym. Eng. Sci.* 44:1036–1045.

CHAPTER 4

Processing of Multifunctional Polymer Nanocomposites

4.1 Synthesis Methods

After the selection of a particular polymer matrix and the appropriate nanoparticles for a specific application, the next challenge is to determine the proper synthesis method to create the desired polymer nanocomposite. Figure 4.1 shows the processing challenge of transforming 8-μm particles into more than 1 million platelets.

In general, for solid thermosetting reactive prepolymers or thermoplastic polymers with solid nanoparticles, the following processing methods are recommended:

- Solution intercalation
- Melt intercalation
- Three-roll milling
- Centrifugal

For liquid thermosetting reactive prepolymers or thermoplastic polymers in solvent with solid nanoparticles, the following processing methods are recommended:

- In situ polymerization
- Emulsion polymerization
- High-shear mixing
- Ultrasonication

Examination of Figs. 4.2 and 4.3 identifies the "real" challenge that confronts the investigator into the chosen polymer. Solution intercalation, mesophase-mediated emulsion or suspension, in situ polymerization, and melt processing are considered convenient methods to disperse layered silicates into nanocomposites in an intercalated or exfoliated state (Fig. 4.3). As mentioned later in this chapter, optimum mechanical properties of the resulting nanocomposite are obtained when the layered silicate is fully exfoliated with polymer inserted into the silicate galleries. Intercalation of layered silicate with polymer is somewhat beneficial in polymer properties enhancement but not as optimum as exfoliated systems [1–10]. An excellent review by Alexandre and Dubois lists recent developments in synthesis,

Challenge
Age-Old Blend Questions:
• Process from Agglomerate to Dispersion: Activation Energy?
• Final Dispersion: Thermodynamic v. Kinetic Stability?
8-μm Particle
>1 Million Particles

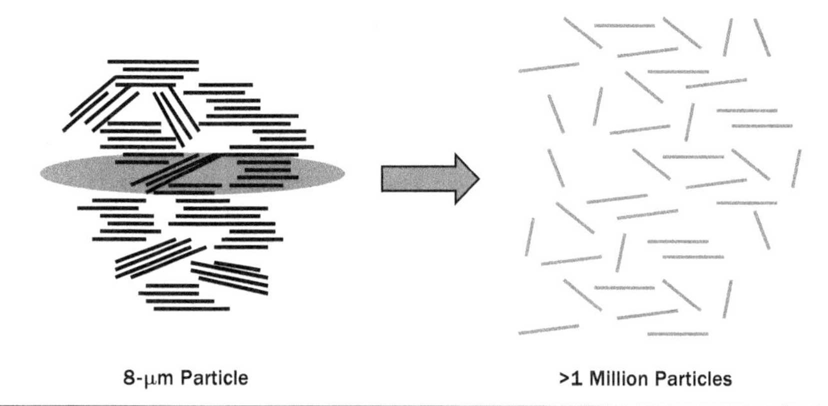

FIGURE 4.1 Processing challenge of layered silicate. (*Courtesy of SCP.*)

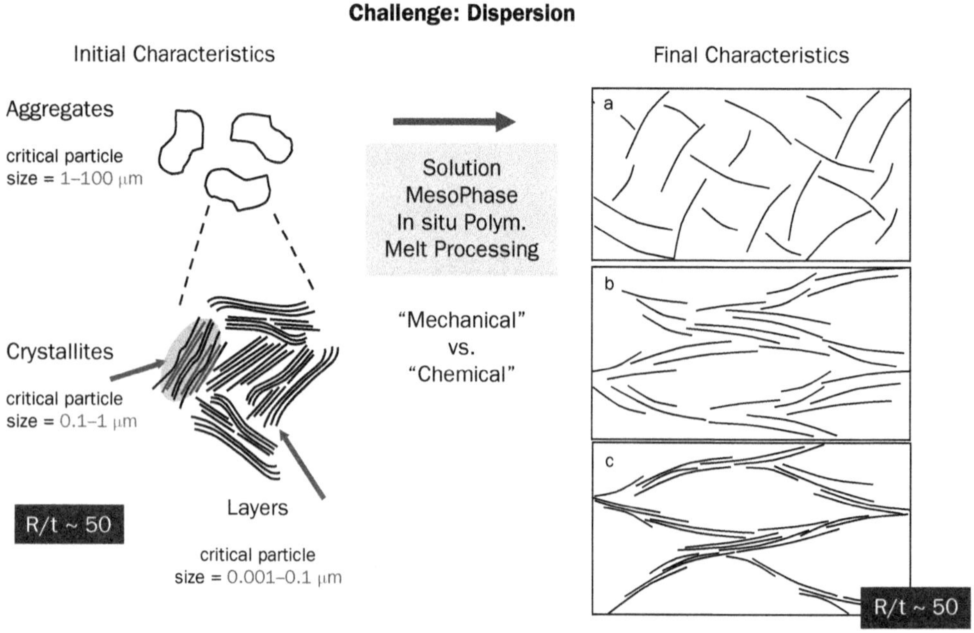

Challenge: Dispersion
Initial Characteristics
Final Characteristics
Aggregates
critical particle
size = 1–100 μm
Solution
MesoPhase
In situ Polym.
Melt Processing
"Mechanical"
vs.
"Chemical"
Crystallites
critical particle
size = 0.1–1 μm
Layers
critical particle
size = 0.001–0.1 μm
R/t ~ 50
a
b
c
R/t ~ 50

FIGURE 4.2 Dispersion of layered silicates using different processing methods. (*Courtesy of R. Vaia.*)

properties, and future applications of polymer-layered silicate nanocomposites [8]. Different approaches and a wide range of polymer matrices (i.e., thermoplastics, thermosets, and elastomers) were considered. Parallel to process selection is the use of WAXD and TEM (Chap. 5) to characterize the uniformity of lack thereof, of the

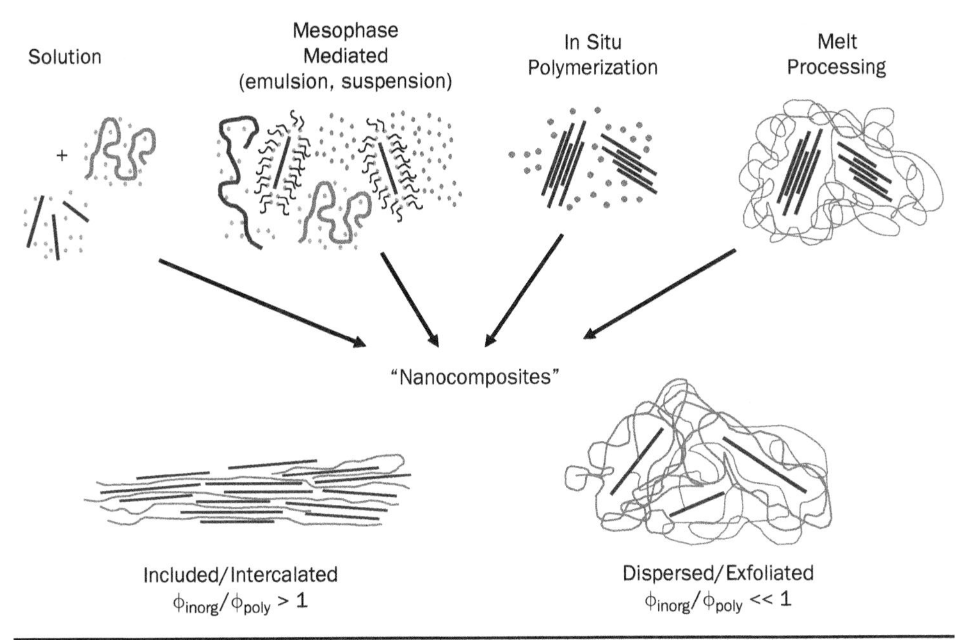

FIGURE 4.3 Various synthesis methods to disperse layered silicate into nanocomposites. (*Courtesy of R. Vaia.*)

dispersion. TEM is vital in appreciation of "viewing" an intercalated or exfoliated nanocomposite system.

Two types of structures are obtained, namely intercalated nanocomposites, where the polymer chains are sandwiched between silicate layers, and exfoliated nanocomposites, where the separated, individual silicate layers are more or less uniformly dispersed in the polymer matrix. This new family of materials exhibits enhanced properties, such as increased Young's modulus and storage modulus, increased thermal stability and gas barrier properties, and good flame retardancy at a low filler level, usually about 5 wt%.

Sanchez et al. surveyed the applications of hybrid organic–inorganic nanocomposites in their recent review [9]. This new generation of hybrid materials is opening many promising applications in several diverse areas: optics, electronics, ionics, mechanics, energy, environment, biology, medicine, functional smart coatings, fuel and solar cells, catalysts, and sensors, among others. These hybrid materials will be described in Chap. 12 under new trends and opportunities.

The following sections show how to use these seven synthesis approaches and different types of nanoparticles to form polymer nanocomposites.

4.2 Solution Intercalation

The layered silicate is exfoliated into single layers using a solvent in which the polymer (or a prepolymer, in the case of insoluble polymers, such as polyimide) is soluble. It is well known that such layered silicates, due to the weak forces that stack the layers together, can be easily dispersed in an adequate solvent. The polymer then absorbs onto the delaminated sheets, and when the solvent is evaporated (or the mixture precipitated), the sheet reassembles, sandwiching the polymer to form an ordered, multilayered structure.

The layered silicate is swollen using a solvent in which the polymer is also soluble. By swelling the layers in the nanoparticle, the solvent allows the polymer access to the space in between the silicates' layers and thus intercalates them. The solvent is then removed and the intercalated silicate and polymer structure remains intact.

4.2.1 Solution Intercalation from Polymers in Solution

This technique has been widely used with water-soluble polymers to produce intercalated nanocomposites [11, 12] based on polyvinyl alcohol (PVOH) [13, 14], polyethylene oxide (PEO) [14–18], polyvinylpyrrolidone (PVPyr) [19], or polyacrylic acid (PAA) [18]. When polymeric aqueous solutions are added to dispersions of fully delaminated sodium layered silicate, the strong interaction existing between the water-soluble macromolecules and the silicate layers often triggers the re-aggregation of the layers as it occurs for PVPyr [18] or PEO [14]. In the presence of PVOH, the silicate layers are colloidally dispersed resulting in a colloidal distribution of nanoparticles in PVOH [14].

Polymer intercalation using this technique can also be performed in organic solvents. Polyethylene oxide has been successfully intercalated in sodium montmorillonite (MMT) and sodium hectorite by dispersion in acetonitrile [20], allowing stoichiometric incorporation of one or two polymer chains between the silicate layers and increasing the intersheet spacing from 0.98 to 1.36 and 1.71 nm, respectively. Study of the chain conformation using two-dimensional, double-quantum Nuclear Magnetic Resonance (NMR) on ^{13}C-enriched PEO intercalated in sodium hectorite [21] reveals that the conformation of the –OC–CO– bonds of PEO is 90% ± 5% gauche, including constraints on the chain conformation in the interlayer.

4.2.2 Solution Intercalation from Prepolymers in Solution

The Toyota Research group was the first to use solution intercalation technique to produce polyimide (PI) nanocomposites [22]. The polyimide-MMT nanocomposite has been synthesized by mixing in dimethylacetamide, a modified MMT with the polyimide precursor, that is, a polyamic acid obtained from the step polymerization of 4,4′-diaminodiphenyl ether with pyromellitic dianhydride. The organic-modified MMT was prepared by previous intercalation with dodecylammonium hydrochloride. After elimination of the solvent, an organoclay-filled polyamic acid film was recovered, which was thermally treated up to 300°C to trigger the imidization reaction and to produce the polyimide nanocomposite. The WAXD patterns (Chap. 5) of these filled PI films do not show any diffraction-peak-typical structure of an intercalated morphology. This led Yano et al. [22] to conclude that an exfoliated structure was formed and explained the excellent gas barrier properties of the resultant films. Enhanced properties and excellent gas barrier performance are attributable to the exfoliated nanocomposite structure. This experiment has been extended to other layered silicates (hectorite, saponite, and synthetic mica) with different aspect ratios, as shown in Table 4.1 [22].

X-ray diffraction (see Chap. 5, where the absence of a peak for layered silicates is strong evidence for exfoliation) data of the resulting polyimide-based nanocomposites show no noticeable peak, indicating a possible exfoliated structure for the MMT and the synthetic mica. For both hectorite and saponite, a board peak, centered on a value of 15 Å, is observed, indicating that for these layered silicates, polymer intercalation occurs,

Table 4.1 Nature, CEC, and Length of the Layered Silicates Used in the Synthesis of Polyimide

Layered silicate	CEC (meq/100 g)	Length of dispersed particles (angstroms)*
Hectorite Na^+	55	460
Saponite Na^+	100	1650
Montmorillonite Na^+	119	2180
Synthetic mica Na^+	119	12,300

*Longer particle dimension as determined by TEM observation.

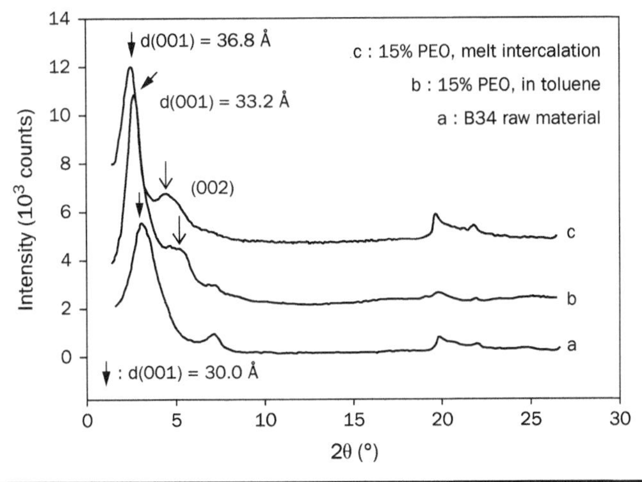

Figure 4.4 WAXD patterns for the PEO/B34 system: (a) pure B34, (b) 15% PEO via solution intercalation in toluene, and (c) 15% PEO via melt interaction.

probably mixed with some exfoliation. For saponite, the measured interlayer spacing is even smaller than the value measured for the starting organically modified clay (18 Å), suggesting that the organic cation may have been expelled from the clay interlayer during the imidization reaction.

Shen et al. [23] used solution intercalation and melt intercalation to create a polymer nanocomposite using polyethylene oxide (PEO) and two nanomaterials: Na^+-MMT and organo-modified bentonite (B34). They used wide-angle x-ray diffraction (WAXD) and Fourier transform infrared (FTIR) analyses to characterize their polymer nanocomposites. The researchers did not perform TEM analysis of their samples, so it is impossible to definitively say which processing method resulted in better dispersion. Figure 4.4 shows the WAXD patterns they obtained for three of their samples [23]. It shows that neither the melt intercalation nor the solution intercalation in toluene dispersed the B34 in the PEO.

Chen and Qu [24] used solution intercalation to exfoliate polyethylene-grafted maleic anhydride (PE-g-MA) into organo-modified MgAl layered double hydroxide (OMgAl-LDH) under reflux in xylene. They found that they were able to create exfoliated nanocomposites using solution intercalation.

In summary, solution intercalation is a good processing technique because of its relative simplicity. It doesn't require large amounts of expensive equipment to perform. It is also a good technique for producing thin films of polymer nanocomposites with oriented clay layers. However, because it does require the use of solvents, many of which are hard to deal with, it is not considered an environmentally friendly method. For this same reason, it is not easily scalable because dealing with large amounts of organic solvents is cost prohibitive on a commercial scale.

4.3 Melt Intercalation

The layered silicate is mixed with the solid polymer matrix in the molten state. Under these conditions, and if the layer surfaces are sufficiently compatible with the selected polymer, the polymer can be inserted into the interlayer space and form either an intercalated or an exfoliated nanocomposite. This melt intercalation method does not rely on a solvent to disperse the nanomaterials, but rather uses heat and kinetic energy to break apart the nanomaterial into individual nanoparticles while allowing the polymer to insert itself between the nanoparticles. In addition to this method being compatible with a number of thermoplastic polymers, melt intercalation has the added benefit of being environmentally friendly. This is due to the fact that melt intercalation does not require a solvent like solution intercalation does, and these solvents used in solution intercalation can be harmful to humans and environmentally damaging. This leads to added costs of solution intercalation, especially when done on a large scale, to properly dispose of the solvents used in the process.

As mentioned earlier, melt intercalation in addition to heat uses kinetic energy to disperse the nanomaterials within the polymer matrix. One (single-screw extruders) or two (twin-screw extruders) screws with varying configurations usually provide this kinetic energy. In a single-screw extruder, the screw has a deep-channeled feed zone that transports solids from the feed throat and compresses them; a transition or compression zone where the channels become shallower and the polymer is melted; and a shallow-channeled metering zone where the melt is conveyed to the adapter and die (Fig. 4.5). Single-screw extruders are widely used for extruding film, sheet, and other applications that use pre-compounded pellets.

Several different types of twin-screw extruders are used for different applications. Parallel twin-screw extruders have two screws that rotate in the same direction (co-rotating) or opposite direction (counter-rotating) with non-intermeshing or fully intermeshing flights. The size of twin-screw extruders can vary considerably depending on how much polymer they need to process and under what kind of shear rates. Figure 4.6 shows a commercial grade twin-screw extruder.

While single-screw extruders are typically flood-fed, with the feed filling the screw channel, twin-screw extruders and extruder equipment are typically starve-fed, with only a partially filled screw channel. Because the screw channel is not completely filled, downstream feed ports or vents can be easily added. Compared to single-screw extruders, twin-screw extruders are better at feeding and conveying, especially of powdered material or large-amount fillers, and typically created more dispersive mixing.

4.3.1 Thermoplastic Nanocomposites

Vaia and Giannelis [25] have studied polystyrene (PS) as the matrix for dispersing different types of clays. Li-fluorohectorite [cation exchange (CEC) = 150 meq/100 g],

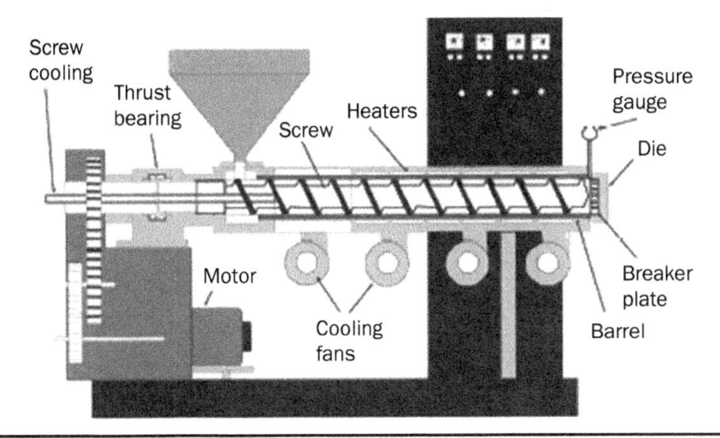

Figure 4.5 Single screw extruder.

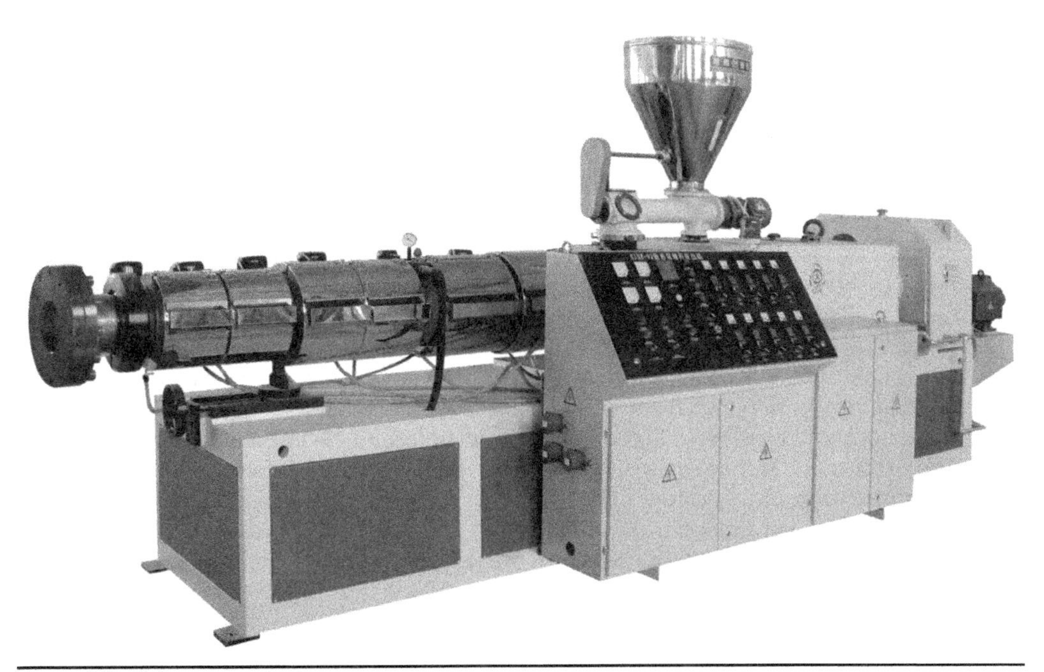

Figure 4.6 Twin-screw extruder.

saponite (100 meq/100 g), and sodium MMT (80 meq/100 g) were modified accordingly, using various ammonium cations: dioctadecyldimethylammonium, octadecyltrimethylammonium, and a series of primary alkylammonium cations with carbon chains of 6, 9–16, and 18 carbon atoms. The nanocomposites were synthesized by statically annealing (without any mixing or shearing) a pelletized intimate mix of the modified silicate in PS (M_w = 30,000, M_w/M_n = 1.06) under vacuum at 170°C, a temperature well above the PS glass transition (Table 4.2) [25].

Table 4.2 Characteristics of Polystyrene Melt Intercalation within Octadecylammonium Modified Clays

Entry	Clay (CEC in meq/100 g)	Ammonium Cation	Initial Gallery Height (nm)	Final Gallery Height (nm)	Net Change (nm)
1	M (80)	PODA	0.75	0.75	0
2	S (100)	PODA	0.83	0.83	0
3	F (120)	PODA	1.33	2.16	0.83
4	F (120)	QODA	1.57	2.69	1.12
5	M (80)	DODMDA	1.43	2.25	0.82
6	S (100)	DODMDA	1.50	2.35	0.85
7	F (120)	DODMDA	2.85	2.85	0

F, fluorohectorite; M, montmorillonite; S, saponite; DODMDA, dioctadecyldimethylammonium; PODA, primary octadecylammonium; QODA, quaternized octadecylammonium.

Comparison of the first three entries indicates that, for a given alkyl surfactant, increasing the cation exchange from 80 to 120 meq/100 g allows for PS intercalation to occur. At a low CEC and for single alkyl chain buildup cations (entries 1 and 2), no intercalation is observed. Under these conditions, the aliphatic alkyl chain of the organic cation adopts a pseudo-monolayer arrangement characterized by low gallery height. Intercalation can take place at lower CECs when the clay surface possesses an ammonium cation bearing two long alkyl chains (entries 5 and 6). However, excessive packing of the chains all along the layer surface (high CEC and two long alkyl chains per cationic head, entry 7) leads to the formation of a non-intercalated structure, as predicted by the theory introduced by Vaia et al. [26]. Vaia et al. also found that polymer intercalation depends on the length of the exchange ammonium cation as well as on the annealing temperature.

Dennis et al. [27] did a study to see how different extruder setups would affect the processing of a polymer nanocomposite. These researchers studied the polyamide 6 polymer (from Honeywell) in conjunction with both Cloisite 15A and Cloisite 30B (from BYK Additives). To perform their experiment, they used four different extruders in conjunction with multiple-screw designs. For this polymer nanocomposite system, they found that the non-intermeshing twin-screw extruder performed the best in terms of delamination and dispersion. However, the researchers noted that it is very likely that co-rotating and counter-rotating twin-screw extruders can provide excellent results if a fully optimized screw configuration is used.

Figure 4.7 shows TEM images of some of the processed polymer nanocomposites that these researchers produced [27]. It is clearly evident that the counter-intermeshing, medium-shear image shows the best dispersion. Table 4.3 shows the mechanical properties of all of the different test samples [27]. In this table, it is evident that the counter-rotating non-intermeshing extruder does provide very good results. It is also interesting to note that the two high-shear samples had very high percentage of elongation at break results—even higher than, or as high as, the neat polymer.

4.3.2 Elastomer Nanocomposites

Burnside and Giannelis [28] have described the two-step preparation of silicon-rubber–based nanocomposites. First, silanol-terminated polydimethylsiloxane (PDMS,

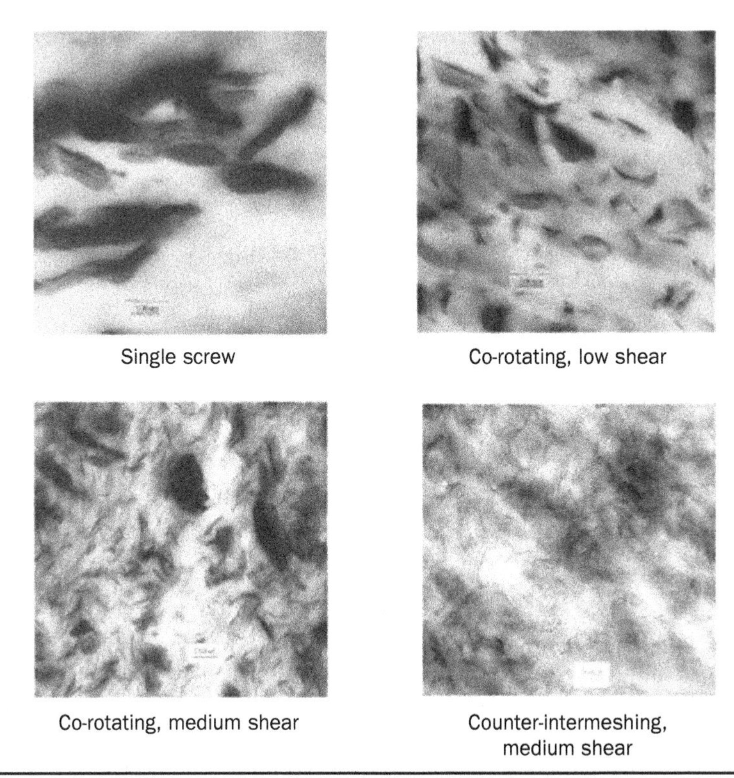

FIGURE 4.7 TEM images of polymer nanocomposites (polyamide 6 and Cloisite 15A) using various extruders.

TABLE 4.3 Montmorillonite Feed Ratios, PMMA Contents in Nonextracted and Extracted Composites, Average Molecular Weights, and Polydispersities of Extracted PMMAs

Sample	Feed ratio of PMMA/ clay (g/g)	PMMA content (wt%)* Nonextracted	PMMA Content (wt%)* Extracted†	$M_n \times 10^{-3}$ (g/mol)	$M_w \times 10^{-3}$ (g/mol)	M_w/M_n
PMMA	100/0	–	–	23	160	6.6
PMMA10	100/10	87.4	58.7	44	250	5.8
PMMA20	100/20	79.3	49.6	60	200	3.4
PMMA30	100/30	60.4	33.4	63	150	2.4
PMMA40	100/40	58.6	22.8	82	390	4.8
PMMA50	100/50	46.1	18.4	38	290	7.6

*As determined by TGA.

†Composite recovered after Soxhlet extraction in toluene for 5 days.

M_w = 18,000) was melt blended at room temperature with dimethyl ditallow ammonium–exchanged MMT. Then the silanol end groups were cross-linked with tetraethyl orthosilicate (TEOS) in the presence of tin bis(2-ethylhexanoate) as a catalyst at room temperature. To obtain exfoliated nanocomposites (as determined by featureless XRD patterns; see Chap. 5), several conditions were required, such as mixing the modified clay and PDMS under sonication (for better mixing) and with the addition of a small amount of water extending/cross-linking TEOS end groups. The nature of both the silicone matrix and the clay modifier plays an important role. Neither exfoliation nor intercalation can occur if MMT is modified with a different surfacing agent, such as benzyl dimethyloctadecylammonium cation, or if too much water is added. However, only intercalation was observed when a PDMS-polydiphenylsiloxane random copolymer containing 14–18 mol% diphenylsiloxane units was used. The results stress the key importance of getting the right match between the matrix and the organoclay to optimize the layered silicate exfoliation.

Li and Shimizu [29] used melt intercalation to fabricate a multiwall carbon nanotube (MWNT)–thermoplastic elastomer nanocomposite. In their research, Li and Shimizu mixed poly[styrene-b-(ethylene-co-butylene)-b-styrene] triblock copolymer (SEBS) from Asahi Kasei Corporation (Japan) with high-purity MWNTs from Nikkiso Co., Ltd. (Japan). The aim of the researchers was to create an elastic, conductive nanocomposite. They found that they were able to disperse the MWNTs uniformly throughout the SEBS matrix using high-shear melt intercalation. They used a HSE3000 mini high-shear extruder (Imoto, Japan) at a speed of 1,000 rpm.

Elastomer nanocomposites and thermoplastic elastomer nanocomposites that possess enhanced ablation and fire retardancy characteristics were developed by the Air Force Research Laboratory/Polymer Working Group at Edwards Air Force Base, California [30–35]. The TPSiV X1180 thermoplastic elastomer (TPE) is a polyamide-based, vulcanized silicone thermoplastic resin manufactured by Dow Corning, which was selected as a potential replacement material as insulative materials in solid rocket propulsion systems. Three formulations, TPSiV X1180:Ph12 T12 POSS® in weight ratio of 80:10, TPSiV X1180:Cloisite 30B in weight ratio of 92.5:7.5, and TPSiV X1180:PR-24-PS carbon nanofibers (CNFs) in weight ratio of 85:15, were processed by twin-screw extrusion. None of these nanoparticles was compatible with the resin system as observed by TEM analysis. The above blends with the baseline Kevlar-filled EPDM rubber were tested for ablation performance using a subscale solid rocket motor. All three TPEs were outperformed by the baseline material. These observations clearly demonstrated that if the nanoparticles are poorly dispersed in the polymer matrix, no thermal performance improvement can be expected [35].

Overall, melt intercalation is a very powerful technique of processing polymer nanocomposites. This technique is environmentally friendly, scalable, and effective at dispersing a variety of nanomaterials in polymer matrices.

4.4 Three-Roll Milling

Three-roll milling is considered as low-shear mixing for incorporating solid nanoparticles into a liquid polymer, as compared to high-shear mixing, which will be discussed in Sec. 4.6. This process uses multiple cylinders rotating in different directions to mix nanomaterials with a given polymer. Figure 4.8 shows how a three-roll mill looks like and how the various cylinders rotate in relation to each other [36]. The three-roll mill,

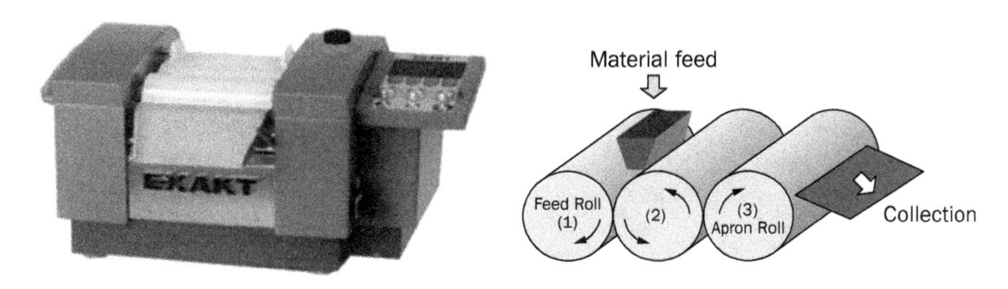

FIGURE 4.8 Three-roll milling equipment and schematic.

like melt intercalation, uses shearing to disperse nanomaterials throughout a polymer matrix. Three-roll milling is an effective processing technique for some polymer nanocomposites. Like melt intercalation it is environmentally friendly in that it uses no solvents. Three-roll milling also has the advantage of being relatively low-shear devices, so this processing technique does not damage CNTs, which is a large benefit.

Polymer nanocomposites were developed by incorporating different POSS nanostructured chemicals in Derakane™ Momentum™ 441-400 epoxy vinyl ester (EVE) resin [36]. The various POSS polymers and monomers used in this study were polyphenyl silsesquioxane uncured (phenyl resin), polyvinyl silsesquioxane uncured (vinyl resin), polyvinyl silsesquioxane fully cured (Firequench®), and vinyl-POSS-cage mixture (vinyl POSS). These formulations were prepared by blending POSS (15 wt%) into the EVE resin using three-roll milling, low-shear mixing, and solvent-assisted blending synthesis methods. Three-roll milling is the preferred method for this study. It is very fast and cost-effective, and results in the optimum POSS dispersion in the EVE resin. The E-glass-reinforced EVE-POSS® nanocomposites exhibit better flammability properties than pure E-glass-reinforced Derakane Momentum 441-400 resin, with reduced smoke, a reduced heat release rate, and an increased ignition time using cone calorimetry. The EVE-POSS nanocomposites were reported to have equivalent or better mechanical, thermal, cure, and processing characteristics. The properties were also found to be comparable or better than the halogenated EVE resin (Derakane Momentum 510A-40). The nano-reinforced polymers acquired fire retardancy via two major mechanisms: (a) reduced volatilization of fuel (organic monomer/polymer) and (b) the formation of thermo-oxidative, stable, and nonpermeable surface chars.

A diglycidyl ether of bisphenol A (DGEBA) epoxy resin cured with methyl tetrahydrophdralic anhydride hardener was blended with Nanomer I.28E using three-roll milling [37]. Good dispersion was reported by Yasmin et al. [37] using this mixing technique.

Rosca and Hoa [38] used three-roll milling to disperse multiwall carbon nanotubes in epoxy. These researchers found that for higher aspect ratio MWNTs (NLIG, NL15L, NL15S) shear alignment occurred. The shearing forces from the three-roll milling equipment forced the MWNTs into lower energy configuration. This did not happen for all MWNTs in their experiments, which Rosca et al. hypothesized was either because in order for this to occur the MWNTs need to be of a certain length as the viscosity of some mixtures was so high that the MWNTs could not move within it.

Yasmin et al. [39] used three-roll milling to disperse clay nanoparticles (Cloisite 30 B) in an epoxy matrix. They found that the three-roll mill yielded good dispersion of clay particles within a short period of time. Figure 4.9 shows TEM images of their samples.

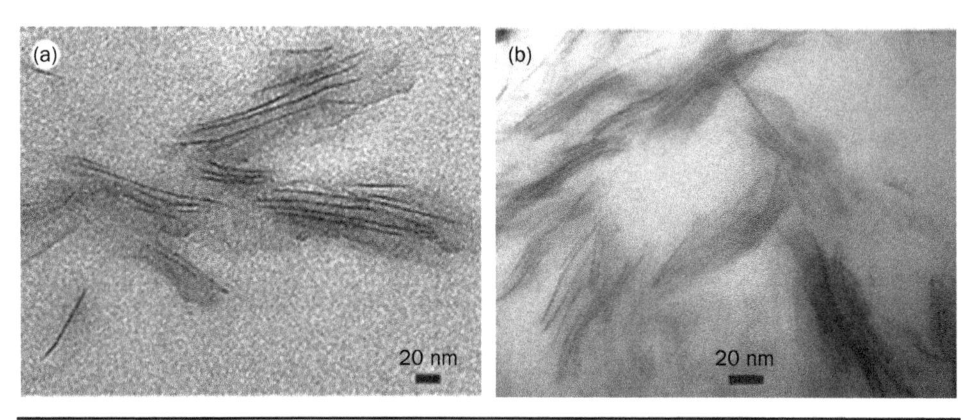

FIGURE 4.9 TEM images of clay nanocomposites: (a) 5 wt% and (b) 10 wt%.

It does appear they have broken down most large agglomerates, but most of the clay platelets are still in tactoids [39].

Seyhan et al. [40] used three-roll milling to disperse MWNTs and double-wall carbon nanotubes (DWNTs) within a styrene-free resin (POLIYA 420) and a vinyl-ester-epoxy resin (POLIPOL 701). They noted that they chose the three-roll milling technique to disperse the CNTs due to the fact that three-roll milling does not damage or rupture CNTs like other processing techniques. They also note that the three-roll mill they used, much like the equipment shown in Fig. 4.8, produces almost entirely shear loads with very little compressive loads.

4.5 Centrifugal Processing

Centrifugal processing is used primarily in powder-to-powder processing. It can also be used to mix nanoparticles with aqueous solutions. Materials are deposit into the container directly and rotation axis has the angle of 45 degrees to revolution. Defoaming by 400G acceleration is generated from 2,000 rpm of revolution. Mixing by centrifugal force from rotation can be applied to materials simultaneously. As a result, this equipment manufactured by Thinky Corporation (Tokyo, Japan, website: www.thinky.com) can be used for blending and degassing at the same time. Figure 4.10 shows the principles of this centrifugal processing technique. Figure 4.11 shows the Thinky ARV-310 mixer with different types of adapters for a variety of applications. When using standard 300-mL container and vacuum state, it has a maximum mixing capacity of 200 mL and a maximum mixing weight of 310 g (gross), and the revolution speed of 2,000 rpm (max) and rotation speed of 1,000 rpm (max). Both mixing and defoaming abilities are proved to be superior.

Thinky Corp. also has some unique and high-quality products for nanodispersion. Figure 4.12 shows Thinky's new product, Nano Premixer PR-1. It is designed to highly reproduce the dispersion of nanomaterials, such as CNTs, while operating safely by keeping samples in an enclosed container (or test tube–shaped vial). Figure 4.13 shows that the dual-sonic technology (patented) rotates the container at a high speed while ultrasonic technology irradiates from the bottom and side of the ultrasonic bath. Rotating the container at an angle of 45 degrees causes convection in the materials,

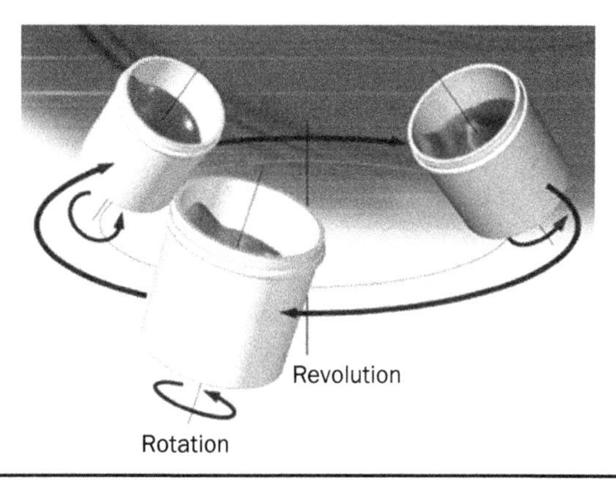

Figure 4.10 Schematics showing the principle of the centrifugal processing. (*Courtesy of Thinky.*)

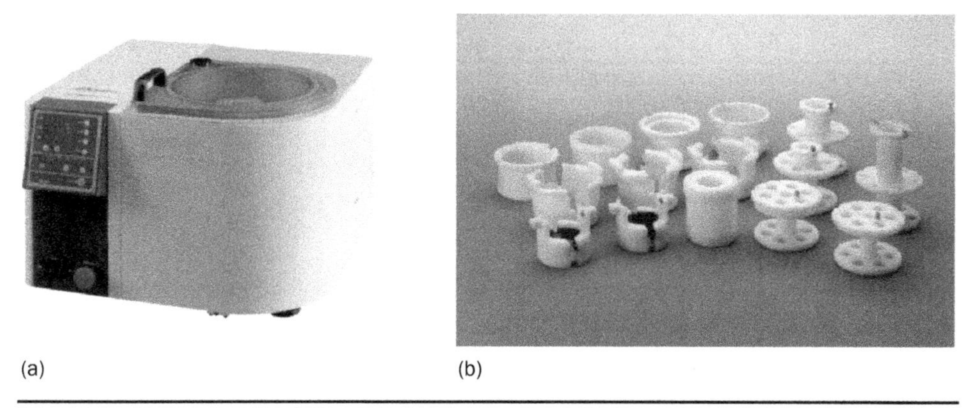

Figure 4.11 Thinky (a) ARV-310 mixer and (b) adaptors for different type of experiments. (*Courtesy of Thinky.*)

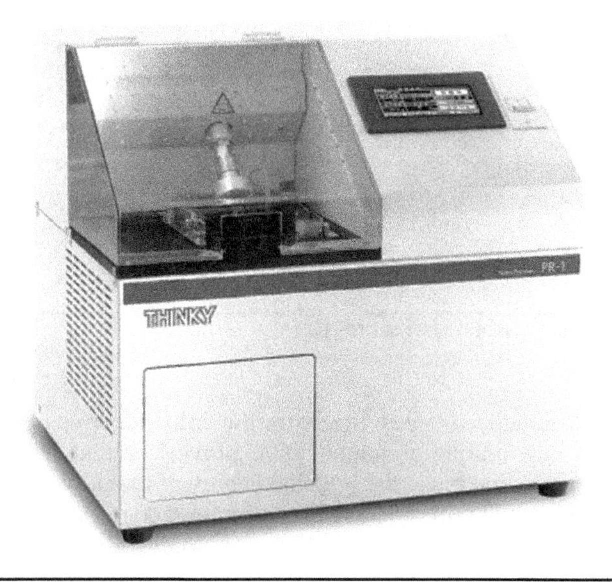

Figure 4.12 Thinky Nano Premixer PR-1.

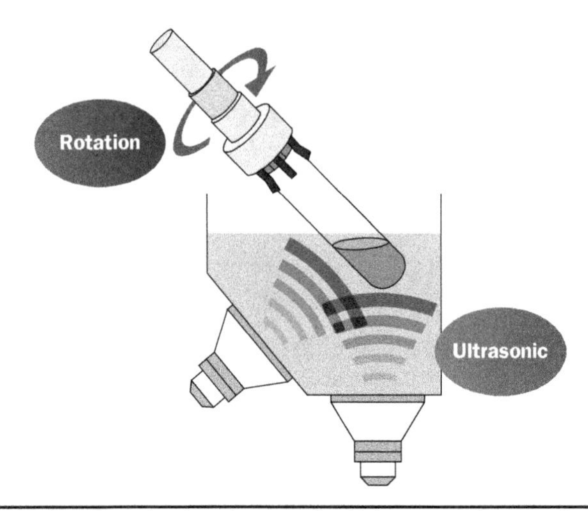

Figure 4.13 Thinky dual-sonic technology (patented) which rotates the container at a high speed while ultrasonic technology irradiates from the bottom and side of the ultrasonic bath.

Figure 4.14 Thinky Nano Pulverizer NP-100.

and thus ultrasonic technology irradiates the entire sample. Nano Pulverizer NP-100 (Figure 4.14) is a unique and innovative pulverizer that has an optimal rotation-revolution ratio to maximize the collision energy of the pulverizing medium. It allows ultralow volume (100 mg) process, fine crushing (100 nm), and rapid (2 minutes) pulverization, which used to be regarded as impossible. Moreover, the mounted cooling

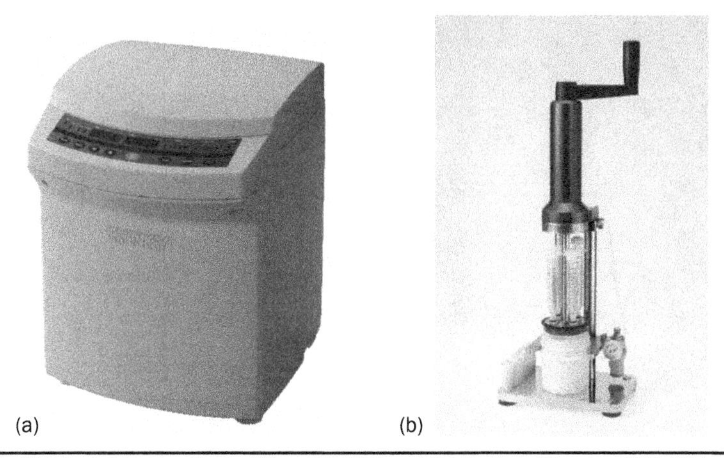

FIGURE 4.15 Thinky (a) Solder Paste Mixer SR-500 and (b) Syringe Charger ARC-40H.

mechanism can decrease the temperature inside the chamber to –20°C, reducing the risk of contamination and keeping the crystal structure of compounds with low melting point during nano pulverization. Thinky Corp. also manufactures other signature productions, such as Solder Paste Mixer and Springe Charger.

Figure 4.15a shows the Solder Paste Mixer SR-500. This mixer can optimize temperature and viscosity of solder paste at the same time. Off-the-shelf 500-g solder paste containers can be set directly into the machine just as they are. SR-500 eliminates the fine air bubbles that are believed to cause solder ball scattering. Figure 4.15b shows the Syringe Charger ARC-40H, which realizes an easy, quick, and bubble-free charging into a syringe after mixed, defoamed, and dispersed by Thinky mixer. Small syringes such as 3, 5, and 10 mL can be used. Figure 4.16 shows the automatic control system vacuum filler ARC-600TWIN which can simultaneously fill a maximum 16 syringes with material. It contributes greatly to reduce costs by streamlining the work that requires time and great care when manually operated.

As an example, powder-to-powder blending using a rotation and revolution Thinky® ARE-310 mixer was used to combine PA11 polymer and MWNTs via centrifugal processing [41]. The PA11 powder was mixed gradually in three steps of 30, 30, and 39 g to 1 g of MWNT (based on the desired final loading of MWNTs, where the total mixture is 100 g in this example), each time rotating at 2,000 rpm for 10 seconds, resulting in PA11-1wt% MWNT mixtures. Substantial improvements in electrical conductivity were observed with increasing loading of MWNT.

Lu et al. [42] processed and fabricated poly(3,4 ethylenedioxythiophene)-poly(4-styrene sulfonate) (PEDOT-PSS)/nickel nanostrands nanocomposites using the centrifugal mixing method. The electrical properties of the nanocomposites were studied and the addition of nanostrands into the intrinsically conductive polymer system with shorter mixing time did not improve the electrical conductivity, since there was no conductive network of nanofillers formed. With longer mixing time, better dispersion of the nanostrands was achieved and a highly significant, two-orders-of-magnitude improvement in electrical conductivity was observed with the 10% nanostrands loading. Additional processing parameters (mixing speed) of the centrifugal mixing method

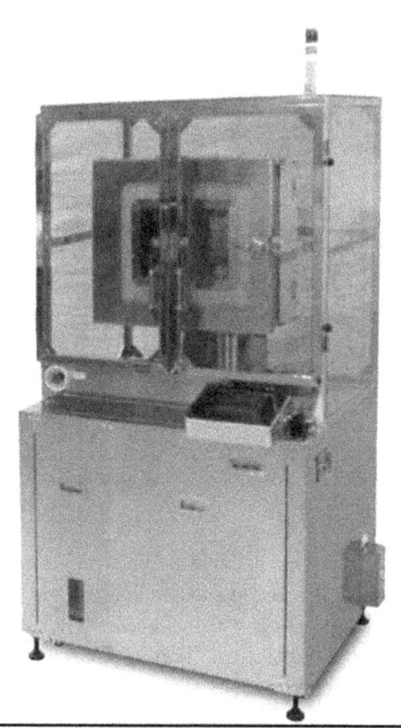

FIGURE 4.16 Thinky automatic control system vacuum filler ARC-600TWIN which can simultaneously fill a maximum 16 syringes with material.

were explored by Lu et al. to obtain the optimum conditions for the processing of the nanocomposites.

Hansen et al. [43] used a desired volume fraction of nickel nanostrands added to a polymer sample, then "wet" by manual mixing methods. Dispersion was achieved by using a Thinky AR-25 centrifugal double planetary mixer. A mixing duration of 20 seconds at 2,000 rpm was used for baseline samples. This polymer-nanostrand composite was then used as a conductive coating. Coating samples were screened through a 50-woven stainless mesh before spray application. Polymer-nanostrand systems typically are visibly percolated at volume fractions of nanostrands above 0.5%, and imaging of sample with this loading level is challenging. However, the electrical conductivity of nanocomposite at these levels is very close to the percolation threshold, and results can be highly sensitive with respect to differences in conductor volume fraction. Test specimens of nanostrand-polymer mixtures were sprayed to an electrically insulating Kapton® polyimide film substrate, and tested for electrical resistivity. Figure 4.17 shows the electrical resistivity of nanostrands in water-based acrylic/urethane polymer films [43].

In general, centrifugal processing is a very effective powder-to-powder and liquid-to-liquid mixer and also an environmentally friendly method. Thinky Corp. has a total product line from research and development (R&D) to production of mixer for larger volume (Fig. 4.18).

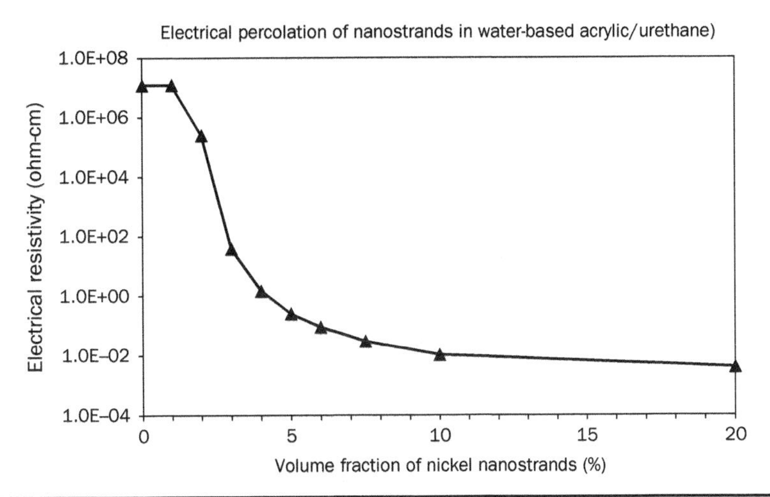

Figure 4.17 Electrical resistivity of nanostrands in water-based acrylic/urethane polymer films.

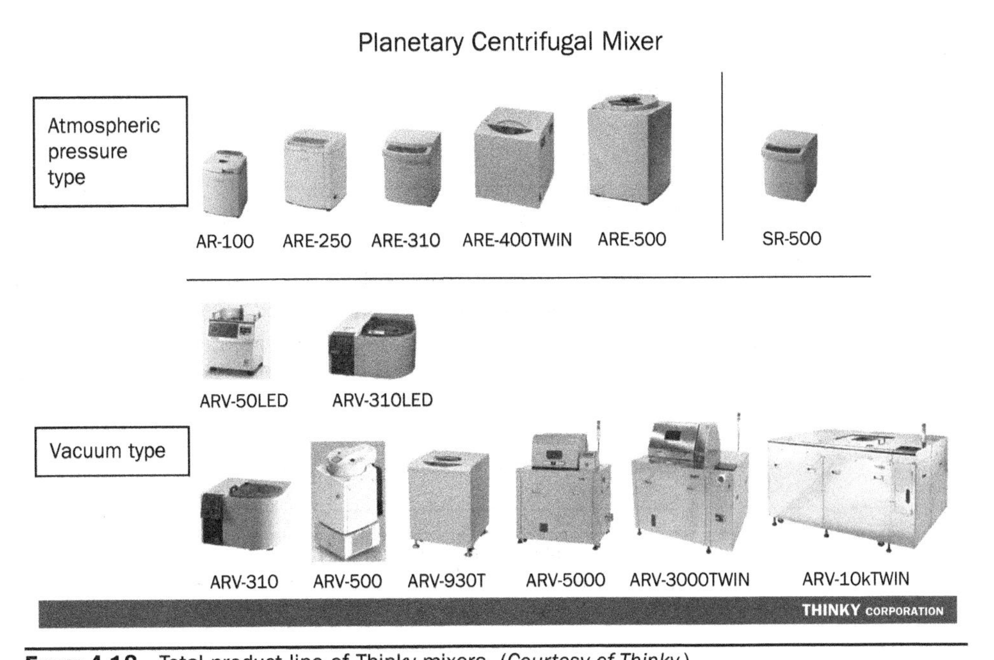

Figure 4.18 Total product line of Thinky mixers. (*Courtesy of Thinky.*)

4.6 In Situ Polymerization

The layered silicate is swollen within the liquid monomer (or a monomer solution) so that the polymer formation can occur between the intercalated sheets. Polymerization can be initiated by different polymerization methods, such as heat or radiation, diffusion of a suitable initiator, or an organic initiator or catalyst fixed through cationic exchange inside the interlayer before the swelling step by the monomer. The in situ polymerization technique initially mixes a monomer with the nanoparticle being used. The addition of the monomer swells the layered silicate creating small openings between the layers of the nanomaterial. The monomer is then polymerized through a variety of methods. During the polymerization the growing polymer uses the small openings to grow into the nanomaterial, thus intercalating it. As this progress and more polymer grows in between layers of the nanomaterial, the nanomaterial becomes more and more dispersed. The polymerization reaction can be driven by a variety of factors including heat, radiation, or catalysts.

4.6.1 Thermoplastic Nanocomposites

Many "interlamellar" polymerization reactions were studied by various research teams in the 1960s and 1970s using layered silicate [12, 44, 45], but it is with the work initiated by the Toyota research team [46, 47] about 20 years ago that the study of polymer-layered silicate nanocomposites became recognized. The Toyota team studied the Na^+-MMT organically modified by protonated α,ω-amino acid [$^+H_3N–(CH_2)_{n-1}–COOH$, with $n = 2, 3, 4, 5, 6, 8, 11, 12, 18$] to be swollen by the ε-caprolactam monomer (melting temperature is 70°C) at 100°C. Subsequently, the nano-modified caprolactam was transformed by ring opening to nylon 6–based nanocomposites [48–50]. A clear difference occurs in the swelling behavior between the MMT with relatively short ($n < 11$) and longer alkyl chains, as depicted in Table 4.4, indicating that a larger amount of monomer can be intercalated for longer alkyl chains.

Montmorillonite modified by 12-aminolauric acid ($n = 12$) was selected during the intercalative ring opening polymerization of ε-caprolactam [49]. The modified MMT (12-Mont) was mixed with the monomer containing a small amount of 6-aminocaproic acid as a polymerization accelerator. The mixture was heated first at 100°C for 30 minutes, then at 250°C for 6 hours. The cooled and solidified product was crushed, washed with water at 80°C, and then dried. Depending on the amount of 12-Mont introduced, either exfoliated (for less than 15 wt%) or delaminated structure (from 15 to 70 wt%) was obtained, as evidenced by wide-angle x-ray diffraction (WAXD) and transmission electron microscopy (TEM) measurements (Chap. 5). Comparison of the titrated

Table 4.4 Basal Spacing on Organo-Modified Montmorillonite in the Presence of ε-caprolactam (ε-CLa) at 100°C

$^+H3N–(CH_2)_{n-1}–COOH$ (n)	Interlayer spacing of the modified clay (Å)	Interlayer spacing when swollen by (ε-CLa) at 100°C (Å)
2	12.7	14.1
3	13.1	19.7
4	13.2	19.9
5	13.2	20.4
6	13.2	23.4
8	13.4	26.4
11	17.4	35.7
12	17.2	38.7
18	28.2	71.2

amounts of COOH and NH_2 end groups present in the synthesized nanocomposites with given values, such as the cation exchange (CEC) of the MMT used (119 meq/100 g), led to the conclusion that the COOH end groups present along the 12-Mont surface are responsible for the initiation of polymerization. Nearly all of the $^{+}NH_3$ end groups present in the matrix should be interacting with the MMT anions. Finally, the ratio of bonded to nonbonded polymer chains increased with the amount of incorporated MMT (from 32.3% of bonded chains for 1.5 wt% MMT to 92.3% of bonded chains for 59.6 wt% clay).

Maneshi et al. [50] used in situ polymerization to produce polyethylene-clay nanocomposites. They used Cloisite 93A clay (BYK) as their nanomaterial. To facilitate the polymerization reaction the researchers used bis(cyclopentadienyl)zirconium dichloride (Cp_2ZrCl_2, Aldrich). They noted that the larger the amount of Zr catalyst used in the reactor, the higher polymerization activities occurred and better clay exfoliation was visible.

Namazi et al. [51] used in situ polymerization to create starch-g-polycaprolactam (Starch-g-PCL). They used Cloisite 15A (BYK Additives) along with potato starch from Merck. These researchers were mainly interested in comparing the results of creating this polymer nanocomposite through in situ polymerization and solution intercalation. They found that the best diffusion and intercalation of the polymer occurred through the solution intercalation method.

4.6.2 Thermoset Nanocomposites

In situ polymerization has also been explored to create thermoset-based nanocomposites. This method has been extensively used for the production of both intercalated and exfoliated epoxy-based nanocomposites. Messersmith and Giannelis [52] have analyzed the effects of different curing agents and conditions in the formation of epoxy nanocomposites based on the diglycidyl ether of bisphenol A (DGEBA) and a MMT modified by bis(2-hydroxyethyl)methyl hydrogenated tallow alkyl ammonium cation. They observed increase in viscosity at relatively low-shear rates and the transition of opaque to semitransparent suspension indicated that the modified MMT clay dispersed readily in DGEBA when sonicated for a short period of time. The increase in viscosity was attributed to the formation of a *house-of-cards* structure, in which edge-to-edge and edge-to-face interactions between dispersed layers form percolation structures. WAXD patterns of uncured clay-DGEBA also indicate that intercalation occurred during mixing and that this intercalation improves as the temperature is increased from room temperature to 90°C.

4.6.3 Rubber-Modified Epoxy Nanocomposites

Lan and Pinnavaia [53] examined the formation of nanocomposites with a rubber-epoxy matrix obtained from DGEBA derivative (Epon 828) and cured with a polyether diamine (Jeffamine D2000) to obtain low glass transition temperature materials. Two MMTs modified by the protonated *n*-octylamine and the protonated *n*-octadecylamine, respectively, were used in this study. It was shown that depending on the alkyl chain's length of the modified clays, an intercalated and partially exfoliated (*n*-octyl) or a totally exfoliated (*n*-octadecyl) nanocomposite can be obtained.

Lan et al. have studied other parameters, such as the nature of alkylammonium cations present in the gallery and the effect of the cation-exchange capacity of the clay [54], when Epon 828 was cured with *m*–phenylene diamine. It was demonstrated that when mixed with the epoxide and any length of the protonated primary alkylamine (with 8, 10, 12, 16 or 18 carbons), the modified clays adopt a structure where the carbon chains are fully extended and oriented perpendicularly to the silicate plane incorporating a maximum of one monolayer of epoxide molecules.

In summary, in situ polymerization is an effective method of creating certain polymer nanocomposites. It has the advantages of being able to be scaled up for large volumes of production. However, it does not always provide the necessary amount of dispersion of nanoparticles due to no shearing force of theirs forcing the tightly bound layers apart.

4.7 Emulsion Polymerization

In a manner analogous to the solution intercalation technique (Sec. 4.2), the layered silicate is dispersed in the aqueous phase, and the polymer nanocomposites are formed. Emulsion polymerization process is also somewhat of a mix of in situ polymerization (Sec. 4.6) and solution intercalation. In emulsion polymerization, a monomer is added to an emulsion, which contains the nanoparticles. The nanoparticles are swollen by the emulsion, which allows the monomer, when an appropriate initiator is added, to grow into a polymer. The growing polymer is then able to intercalate itself throughout the nanoparticle layers that are present.

The emulsion polymerization technique has been used to study the intercalation of water-insoluble polymers within Na^+-MMT, which is well known to readily delaminate in water [55–57]. Polymethyl methacrylate (PMMA) was first tested using this technique [55]. Emulsion polymerization was carried out in water in the presence of various amounts of MMT clay. Distilled methyl methacrylate monomer (MMA) was dispersed in the aqueous phase with the aid of sodium lauryl sulfate as a surfactant. Polymerization was conducted at 70°C for 12 hours, using potassium persulfate as the free-radical initiator. The resultant latex is then coagulated with an aluminum sulfate solution, filtered, and dried under reduced pressure. The obtained composites were extracted with hot toluene for 5 days (Soxhlet extraction). The amounts of intercalated polymer were determined for both extracted and nonextracted materials and are presented in Table 4.5, with the molecular weights and polydispersities of the extracted polymers.

These results demonstrate that part of the PMMA chain remains immobilized onto and/or inside the layered silicates and is not extracted. This finding is further confirmed by FTIR spectroscopy of the extracted composite, which shows the absorption bands typical of PMMA chains. It can be observed that the relative content of clay does not substantially modify the PMMA molecular weights (M_w), the value of which is quite comparable to M_w of PMMA polymerized in the absence of clay (entry 1, Table 4.5). The presence of layered silicates does not seem to perturb the free-radical polymerization. Intercalation is evidenced by x-ray diffraction (WAXD) (Chap. 5), where an increase of about 5.5 Å is observed for PMMA 10, 20, and 30. This increase correlates relatively well with the thickness of the polymer chain in its extended form. Differential scanning calorimetry (DSC) data obtained for the extracted nanocomposites did not show any glass transition, in accordance with what is usually observed for intercalated polymers. Ion–dipole interactions are believed to be the driving force for the immobilization of the organic polymer chains that are flat on the layered silicate surface.

Table 4.5 Summary of Mechanical Properties of Polymer Nanocomposites [57]

Extruder and screw type	XRD basal spacing (A)	XRD area under curve	TEM platelets or intercalates per 6.25 cm^2 (at 130, 500×)	Extruder mean residence time (s)	Normalized variance (O'^2)
Single-screw extruder					
30B	30.9	120	13	141	0.0049
15A	32.2	82.5	4	141	0.0049
Twin-screw extruder					
Co-rotating intermeshing					
Low-shear 15A	36.2	382	7	67	0.090
Medium-shear 15A	37.7	146	16	153	0.113
Counter-rotating intermeshing					
Low-shear 10A	34.4	263	8	47	
Medium-shear 15A	38.0	106	14	102	0.0557
Medium-shear 30B	No Peak	No Peak	35	102	0.0557
High-shear 15A	37.9	164	10	117	0.049
Counter-rotating non-intermeshing					
Low-shear 15A	34.7	581	11	108	0.0895
Medium-shear 15A	No Peak	No Peak	27	162	0.0653
High-shear 15A	37.9	277	20	136	0.0579

Zhang et al. [58] used emulsion polymerization to create a polystyrene-SiO_2 microsphere polymer nanocomposite. In their experiment, the researchers actually dispersed the SiO_2 ultrasonically before adding them to the oil phase created dissolving AIBN in styrene. To complete the polymerization reaction, they agitated the mixture (50 rpm) and kept it at 78°C for 24 hours.

Meneghetti and Qutubuddin [59] studied the properties of PMMA and a MMT clay nanocomposite. Both nanocomposites begin breaking down at higher temperatures; however, it appears the PMMA/clay nanocomposite created using in situ polymerization is more effective than the polymer nanocomposite created using emulsion polymerization.

4.8 High-Shear Mixing

The solid or liquid nanoparticles are mixed with the liquid polymer matrix using high-shear equipment (e.g., an IKA® mixer, IKA, Wilmington, NC) (Fig. 4.19). Under these conditions and if the surface-treated nanoparticles are compatible with the selected polymer, the high-shear mixing will disrupt the nanoparticle aggregates and disperse the polymer matrix into the nanoparticle layers. An intercalated or an exfoliated nanocomposite will result. This technique may or may not (especially in the case of water-based polymer) require solvent.

Figure 4.19 IKA colloid mill: (a) high-shear mixer overall setup, (b) mixing head stator top (from underside up), and (c) mixing head rotor bottom on pump housing and collector plate.

In this process, the liquid polymer mixes with the nanomaterial and then runs through a high-shear mixer to disperse and exfoliate the nanomaterial. The mixer does this by shearing the nanomaterial into platelets, which the polymer can then incorporate into its matrix. An example of an industrial-scale high-shear mixer is that manufactured by Ross (Hauppauge, NY, website: www.mixers.com). Two rotor configurations that physically shear the polymer-nanoparticle mixer are (a) Quad slot mixing head and (b) MegaShear mixing head. The mixing process is done in a loop so that the polymer-nanoparticle liquid can be processed as many times as is needed.

Frohlich et al. [60] used a mixture of two epoxy resins, tetraglycidyl 4,4′-diaminodiphenyl methane (TGDDM) and bisphenol-A-diglycidylether (BADGE), cured with 4,4′-diaminodiphenyl sulfone (DDS), as the matrix material for high-performance hybrid nanocomposites containing organophilic layered silicate (OLS, synthetic fluorohectorite) and functional six-arm star poly(propylene oxide-block-ethylene) (PPO) as a toughening agent. The hybrid nanocomposites were composed of OLS as well as separated liquid PPO spheres in the epoxy matrix. Synthetic fluorohectorite (Somasif MX-100) was supplied by Unicoo Japan Ltd. (Tokyo, Japan) and octadecylamine (ODA) was supplied by Fluka (Buchs, Switzerland). The liquid PPO was supplied by Bayer AG (Leverkusen, Germany).

To prepare the epoxy hybrid nanocomposites, mixtures of equal amounts of BADGE (Aradite GY 250) and TGDDM (Araldite MY 720) were used as epoxy components. BADGE (130.5 g) and TGDDM (130.5 g) were mixed together with 20 g of the modified PPO with stearate end groups for tailored polarity as well as phenol end groups for covalent bonding to epoxy resin at 80°C and 8 hPa pressure over a period of 30 minutes with a Molteni Labmax high-shear mixer (Rheinfelder, Switzerland). Somasif/ODA (10 g) was then added to the liquid prepolymers and dispersed therein using an IKA Ultra-Turrax T25 basic high-performance disperser (IKA-Werke, Staufen, Germany) for 6 minutes. The organophilic layered silicate was swollen in the resin for 3 hours at 90°C under reduced pressure with the Labmax mixer. Then 109 g DDS (Hardener HT 976) was dissolved in the reaction mixture at 135°C over a period of 25 minutes of mixing at 8 hPa.

The epoxy resin was then poured into a mold and cured at 140°C for 3 hours and at 220°C for 7 hours in a vented oven to produce epoxy hybrid nanocomposites containing 5 wt% of PPO liquid polymer and 2.5 wt% of Somasif/ODA OLS. Various other hybrid nanocomposites with different PPO and OLS were produced by this method. Controls containing either liquid PPO toughening agent or surface-treated clay (OLS) were also prepared for comparison.

Various analytical procedures, such as gas permeation chromatography (GPC), differential scanning calorimetry (DSC), transmission electron microscopy (TEM), dynamic mechanical analyzer (DMA), linear coefficient of thermal expansion (CTE), tensile properties, and fracture toughness were determined. To evaluate the effect of morphology on the fracture mechanical response, the surface of broken compact-tension specimens was examined using scanning electron microscopy (SEM).

The hybrid nanocomposites possessed high glass transition temperature of 220°C. The T_g values were elevated compared to those of the neat resin (212°C) by the addition of the modified liquid PPO. The CTE of the composite material was increased by the addition of liquid polymer toughening agent. The stiffness was only slightly increased, and the strength of the hybrid composites was improved by 20%, mainly because of the addition of the layered silicates. The K_{Ic} was decreased by 15% to values of 1.15 MPa $m^{-1/2}$, compared to 1.35 MPa $m^{-1/2}$ for the neat resin. Hence, the extensive matrix yielding of the neat resin with the generation of shear ribs was the most effective failure mode. The predominant failure mode of the hybrid nanocomposites initiated by the silicate particles, crack bifurcation and branching, was not effective. Crack pining by the liquid PPO spheres could be resolved only when the modified polymer was used as a single additive and afforded the lowest K_{Ic} values.

The Koo Research Group has been successfully using high-shear mixing techniques to incorporate layered silicates, nanosilicas, CNFs, and POSS to form polymer nanostructured materials in several of its research programs [61–67]. Layered silicates, CNFs, and POSS were incorporated separately into resole phenolic (Hexion SC-1008) using high-shear, nonsparking paint mixing equipment to form different types of polymer nanocomposites [61–67]. Neat resin castings (without fiber reinforcement) were made for WAXD and TEM analyses (Chap. 5). The polymer nanocomposites were then impregnated with carbon fabric to form polymer matrix composites. These nanocomposite rocket ablative materials (NRAMs) were tested under a simulated solid rocket motor environment. Ablation and heat transfer characteristics of these NRAMs will be discussed in more detail in Chap. 9.

The Koo Research Group [65–67] also incorporated layered silicates, CNFs, and POSS into resole phenolic (Hitco 134A) and cyanate ester (Lonza PT-15 and PT-30) using high-shear mixing equipment to form polymer nanocomposites. These polymer nanocomposites were used to create a new type of nano-modified carbon-carbon composite (NCCC). WAXD and TEM analyses (Chap. 5) were used to determine the degree of dispersion. Six resin/nanoparticle materials were selected to produce prepregs. These prepregs were fabricated into composites for carbonization and densification to produce carbon-carbon composites. Heat aging was conducted for 24 hours for the six candidates and the baseline Hitco CC139 materials. Mechanical properties such as the tensile, compressive, and flexural strengths and moduli were determined and compared with the baseline CCC [65, 66].

The incorporation of layered silicates, nanosilicas, and CNFs into a high-temperature, damage-tolerant epoxy resin system (CYCOM® 977-3, from Cytec-Solvay Group) was reported by the Koo Research Group [67]. WAXD and TEM were used to determine the degree of dispersion. Dynamic mechanical thermal analysis (DMTA) was used to determine the T_g and complex modulus of the polymer nanocomposites. The TEM analyses indicated the layered silicates, nanosilicas, and CNFs dispersed very well in the epoxy resin system. High T_g and complex modulus values from DMTA for the nano-modified materials are presented as evidence for nanophase presence in the epoxy system as compared to lower T_g and complex moduli for the epoxy resin control. The DMTA data of the neat epoxy nanosilica nanocomposite (2 wt% Aerosil® R202) show the high T_g (258°C) and the highest complex modulus (964 MPa).

Wang et al. [68] used a high-shear mixer to help fabricate a nano-$CaCO_3$/polypropylene composite. The mixer they used had a top speed of 6,000 rpm and was outfitted with special rotors for nanocomposites. The researchers did not identify the manufacturer of high-shear mixer used in their study. In the experiment, the researchers only used the high-shear mixer to break down the $CaCO_3$ into nano-sized particles before using melt intercalation to actually produce the polymer nanocomposite. The researchers found that their final polymer nanocomposites were much better dispersed and had better mechanical properties when the $CaCO_3$ was first processed by the high-shear mixer.

Zunjarrao et al. [69] used high-shear mixing and ultrasonication to create a composite out of onium ion surface modified MMT clay and epoxy resin (DEGBF). They operated their high-shear mixer at 15,000 rpm for 30 minutes and kept the mixture at 65°C using an ice bath. They then used a vacuum chamber for 12 hours to degas the composite. The researchers found that the high-shear mixing samples showed a higher degree of dispersion than those created using ultrasonication. However, the researchers found that at low clay loadings both processing techniques yielded exfoliated clay particles.

In summary, high-shear mixing is an effective processing technique of polymer nanocomposites. It is very effective in exfoliating nanoparticles within polymers due to the extremely high-shear rates involved. It is relatively scalable, but the number of times you can cycle a polymer-nanomaterial mixture through is limited due to viscosity increase of the mixture.

4.9 Ultrasonic Mixing

The ultrasonic mixing is used to disperse nano-sized particles into liquids, such as water, oil, solvents, or resins. The application of ultrasonic to nanomaterials has multiple effects. The most obvious is the dispersing of nanomaterials in liquids to break particle agglomerates. Another process is using ultrasound during particle synthesis or precipitation. Nanomaterials—nanoclays, carbon nanotubes, graphenes, or metal oxides—tend to be agglomerated when mixed into a liquid. Effective means of de-agglomerating and dispersing are needed to overcome the bonding forces after wetting the powder. The ultrasonic breakup of the agglomerated structures in aqueous and nonaqueous suspensions will utilize the full potential of nanomaterials.

The ultrasonic electronic generator transforms AC line powder to a 20-kHz signal that drives a piezoelectric convertor/transducer (Fig. 4.20a) (Qsonica, LLC, Newtown, CT, website: www.sonicator.com). This electrical signature is converted by the transducer to a mechanical vibration due to the characteristics of the internal piezoelectric crystals. The vibration is amplified and transmitted down the length of the horn/probe where the tip longitudinally expands and contracts. The distance the tip travels is dependent on the amplitude selected by the user through the amplitude control knob. In liquid, the rapid vibration of the tip causes cavitation, the formation and violent collapse of the microscopic bubbles (Fig. 4.20b). The collapse of thousands of cavitation

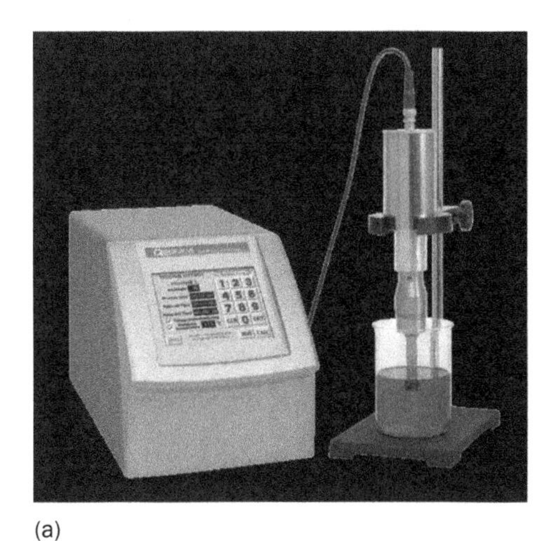

(a)

(b)

Figure 4.20 (a) A typical probe sonicator Q700 manufactured by QSconica and (b) the probe sonicator creating cavitation field in a liquid. (*Courtesy of QSonica.*)

bubbles releases tremendous energy in the cavitation field. Objects and surfaces within the cavitation field are processed. The probe tip diameter dictates the number of samples that can be effectively processed. Smaller tip diameter delivers high-intensity sonication, but the energy is focused within small, concentrated area. Large tip diameters can process larger volumes but offer lower intensity. The choice of a generator and horn/probe types are matched to the volume, viscosity, and other parameters of the particular application.

The key when considering a sonicator is the sample volume range. The sample volume indicates which size probe is required for the experiment. Probe sonicators are more powerful and effective than ultrasonic cleaner baths. A cleaner bath takes hours to accomplish what a probe sonicator can do in minutes. It is clear that MWNTs are not completely soluble in water by using bath sonicator for 8 hours. There are much sedimentation of MWNTs at the bottom of a small bottle (Fig. 4.21a). Upon operating 20-kHz applied by a probe sonicator, the MWNTs are entirely dispersed in aqueous solution, forming a homogeneous free solution (Fig. 4.21b). Remarkably, there is no sedimentation observed even after 4 months of sitting at room temperature (Fig. 4.21c). The concentration of MWNTs is 2,500 mg/L and the MWNTs/SDS ratio is 1:10. Figure 4.21d shows that MWNTs of Fig. 4.21c was diluted to 25 mg/L with deionized water.

The dispersion of nano–zirconium dioxide (ZrO_2) and its effect on the mechanical properties were investigated by Halder et al. [70]. They fabricated ZrO_2-epoxy nanocomposites via mechanical mixing (MM) and ultrasonic dual-mode mixing (UDMM) methods. The mechanical mixing of ZrO_2 nanoparticles in epoxy resin was employed using glass rod stirring, and the ultrasonic dual mode mixing was employed by ultrasonic vibration along with magnetic stirring to produce ZrO_2-epoxy nanocomposite. Micrographs obtained using a field emission scanning electron microscope (FESEM) revealed an improved dispersion quality of ZrO_2 nanoparticles especially for the UDMM method. A series of FESEM images of ZrO_2-epoxy nanocomposite processed by UDMM in which nanoparticles of 5, 10, and 20 wt% revealed good dispersions of ZrO_2 in the epoxy resin at low, high, and very high magnifications are shown in [70]. The improvement in dispersion was reflected in much improved tensile and fracture properties of the nanocomposite.

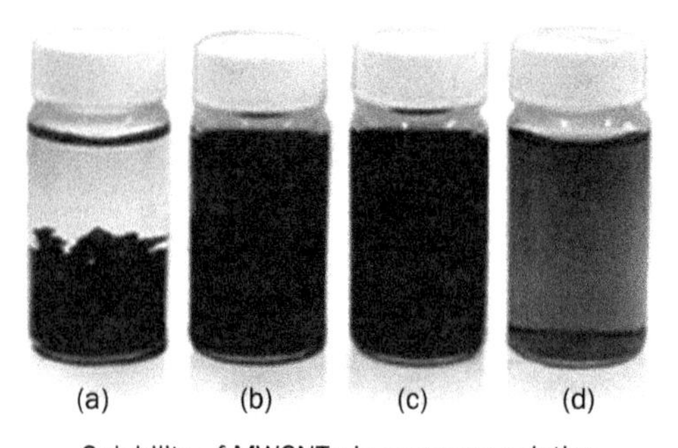

Figure 4.21 Solubility of MWNTs in aqueous solution. (*Courtesy of QSonica.*)

Heidarian and Shishesaz [71] investigated the effect of duration of ultrasonication process on structural characteristics and barrier properties of solvent-free castor oil–based polyurethane (PU)/organically modified montmorillonite (OMMT) nanocomposites. A series of PU/OMMT composites were synthesized by in situ polymerization technique through an ultrasonication-assisted process at various processing durations. Effect of ultrasonication duration on de-agglomeration of clay stacks in castor oil dispersions was evaluated by optical microscopy, sedimentation test, and viscosity measurement. Figure 4.22 shows the suspension state of castor oil/OMMT mixtures, prepared after 2-hour mechanical agitation and 15-, 30-, and 60-minute sonication [71]. Suspensions were placed at 50°C for 2 weeks after preparation. WAXD and FTIR spectroscopy techniques were employed to investigate the effect of processing time on degree of delamination of clay platelets and interfacial strength between clay layers and PU matrix. The surface morphology of the nanocomposites was analyzed by atomic force microscopy (AFM). The topographic morphology of surface of molded specimens was scanned by AFM. Figure 4.23 reveals the representative AFM pictures [71]. The results showed that by increasing the ultrasonication time up to 60 minutes, the size of clay agglomerates decreased and the interlayer spacing of clay platelets increased. To evaluate the effect of ultrasonication duration on transport properties of the PU/OMMT composites, diffusion coefficient and permeability were determined through water uptake test. Electrochemical impedance spectroscopy was carried out to analyze the barrier properties and to evaluate the corrosion performance of these composite coatings on carbon steel panels. It was found that by increasing sonication time, the barrier property of nanocomposites against diffusion of water molecules improved, which is due to further separation of clay platelets, enhancement of the traveling pathways for water molecules, and improvement of interactions between the two components.

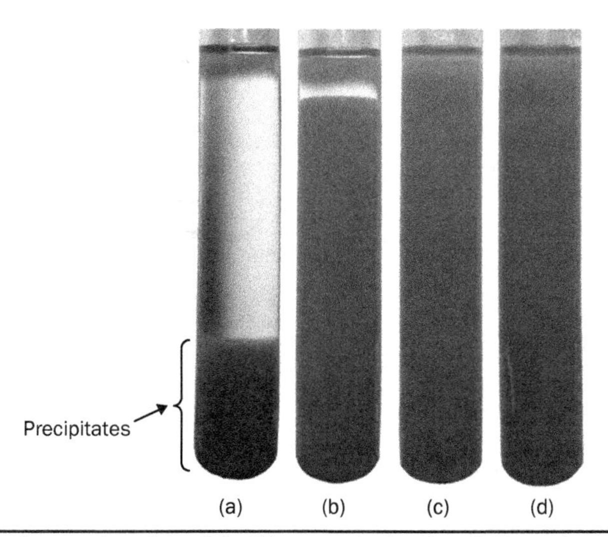

Figure 4.22 Suspension state of 3 wt% organoclay/castor oil dispersions: (a) after 2-hour mechanical agitation, (b) after 15-minute sonication, (c) after 30-minute sonication, and (d) after 60-minute sonication.

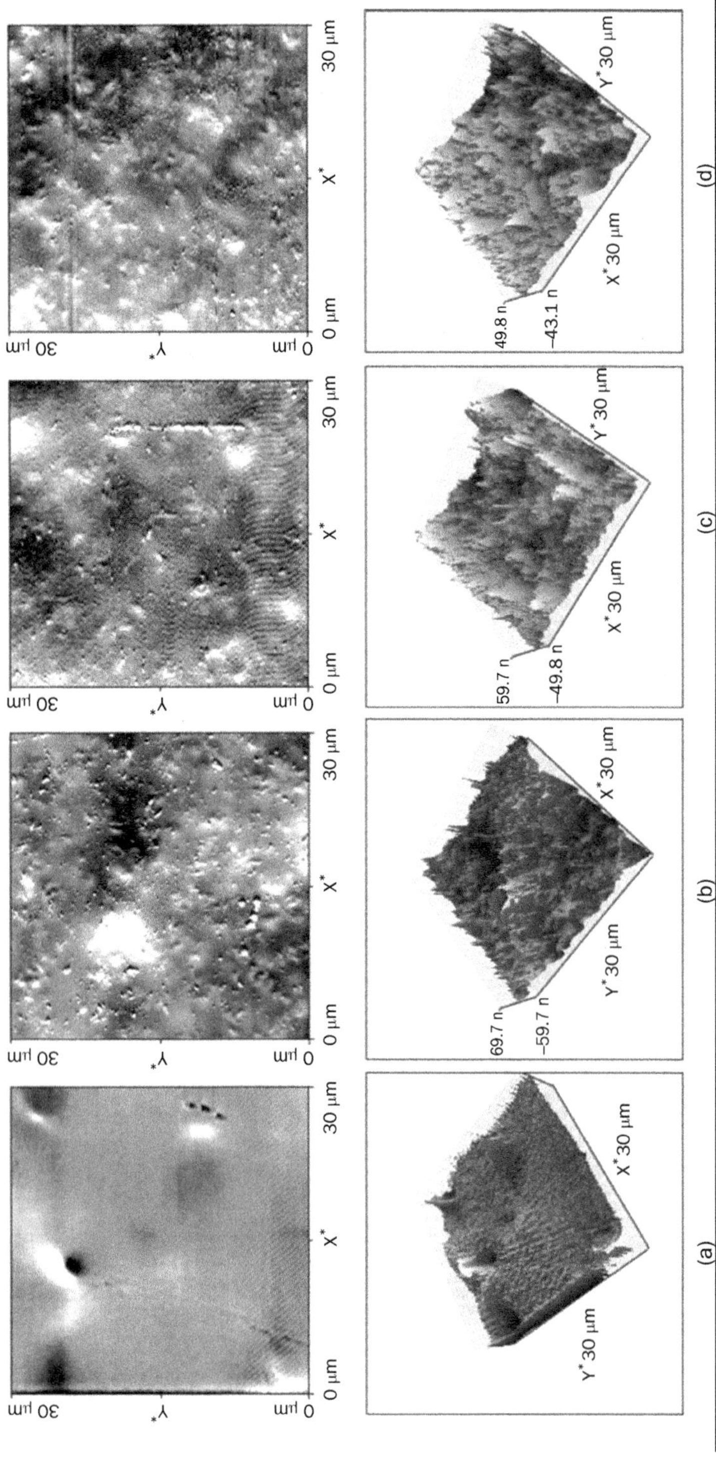

Figure 4.23 Typical topographic images of (a) pristine PU surface and the surfaces of 3 wt% PU/OMMT composites premixed by (b) 15-minute sonication, (c) 30-minute sonication, and (d) 60-minute sonication.

Polystyrene (PS) and multiwalled carbon nanotube (MWNT) nanocomposites were synthesized by Zhang et al. [72] via an in situ bulk polymerization by employing an ultrasonicator without adding an initiator. They found that the ultrasonication technique produced well-dispersed MWNT in the PS matrix. Morphology of the as-synthesized PS/MWNT nanocomposite was investigated by both SEM and TEM. Figure 4.24a shows several SEM images of PS and Fig. 4.24b–d depicts PS/MWNT nanocomposites [72]. Figure 4.25a,d shows photographs of PS/MWNT nanocomposites at 20°C without ultrasonication; Fig. 4.25c,f, at 20°C without ultrasonication; and Fig. 4.25b,e, at 20°C with ultrasonication [72]. Electrical conductivity of the PS/MWNT nanocomposite film fabricated by a solvent casting method was also examined to be enhanced with MWNT content, while average molecular weights of the synthesized PS in the PS/MWNT nanocomposites analyzed by a gel permeation chromatography increased and then saturated at 2 wt% MWNT. Rheological properties of MWNT containing PS were enhanced because of improved dispersion of the MWNT through an interaction between MWNT and PS.

Ultrasonic processors and flow cells for de-agglomeration and dispersion from laboratory to production level are manufactured by Hielscher Ultrasonics GmbH (Teltow, Germany, website: www.hielscher.com). Hielscher offers a broad range of ultrasonic devices and accessories for dispersion of nanomaterials. Compact laboratory devices of up to 400 W power are mainly used for sample preparation or initial feasibility studies. Larger devices such as 500-, 1,000-, and 2,000-W ultrasonic processors with flow cell and various booster horns and sonotrodes can process larger volumes. Ultrasonic processors of 2-, 4-, 10-, and 16-kW and larger clusters can process production volumes at almost any level. The industrial systems can be retrofitted to work inline.

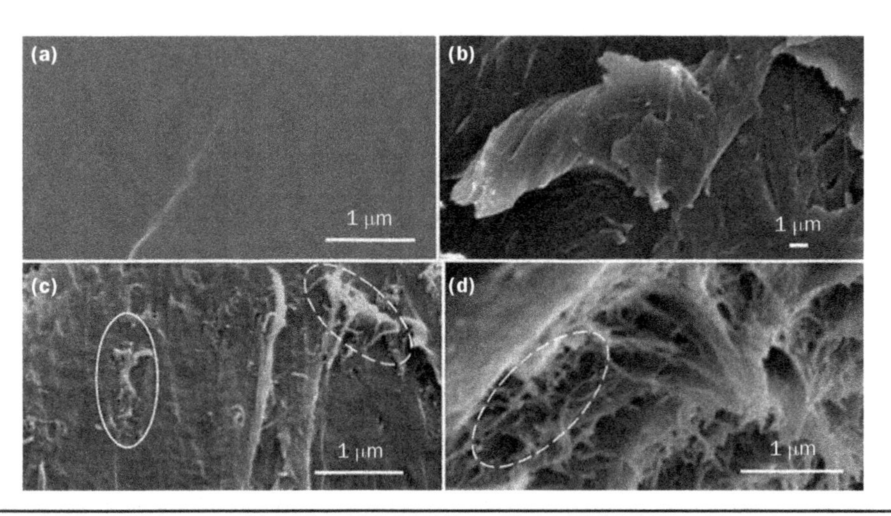

Figure 4.24 SEM images of (a) PS, (b) PS/MWNT 1 wt%, (c) PS/MWNT 2 wt%, and (d) PS/MWNT 4 wt%.

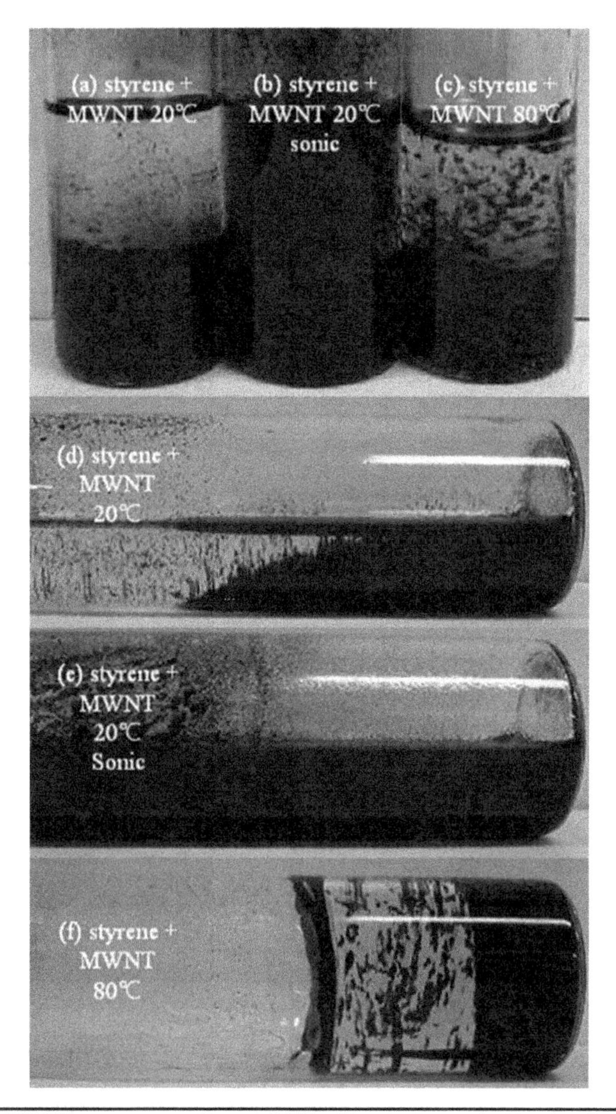

FIGURE 4.25 PS/MWNT nanocomposites at 20°C without ultrasonication (a, d), at 80°C without ultrasonication (c, f), and at 20°C with ultrasonication (b, e).

4.10 Summary

In this chapter, a brief overview of seven popular synthesis approaches to incorporating different types of nanoparticles into a variety of polymer matrices was presented. The fundamentals behind each technique were examined as well as their advantages and disadvantages. Selected literature examples were used to demonstrate these synthesis methods employed successfully by researchers. Selective examples will be described in more detail in Chaps. 7 through 11, and will emphasize the properties and performance of these polymer nanostructured materials.

4.11 Study Questions

4.1 Besides the processing techniques described in this chapter, identify other common processing techniques used to formulate polymer nanocomposites and discuss the advantages and disadvantages of these processing techniques. Present your answers in the format used in this chapter.

4.2 Why a longer curing cycle usually means a better final polymer?

4.3 What are common catalysts used in the curing cycle for thermosets; provide several examples?

4.4 How does the density of the polymer may affect the dispersion and processing techniques?

4.5 Using Fig. 4.2, describe the methods that were used to produce the final characteristics as shown in parts a–c?

4.6 Why did the Toyota Research Group (Sec. 4.2.2) select polyamide nanocomposites for their applications?

4.7 What is solution intercalation? How can one tell if the nanoparticles are exfoliated if only x-ray diffraction is used to evaluate the state of nanoparticle dispersion within the nanocomposite?

4.8 What is melt intercalation? Provide some examples. What are some advantages of melt intercalation method as compared to solution intercalation?

4.9 What are some nanoparticles which have been dispersed with three-roll milling, emulsion polymerization, and in situ polymerization? Which of these three methods seems more versatile?

4.10 Describe high-shear mixing and the nanoparticles which have been successfully dispersed using this method. Is there any limitation to this method?

4.11 Is the solution intercalation from polymers in solution widely used with some non-water-soluble polymer? If so, how does it work?

4.12 Can we use the solution intercalation method to prepare the solvent-soluble polymer? Why or why not?

4.13 Which test method do we usually use to characterize the polymer prepared from the solution intercalation from polymers in solution?

4.14 What are the property differences (e.g., mechanical properties, electrical properties) among the polymers of thermoset-based nanocomposites, thermoplastic-based nanocomposites, and elastomer-based nanocomposites?

4.15 What are some of the processing methods for dispersing nanoparticles in polymers? How can we tell whether or not the nanoparticles are properly dispersed?

4.16 Provide three examples of water-soluble polymers used in solution intercalation.

4.17 Describe the melt intercalation process and how surface modification affects the nanoparticle dispersion.

4.18 Is it expected to find a glass transition temperature (Tg) when observing the DSC data of an intercalated polymer?

4.19 Why are solution intercalation, mesophase mediated, in situ polymerization, and melt processing considered convenient processing techniques?

4.20 What are the water-soluble polymers that are used with the solution intercalation technique?

4.21 For solution intercalation from prepolymers in solution, how is the solvent eliminated?

4.12 References

1. Tomanek, E., and Enbody, R. J., eds. (2000). *Science and Application of Nanotubes*. New York, NY: Kluwer Academic/Plenum Publishers.
2. Pinnavaia, T. J., and Beall, G. W., eds. (2000) *Polymer-Clay Nanocomposites*. New York, NY: John Wiley & Sons.
3. Krishnamoorti, R., and Vaia, R. A., eds. (2001) *Polymer Nanocomposites: Synthesis, Characterization, and Modeling*. ACS Symposium Series 804. Washington, DC: ACS.
4. Wang, Z. L., Liu, Y., and Zhang, Z., eds. (2003) *Handbook of Nanophase and Nanostructured Materials, Vol. 4: Materials Systems and Applications (II)*. New York, NY: Kluwer Academic/Plenum Publishers.
5. Cao, G. (2004) *Nanostructures & Nanomaterials Synthesis, Properties & Applications*. London, England: Imperial College Press.
6. Di Ventra, M., Evoy, S., and Heflin Jr., J. R., eds. (2004) *Introduction to Nanoscale Science and Technology*. New York, NY: Kluwer Academic Publishers.
7. Schulz, M. J., Kelkar, A. D., and Sundaresan, M. J., eds. (2006) *Nanoengineering of Structural, Functional, and Smart Materials*. Boca Raton, FL: CRC.
8. Alexandre, M., and Dubois, P. (2000) Polymer-layered silicate nanocomposites: preparation, properties, and uses of a new class of materials. *Mater. Sci. Eng.*, R28:1–63.
9. Sanchez, C., Julian, B., Belleville, P., and Popall, M. (2005) Applications of hybrid organic-inorganic nanocomposites. *J. Mater. Chem.*, 15:3559–3592.
10. Lerner, M., and Oriakhi, C. (1997) *Handbook of Nanophase Materials*. New York, NY: Mekker Decker.
11. Lagaly, B. (1999) Introduction from clay mineral-powder interactions to clay mineral-polymer nanocomposites. *Appl. Clay Sci.*, 15:1–9.
12. Greenland, D. J. (1963) Adsorption of polyvinylalcohols by montmorillonite. *J. Colloid Sci.*, 18:647–664.
13. Ogata, N., Kawakage, S., and Orgihara, T. (1997) Poly (vinyl alcohol)-clay and poly (ethylene oxide)-clay blend prepared using water as solvent. *J. Appl. Polym. Sci.*, 66:573–581.
14. Parfitt, R. L., and Greenland, D. L. (1970) Absorption of poly (ethylene glycols) on montmorillonites. *Clay Mineral.*, 8:305–323.
15. Zhao, X., Urano, K., and Ogasawara, S. (1989) Adsorption of polyethylene glycol from aqueous solutions on montmorillonite clays. *Colloid Polym. Sci.*, 267:899–906.
16. Ruiz-Hitzky, E., Aranda, P., Casal, B., and Galvan, J. C. (1995) Nanocomposite materials with controlled ion mobility. *Adv. Mater.*, 7:180–184.
17. Billingham, J., Breen, C., and Yarwood, J. (1997) Adsorption of polyamide, polyacrylic acid and polyethylene glycol on montmorillonite: an in situ study using ATR-FTIR. *Vibr. Spectosc.*, 14:19–34.

18. Levy, R., and Francis, C. W. (1975) Interlayer adsorption of polyvinylpyrrolidone on montmorillonite. *J. Colloid Interface Sci.*, 50:442–450.
19. Wu, J., and Lerner, M. M. (1993) Structural, thermal, and electrical characterization of layered nanocomposites derived from sodium-montmorillonite and polyethers. *Chem. Mater.* 5:835–838.
20. Harris, D. J., Bonagamba, T. J., and Schmidt-Rohr, K. (1999) Conformation of poly (ethylene oxide) intercalated in clay and MoS_2 studied by two-dimensional double-quantum NMR. *Macromolecules*, 32:6718–6724.
21. Yano, K., Usuki, A., Okada, A., Kurauchi, T., and Kamigaito, O. (1993) Synthesis and properties of polyimide-clay hybrid. *J. Polym. Sci. Part A: Polym. Chem.*, 31:2493–2498.
22. Yano, K., Usuki, A., and Okada, A. (1997) Synthesis and properties of polyimide-clay hybrid films. *J. Polym. Sci. Part A: Polym. Chem.*, 35:2289–2294.
23. Shen, Z., Simon, G. P., and Cheng, Y. B. (2002) Comparison of solution intercalation and melt intercalation of polymer–clay nanocomposites. *Polymer*, 43:15:4251–4260.
24. Chen, W., and Qu, B. (2003) Structural characteristics and thermal properties of Pe-G-Ma/Mgal-Ldh exfoliation nanocomposites synthesized by solution intercalation. *Chemistry of Materials*, 15:16:3208–3213.
25. Vaia, R. A., and Giannelis, E. P. (1997) Polymer melt intercalation in organically modified layered silicates: model predictions and experiment. *Macromolecules*, 30:8000–8009.
26. Vaia, R. A., Ishii, H., and Giannelis, E.P. (1993) Synthesis and properties of two-dimensional nanostructures by direct intercalation of polymer melts in layered silicates. *Chem. Mater.*, 5:1694–1696.
27. Dennis, H. R., Hunter, D. L., Chang, D., Kim, S., White, J. L., Cho, J. W., and Paul, D. R. (2001) Effect of melt processing conditions on the extent of exfoliation in organoclay-based nanocomposites. *Polymer*, 42(23):9513–9522.
28. Burnside, S. D., and Giannelis, E. P. (1995) Synthesis and properties of new poly (dimethylsiloxane) nanocomposites. *Chem. Mater.*, 7:1597–1600.
29. Li, Y., and Shimizu, H. (2009) Toward a stretchable, elastic, and electrically conductive nanocomposite: morphology and properties of poly [styrene-B-(ethylene-Co-butylene)-B-styrene]/multiwalled carbon nanotube composites fabricated by high-shear processing. *Macromolecules*, 42(7):2587–2593.
30. Blanski, R., Koo, J. H., Ruth, P., Nguyen, N., Pittman, C., and Phillips, S. (2004). Polymer nanostructured materials for solid rocket motor insulation–ablation performance. *Proc. 52nd JANNAF Propulsion Meeting*, CPIA, Columbia, MD.
31. Koo, J. H., Marchant, D., Wissler, G., Ruth, P., Barker, S., Blanski, R., and Phillips, S. (2004) Polymer nanostructured materials for solid rocket motor insulation—processing, microstructure, and mechanical properties. *Proc. 52nd JANNAF Propulsion Meeting*, CPIA, Columbia, MD.
32. Ruth, P., Blanski, R. and Koo, J. H. (2004) Preparation of polymer nanocomposites for solid rocket motor insulation. *Proc. 52nd JANNAF Propulsion Meeting*, CPIA, Columbia, MD.
33. Blanski, R., Koo, J. H., Ruth, P., Nguyen, N., Pittman, C., and Phillips, S. (2004) Ablation characteristics of nanostructured materials for solid rocket motor insulation. *Proc. National Space & Missile Materials Symposium*, Seattle, WA, June 21–25.

34. Koo, J. H., Marchant, D., Wissler, G., Ruth, P., Barker, S., Blanski, R., and Phillips, S. (2004) Processing and characterization of nanostructured materials for solid rocket motors. *Proc. National Space & Missile Materials Symposium*, Seattle, WA, June 21–25.
35. Marchant, D., Koo, J. H., Blanski, R., Weber, E. H., Ruth, P., Lee, A., and Schlaefer, C. E. (2004) Flammability and thermophysical characterization of thermoplastic elastomer nanocomposites. *ACS National Meeting, Fire & Polymers Symposium*, Philadelphia, PA, August 22–26.
36. Gupta, S. K., Schwab, J. J., Lee, A., Gu, B. X., and Hsiao, B. S. (2002) POSS® reinforced fire retarding EVE resins. *Proc. SAMPE 2002 ISSE*, SAMPE, Covina, CA, pp. 1517–1526.
37. Yasmin, A., Abot, J. L., and Daniel, I. M. (2003) Processing of clay/epoxy nanocomposites with a three-roll mill machine. *Mater. Res. Soc. Symp. Proc.*, 740:75–80.
38. Rosca, I. D., and Hoa, S.V. (2009) Highly conductive multiwall carbon nanotube and epoxy composites produced by three-roll milling. *Carbon*, 47(8):1958–1968.
39. Yasmin, A., Abot, J. L., and Daniel, I. M. (2003) Processing of clay/epoxy nanocomposites by shear mixing." *Scripta Materialia*, 49(1):81–86.
40. Seyhan, A. T., Tanoğlu, M., and Schulte, K. (2009) Tensile mechanical behavior and fracture toughness of MWCNT and DWCNT modified vinyl-ester/polyester hybrid nanocomposites produced by 3-roll milling. *Materials Science and Engineering: A*, 523(1):85–92.
41. Yuan, M., Johnson, B., Koo, J. H., and Bourell, D. (2013) Polyamide 11-MWNT nanocomposites: thermal and electrical conductivity measurements. *Journal of Composite Materials*. https://doi.org/10.1177/0021998313490975.
42. Lu, C., Krifa, M., and Koo, J. H. (2013) Conductive poly (3,4 ethylenedioxythiophene):poly(4-styrene sulfonate) (PEDOT:PSS)/nickel nanostrands nanocomposites. *Proc. SAMPE 2013 ISSE*, SAMPE, Covina, CA.
43. Hansen, N., Adams, D. O., and Fullwood, D. T. (2012) Quantitative methods for correlating dispersion and electrical conductivity in conductor-polymer nanostrand composites. *Composites: Part A*, 43:1939–1946.
44. Lagaly, G. (1999) Introduction: from clay mineral-polymer interactions to clay mineral-polymer nanocomposites. *Appl. Clay Sci.*, 15:1–9.
45. Eastman, M. P., Bain, E., Porter, T. L., Manygoats, K., Whitehorse, R., Parnell, R. A., and Hagerman, M. E. (1999) The formation of poly(methyl-methacrylate) on transition metal-exchanged hectorite. *Appl. Clay Sci.*, 15:173–185.
46. Fukushima, Y., Okada, A., Kawasumi, M., Kurauchi, T., and Kamigaito, O. (1988) Swelling behavior of montmorillonite by poly-6-amide. *Clay Mineral*, 23:27–34.
47. Usuki, A., Kojima, Y., Kawasumi, M., Okada, A., Fukushima, Y., Kurauchi, T., and Kamigaito, O. (1993) Synthesis of nylon-6-clay hybrid. *J. Mater. Res.*, 8:1179–1183.
48. Usuki, A., Kawasumi, M., Kojima, Y., Okada, A., Kurauchi, T., and Kamigaito, O. (1993) Swelling behavior of montmorillonite cation exchanged for ω-amino acid by ε-caprolactam." *J. Mater. Res.*, 8:1174–1178.
49. Kojima, Y., Usuki, A., Kawasumi, M., Okada, A., Kurauchi, T., and Kamigaito, O. (1993) Synthesis of nylon-6-clay hybrid by montmorillonite intercalated with ε-caprolatam." *J. Polym. Sci. Part A: Polym. Chem.*, 31:983–986.
50. Maneshi, A., Soares, J. B. P., and Simon, L. C. (2011) An efficient in situ polymerization method for the production of polyethylene/clay nanocomposites: effect of polymerization conditions on particle morphology. *Macromolecular Chemistry and Physics*, 212:2017–2028.

51. Namazi, H., Mosadegh, M., and Dadkhah, A. (2009) new intercalated layer silicate nanocomposites based on synthesized starch-G-Pcl prepared via solution intercalation and in situ polymerization methods: as a comparative study. *Carbohydrate Polymers*, 75(4):665–669.
52. Messersmith, P. B., and Giannelis, E. P. (1994) synthesis and characterization of layered silicate-epoxy nanocomposites. *Chem. Mater.*, 6:1719–1725.
53. Lan, T., and Pinnavaia, T. J. (1994) Clay-reinforced epoxy nanocomposites. *Chem. Mater.*, 6:2216–2219.
54. Lan, T., Kaviratna, P. D., and Pinnavaia, T. J. (1995) Mechanism of clay tactoid exfoliation in epoxy-clay nanocomposites. *Chem. Mater.*, 7:2144–2150.
55. Lee, D. C., and Jang, L. W. (1996) Preparation and characterization of PMMA-clay hybrid composite by emulsion polymerization. *J. Appl. Polym. Sci.*, 62:1117–1122.
56. Lee, D. C., and Jang, L. W. (1998) Characterization of epoxy-clay hybrid composite prepared by emulsion polymerization. *J. Appl. Polym. Sci.*, 68:1997–2005.
57. Noh, M. W., and Lee, D. C. (1999) Synthesis and characterization of PS-clay nanocomposite by emulsion polymerization. *Polym. Bull.*, 42:619–626.
58. Zhang, K., Wu, W., Meng, H., Guo, K., and Chen, J. F. (2009) Pickering emulsion polymerization: preparation of polystyrene/nano-SiO_2 composite microspheres with core-shell structure. *Powder Technology*, 190(3):393–400.
59. Meneghetti, P., and Qutubuddin, S. (2006) synthesis, thermal properties and applications of polymer-clay nanocomposites. *Thermochimica Acta*, 442(1):74–77.
60. Frohlich, J., Thomann, R., Gryshchuk, O., Karger-Kocsis, J., and Mulhaupt, R. (2004) High-performance epoxy hybrid nanocomposites containing organophilic layered silicates and compatibilized liquid rubber. *J. Appl. Polym. Sci.*, 92:3088–3096.
61. Koo, J. H., Stretz, H., Bray, A., Wootan, W., Mulich, S., Powell, B., Weispfenning, J., et al. (2002) Phenolic-clay nanocomposites for rocket propulsion system. *Proc. SAMPE 2002 ISSE*, SAMPE, Covina, CA.
62. Koo, J. H., Stretz, H., Bray, A., Weispfenning, J., Luo, Z. P., and Wootan, W. (2003) Nanocomposites rocket ablative materials: processing, characterization, and performance." *Proc. SAMPE 2003 ISSE*, SAMPE, Covina, CA, pp. 1156–1170.
63. Koo, J. H., Chow, W. K., Stretz, H., Cheng, A. C-K., Bray, A., and Weispfenning, J. (2003) Flammability properties of polymer nanostructured materials. *Proc. SAMPE 2003 ISSE*, SAMPE, Covina, CA, pp. 954–964.
64. Koo, J. H., Stretz, H., Weispfenning, J., Luo, Z. P., and Wootan, W. (2004) Nanocomposite rocket ablative materials: subscale ablation test. *Proc. SAMPE 2004 ISSE*, SAMPE, Covina, CA.
65. Koo, J. H., Pittman, Jr., C. U., Liang, K., Cho, H., Pilato, L. A., Luo, Z. P., Pruett, G., et al. (2003) Nanomodified carbon/carbon composites for intermediate temperature: processing and characterization. *Proc. SAMPE 2003 ISTC*, SAMPE, Covina, CA.
66. Koo, J. H., Pilato, L. A., Winzek, P., Shivakumar, S., Pittman, Jr., C. U., and Luo, Z. P. (2004) Thermo-oxidative studies of nanomodified carbon/carbon composites." *Proc. SAMPE 2004 ISSE*, SAMPE, Covina, CA.
67. Koo, J. H., Pilato, L. A., Wissler, G., Lee, A., Abusafieh, A., and Weispfenning, J. (2005) Epoxy nanocomposites for carbon fiber reinforced polymer matrix composites. *Proc. SAMPE 2005 ISSE*, SAMPE, Covina, CA.
68. Wang, G., Chen, X. Y., Huang, R., and Zhang, L. (2002) Nano-$CaCO_3$/polypropylene composites made with ultra-high-speed mixer. *Journal of Materials Science Letters*, 21(13):985–986.

69. Zunjarrao, S.C., Sriraman, R., and Singh, R.P. (2006) Effect of processing parameters and clay volume fraction on the mechanical properties of epoxy-clay nanocomposites. *Journal of Materials Science*, 41(8):2219–2228.
70. Halder, S., Ghosh, P. K., and Goyat, M. S. (2012) Influence of ultrasonic dual mode mixing on morphology and mechanical properties of ZrO_2-epoxy nanocomposite. *High Performance Polymers*, 24(4):331–341.
71. Heidarian, M., and Shishesaz, M. R. (2012) Study on effect of duration of the ultrasonication process on solvent-free polyurethane/organoclay nanocomposite coatings: structural characteristics and barrier performance analysis. *Journal of Applied Polymer Science*, 126:2035–2048.
72. Zhang, K., Lim, J. Y., Choi, H. J., Lee, J. H., and Chio, W. J. (2013) Ultrasonically prepared polystyrene/multi-walled carbon nanotube nanocomposites. *J. Mater. Sci.*, 48:3088–3096.

CHAPTER 5

Structure and Property Characterization

5.1 Global Characterization Methods

There are three key steps in the development of polymer nanocomposites (PNCs): (a) materials preparation, (b) property characterization, and (c) material performance or device fabrication. The materials preparation involved in the processing of nanoparticles with polymer matrix into a PNC has been discussed in Chap. 4. The next challenge is to determine the degree or quality of dispersion of the nanoparticles in the polymer matrix. Characterization involves two main processes: structure or morphological analysis and property measurements. Structure analysis is carried out using a variety of microscopy and spectroscopy techniques, while property characterization is rather diverse and depends on the individual application.

Due to the high selectivity of the size and structure of the PNCs, their physical properties can be quite diverse. It is known that the physical properties of nanostructures depend strongly on their size and shape. The properties measured from a large quantity of nanomaterials could be an average of the overall properties, so that the unique characteristics of individual nanostructure could be embedded. An essential task as one develops capability in the preparation of nanomaterials is property characterization of an individual nanostructure with a well-defined atomic structure.

Characterizing the properties of individual nanoparticles presents a challenge to many existing testing and measuring techniques because of the following constraints. First, the size (diameter and length) is rather small, prohibiting the application of well-established techniques. Second, the small size of nanostructures makes their manipulation rather difficult, and specialized techniques are needed for identifying and analyzing individual nanostructures. Finally, new methods and analytical tools must be developed to quantify the properties of individual nanostructures. In this chapter, several commonly used structural characterization techniques are briefly discussed, along with examples demonstrating the techniques. More detailed characterization of nanophase materials can be found elsewhere [1–5]. Property characterization for the individual PNCs is also briefly discussed. It is also necessary to employ more than one characterization technique in order to accurately characterize the nanomaterials and PNCs. The following section shows an overview of some commonly used material characterization techniques. It is divided into two general groups: the first group is used for morphology characterization and the second group is used for material property characterization.

- Morphology Characterization:
  - Optical microscopy
  - Wide-angle x-ray diffraction (WAXD)
  - Scanning electron microscopy (SEM)
  - Transmission electron microscopy (TEM)
  - Energy-dispersive x-ray spectroscopy (EDS or EDX)
  - Small angle x-ray scattering (SAXS)
  - Scanning tunneling microscopy (STM)
  - Atomic force microscopy (AFM)
  - Raman spectroscopy
  - X-ray photoelectron spectroscopy (XPS)
  - Others
- Global Methods for Polymer Nanocomposite Property Characterization:
  - Mechanical properties (tensile, flexural, and compressive strength and modulus, creep, shear, and impact)—see Chap. 6
  - Thermal and rheology properties [thermogravimetric analysis (TGA), differential scanning calorimetry (DSC), dynamic mechanical thermal analysis (DMTA), and thermal conductivity]—see Chap. 7
  - Flammability properties (cone calorimetry, mass loss calorimetry, microscale combustion calorimetry, limiting oxygen index, UL 94, and Steiner tunnel test)—see Chap. 8
  - Ablation properties [solid rocket motor simulation (SSRM), oxyacetylene test bed (OTB), in situ ablation recession and thermal sensor, and char strength sensor]—see Chap. 9
  - Electrical properties—see Chap. 10
  - Other properties—see Chap. 11

Sections 5.2–5.9 discuss the techniques that are commonly used to characterize the morphology of the microstructures and nanostructures of PNCs.

5.2 Optical Microscopy

Optical microscopy is the most inexpensive imaging technique to inspect the dispersion of nanomaterials in a PNC. If the nanomaterials are not uniformly dispersed in a specific resin matrix, aggregates of nanomaterials can easily be seen using this optical technique before moving to the high power imaging instruments, such as SEM, STM, and TEM.

5.3 X-Ray Diffraction

Wide-angle x-ray diffraction (WAXD) is the most commonly used technique to characterize the degree of nanodisperions of MMT organoclay in a specific polymer. Wide-angle x-ray diffraction measures the spacing between the ordered crystalline layers of the organoclay. By using Bragg's law:

$$\sin\theta = n\lambda/2d \tag{5.1}$$

where d is the spacing between atomic planes in the crystalline phase and λ is the x-ray wavelength. The intensity of the diffracted x ray is measured as a function of the diffraction angle 2θ and the specimen's orientation. This diffraction pattern is used to identify the specimen's crystalline phases and to measure its structural properties. WAXD is nondestructive and does not require elaborate sample preparation which partly explains the wide usage of this technique in materials characterization. Spacing change (increase or decrease) information can be used to determine the type of PNC formed, such as the following:

- Immiscible (no d-spacing change)
- Decomposed/deintercalated (d-spacing decrease)
- Intercalated (d-spacing increase)
- Exfoliated (d-spacing outside of wide-angle x-ray diffraction, or expansive and disordered as to give a signal)

However, WAXD can be affected by many parameters:

- Sampling (powder vs. solids, alignment of clay platelets)
- Experimental parameters (slit width, count time, angle step rate)
- Layered silicate order (disordered/amorphous materials exhibit no signal by WAXD)
- WAXD measures d-spacing, not overall (global) clay dispersion in the sample

To reveal the capabilities of a modern x-ray characterization of nanoparticles and the fundamentals of WAXD, the reader should refer to other sources [3, 5]. Figure 5.1 displays the WAXD data showing the intensity of various wt% (5, 10, and 15) of SCP's Cloisite® 30B MMT organoclay dispersing in a resole phenolic (Hexion SC-1008). The fact that no peaks were shown around 2.5 for all three wt% indicates that all three formulations were possibly in an exfoliated state. Unfortunately, in this study WAXD provided the wrong diagnostics, which became evident when TEM analysis was done on the same three cases. It is discussed in Sec. 5.4.2 and more details are provided in Chap. 9.

5.4 Electron Microscopy and Spectroscopy

5.4.1 Scanning Electron Microscopy (SEM)

A scanning electron microscope scans electron beam across the sample surface and collects scattered electrons for imaging. Because the image is formed using backscattered signals instead of forward-transmitted signals, the electron beam energy does not need

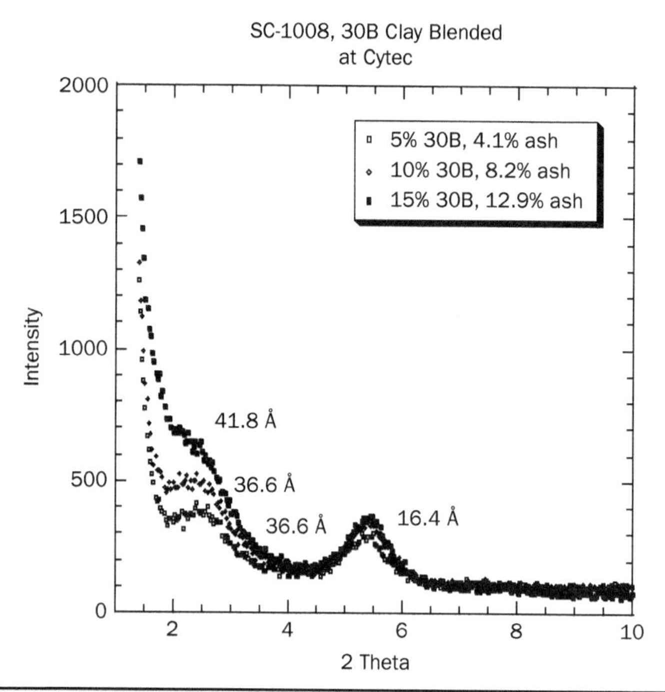

FIGURE 5.1 WAXD data showing the intensity of 5%, 10%, and 15% of Cloisite 30B MMT nanoclay in resole phenolic.

to be high (less than 40 keV) and the sample also does not require electronic transparency. It has only a conductive requirement to prevent charging. A high-powered electron beam originating from an electron gun near the top of the column is accelerated down a column through sets of collimators and magnetic focusing lenses. Detectors are placed near and around a sample. Many kinds of 2-degree signals are collected and assimilated. If the SEM has EDX capability, in situ elemental analysis is also conducted [6]. Figure 5.2 shows a FEI Quanta 650 SEM, capable of performing chemical analysis by using a Bruker EDX system. The resolution of this instrument is less than 1 nm. Magnification is on the order of 2 million. The resolution is 0.6 nm at 5 kV and 1.0 nm at 1 kV. Magnification capability ranges from 25 to 2 million. Accelerating voltage ranges 0.5 kV to approximately 30 kV.

At least four different kinds of signals are scattered backwards that can potentially be used: secondary electrons, backscattered electrons, characteristic x ray, and Auger electrons. The most popular SEM imaging method is secondary electron imaging (SEI). SEI can produce good resolution down to 1–5 nm with a large depth of field; this feature helps reveal much information about surface topography. However, backscattered electrons are often scattered by samples elastically and thus is strongly related to atomic number (Z) near the surface as well as crystallographic orientation of the surface. Therefore, backscattered electrons are often used to identify chemical composition together with characteristic x rays. Backscattered electrons are also used to image the distribution of elements with significant difference in weight. Characteristic x rays have

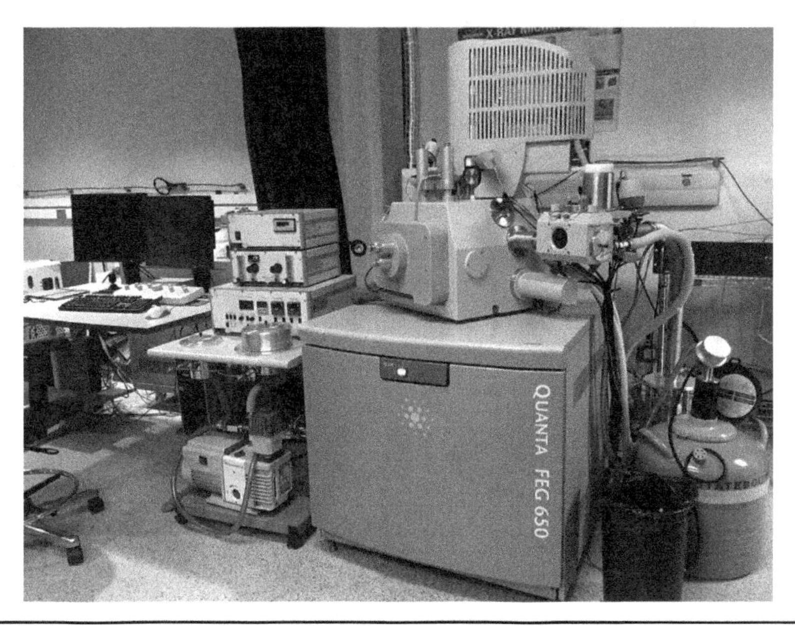

FIGURE 5.2 A FEI Quanta 650 SEM.

also been widely used in SEM to identify chemical composition of materials. This is called energy-dispersive x-ray spectroscopy (EDX or EDS) (this technique is discussed later in this chapter).

5.4.2 Transmission Electron Microscopy (TEM)

One of the characteristics of polymer nanostructured materials is their extremely small particle size of the added nanomaterials. Although some structural features can be revealed by x-ray and neutron diffraction, direct imaging of individual nanomaterials is only possible using TEM and scanning probe microscopy. TEM is unique because it can provide a real space image on the atom distribution in the nanocrystal and on its surface [1, 5, 7]. Today's TEM is a versatile tool that provides not only atomic resolution lattice images, but also chemical information at a spatial resolution of 1 nm or better, allowing direct identification of the chemistry of a single nanocrystal. With a finely focused electron probe, the structural characteristics of a single nanomaterial can be fully characterized. To reveal the capabilities of a modern TEM and the fundamentals of TEM and its applications in characterization of nanophase materials, the reader should refer to other sources [1–9].

TEM has resolving powder in the range of 0.05 nm. TEMs are also equipped with in situ EDX and electron diffraction capability. Figure 5.3 shows a FEI Tecnai Transmission Electron Microscope. This TEM is equipped with a 20–120 kV FEG (field emission gun). The chemical analysis system involves the EDS and electron energy-loss spectroscopy (EELS); these can be used in a complementary manner to quantify the chemical composition of the specimen. EELS can also provide information about the electronic structure of the specimen. EELS can also provide information related to the electronic structure of the specimen. The element in the electron gun is ZrO/W and its point-image resolution

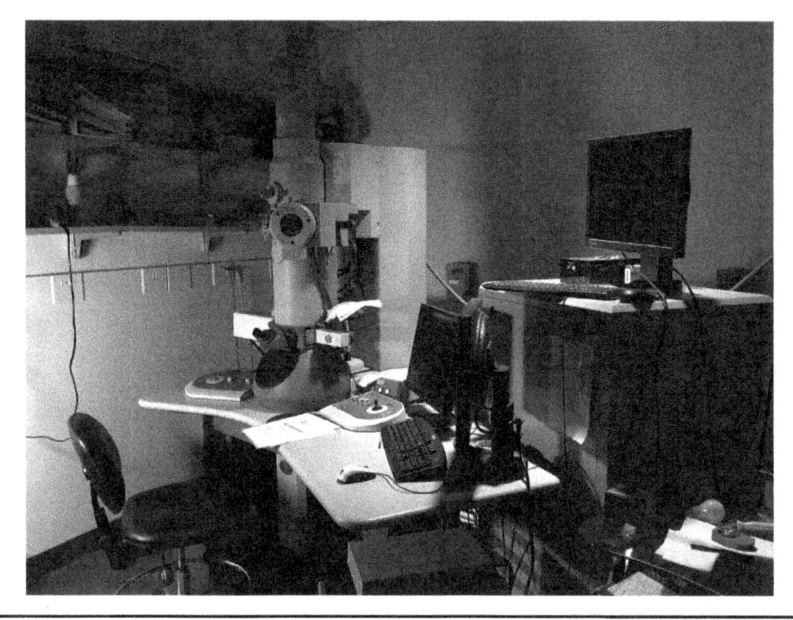

FIGURE 5.3 A FEI Tecnai Transmission Electron Microscope.

capability is 0.17 nm. Beam spot size can be focused down to 0.4 nm in diameter. Magnification ranges from 100 to 1.5 million.

TEM sample preparation is of utmost importance in obtaining TEM with good resolution. Figure 5.4 illustrates the sequence of TEM sample preparation that the author followed in his study. The basic requirements are that the specimen has to be thin enough to be transparent to the electron beam, must be clean, without much damage or contamination. There are several common methods for TEM specimen preparation:

- *Ion-milling:* almost for all kinds of materials
- *Electropolishing:* for conductive bulk materials
- *Crushing powders:* the simplest way to prepare for TEM specimens, but microstructural details may be lost

The bulk sample is first cut into thin slices about 0.6 mm thick. The thin sample slices are then either core-drilled or punched to obtain a 3-mm-diameter disk with a thickness of about 60 μm. The disk is then dimpled until the thickness in the center reaches about 10–20 μm and the specimen is then ion-milled until it is penetrated.

Figure 5.5 shows TEM micrographs of three types of nanoparticles in isopropyl alcohol (IPA) solvent as a 99.5% solution in IPA [10]. The unit bar is 500 nm in all three micrographs. Cloisite 30B in Fig. 5.5a shows intercalated nanoclay. Figure 5.5b shows PR-24-PS carbon nanofibers (CNFs) entangled in bundles. Figure 5.5c shows TriSilanolPhenyl-POSS®-SO1458 ($C_{48}H_{38}O_{12}Si_7$) and Fig. 5.5d shows polyvinyl silsesquioxane uncured PM1285 $(C_2H_3O_{1.5}Si)n$ in micrometer sizes.

- *Basic requirements:*
  - Thin enough to be transparent to the electron beam
  - Clean without much damage or contamination

- *Several common methods:*
  - *Ion-milling* – almost for all kinds of materials
  - *Electropolishing* – for conductive bulk materials
  - *Crushing powders* – the simplest way to prepare the TEM samples but microstructural details may be lost

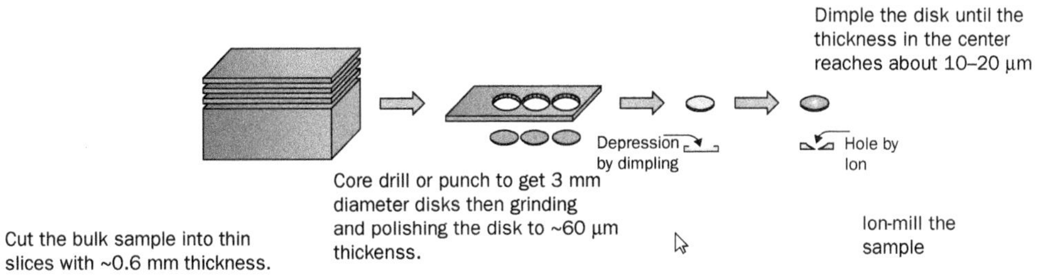

FIGURE 5.4 TEM sample preparation process by ion-milling.

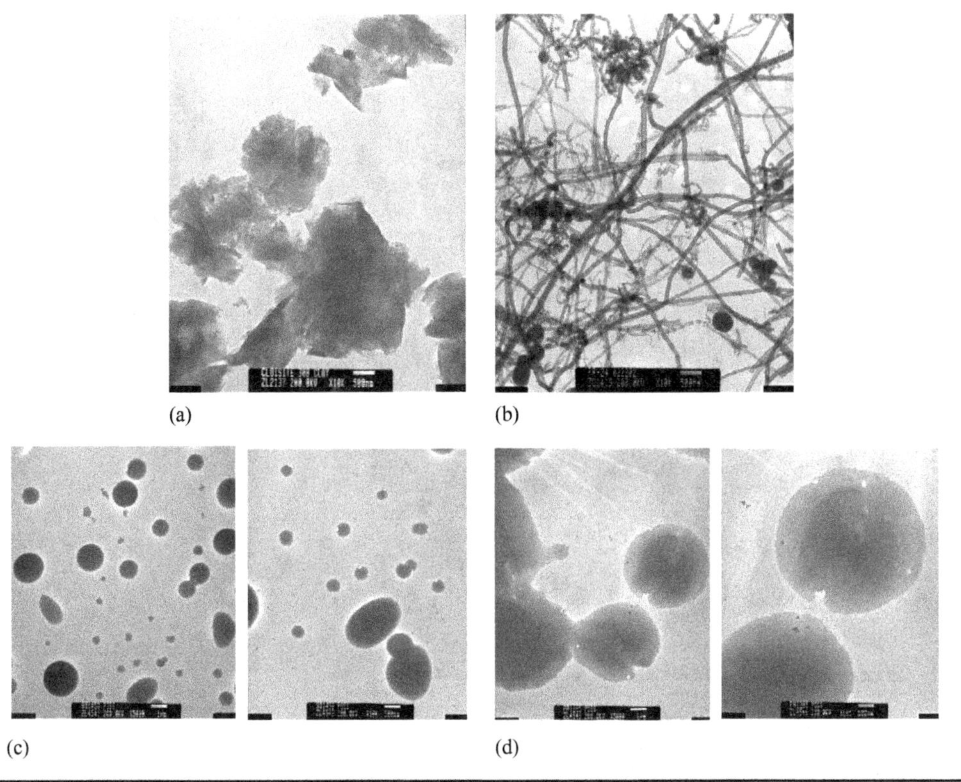

FIGURE 5.5 TEM micrographs of nanoparticles in IPA solvent: (a) Cloisite 30B, (b) PR-24-PS CNFs, (c) TriSilanolPhenyl-POSS-SO1458, and (d) polyvinyl silsesquioxane uncured—PM1285.

Figure 5.6 shows the TEM images of (a) the 5 wt% Cloisite 30B:95 wt% SC-1008 (Fig. 5.6a), (b) 10 wt% Cloisite 30B:90 wt% SC-1008 (Fig. 5.6b), and 15 wt% Cloisite 30B:85 wt% SC-1008 (Fig. 5.6c). The TEM images indicated intercalation and exfoliation of the nanoclay before committing a 20-lb run at Cytec Solvay Group of these nanoclay-resin mixtures for the preparation of prepregs. These TEM results verified the WAXD results, as shown in Fig. 5.1, to be incorrect. This proved to be a very cost-effective and efficient technique for screening different formulations on a small scale.

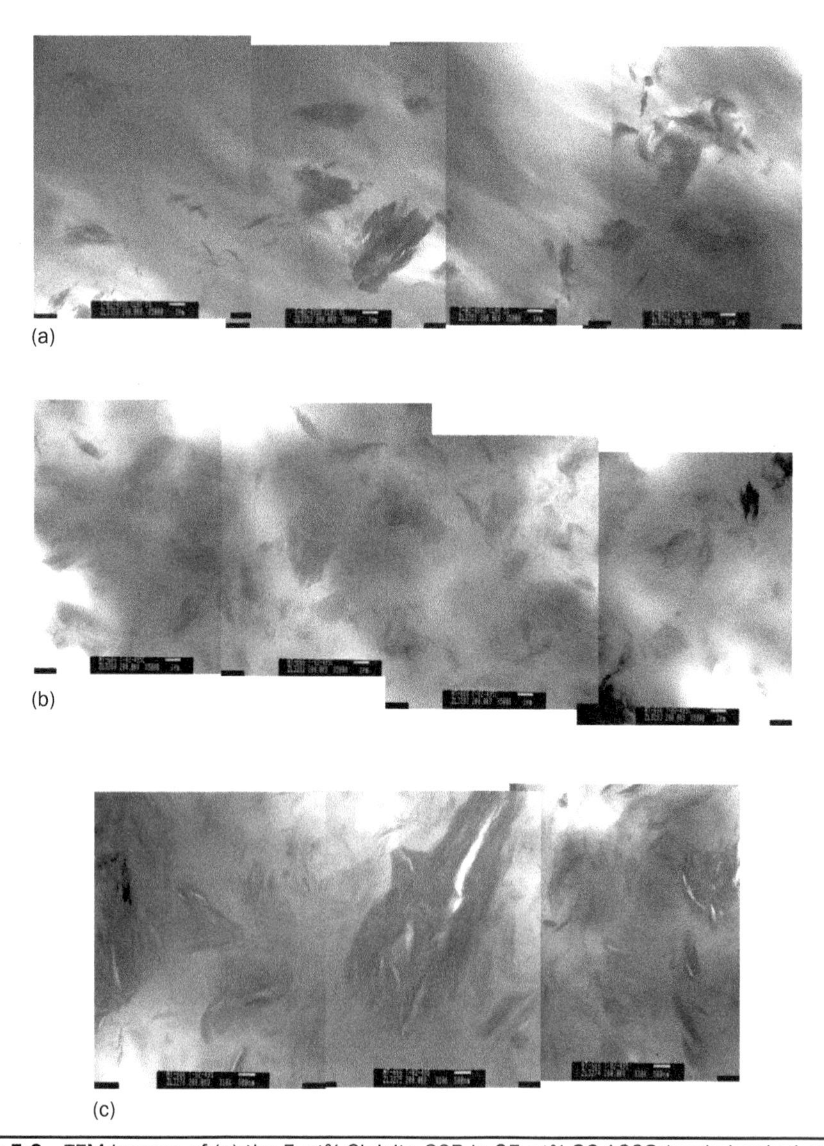

FIGURE 5.6 TEM images of (a) the 5 wt% Cloisite 30B in 95 wt% SC-1008 (scale bar is 1 μm), (b) 10 wt% Cloisite 30B in 90 wt% SC-1008 (scale bar is 1 μm), and (c) 15 wt% Cloisite 30B in 85 wt% SC-1008 (scale bar is 500 nm).

Transmission electron microscopy allows the observation of the overall organo-clay dispersion in the PNC sample. Clay dispersion and structure observed under the microscope can determine the nature of a clay nanocomposite as follows:

- *Immiscible:* usually large clay tactoids, undispersed clay particles
- *Intercalated:* clay layers in ordered stacks can be observed
- *Exfoliated:* single clay layers can be observed

Global microscale dispersion, as well as nanoscale dispersion and structure, can be determined by TEM.

Several limitations and drawbacks of the TEM procedure include the following:

1. Extensive sample preparation is time-consuming and this limits the number of samples that can be imaged and analyzed.
2. In addition to the small physical size of samples, the field of view is relatively small as well. For example, the section under analysis may not be representative of the sample as a whole.
3. Sample structure and morphology have been known to change drastically under exposure to electrons with extremely high energy, and damage to biological samples in particular can occur.
4. TEM cannot measure *d*-spacing of clay; therefore, it cannot easily determine difference between intercalated clay nanocomposite and well-dispersed, immiscible nanocomposite.
5. TEM is a high-vacuum instrument that is costly to operate and to maintain, certainly playing a role in the high barrier entry required to conduct nanoscience research.

However, the advantages of the TEM technique are numerous. Any type of sample, whether electrically insulating, semiconducting, or conducting, is able to be imaged by TEM. The incredible resolution capability allows for atomic-level inspection. TEM measures overall *characteristics* of clay, such as uniformity, tactoids, or nonuniformity of the clay within the material. It is recommended that TEM measurements should be combined with WAXD data for thorough analyses of the nanomaterials.

5.4.3 Energy-Dispersive X-Ray Spectroscopy (EDS or EDX)

X rays emitted from atoms are characteristics of the elements, and the intensity distribution of the x rays represents the thickness-projected atom densities in the specimen. This is known as *energy-dispersive x-ray spectroscopy* (EDS or EDX), which has played an important role in microanalysis, particularly for heavier elements. EDS is a key tool for identifying the chemical composition of a specimen. A modern TEM is capable of producing a fine electron probe smaller than 2 nm, allowing direct identification of the local composition of an individual element. Examples will be provided in later chapters. When POSS materials were incorporated into the cyanate ester resin, EDS was used to demonstrate that molecular dispersion was achieved by the PT-15 cyanate ester-Trisilanolphenyl-POSS nanocomposite. Figure 5.7 shows the

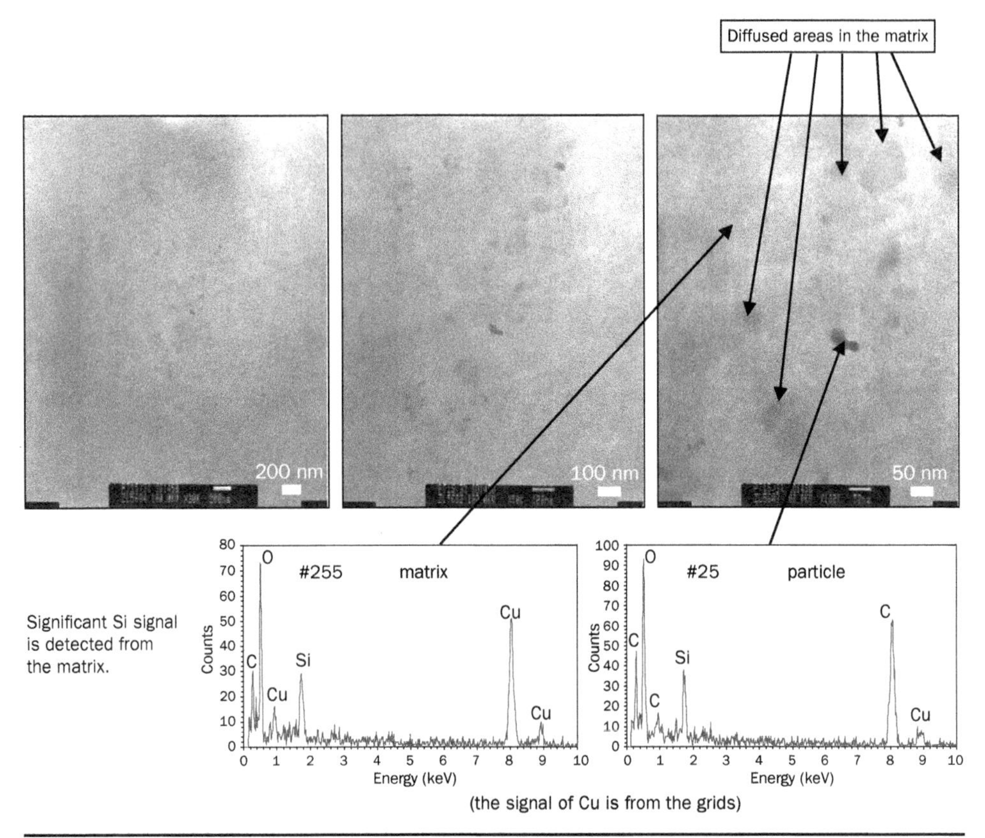

FIGURE 5.7 TEM micrographs of PT-15/SO1458 POSS in progressive magnification; EDS shows Si is dispersed very well in the PT-15 resin system.

TEM micrographs of PT-15/SO1458 POSS in progressive magnification. EDS shows Si is dispersed very well in the PT-15 resin system. Si counts were observed in the particle and matrix areas [10]. Thus, molecular dispersion of POSS particle in the PT-15 CE resin is evident.

5.5 Small-Angle X-Ray Scattering (SAXS)

Small-angle x-ray scattering (SAXS) is an analytical x-ray application technique for the structural characterization of solid and fluid materials in the nanometer range. In SAXS experiments, the sample is irradiated by a well-defined, monochromatic x-ray beam. When a nonhomogeneous medium, for example, proteins in water, is irradiated, structural information about the scattering particles can be derived from the intensity distribution of the scattered beam at very low scattering angles.

With SAXS it is possible to study both monodisperse and polydisperse systems. In the case of monodisperse systems, one can determine size, shape, and internal structure of the particles. For polydisperse systems, a size distribution can be calculated under the assumption that all particles have the same shape. SAXS is applied to investigate structural details in the 0.5–50 nm size range in materials such as the following:

- Particle sizing of suspended nanopowders
- Materials in life science and biotechnology (proteins, viruses, DNA complexes)
- Polymer films and fibers
- Catalyst surface per volume
- Microemulsions
- Liquid crystals

Small-angle x-ray scattering can provide the following information:

- Lamella repeat distance
- Radius of gyration
- Particle size and shape
- Large-scale structure and long-range order

Readers should refer to existing texts describing the SAXS technique [4, 11, 12].

5.6 Scanning Probe Microscopy (SPM)

Scanning probe microscopes were developed as an important technique to characterize the surface properties of materials, especially nanomaterials. Scanning probe microscopes are capable to create detailed three-dimensional images of specimen surfaces with atomic resolution [5]. Two basic types of scanning probe microscopes are used in nanoscience. In 1981, Binnig et al. developed the scanning tunneling microscopy (STM) technique [13, 14]. In the STM technique, the magnitude of the tunneling current between the probe tip and the atoms of a substrate surface is monitored. STM samples must be electrically conducting. Atomic force microscopy (AFM) and its derivatives comprise a second-class SPM. In AFM, the size of the force between a probe tip and the atoms of substrate surface is monitored. The AFM technique was also developed by Binning and colleagues in 1985 especially for insulating materials [14–16]. The AFM technique is able to measure forces on the order of 1 μN and less. There are many similarities between STM and AFM. Both are equipped with a probe tip fastened to a cantilever, a scanning (motion) mechanism, and a detector system. Bhushan and Marti's chapter on "Scanning Probe Microscopy—Principle of Operation, Instrumentation, and Probes" should be an excellent review article for the readers [14]. Four achievements that contributed to the development of scanning probe methods are summarized by Hornyak et al. [5] in the following:

1. Physical cushioning mechanisms, motion-damping eddy currents induced by magnets and copper plates, and nitrogen-gas–regulated suspension systems serve to isolate the SPM from external vibration sources. Protective enclosures shield the microscope from room drafts. If extremely high-resolution study is required, cryogenic temperatures eliminate atomic and molecular movement.
2. Computerized feedback system control of piezoelectric devices with nanometer level precision guides the probe during its descent to the surface—less than a

nanometer above where van der Waals attraction exerts force on the probe tip (in AFM) or tunneling current to be able to flow (in STM).

3. Computerized control of piezoelectric scanners moves the probe tip across the sample surface with nanometer precision.
4. The fabrication of the sharper probes allows for better resolution of surface features. For example, carbon nanotube probes with sharpness of less than 1 nm have shown promise in the past few years.

5.6.1 Scanning Tunneling Microscopy (STM)

The fundamental principle of STM is quantum tunneling, as electrons can tunnel through a finite energy barrier if the physical distance of the barrier is small enough. Electron (or quantum) tunneling is attained when a particle (an electron) with lower kinetic energy is able to exist on the other side of an energy barrier with higher potential energy, thus disobeying a fundamental law of classical mechanics. Tunneling is the penetration of an electron into a classically forbidden region [17]. Electrons exhibit wave behavior and their position is represented by a wave (probability) function. The wave function represents a finite probability of finding an electron on the other side of the potential barrier. Since the electron does not possess enough kinetic energy to overcome the potential barrier, the only way the electron can appear on the other side is by tunneling through the barrier. The principle of quantum tunneling was also used to explain exponential radioactive decay rates. In quantum mechanics, a finite probability is allocated to tunneling and hence nuclear decay is allowed.

STM relies on an electronic signal to relay information about a sample—the strength of a tunneling current potential that exists between the probe tip and the substrate surface. Small changes in the distance between the probe tip and the substrate surface translate into large changes in the tunneling current. By this phenomenon, atomic-scale resolution by STM is possible in the x, y, and z directions. The density of states (DoS) of solid-state materials can be also mapped by the technique called *scanning tunneling spectroscopy* (STS). Chemical reactions induced and oriented by the STM probe are also available by means of this technique.

5.6.2 Atomic Force Microscopy (AFM)

Atomic force microscopy relies on the mechanical deflection of a cantilever to replay information about the contour of a sample surface. An atomically sharpened probe tip (20–50 nm or less) descends perpendicularly from the distal end of a cantilever. The tip-to-sample distance is fixed by means of a feedback mechanism that maintains constant force between them (i.e., constant height). A focused beam is reflected off the back of the cantilever equipped with a sharpened probe and into a photodiode detector [18]. Any deflection of the beam is translated into a topographical feature. Rastering of the tip over the surface produces a topographical image. The photodiode is split into two compartments (or vice versa) and is able to detect differences in beam position to the level of a nanometer. The Agilent 5500 Atomic Force Microscope is shown in Fig. 5.8.

The AFM technique relies on a balance between attractive van der Waals and repulsive electrostatic forces between the probe tip and the surface. The net force

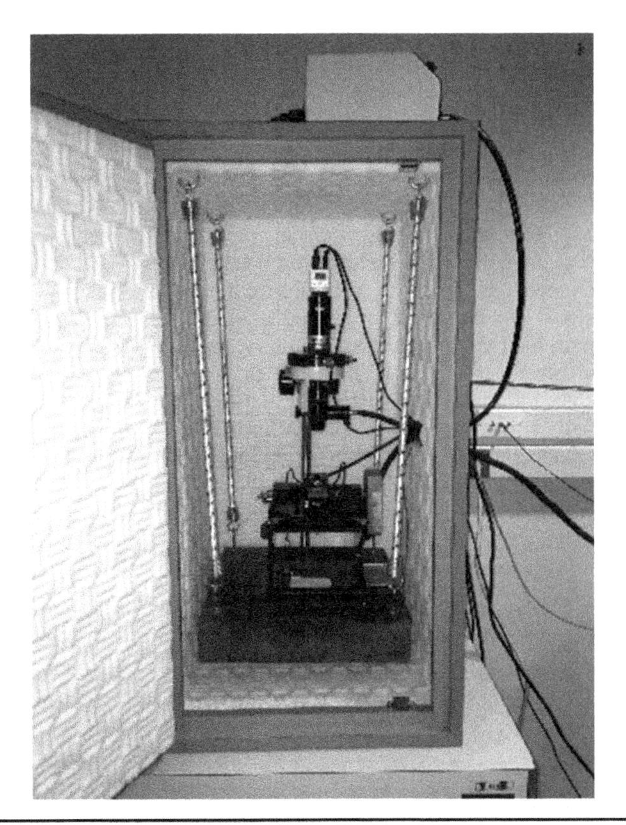

FIGURE 5.8 The Agilent 5500 Atomic Force Microscope.

is a function of the distance between the two. The force is also a function of the dielectric constant ε of the medium. Hand-in-hand with atomic-order resolution, the theoretical magnification potential of the AFM technique is on the order of 10^9 [5, 14–16]. The versatility of AFM is demonstrated further by its ability to image large objects that are tens of micrometers or more in size. There are three major mode of operations of AFM: contact, tapping, and noncontact (Table 5.1), as summarized by Hornyak et al. [5].

Different cantilever-probe tip combinations are available for different types of AFM technique. Cantilevers are fabricated from single crystal silicon. The length of AFM contact and tapping mode cantilevers is between 100 and 500 μm; the width is between 25 and 40 μm, and the thickness is between 1 and 10 μm. The resonant frequency of these cantilevers dwells between 10 and 300 kHz and the spring constant between 0.1 and 50 $N \cdot m^{-1}$. The probe tip diameter ranges from a few nanometers for finer resolution to 20 nm for an average AFM to as large as 50 nm. Noncontract AFM mode cantilevers are made of highly doped single crystal silicon.

Nakajima et al. [19] investigated the surface mechanical properties including Young's modulus, adhesion, viscoelasticity, and energy dissipation during probe–sample interactions by employing AFM nanomechanical mapping. The ability of

Table 5.1 Atomic Force Microscopy (AFM) Primary Modes

AFM mode	Configuration	Capability
Contact mode	The probe-cantilever assembly applies a constant force to the surface; the force constant is $<1\ \text{N} \cdot \text{M}^{-1}$. The deflection of the cantilever is due to topographical changes characteristic of the surface. The mechanical deflection is translated into an optical signal by reflection of an aligned laser light from the top surface of the cantilever into an aligned-calibrated dual photodetector collector. Due to intimate contact with the surface, large lateral forces can influence the probe. Surface contamination, electrostatic forces, and heterogeneous surfaces are able to impact the action of the probe. AFM is capable of imaging nanomaterials in air, vacuum, and liquids.	Samples with hard surfaces are appropriate for contact mode AFM analysis. AFM is capable of imaging insulator, semiconductor, and conductor surfaces. Qualitative information includes 3D visualization and material sensing by phase contrast. Quantitative information includes topographic mapping, particle and pore size, particle and pore morphology, surface roughness and texture, particle count, size distribution, and volume-mass distribution.
Tapping mode	• Strong repulsive regime • Oscillating probe (>100 kHz) • Intermittent contact with surface • Lateral forces reduced significantly • Amplitude and phase imaging	• Three-dimensional topography • Use for soft samples, or weakly bound to surface • Biological materials • DNA • Carbon nanotubes
Noncontact mode	• Weak-attractive regime • Oscillating probe • Can be applied with water layer	Soft surfaces

nanomechanical mapping for identifying the composition and evaluating the mechanical properties at nanoscale will open a new way to study surface properties and explore the *microstructure–properties* relationship in a large range of polymeric and biological materials.

5.7 Raman Spectroscopy

Far-infrared, infrared, near-infrared, Raman spectroscopy, and surface-enhanced Raman spectroscopy (SERS) are extremely relevant methods used to characterize nanomaterials. These methods provide information about vibrational energy: IR methods rely on asymmetrical (dipolar) vibrations and Raman methods on symmetrical (polarizable) vibrations. In this way, the two methods complement each other.

C. V. Raman and K. S. Krishnan in 1928 discovered that light scattered off certain molecules changed the polarization state of the molecules. The Raman Effect is a scattering phenomenon that links the vibrational frequencies of the molecule to the energy difference between the incident and scattered light. However, only molecules with symmetric vibrational (polarization) modes and transitions were amenable to Raman spectroscopic analysis. The SERS process depends on nanomaterial facets or particles that are able to enable signal intensity by multiple orders of magnitude—an example of a spectroscopic method enabled by nanotechnology.

Two types of photon scattering exist—elastic scattering and inelastic scattering. Elastic scattering, in which there is no energy loss, is called Rayleigh (Rayleigh–Debye) or Mie scattering. Such scattering occurs when particle dimensions are on the order of or smaller than the wavelength of the incident radiation. Remarkably, then scattering is a true nanoscale phenomenon. If particle size does not conform to this limit (e.g., larger materials), then reflection and refraction also occur. Rayleigh scattering (elastic) involves no change in the polarizability of the molecule. However, a small portion of the incident photons is scattered at frequencies that are different form the incident light. This is called *inelastic scattering* and the result is exhibited in the form of vibrational, rotational, or electronic energy changes in the molecule. Another form of Raman spectroscopy called *resonant Raman spectroscopy* (RRS) occurs when there is resonance between the incident radiation and an electronic transition of the molecule.

SERS is a technique that exhibits incredible sensitivity and has great importance to nanoscience. Substrates in the SERS technique are noble metals, such as Au and Ag that happen to be in nanoparticle form. The primary mechanism of SERS relies on the surface plasmon resonance of nanometal particles. An electromagnetic enhancement on the order of 10^6 of the vibrational signal from an adsorbed species is imparted to the spectrum. Analysis of single-wall nanotubes by SERS is diagnostic of their presence. Single-molecule detection is also possible with this method. Garea and Iovu [20] investigated the synthesis of nanocomposites based on POSS and polymer matrix using Raman spectroscopy and x-ray photoelectron spectroscopy.

5.8 X-Ray Photoelectron Spectroscopy (XPS)

An x-ray photoelectron spectroscope is considered a modern tool for characterization of carbonaceous materials to determine the chemical composition, impurity presence, and nature of chemical bonds. The most important application of XPS in nanocomposites based on polymer-CNT is to identify the chemical composition of the modified carbon nanotube surface in order to confirm whether the surface modification has occurred or not. The XPS analysis performed on the CNT surface was focused on measuring the binding energy of photoelectrons ejected when CNTs are irradiated with different agents pointed out by XPS analysis [20].

5.9 Other Techniques

A variety of other morphological characterization techniques can be found in [1–20].

Sections 5.10–5.16 discuss the global methods that are commonly used to characterize the bulk material properties, such as mechanical, thermal, flammability, ablation, electrical, optical, and other properties, of the PNCs. These material properties are discussed in detail in Chaps. 6–11.

5.10 Mechanical Properties

When nanomaterials are added to neat polymers to enhance specific properties, the most important task is to maintain the mechanical properties of these PNCs. The mechanical properties, among all the properties of these PNCs, are often the most important properties because virtually all service conditions are the majority of end-use applications involving some degree of mechanical loading. The basic understanding of stress–strain behavior of polymers is of utmost importance to design engineers. A typical stress–strain diagram is shown in Fig. 5.9. For a better understanding of the stress–strain curve, a few basic terms associated with the stress–strain diagram are defined below:

- *Stress:* The force applied to produce deformation in a unit area of a test specimen. Stress is a ratio of applied load to the original cross-sectional area expressed in kg/m^2 ($lb/in.^2$).
- *Strain:* The ratio of the elongation to the gauge length of the test specimen, or simply stated, change in length per unit of the original length ($\Delta l/l$). It is expressed in a dimensionless ratio.
- *Elongation:* The increase in length of a test specimen produced by a tensile load.
- *Yield point:* The first point on the stress–strain curve at which an increase in strain occurs without the increase in stress.

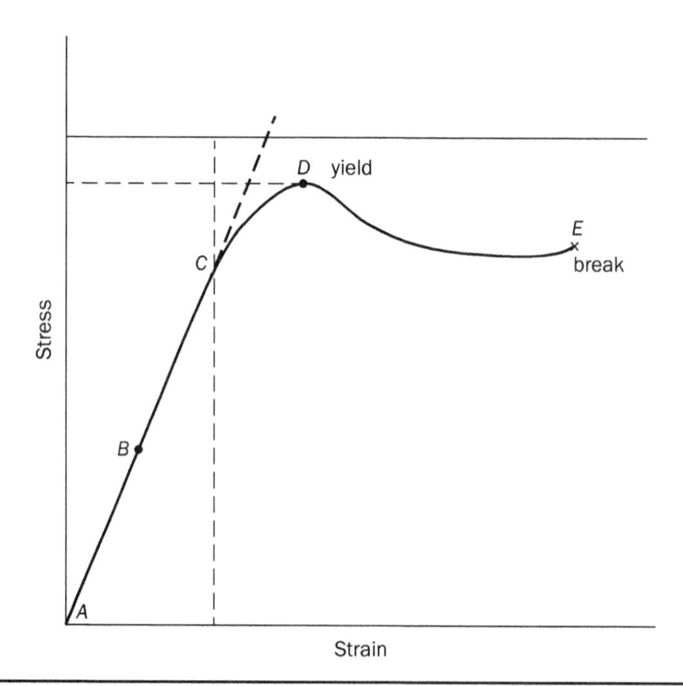

Figure 5.9 A typical stress–strain curve.

- *Yield strength:* The stress at which a material exhibits a specified limiting deviation from the proportionality of stress to strain. This stress will be at the yield point.
- *Proportional limit:* The greatest stress at which a material is capable of sustaining the applied load without any deviation from proportionally of stress to strain (Hooke's law), expressed in lb/in.^{2}.
- *Modulus of elasticity:* The ratio of stress to corresponding strain below the proportion limit of a material. It is expressed in F/A, usually in kg/m^2 (lb/in.^{2}). This is also known as Young's modulus. A modulus is a measure of the material's stiffness.
- *Ultimate strength:* The maximum unit stress a material will withstand when subjected to an applied load in compressing, tension, or shear, expressed in kg/m^2 (lb/in.^{2}).

Some commonly used mechanical property tests are as listed below [21]:

- Tensile Properties Test (ASTM D 638, ISO 527-1)
- Flexural Properties Test (ASTM D 790, ISO 178)
- Compressive Properties Test (ASTM D 695, ISO 75-1 and 75-2)
- Creep Properties Test
- Impact Properties Test (Izod and Charpy impact tests)
- Shear Properties Test (ASTM D 732)
- Other tests, such as abrasion, fatigue resistance, and hardness tests.

5.11 Thermal Properties

The thermal properties of PNCs are equally as important as the mechanical properties. Unlike metals, polymers are extremely sensitive to changes in temperature. Four commonly used techniques—thermogravimetric analysis, differential scanning calorimetry, dynamic mechanical and thermal analysis, and thermal conductivity—are discussed in this section.

5.11.1 Thermogravimetric Analysis (TGA)

Thermogravimetric analysis is a technique in which the mass of a substance is monitored as a function of temperature or time when the sample specimen is subjected to a controlled temperature program in a controlled atmosphere. TGA is a technique in which, upon heating a material, its weight increases or decreases. A TGA consists of a sample pan that is supported by a precision balance. The pan resides in a furnace and is heated or cooled during the experiment. The mass of the sample is monitored during the experiment. A sample purge gas controls the sample environment. This gas may be inert or a reactive gas that flows over the sample and exits through an exhaust. TGA can determine the thermal stability of the PNCs compare to the base polymer.

Lao et al. studied the thermal and flammability properties of polyamide 12 (PA12) with different wt% loadings of Cloisite 30B nanoclay (NC2) and PR-19-PS CNFs [22]. They performed TGA on neat and nanoparticle-reinforced PA12 (Fig. 5.10) under

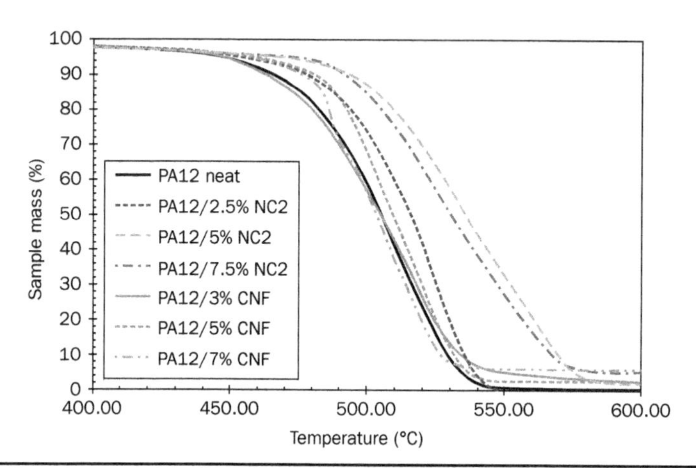

FIGURE 5.10 Sample mass (%) of neat and nanoparticle-filled PA12 as a function of temperature (TGA, scan rate 10°C/min).

nitrogen using scan rates of 10°C/min and 20°C/min [22]. The test could not be performed under air since the intumescent flame retardant (FR) additives would expand at high temperature under an air environment resulting in erratic results. Since lower and higher scan rates showed similar trend of curve shifts with the lower scan rate exhibiting a better separation of curves, only the TGA data obtained using the scan rate of 10°C/min was shown. From these raw data, the decomposition temperatures for both 10% and 50% mass loss, $T_{10\%}$ and $T_{50\%}$, were inferred and summarized [22]. The $T_{10\%}$ and $T_{50\%}$ of the PA12/NC2 blends were higher than those of neat PA12, while those for the PA12/CNF blends were about the same as those of neat PA12 (Fig. 5.10).

5.11.2 Differential Scanning Calorimetry (DSC)

Vaia et al. [23] reported that the modification chains interacted in the filler interlayers exist in states with varying degrees of order. In general, as the interlayer packing density or the chain length decreases (or the temperature increases), the intercalated chains adopt a more disordered, liquid-like structure resulting from an increase in the gauche/trans conformer ratio. In order to quantify such phase behavior transitions and chain dynamics for organic monolayers immobilized on the surface of montmorillonites, DSC analysis is performed.

5.11.3 Dynamic Mechanical Thermal Analysis (DMTA)

The introduction of nanomaterials into a polymeric matrix causes difficulties in processing. The mobility of the nanomaterials is reflected by viscosity, viscoelasticity, moduli, etc. Therefore, understanding the rheological properties of such nanocomposites is of significant importance for optimizing the process parameters that lead to controlled microstructure, which enables the design of the properties of the PNCs [24, 25]. Rheology has been extensively used in the study of nanocomposite system in conjunction with basic characterization techniques, such as WAXD, SEM, and TEM [26]. A large number of rheological studies on PNCs have been published in the past 20 years. Rheology offers an effective way to assess the state of nanomaterial dispersion and to detect the presence of interconnecting microstructure of the nanocomposites in molten

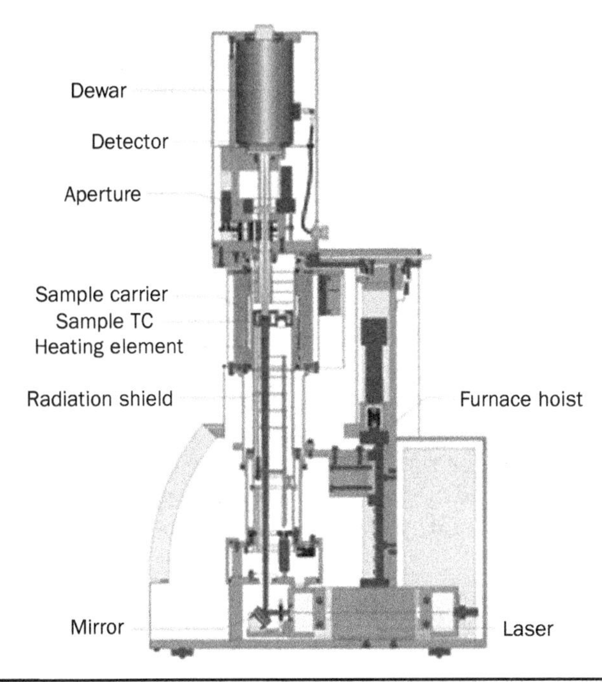

FIGURE 5.11 Netzsch LFA 457 Microflash for thermal diffusivity measurements.

state directly. Song and Jin [27] reported the characterization of rheological properties of PNCs. They provided the fundamental rheological theory for studying PNCs.

Krishnamoorti et al. [28] described the steady shear rheological behavior of intercalated or exfoliated nanocomposites based on poly(dimethyldiphenyl siloxane) (PDS) and layered silicate. The viscosity systemically increases with silicate loading but still obeys the Newtonian behavior in the lower shear rate region, even at the highest silicate loadings.

5.11.4 Thermal Conductivity

The thermal conductivity property is commonly characterized according to ASTM E1461 [29]. A Netzsch LFA 457 Microflash is usually used, as shown in Fig. 5.11, for thermal diffusivity (α) measurements. Here, thermal diffusivities of samples 8 × 8 mm^2 will be measured at a temperature range from 30°C to 175°C. Two from each set of NGP and n-Al loadings of the samples will be simultaneously tested with a reference sample of either graphite or Pyrex. After diffusivity results are obtained, the specific heat (C_p) of each sample will be calculated through the use of LFA software. This is accomplished by using the comparative method of contrasting the change in temperature curves used to calculate the thermal diffusivity of the test sample with those of the reference samples. The reference samples possess a known specific heat while the specific heat of the test samples is determined. Gathering the diffusivity results along with the measured density (ρ) and C_p, using Eq. (5.2), the thermal conductivity (K) of the samples will be determined:

$$K = \alpha \rho C_p \tag{5.2}$$

5.11.5 Other Thermal Properties

Besides the above thermal properties, other thermal properties are also characterized [30], such as (a) heat deflection temperature (HDT) measured by ASTM D 648, ISO 75-1 and 75-2, (b) Vicat softening temperature measured by ASTM D 1525, ISO 306, (c) torsion pendulum test using ASTM D 2236, (d) long-term heat-resistance test using ASTM D 794, (e) UL temperature index, (f) creep modulus/creep rupture tests, (g) coefficient of linear thermal expansion, and (h) brittleness temperature measured by ASTM D 746, ISO 974.

5.12 Flammability Properties

There are numerous techniques to evaluate the thermal, flammability, and smoke properties of PNCs, depending on the type of information desired. A brief discussion of the techniques is not intended to be exhaustive or authoritative, but only as an aide to the reader in identifying the test methods used worldwide; the reader should refer to the original ASTM, UL, and other organizations for more detailed information. The reader is referred to Troitzsch [31], Hirschler [32], and Shah [33] for an overview of all flammability test methods.

5.12.1 Cone Calorimeter (CC)

The potential fire hazard of any material depends on many factors, including ignitability, rate of surface flame spread, heat release rate (peak, average, and total), mass loss rate, smoke evolution, and evolution of toxic gases (e.g., CO). The cone calorimeter (CC) has been regarded as the most significant bench scale instrument for evaluating reaction-to-fire properties of materials. CC data address most of these factors, either directly [e.g., heat release rate (HRR)] or indirectly (e.g., time to ignition and HRR as functions of external heat flux as indicators of flame spread potential). The data obtained from the CC test are available in engineering units that are conducive to extrapolation to other fire sizes and scenarios. This section briefly discusses how does the CC operates. Readers should refer to Babrauskas' excellent description of the CC apparatus in [34], the ASTM E 1354 [35], or the ISO 5660 [36] for more details of the CC apparatus.

A conceptual view of the CC apparatus is shown in Fig. 5.12. The CC apparatus is a fire test instrument based on the principle of oxygen consumption calorimetry. This empirical principle is based on the observation that, generally, the net heat of combustion of any organic materials is directly related to the amount of oxygen required for combustion. Approximately 13.1 MJ of heat are released per kilogram of oxygen consumed. At the core of the instrument is a radiant electrical heater in the shape of a truncated cone (hence the name). This heating element irradiates a flat horizontal sample 100 mm × 100 mm and up to 50 mm thick, placed beneath it, at a preset heat flux of up to 100 kW/m^2. The sample is placed on a load cell for continuous monitoring of its mass as it burns. Ignition is provided by an intermittent spark igniter located 13 mm above the sample.

The gas stream containing the combined combustion products is captured through an exhaust duct system, consisting of a high temperature centrifugal fan, a hood, and an orifice-plate flow meter. The typical air flow rate is 0.024 m^3/s. Oxygen concentration in the exhaust stream is measured with an oxygen analyzer capable of an accuracy

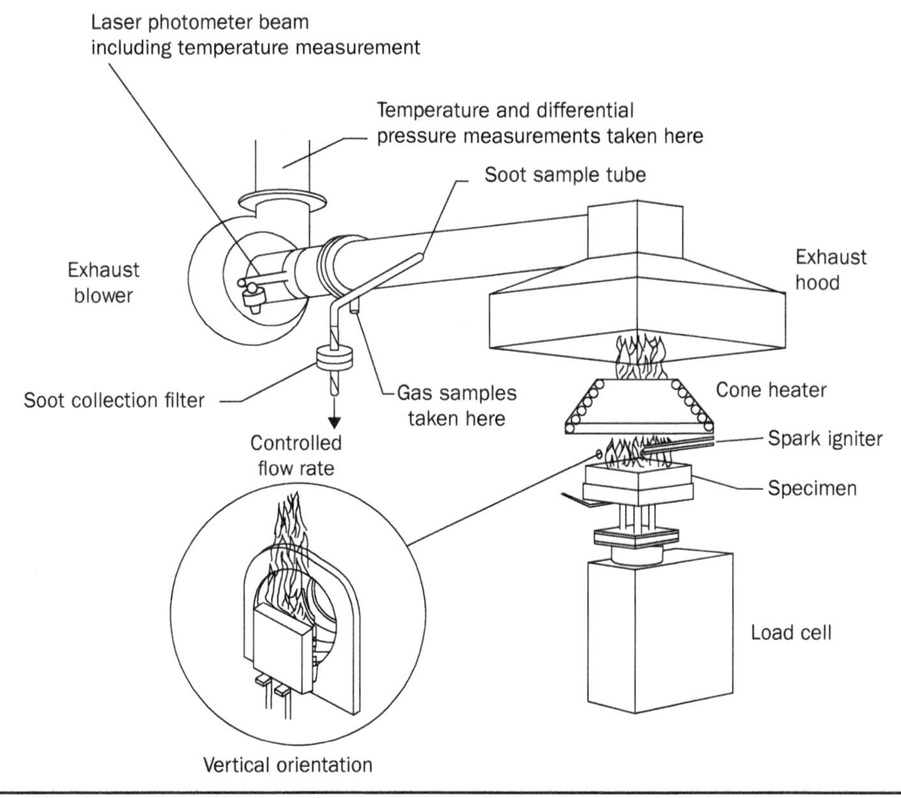

Figure 5.12 Schematic view of the cone calorimeter.

of 50 ppm and the heat release rate is determined by comparing the oxygen concentration with the value obtained when no sample is burning.

Smoke obscuration measurement is made in the exhaust duct by a helium-neon laser, with silicon photodiodes as the main beam and reference detectors, and appropriate electronics to derive the extinction coefficient and set the zero reading. Locations are also provided in the exhaust duct for additional sampling probes to determine concentrations of other combustion products such as carbon dioxide and carbon monoxide. All data are collected with a PC which records data continuously at fixed intervals of a few seconds during the test.

The CC apparatus is used to determine the following principal flammability properties:

- Heat release rate per unit area (kW/m^2)
- Cumulative heat released (kW)
- Effective heat of combustion (MJ/kg)
- Time to ignition (s)
- Mass loss rate (kg/s)
- Total mass loss (kg)
- Smoke density (m^2/kg)

Ignitability is defined as the propensity to ignition, as measured by the time in seconds to sustain flaming (the existence of flame on or over most of the specimen surface for periods of at least 4 seconds) at a specified heat flux. The ignitability is thus a measure of how quick the material will ignite in a fire situation.

Heat release is defined as the heat generated in a fire due to various chemical reactions occurring within a given weight or volume of material. The major contributors are those reactions where carbon monoxide and carbon dioxide are generated and oxygen is consumed. Heat release data provide a relative fire-hazard assessment for materials. Materials with low heat release per unit weight or volume will do less damage to the surroundings than materials with high heat release rate. The heat release rate, especially the peak amount, is the primary characteristic determining the size, growth, and suppression requirements of a fire environment. The rate of heat release is determined by measurement of the oxygen consumption as determined by the oxygen concentration and the flow rate in the exhaust product stream.

Effective heat of combustion, mass remaining, smoke ratio, carbon dioxide, and carbon monoxide are the additional CC data that are obtained during the experiment.

5.12.2 Mass Loss Calorimetry (MLC)

The mass loss calorimeter per ASTM E2102 [37] or ISO 13927 is used to determine mass loss rate and ignitability at any heat flux condition defined within the CC fire model of ISO 5660 or ASTM E1354. This instrument, manufactured by Fire Testing Technology Ltd. (FTT), West Sussex, the United Kingdom, is a complete assembly that is also suitable to replace the cone heater/load cell assembly inside a CC apparatus. The MLC, as a stand-alone instrument, serves as a quality control tool. The instrument is supplied as two-part component system: the cone heater assembly (with conical heater, shutter mechanism, spark ignition, load cell) and the control. It can also be provided as an enclosed system in order to measure combustion products and toxicity yields or to use with low oxygen conditions. Heat release rate instrument can be used to determine the effect of combustion, as a ratio of heat release rate and mass loss rate. Thus, if the effective heat of combustion is known, mass loss rate data can be used to calculate approximate values of heat release rate. The MLC is an inexpensive way of assessing fire properties of materials. Heat release is calculated using thermopile in the MLC.

5.12.3 Microscale Combustion Calorimetry (MCC)

The microscale combustion calorimeter was developed by Federal Aviation Administration (FAA) to offer industry a research tool to assist FAA in its mandate to dramatically improve the fire safety of aircraft materials. It is a standardized ASTM D 7309 test method [38] (Fig. 5.13). The MCC instrument is becoming a useful research tool due to its ability to obtain meaningful test data with a sample size in the range of 0.5–50 mg. Potential measurements carried out using this tool are, for example, heat of combustion, ignition temperature, heat release, heat release capacity, and flame resistance.

5.12.4 Oxygen Index—Limiting Oxygen Index (LOI)

Limiting oxygen index (LOI) is another test commonly used in the R&D laboratory; it is also known as the *oxygen index* (OI). It is included in some national or international standards, such as ASTM D2863-12el [39] or BS ISO 4589-2. The specimen size and

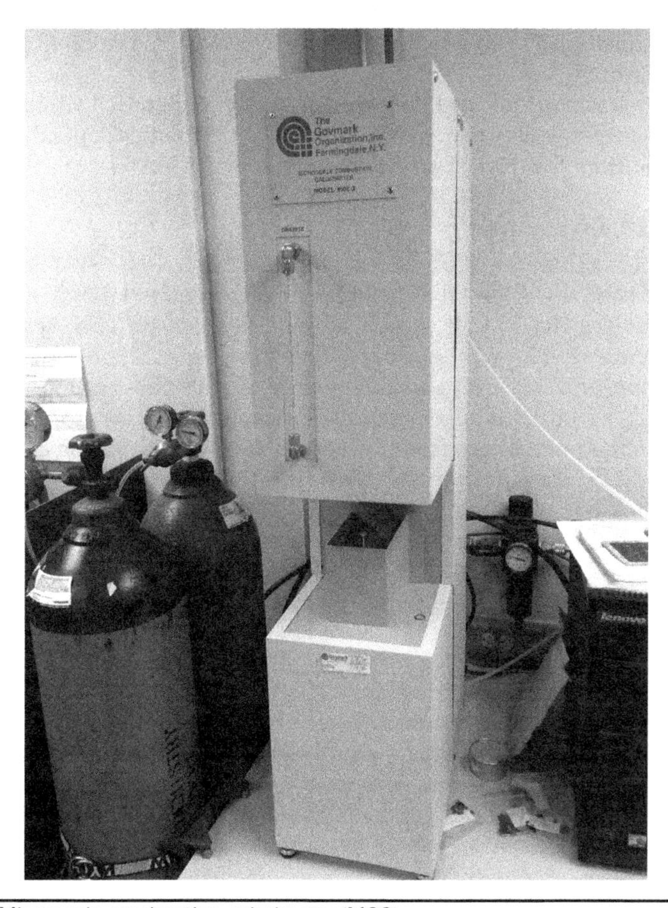

Figure 5.13 Microscale combustion calorimetry (MCC).

shape is not as strictly specified as the UL 94 test, but usually bars of about 100 × 65 × 3 mm are used in testing of rigid plastics. The specimen is positioned in a candle-like position vertically at the exit of a glass chimney and clamped at the bottom. The chimney is continuously fed with a controlled mixture of nitrogen and oxygen. The flame of a small igniting burner is applied to the top of the specimen until the entire top surface is ignited. If the specimen does not ignite or extinguishes in less than 30 seconds, the concentration of oxygen is then increased, in steps with repeated trial ignitions, until the specimen shows stable candle-like combustion for more than 3 minutes after removal of the ignition source or if more than 5 cm length of the sample is consumed, a new specimen should be put in place and tested at a slightly lower oxygen concentration. By iteration of these steps, finally the LOI value is determined; this is the highest concentration of oxygen at which the tested sample self-extinguished in less than 3 minutes and less than 5 cm of the material is consumed. The LOI test does not represent any real fire scenario, but it is quite reproducible, and fairly insensitive to sample thickness. It is often use as a screening, research statistical design tool, because it gives a numerical value typically to two significant figures, instead of a mere classification as given by the UL 94. It can even be run on films and textiles with suitable clamping.

Materials which melt and flow readily will give high LOI values, and in the case of nylon, it is even possible to arrive two different LOI values depending on whether they are reached from the oxygen side or the high oxygen side.

5.12.5 UL 94

Underwriters Laboratories UL 94 test [40] (also known as ASTM D 3801 and ISO 1210) is designed for assessing "flammability of plastic materials for parts in devices and appliances" (see Fig. 5.14). It is run on a sample of plastic itself, at specific dimensions. This test measures how long a bar of polymer burns after exposure to a small gas burner flame, and it is a reasonable measure of material's response to a small ignition source, such as a candle or a match. It is accepted for standardization internationally. Five different classifications from the vertical burn configuration are included, but only V-0, V-1, and V-2 classifications are introduced because these are most often cited in the literature.

To assign this set of classifications, a plastic bar of 125 mm length × 13 mm width, with smooth edges, is positioned vertically and held from the top. Depending on the application of the plastic, bars may have a thickness of 3.2, 1.6, or 0.8 mm. Thinner specimens are usually more flammable. A wad of surgical cotton is placed 300 mm below the specimen to detect flaming drips which will ignite the cotton. The Bunsen burner flame (about 19 mm high, fed with methane) is applied twice to the test bar (10 seconds each). After each application the burning time is recorded. The second application of the flame should follow immediately after the bar ceases to burn from the first application. A V-0 classification is given when the five testy bars extinguish in less than 10 seconds after any flame application. The average burning time for 5 bars tested (10 flame applications) should not be greater than 5 seconds and there should be no burning drips that ignite the cotton. The individual specimen afterflame plus afterglow must not exceed

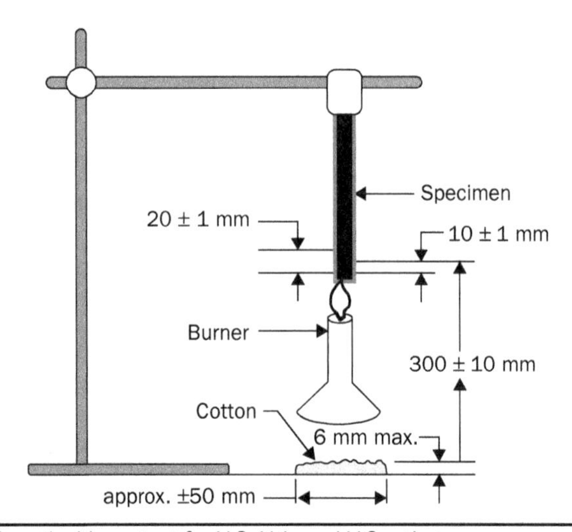

FIGURE 5.14 UL 94 vertical burn test for V-0, V-1, and V-2 ratings.

30 seconds. A V-1 classification is given to samples with maximum total combustion time less than 50 seconds and average burning time for five specimens less than 25 seconds. The individual specimen afterflame plus afterglow must not exceed 60 seconds. There should be no flaming drips. The sample is classified V-2 if it meets the flaming time criteria of V-1 or V-0 but has flaming drips that ignite the cotton.

5.12.6 Steiner Tunnel Test (ASTM E 84)

This Stein tunnel test (ASTM E 84) [41] is used mainly for building materials in the United States and is written in various building codes. The test apparatus is a tunnel measuring 8.7 × 0.45 × 0.31 m (25 × 1.5 × 1 ft) with viewing windows along its length. The sample to be tested is 7.6 m (24 ft) long and 0.51 m (1.67 ft) wide, mounted on the ceiling position. It is exposed from 10 minutes at one end of the tunnel to a 88-kW (5000 Btu/min) gas burner, with a forced draft through the tunnel, from the burner end, at an average initial air velocity of 1.2 m/s (240 ft/min). The flame spread index (FSI) is calculated on the basis of the area under the curve of the flame tip location as a function of time, compared to assigned values of 0 for an inert board and 100 for a standard red oak flooring. The smoke developed index (SDI) is calculated from the area under the curve of the light obscuration versus time as measured at the tunnel outlet, and is compared to the reference values of 0 for inert board and 100 for the red oak. Some codes require that the test be run on the minimum and maximum thickness and with construction similar to the manner in which the laminate will be used.

5.13 Ablation Properties

The testing of ablative materials involves simulating the extreme conditions they are exposed to in actual applications, while measuring various metrics. Additional measurements can be made after screening test is completed. The specific experimental inputs for ablation testing of a sample include extremely high heat flux, high temperature, and typically surface erosion from high-velocity fluid or particulate matter. Some common metrics for ablative performance measurement include mass loss, recession and/or recession rate, front-face temperature, back-face temperature, in-depth temperature, char thickness, and off-gas composition. The metrics hold different weights depending on the specific application of the ablative material. The material's resistance to heat is determined via measurements of the back-face, front-face, and heat soak (internal) temperatures during and after the testing procedure. These measurements show how well the material inhibits heat flow and can determine whether it fails to meet specific requirements regarding back-face temperature. For example, if the back-face temperature is too high, it may result in failure of the adhesive bonding of the ablative material to the supporting structure, which would result in a failure of the overall material. In addition, posttest measurements of the char thickness and material erosion can provide additional insight into the resistant capabilities of the material relative to exposure time. If the reaction progresses too quickly through the material or the protective char layer erodes too quickly due to surface shear, the material may not provide thermal protection for a long enough time period.

There are no commercially available ablation testing systems, but a number of independently designed and managed apparatus are available [42]. These can be roughly split into two classes of apparatus: large scale and small scale. Large-scale ablation testing is usually done in the final stages of development for ablative materials and it is intended to actually predict and rate how the material will perform in its real-world application. They are used to validate any tests or models employed in development and to determine whether materials qualify for their intended use. Because of this, large-scale testing setups are typically fine-tuned to very accurately recreate the conditions that thermal protection system (TPS) materials will experience in application and are very adjustable so that a variety of conditions can be simulated. As a result, large-scale ablation testing is very expensive and time consuming to perform, and there are a limited number of facilities that can reliably do it. The VKI Plasmatron developed and operated by the von Karman Institute for Fluid Dynamics in Belgium and the Arc Jet Complex at NASA's Ames Research Center are two examples of large-scale ablation testing facilities [43].

Ablation and thermal performance of the resulting nanostructured materials for propulsion and reentry applications are evaluated using established laboratory devices, such as simulated solid rocket motor (SSRM), subscale solid rocket motor (char motor), and oxyacetylene test bed (OTB). These three ablation testing devices are briefly discussed in this section. Besides these three laboratory devices, two other capabilities relating to ablation research are included in this section, namely the in situ ablation recession and thermal sensor and a device to measure the char strength of posttest PNC ablatives.

5.13.1 Simulated Solid Rocket Motor (SSRM)

The SSRM is an established testing approach developed by Koo et al. in the 1990s [41] that has been used extensively in the development of ablative materials [44–54]. It is a small-scale, supersonic, liquid-fueled rocket motor burning a mixture of kerosene and oxygen, as shown in Fig. 5.15. Aluminum oxide particles are injected into the plume to simulate the particle-laden flow of solid rocket exhaust plumes. The SSRM is a controlled laboratory device capable of producing a particle-laden exhaust environment with measured heat fluxes from 40 to 1,250 Btu/ft^2-s (454 to 14,200 kW/m^2). The flame

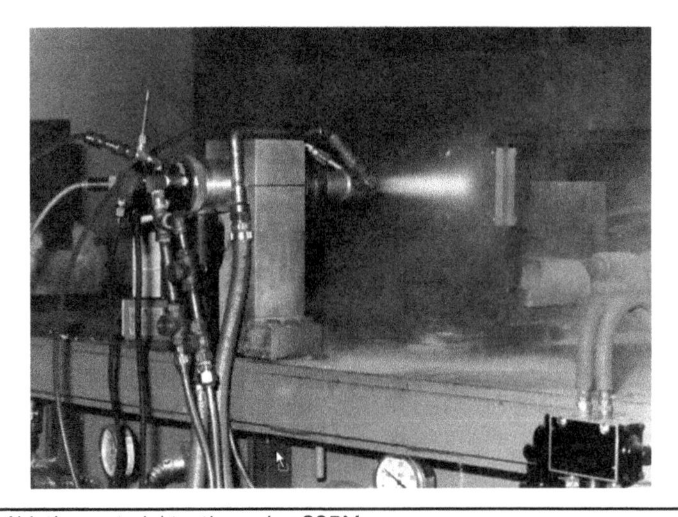

Figure 5.15 Ablative material testing using SSRM.

temperature is approximately 3,992°F (2,200°C), and the velocity of the particle-laden exhaust is approximately 2,000 m/s. Calibration of the SSRM was performed using a Medtherm water-cooled heat flux gauge. The highest heat flux is 1,250 Btu/ft^2-s (14,200 kW/m^2) at 2 in. (5.08 cm) from the nozzle exit, and the lowest heat flux is about 40 Btu/ft^2-s (454 kW/m^2) at 14 in. (35.56 cm) from the nozzle exit. The standard sample size was 4 by 4 by 1/4 or 1/2 in. thick (10.2 by 10.2 by 0.64 or 1.28 cm thick). The composite materials were bonded to 4 by 4 by 1/8 in. (10.2 by 10.2 by 0.32 cm) steel substrate. A narrow slot was machined into the bondline side of the steel substrate. The thermocouple (TC) was embedded into this slot. The bead of the TC was placed at the center point of the plate. The temperature history of the bondline was recorded with this TC. C-clamps were used to clamp the ablatives to the steel substrate during testing. The test samples were placed in a fixture downstream from the SSRM nozzle exit. The fixture had adjustments in axial distance and impingement angle with reference to the plume centerline. A normal 90° impingement was used for this study. Three axial distances were used to correspond with heat fluxes of 250, 625, and 1,000 Btu/ft^2-s (2,838, 7,094, and 11,345 kW/m^2). These distances were selected to simulate low, medium, and high heating conditions. Prior to the motor burn, the bondline temperature was recorded. Peak erosion was determined by pre- and posttest measurements using a pencil-point dial indicator. Mass loss was determined by pre- and posttest weight loss measurements. Consistent data sets were averaged for this study. Data on the following parameters were acquired during the experiment:

- Peak erosion (mm)
- Residual mass (%)
- Maximum backside heat-soaked temperature (°C)
- Time to reach maximum heat-soaked temperature (minutes)
- Surface temperature (°C)
- Microstructures of posttest specimens using SEM

Koo and colleagues used this device to study a family of nanocomposite rocket ablative materials (NRAMs). Three nanomaterials, MMT nanoclays, CNFs, and POSS, were incorporated into a phenolic resole resin (Hexion SC-1008) and were impregnated into carbon fibers to form carbon fiber–reinforced composites [10, 50–52]. NRAMs of Koo and colleagues were compared with Cytec Solvay Group's industry standard MX-4926 carbon/phenolic ablative. More detailed discussions are provided in Chap. 9 on ablation properties (see the section "Phenolic Nanocomposite Studies by The University of Texas at Austin, United States").

5.13.2 Subscale Solid Rocket Motor (Char Motor)

At Edwards Air Force Base, California, AFRL/PRSM has developed a very cost-effective char motor to screen solid rocket motor (SRM) insulation materials [55–57]. A schematic of the Pi-K SRM experimental setup is illustrated in Fig. 5.16. The motor is operated and fired in the upward direction. It usually contains up to 2 lbs (4.4 kg) of solid rocket propellant in an end-grain burning configuration. Both non-aluminized and aluminized propellants have been used for insulation material testing. The average chamber pressure is typically about 800 psia (5.5 MPa), and the test duration is about 8 seconds. For insulation material testing, a cylindrical cone made of glass/phenolic

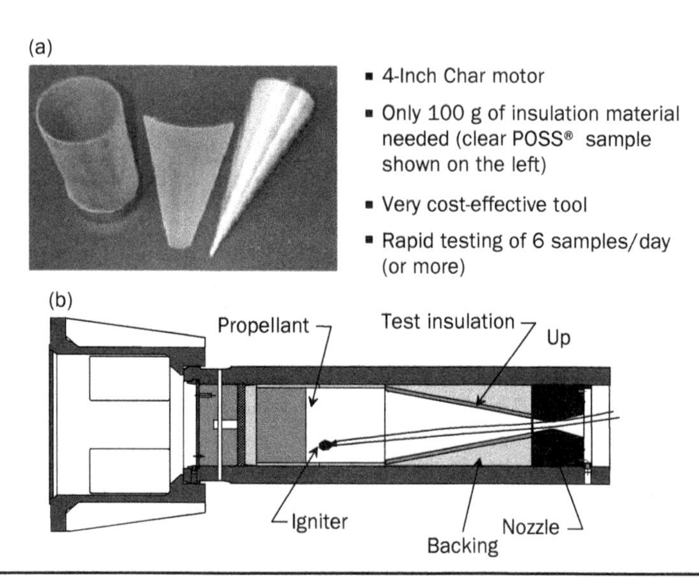

Figure 5.16 Schematic of the AFRL/Pi-K SRM experimental setup for ablation testing. Both non-aluminized and aluminized propellants can be used. Average chamber pressure is about 800 psia and test duration is about 8 seconds when an end-burning propellant grain was used. NRAMs will be tested at the nozzle section.

material of different area ratio is used inside the Char motor. This configuration is used to simulate a range of Mach numbers that the insulation materials are exposed during to the combustion of gaseous products. Rapid testing of six samples or more per day can be achieved. For nozzle material testing, the cylindrical cone is removed, and the whole nozzle assembly at the back end of Pi-K SRM is replaced by NRAM candidate materials. The whole assembly is a convergent/divergent nozzle, and the nozzle throat is a machined graphite nozzle insert, as shown in Fig. 5.17. The graphite nozzle insert piece prevents excess erosion in the nozzle throat area. This configuration enables the evaluation of candidate materials in the entrance region and exit cone of the nozzle of SRM. Four successful firings were conducted and erosion data at different area ratios were collected. Posttest specimens of selective regions of the nozzle assembly were sectioned and analyzed using SEM. More detailed study using this device is presented in Chap. 9 on ablation properties (see the section "Phenolic Nanocomposite Studies by The University of Texas at Austin, United States").

5.13.3 Oxyacetylene Test Bed (OTB)

Small-scale ablation testing is used in the early stages of ablative material development program to determine the relative worth of a potential ablative through comparison to well-performing ablatives. They recreate the ablative environment to a degree that the material will undergo ablation similar to what it would see in the real-world, but do not control or account for certain variables in order to allow for lower cost and easier testing. These inaccuracies are why small-scale testing is only used in early stages as screening tool, and large-scale testing must be used to validate these early test results. However, the primary advantage of small-scale testing is in cost and speed of testing. Two instances of small-scale ablative testing systems are the apparatus developed by

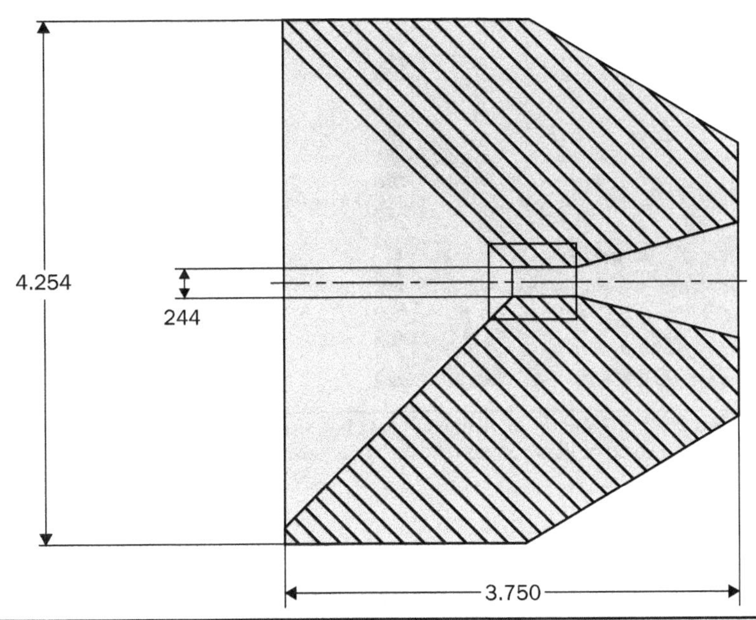

Figure 5.17 Cross-sectional diagram of the motor nozzle assembly showing the graphite insert at the nozzle throat of the NRAM material. This nozzle assembly fits into the back end of the Pi-K motor. The direction of the combustion gaseous products from the rocket chamber is moving from right to left. All dimensions are in inches.

M. Natali et al. [58] of the University of Perugia in Terni, Italy and by G. Pulci et al. [59] of the Sapienza University of Rome.

The ablation testing setup used in Koo Research Group described in this section is a small-scale setup developed and managed at The University of Texas at Austin [60, 61]. It is known as The University of Texas Oxyacetylene Test Bed (UT-OTB or OTB). M. Natali provided insight and advice aiding in the development of the OTB. The OTB is based on an oxyacetylene torch that is used to create the very high temperatures needed for ablation and which has its flame concentrated on a small area, such that very high heat flux values are also produced. The OTB is a small-scale apparatus, so high temperature and heat flux are the only two experimental conditions of an ablative environment intended to be accurately simulated. Acceptable values for these parameters were determined to be a temperature of at least 2,200°C and a heat flux in excess of 1,400 W/cm^2. Small sample size is also desirable as it reduces the cost of each test and enables a greater number of tests to be performed.

Once the OTB was completed, initial testing was done to confirm that the experimental conditions met the required parameters. A literature search of the stoichiometric combustion temperature of acetylene with oxygen confirms that our oxyacetylene torch will produce a temperature in excess of 3,000°C, much greater than the 2,200°C that we required. We were also able to design our setup such that the area of the sample exposed to the ablation conditions is roughly 10 mm in diameter.

The OTB consists of a lathe chuck mounted on two perpendicular rail systems, as shown in Fig. 5.18a [61]. The chuck can hold test samples of different dimensions and geometries. The first rail allows the sample to be exposed to the front of the torch tip,

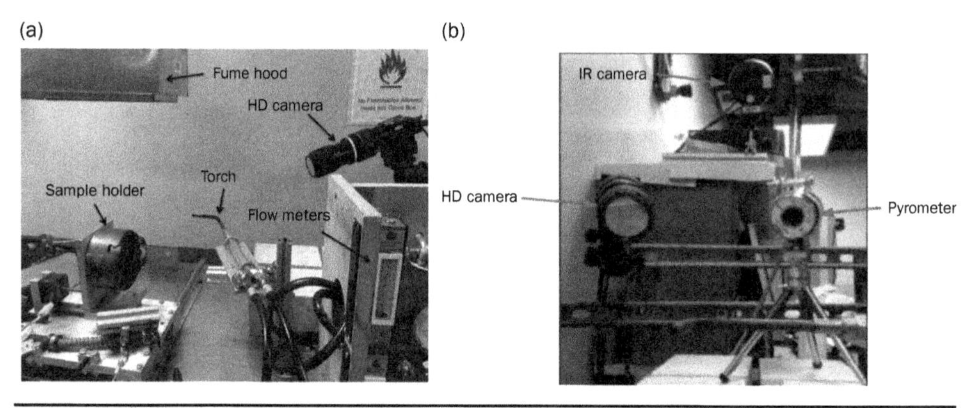

FIGURE 5.18 (a) UT Austin oxyacetylene test bed experimental setup and (b) advanced diagnostics (two-color IR pyrometer, IR video camera, and HD video camera).

while the second rail controls the distance from the torch tip. The torch is rigidly mounted to the table and the flow rate of each gas is controlled by a flow meter. While a test is being performed, it is monitored by a two-color IR pyrometer, an IR video camera, and an HD video camera (Fig. 5.18b). The two-color IR pyrometer measures the surface temperature of a single point, in this case the center of the sample. The IR video camera also measures surface temperature, but of the entire sample. The HD video camera is used to view the evolution of surface features, and provides an optical ablation rate.

The heat flux experienced by the sample is a function of standoff distance to the torch tip, fuel ratio, and the flow rates of each of the gasses. In order to ensure the correct heat flux is being used, a slug calorimeter in which a copper slug of known mass is placed in the sample holder and exposed to the oxyacetylene torch flame [62] or a Gardon heat flux transducer (manufactured by either Medtherm Corp., Huntsville, AL or Vatell Corp., Christiansburg, VA) is placed in the sample holder and exposed to the oxyacetylene torch flame. The Gardon gauge is a water-cooled copper cylinder with a TC embedded in the center, surrounded by a sleeve of constantan. A high-emissivity paint covers the center to absorb the incident radiation. The TC signal is sent to a data acquisition system (DAQ) where it is then translated into a heat flux by a LabVIEW program. After adjusting the various parameters, the desired heat flux is attained, and each sample can be tested under the same conditions. Depending on the standoff distance from the sample to the torch, the OTB is capable of producing heat flux values in excess of 1,400 W/cm^2 (Fig. 5.19).

Lee and colleagues used the OTB to conduct high heat flux ablation experiments [63–66]. The thermoplastic polyurethane elastomer nanocomposites (TPUNs) containing different loading MMT nanoclays, CNFs, and MWNTs were processed via twin-screw extrusion test specimens fabricated by injection-molding. These TPUNs were exposed to the hottest portion of the flame, with a 7-mm flame diameter, for 25 seconds. For the materials tested in the above study, specific settings within the OTB's range of operation were selected and all the material formulations were tested at these conditions. The nanocomposite formulations were cut into 1.27 cm × 1.27 cm × 5 cm (0.5 in. × 0.5 in. × 2 in.) samples for testing with three samples tested for each formulation to increase the statistical confidence in the results. The square face of the samples was exposed to the

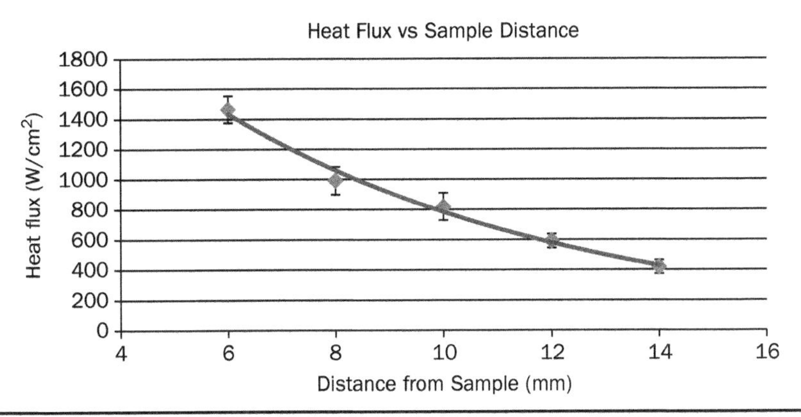

Figure 5.19 Calibration of heat flux for OTB.

oxyacetylene torch at a calibrated heat flux of approximately 1,000 W/cm^2 for a duration of 25 seconds. The OTB facilitates three K-type TC measurements during testing, and for our study these TCs were placed within the samples at distances of 5, 10, and 15 mm from the exposed surface with a heat shield in place to protect the protruding wires. In addition to these in situ temperature curves, the mass loss, recession, and char thickness of each sample were determined based on pre- and posttest measurements. For more detailed discussions, refer to Chap. 9 on ablation properties (see the section "TPUN Studies by The University of Texas at Austin, United States").

5.13.4 Char Strength Sensor

In order to maximize the payload of SRMs, ethylene-propylene-diene-monomer (EPDM) rubber has shown to be an ideal candidate matrix for internal SRM insulation and it practically replaced all other elastomers, such as nitrile and silicone as insulators in advanced SRMs [67–71]. Among fibrous reinforcements, aramid fibers (Kevlar®) in the form of fibers or pulp have been incorporated in the EPDM rubber because of their high thermal capacity, chemical stability, fire resistance, low thermal conductivity, high ablation resistance, and high physical adhesion [67, 68].

The evaluation of the mechanical properties of the charred layer is very important, since it can enable the possibility to better tune the TPUNs, as discussed by Koo and colleagues [72–75] and as discussed in Chap. 9 of this book on ablation properties. This char strength property also will produce a more comprehensive screening of the TPUN materials. However, due to the fragile nature of the char, ordinary testing method, such as Rockwell hardness is not feasible, thus requiring the development of a new testing protocol and sensor.

In 2009, a team from The University of Texas at Austin contributed toward the final success of this new char strength sensor. The team, Redondo et al. [76], provided background and evaluated available testing methods. The preferred testing method selected was to crush the char samples, thus a prototype was created. However, the data obtained from the prototype proved unreliable due to poor structural tolerances, mechanical vibration, and signal noise. As a result of the prototype not being able to complete the task, it was used as a *proof of concept* allowing for complete redesign. Lastly, this research utilizes information from the Reshetnikov et al. study which details another compression testing method used to test the strength of intumescent char. These Russian scientists cited various ways to test char material, such as centrifugal shear, compression, and fluid flow shear [77].

Air Force Research Laboratory (AFRL)/Edwards Air Force Base, California currently uses a method to test char strength. However, the test is entirely qualitative and subjective. The tester brushes a charred sample with a wire brush and determines the relative strengths of different samples visually by the remaining thickness [78]. The lack of a repeatable, quantitative test prompted the need for a scientific testing method, since the brush test contains significant human error in the applied force. Two sensors were developed: a compression sensor and a shear senor, and are discussed briefly in this section.

Compression Sensor

The schematic of the compression char strength tester is shown in Fig. 5.20 [79–81]. It consists of three main components: the sensor, the power supply, and the stand to hold the sensor. Figure 5.21 shows the compression char strength tester and a technical drawing of it [81]. In order to generate a correlation for the LabVIEW program to be used, the potentiometer and force sensor are calibrated. The potentiometer is calibrated to give a correlation of the actuator's probe tip position as a function of voltage across the potentiometer. The force sensor inside the probe tip is calibrated to determine the force output as a function of voltage. The potentiometer calibration data are obtained using calipers to measure the distance traveled by the actuator and using a multimeter to measure the voltage drop across its potentiometer. The force sensor is calibrated using an apparatus made up of four springs sandwiched by two metal plates. As the probe presses down on the top metal plate, the four springs are compressed, and the force sensor inside the probe tip sends a voltage signal to the control program in LabVIEW. Using the potentiometer in the linear actuator, the distance that the top plate was pressed down is known. Using the distance that the springs are compressed and the stiffness of the springs, the force can then be calculated. The final step is then to generate a correlation of force as a function of voltage. Sample data of the calibration are shown in Fig. 5.22 [81].

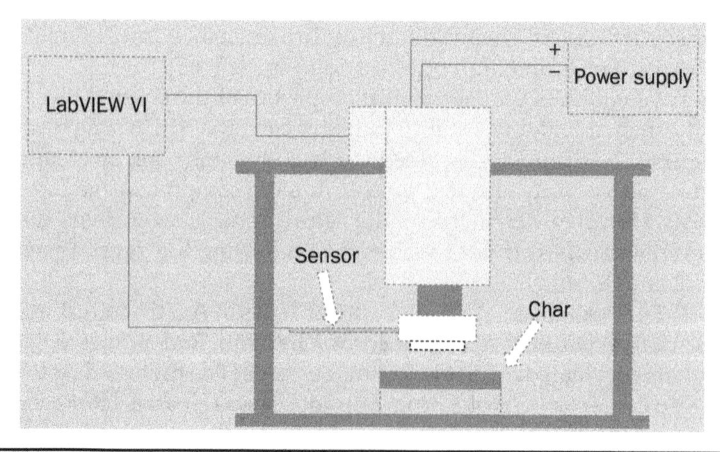

FIGURE 5.20 Char compressor schematic diagram showing the power supply, LabVIEW VI, sensor, and the char specimen.

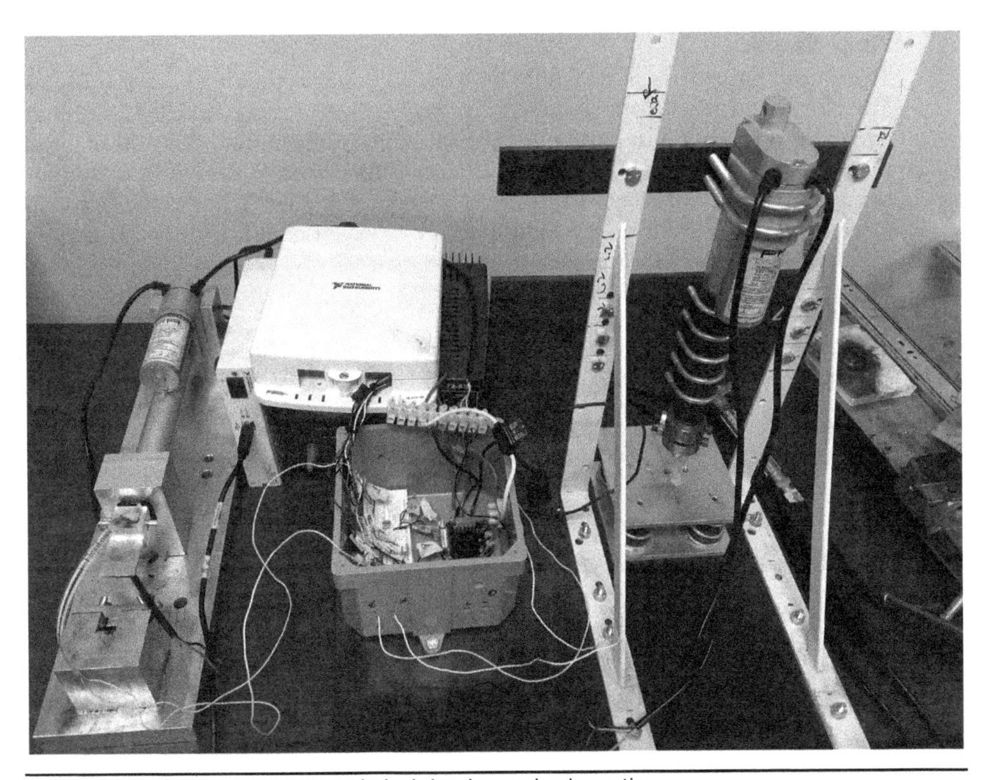

FIGURE 5.21 Char compressor technical drawing and schematics.

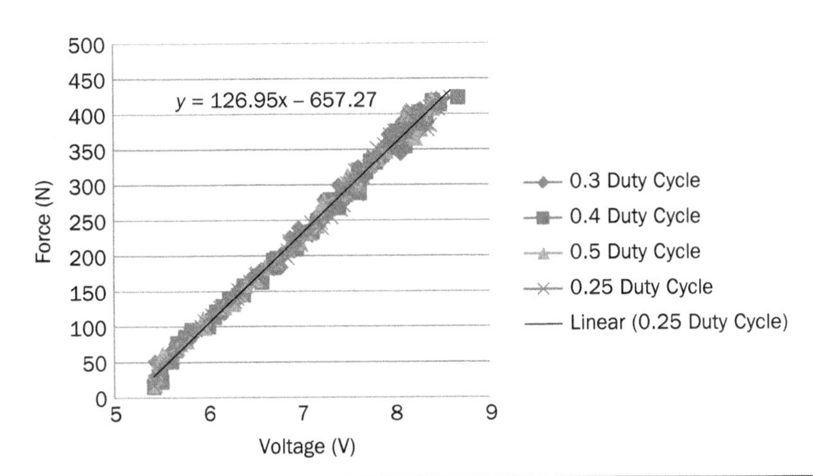

FIGURE 5.22 Sensor force versus voltage using several duty cycles with correlation equation.

Shear Sensor

The shear sensor uses a horizontally oriented linear actuator to shear a char sample. The shear sensor uses the same control box, power supply, and DAQ system as the compression sensor, but a different LabVIEW control program. The shear sensor schematic is shown in Fig. 5.23. Figure 5.24 shows the shear sensor without all the control components (control box, DAQ, power supply, etc.) [80–85]. The shear sensor uses the same type and brand load cell as the compression sensor. It is calibrated using the same method as described for the compression sensor.

Analysis Method

Since the criteria for the strength of the char material could consist of a variety of factors, the sensor gathers force and distance (depth) data as the probe compresses or shears off

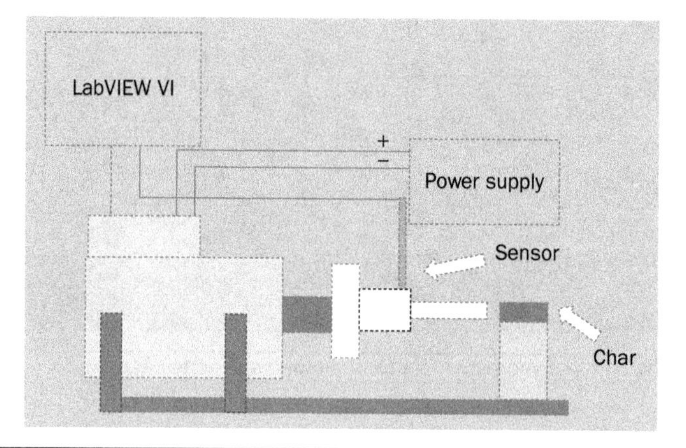

Figure 5.23 Char shear schematic diagram showing the power supply, LabVIEW VI, sensor, and the char specimen.

Figure 5.24 Photograph of shear strength sensor.

the charred material. Using these data, two methods were developed to determine the strength of the char.

In the first method, the force data can be used to observe the yield strength of the char. As the piezoresistive load cell implemented in the design changes resistance in response to pressure, the applied force is recorded. A peak in the force followed by a significant decrease in applied force on the char will indicate that the ultimate strength has been reached, revealing the maximum allowable stress for the material. Using this method, one can rank the strength of several samples, the strongest being the sample that has the largest ultimate stress.

The second method to determine the strength of the char material is to generate a plot of force versus distance (depth). If one integrates the force versus distance plot to find the area under the curve (F*D = W), one can determine the energy required to destroy the char. The strongest char will produce the highest energy of destruction or energy dissipated.

Compression Char Strength Results The average energy dissipated results depicted in Fig. 5.25 show that TPUN-10 wt% MWNT ranks higher with a value of 703 ± 335 N-mm, followed by Kevlar-filled EPDM with a value of 466 ± 278 N-mm, and lastly TPUN-7.5 wt% MWNT ranks lowest with a value of 396 ± 254 N-mm [81]. The average maximum force results of Fig. 5.26 show that TPUN-10 wt% MWNT ranks higher with a value of 170 ± 53 N, followed by Kevlar-filled EPDM with a value of 136 ± 62 N, and lastly TPUN-7.5 wt% MWNT again ranks lowest with a value of 112 ± 66 N [81]. As it can be seen from Figs. 5.25 and 5.26, TPUN-10 wt% MWNT has the potential to replace Kevlar-filled EPDM, since it is shown to rank higher in average energy dissipated and average maximum force than Kevlar-filled EPDM if no other material properties are considered [81].

Shear Char Strength Results Although it was determined from the compressive strength tests conducted in this study that MWNT was the strongest TPUN, Figs. 5.27 and 5.28 show that MMT is not that far behind in strength. To determine which of the two materials is stronger, and therefore warrant further testing to determine if it may replace Kevlar-filled EPDM, shear strength testing was conducted on Kevlar-filled EPDM, TPUN-7.5 wt% MMT, and TPUN-7.5 wt% MWNT [84]. The final average energy dissipation results are shown in Fig. 5.29 [84]. TPUN-7.5 wt% MWNT ranks higher with

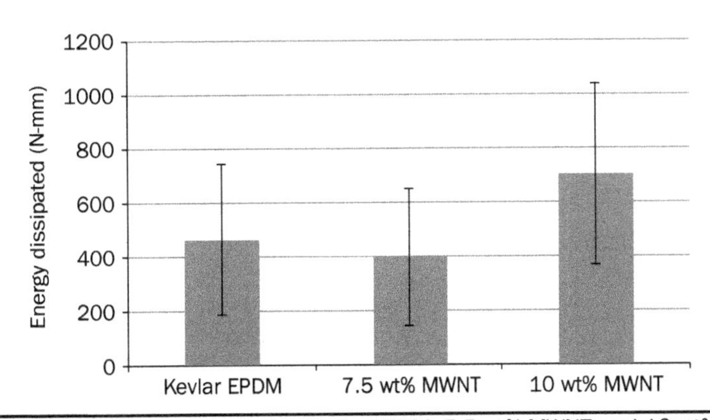

Figure 5.25 Average energy dissipated of Kevlar EPDM, 7.5 wt% MWNT, and 10 wt% MWNT.

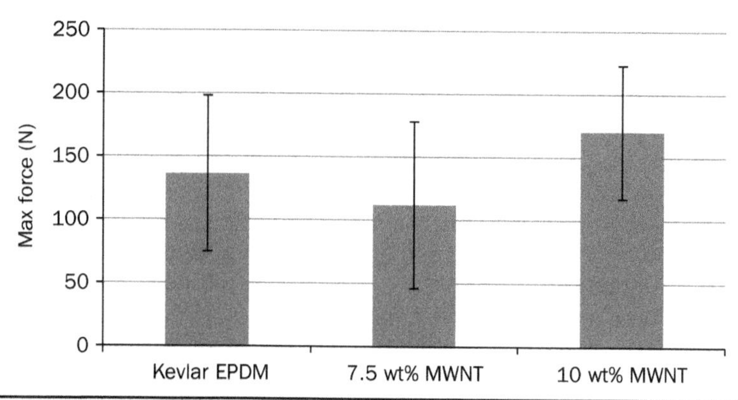

Figure 5.26 Average maximum force of Kevlar EPDM, 7.5 wt% MWNT, and 10 wt% MWNT.

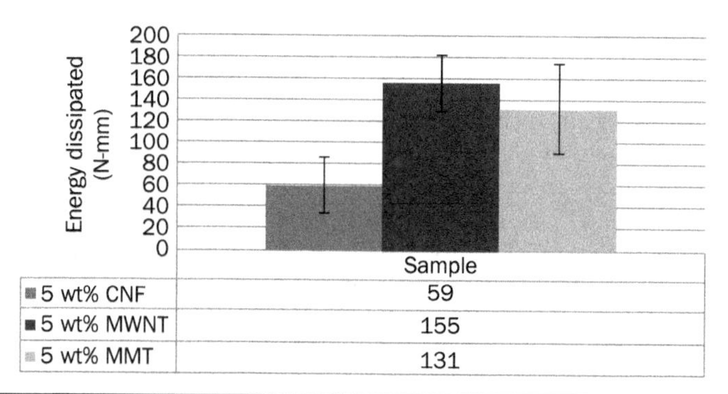

Figure 5.27 Average energy dissipated of each type of thermoplastic elastomer nanocomposites.

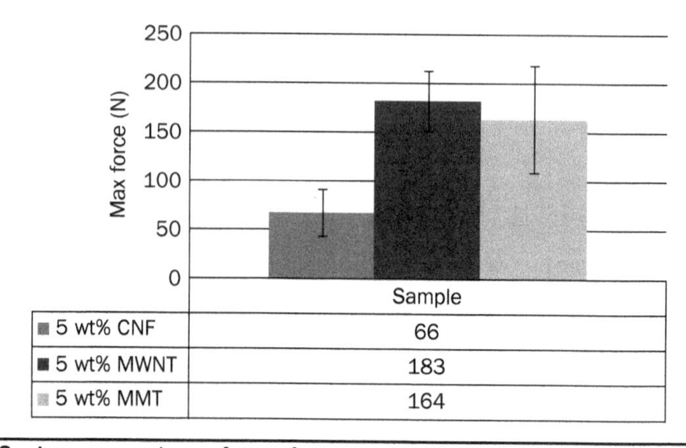

Figure 5.28 Average maximum force of each type of thermoplastic elastomer nanocomposites.

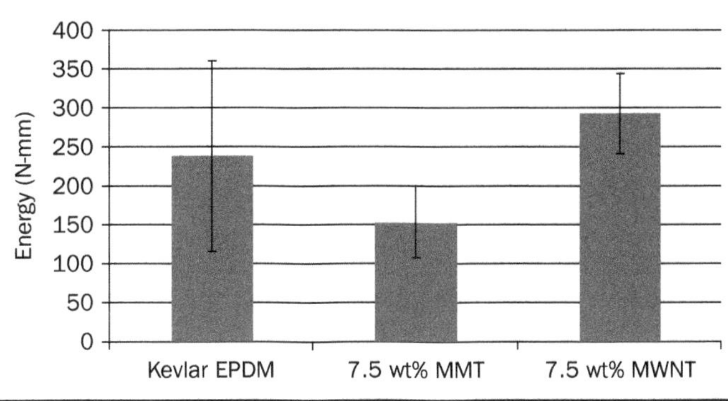

FIGURE 5.29 Average energy dissipated in shear of Kevlar-filled EPDM, 7.5 wt% MMT, and 7.5 wt% MWNT.

a value of 292 ± 51 N-mm, followed by Kevlar-filled EPDM with a value of 239 ± 122 N-mm, and lastly TPUN-7.5 wt% MMT ranks lowest with a value of 154 ± 45 N-mm.

Average maximum force analysis was not conducted on the shear tests because the force versus distance plots on shear tests contain several peaks that could potentially be the yield strength. For detailed description of the compression and shear sensors and additional results refer to [81, 84, 85].

5.13.5 In Situ Ablation Recession and Thermal Sensors

Two types of in situ ablation recession and thermal sensors were developed by Koo and colleagues [86–100] based on break-wire concept using ultra-miniature TCs: one sensor was based on 0.25-mm-diameter TCs [86–89, 100] and another sensor was based on 0.55-mm-diameter TCs [89–100]. These sensors were demonstrated using low-density carbon/carbon (LDCC) and high-density carbon/carbon (HDCC) composites and carbon/phenolic composite. Sensors with different configurations—four-level (four TCs) sensors, eight-level (eight TCs) sensors, and nine-level (nine TCs) sensors—were designed, fabricated, and tested under high heat flux (1,000 W/cm^2) via the OTB. Representative results of this novel ablation and thermal sensing technique are discussed in this section.

The 0.25-mm-Diameter TC Ablation Sensor

A break-wire–like ablation recession sensor based on commercial, ultrafine TCs (250 μm diameter) was designed and tested [86–89, 100]. A series of TCs are embedded in the TPS, perpendicularly to the ablator surface (Fig. 5.30). Each sensing head of the TC is positioned at a well-established depth from the surface. During the heating of the ablator, the TCs would first work as a temperature sensor providing invaluable information about the state of the TPS. When the temperature rises above the melting point of the metal sheath and of the Seebeck junction, the TC would break. Due to this dual nature of the sensing heads—as a traditional Seebeck junction and as a position marker—it would be possible to obtain a wide range of data on the recession state of the TPS. This ablation recession sensor was tested on carbon/carbon composite (CCC) under a severe hyperthermal environment produced by an OTB. Two types of CCC materials with different densities (LDCC = 1.34 g/cc and HDCC = 1.70 g/cc) were considered. At the

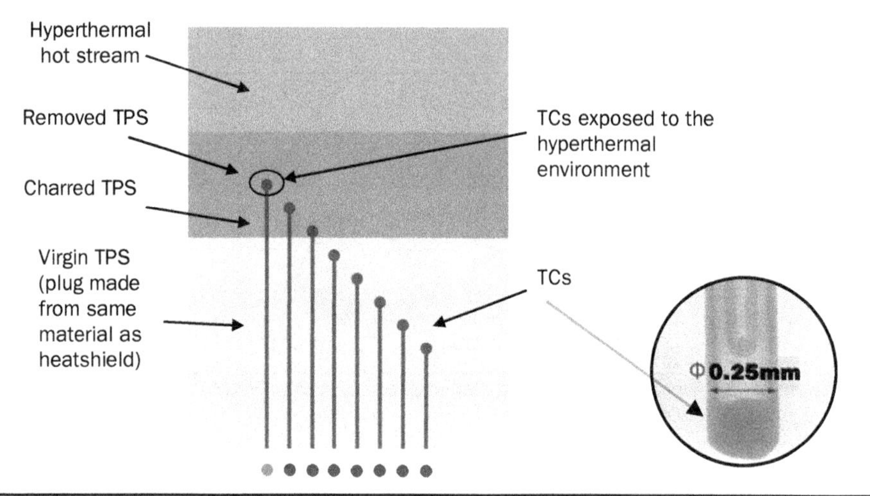

Figure 5.30 Scheme of the proposed thermocouple-based recession sensor.

same time, two different configurations were tested (four levels and eight levels). The obtained results were quite encouraging, showing the TC-based approach can provide accurate data on the recession rate of the TPS material. Due to the uniqueness of the proposed sensing technique, it can be applied both on TPS of spacecraft as well as on rocket nozzles. Moreover, the proposed approach uses low-cost, commercially available TCs, and industry-standard processing techniques.

Materials The sensing technique was tested on 15 × 15 × 15 mm carbon/carbon composite cubes (CCCs). Two types of carbon/carbon materials were considered: a low-density material (LDCC) and a high-density material (HDCC). These cubes were machined with a waterjet cutter starting from a slab of carbon/carbon composite. Miniature K-type TCs having a stainless steel sheath and with an outer diameter of 0.25 mm were chosen.

Production of the Sensor Plug First of all, it was necessary to identity a suitable technology to drill the CCC plug (a cube) to produce a series of blind holes at an increasing depth from one of the faces of the cube. Using blind holes and knowing the depth at which the TC must be inserted, a defined hole arrangement can be produced (Fig. 5.31). The drilled samples were produced using two techniques: a laser source and electrical discharge machining (EDM).

In the first part of the research, in order to keep down the costs, only four levels of depth were considered. This means that only four TCs were embedded in the CCC plug. Figure 5.31 shows a detailed technical drawing of the produced plug. Compared to the laser drilling, the EDM produced holes with a superior quality. Figure 5.32 shows an SEM image of a CCC drilled by EDM. The figure clearly shows the high quality of the produced holes. After receiving the sample from the drilling company, the actual depth of the produced holes was checked by means of a caliper coupled with gage pins. On the basis of the experimental results it was concluded that the produced holes had the depths which are reported in Table 5.2.

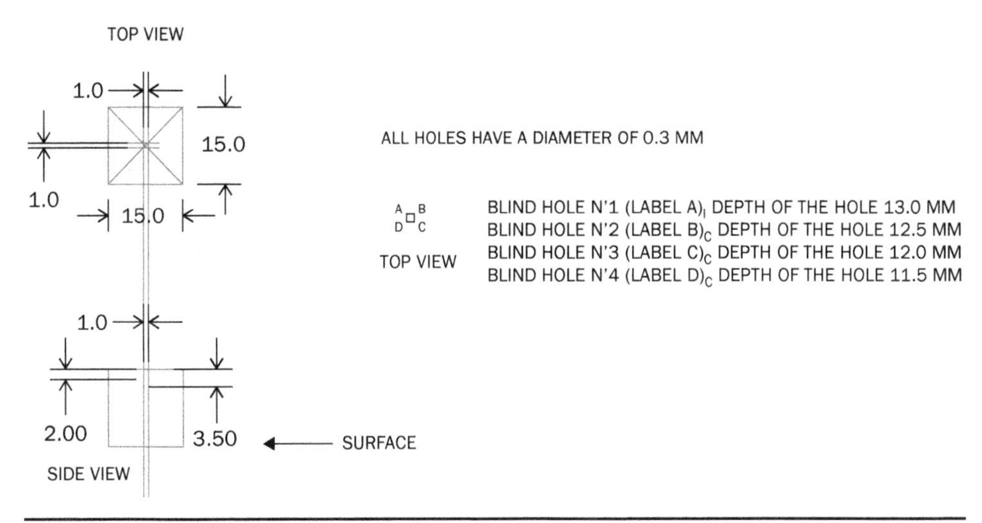

FIGURE 5.31 Drilling layout of the CCC plug.

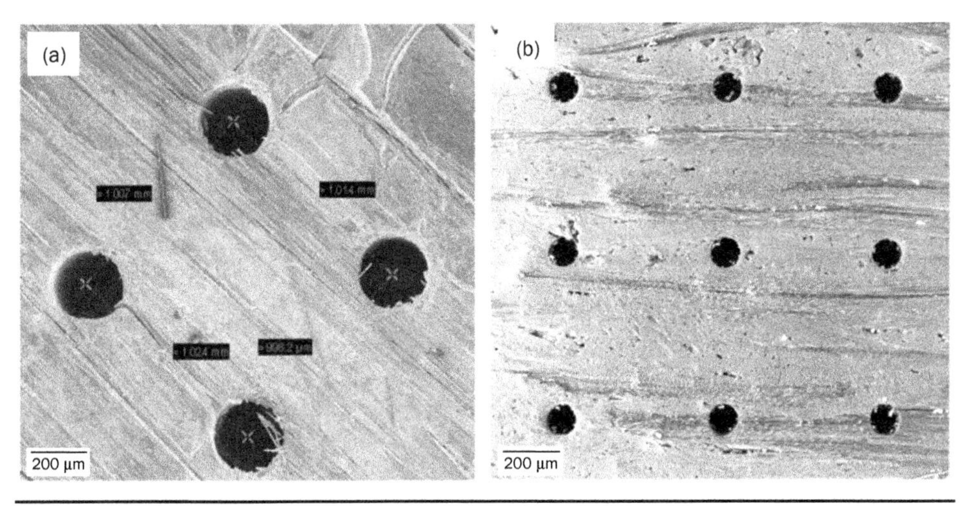

FIGURE 5.32 SEM of EDM drilled holes of the (a) four-hole configuration and (b) nine-hole configuration of the carbon/carbon composites.

TABLE 5.2 Theoretical and Real Depths of the Produced Blind Holes

Hole tag	Theoretical depth of hole (mm)	Actual depth (mm)
1	13.00	13.15 ± 0.05
2	12.50	12.55 ± 0.05
3	12.00	12.15 ± 0.05
4	11.50	12.20 ± 0.05

Table 5.2 shows the theoretical and actual holes 1 to 3 matched; the actual depth of hole number 4 did not match the required depth. Since most of the holes in the sample were drilled with the right depth, and due to economic considerations, the sample was used: the CCC was considered adequate to test the sensing technique. Once the holes were drilled in the cubic sample, the TCs were glued into CCC plug using a high-temperature adhesive suitable for carbon/carbon composites [86, 87].

A special sample holder designed to ensure a correct positioning of the TCs during the curing of the adhesive was designed and produced. Figure 5.33a shows the fully assembled CCC sensor. Also, the details of the cold face of the plug are shown in Fig. 5.33b.

Sensor Calibration Results Prior to testing the recession-sensing technique with the OTB, a nondestructive test was carried out in order to obtain some preliminary data on the response of the CCC plug. For this test, the hot face of the sample was heated with a low-power burner. Figure 5.34 represents a typical temperature profile obtained with

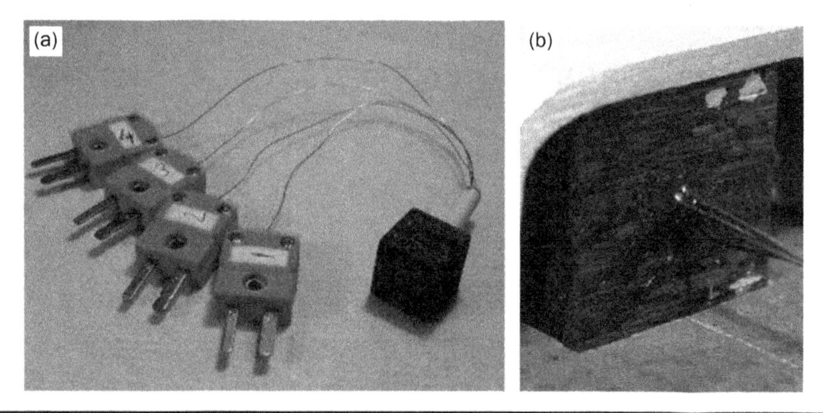

Figure 5.33 Fully assembled CCC sensor plug (a) and details of the cold face of the CCC plug (b).

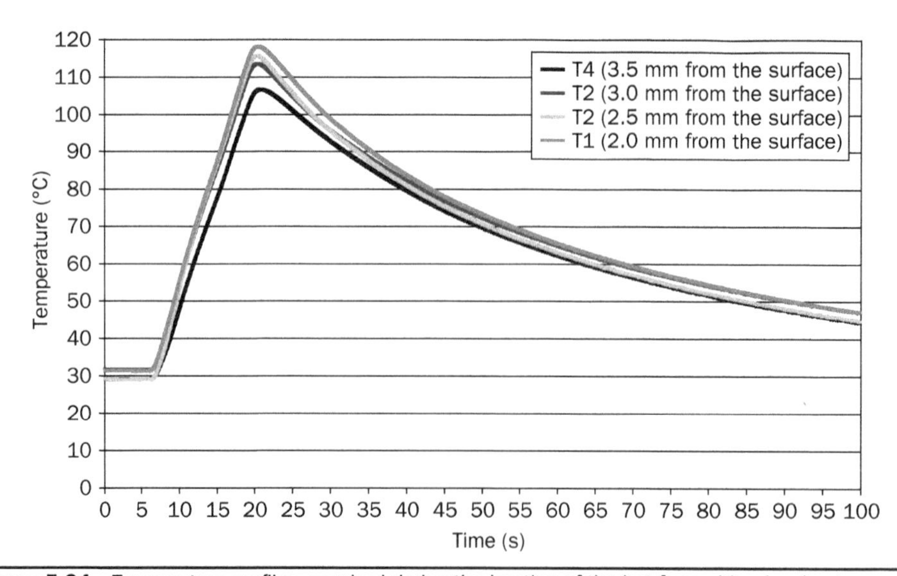

Figure 5.34 Temperature profiles acquired during the heating of the hot face with a low heat source.

this heat source. According to the experimental results, the TC embedded in the deepest hole (closer to the exposed surface) exhibited the higher temperature. In general, during the heating up of the sample, the hierarchy of the displayed temperature profiles exhibited a full dependence on the hole depth: the greater the depth of the hole, the higher the temperature measured.

Ablation Test Results Figure 5.35a shows the test of the CCC plug and Fig. 5.35b shows the post-burning appearance of the sensor. The temperature profiles acquired during the experiment are shown in Fig. 5.36. All temperature patterns displayed a behavior

FIGURE 5.35 (a) CCC tested on the OTB. (b) Burnt CCC immediately after the test.

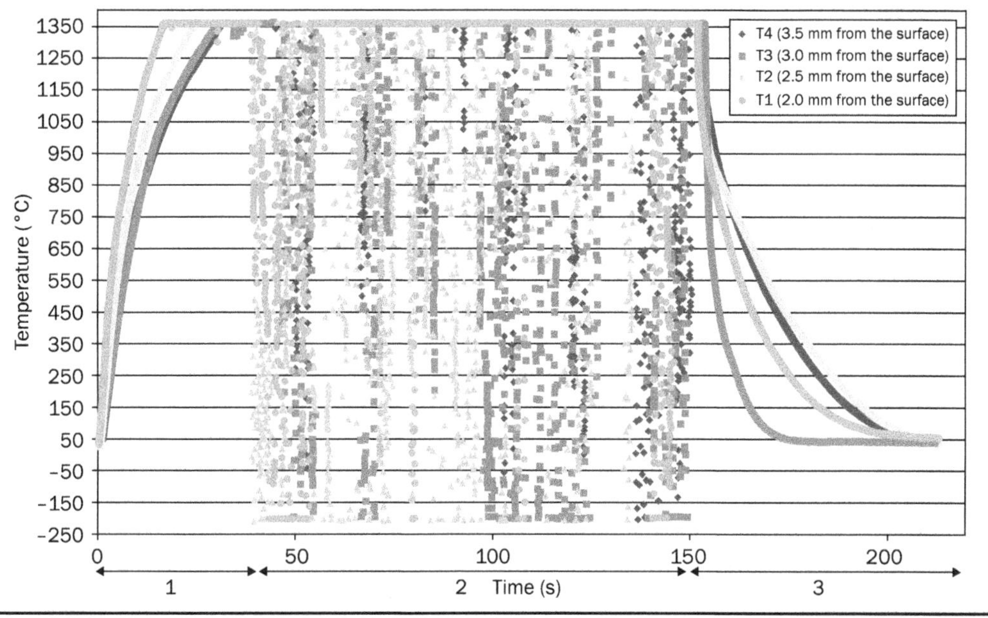

FIGURE 5.36 Temperature profiles of typical four-level ablation sensor.

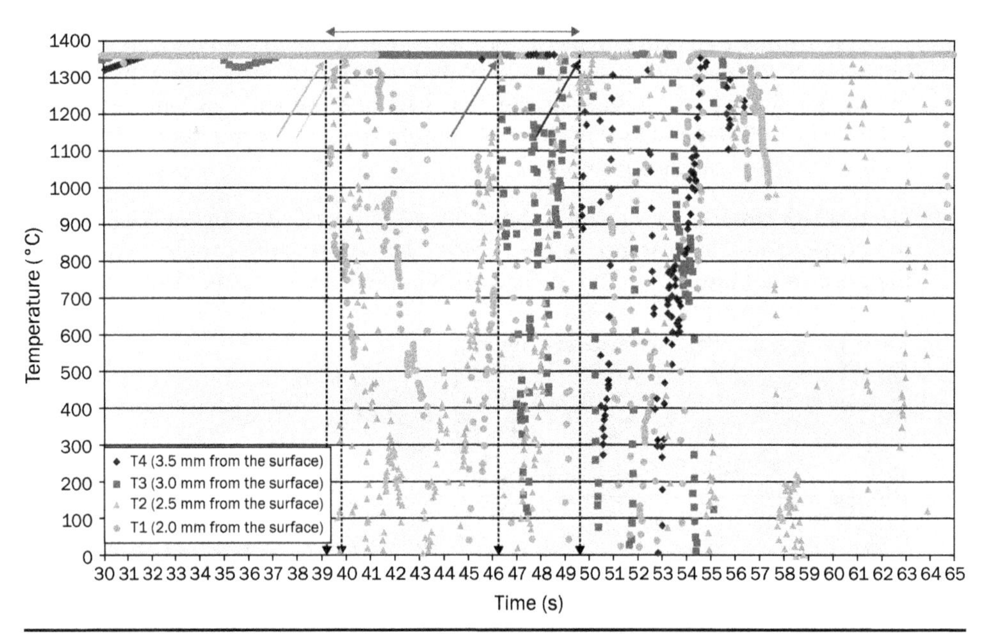

Figure 5.37 Detail of the temperature profiles in the range of time in which all TCs experienced the first break.

characterized by three common steps. In this first part (Fig. 5.37), all TCs properly worked as a temperature sensor. Similar to results obtained during the preliminary study, the smaller the distance of the TC from the surface, the higher the temperature displayed. The data in this first step of the temperature profiles showed that the signals of TC3 and TC4 were very similar: this evidence could confirm that the depth of these two holes was very similar as evidenced by the measure of the hole depths reported in Table 5.2. Unfortunately, once a temperature of 1,360°C was reached by each TC, the DAQ cut the data, which did not allow continued recording of the signal from the TCs. After about 40 seconds, the Seebeck junction SJ of the first TC experienced a break: after this period of time, the steel sheath of the TC melted and then the junction directly exposed to the flame, lost its integrity, causing a loss of electric contact between the two terminals of the TCs. Consequently, this point marked the exact time at which TC1 was broken.

The strong discontinuity in the temperature profile clearly marked this event. Figure 5.37 shows the detail of the temperature profiles in the range of time in which all TCs experienced the first break. Starting from the first break of the TC1, the Seebeck junction started to work improperly, providing an intermittent signal. This phenomenon can be explained as follows: due to the presence of the flame, after the TC experienced the first loss of electrical contact, the terminals of the TC continued to melt. These droplets of melt metal were able to close the electrical contacts between terminals, producing a detectable signal. Also, the steel sheath of the TC provided a source of liquid metal able to close the electrical loop. This process was regulated by several uncontrollable factors producing a randomly distributed opening and closing of the junction.

The above-described mechanism occurred for all four TCs but, due to their different displacement (depth) in the CCC plug, the first break of each TC happened at different times. The working principle of this recession sensor is exactly based on this behavior. The Seebeck junction of the first TC exhibited the loss of integrity after about 39.3 seconds from the beginning of the experiment. TC2 broke less than one second later than TC1. TC3 and TC4 broke at about 46.5 seconds and 50 seconds, respectively, from the beginning of the test. A possible reason that the first two TCs broke at nearly the same time could be that the Seebeck junctions embedded in the metal sheath had a different position with respect to its own outer case. In this case, even if the two TCs were placed at exactly the desired depth into the plug, the actual position of the junctions with respect to the surface of the CCC may not be correct. If this hypothesis is correct, the two TCs would break at nearly the same time. In any case, since commercial TCs were used, there was no way to control the positioning of the Seebeck junction in the sheath of the TC: in fact, this parameter strictly depends on the manufacturing process of the TCs.

However, this situation does not represent a problem, since the proposed recession-sensing technique is based on the use of many TCs placed at different depths. Moreover, placing different sensing plugs in TPS could be possible to minimize the problem related to the manufacturing tolerance of the TCs. Another possible reason that the first two TCs broke at nearly the same time could be related to local nonuniformity of the erosion. In any case, even considering the response of the first two TCs that broke very close one another, the system was shown to work properly. In fact, the TCs generally broke accordingly to their distance from the surface marking the recession layer of the material. The first break of each TC precisely indicated the recession.

In order to further understand the real behavior of the employed TCs embedded in the CCCs when exposed to a severe hyperthermal environment, the exposure time was extended up to 150 seconds. According to Fig. 5.37, it is possible to see that once the TCs experienced the first break, all of them continued to display the previously mentioned intermitting behavior (second step of the graph in Fig. 5.37). Finally, in the third step of Fig. 5.37, the TCs display the cooling down of the system. However, once broken and open, each original Seebeck junction was altered by the presence of some unknown material between the two terminals of the TC [original composition of the K-TC: chromel (90% nickel and 10% chromium) and alumel (95% nickel, 2% manganese, 2% aluminum, and 1% silicon)], thus leading to a practically meaningless reading of the temperature.

In order to get a quantitative evaluation of the real erosion rate provided by the sensor, the actual position of the spatial markers—the TC heads—was plotted as a function of time in which it experienced the first break. This calculation was carried out in two ways. In the first method, the four positions of each TC were considered and correlated with the time of the first break (Fig. 5.38). Using this method, the sensor provided a measured recession rate of 0.07128 ± 0.00548 mm/s. Comparing this data with the value measured on the neat CCC samples, it was possible to establish that the sensor provided a very good indication of the actual recession rate of the material. In fact, the erosion rate estimated on the neat CCCs was 0.056 ± 0.007 mm/s In the second approach, the recession rate (RR) was calculated considering the (0,0) point—in other words, as if there was a virtual TC placed on the surface of the plug experiencing a breaking at the time 0 second. In this case, the plot indicates that the recession rate was about 0.05751 ± 0.00124 mm/s (Fig. 5.39) with an impressive agreement with the value

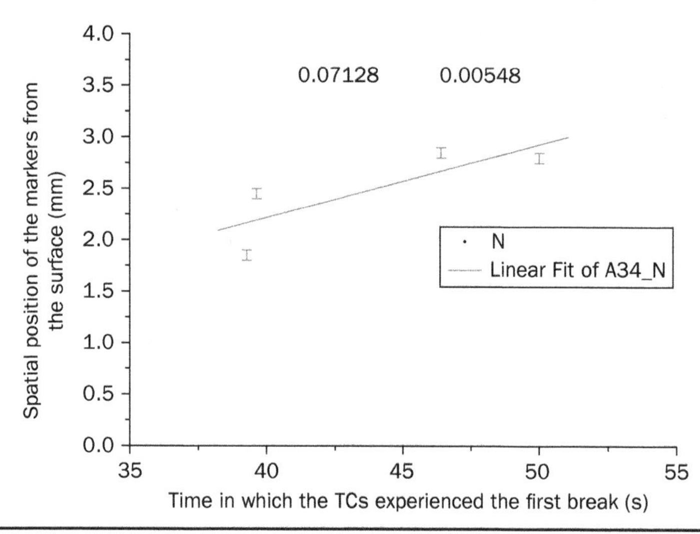

Figure 5.38 Position of the markers versus time of the first break of each marker.

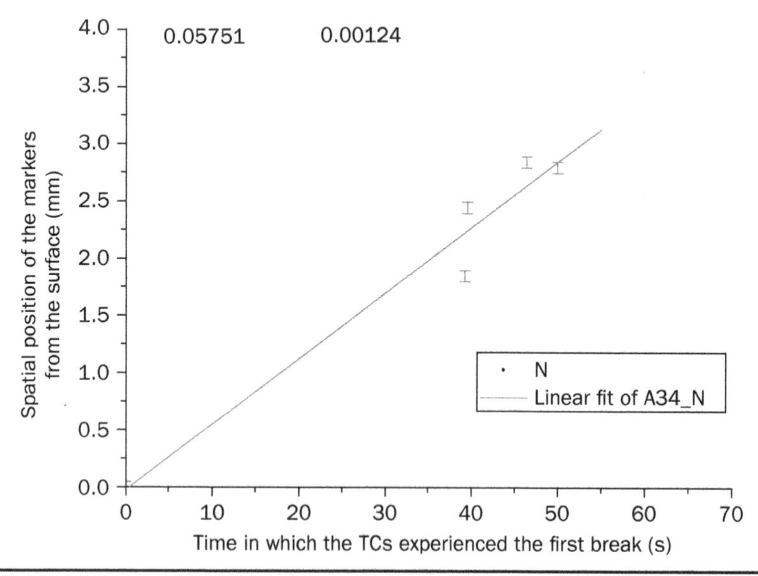

Figure 5.39 Position of the markers versus time of the first break of each marker including (0,0) point.

found for the neat CCCs. In any case, even considering the limited number of the available experimental points, both approaches showed that the sensor was able to provide a good evaluation of the erosion rate.

After the ablation testing of the sample was conducted, visual inspection of the tested CCC clearly showed the presence of the TC heads on the bottom of the crater produced by the torch (Fig. 5.40). SEM images were taken of the deepest part of the erosion

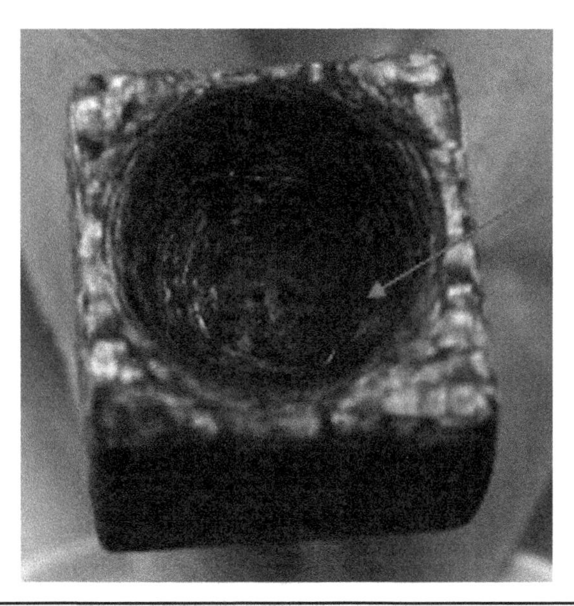

Figure 5.40 Post-burning image of the CCC plug: the arrow indicates the heads of the four TCs.

crater on the CCC sample (Fig. 5.41). The first picture (top) clearly shows that each TC head was covered by a layer of material probably made of the steel of the sheath. However, by increasing the magnification (bottom), it was possible to see that the cover of each TC was composed of some metal blend in which some chopped carbon fibers were also embedded. Such a layer of electrically conductive material should be responsible for the electrical contact between the two TC terminals.

After all these tests, the adhesion of the burnt TCs on the CCC was also qualitatively evaluated. This test was carried out considering the pull out resistance of the TCs. Preliminary results indicated that the TCs exhibited an appreciable pull out resistance. This indication is very important since, during the atmospheric reentry, the role played by the TPS cannot be compromised by the presence of the TCs. In any case, the pull down resistance of the burnt TCs can be increased with some dedicated designing criteria and a proper holder of the TCs on the cold side of the heat shield. Additional test results of this sensor using different types of ablatives can be found in [89, 91–96] and in Sec. 9.4.

Summary and Conclusion The proposed in situ ablation sensing system that can monitor the ablation recession rate of a TPS employs a simple working principle, combined with an inexpensive and commercially available raw materials, and processing techniques. The in situ ablation sensor provides a real-time monitoring of the ablation rate of an ablative material and even though the method here considered is an intrusive system, the use of ultrafine TCs (250 μm) requires a limited modification of the TPS. To date, due to the selected drilling technique (EDM) used to produce the blind holes in which the TCs are hosted, the proposed approach can be applied only to electrically conductive materials, such as carbon/carbon composite due to the nature of the EDM drilling technique.

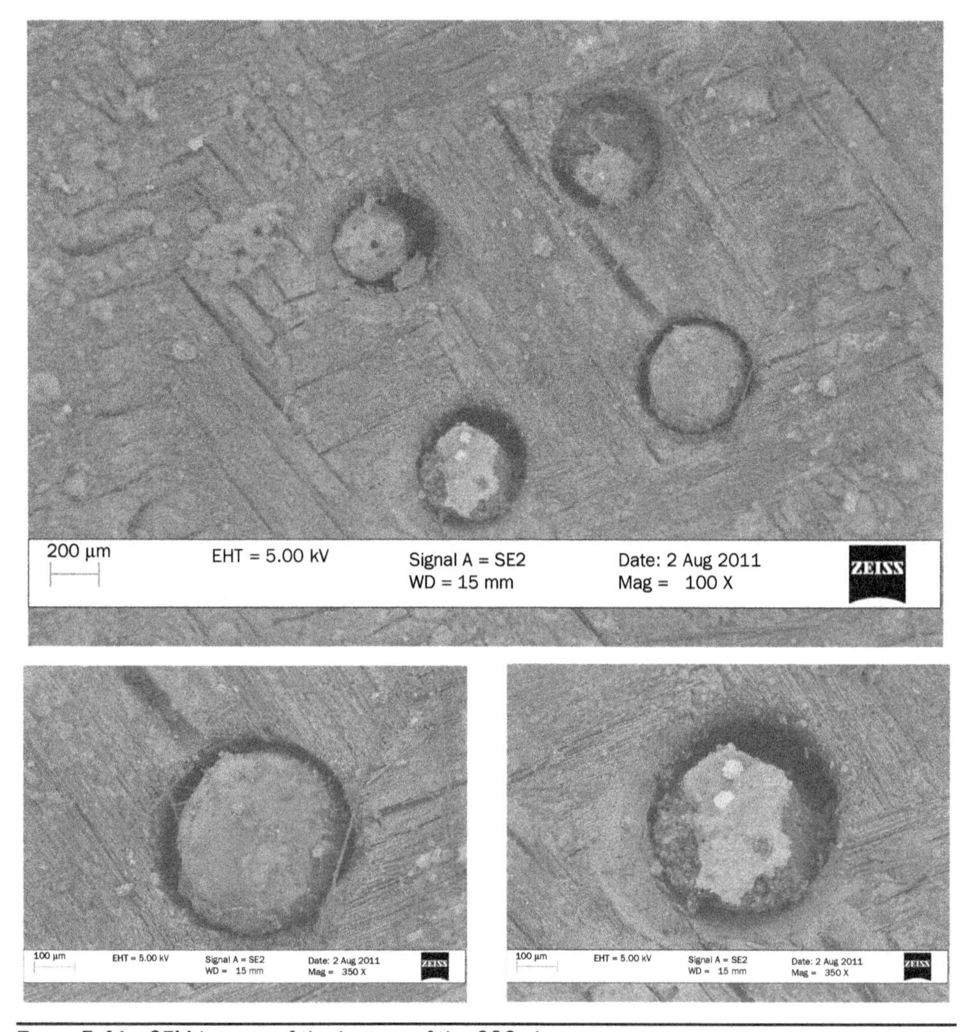

Figure 5.41 SEM images of the bottom of the CCC plug.

During exposure to the torch, the TCs experienced multiple electrical breaks. The detection of the ablation recession can be strictly related to the first break of each Seebeck junction, which depends on the depth at which the TC is embedded. After this event, the electrical signal provided by the broken TC can be filtered by using some dedicated software able to neglect the following breaks. Table 5.3 summarizes the results of all tests performed on the carbon/carbon plugs. It can be seen that both the LDCC- and HDCC-based plugs using this technique was able to provide a very good real-time evaluation of the ablation rate [87, 89]. These data also indicated that the technology is easily scalable and extremely flexible.

Once combined with a proper hardware/software, this proposed system can provide precious data for ground testing as well as obtaining flight data used as a feedback

TABLE 5.3 Final Results for the Four- and Eight-Level Sensors

Plug tag	# TCs	Material	First approach				Second approach				MARR***/° (mm/s)	*SD* (mm/s)	Notes
			RR* (mm/s)	mean (mm/s)	*SD* (mm/s)	*mSD* (mm/s)	RR** (mm/s)	mean (mm/s)	*SD* (mm/s)	*mSD* (mm/s)			
ccold1	4	LDCC	0.0713	0.0645	*0.0055*	*0.0038*	0.0575	0.0576	*0.0012*	*0.0012*	0.056	*0.007*	°
ccold3	4	LDCC	0.0611		*0.0028*		0.0552		*0.0011*				
ccold4	4	LDCC	0.0610		*0.0030*		0.0602		*0.0012*				
ccnew1	4	HDCC	0.0137	0.0262	*0.0008*	*0.0013*	0.0229	0.0251	*0.0005*	*0.0005*	0.023	*0.005*	°
ccnew2	4	HDCC	0.387		*0.0017*		0.0273		*0.0005*				
34	8	HDCC	0.0510	0.0479	*0.0008*	*0.0024*	0.0336	0.0327	*0.0004*	*0.0004*	0.043	*0.003*	§
42	8	HDCC	0.0421		*0.0057*		0.0308		*0.0003*				
44	8	HDCC	0.0506		*0.0008*		0.0335		*0.0004*				

* Erosion rate calculated with the first approach
** Erosion rate calculated with the second approach
*** MARR: mean erosion rate measured as a function of the final depth of the crater
° Measured on a 4 mm depth crater
§ Measured on a 8 mm depth crater

for reentry TPS application. Moreover, the acquired data could be used to improve the comprehension of ablation mechanism of the ablative material development for ground testing, especially when combined with the data provided by other in situ sensors. Finally, the proposed technique can also be exploited in other fields of the aerospace industry: as an example, it can be employed to monitor the erosion rate of the nozzle throats and nozzle entrance region and exit cone of solid rocket motors. More detailed results can be found in [87, 89, 100] and in Sec. 9.4.

The 0.5-mm-Diameter TC Ablation Sensor

Two types of in situ ablation recession and thermal sensor were developed by Koo and colleagues [86–100] based on break-wire concept using ultra-miniature TCs: one sensor was based on a 0.25-mm-diameter TC [86, 87, 100] and the other sensor was based on a 0.55-mm-diameter TC [88–100]. Earlier in situ ablation sensor designs developed by Natali and colleagues involved the use of four 0.25-mm-diameter TCs and EDM drilling technique [86, 87, 100]. Later designs used four 0.55-mm TCs and conventional micro-drilling in order to significantly reduce the cost and the lead time of manufacturing each sensor. Further designs involved the use of nine 0.55-mm TCs in addition to reduced distances between TCs. After manufacturing three samples, testing using the OTB gave results which were compared with data obtained using the 0.25-mm micro-TC design. The results showed that unlike the 0.25-mm micro-TC, the 0.55-mm TC existed well outside of the ablative material and did not disintegrate immediately upon exposing to the flame. Statistical analysis was used to determine the approximate break temperature for each TC. The break temperature of 1,432°C was determined to match well with the sensor-calculated ablation rates and the manually measured average ablation rates. Three more samples were constructed and tested in order to confirm that 1,432°C was a suitable break temperature in order to achieve repeatable results with the carbon/carbon composite sensor plugs. It was determined that through the 0.25-mm micro-TC, crisp data resolution was provided, since the TCs broke upon exposure to the environment. The 0.55-mm TCs utilizing break temperature analysis provided the ability for the larger TCs to obtain higher-resolution recession rates. Three more samples using nine TCs were fabricated and tested in order to determine if a higher TC count would compromise the structural integrity of the plug. Additionally, four samples were fabricated and tested using carbon/phenolic as the base ablative material in place of carbon/carbon composite. In this section only the second round of four levels CCC sensor results are presented, additional results can be found in [86, 87] and in Chap. 9.

Design Specifications The sensor design was intended to be similar enough to the setup used by Natali and Koo [86, 87], so that the data gained from testing would be easily comparable. The ablation sensor developed by Natali and Koo consisted of four ultra-miniature TCs installed at staggered depths from the back of the CCC plug [86, 87]. As the plug is heated the TCs provide data about the temperature distribution within the plug, and finally break when the ablation front reaches the TC and destroys the junction in the tip of the TC. By knowing the depth of the TC and the time at which it broke, the position of the ablation front and rate of ablation can be calculated.

The CCC plugs have dimensions of 15 mm^3, so that they could be mounted and used with the OTB. The TCs had to be larger than 0.25 mm but small enough to fit four within the area under the flame generated by the OTB's flame envelope.

The deepest TC hole to be drilled was 13 mm deep, 2 mm from the front surface of the plug, and each following hole was to be shallower by 0.5 mm. The holes which house the TCs had to be created through conventional drilling. All parts had to be commercially available. The primary purpose of these constraints was to determine whether the structural integrity of the CCC sensor would be significantly affected during testing by the increase in TC diameter from the 0.25-mm version. Additionally, the larger TC size and manufacturing requirements would greatly reduce the cost of producing the sensor plugs.

Production of 0.55-mm TC Sensor All production had to be accomplished in-house at The University of Texas at Austin (UT) for cost saving and shorter lead time. The actual manufacturing of the sensor systems consisted of two major steps: using subtractive manufacturing processes to prepare the thermal protection material to hold the TCs at the proper depth and installing the TCs with the use of a durable high-temperature adhesive.

Design The spacing between each drilled TC hole must be small enough so that during testing the oxyacetylene test bed applies its heat flux on each TC equally. The OTB torch acts on a very fine area that is roughly 3 mm in diameter. As seen in Fig. 5.42, the oxyacetylene torch acts upon the sample in a cone. By keeping the holes closer together, the same heat flux is applied to all four TCs and the TCs remain at the deepest point of the ablation area.

However, the holes cannot be spaced too close together for two important reasons. First, the hole spacing must reflect the increasing TC size. If the holes are too close, the structural strength of the ablative material could be weakened enough such that the walls between the TCs collapse, resulting in a significant weak point that may be blown through by the environment. Second, the hole spacing must be large enough to prevent TC cross-talk within the conductive CCC sample in order to reduce signal noise during testing. After testing, the shield for each TC is constituted by a metal blend embedded with the carbon fibers of the sample. Since CCC is a conductive material, the effectiveness of the shielding is reduced. By maintaining enough distance between each TC, cross-talking between each TC will not occur.

This design will maintain similar dimensions as the EDM-drilled sample by Natali and Koo [86, 87]. With the larger 0.55-mm TCs, this resulted in hole spacing of roughly 2 mm. The drill size and hole diameter were determined by the amount of clearance needed between the TC and the hole wall. Figure 5.42 shows the dimensions of the final working print for the sensor plugs utilizing four TCs.

Originally the holes in the CCC samples were to be drilled by EDM, but microscale drilling was used instead due to longer lead time and additional funds required in offsite EDM drilling. Having already determined the TC hole size for the samples, a methodology for manufacturing the test samples in a fast and reliable way has to be developed.

Fabrication The CCC acquired from South Korea was extremely tough and had an average density of 1.7 g/cm^3, which is roughly 50% higher than the CCC used in the original micro-TC tests by Natali and Koo [86, 87]. In order to improve manufacturing speeds, improve accuracy, and reduce the risk of drills breaking down-hole, a three-step process was developed to produce the holes.

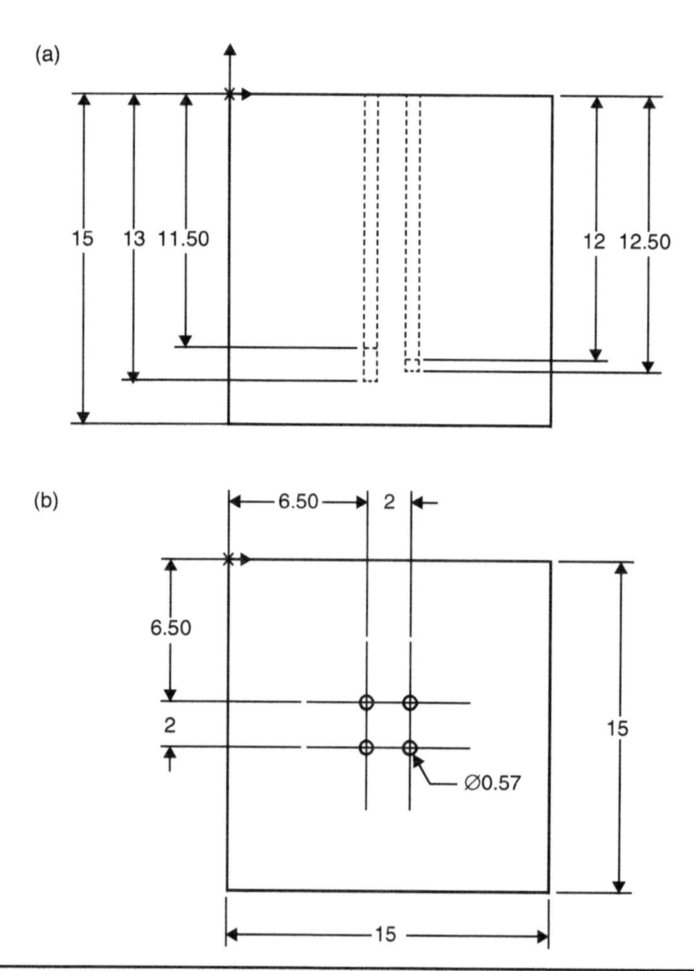

FIGURE 5.42 Side (top) and top (bottom) views of the sample. All dimensions are in millimeters for 0.50-mm-diameter TC CCC ablation recession sensor.

The carbon/phenolic was significantly more difficult to drill because the phenolic would melt slightly during drilling. Melted phenolic would stick to the drill bit, resulting in excessive torque on the drill bit and breakage when attempting to drill deeper than 13 mm. Spindle speeds and feed speeds were reduced in an attempt to prevent the heating of the carbon/phenolic, though this did not prove successful and drill bits continued to break inside the samples. The TC depths were reduced by 0.5 mm each on the carbon phenolic samples in order to carry on with the tests.

Experimental Setup The burn tests involved using the UT OTB to generate a heat flux of 940 W/cm^2 to erode the CCC plug. The heat flux of 940 MW/m^2 was used by Natali in his original tests, so the tests were done at the same heat flux in order to make sure that the data would be easily comparable with the Italian team's test results. The four TCs were connected to a DAQ, which in turn interfaced with a LabView program on a laptop. The temperatures of the TCs were recorded over the length of the burn test, which ranged from 90 to 180 seconds.

The OTB torch was held at 6 mm from the sample in order to reach the proper heat flux as determined by calibration performed by Allcorn et al. [73]. A vice holds the brass sample holder, which in turn holds the CCC sample in place during the test. The brass sample holder was originally designed for use with cylindrical samples where the TCs were not fixed inside the test sample, such as the copper slugs used to calibrate the heat flux. The holder was modified by cutting a slot into the side so that the TCs on the completed CCC samples could be properly inserted behind the sample. Additional holes were drilled and tapped so that cube-shaped samples could be securely held in place.

Ninety-second burn tests were performed on two virgin CCC samples in order to ensure that the OTB was calibrated to have the same heat flux as the OTB that Natali's team used. The first sample was composed of the same CCC that Natali used, while the second sample was made of the high density CCC from South Korea. After the samples were burned, the crater depths were measured and the average ablation rates were calculated. The ablation rate of the first sample, 0.054 ± 0.007 mm/s, matched the ablation rate of Natali's test sample, 0.057 ± 0.007 mm/s, within experimental uncertainty [86, 87]. The ablation rate of the second sample was found to be 0.032 ± 0.007 mm/s. This recession rate was used to determine a test time of 180 seconds based on the recession rate times the desired burn depth of at least 3.5 mm, which was necessary to burn all four TCs and show that the TCs could act in a similar capacity to break wires.

Results and Discussion Three rounds of three tests each were conducted using the testing procedures described above in order to demonstrate repeatability in using the TC-based ablation sensor. The results of the first round were not as expected and so a second round of testing was performed using the same setup. Between the first six tests, a consistent method for using the TC-based in situ ablation sensor was demonstrated. The last three tests involved the use of plugs with nine TCs instead of four, and focused primarily on determining whether or not increasing the number of TCs would have an adverse effect on the structural integrity of the CCC plug. In this study, Phase I testing refers to tests utilizing the four TC configuration, while Phase II testing refers to the nine TC configuration. Only the test results of Tests 4, 5, and 6 are discussed in this section; more detailed results can be found in [88, 89] and also in Chap. 9.

Results of Tests 4, 5, and 6 In the second round (Phase I) of testing, the precautions listed above were taken into account. Despite these adjustments, the data were found to be similar to the first round of testing. The data from the second round of testing are shown in Fig. 5.43 (Test 4). The average recession rate was calculated both with the TCs and manually using the same method as round one. The recession rates as calculated by the TCs are shown in Fig. 5.44 (Test 4). As with the first round of testing, it can be seen that the average recession rates calculated using both methods agree within uncertainty (Fig. 5.45).

These results further demonstrate the validity of using a break temperature of 1,432°C. The results presented above demonstrate that using the TC ablation sensor design with the break temperature method should yield consistent and reliable measurement of ablative recession rates. To demonstrate the use of this method in practice the research team prepared a graph of the average recession rate calculated at each TC, as shown in Fig. 5.46.

In the real-world application, this ablation sensor and a computer can produce a graph of the average recession rate to make in-flight corrections using current state of

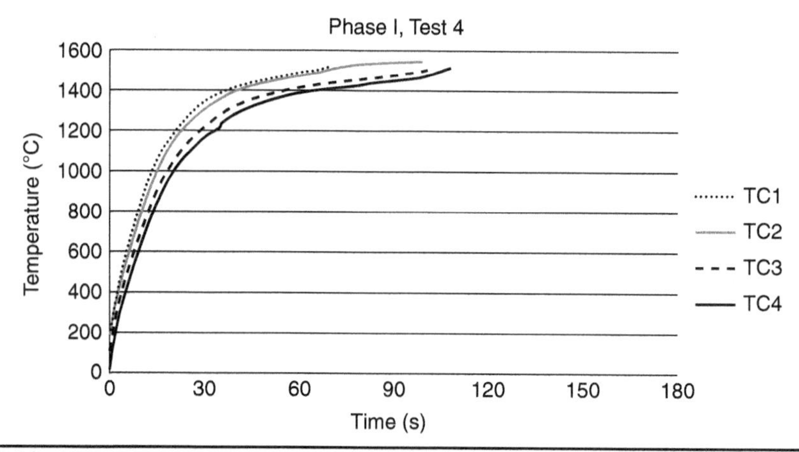

FIGURE 5.43 Test 4 temperature profile for 0.50-mm TC ablation CCC sensor.

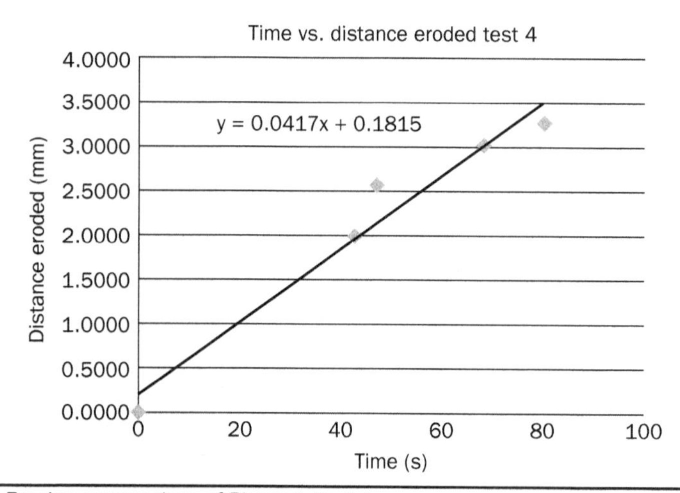

FIGURE 5.44 Erosion versus time of Phase I, Test 4.

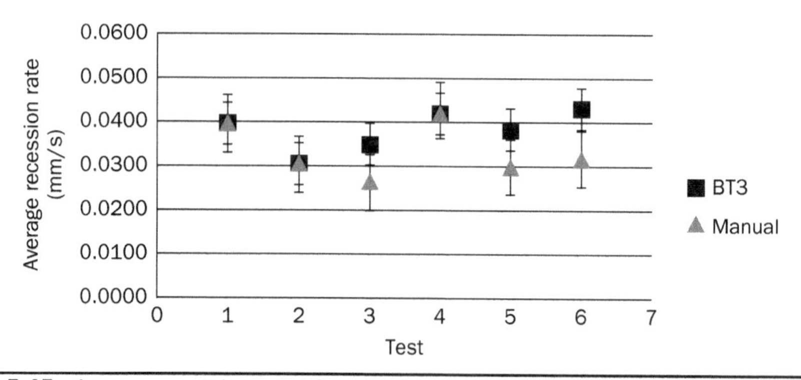

FIGURE 5.45 Average recession rate of samples at varying break temperatures.

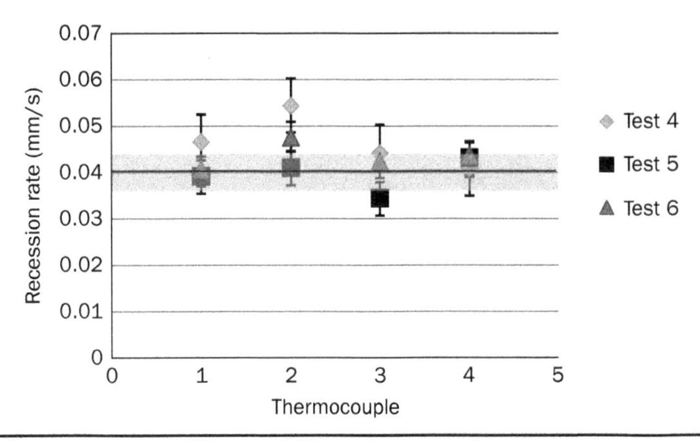

FIGURE 5.46 Recession rate at TC at 1,432°C for Tests 4 to 6.

ablation of the thermal protection system. Furthermore, it is believed that the TC-based ablation sensor with the break temperature method would work with a large variety of TPS materials as long as the TC remains constant, as the break temperature method only relies on the melting temperature of the TC.

Summary and Conclusion The four 0.50-mm-diameter TC sensor design is an extremely cost-effective way to accurately measure the in situ recession rate of a material exposed in a hyperthermal environment when used with the cutoff temperature methodology developed and the sensor does not compromise the structural strength of the original ablative material. Cross-talk between TCs is not an issue within conductive materials when shielded TCs are used. While utilizing more TCs may increase the resolution of measurement, it also significantly affects the structural integrity of the material that it is embedded. A different design will need to be developed in order to properly utilize more TCs without creating weak points in the ablative. Additionally, a different DAQ should be used to take measurements from large numbers of TCs, so as to prevent major signal noise problems. Additional test results of this sensor technology applying to PICA, AVCOAT, SiC/glass ceramic matrix composites, and 3D carbon/phenolic can be found in [83–100] and also in Chap. 9.

5.14 Electrical Properties

Electrical conductivity of PNC samples is measured typically by four-point probe method. A commercial electrical conductivity setup, such as Hioki Megaohmmeter or specialized multimeter from Keithley, a U.S.-based company, can be used to measure electrical conductivity. For EMI shielding characterization, ASTM D4935-10 [101] is used for most cases using network analyzer. However, the limitation for this method is that it required large sample size, which is a circular plate having a diameter of 6 in. (15.24 cm). The amplitudes of incident wave, transmitted wave, and reflected wave were used to determine the effectiveness of EMI shielding.

Additional test results of this technology are presented in Chap. 10 on electrical properties of PNCs. A few conventional electrical properties for polymers are discussed by Shah [102], such as (a) dielectric strength (ASTM D 149, IEC 243-1), (b) dielectric constant and dissipation factor (ASTM D 150, IEC 250), (c) electrical resistance tests—volume resistance and surface resistance (ASTM D 257, IEC 93), (d) ARC resistance (ASTM D 495), and (e) EMI/RFI shield.

5.15 Other Properties

Tribological property characterization, such as abrasion, wear, and scratch resistance characterization, and permeability property characterization are discussed in Chap. 11. Besides the above properties and characterization techniques, other properties, such as weathering properties, chemical properties, analytical tests, conditioning properties, etc., can be found in Shah's *Handbook of Plastics Testing and Failure Analysis* [103].

5.16 Summary, Future Needs, and Assessments

No one "single" technique provides sufficient analytical information related to the structural characteristics of a PNC. More than one technique is recommended to understand the structures and properties of the PNCs. It is time consuming and costly to perform all the necessary experiments and to accumulate all of the data to verify the development of each PNC formulation. One needs to be selective in the choice of analytical tools to be used in their studies.

There are other techniques that show promise for PNC analysis but that need to be validated, such as the following:

- Neutron scattering: it has similar issues with WAXD and also limited access.
- Nuclear magnetic resonance (NMR): currently only works for clays with iron in the clay structure. This technique would need to be adapted for synthetic clays with no iron!

Polymer nanocomposites present a complex analytical problem. Properties of PNCs are dependent on many factors. No one analytical technique can analyze all of these factors. TGA, WAXD, and TEM are currently the major tools to characterize these materials. Combining these techniques will give us a better description and analysis of these PNCs. Each technique has its own advantages and shortcomings. To shorten the time to complete analysis on PNCs, new analytical techniques are necessary to facilitate analyses more rapidly to reduce the time lag for complete analyses of PNCs. Furthermore, techniques that probe polymer-nanoparticle/polymer-organic treatment interactions are warranted.

5.17 Study Questions

5.1 Besides the characterization methods described in this chapter, identify other common characterization techniques used to evaluate nanomaterials and polymer nanocomposite properties and discuss the advantages and disadvantages of these characterization methods. Present your answers in a similar format used in this chapter.

5.2 Describe the different characterization methods that are commonly used to characterize the degree of dispersion of nanomaterials in a polymer nanocomposite material. Provide examples. Compare these characterization methods and discuss the advantages and disadvantages of these methods.

5.3 What are the limitations and benefits of each scanning method, such as optical microscopy, wide-angle x-ray diffraction (WAXD), scanning electron microscopy (SEM), transmission electron microscopy (TEM), and scanning tunneling microscopy (STM)?

5.4 What are some advantages and disadvantages of WAXD method? How can *d*-spacing be used to determine the structural characteristics of polymer nanostructured materials?

5.5 Why is transmission electron microscopy (TEM) such a useful technique compared to other characterization methods? What are some methods for preparing TEM samples?

5.6 How is energy-dispersive x-ray spectroscopy used in conjunction with TEM?

5.7 What are some advantages that small-angle x-ray scattering (SAXS) provide in comparison to other methods?

5.8 Briefly describe how cone calorimetry is performed. How are the properties measured by the cone calorimeter to characterize the flammability properties of polymer nanocomposite materials?

5.9 Are there correlations and general trends for these industry test methods, such as cone calorimeter, microscale combustion calorimeter, limiting oxygen index (LOI), UL 94, and Stein tunnel test, for flammability properties of polymer nanocomposites?

5.10 Briefly describe the working principle of the in situ ablation recession and thermal sensor. How many types of ablative materials have this technique demonstrated?

5.11 Briefly describe the working principle of the char strength sensor to evaluate polymer nanocomposite chars?

5.18 References

1. Wang, Z. L., ed. (2000) *Characterization of Nanophase Materials.* Weinheim, Germany: Wiley-VCH, pp. 37–80.
2. Yan, N., and Wang, Z. L., eds. (2005) *Handbook of Microscopy for Nanotechnology.* Boston, MA: Kluwer Academic Publishers.
3. Wang, Z. L., ed. (2000) *Characterization of Nanophase Materials.* Weinheim, Germany: Wiley-VCH, pp. 13–36.
4. Cao, G. (2004) *Nanostructures and Nanomaterials: Synthesis.* London, England: Properties & Applications, Imperial College Press, pp. 329–390.
5. Hornyak, G. L., Tibbals, H. F., Dutta, J., and Moore, J. J. (2009) *Introduction of Nanoscience & Nanotechnology.* Baca Raton, FL: CRC Press, pp. 107–175.
6. Crewe, V. (1970) The current state of high resolution scanning electron microscopy. *Quarterly Review of Biophysics,* 3(I):137–175.
7. Buseck, P., Cowley, J. M., and Eyring, L., eds. (1988) *High Resolution Transmission Electron Microscopy and Associated Techniques.* New York, NY: Oxford University Press.

8. Browing, N. D., Chisholm, M. F., and Pennycook, S. J. (1993, November) Atomic-resolution chemical analysis using a scanning transmission electron microscope. *Nature,* 366:143–146.
9. Hobbs, S. Y., and Watkins, V. H. (2000) Morphology characterization by Microscopy Techniques. In *Polymer Blends, vol. 1: Formulation,* Paul, D. R., and Bucknall, C. B., eds. New York, NY: John Wiley & Sons, pp. 239–289.
10. Koo, J. H., Stretz, H., Bray, A., Weispfenning, J., Luo, Z. P., and Wootan, W. (2004) Nanocomposite rocket ablative materials: processing, microstructure, and performance. AIAA-2004-1996 paper, *44th AIAA/ASME/ASCE/AHS Structures, Structural Dynamics, and Materials Conference.* Palm Springs, CA. April 19–22.
11. Guinier, A., and Fournet, G. (1955) *Small-Angle Scattering of X-Rays.* New York, NY: Wiley.
12. Glatter, O., and Kratky, O. (1982) *Small-Angle X-Ray Scattering.* London, England: Academic Press.
13. Binnig, G., Rohrer, H., Gerber, C., and Weibel, E. (1982) Surface studies by scanning tunneling microscopy. *Phys. Rev. Lett.,* 49:57–61.
14. Bhushan, B., and Othmar Marti, O. (2017) Scanning probe microscopy—principle of operation, instrumentation, and probes. In *Nanotribology and Nanomechanics,* Bhushan, B., ed. Springer International Publishing AG, pp. 33–93. doi: 10.1007/978-3-319-51433-8_233.
15. Binnig, G., Quate, C. F., and Gerber, C. (1986) Atomic force microscope. *Phys. Rev. Lett.,* 56:930–933.
16. Binnig, G., Gerber, C., Stoll, E., Albrecht, T. R., and Quate, C. F. (1987) Atomic resolution with atomic force microscope. *Europhys. Lett.,* 3:1281–1286.
17. Levine, A. (1991) *Quantum Chemistry,* 4th ed. Upper Saddle River, NJ: Prentice Hall.
18. Jalili, N., and Laxminarayana, K. (2004) A review of atomic microscopy imaging systems: application to molecular metrology and biological sciences. *Mechatronics,* 14:907–945.
19. Nakajima, K., Wang, D., and Nishi, T. (2012) AFM characterization of polymer nanocomposites. In *Characterization Techniques for Polymer Nanocomposites,* Mittal, V., ed. Weinheim, Germany: Wiley-VCH, pp. 185–228.
20. Garea, S. A., and Iovu, H. (2012) Following the nanocomposites synthesis by Raman spectroscopy and x-ray photoelectron spectroscopy (XPS). In *Characterization Techniques for Polymer Nanocomposites,* Mittal, V., ed. Weinheim, Germany: Wiley-VCH, pp. 115–142.
21. Shah, V. (2007) *Handbook of Plastics Testing and Failure Analysis.* Hoboken, NJ: Wiley & Sons, pp. 17–93.
22. Lao, S. C., Yong, W., Nguyen, K., Moon, T. J., Koo, J. H., Pilato, L., and Wissler, G. (2010) Flame-retardant polyamide 11 and 12 nanocomposites: processing, morphology, and mechanical properties. *Journal of Composite Materials,* 44(25):2933–2951.
23. Vaia, R. A., Teukolsky, R. K., and Giannelis, E. P. (1994) Interlayer structure and molecular environment of alkylammonium layer silicates. *Chem. Mater.,* 6:1017.
24. Eslami, H., Grmela, M., and Bousmina, M. (2009) A mesoscopic tube model of polymer/layered silicate nanocomposites. *Rheol. Acta,* 48:317.
25. Eslami, H., Grmela, M., and Bousmina, M. (2009) Structure Build-Up at Rest in Polymer Nanocomposites: Flow Reversal Experiments. *J. Polym. Sci.,* 47:1728.
26. Ray, S. S. (2006) Rheology of Polymer/Layered Silicate Nanocomposites. *J. Ind. Eng. Chem.,* 6:811.

27. Song, M., and Jin, J. (2012) Characterization of rheological properties of polymer nanocomposites. In *Characterization Techniques for Polymer Nanocomposites*, Mittal, V., ed. Weinheim, Germany: Wiley-VCH, pp. 251–281.
28. Krishnamoorti, R., Vaia, R. A., and Giannelis, E. P. (1996) Structure and dynamics of polymer-layered silicate nanocomposites. *Chem. Mater.*, 8:1728.
29. *Standard Test Method for Thermal Diffusivity by the Flash Method* (ASTM E1461-11). Philadelphia, PA: American Society for Testing and Materials.
30. Shah, V. (2007) *Handbook of Plastics Testing and Failure Analysis*. Hoboken, NJ: Wiley, pp. 94–116.
31. Troitzsch, J., ed. (2004) *Plastics Flammability Handbook*, 3rd ed. Cincinnati, OH: Hanser.
32. Hirschler, M. (2008) *NFPA Fire Protection Handbook*, 20th ed. Quincy, MA: NFFA. Chapter 3.
33. Shah, V. (2007) *Handbook of Plastics Testing and Failure Analysis*. Hoboken, NJ: Wiley, pp. 218–250.
34. Babrauskas, V. (1996) The cone calorimeter. In *Heat Release in Fires*, Babrauskas, V., and Grayson, S. J., eds. London: E & FN Spon, pp. 61–91.
35. *Standard Test Method for Heat and Visible Smoke Release Rates for Materials and Products Using an Oxygen Consumption Calorimeter* (ASTM E1354). Philadelphia, PA: American Society for Testing and Materials.
36. Fire Tests—Reaction to Fire—Part 1: Rate of Heat Release from Building Products. ISO DOS 5660. International Organization for Standardization, Geneva, Switzerland.
37. *Standard Test Method for Screening Test for Mass Loss and Ignitability of Materials* (ASTM E2102). Philadelphia, PA: American Society for Testing and Materials.
38. *Standard Test Method for Determining Flammability Characteristics of Plastics and Other Solid Materials Using Microscale Combustion Calorimetry* (ASTM D7309-11). Philadelphia, PA: American Society for Testing and Materials.
39. *Standard Test Method for Measuring the Minimum Oxygen Concentration to Support Candle-Like Combustion of Plastics* (Oxygen Index) (ASTM D2863-12e1). Philadelphia, PA: American Society for Testing and Materials.
40. UL 94, the Standard for Safety of Flammability of Plastic Materials for Parts in Devices and Appliances testing, Underwriters Laboratories, Northbrook, IL.
41. *Standard Test Method for Surface Burning Characteristics of Building Materials* (ASTM E84-13a). Philadelphia, PA: American Society for Testing and Materials.
42. Schmidt, D. L., and Schwartz, H. S. (1963) Evaluation methods for ablative plastics. *SPE Transactions*, 238–250.
43. Botton, B., Chazot, O., Carbonaro, M., Van Der Haegen, V., and Paris, S. (1999, October) The VKI plasmatron characteristics and performance. DTIC Compilation Part Notice ADP010745. Rhode-Saint-Genese, Belgium.
44. Koo, J. H., Kneer, M., and Schneider, M. (1992) A cost-effective approach to evaluate high-temperature ablatives for military applications. *Naval Engineers J.*, 104: 166–177.
45. Miller, M. J., Koo, J. H., and Lin, S. (1993, January) Evaluation of different categories of composites ablative for thermal protection. AIAA-93-0839, 31st AIAA Aerospace Sciences Meeting, Reno, NV.
46. Yang, B.-C., Cheung, F.-B., and Koo, J. (1995, January) Prediction of thermomechanical erosion of high-temperature ablatives in the SSRM facility. AIAA-95-0254, 33rd Aerospace Sciences Meeting, Reno, NV.

47. VanMeter, M., Koo, J. H., Wilson, D., and Beckley, D. A. (1995) Mechanical properties and material behavior of a glass silicone polymer composite. *Proc. 40th International SAMPE Symposium*. SAMPE, Covina, CA, pp. 1425–1434.
48. Koo, J. H., et al. (1998) Effect of major constituents on the performance of silicone polymer composites. *Proc. 30th International SAMPE Technical Conference*. SAMPE, Covina, CA.
49. Koo, J. H., et al. (1999) Thermal protection of a class of polymer composites. *Proc. 44th International SAMPE Symposium*. SAMPE, Covina, CA, pp. 1431–1441.
50. Koo, J. H., Stretz, H., Weispfenning, J., Luo, Z., and Wootan, W. (2004) Nanocomposite rocket ablative materials: subscale ablation test. *Proceedings International SAMPE 2004 Symposium on Disc* [CD-ROM]. Covina, CA: SAMPE.
51. Koo, J. H., Stretz, H., Weispfenning, J., Luo, Z., and Wootan, W. (2004, April) Nanocomposite rocket ablative materials: processing, microstructures, and performance. AIAA-2004-1996, AIAA, Reston, VA.
52. Koo, J. H., Pilato, L., and Wissler, G. (2007) Polymer nanostructured materials for propulsion systems. *Journal of Spacecraft and Rockets*, 44(6):1250–1262.
53. Koo, J. H., Miller, M. J., Weispfenning, J., and Blackmon, C. (2011) Silicone polymer composite for thermal protection of naval launching system. *Journal of Spacecraft and Rockets*, 48(6):904–919.
54. Koo, J. H., Miller, M. J., Weispfenning, J., and Blackmon, C. (2011) Silicone polymer composites for thermal protection system: fiber reinforcements and microstructures. *Journal of Composite Materials*, 45(13):1363–1380.
55. Blanski, R., Koo, J. H., Ruth, P., Nguyen, H., Pittman, C., and Phillips, S. (2004, May) Polymer nanostructured materials for solid rocket motor insulation-ablation performance. *Proc. 52nd JANNAF Propulsion Meeting*, CPIAC, Columbia, MD.
56. Koo, J. H., Marchant, D., Wissler, G., Ruth, P., Barker, S., et al. (2004, May) Polymer nanostructured materials for solid rocket motor insulation–processing, microstructure, and mechanical properties. *Proc. 52nd JANNAF Propulsion Meeting*, CPIAC, Columbia, MD.
57. Ruth, P., Blanski, R., and Koo, J. H. (2004, May) Preparation of polymer nanostructured materials for solid rocket motor insulation. *Proc. 52nd JANNAF Propulsion Meeting*, CPIAC, Columbia, MD.
58. Natali, M., Monti, M., Kenny, J. M., and Torre L. (2011) A nanostructured ablative bulk moulding compound: development and characterization. *Compos. Part A–Appl. S.*, 42(9):1197–1204.
59. Pulci, G., Tirillo, J., Marra, F., Fossati, F., Bartuli, C., and Valente, T. (2010) Carbon-phenolic ablative materials for re-entry space vehicles: manufacturing and properties. *Composites Part A*, 41(10):1483–1490.
60. Allcorn, E., Robinson, S., Tschoepe, D., Koo, J. H., and Natali, M. (2011) Development of an experimental apparatus for ablative nanocomposites testing. AIAA-2011-6050, 47th *AIAA/ASME/SAE Joint Propulsion Conference*, San Diego, CA, August 1–4.
61. Gutierrez, L., Reyes, J., Scott, S., Sada, A., and Koo, J. H. (2015) Design of small-scale Ablative testing apparatus with sample position and velocity control. AIAA-2015-1584, *AIAA SciTech 2015*, Kissimmee, FL, January 5–9.
62. *Standard Test Method for Measuring Heat-Transfer Rate Using a Thermal Capacitance (Slug) Calorimeter* (ASTM E457-08). Philadelphia, PA: American Society for Testing and Materials.

63. Lee, J. C. (2010, December) *Characterization of Ablative Properties of Thermoplastic Polyurethane Elastomer Nanocomposites*. Ph.D. dissertation, The University of Texas at Austin, Austin, TX.
64. Lee, J. C., Koo, J. H., and Ezekoye, O. A. (2011) Thermoplastic polyurethane elastomer nanocomposite ablatives: characterization and performance. AIAA-2011-6051, *47th AIAA/ASME/SAE Joint Propulsion Conference*, San Diego, CA, August 1–4.
65. Lee, J. C., Koo, J. H., and Ezekoye, O. (2009) Thermoplastic polyurethane elastomer nanocomposites: density, hardness, and flammability properties correlations. AIAA-2009-5273, *AIAA Joint Propulsion Conference*, Denver, CO, August 2–5.
66. Lee, J. C., Koo, J. H., Lam, C., Ezekoye, O., and Erickson, K. (2009) Heating rate and nanoparticle loading effects on thermoplastic polyurethane elastomer nanocomposite kinetics. AIAA-2009-4096, *AIAA Thermophysics Conference*, San Antonio, TX, June 22–25.
67. Donskoy, A. (1996) Elastomeric Heat Shielding Materials for Internal Surfaces of Missile Engines. *Int. J. Polym. Mater.*, 31(1):215–236.
68. Solid Rocket Motor Internal Insulation. (1976) *NASA Space Vehicle Design Criteria*. NASA-SP-8093.
69. Bell, M. S., and Tam, W. (1992) *ASRM Case Insulation Design and Development*. NASA-CR-191947.
70. Bhuvaneswari, C. M., Kakade, S. D., Deuskar, V. D., Dange, A. B., and Gupta, M. (2008) Filled ethylene-propylene dieneterpolymer elastomer as thermal insulator for case-bonded solid rocket motors. *Defence Science Journal*, 58(1):94–102.
71. Bhuvaneswari, C. M., Sureshkumar, M. S., Kakade, S. D., and Gupta, M. (2006) Ethylene-propylene Diene Rubber as a futuristic elastomer for insulation of solid rocket motors. *Defence Science Journal*, 56(3):309–320.
72. Allcorn, E., Natali, M., and Koo, J. H. (2011) Ablation performance and characterization of thermoplastic elastomer nanocomposites. *Proc. SAMPE 2011 ISTC*, Fort Worth, TX, October 17–20.
73. Allcorn, E. K., Natali, M., and Koo, J. H. (2013) Ablation performance and characterization of thermoplastic polyurethane elastomer nanocomposites. *Composites: Part A*, 45:109–118. doi: 10.1016/i.compositesa.2012.08.017.
74. Wong, D., Pinero, D., Koo, J. H., Stretz, H., and Ambuken, P. (2013) Thermoplastic polyurethane elastomer nanocomposites: ablation and charring characteristics. *Proc. SAMPE 2013 ISSE*, Long Beach, CA, May 6–9.
75. Wong, D., Pinero, D., Jaramillo, M., Koo, J. H., Ambuken, P., and Stretz, H. (2013) Ablation and combustion characteristics of thermoplastic polyurethane nanocomposites, AIAA-2013-3862. *49th AIAA/ASEM/SAE/ASEE Joint Propulsion Conference*, San Jose, CA, July 14–17.
76. Redondo, H., Atreya, M., Kan, M., and Koo, J. H. (2010, May) Evaluation of char strength of polymer nanocomposites for propulsion systems. *Proc. SAMPE 2010 ISSE* [CD-ROM]. Society for Advancement of Material and Process Engineering, Covina, CA.
77. Reshetnikov, S., Garashenko, A. N., and Strakhov, V. L. (2000) Experimental investigation into mechanical destruction of intumescent chars. *Polymers for Advanced Technologies*, 11:392–397.
78. Nguyen, H. (2012) Air Force Research Laboratory (AFRL), Edwards AFB, CA, private communication.

79. Jaramillo, M., Koo, J. H., Edd, A., and Wells, D. (2011, October) An experimental investigation of char strength of polymer nanocomposites for propulsion applications. *Proc SAMPE 2011 ISTC* [CD-ROM]. Society for Advancement of Material and Process Engineering, Covina, CA.
80. Jaramillo, M., Forinash, D., Wong, D., Natali, M., and Koo, J. H. (2013) An investigation of compressive and shear strength of char from polymer nanocomposites for propulsion applications, AIAA-2013-3864. *49th AIAA/ASEM/SAE/ASEE Joint Propulsion Conference*, San Jose, CA, July 14–17.
81. Jaramillo, M., Koo, J. H., and Natali, M. (2014) Compressive char strength of polyurethane elastomer nanocomposites. *Polymers for Advanced Technology,* 25(77):742–751. doi: 10.1002/pat.3287.
82. Forinash, D. M., Alter, R. J., Clatanoff, S. B., Newman, J. E., Jaramillo, M., and Koo, J. H. (2012, October) Development of an apparatus for measuring the shear strength of charred ablatives. *Proc. SAMPE TECH 2012* [CD-ROM]. Society for Advancement of Material and Process Engineering, Covina, CA.
83. Lewis, J., Koo, J. H., et al. (2015, October) Sensor to measure the shear strength of ablative polymer nanocomposites. *Proc. CAMX 2015* [CD-ROM]. Society for Advancement of Material and Process Engineering, Covina, CA.
84. Lewis, J., Koo, J. H., et al. (2017) Development of a shear char strength sensing technique to study thermoplastic polyurethane elastomer nanocomposites. *Polymer for Advanced Technologies,* 28(12):1707–1725, doi: 10.4044.1002/pat.
85. Mendez, J., Koo, J. H., et al. (2017) Char strength studies of glass/phenolic ablatives and graphite. *SAMPE 2017 ISTC,* Seattle, WA, May 22–25.
86. Natali, M., Koo, J. H., Allcorn, E., and Ezekoye, O. A. (2013) In situ ablation recession sensor based on ultra-miniature thermocouples—part a: 0.25 mm diameter thermocouples. AIAA-2013-3660, *49th AIAA/ASME/SAE/ASEE Joint Propulsion Conference,* San Jose, CA, July 15–17.
87. Natali, M., Koo, J. H., Allcorn, E., and Ezekoye, O. A. (2014) An in situ ablation recession sensor for carbon/carbon ablatives based on commercial ultra-miniature thermocouples. *Sensors and Actuators B: Chemical,* 196:46–56, doi: 10.1016/j.snb.2014.01.022.
88. Yee, C., Ray, M., Tang, F., Wan, J., Koo, J. H., and Natali, M. (2013) In situ ablation recession sensor based on ultra-miniature thermocouples—part B: 0.50 mm diameter thermocouples. AIAA-2013-3659, *49th AIAA/ASME/SAE/ASEE Joint Propulsion Conference,* San Jose, CA, July 15–17.
89. Yee, C., Ray, M., Tang, F., Wan, J., Koo, J. H., and Natali, M. (2014) In situ ablation recession and thermal sensor based on ultra-fine thermocouples. *Journal of Spacecraft and Rockets,* 51(6):1789–1796.
90. Lisco, B., Yao, E., Pinero, D., and Koo, J. H. (2014) In situ ablation recession and thermal sensors for low density ablators—revisited. *Proc. CAMX 2014,* Orlando, FL, October 13–16.
91. Cameron, S., Astley, A., Leggett, S., Sirgo, G., and Koo, J. H. (2015) In situ ablation recession and thermal sensor based on ultra-fine thermocouples. *Proc. SAMPE 2015 ISTC,* Baltimore, MD, May 18–21.
92. Koo, J. H., Natali, M., et al. (2015) A versatile in situ ablation recession and thermal sensor adaptable for different ablatives. AIAA-2015-1122, *AIAA SciTech 2015,* Kissimmee, FL, January 5–9.

93. Grantham, T. S., Tanner, G., Molina, R., Duong, N. M., and Koo, J. H. (2015) Ablation, thermal, and morphological properties of SiC fiber–reinforced ceramic matrix composites. AIAA-2015-1581, *AIAA SciTech 2015*, Kissimmee, FL, January 5–9.
94. Koo, J. H., et al. (2015) A versatile in situ ablation recession and thermal sensor based on ultra-fine gage thermocouples for ablative TPS materials. *Proceedings of National Space & Missile Materials Symposium (NSMMS)*, Chantilly, VA, June 22–25.
95. Koo, J. H., and Natali, M. (2018) In situ ablation recession and thermal sensor for thermal protection systems. *Journal of Spacecraft and Rockets*, 55(4):783–796.
96. Lisco, B., Koo, J. H., et al. (2017) In-situ ablation and thermal sensing of a 3-dimension woven carbon/phenolic composite for computer modeling and simulation. AIAA-2017-0356, *2017 AIAA SciTech Forum*, Grapevine, TX, January 9–13.
97. Sammak, J., Celler, J., Guarino, M., Schlomer, D., and Koo, J. H. (2017) In-situ ablation sensor and numerical modeling to study three-dimensional woven carbon/phenolic ablative. *SAMPE 2017 ISTC*, Seattle, WA, May 22–25.
98. Berdoyes, M. (2107) Ablation experimental characterization and numerical investigation of a 3-dimensionally built carbon/phenolic composite for aerospace applications. *68th International Astronautical Congress (IAC)*, Adelaide, Australia, September 25–29.
99. Menz, R., Koo, J. H., Lisco, B. E., Sammak, J., Mendez, J., Berdoyes, M., and Castex, J. (2018) Characterization of 3-D woven carbon/phenolic using in-situ ablation sensing, video imaging, and numerical simulation. AIAA-2018-0098, *2018 AIAA SciTech Forum*, Kissimmee, FL, January 8–12.
100. Koo, J. H., Natali, M., et al. (2018) In situ ablation recession and thermal sensor for thermal protection systems. *Journal of Spacecraft and Rockets*, doi: 10.2514/1.A33925.
101. ASTM D4935—10 *Standard Test Method for Measuring the Electromagnetic Shielding Effectiveness of Planar Materials*. Philadelphia, PA: American Society for Testing and Materials.
102. Shah, V. (2007) *Handbook of Plastics Testing and Failure Analysis*. Hoboken, NJ: Wiley, pp. 157–175.
103. Shah, V. (2007) *Handbook of Plastics Testing and Failure Analysis*. Hoboken, NJ: Wiley.

PART 2

Multifunctional Properties of Polymer Nanocomposites

CHAPTER 6

Mechanical Properties of Polymer Nanocomposites

6.1 Introduction

In the previous chapters, the three key steps in the development of polymer nanocomposites—(a) nanomaterial selection, (b) polymer nanocomposite processing, and (c) polymer nanocomposite property characterization and performance or device fabrication—were discussed. In the following chapters, characterization of specific polymer nanocomposite properties will be investigated, including mechanical, thermal, flammability, ablation, electrical, and other properties.

This chapter is organized into three sections to discuss the mechanical properties of (a) thermoplastic-based, (b) thermoset-based, and (c) thermoplastic elastomer–based nanocomposites.

6.2 Thermoplastic-Based Nanocomposites

In this section, examples provided are based on the research conducted to analyze the enhancements of mechanical properties of thermoplastic-based nanocomposites. They are grouped into nanoclays, carbon-based nanocomposites, and other nanomaterials.

6.2.1 Nanoclay-Based Thermoplastic Nanocomposites

Montmorillonite (MMT) clay platelets are probably the most widely researched nanoparticles. These nanoparticles have unique effects on the mechanical properties of the polymer, which are influenced by the type of clay, surface treatment of the clay, and how the clay is processed in the polymer matrix.

Glass Fiber/MMT Thermoplastic Bionanocomposites

Kord reported a study conducted on glass fiber–reinforced high-density polyethylene (HDPE) [1]. In the study the fibers were to be reinforced by rice husk flour, and Cloisite 30B clay platelets were added to the mixture to increase mechanical and water absorption properties. The purpose of this study was to see if the clay platelets could be added effectively and if their benefits justified the further studying of this process.

The raw materials used in the experiment included HDPE at 0.954 g/cm^3, rice husk flour with a particle size of around 100 meshes, glass fibers, and Cloisite 30B clay platelets. During mixing, maleic anhydride was used as a coupling agent. The proportion in

Table 6.1 Compositions Used in the Making of the Glass Fiber/MMT Nanocomposite [1]

High-density polyethylene (wt%)	Rice husk flour (wt%)	Glass fiber (wt%)	Nanoclay (phc)	Coupling agent (phc)
50	50	0	0	2
50	45	5	2	2
50	40	10	4	2
50	35	15	6	2

phc, per hundred compounds

which the materials were mixed in the experiment are presented in Table 6.1 [1]. Before mixing, the rice husk flour was dried at 65°C for 24 hours. The mixing was conducted using a Hake internal mixer at melting temperature for the HDPE. After the polymer was melted, MMT and the coupling agent were added. After 5 minutes, the glass fibers were added along with the rice husk flour. The total mixing time was around 13 minutes. Samples for tensile testing were made by injection molding. The flexural and tensile tests were conducted at 2 mm/min under ASTM D790-03 and D638-03 standards, respectively. There were no TEM images presented to characterize the nanocomposite, but Kord did mention that the particles did not disperse as well as they would have liked. This uneven dispersion causes degradation in properties such as the impact strength.

According to the results that follow, the mechanical properties generally increased with increased loadings of glass fibers and MMT. The data for tensile strength and flexural strength seem to peak at 4 wt% nanoclay loading (Fig. 6.1a,b) [1]. The nanoclay increases both the flexural strength and modulus but decreases the impact strength of the material (Fig. 6.1c) [1]. This follows the typical trend that anytime a nanoclay is added, the elongation or elasticity of the nanocomposite is decreased due to a more rigid matrix.

Kord concluded that this method did indeed enhance certain mechanical properties. It also, however, had some negative affects at higher loadings [1]. For 4 wt% MMT loading, the properties seemed to be optimal with reasonable loss to impact strength. The study also indicates that the assimilation of the clay platelets into the polymer matrix may produce better results with the addition of a compatibilizer.

6.2.2 Carbon-Based Thermoplastic Nanocomposites

Forming Multiwalled Carbon Nanotubes into Buckypaper

Fernandez-d'Arlas et al. [2] include in their study a report by several scientists working in Spain that focused on two different processing methods for mixing multiwalled carbon nanotubes (MWNTs) into a polyurethane matrix. In the experiment, the scientists looked specifically at the mechanical effects of using either a solvent processing method or implementing buckypaper into the matrix. The buckypaper was theorized to help the strengthening process of the MWNT and create a consistent network within the polymer matrix [2].

In the case of the solvent method, the scientists chose to prepare the MWNT with a combination of dimethylformamide and tetrahydrofuran at a 1-to-1 ratio. These

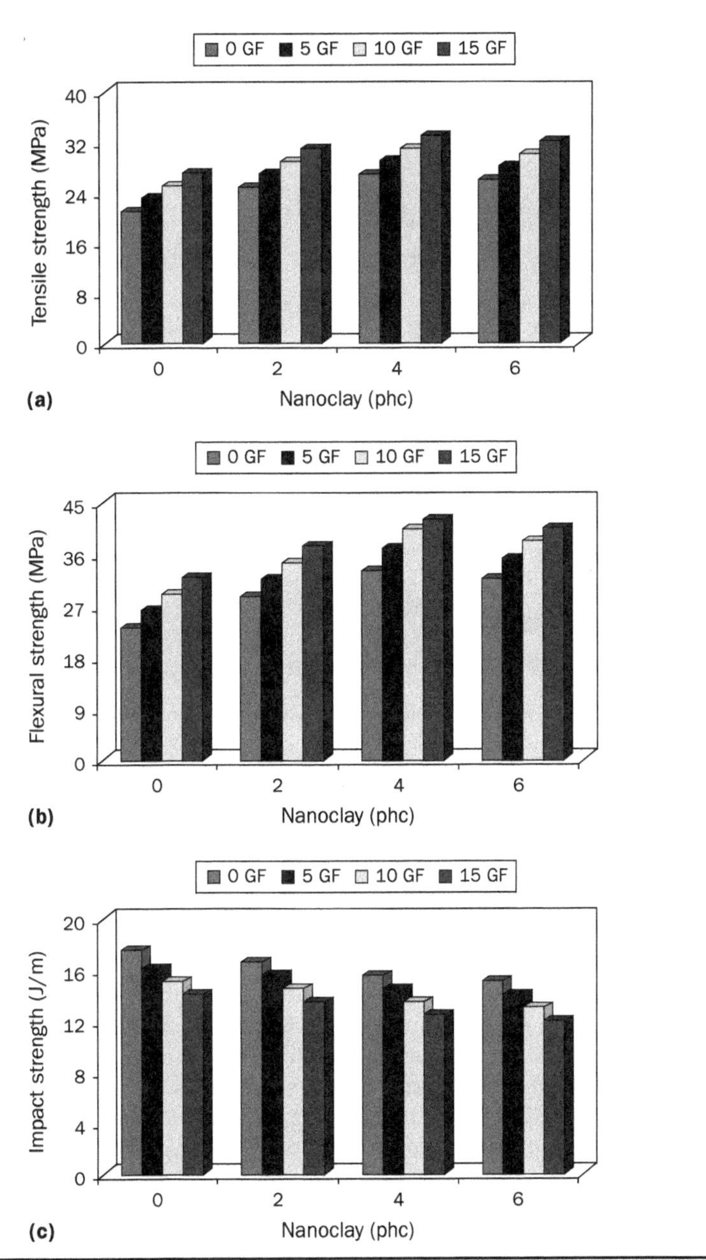

Figure 6.1 Effect of different glass fiber/MMT loadings on (a) tensile strength, (b) flexural strength, and (c) impact strength [1].

solvents were chosen because of their low boiling points, meaning that it would be easy for them to be evaporated out. The MWNTs were placed in a 6 mL solution of the solvent and mixed with a sonication tip. Then the polymer was pipetted from the same solvent solution and combined to make a total composite weight of 300 g [2]. However, the buckypaper method started by acid treating the MWNT in a dimethylformamide

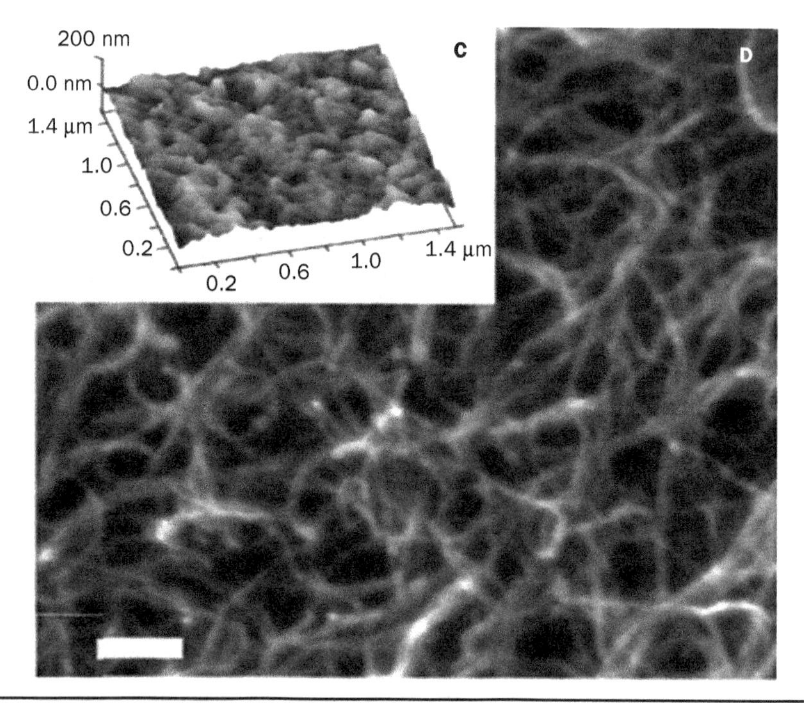

Figure 6.2 TEM image of the buckypaper network in the polyurethane matrix [2].

solution. The buckypaper is created by forming a network of MWNT that is very porous in nature. That way when the polymer is added, it creates passages for it to seep through. The results for the buckypaper method were as predicted, reducing the porosity of the overall mixture and creating a nanocomposite with enhanced mechanical and electrical properties [2]. The buckypaper network was formed in the polymer [2], as shown by the TEM image in Fig. 6.2.

Samples of the nanocomposite were put through stress and strain tests in order to see what the enhancements were to the mechanical toughness of the polymer. Figure 6.3 shows the mechanical properties of the polymer nanocomposites [2]. The higher loading of MWNTs causes the yield stress and modulus to increase (Fig. 6.3). However, around 1 wt% loading, the maximum strain of the material begins to degrade. This is due to the carbon nanotube network taking over and becoming the dominant component. The MWNTs that make up the buckypaper are brittle and as more of it is added to the polymer, the buckypaper begins to feel more of the stress [2].

Graphene-Reinforced Thermoplastics for Strain Sensors in Structural Health Monitoring

Throughout the years smart materials have been researched in order to provide better response time for sensory equipment in the medical field. Recently, a research group [3] published an article about their study conducted using graphene-reinforced polyvinylidene fluoride (PVDF) for structural strain sensing.

To synthesize the nanocomposite, a simple solvent casting technique was used. Initially, the graphene was placed in the dimethylformamide (DMF) solvent and stirred

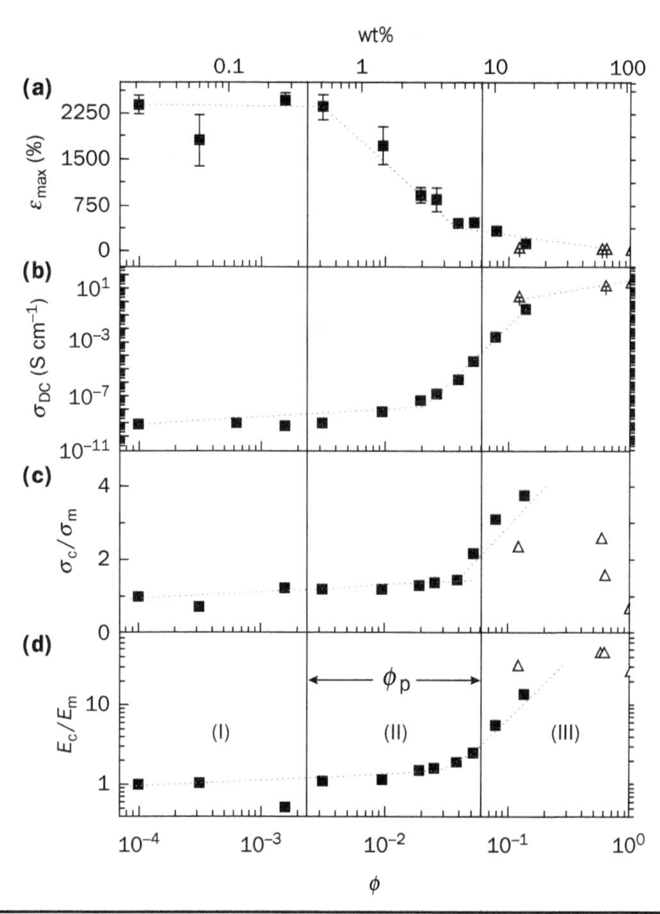

FIGURE 6.3 Mechanical properties of the new nanocomposite [2].

for 60 minutes. The same process was repeated to a specific amount of PVDF polymer in a magnetic stirrer for 30 minutes. Afterward, the two solutions were mixed together by means of an ultrasonicator for a period of 1 hour and then moved to a shear mixer to mix at room temperature for 2 hours and then at 80°C for 30 minutes at 4,000 rpm. The TEM image shows the graphene sheets arranged themselves in a random crumpled order, creating a disordered solid [3].

The TEM image shows good dispersion of graphene within the polymer matrix [3]. The samples made for testing were cut into thin films (containing 2% wt of graphene) and adhered to the back of a piece of aluminum using epoxy glue. The tensile, compression, and ramping load tests were conducted using a servo hydraulic test machine. A constant electrical current ran through the material when it was being tested. The samples containing the graphene sheets reacted very dramatically under strain [3]. For a 0.1% increase in strain, the voltage changed by 20 mV, which is high.

In conclusion, graphene proved to be a viable additive to increase the mechanical sensitivity of the PVDF polymer [3]. Because the results were positive, for the material's response it is reasonable to use it in a medical application where stress must be measured.

6.2.3 Other Nanomaterial-Based Thermoplastic Nanocomposites

So far only two different types of nanofillers—MMT clay and carbon nanotubes/graphene sheets—have been discussed. Materials, such as clay and graphite are strong and helpful in boosting mechanical properties in polymers. However, there are other types of nanomaterials that may have a similar effect. When speaking of mechanical strength, exotic materials, such as diamond are at the peak of the spectrum. Can it be used effectively and properly though to make a viable nanocomposite? Also, after the nanocomposite is formed, what shapes can we make while still ensuring good dispersion? These are some of the topics discussed in the following sections.

Using Electrospinning to Process the Mixed Nanocomposite into Fibers

There are many different shapes that can be molded using nanocomposites after it is mixed. A group of scientists in Beijing [4] conducted a study on preparing polymer/nanoparticle fibers for insertion into a polyurethane matrix by electrospinning.

The polyurethane was solvent surface treated in a dimethylformamide solution. The idea was to create small composite fibers that would reinforce the nanocomposites that were added and thus enhancing the mechanical properties [4]. The electrospinning process works by pressing the composite through a spindle hooked to a voltage source. As the liquid is released, the charged solution is stretched, and a long fiber is formed. In this case the experiment was done at a constant 24 kV and the solution was passed through the spindle at 0.3 mL/h [4]. A set of TEM images of the resulting mixture is presented in Fig. 6.4. It is shown in these images that the silica particles are

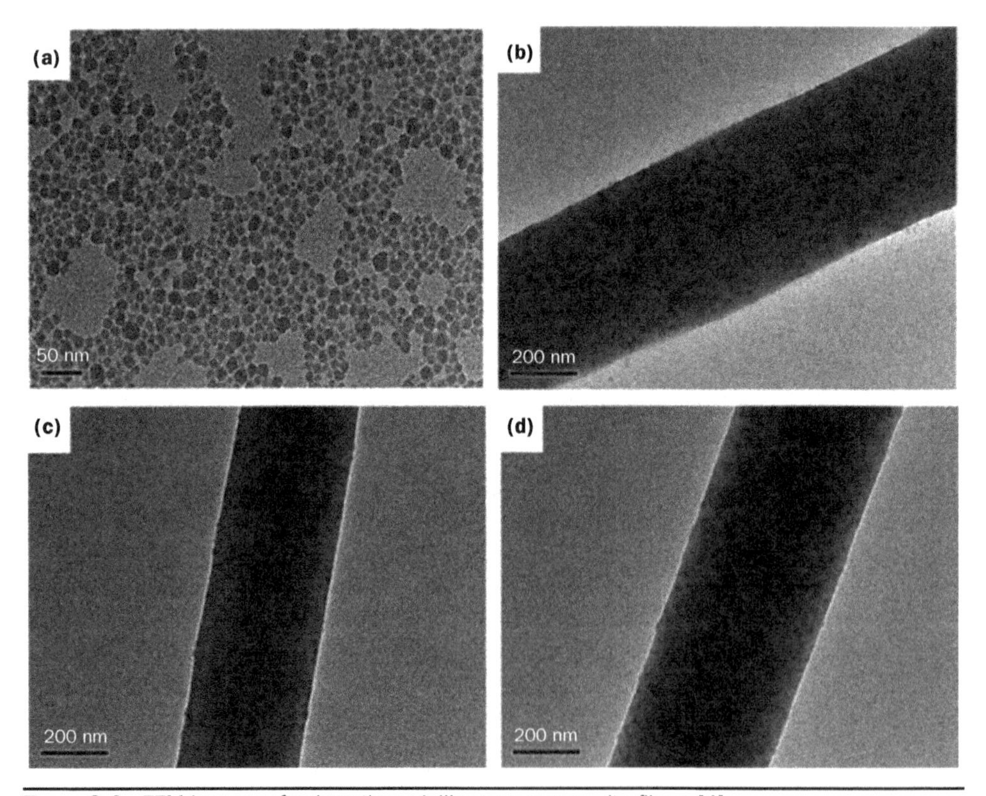

Figure 6.4 TEM images of polyurethane/silica nanocomposite fibers [4].

well dispersed throughout the fibers. Figure 6.4 has four TEM images (a, b, c, d) at three different silica loadings [4]. Figure 6.4b–d shows fibers that are made up of three different silica loadings containing 1%, 5%, and 10% silica. The mechanical properties of the electrospun fiber were calculated using a variety of mechanical testing techniques including tensile tests on some samples [4]. The stress–strain plots for each of the separate loadings are shown in Fig. 6.5 [4].

The stress–strain plot shows that as the silica loading increases, the material is able to take on more stress before yielding. It also experiences a large amount of strain reaching values of up to almost 1,500%. Surprisingly, the strongest material (the 10% load) experiences almost as much strain as the 1% loaded case. It is depicted in the TEM images that all the cases studied showed even dispersion within the fibers [4].

Thermoplastics Modified with Nanodiamonds

From a mechanical property standpoint, diamonds possess desirable attributes in both strength and hardness. A group of researchers in Russia are experimenting by incorporating nanodiamonds (NDs) into several thermoplastic matrices. The two thermoplastics used in their experiments were an amorphous styrene-acrylonitrile copolymer and an amorphous polysulfone (PSP). Another main goal of the project was to experiment with processing conditions. The differences in microstructure and properties were examined when applying both normal melt mixing and mixing under spurt conditions [5].

The NDs used were a PUOO-CX DND 96 brand that came in the form of gray powder with a density of 3.3 g/cm^3. The nanocomposite was processed by mechanical blending in the molten state with a worm-piston mixer in 4–5 mL samples. The copolymer was mixed at 190–200°C and the PSP was mixed at 270–280°C. Samples were made for tensile testing that contained 0.5, 1, 2.5, and 5 wt% NDs for both thermoplastics [5]. The researchers also experimented with mixing conducted during elastic turbulence. This type of mixing exposes the solution to high degrees of stress which

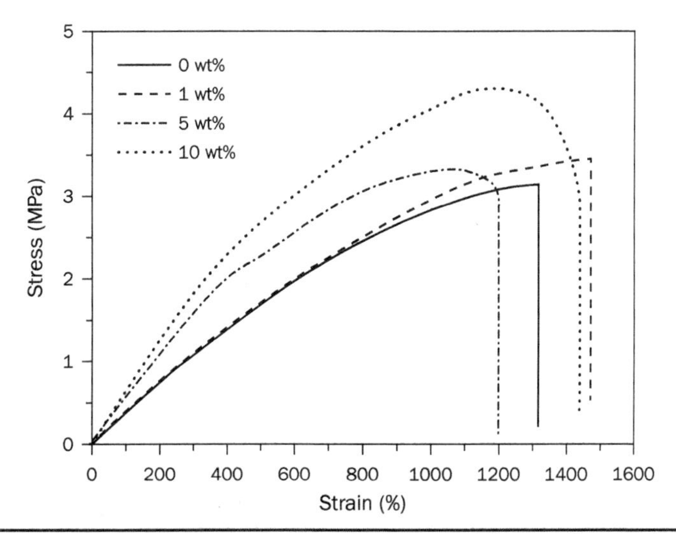

FIGURE 6.5 Stress–strain results for nanocomposite fibers at different silica loadings [4].

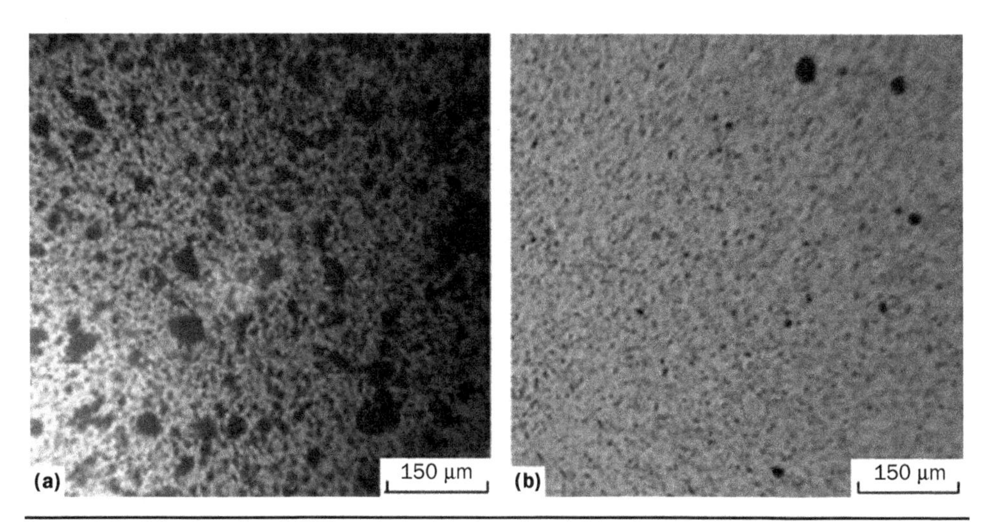

FIGURE 6.6 Micrographs of PNC made by (a) normal melt blending and (b) spurt/elastic conditions [5].

in theory is supposed to break open NDs even further in order to obtain smaller particles. In Fig. 6.6, the micrographs indicated that the elastic mixing procedure did help to produce finer ND particles (Fig. 6.6b) compared to the normally mixed solution (Fig. 6.6a) [5].

Figure 6.7a (left) shows the ratio between the elastic moduli of copolymer-ND composites and the modulus of the unfilled polymer versus the loading of NDs from 0 to 5 wt%. Figure 6.7b (left) shows the ratio between the tensile strengths of copolymer-ND composites and the unfilled polymer versus the loading of NDs from 0 to 5 wt%. Composites were prepared through (1, 3) blending in the spurt mode and (2, 4) by conventional normal melt. For the copolymer prepared by the spurt method, as the content of NDs is increased, the elastic modulus significantly increases, while the elongation at break for the neat polymer remains unchanged [Fig. 6.7a (left)]. For the systems prepared through the conventional normal melt method, the main mechanical parameters tend to decrease as the loading of the filler increases [Fig. 6.7b (left)].

Figure 6.7a (right) and 6.7b (right) present similar dependences of the PSP-based systems. The mechanical properties of PSP before and after treatment in mixer remain invariable. From these data, blending in the spurt mode provides higher strength parameter than the conventional melt blending. The elastic modulus increased with the increase of ND loading [Fig. 6.7a (right)]. The modifying effect of NDs for the composites prepared using the conventional melt method is very statistically insignificant and within the scatter of the experimental data [Fig. 6.7a (right)]. The modulus and max stress ratios shown in Fig. 6.7b (right) are higher for the samples mixed under spurt/elastic conditions [5]. The researchers also found that for the spurt samples, the elongation at break remained the same as they were in the neat polymer. However, all properties degrade under small loadings with conventional normal mixing.

This was explained by the fact that the larger ND particles in the normally mixed polymers acted as defects in the matrix. Furthermore, the spurt conditions seemed to help the overall assimilation by creating nonuniform shearing rates within the polymer.

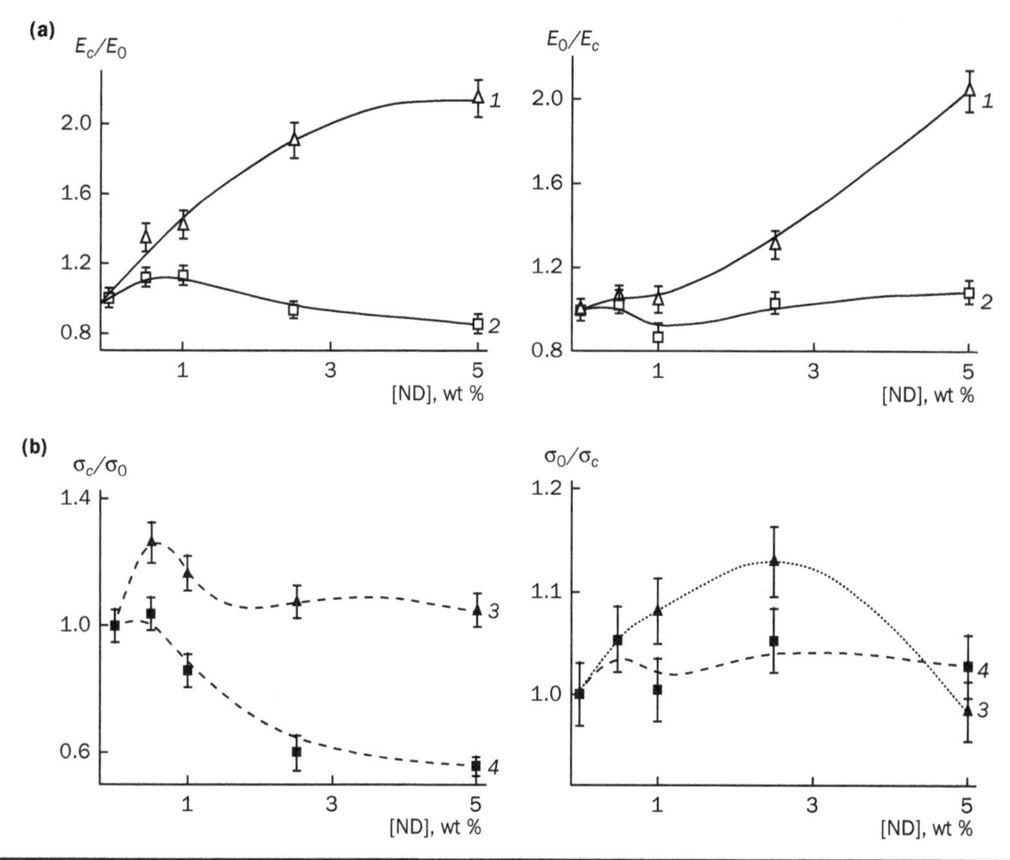

Figure 6.7 (a) Modulus ratios for PSP (right) and the modulus ratios for the copolymer (left); (b) maximum stress ratios for both PSP (right) and copolymer (left) (order is the same as modulus). Note: Samples 1 and 3 were blended through spurt/elastic conditions and Samples 2 and 4 by conventional normal melt conditions [5].

These conditions allowed the NDs to assimilate into weak spots within the polymer matrix. Overall, the shearing seemed to create a better dispersion of nanoparticles [5].

6.2.4 Summary of Thermoplastic-Based Nanocomposites

The mechanical properties of polymers can be strengthened dramatically by adding nanoparticles if the right process and materials are chosen. Incorrect combinations may produce no changes, or, in some cases, even adverse effects. Also, it is not only the process that matters; the fabrication of nanocomposite parts is also important.

6.3 Thermoplastic Elastomer-Based Nanocomposites

This section reviews thermoplastic elastomer–based nanocomposites, subdivided into thermoplastic elastomer-clay nanocomposites, thermoplastic elastomer-carbon–based nanocomposites, and thermoplastic elastomer with other nanomaterials research.

6.3.1 Nanoclay-Based Thermoplastic Elastomer Nanocomposites

High-Performance Elastomeric Nanocomposites via Solvent-Exchange Processing

Liff et al. examined a thermoplastic elastomer nanocomposite formed from Elasthane 80A, a commercially available thermoplastic polyurethane, and unmodified Lamponite RD, a synthetic smectic clay [6]. Nanocomposites with weight loading of 2, 4, 10, 15, and 20 wt% Lamponite RD were prepared via a "novel solvent exchange process." This study describes this process as using two separate solvents to first dissolve the polymer and nanomaterial separately, and then to combine the resulting solutions. The solvents are then extracted, and the remainder is a well-dispersed nanocomposite.

The solution exchange method employed in this study achieved good dispersion and exfoliation of the Lamponite platelets (Fig. 6.8) [6]. The clay had a noticeable effect on the mechanical properties of the polyurethane. Figure 6.9 shows that as clay loading

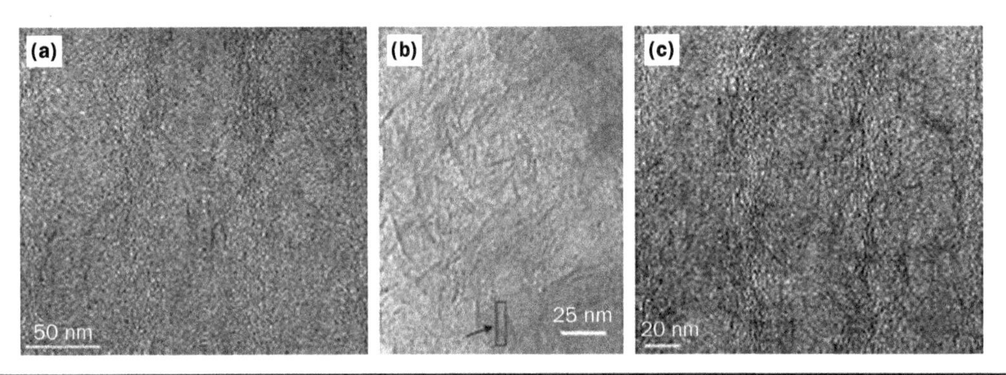

FIGURE 6.8 TEM images of nanocomposites containing (from left to right) 4, 10, and 20 wt% Lamponite RD [6].

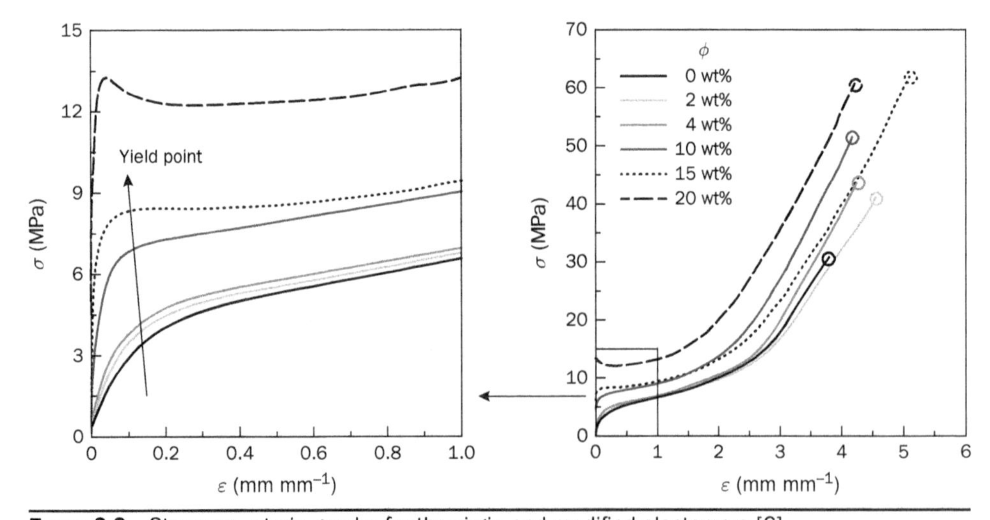

FIGURE 6.9 Stress vs. strain graphs for the virgin and modified elastomers [6].

increased, so did the modulus of elasticity and the ultimate tensile strength (UTS) [6]. The nanocomposite also developed a marked yield point at 20 wt% clay and the UTS doubled compared to virgin polymer. Elongation at break seemed slightly improved by the addition of nanoclays, although the 15 wt% loading had markedly superior performance compared to the other weight loads. The 15% weight load also exhibited the highest UTS. The study did not explore if this was an artifact of testing or a more optimal loading [6].

Morphology and Properties of Thermoplastic Polyurethane Nanocomposites

Chavarria and Paul [7] examined how varying the structure of organoclays would affect the properties of the resulting polymer. They produced nanocomposites from five varieties of MMT clays from Southern Clay Products (now known as *BYK Additives*). The clays varied in how they had been surface-treated by the supplier, which affected the surface structures of the clay platelets. Two varieties of polymer were also utilized, one medium hardness and the other high hardness. The nanocomposites were prepared in a twin-screw, co-rotating, and intermeshing extrusion machine to a variety of clay loading levels.

As depicted in the TEM micrographs in Figs. 6.10 and 6.11, some of the nanoclays dispersed better than others [7]. In most cases, the addition of nanoclays seemed to have significantly increase the modulus of elasticity, also to have increased the upper

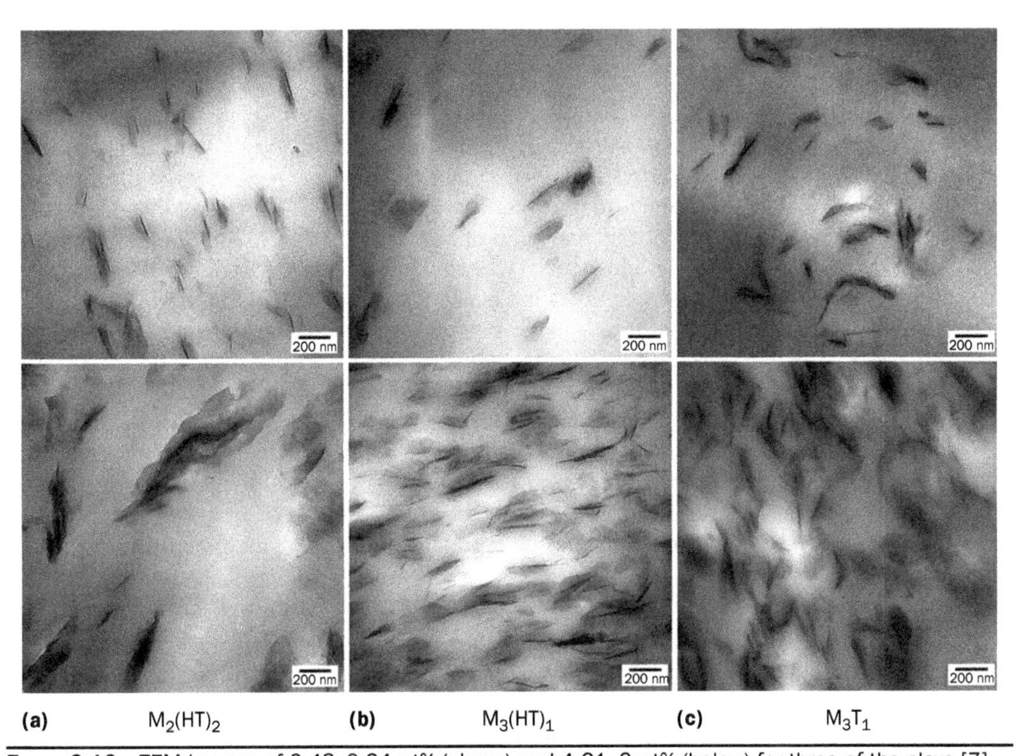

Figure 6.10 TEM images of 0.48–0.94 wt% (above) and 4.61–6 wt% (below) for three of the clays [7].

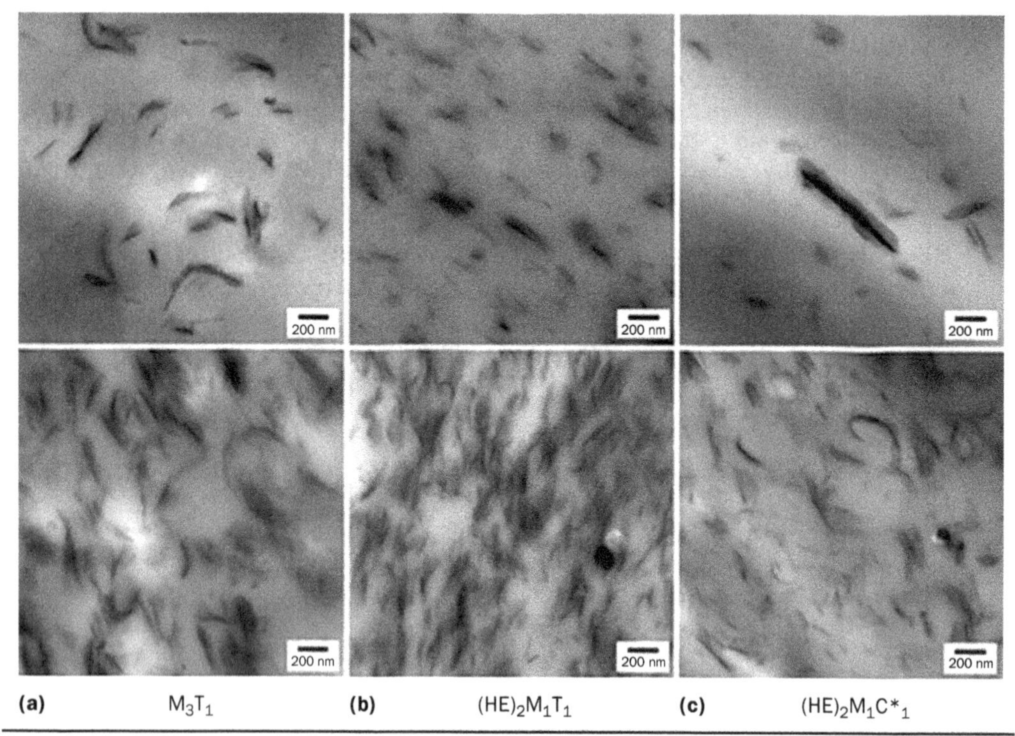

FIGURE 6.11 TEM images of 0.9–0.95 wt% (above) and 5.46–6 wt% (below) for three of the clays [7].

tensile strength. The stress versus strain graphs in Fig. 6.12 show that more clay did not necessarily yield greater improvements in qualities [7]. Most cases showed diminishing returns, or an outright reduction, in upper tensile strength as loading of clay increased. $M_2(HT)_2$-treated clays did not increase the upper tensile strength of the resulting polymer, but instead made it progressively worse as more clay was added. Elongation at break is not explored greatly here, as the tests only ran to 400% elongation, which evidentially was less than what the nanocomposites could withstand. The 5.5 wt% MMT of $(HE)_2M_1T_1$ was noted to have an ultimate shear of less than 400% in medium hardness polyurethane, when it broke at approximately 275%. The high harness polyurethane became much more brittle as more $(HE)_2M_1T_1$ was added. The 3 wt% MMT loading had an ultimate shear one-third of what was to be expected from virgin high-hardness polyurethane. As can clearly be seen here, the surface treatments of the nanomaterial have a tremendous influence on the behavior of the resulting nanocomposite. This research showed impressive depth and examined a great deal of material behavior [7].

6.3.2 Carbon-Based Thermoplastic Elastomer Nanocomposites

Nanostructured Carbon-Reinforced Polyisobutylene-Based Thermoplastic Elastomer

Thermoplastic elastomers have a great track record of being used as biomaterials, so it is only natural that work should be done on exploring thermoplastic elastomer nanocomposites to use inside a living body. Puskas et al. explored the properties of a carbon black nanoparticle reinforced copolymer [8]. The copolymer was formed by

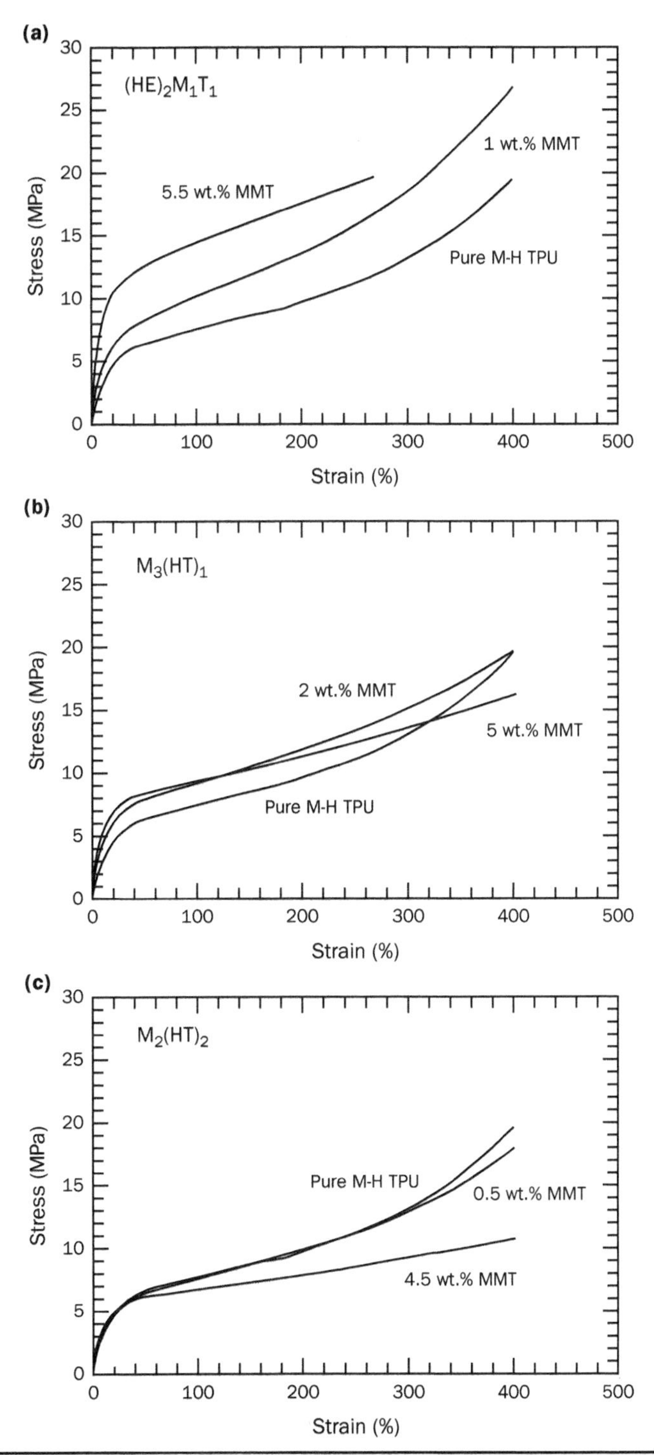

FIGURE 6.12 Stress vs. strain curves for three different nanoclay composites. M-H denotes medium-hardness polyurethane [7].

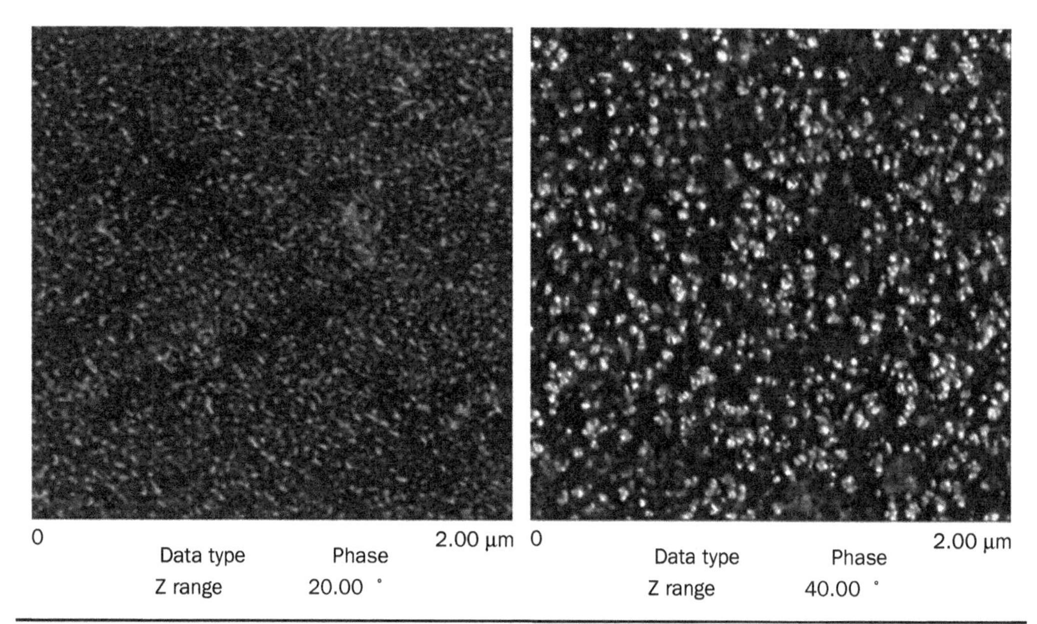

Figure 6.13 AFM of virgin copolymer (left) and copolymer with carbon black added (right). Dark material is elastomeric phase; light material is carbon black or plastic phase [8].

Table 6.2 Experimental Copolymer Tested Ultimate Tensile Strength (UTS) and Ultimate Strain (US) Compared with Silicon and Two Biomaterials [8]

ID	Shore A	Modulus at 100% strain (MPa)	UTS (MPa)	US (%)
D_IB(OH)-MS	52	0.5	6.8	660
D-IB(OH)-MS_CB	56	2.3	13.8	410
SIBS	–	0.5	7.6	650
Silicone	80	1.5	10.2	850
Hexsyn	60	3.4	13.8	314

polymerizing polyisobutylene and poly(isobutylene-co-para-methlystyrene) together in the lab, rather than being purchased commercially. Two batches were prepared, one that was stock polymer and the other being loaded with 37.5 wt% carbon black nanoparticles. The nanoparticles conformed to ASTM N234, indicating a particle size of less than 100 nm. Blending was performed in a Brabender mixer, a laboratory-sized internal mixer. The finished polymers were denoted D_IB(OH)-MS and D_IB(OH)-MS_CB, the latter of which contained the carbon black. Silicone, a thermosetting elastomer, is commonly used in bioimplantation applications, so it served as the control for this experiment.

The carbon black seems to have been fairly well dispersed into the copolymer matrix (Fig. 6.13) [8]. The resulting nanocomposite seems to be quite homogenous, with little agglomeration of the 100 nm or less carbon black particles. Table 6.2 and Fig. 6.14

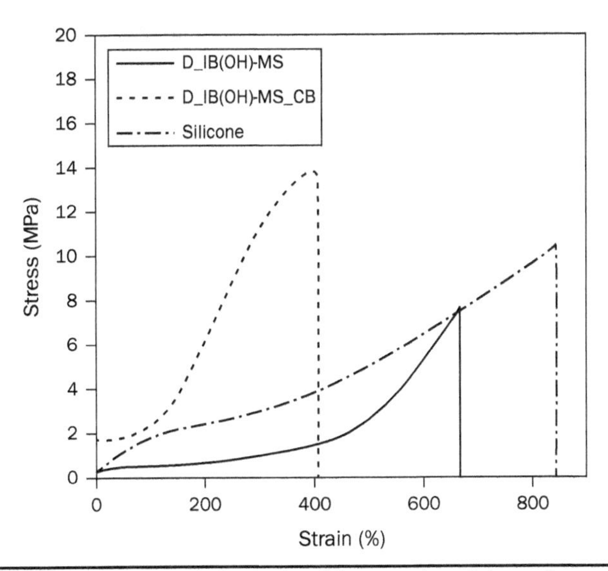

FIGURE 6.14 Graph of stress vs. strain for tests of the experimental copolymer and silicone [8].

show that in testing, the carbon black dramatically doubled the UTS of the experimental copolymer, but reduced its ultimate shear (elongation at break) by approximately 40% [8]. Even so, the heavily loaded copolymer (37.5 wt% carbon black) still stretched four times its length before breaking. The modulus of elasticity of the nanocomposite was also observed as being more than four times that of the unmodified copolymer [8].

Carbon Nanotube-Reinforced Polyurethane Composite Fibers

Chen et al. were interested in other work carried out with nylon-6 and carbon nanotubes [9]. Those studies had shown dramatic increase in the mechanical properties of thermoplastic polymers with the addition of nanotubes, and Chen et al. were curious how thermoplastic elastomers performed with similar additions of carbon nanotubes. Polyurethane was chosen for this study's elastomer matrix, along with MWNTs with diameters between 40 and 60 nm. The MWNTs were chemical functionalized by washing them in an acid, then washing the acid away with water. After washing, the MWNTs were dispersed into the polyurethane in a counter-rotating twin-screw extruder that then extruded 1.75-mm fibers out through a die. These fibers were then melted down to final diameters of about 200 μm. Batches were prepared with 5.6, 9.3, and 17.7 wt% MWNTs.

The combination of acid washing and twin-screw extrusion did a good job of dispersing the carbon nanotubes into the polyurethane matrix. Figure 6.15 shows that the MWNTs were dispersed randomly and seemingly homogenously at the lower two loadings, but at the 17.7 wt% loading, fiber crowding led to some low-level agglomeration [9]. Mechanical testing resulted in promising data, which can be seen in Fig. 6.16 [9]. Additional MWNT loading led to an increase in Young's modulus, UTS, and elongation at break. The increase in Young's modulus was noted with all loadings, but the increase in tensile strength and elongation at break had seemingly optimal loading levels. UTS was most greatly improved at the 5.6 wt% level and

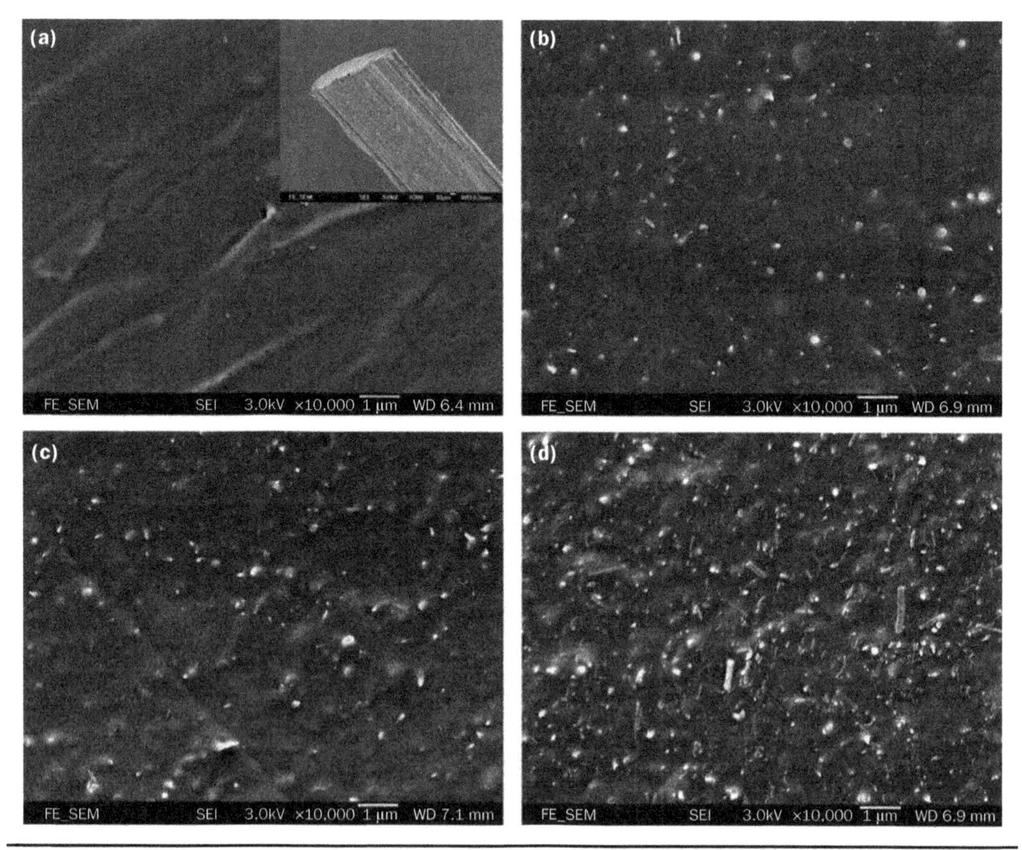

FIGURE 6.15 SEM images of neat polymer and inset of (a) cross-sectional composite fiber, (b) 5.6 wt% MWNT composite, (c) 9.3 wt% MWNT composite, and (d) 17.7 wt% MWNT composite [9].

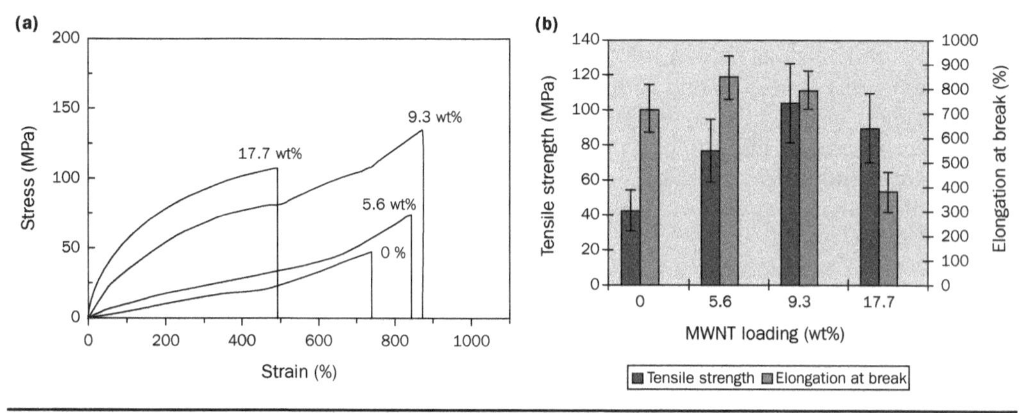

FIGURE 6.16 Stress vs. strain graph of the nanocomposites (left) and a bar graph summarizing the mechanical test results (right) [9].

elongation at break was the highest recorded at the 9.3 wt% level. The 17.7 wt% was noticeable stronger than the virgin polymer, but had a marked decrease in elongation at break. The bar graph in Fig. 6.16 indicates that an overall optimal balance of increased tensile strength and elongation at break seems to exist relatively close to 9.3 wt% MWNT [9]. Chen et al. note that, to their knowledge, simultaneous improvement in Young's modulus, tensile strength, and elongation at break in polyurethane had not been addressed before [9].

6.3.3 Other Nanomaterial-Based Thermoplastic Elastomer Nanocomposites

Mechanical and Fracture Behaviors of Elastomer-Rich Thermoplastic Polyolefin/SiC Nanocomposites

Liao and Tjong explored how the addition of 45–55 nm silicon carbide (SiC) nanoparticles to an elastomer-rich thermoplastic polyolefin (ETPO) affected fracture behavior [10]. The thermoplastic polyolefin utilized was a copolymer of 30 wt% commercial polypropylene and 70 wt% commercial poly(styrene-ethylene-butylene-styrene) grafted with malic anhydride. The copolymer was prepared in a twin-screw extruder where silicon carbide nanoparticle loadings of 1, 3, and 5 wt% were introduced. Izod impact tests were also conducted, in addition to traditional tensile tests, at two different crosshead speeds.

Figure 6.17 shows that the SiC nanoparticles were left behind in groups after etching [10]. The study notes that the nanoparticles were not well dispersed, and tended to agglomerate within the polymer matrix, and that a sizeable percentage of the SiC particles could not be seen in the SEM images, as they were removed along with the elastomer during etching. Table 6.3 and Fig. 6.18 lists and charts, respectively, the mechanical properties of the ETPO-SiC nanocomposites [10]. As SiC loading increased, the Young's modulus and tensile strength at a given strain or strain rate decreased slightly. Elongation at fracture was not explored in Liao and Tjong's work, but all

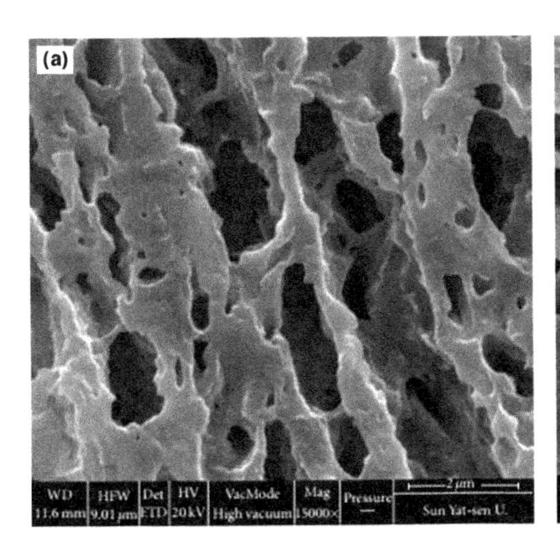

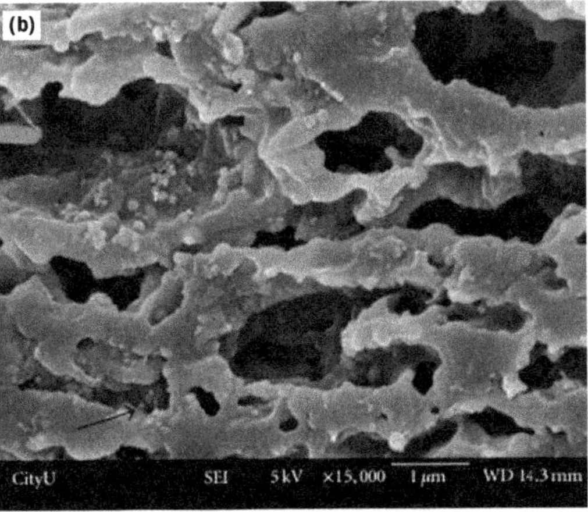

FIGURE 6.17 SEM images of etched virgin ETPO (left) and etched ETPO with 3 wt% SiC (right). The etchant removed the elastomer and left the polypropylene behind. The small light color particles in the right image are SiC nanoparticles [10].

Table 6.3 Mechanical properties of ETPO and SiC nanocomposites [10]

Sample	HDT (°C)	Young's modulus (MPa) 10 mm/min	Tensile strength (MPa)				Impact strength (kJ/m²)
			10 mm/min		300 mm/min		
			50%	300%	50%	300%	
ETPO	53.64	269 ± 5	6.92 ± 0.02	8.00 ± 0.11	8.57 ± 0.10	9.30 ± 0.05	2.84 ± 0.02
ETPO-1 wt% SiC_p	54.32	255 ± 7	6.61 ± 0.22	7.80 ± 0.02	8.30 ± 0.12	9.18 ± 0.14	2.90 ± 0.02
ETPO-3 wt% SiC_p	55.28	236 ± 10	6.21 ± 0.13	7.54 ± 0.16	7.81 ± 0.09	8.96 ± 0.13	2.92 ± 0.02
ETPO-5 wt% SiC_p	56.57	238 ± 12	6.22 ± 0.08	7.86 ± 0.09	7.92 ± 0.21	9.10 ± 0.07	3.11 ± 0.25

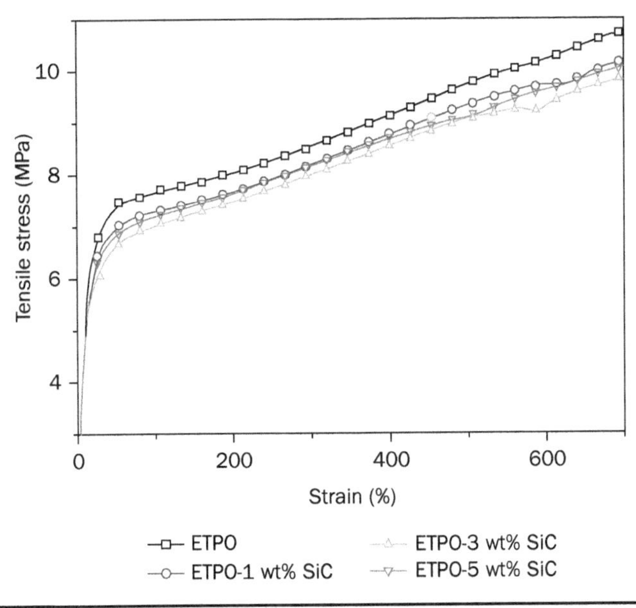

FIGURE 6.18 Stress vs. strain graph for ETPO-SiC nanocomposites [10].

nanocomposite compositions demonstrated capacities to withstand strains in excess of 600%. A modest increase in impact strength can be seen with increased SiC nanoparticle content, and the article by Liao and Tjong notes that the fracture toughness of the material was five times higher in the 5 wt% SiC nanocomposite compared to the virgin copolymer [10]. They made no mention of this, but it may be possible that the reduction in tensile strength may have been related to the poor dispersion of the SiC nanoparticles within the polymer matrix.

Surface Modification Effects on the Structure and Properties of Nanosilica-Filled Thermoplastic Elastomer

Some of the previous studies have shown the importance of surface treatments for organic nanoclays in thermoplastic elastomers, and Aso et al. explored the same question for nanosilica [11]. The study examined nanocomposites formed by dispersing unmodified or surface-modified nanosilica into Hytrel (a DuPont brand name of a thermoplastic copolyetherester elastomer) using a twin-screw extruder. The unmodified nanosilica particles had an average size of 16 nm, while the modified particles averaged 12 nm in size. Composites of 0.5, 1, 2, 3, and 6 wt% nanosilicates were prepared. The XRD patterns of neat Hytrel, 6 wt% modified nanosilicate, and unmodified nanosilicates are shown in [11].

Even without the modifier, the nanosilica was fairly well dispersed, as can be seen in Fig. 6.19, but the structures formed were as large as 1 μm in some cases [11]. However, the surface modification helped reduce the size of these agglomerates, as can be seen by comparing Fig. 6.19b with Fig. 6.19d. The surface modification reduced most of the nanosilica agglomerates to between 15 and 150 nm. This difference in agglomerate size likely contributed to the resulting differences in mechanical

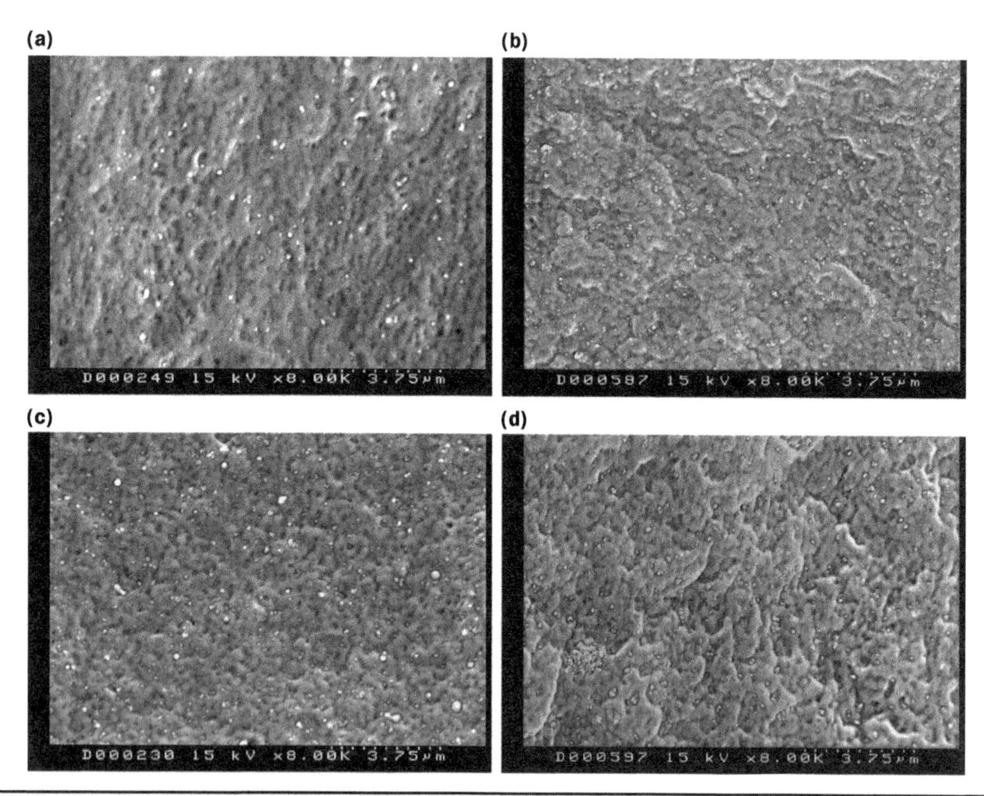

Figure 6.19 SEM of modified nanosilicate composite: (a) 1 wt%, (b) 3 wt%, (c) 6 wt%, and (d) 3 wt% unmodified nanosilicate [11].

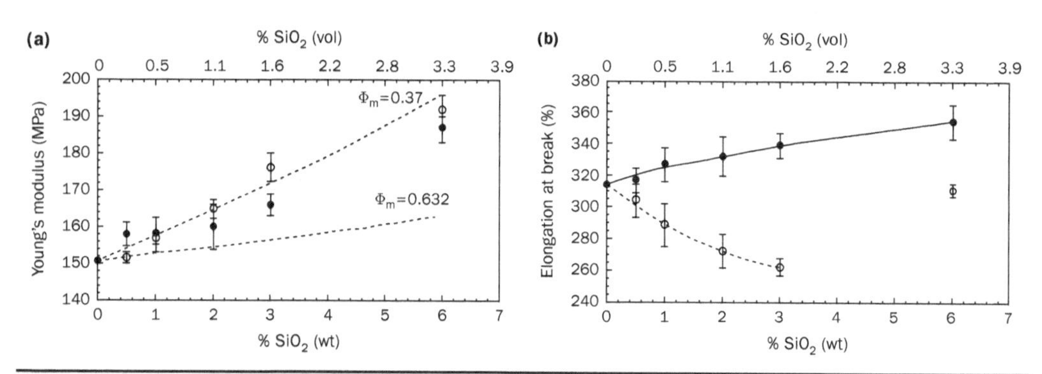

Figure 6.20 Graph of Young's modulus vs. wt% nanosilica (left) and elongation at break vs. wt% nanosilica (right). Solid dots are modified nanosilica composites; hollow circles are unmodified nanosilica [11].

properties seen in the nanocomposites. Figure 6.20 shows that the Young's modulus increased as more modified or unmodified nanosilica was blended into the Hytrel matrix, but this was not the case with elongation at break [11]. The addition of modified nanosilica, with its smaller agglomerates, led to a modest increase in elongation at break. Conversely, the addition of unmodified nanosilica reduced the Hytrel's elongation at break to a greater degree than the modified nanosilica increased it. The paper also makes note that the addition of modified nanosilica improved the ability of the polymer to resist creep and the addition of unmodified nanosilica improved creep resistance even more so [11]. Clearly, just as with clays, the surface modification applied to nanosilica plays a great deal of importance when determining a nanocomposite's mechanical properties.

6.3.4 Summary of Thermoplastic Elastomer-Based Nanocomposites

The general observation that can be derived from this review is that when adding a nanomaterial to a thermoplastic elastomer, it is crucial that the surface of the nanomaterial should be compatible with the polymer. If not, the nanomaterial will just agglomerate into micrometer-sized structures. The method of processing also must suitably disperse the nanomaterial, or else, again, it will just agglomerate. If sufficiently dispersed, the material will almost always display an increased Young's modulus and will frequently have a higher UTS. The amount of elongation at break is typically negatively impacted by the addition of nanomaterials, but in certain cases it can actually be improved along with Young's modulus and UTS.

As the base polymers are elastomers, the final elongation at break may still be considerable even after being reduced by adding a nanomaterial. Also, the same nanomaterial may perform differently in different thermoplastic elastomer, so it is crucial that perspective formulators of nanocomposites should be aware of what is possible from what combinations of nanomaterials and polymers. Hopefully researches will continue to explore thermoplastic elastomer nanocomposites because there certainly seems to be a great deal of promise there.

6.4 Thermoset-Based Nanocomposites

Thermoset nanocomposites are widely used in many applications, including the aerospace industry. For the past few decades, composite materials have come into use in defense applications, due to the improved efficiencies caused by the weight reduction compared to their metallic counterparts. The higher performance demands of new applications require composites with improved material properties. Composites enhanced with nanomaterials serve to improve properties, such as Young's modulus, toughness, and yield strength, while still maintaining the weight advantage these materials offer. In this section, examples provided are based on the research conducted to analyze the enhancements of mechanical properties of epoxy-based nanocomposites: nanoclay, carbon-based, nanosilica, nano–titanium dioxide, and nano–zirconium dioxide.

6.4.1 Epoxy Nanocomposites

Epoxy-Clay Nanocomposites

Research carried out by Kinloch and Taylor [12] tested the mechanical properties and fracture toughness of DGEBA, a diglycidyl ether of bisphenol A modified with MMT clay. A comparison to theoretical mechanics models was included in the research. The types of clay used were Cloisite 25A, Cloisite 30B, Cloisite Na^+, and Nanomer I30E. Plates of epoxy composite were produced by stirring the clay modifier into the epoxy and then baking in a vacuum oven at 75°C. The vacuum was removed, and the plates

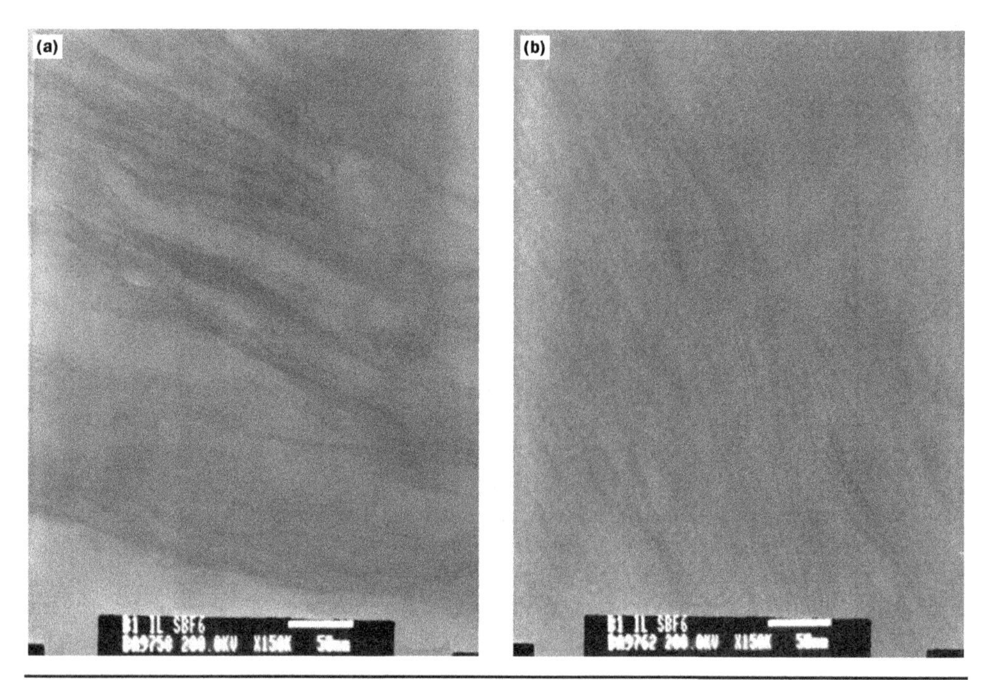

FIGURE 6.21 Transmission electron micrograph of epoxy nanocomposite with 5 wt% Cloisite 25A (left) and 5 wt% Nanomer 130E (right) [12].

were baked at this temperature for another 24 hours. Hardener was added to the mixture, stirred, cured for 3 hours at 75°C, followed by another round of curing for 12 hours at 110°C.

Characterization of the dispersion of the nanomaterial was performed using wide-angle x-ray diffraction (WAXD) and visual inspection via transmission electron microscopy (TEM) analyses. Mechanical testing was performed at a displacement rate of 1 mm/min and the Young's modulus was determined.

Several analytical models were employed for comparison to experimental data. For the prediction of the tensile modulus, the modified rule of mixtures, the Halpin–Tsai model, and the Mori–Tanaka model were utilized. For the prediction of the fracture energy, the Farber and Evans analysis was utilized.

WAXD data showed differing levels of dispersion of the nanoparticles. A loading of 5 wt% I30E reveals intensity with no "001" peak, an indication of an exfoliated nanostructure. A loading of 5 wt% Cloisite 25A and Cloisite 30B revealed a "001" peak that indicated an intercalated nanostructure, with darker areas representing the agglomerated clay. TEM analysis confirmed these observations, further adding that the exfoliated structure formed in the Nanomer I30E composite was of an ordered nature (Fig. 6.21) [12].

The tensile modulus experiments show that as the mass of the modifier increases, the Young's modulus increases [12]. The nanoparticle that was considered the best modifier of the Young's modulus was Nanomer I30E. However, at 10 wt%, the I30E mixture is too viscous and reaches the maximum amount of modifier that can be added to the filler material. The other modifiers follow a similar modulus trend as nanomaterial is

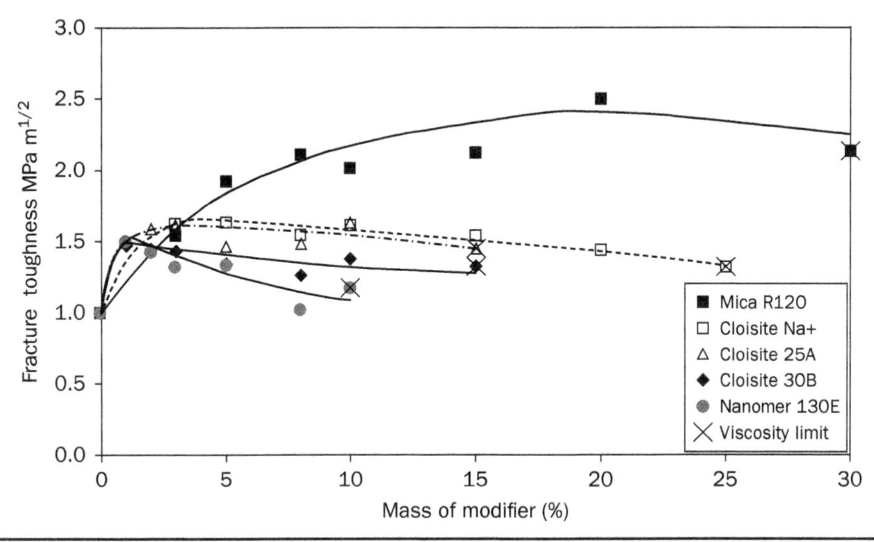

FIGURE 6.22 Fracture toughness of modified epoxy nanocomposites [12].

added to the epoxy, but reach a viscosity limit at 15 wt%. The exception to this trend is Cloisite Na^+, which reaches a viscosity limit of 25 wt% and a modulus improvement of 2.5 GPa (unmodified epoxy is 3 GPa).

The fracture toughness experiments show an initial increase of the toughness at the smallest amount of nanomaterial added, but further additions of material eventually reveal a peak in the toughness followed by a decrease which occurs roughly around 2–4 wt% loading, with the exception of Cloisite 25A (about 10%) (Fig. 6.22) [12].

The analytical models for the tensile modulus agreed fairly well with the tensile data, specifically the Mori–Tanaka model. The model used to predict fracture toughness did not produce acceptable correlation with the experimental data. This is thought to be caused by a fracture mechanism dominated by cavities formed by plastic deformation, as opposed to a crack deflection mechanism described by the Farber and Evans model [12].

Epoxy Carbon-Based Nanocomposites

A study carried out by Gojny et al. [13] explores the influence of different kinds of carbon nanotubes on the mechanical properties of epoxy matrix composites. The epoxy matrix used in their study was a DGEBA epoxy resin modified with single-walled nanotubes (SWNTs) with an average diameter of 2 nm, double-walled nanotubes (DWNTs) with an average diameter of 2.8 nm, and multiwalled nanotubes (MWNTs) with an average diameter of 15 nm. Additionally, DWNTs and MWNTs functionalized with ammonia were used as resin nanomodifiers ($-NH_2$). Resin samples were prepared with 0.1, 0.3, and 0.5 wt% nanomaterial for each of the aforementioned materials.

The epoxy resin was prepared by coarsely mixing the nanotubes into the resin without hardener, followed by high-shear mixing with a gap space of 5 μm and speed ranging from 20 to 180 rpm. Following the mixing, a hardener was added to the mixture and then cured for 24 hours at room temperature, followed by additional curing at 60°C for 24 hours.

Mechanical characterization was performed according to DIN EN 527.1/2 with dog bone specimens using a Zwich universal tensile tester at a crosshead speed of 1 mm/min. Fracture toughness testing was performed according to ASTM standards and at a crosshead speed of 1.3 mm/min.

TEM images revealed adequate dispersion for all samples used. MWNT samples showed the best dispersion of the modifier, with individual nanotubes spaced evenly throughout the TEM sample (Fig. 6.23) [13]. Both SWNTs and DWNTs showed a mixture of small broken-up agglomerates and individual nanotubes. The

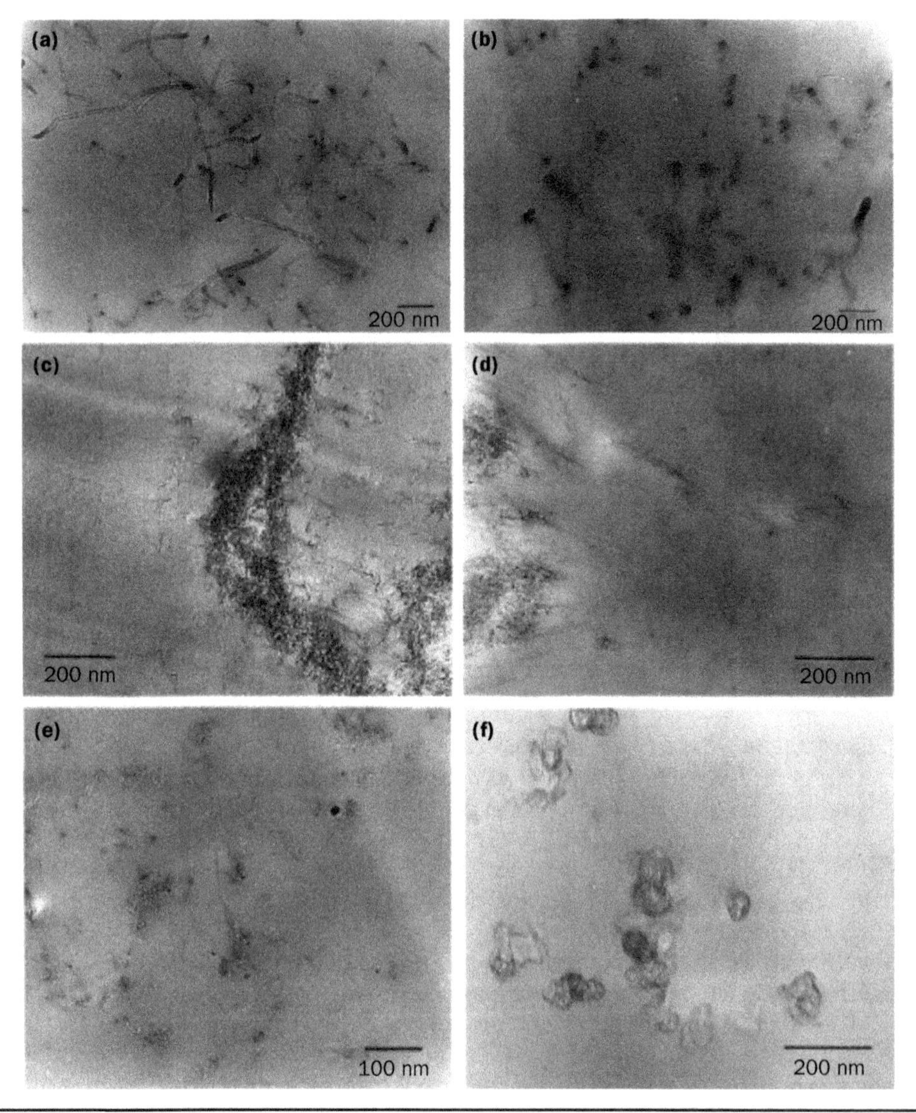

FIGURE 6.23 TEM images of nanocomposites: (a) epoxy/MWNTs, (b) epoxy/MWNTs–NH_2, (c) epoxy/DWNTs, (d) epoxy/DWNTs–NH_2, (e) epoxy/SWNTs, and (f) epoxy/CB [13].

ammonium-functionalized MWNTs and DWNTs showed better dispersion of the nanotubes and an overall smaller size agglomerate diameter when compared to their non-functionalized counterparts.

In general, the addition of carbon nanotubes to the epoxy resin increased the Young's modulus, UTS, and fracture toughness for all types of carbon nanotube and weight percentage. The experimental data are shown in Table 6.4 [13].

The most notable property improvement is the functionalized DWNT, resulting in 14% improvement in Young's modulus, an 8% increase in UTS, and a 90% improvement in fracture toughness. In general, fracture toughness was greatly enhanced for all modifiers and weight percentages when viewed on a percentage basis. Mechanical properties improve as weight percentage increases from 0.1% to 0.3% for all nanomodifiers, but decreases occur at 0.5 wt% for unfunctionalized DWNT and MWNT modifiers and in fracture toughness for SWNT nanomaterial.

Table 6.4 Mechanical Properties of Epoxy Nanocomposites [13]

	Filler type/ content (wt%)	Young's modulus (MPa)	Ultimate tensile strength (MPa)	Fracture toughness K_{Ic} (MPa $m^{1/2}$)
Epoxy	**0.0**	**2599** (±**81**)	**63.80** (±**1.09**)	**0.65** (±**0.062**)
Epoxy/CB	0.1	2752 (±144)	63.28 (±0.85)	0.76 (±0.030)
	0.3	2796 (±34)	63.13 (±0.59)	0.86 (±0.063)
	0.5	2830 (±60)	65.34 (±0.82)	0.85 (±0.034)
Epoxy/SWCNTs	0.05	2681 (±80)	65.84 (±0.64)	0.72 (±0.014)
	0.1	2691 (±31)	66.34 (±1.11)	0.80 (±0.041)
	0.3	2812 (±90)	67.28 (±0.63)	0.73 (±0.028)
Epoxy/DWCNTs	0.1	2785 (±23)	62.43 (±1.08)	0.76 (±0.043)
	0.3	2885 (±88)	67.77 (±0.40)	0.85 (±0.031)
	0.5	2790 (±29)	67.66 (±0.50)	0.85 (±0.064)
Epoxy/DWCNTs–NH_2	0.1	2610 (±104)	63.62 (±0.68)	0.77 (±0.024)
	0.3	**2944** (±**50**)	**67.02** (±**0.19**)	**0.92** (±**0.017**)
	0.5	**2978** (±**24**)	**69.13** (±**0.61**)	**0.93** (±**0.030**)
Epoxy/MWCNTs	0.1	2780 (±40)	62.97 (±0.25)	0.79 (±0.048)
	0.3	2765 (±53)	63.17 (±0.13)	0.80 (±0.028)
	0.5	2609 (±13)[a]	61.52 (0.19)[a]	[a]
Epoxy/MWCNTs–NH_2	0.1	2884 (±32)	64.67 (±0.13)	0.81 (±0.029)
	0.3	2819 (±45)	63.64 (0.21)	0.85 (±0.013)
	0.5	2820 (±15)	64.27 (±0.32)	0.84 (±0.028)

[a] High viscosity disabled degassing – composite contained numerous voids.

In summary, the study revealed that the addition of carbon-based nanomaterials to the DGEBA epoxy matrix improved Young's modulus, and UTS improved while increasing the toughness of the material.

Research performed by Zheng et al. [14] explored the properties of epoxy resin TDE-85 modified with MWNTs. Two types of MWNTs were utilized for dispersion: untreated MWNTs and amino-functionalized carbon nanotubes (MWNTs–NH_2). This chemical modification is thought to enhance the compatibility of the particle with the matrix. The diameter of the nanotubes ranged from 30 to 100 nm. The MWNTs were heated and sonicated for 30 minutes in a triethylenediamine ethanol solution. The nanotubes were then added to the epoxy resin at 120°C to lower the viscosity of the resin. The nanocomposite slurry was then dispersed in a high-speed homogenizer of 20,000 rpm for 20 minutes and degassed at 90°C for 30 minutes. The epoxy resin was mixed with 0.2%, 0.4%, 0.6%, and 1.0% of MWNTs and MWNTs–NH_2 separately.

The chemical modification of the carbon nanotubes proved to be an effective way to improve dispersal. TEM images of the epoxy resins showed the improvement in the dispersal of the carbon nanotubes, indicated by the appearance of individual nanotubes in the chemically modified resin. The experimental data from impact strength, bending modulus, and bending strength show improved performance for the chemically modified MWNTs, further indicating enhanced properties due to improved particle dispersion. The mechanical properties discussed later will address the improvements in performance from the inclusion of the chemically modified MWNTs. The TEM image of dispersion is shown in Fig. 6.24 [14].

The addition of MWNTs to the neat resin showed an increase of impact strength, bending modulus, and bending strength up to 0.6 wt% carbon nanotube. The peak of property performance occurs at 0.6 wt% for the aforementioned properties. A maximum enhancement of 80%, 50%, and 100% were observed for impact strength, bending modulus, and bending strength, respectively. It was observed that the addition of carbon nanotubes past 6 wt% resulted in a decrease of the aforementioned properties (up to 1 wt%). Additionally, it should be noted that the addition of MWNT to the neat

Figure 6.24 Transmission electron microscopy image of (a) unmodified MWNTs and (b) MWNTs-NH_2 [14].

resin increased the density with greater wt%, but this increase in density is minimal. Experimental data for the above observations are shown in Fig. 6.25 [14].

Research conducted by M. A. Rafiee and J. Rafiee [15] considered the previously mentioned MWNTs and SWNTs, and also included graphene platelets (GPL) as a modifier in an epoxy matrix. Graphene platelet consists of multiple stacked sheets of single-atom thick graphene. Young's modulus, tensile strength, and fracture toughness were measured in a series of experiments. The GPL, SWNTs, and MWNTs were dispersed in a bisphenol A epoxy. The sample preparation is as follows: nanoparticles are sonicated in a solvent and stirred with a Teflon magnet. The solvent is removed, and the mixture is processed in a high-shear mixer. The resulting mixture is then degassed under a vacuum and cured in the final part molds. All mixtures prepared contained 0.1 wt% of nanomaterials. TEM images revealed that the preparation described above produced a high level of dispersion of the nanomaterials within the epoxy matrix.

Tensile strength tests revealed that all nanoparticle additives improved the property performance when compared to the pristine epoxy. The GPL were shown to improve the performance the most, followed by MWNTs and SWNTs. The GPL improved the pristine epoxy tensile strength performance by 45% (Fig. 6.26) [15].

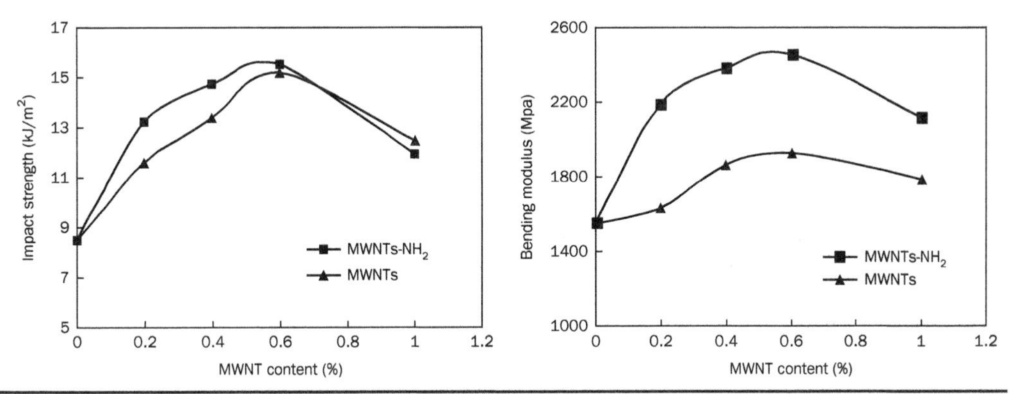

FIGURE 6.25 Plots of mechanical properties for modified and unmodified MWNTs [14].

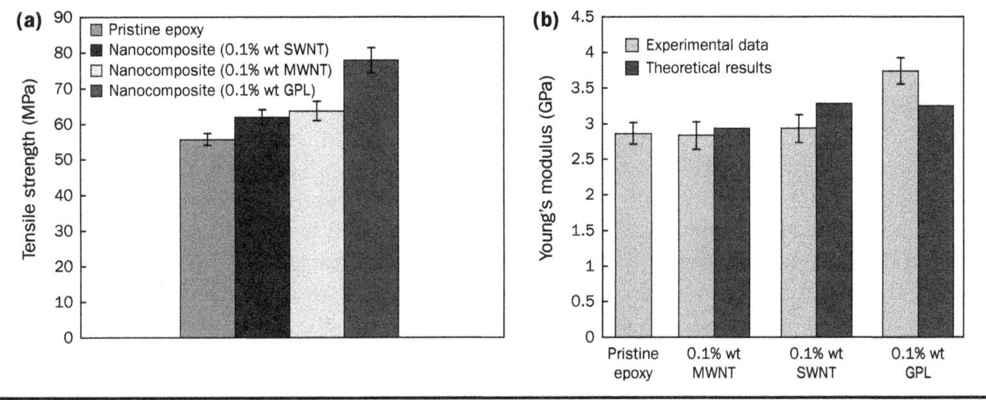

FIGURE 6.26 Tensile strength and Young's modulus plots for carbon-modified epoxies [15]. (*Courtesy of ACS.*).

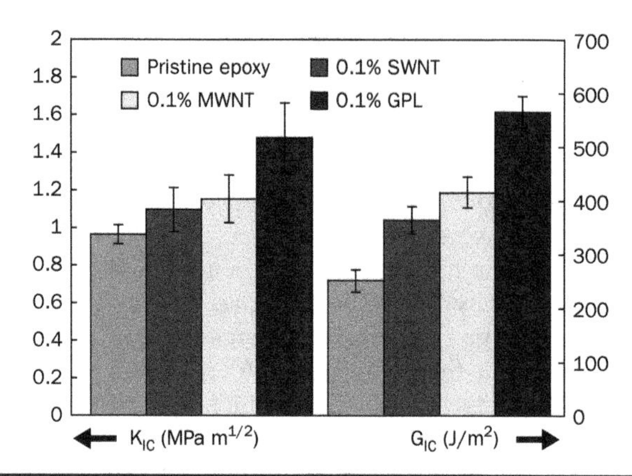

FIGURE 6.27 Fracture toughness and fracture energy plots for carbon-modified composites [15]. (*Courtesy of ACS.*)

The Young's modulus was also shown to improve the performance of the pristine epoxy, with the GPL leading to the greatest improvement, followed by SWNTs and MWNTs. The GPL improved the pristine epoxy modulus by about 30%. The results for these experiments are shown in Fig. 6.27 [15].

The fracture toughness and fracture energy showed similar trends seen in the tensile strength tests. The GPL performed the best, followed by MWNTs and SWNTs. All nanoparticles improved the pristine epoxy performance, with GPL improving the fracture toughness by 50% and the fracture energy by 125%. These results of the fracture toughness tests are shown in Fig. 6.27 [15].

Epoxy with Other Nanomaterials

Blackman et al. [16] explored the fracture and fatigue behavior of diglycidyl ether of bisphenol A (DGEBA) epoxy resin modified with silica (SiO_2) nanoparticles. The nanoparticles are dispersed via the sol-gel method and cured for 1 hour at 90°C followed by 2 hours at 160°C. Specimens were prepared for 4.0, 7.8, 14.8, 20.2 wt% silica. Atomic force microscopy (AFM) characterized the level of dispersion of the nanoparticles and revealed good dispersion of the nanomaterial. An atomic force microscope AFM image of the dispersion is shown in Fig. 6.28 [16].

Glass temperature, tensile, and fracture toughness tests were applied to the silica-enhanced specimen. Tests revealed that there was no substantial increase in glass transition temperature with increasing amounts of weight percent silica. The tensile tests revealed that increasing weight percent of silica led to an increase in Young's modulus. Furthermore, the addition of silica nanoparticles was positively correlated to fracture toughness. A 20.2 wt% silica addition to the neat resin led to a 30% increase in modulus and a 70% in fracture toughness. An analysis of the crack growth rate per cycle revealed that the value of the range of the applied stress-intensity factor at threshold increases steadily as the fracture toughness increases, or as the concentration of silica increases. Table 6.5 shows the experimental data of these epoxy-SiO_2 nanocomposites [16].

TABLE 6.5 Mechanical Properties of Nanosilica-Modified Composites [16]

wt% nano-SiO_2	T_g (°C)	E (GPa)	K_{Ic} (MPa $m^{1/2}$)
0.0	153	2.96	0.51
4.0	152	3.20	0.65
7.8	154	3.42	0.79
14.8	152	3.60	0.83
20.2	150	3.85	0.88

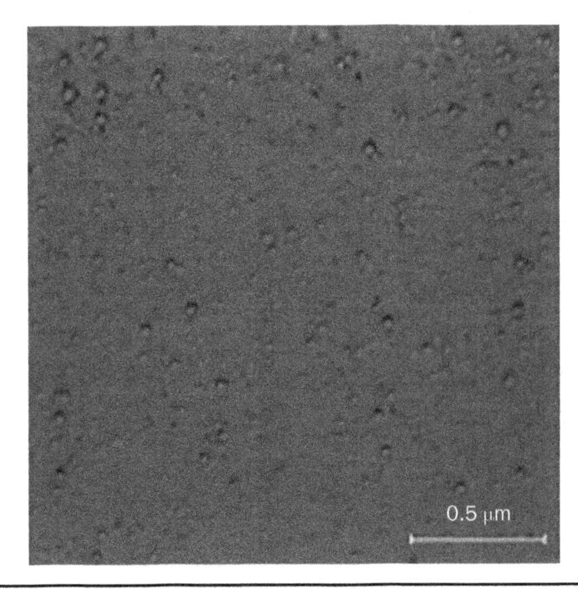

FIGURE 6.28 Atomic force microscopy image showing the dispersion of SiO_2 nanoparticles in the epoxy matrix [16].

Al-Turaif studied the effect of nano–titanium dioxide (n-TiO_2) on the mechanical properties of cured epoxy resins [17]. Particles of titanium oxide with sizes 220, 50, and 17 nm and at 1, 3, 5, and 10 wt% were dispersed into a DER331 epoxy, an undiluted DGEBA-based liquid epoxy resin. The composite samples were prepared by using a moderate speed paddle mixer for 30 minutes followed by a 6-hour degassing at 80°C and 700 mbar.

Tensile tests were conducted on the Instron 5568 using ASTM standard samples and an extension rate of 0.5 mm/min. The tensile tests provided data to determine toughness, UTS, ultimate elongation at break, and Young's modulus. A test was performed to determine the flexural stress of the material according to ASTM D-790 standards and at a crosshead rate of 2.5 mm/min.

The tensile tests revealed that at 1 and 3 wt% smaller particle sizes resulted in improved UTS. Conversely, at 5 and 10 wt%, larger particle sizes resulted in improved

UTS. The peak in UTS was observed to occur at 3 wt% for the two smaller particle sizes and at 5 wt% for the larger 220-nm particle size. The maximum value of UTS occurred at 17 nm at 3 wt% with a value that was 14% greater than the unfilled epoxy matrix. For Young's modulus, a smaller particle size resulted in the largest value. Like in the UTS data, a peak occurred at 3 wt% for all particle sizes. The maximum value for the modulus was observed to occur at 3 wt% for the smallest particle size with a value that was 24% greater than the unfilled epoxy matrix. There was no noticeable correlation between particle size and elongation at break. However, the largest value of elongation occurred at 3 wt% and smallest overall modified value occurred at 1 wt%, a trend similar to UTS and Young's modulus data. The maximum value of elongation at break occurred for 17-nm particles at 3 wt% with a value that was 70% greater than the unfilled epoxy matrix (Fig. 6.29) [17].

Toughness tests continued the general trend of smaller particle size leading to improved material toughness. The common trend of peak values occurring at 3 wt% for all particle sizes was also observed. The maximum value of toughness occurred at 17 nm and 3 wt% with a value that was 100% greater than the unfilled epoxy matrix [17].

Flexural stress also showed a similar trend, but reflected for the desire to minimize the stress values. The lowest values for stress occurred at the larger particle sizes and increased as the particle size decreased. The lowest values for the stresses were observed at 3 and 5 wt% for the 220 and 50 nm particle size and at 5 wt% for the 17 nm particle size. Overall, the flexural stress for all modified composites was larger than the unfilled epoxy matrix. The smallest flexural stress increase occurred at 3 and 5 wt% for the 220 nm particle size with a value that was 14% larger than the unfilled epoxy. The largest increase occurred at 1 wt% for the 17 nm particle size with a value that was 23% larger than the unfilled epoxy (Fig. 6.29) [17].

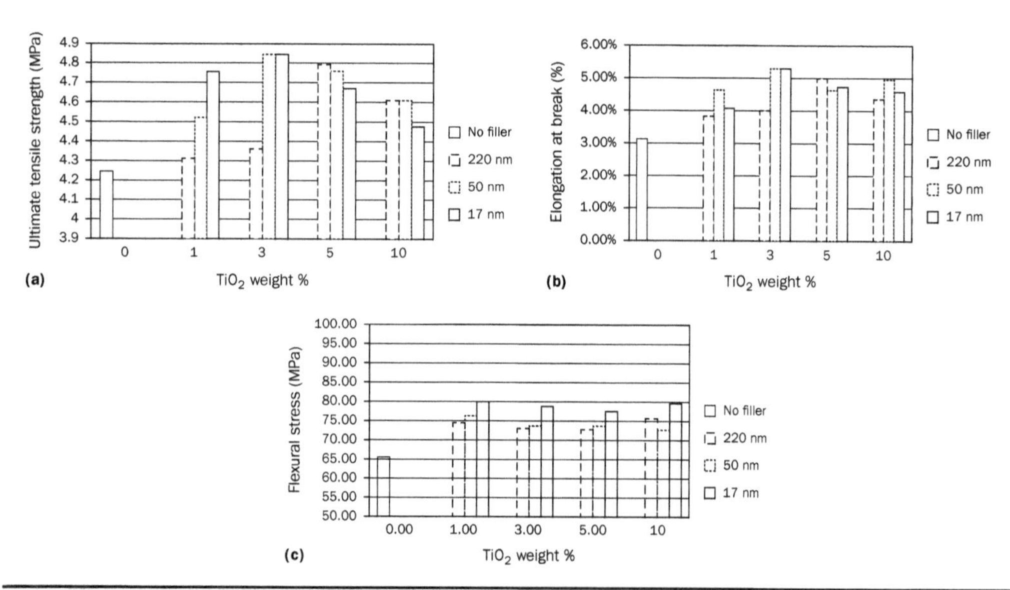

FIGURE 6.29 Mechanical property plots of titanium oxide–modified composites with varying particle size [17].

It should be noted that even though this study describes the use of x-ray photoelectron spectroscopy (XPS) to confirm the dispersion of the nanoparticles within the epoxy matrix, no data are presented displaying the actual dispersion that was achieved in the samples.

A study done by Cheema et al. explored the mechanical properties of a bisphenol A–based epoxy resin modified with ZrO_2 nanoparticles [18]. The size of the nanoparticles is 10 nm, and the epoxy resin is based on bisphenol A (RIMR 135i) and cured with a mixture of two amine-based curing agents (RIMH 134 and RIMH 137). The nanomaterial chosen for this polymer matrix is treated with sorbic acid (SA) and diphenyl hydrogen phosphate (DPP) ligands because of the different anchoring abilities to the resin matrix, resulting in high dispersibility.

A loading of 2 wt% ZrO_2 was added to the epoxy resin and mixed on a magnetic stirrer for 15 minutes at ambient conditions. The mixture was then treated using a vacuum dissolver system fitted with a d25-mm dissolver disk and operated at 3,000 rpm for 2 hours at 50°C. The mixture was cured with the aforementioned curing agents and placed in an oven for 3 hours at 65°C, cooled to room temperature then tempered for 8 hours at 80°C. Mechanical properties such as maximum tensile strength and Young's modulus were determined from tensile testing.

The nanocomposites were characterized by TEM to determine the particle dispersion in the polymer matrix. TEM images, transmittance measurements, and small-angle x-ray *scattering* (SAXS) indicated a high level of dispersion of both DPP- and SA-modified ZrO_2 nanoparticles within the polymer matrix. The mechanical properties of the nanocomposites are summarized in Table 6.6 [18].

Both the treated SA and DPP nanoparticles improved the pure polymer maximum tensile strength by 9.3% and 11.9%, respectively. Both treated nanoparticles improved the Young's modulus of the pure polymer as well, with the DPP-treated particle improving the pure polymer by 12.5% and the SA particle improving the polymer by 0.4%. It should be noted that the SA improvement is within the margin of error for this set of tests. The DPP tests also exhibited larger margins of error, hinting that the trials for the tensile testing resulted in relatively high variance. In summary, the improvement in the mechanical properties is thought to be caused by the enhanced polymer–particle interactions caused by the treatment of the nanoparticles.

6.4.2 Special Types of CNT-Based Thermoset-Based Nanocomposites

Chou et al. recently published an excellent review article that examines the recent advancements in the science and technology of carbon nanotube (CNT)–based fibers and composites [19]. The assessment is based on the hierarchical structural levels of CNTs used in composites, from 1D to 3D. Chou et al. defined the 1D level because

Table 6.6 Mechanical Properties of Zirconium Oxide–Modified Epoxy Resin [18]

Sample	Maximum tensile strength (MPa)	Relative change (%)	Young's modulus (MPa)	Relative change (%)
Pure polymer	69 ± 0.9	–	2630 ± 80	–
NC-ZrO_2-SA	75 ± 0.5	9.3	2640 ± 60	0.4
NC-ZrO_2-DPP	77 ± 3.8	11.9	2960 ± 120	12.5

fibers are composed of pure CNTs or CNTs embedded in a polymer matrix. The 2D level is defined as the CNT-modified advanced fibers, CNT-modified interlaminar surfaces, and highly oriented CNTs in planar form. At the 3D level, examples of mechanical and physical properties of the CNT/polymer composites, CNT-based damage sensing, and textile assemblies of CNTs are included. Chou et al. also discuss the opportunities and challenges in basic research at these hierarchical levels. The 1D level of CNT fibers is beyond the scope of this text; the discussions on 2D level of CNT-modified interlaminar surfaces and highly oriented CNTs in planar form studies are included in this section. The main focus of this text includes the 3D level describing the mechanical and other material properties of the CNT/polymer composites (CNT-based polymer nanocomposites) (see Chaps. 7 to 11).

CNT-Modified Interlaminar Surfaces

Recently, many attempts have been made by researchers in modifying the interlaminar surface by CNTs for improved mechanical and transport properties. Wardle's group at MIT has made significant advancement and contribution in this research area [20–28]. Their recent approach described by Garcia et al. [20, 21] is particularly interesting that the group successfully integrated aligned CNTs with existing carbon fiber prepregs and processing. This is accomplished by first growing a vertically aligned CNT forest at high temperature. The CNTs are then transferred using a rolling transfer scheme to prepregs at room temperature, taking advantage of the tack of the prepregs. The aligned CNT forest can readily draw up the polymer through capillary action when laminated and form a bond with the polymer matrix (Fig. 6.30).

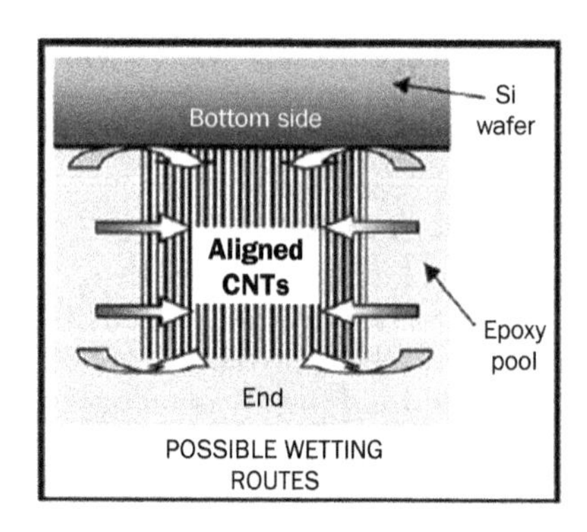

- Identification of wetting route
  - Highly preferred along CNT axis
  - Secondary wetting perpendicular to CNTs

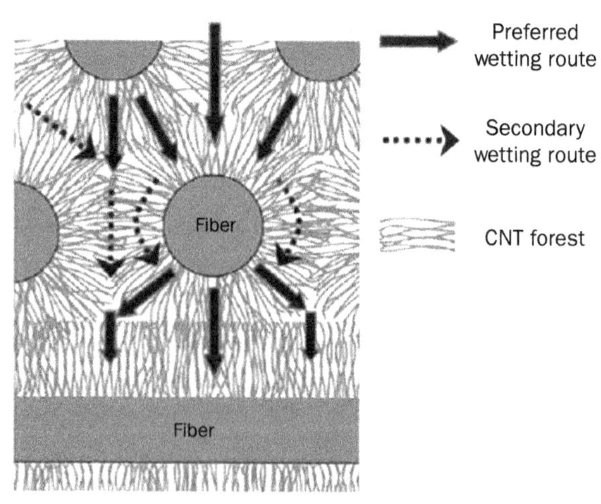

Figure 6.30 Processing nano-engineered composites enabled via capillary-induced wetting of aligned CNTs. (*Courtesy of Prof. Wardle.*)

Garcia et al. [20] claimed that the ideal aligned CNT reinforcement would be comprised of short forest (less than 20 μm). The CNTs penetrate each ply by about 10 μm. As a result, an about 20 μm forest would introduce no additional thickness to the ply interlayer, which can be detrimental to in-plane properties. Figure 6.31a shows the *transfer-printing* process, where the prepreg is attached to a cylinder that is rolled across the Si substrate containing the CNT forest [20]. Figure 6.31b shows the fully transferred CNT forest on a carbon fiber/epoxy prepreg [20]. The CNT forest maintains vertical alignment and is neither broken nor buckled (Fig. 6.31c,d) [20]. The addition of a second ply of prepreg then creates two-ply laminate with an aligned CNT interlayer. Two aerospace-grade unidirectional laminates with CNT interlayer in Mode I and Mode II fracture properties were characterized. In their initial testing, the CNT-modified interface was observed to increase fracture toughness 1.5–2.5 times in Mode I and three times in Mode II. Evidence of CNT bridging was observed in fracture micrographs. Figure 6.32 shows an illustration of ideal hybrid interlaminar architecture showing the aligned CNTs placed in between two plies of a laminated composite [20].

Wicks et al. [26] further explored the 3D reinforcement of woven advanced polymer-matrix composites using CNTs experimentally and theoretically. Aligned CNTs are introduced in both the interlaminar and intralaminar regions of the

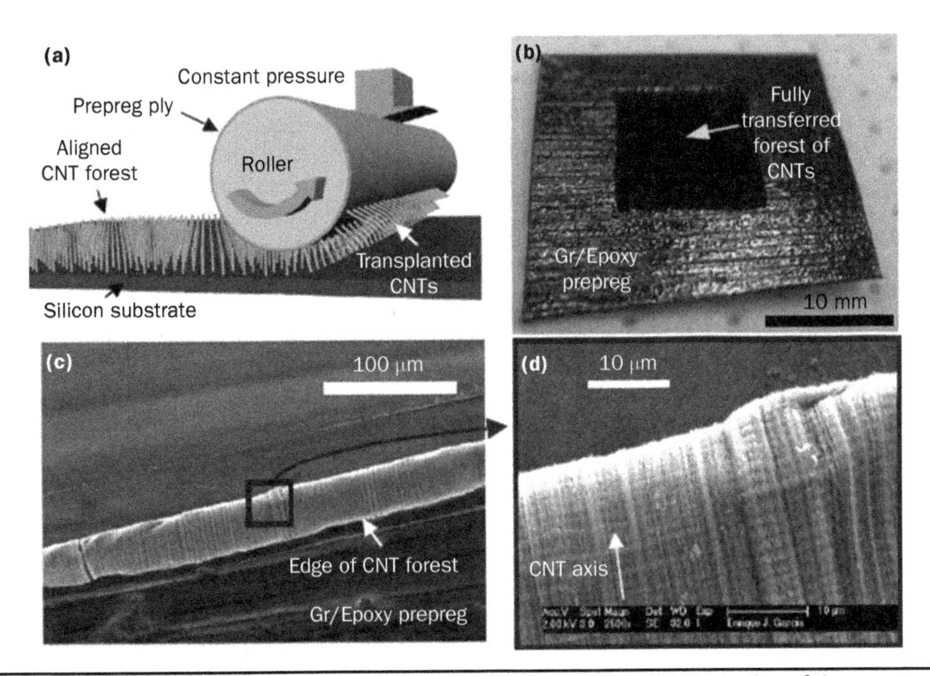

FIGURE 6.31 Transfer-printing of vertically aligned CNTs to prepreg: (a) illustration of the "transfer-printing" process; (b) CNT forest transplanted from its original silicon substrate to the surface of a Gr/Ep prepreg ply; (c,d) SEM micrographs of the CNT forest, showing CNT alignment after transplantation [20].

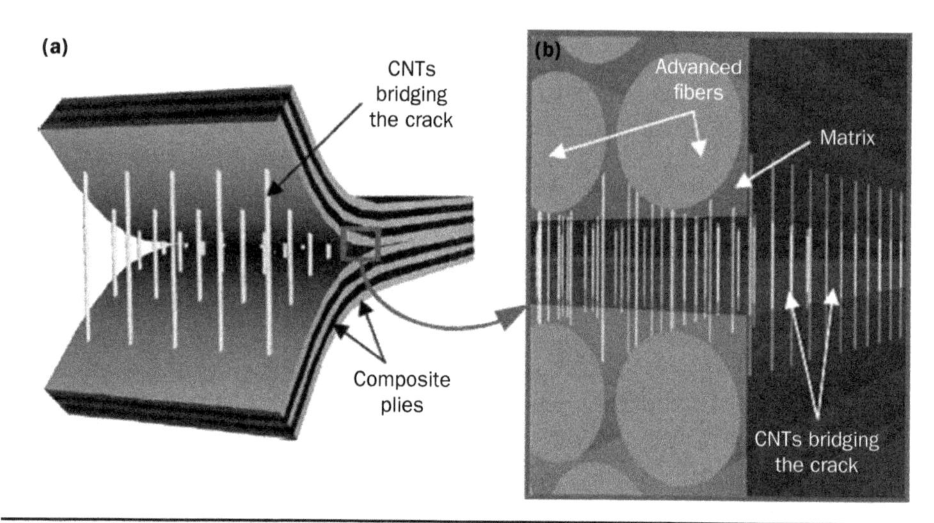

FIGURE 6.32 Illustration of the ideal hybrid interlaminar architecture: (a) CNTs placed in between two plies of laminated composite and (b) close-up of the crack, showing CNTs bridging the crack between the two pliers. Illustrations are not to scale [20].

laminate, providing 3D reinforcement (Fig. 6.33) [26]. This architecture has been introduced previously by Garcia et al. in the literature as fuzzy fiber–reinforced plastic (FFRP) laminate [21]. Radially aligned CNTs grown in situ on the surface of fibers in a woven cloth provide significant 3D reinforcement, as measured by Mode I interlaminar fracture testing and tension-bearing experiments. Aligned CNTs bridge the ply interfaces enhancing both initiation and steady-state toughness, and improving the already tough system by 76% in steady state (more than 1.5 kJ/m^2 increase). CNT pullout on the crack face is the observed toughening mechanism, and an analytical model is correlated to the experimental fracture data. In the plane of the laminate, aligned CNTs enhanced the tension-bearing response with increase of 19% in bearing stiffness, 9% in critical strength, and 5% in ultimate strength accompanied by a clear change in failure mode from shear-out failure (matrix dominated) without CNTs to tensile fracture (fiber dominated) with CNTs.

As discussed by Wicks et al. [26], areas for future research in aligned-CNT reinforced advanced composites are numerous and can include (a) additional mechanical characterizations, especially toughness (Mode II, where z-pinning and stitching are particularly ineffective), impact resistance, and damage tolerance; (b) determination of any additional toughening mechanisms at the interface beyond CNT bridging; (c) modeling of the multi-scale elastic and damage response for these systems [26]; (d) multifunctional property testing, such as electrical and thermal conductivities [28–30]; (e) development of other fabrication routes beyond hand layup for the FFRP system; (f) replication of the FFRP architecture using carbon fibers, where CNT growth on the carbon fibers must be shown to maintain fiber tensile properties; (g) continuous rather than batch production of aligned CNTs; and (h) particle exposure studies during the fabrication and machining of CNT-containing laminates, such as the FFRP [31–33].

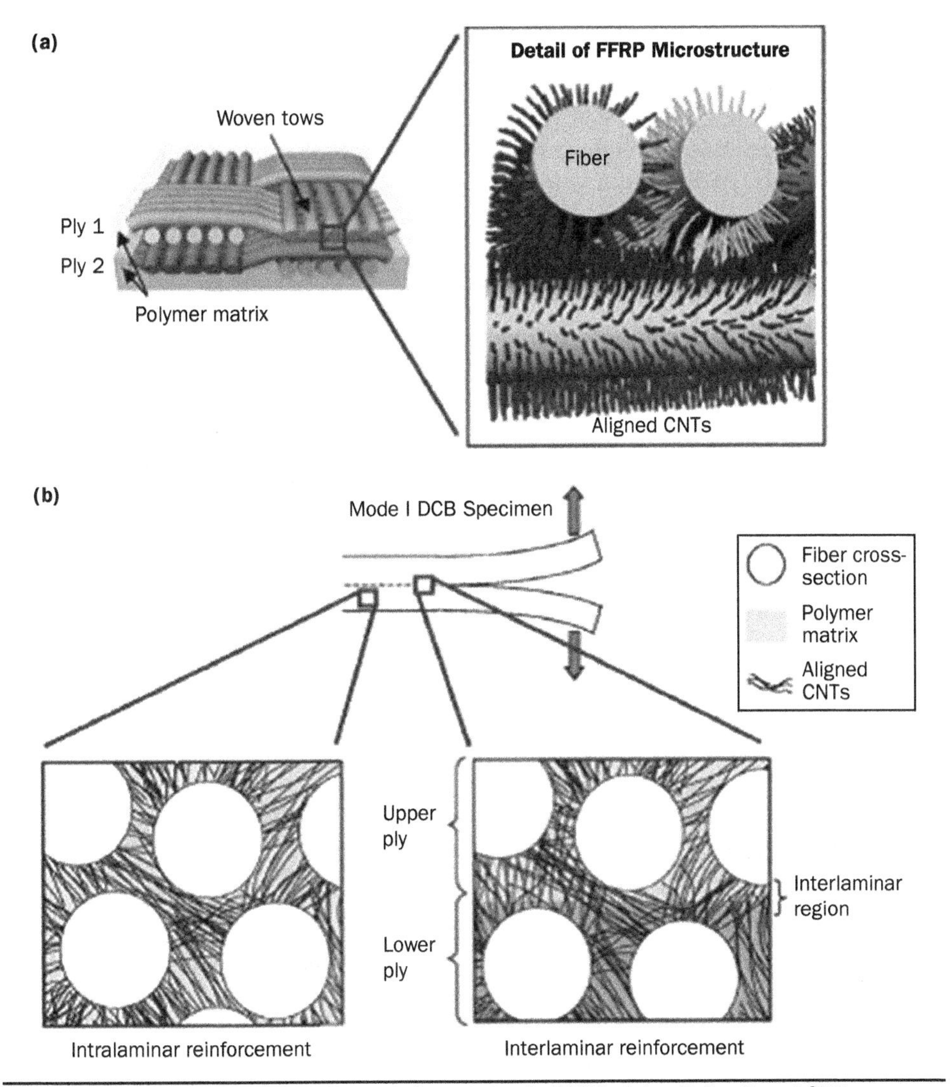

FIGURE 6.33 Illustration of fuzzy fiber–reinforced plastic (FFRP): (a) radially aligned CNTs grown on advanced fiber cloth; (b) CNT intralaminar and interlaminar reinforcement [26].

6.4.3 Summary of Thermoset-Based Nanocomposites

The addition of clay, SWNTs, DWNTs, MWNTs, GPL, SiO_2, TiO_2, and ZrO_2 nanoparticles improved the mechanical properties, such as Young's modulus, UTS, and toughness of the different epoxy resin systems. WAXD, TEM, and AFM techniques were used to determine the degree of nanoparticle dispersion in the resin matrix. The experimental data from impact strength, bending modulus, and bending strength show improved performance for the chemically modified MWNTs, further indicating enhanced properties due to improved particle dispersion. The analytical models for the tensile modulus agreed fairly well with the tensile data, specifically the Mori–Tanaka model.

The model used to predict fracture toughness did not produce acceptable correlation with the experimental data.

Overall, values for UTS, tensile modulus, and work to break were improved. Aligned CNTs are introduced in both the interlaminar and intralaminar regions of the laminate, provide 3D reinforcements, and enhance thermal and electrical conductivity values of the prepreg laminates.

6.5 Overall Summary

The mechanical properties of the thermoplastics, thermoplastic elastomers, and thermosets can be strengthened dramatically by adding a small amount of nanoparticles if the right processing techniques and conditions are selected. Incorrect combinations may produce no change at all or, in some cases, adverse effects. Studies that relied on WAXD techniques to determine the dispersion of the nanoparticles within the polymer matrix can be further investigated by obtaining TEM micrographs of the sample to gauge whether the degree of dispersion within the polymer matrix is optimal. It is highly recommended to assess nanoparticle dispersion first, before launching into the expensive and lengthy material properties characterization.

However, all the studies demonstrated that even at low-loading percent of nanoparticles, significant improvements in mechanical properties were observed. Also, it is not only the process that matters; the fabrication of the polymer nanocomposite parts is also important to achieve enhanced mechanical properties of polymer nanocomposites. With the possibility of obtaining even more uniformly dispersed nanoparticles, the future application of nanocomposites in terms of cost for the benefit it provides seems promising.

6.6 Study Questions

6.1 How does the nanomaterial dispersion affect the mechanical properties of polymer nanocomposites in general? Provide examples to illustrate your claims.

6.2 How does the mechanical properties of polymer-clay nanocomposites influenced by the type of clay and surface treatment of the clay, and how clay is processed in the polymer matrix? Provide examples.

6.3 Based on the discussions in this chapter on the enhancement of mechanical properties of thermoplastic-based nanocomposites using MMT clay, carbon-based (MWNT, buckypaper, and graphene) nanomaterials, fibers, and nanodiamonds, describe the mechanisms by which each type of nanomaterials provides the enhanced mechanical properties. What other nanomaterials can be used for thermoplastic-based nanocomposites? Provide examples.

6.4 Based on the discussions in this chapter on the enhancement of mechanical properties of thermoplastic elastomer–based nanocomposites using MMT clay, carbon-based (MWNT) nanomaterials, SiC, nanosilica, and TiO_2, describe the mechanisms by which each type of nanomaterials provides the enhanced mechanical properties. What other nanomaterials can be used for thermoplastic elastomer–based nanocomposites? Provide examples.

6.5 Based on the discussions in this chapter on the enhancement of mechanical properties of epoxy-based nanocomposites using MMT clay, carbon-based

(MWNT) nanomaterials, nanosilica, TiO_2, and ZrO_2, describe the mechanisms by which each type of nanomaterials provides the enhanced mechanical properties. What other nanomaterials can be used for thermoset-based nanocomposites? Provide examples.

6.6 The elongation at break is the property of polymer nanocomposites that tends to degrade significantly with the addition of MMT clay and carbon-based nanomaterials. What will be a good strategy to maintain this property in the polymer nanocomposites?

6.7 References

1. Kord, B. (2012) Studies on mechanical characterization and water resistance of glass fiber/thermoplastic polymer bionanocomposites. *Journal of Applied Polymer Science,* 123(4):2391–2396.
2. Fernandez-d'Arlas, B., Khan, U., Rueda, L., Martin, L., Ramos, J. A., Coleman, J. N., González, M. L., et al. (2012) Study of the mechanical, electrical and morphological properties of PU/MWCNT composites obtained by two different processing routes. *Composites Science and Technology,* 72(2):235–242.
3. Eswaraiah, V., Balasubramania, K., and Ramaprabh, S. (2011) Functionalized graphene reinforced thermoplastic nanocomposites as strain sensors in structural health monitoring. *Journal of Materials Chemistry,* 21(34):12626–12628.
4. Zhang, X., Chen, Y., Yu, J., and Guo, Z. (2011) Thermoplastic polyurethane/silica nanocomposite fibers by electrospinning. *Journal of Polymer Science Part B, 49*(23):1683–1689.
5. Karbushev, V., Semakov, A., and Kulichikhin, V. (2011) Structure and mechanical properties of thermoplastics modified with nanodiamonds. *Polymer Science Series A,* 53(9):765–774.
6. Liff, S. M., Kumar, N., and McKinley, G. H. (2007) High-performance elastomeric nanocomposites via solvent-exchange processing. *Nat Mater.,* 6(1):76–83.
7. Chavarria, F., and Paul, D. R. (2006) Morphology and properties of thermoplastic polyurethane nanocomposites: effect of organoclay structure. *Polymer,* 47(22):7760–7773.
8. Puskas, J. E., Foreman-Orlowski, E. A., Lim, G. T., Porosky, S. E., Evancho-Chapman, M. M., Schmidt, S. P., El Fray, M., et al. (2006) A nanostructured carbon-reinforced polyisobutylene-based thermoplastic elastomer. *Biomaterials,* 31(9):2477–2488.
9. Chen, W., Tao, X., and Liu, Y. (2006) Carbon nanotube-reinforced polyurethane composite fibers. *Composites Science and Technology,* 66(15):3029–3034.
10. Liao, C. Z., and Tjong, S. C. (2010) Mechanical and fracture behaviors of elastomer-rich thermoplastic polyolefin/SiC nanocomposites. *Journal of Nanomaterials,*1–9.
11. Aso, O., Eguiazábal, J. I., and Nazábal, J. (2007) The influence of surface modification on the structure and properties of a nanosilica filled thermoplastic elastomer. *Composites Science and Technology.* 67(13):2854–2863.
12. Kinloch, A. J., and Taylor, A. C. (2006) The mechanical properties and fracture behaviour of epoxy-inorganic micro- and nano-composites. *Journal of Materials Science,* 41: 3271–3297.
13. Gojny, F. H., Wichmann, M. H. G., Fiedler, B., and Karl, S. (2005) Influence of different carbon nanotubes on the mechanical properties of epoxy matrix composites—a comparative study. *Composites Science and Technology,* 65:2300–2313.

14. Zheng, Y., Kim, J.-D., Peng, H., Margrave, J. L., Khabashesku, V. N., and Barrera, E. V. (2006) Functionalized effect on carbon nanotube/epoxy nano-composites. *Materials Science and Engineering A*, 435–436:145–149.
15. Rafiee, M. A., and Rafiee, J. (2009) Enhanced mechanical properties of nanocomposites at low graphene content. *ACS Nano* 3(12):3884–3890.
16. Blackman, B. R. K., Kinloch, A. J., Lee, J. S., Taylor, A. C., Agarwal, R., Schueneman, G., and Sprenger, S. (2007) The fracture and fatigue behavior of nano-modified epoxy polymers. *Journal of Materials Science*, 42:7049–7051.
17. Al-Turaif, H. A. (2010) Effect of nano TiO_2 particle size on mechanical properties of cured epoxy resin. *Organic Coatings*, 69:241–246.
18. Cheema, T. A., Lichtner, A., Weichert, C., Böl, M., and Garnweitner, G. (2012) Fabrication of transparent polymer-matrix nanocomposites with enhanced mechanical properties from chemically modified ZrO_2 nanoparticles. *Journal of Materials Science*, 47:2665–2674.
19. Chou, T.-W., Gao, L., Thostenson, E. T., Zhang, Z., and Byun, J.-H. (2010) An assessment of the science and technology of carbon nanotube-based fibers and composites. *Composites Science and Technology*, 70:1–19.
20. Garcia, E. J., Wardle, B. L., and Hart, A. J. (2008) Joining prepreg composite interfaces with aligned carbon nanotubes. *Composites Part A*, 39(6):1065–1070.
21. Garcia, E. J., Wardle, B. L., Hart, A. J., and Yamamoto, N. (2008) Fabrication and multifunctional properties of a hybrid laminate with aligned carbon nanotubes grown in situ. *Composites Science and Technology*, 68:2034–2041.
22. Garcia, E. J., Hart, A. J., and Wardle, B. L. (2008) Long carbon nanotubes grown on the surface of fibers for hybrid composites. *AIAA Journal*, 46(6):1405–1412.
23. Blanco, J., Garcia, E. J., Guzman, R., Villoria, D., and Wardle, B. L. (2009) Limiting mechanisms of mode I interlaminar toughening of composites reinforced with aligned carbon nanotubes. *Journal of Composite Materials*, 43(8):825–841.
24. Yamamoto, N., Hart, A. J., Garcia, E. J., Wicks, S. S., Duong, H. M., Slocum, A. H., and Wardle, B. L. (2009) High-yield growth and morphology control of aligned carbon nanotubes on ceramic fibers for multifunctional enhancement of structural composites. *Carbon*, 47:551–560.
25. Ray, M. C., Guzman de Villoria, R., and Wardle, B. L. (2009) Load transfer analysis in short carbon fibers with radially-aligned carbon nanotubes embedded in a polymer matrix. *Journal of Advanced Materials*, 41(4):82–94.
26. Wicks, S. S., Guzman de Villoria, R., and Wardle, B. L. (2010) Interlaminar and intralaminar reinforcement of composite laminates with aligned carbon nanotubes. *Composites Science and Technology*, 70:20–28.
27. Lachman, N., Wiesel, E., Guzman de Villoria, R., Wardle, B. L., and Wagner, H. D. (2012) Interfacial load transfer in carbon nanotube/ceramic microfiber hybrid polymer composites. *Composites Science and Technology*, 72:1416–1422.
28. Yamamoto, N., Garcia, E. J., Wardle, B. L., and Hart, A. J. (2008) Thermal and electrical properties of hybrid woven composites reinforced with aligned carbon nanotubes. *Proc. 49th AIAA Structures, Dynamics, and Materials Conference*, AIAA-2008-1857, Schaumburg, IL, April 7–10.
29. Vaddiraju, S., Cebeci, H., Gleason, K. K., and Wardle, B. L. (2009) Hierarchical multifunction composites by conformally coating aligned carbon nanotube arrays with conducting polymer. *Applied Materials & Interfaces*, 1(11):2565–2572.

30. Marconnet, A. M., Yamamoto, H., Panzer, M. A., Wardle, B. L., and Goodson, K. E. (2011) Thermal conduction in aligned carbon nanotube-polymer nanocomposites with high packing density. *ACS Nano*, 5(6):4818–4825.
31. Bello, D. B., et al. (2009) Exposures to nanoscale particles and fibers during handling, processing, and machining of nanocomposites and nano-engineering composites reinforced with aligned carbon nanotubes. In *17th International Conference on Composite Materials (ICCM) Proceedings*, Edinburgh, Scotland, July 27–31.
32. Bello, D., Hart, A. J., Ahn, K., Hallock, M., Yamamoto, N., Garcia, E. J., Ellenbecker, M. J., et al. (2008) Particle exposure levels during CVD growth and subsequent handling of vertically-aligned carbon nanotube films. *Carbon*, 46(6):974–977.
33. Bello, D., Wardle, B. L., Yamamoto, N., Guzman deVilloria, R., Garcia, E. J., Hart, A. J., Ahn, K., et al. (2009) Exposure to nanoscale particles and fibers during machining of hybrid advanced composites containing carbon nanotubes. *J. Nanopart Res.*, 11(1):231–249.

CHAPTER 7

Thermal Properties of Polymer Nanocomposites

7.1 Introduction

This chapter discusses recent advancements in the improvement and understanding of the thermal behavior of polymer nanocomposites. Many different kinds of polymer nanocomposite systems were analyzed, focusing on their thermal properties. Most systems analyzed had thermal properties enhanced by the addition of nanoparticles, such as nanoclays or layered double hydroxides. The addition of certain nanoparticles also had a positive impact on mechanical and electrical properties. Some systems, as in the homogeneously nucleated polymethyl methacrylate/polymer single-walled nanotube (PMMA/pSWNT) system, had other interesting thermal behavior. Polymer nanocomposites have found themselves in all kinds of applications. Their light weight offers high specific properties that make them preferable for a wide range of applications. This chapter also focuses on thermal conductivity properties of polymer nanocomposites. The nanocomposites are compared to their neat polymer host and sometimes to other macrocomposites and competing nanocomposites. The composites analyzed are organized by the polymers and the type of nanoparticles used.

7.2 Thermoplastic-Based Nanocomposites

7.2.1 Polypropylene-Clay Nanocomposites

Nanoclay has attracted a lot of attention in polymer nanocomposite due to its relative abundance, good performance, and environmental factors. There are different kinds of commercially available nanoclays. Some are treated to become organophilic in order to be compatible with the polymer matrix. One of the challenges involved in any kind of composite, especially nanocomposites, is dispersing, and exfoliating, nanoparticles in the host material matrix.

Ellis and D'Angelo [1] produced a polypropylene-organophilic montmorillonite clay containing approximately 4 wt% clay. Although the clay was fully dispersed in the polypropylene (PP) matrix, the particles were not fully exfoliated. Figure 7.1 shows that the x-ray diffraction (XRD) pattern of the PP nanocomposite had a peak corresponding to the treated (organophilic treatment) clay [1]. This peak indicates that the clay particles were not fully exfoliated in the PP matrix.

Furthermore, Fig. 7.2 shows transmission electron microscopy (TEM) images of the PP nanocomposite at two different magnifications [1]. The image on the left shows that the particles were successfully dispersed in the polymer matrix. However, the image on the right shows that the particles were not fully exfoliated, they are in an intercalated state.

The PP nanocomposite had overall better mechanical properties than the neat PP. The thermal properties of the composite were relatively unaffected. The percent ash produced was slightly higher, but still lower than a conventional talc macrocomposite. The coefficient of thermal expansion (CTE) was slightly improved. In general, the use of organophilic clay in a PP matrix may lead to increased mechanical properties and slightly improved thermal properties. Due to the intercalated nature of the nanocomposite produced, there is reason to believe that even better results can be obtained from a fully exfoliated nanocomposite [1].

7.2.2 PEEK-Carbon Nanofiber Nanocomposites

Sandlera et al. [2] produced a carbon nanofiber–reinforced poly(ether ketone) (PEEK) nanocomposite. PEEK is an important material that is already being used in a variety of applications, such as seals in the oil industry. The carbon nanofibers (CNFs) in this composite were successfully dispersed and exfoliated. Figure 7.3 shows the fracture

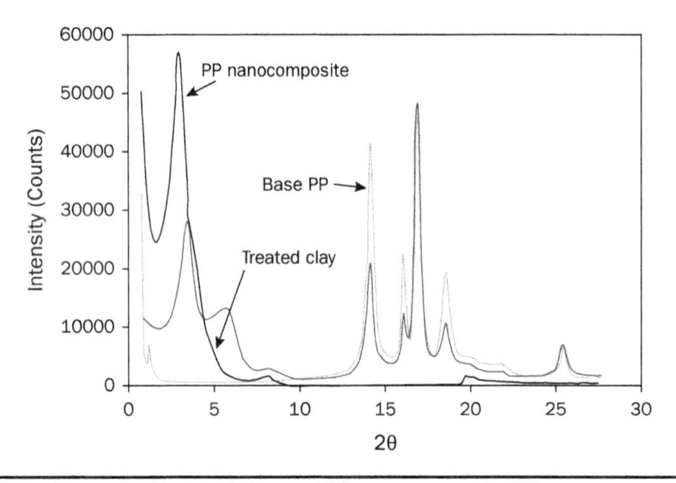

FIGURE 7.1 XRD pattern of PP, treated clay, and PP nanocomposite [1].

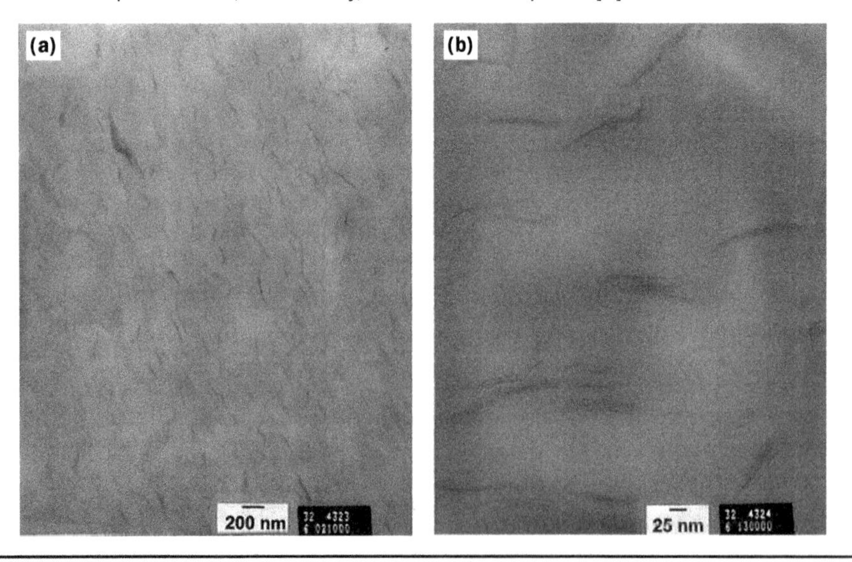

FIGURE 7.2 TEM images of PP-clay nanocomposite [1].

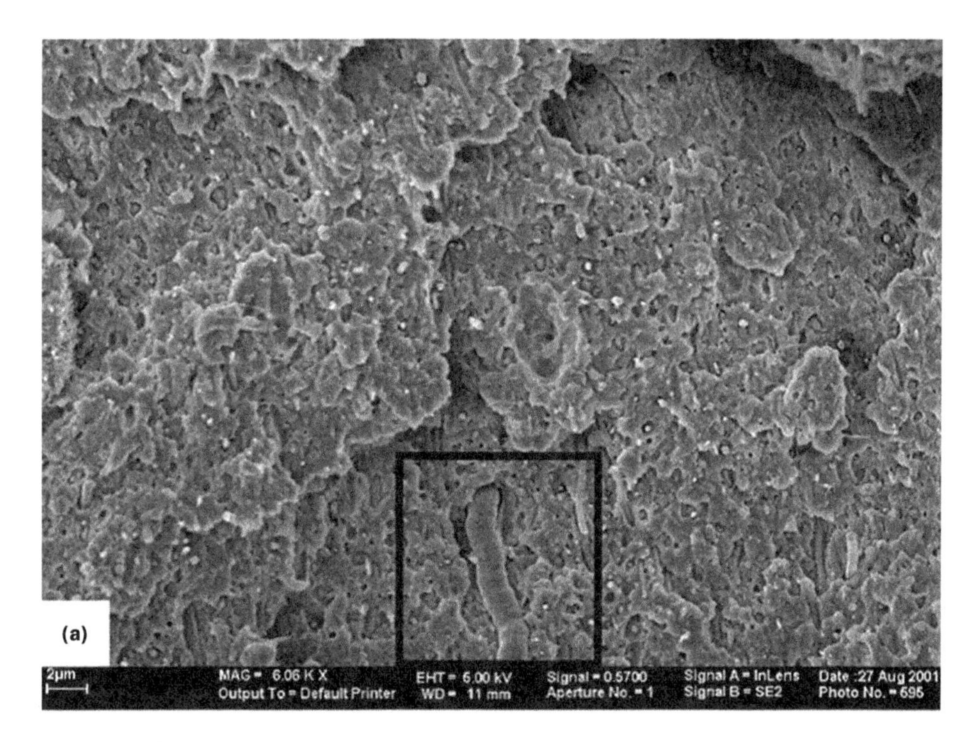

Figure 7.3 SEM micrographs of the fracture surface of (a) 10 wt% and (b) 15 wt% vapor-grown CNFs. The squares represent polymer coated CNFs [2].

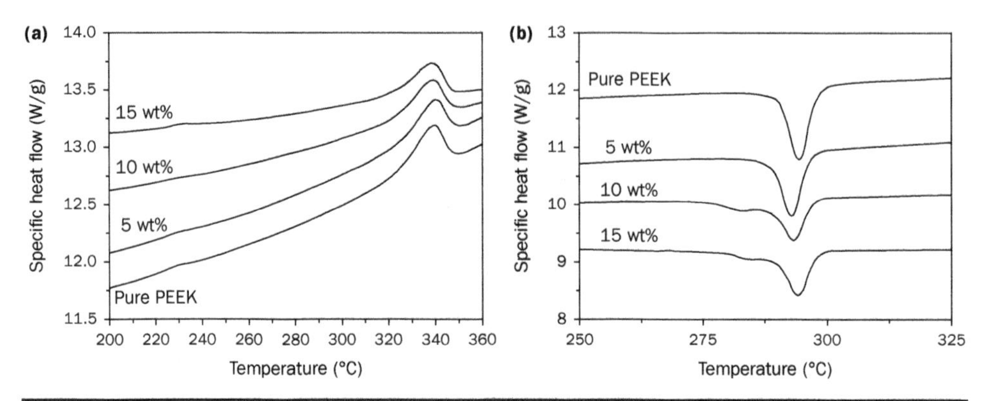

Figure 7.4 Specific heat flow recorded for the PEEK nanocomposites as a function of temperature during an isothermal DSC runs: (a) heating curves and (b) cooling curves. Individual curves are shifted vertically for clarity [2].

surface of the PEEK-CNF nanocomposite [2]. The images show that the CNFs did not agglomerate and were coated by the polymer.

The PEEK-CNF nanocomposite had improved mechanical properties while maintaining good thermal properties. The glass transition temperature, among other thermal properties, was not significantly changed. However, Fig. 7.4 shows interesting crystallization behavior [2]. Lozano and Barrera previously reported that vapor-grown CNFs in an isotactic polypropylene homopolymer served as nucleation sites [3]. However, the CNFs had no significant impact in the crystallization of the composite. Nucleation started homogeneously in the PEEK matrix.

The PEEK-CNF nanocomposite showed improved mechanical properties while maintaining good thermal properties. Unlike as reported before, the vapor-grown CNFs in this nanocomposite did not function as nucleation sites during crystallization. Homogeneous nucleation of this nanocomposite is an important feat that becomes even more important as the size of the filler material further decreases (if, for example, carbon nanotubes were to be used) [3].

7.2.3 PVC-Layered Double-Hydroxide Nanocomposites

Liu et al. prepared a PVC [poly(vinyl chloride)]-layered double hydroxide (LDH) using dodecyl sulfate (DS^-) or stearate anion [4]. The polymer nanocomposite showed significantly improved thermal stability when compared to pure PVC. XRD patterns [4] shown no significant difference between the peaks displayed in pure PVC and the different LDH composites, indicating that the ordered layered structure had been successfully destroyed and the nanoparticles dispersed in the PVC matrix.

Furthermore, TEM images (Fig. 7.5) indicate that the LDHs were successfully exfoliated in the PVC matrix [4]. The results from XRD patterns and TEM images indicate that organo-LDHs containing DS^- and stearate anions can be exfoliated into nanometer scale and homogeneously dispersed in PVC through the solution intercalation procedure.

The thermal stability of PVC-LDH nanocomposites was evaluated by conventional Congo Red test. Figure 7.6 compares the thermal stability of pure PVC, a previously

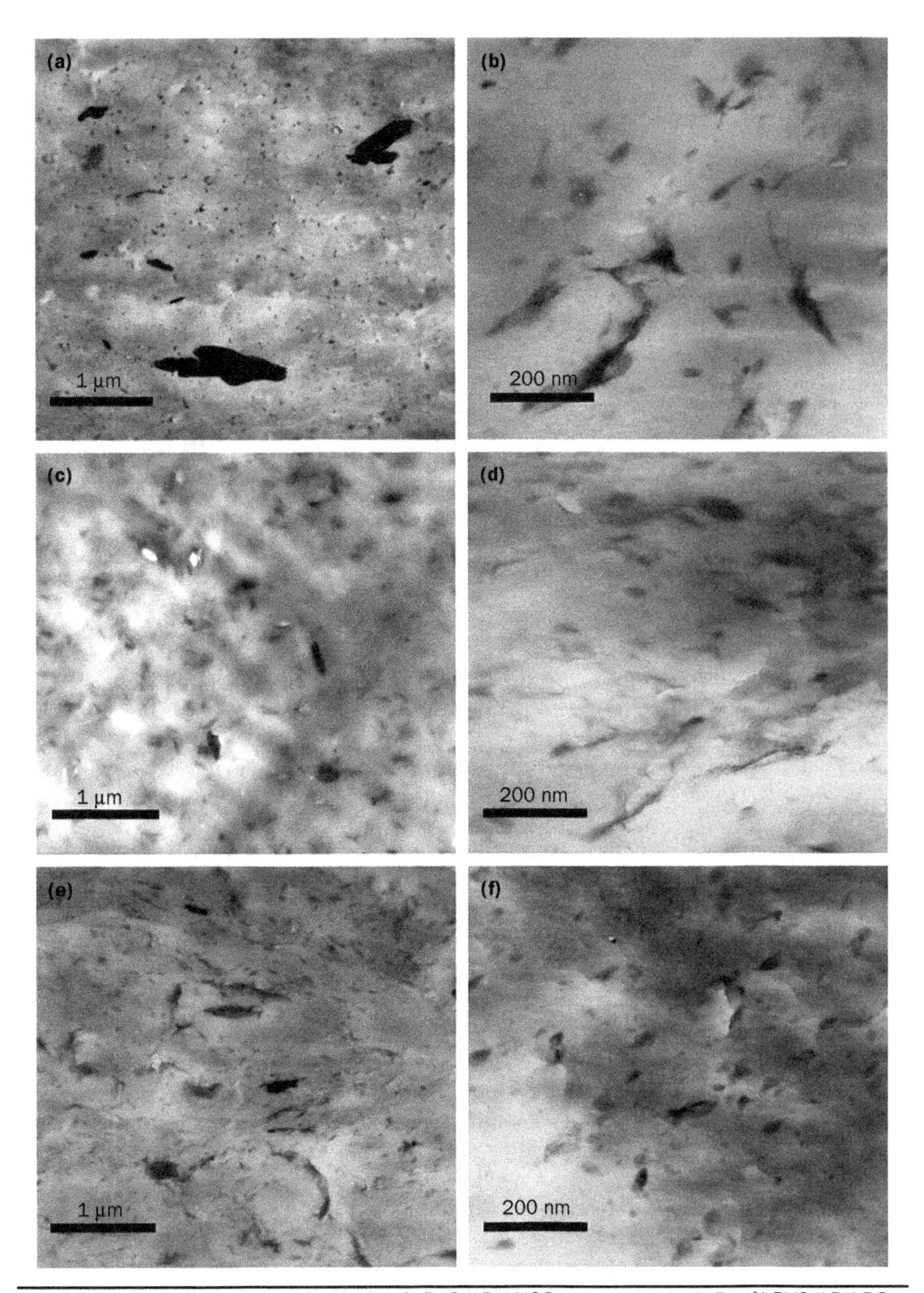

Figure 7.5 TEM micrographs of (a,b) 5 wt% PVC/LDH-NO3 composite, (c,d) 5 wt% PVC/LDH-DS nanocomposite, and (e,f) 5 wt% PVC/LDH-stearate nanocomposite [4].

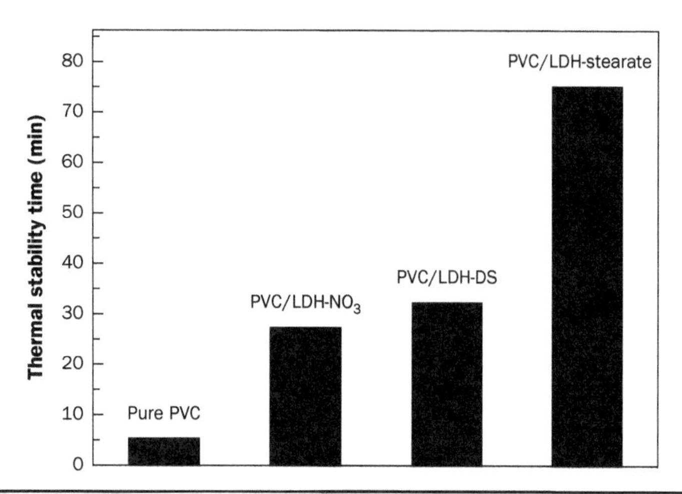

Figure 7.6 Thermal stability time of PVC, PVC/LDH-NO_3 composite, PVC/LDH-DS nanocomposite, and PVC/LDH-stearate nanocomposite measured by Congo Red test [4].

developed PVC-LDH composite, and the newly developed LDH composites [4]. The thermal stability of LDH-DS is slightly better than the previously developed LDH composite but significantly better than pure PVC. Also, LDH-stearate is significantly better than pure PVC and more than twice as good as the previously developed LDH composite.

Furthermore, thermal stability as a function of LDH loading was also analyzed. Figure 7.6 shows the effect of LDH loading for a conventional fire-retardant LDH-AP composite and the new LDH-stearate composite [4]. The stearate hydroxide loading has much bigger effect than the conventional fire-retardant composite in thermal stability and material loading as high as 10 wt% can be used to improve thermal stability.

The PVC-LDH nanocomposites developed by solution intercalation (DS^- and stearate) exhibit significantly improved thermal stability. This composite can be further developed and adapted to promote PVC resins with improved thermal stability, flame retardancy, and smoke suppression properties [4].

7.2.4 Hybrid Systems

Piszczyk et al. prepared rigid polyurethane-polyglycerol-clay nanocomposites using different kinds of polyglycerols and layered silicate nanoclays (montmorillonite) [5]. A polymeric foam-nanoclay composite with improved thermal properties was created. Figure 7.7 shows a set of scanning electron microscopy (SEM) micrographs of the polyurethane-polyglycerol foam with different kinds of clays [5]. The cell structure is maintained by the addition of nanoclay with improved mechanical properties.

The XRD pattern of pure foams and nanocomposite foams presented in [5] shown that the nanocomposites did not show diffraction maximums corresponding to clay, indicating a good exfoliation of nanoclays in the foams.

Thermogravimetric analysis (TGA) data of the nanocomposites are represented in Table 7.1 [5]. Cloisite 30B is clearly the best clay in terms of mass loss. Table 7.1 includes the temperature at which the nanocomposites experience 5%, 10%, and 50% weight loss.

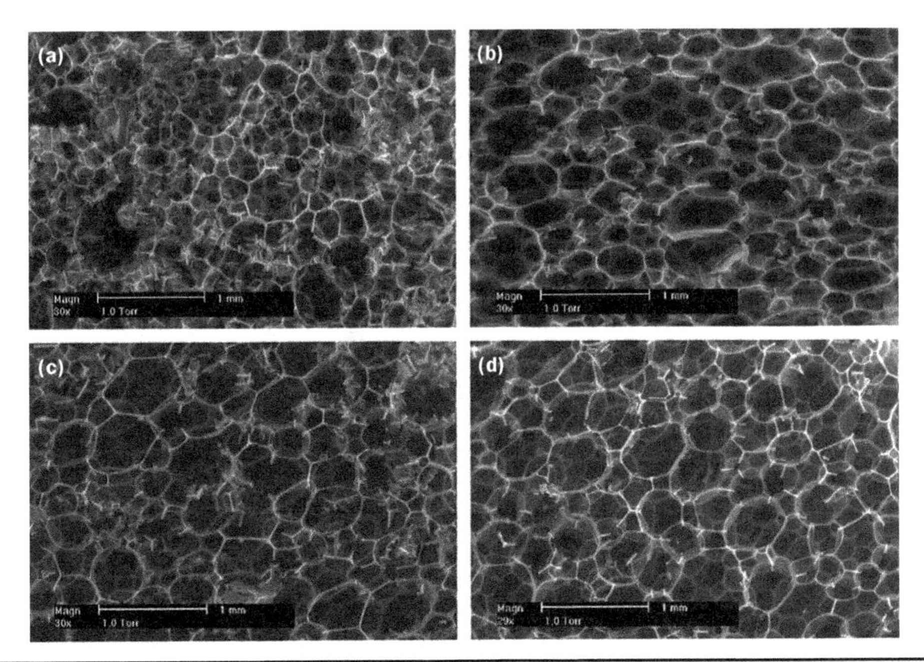

FIGURE 7.7 SEM micrographs showing cell structure of (a) polyurethane–polyglycerol foams and nanocomposite foams with (b) Cloisite 30B, (c) Laponite RD, and (d) Bentonite [5].

TABLE 7.1 Thermogravimetric Analysis of Rigid Polyurethane-Polyglycerol Nanocomposite [5]

Clay in foam	RF 551: Polyglycerol	Weight loss (%) Temperature (°C) 5	10	50
None	100: 0	286	316	480
Cloisite 30B		302	328	533
Laponite RD		300	318	410
Bentonite		280	298	392
None	65:35	271	316	433
Cloisite 30B		289	325	430
Laponite RD		287	306	448
Bentonite		279	312	418
None	30:70	237	285	450
Cloisite 30B		272	326	442
Laponite RD		276	316	439
Bentonite		269	309	441

Three different polyurethane-polyglycerol ratios were analyzed, and three different kinds of clays were used for each kind of foam. The best thermal stability was obtained for foams containing Cloisite 30B. The bio-based polyglycerol was successfully used to manufacture rigid polyurethane foams. The addition of nanoclay, especially Cloisite 30B, not only increased the mechanical performance (compressive strength), but also the thermal stability of the polyurethane-polyglycerol foams [5].

Shen et al. have tested the effects of nanoclay on a woven glass fiber–reinforced polyamide 6 (PA6) nanocomposite [6]. The XRD pattern in [6] shows that the clay was fully dispersed in the PA6 matrix [6]. Cone calorimetry was also used to evaluate combustion performance of composites [6]. The addition of clay, especially at amounts higher than 4 wt%, had a significant positive effect on time to ignition and time of peak HRR, but not significant effect on peak HRR value.

The heat distortion temperature (HDT) is shown in Fig. 7.8 and indicates that the addition of clay to the polymer matrix had no significant impact on laminate composite HDT [6]. The glass transition temperatures of the laminate composites are recorded in Table 7.2 [6]. The glass transition temperature of the composites was significantly enhanced by the addition of clay (27% increase at 6 wt% clay).

Woven glass fiber–reinforced laminates using a PA6/clay polymer nanocomposite matrix were successfully fabricated. The addition of nanoclay delayed the time to ignition and peak heat release rate by 55% and 120%, respectively. The glass transition temperature

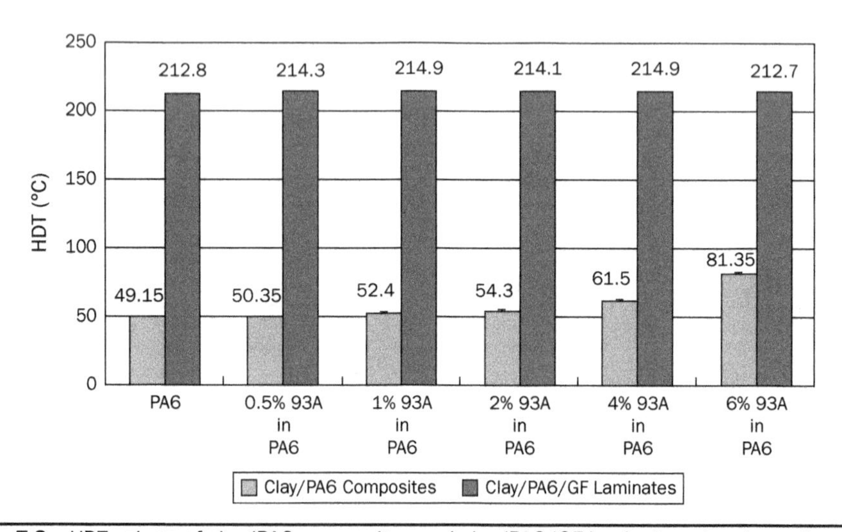

Figure 7.8 HDT values of clay/PA6 composites and clay/PA6/GF laminates at various clay loadings (wt%) in PA6 [6].

Table 7.2 Glass Transition Temperatures of Clay/PA6/GF Laminates [6]

Sample	PA6	+0.5 wt% Clay 93A	+1 wt% Clay 93A	+2 wt% Clay 93A	+4 wt% Clay 93A	+6 wt% Clay 93A
T_g (°C) of laminates	56	60	63	65	68	71

was also increased by 27% at a clay loading of 6 wt%. It should also be mentioned that mechanical properties were generally slightly enhanced by the addition of clay [6].

7.2.5 Summary of Thermal Properties of Thermoplastic-Based Nanocomposites

The thermal performance of PP, PEEK, PVC, and PA6/glass fiber–reinforced polymer nanocomposite systems with clay, CNF, LDH, and clay/glass fiber were analyzed. In general, their thermal performance can be enhanced by the addition of nanoparticles. In particular, the clay/PA6/glass fiber laminates have improved time to significant positive effect on time to ignition and time of peak HRR, but not significant effect on peak HRR value, higher HDT, and improved glass transition temperature. The mechanical and electrical performance of the polymeric matrix can also be improved by the addition of nanoparticles. Sometimes simply maintaining good thermal performance while improving other properties is a significant feat.

It has been found that the performance of nanoparticles on the basis of improved properties is highly dependent on the dispersion and exfoliation of the nanoparticles in the polymeric matrix. Certain surface treatments to nanoparticles and specialized processing techniques can help the dispersion and exfoliation of nanofibers in the composite. Polymer nanocomposite is a promising and growing field of study. It has already led to important advancements and continues to be in high demand in the industry.

7.3 Thermoplastic Elastomer–Based Nanocomposites

Over the years, research into the development, testing, and implementation of polymer nanocomposites has gained both interest and attention. The reinforcement of commodity polymers has allowed for the creation of new materials with enhanced properties to fulfill market demands.

Among the many polymers currently being explored, the class of thermoplastic elastomers (TPEs) offers a broad range of attractive properties. TPEs are a physical mixture of polymers, referred to as copolymers that behave as both thermoplastics and elastomers. This means that they can be used in injection molding, along with other processing methods that exclude elastomeric thermosets. This is because the cross-linking in TPEs occurs via weaker, physical bonds rather than the chemical bonds that form during the curing process of a thermoset. As such, they are prime candidates for reinforcement by nanomaterials. TPEs are replacing vulcanized rubber due to favorable mechanical properties and recyclability. Properties of TPEs are similar to rubber in softness, flexibility, extensibility, and resilience and are more easily processed by the conventional thermoplastic processes of extrusion and injection molding.

Developmental TPEs must meet three criteria: clarity, thermal resistance, and softness. These properties must be idealized for TPEs for their use as thermal interface materials for heat dissipation in electronic devices. These properties are important because heat produced during the operation of electronics is generally dissipated by heat conduction. To transfer this heat away from the electronics, heat sinks are generally used; however, heat dissipation capacity of a heat sink decreases due to the interfacial thermal resistance arising from the mismatch of surface roughness of both the device and the heat sink from a lack of a good thermal contact. Therefore, TPEs can act as thermal interface materials to fill the remaining gaps at this interface [7].

This section provides an overview of recent developments in the field of TPE nanocomposites from the perspective of thermal properties including thermal stability, glass transition temperature, crystallization temperature, thermal expansion, and thermal conductivity. New ways to improve the thermal properties of TPEs by using nanoadditives focusing primarily on increasing thermal conductivity and decreasing the CTE [8] are discussed.

Sections 7.3.1 to 7.3.5 describe research involved in enhancement of thermal stability, glass transition temperature, and crystallization property and Secs. 7.3.6 to 7.3.8 describe research involved in enhancement of thermal expansion and thermal conductivity properties. Step-by-step review of a set of recent experiments and research and their results as far as thermal properties are concerned are presented along with a summary at the end of this section.

7.3.1 Thermoplastic Polyurethane-Montmorillonite Clay

The effects of organically modified montmorillonite (OMMNT) on thermoplastic polyurethane (TPU)–based nanocomposites were studied by the Rubber Technology Centre of the Indian Institute of Technology [9]. The composite was prepared by melt intercalation using the Thermo Scientific HAAKE PolyLab OS Rheomox at 185°C and 100 rpm for 6 minutes. Cloisite 15A was the nanoclay used in the following amounts: 1, 3, 5, 7, and 9 wt% (TPU1A, TPU3A, TPU5A, TPU7A, and TPU9A). Wide-angle x-ray diffraction (WAXD) showed no evidence of d-spacing, which suggests that the nanoparticle may be well dispersed; the TEM analysis showed that there was a mixed morphology of partially exfoliated and intercalated structures. Although the group was unable to thoroughly exfoliate the clay, there is still significant interaction between the TPU and the OMMNT [9].

TPU is a copolymer and there are two different segments. The hard segment is comprised of alternating diisocyanate and diols or diamine units while the soft segment is comprised of linear long-chain diols. Due to the organic modifier on the nanoclay, there is a strong interaction between the highly polar soft segments. There is hydrogen bonding that occurs between the OMMNT and the TPU, and the clay particles appear to preferentially attract to the soft segments.

TGA was performed to compare the residual weight of the pure TPU and the composites (Fig. 7.9a) [9]. Note that there are two distinct peaks in the derivative of the TGA curve, which shows that there are two separate stages of degradation for the hard and soft segments, respectively (Fig. 7.9b). The temperature at the onset of each stage is denoted as Td1 and Td2 in Table 7.3 [9].

Increasing the amounts of Cloisite 15A shifts the peak degradation temperatures (DTG1max, DTG2max) higher, which shows that the TPU-OMMNT composite has better thermal stability than the pure TPU polymer (Table 7.3). These two temperatures correspond to the peak rate of degradation of the hard segment and the soft segment, respectively. At DTG2max, the degradation of the base polymer is dominant [9]. Although the peak degradation temperatures are a maximum at 5 wt%, the residual mass at 500°C (500 wt%) still increases almost linearly. While the clay itself does not burn, the organic modifier is combustible and accounts for some of the mass loss, which explains why the residual mass is less than the amount of Cloisite 15A added.

Barick and Tripathy explain the effect of the nanoclay by the "Labyrinth barrier effect," which is caused by the anisotropic silicate platelet mesh that impedes the diffusion of volatile degradation products [9]. Since the structure of TPU has two distinct hard and soft segments, two melting temperatures are expected. However, in the

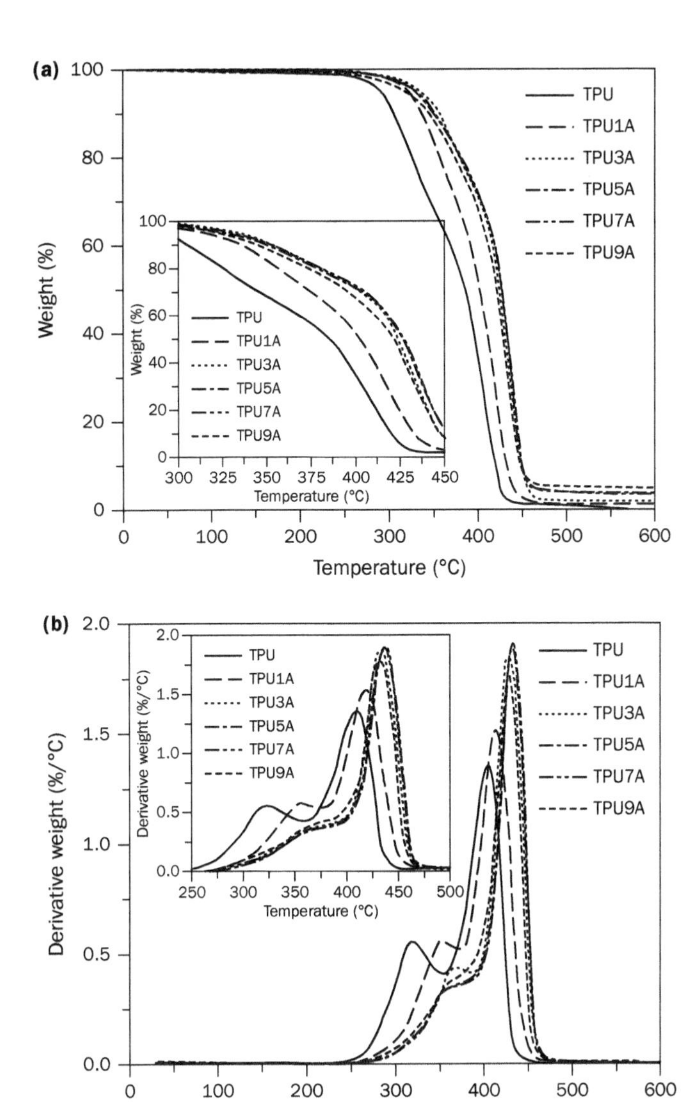

Figure 7.9 (a) Residual mass TGA curve and (b) its derivative of TPU and TPU composites [9].

Table 7.3 TGA Results of Neat TPU and their Composites [9].

Sample codes	DTG_{1max} (°C)	DTG_{2max} (°C)	T_{50} wt% (°C)	T_{d1} (°C)	T_{d2} (°C)	T_{95} wt% (°C)	r_{500} wt% (%)
TPU	321.37	408.65	386.73	297.31	356.85	425.58	0.801
TPU1A	352.21	418.03	407.52	326.21	380.32	433.24	1.451
TPU3A	365.19	432.26	425.33	336.99	386.38	455.54	2.119
TPU5A	373.29	436.65	427.33	332.63	390.41	462.07	4.132
TPU7A	362.16	436.05	427.63	329.03	378.38	461.34	4.209
TPU9A	362.14	432.39	421.81	320.64	376.29	563.67	5.346
Cloisite 15A	—	316.65	—	253.67	—	—	—

differential scanning calorimetry (DSC) curve, only one endothermic phase is detected, which is where the differential heat flow is at a local minimum [9]. It is speculated that the reason for detecting only one melting point is that there is limited movement of the hard segment. Additionally, the hard segment has only a slight change in specific heat and is uniformly dispersed throughout the composite matrix [9]. As a result, all of the data acquired by means of DSC represents the soft portion of the composite [9].

It was observed that increasing amounts of Cloisite 15A slightly increased the glass transition temperature for the soft segment. Barick and Tripathy concluded that the small amount of dispersed nanoclay is insignificant to produce significant changes in T_g of the composite due to the TPU comprising most of the bulk-free volume of the composite and the interactions between the organic and inorganic parts. Consequently, the soft segments become more rigid and have limited movement, which results in a slight increase in T_g [9].

The change in specific heat of the composite across the glass transition was observed for increasing amounts of nanoclay. It is speculated that the difference in specific heat is the result of the nanoclay layers hindering the motion of polymer chains [9]. As the amount of OMMNT is increased, the difference in specific heat decreases. However, at a certain amount of OMMNT, the difference in specific heat begins to increase but still remains significantly below that of pure TPU. This phenomenon is probably due to the specific heat of the OMMNT being significantly higher than that of the TPU.

The addition of OMMNT increases the melting temperature of the composite, which suggests that the nanoclay alters the crystal structure of the soft segments. The data show that the melting temperature plateaus at about 3 wt%, which is a 2.33% improvement over neat TPU [9]. However, better results may be obtained for higher loadings of OMMNT if the nanoclays can be more uniformly dispersed and fully exfoliated.

It was observed that the heat of fusion increased with OMMNT loading. This value plateaued around 3 wt%, which is 211% greater than neat TPU. Increasing amounts of OMMNT increases the degree of crystallinity of the soft segments, and since the enthalpy change during the phase change is small, it is speculated that only some of the polymer linkages formed phase separation domains. Barick and Tripathy attribute the plateau in the heat of fusion to be caused by filler agglomeration at high concentrations, which reduces the effective interfacial surface area of the filler to the polymer [9].

Barick and Tripathy suggest that the peak shown in the DMA curve is associated with the main glass transition temperature of the system due to the mobilization of the soft segments in the TPU matrix (Fig. 7.10) [9]. As Cloisite 15A was added, the tan(δ) peak shifted to higher temperatures due to resistance in motion of the polymer chains caused by the nanoclay [9]. The value of the glass transition temperature appears to plateau at a certain loading of OMMNT resulting in a maximum of 3.9% difference from neat TPU. After 7 wt% loading, T_g begins to decline, which could be the result of poor dispersion of OMMNT in the TPU matrix.

7.3.2 Thermoplastic Polyurethane–MWNT Nanocomposites

A study carried out by Rice University researchers showed the effects of nanotube orientation on the properties of multiwalled carbon nanotubes (MWNTs). The carbon nanotube (CNT) arrays were created using a chemical vapor deposition process, which grew to lengths of 3.5 mm with an average diameter of 75 nm. Different samples of oriented composites and randomly dispersed composites loaded with 6 wt% MWNT were produced for testing. The base polymer used was TPU [10].

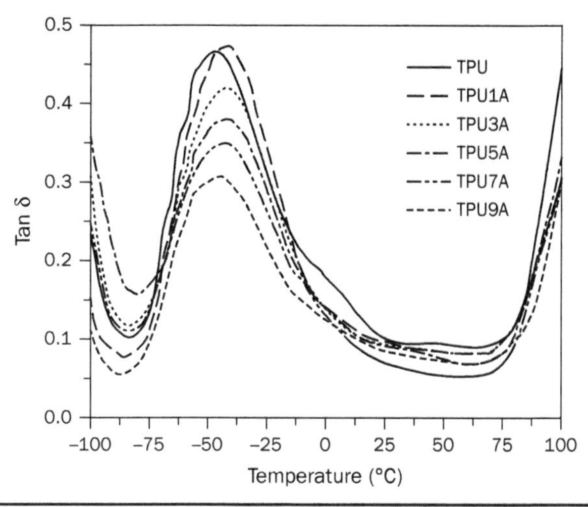

FIGURE 7.10 DMA curve and T_g for TPU and TPU composites [9].

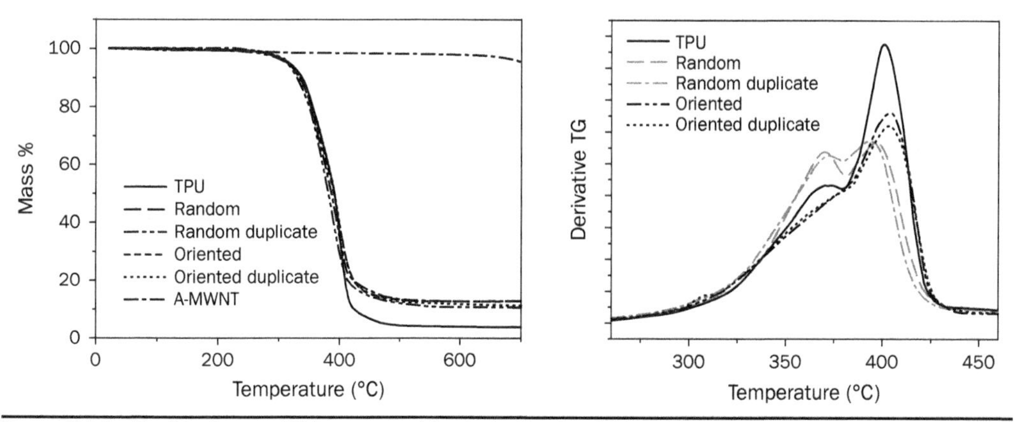

FIGURE 7.11 Residual mass TGA curve and its derivative for TPU and TPU-MWNTs [10].

The TGA was performed in a nitrogen environment, so that the MWNTs do not combust (Fig. 7.11) [10]. Overall, there was virtually no change in the onset of thermal degradation from the derivative TG regardless of MWNT orientation (Fig. 7.11). In general, it appears that the CNTs do not influence the degradation of TPU in an inert environment. The nanotube content was determined from the difference in residual mass at 600°C and there was virtually no change [10]. Unlike OMMNT, the MWNTs do not have the proper aspect ratio to replicate the "Labyrinth effect" discussed earlier (Sec. 7.3.1). Additionally, since they are made of carbon, the MWNTs will probably combust in an oxygen environment and will be detrimental to the degradation of TPU.

Interestingly, the results of the DMA showed that the randomly oriented MWNTs consistently had a higher glass transition temperature than the oriented samples (Fig. 7.12) [10]. This suggests that the random dispersion induced an amorphous phase

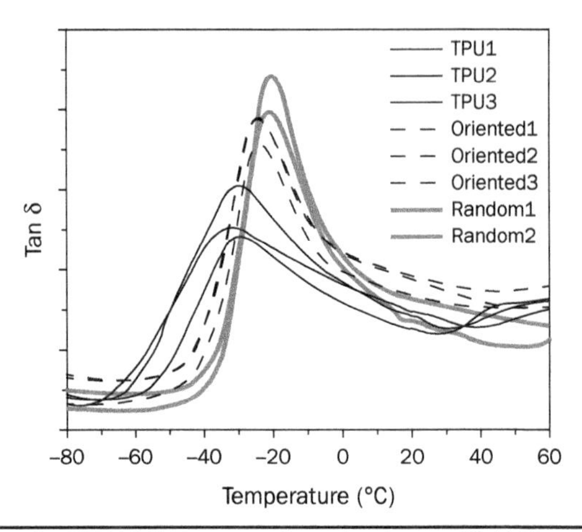

FIGURE 7.12 DMA curve for TPU-MWNT composites showing tan(δ) [10].

that is more closely packed than the oriented samples [10]. The oriented samples had a 2.89% increase in T_g while the randomly dispersed samples had a 4.54% increase.

7.3.3 Thermoplastic Polyurethane Mixed with Laponite and Cloisite

Mishra et al. [11] compared two varieties of clay in preparing TPU-based nanocomposites. The neat resin was Desmopan KU2 8600E, which is a polyether-based TPU prepared from MDI, polytetramethylene glycol, and 1,4-butanediol. The nanoclays were Laponite RD and Cloisite 20A obtained from Rockwood Additives Ltd. (the United Kingdom). Laponite RD was modified with dodecylamine hydrochloride in order to assist in dispersal via a solvent (named L in the figures). Cloisite 20A was modified by dimethyl dehydrogenated tallow quaternary ammonium ion (C in figures). Samples were thus prepared using 20% solution of the TPU in tetrahydrofuran (THF) and mixing the calculated amounts of Laponite/Cloisite while stirring and sonicating. For final preparation, samples were then held in a vacuum oven at 70°C.

Mishra et al. used WAXD, SEM, TEM, and atomic force microscopy (AFM) to characterize the material morphology. Fourier-transform infrared spectroscopy (FTIR) was used to differentiate between the modified and unmodified nanoclays. WAXD results for neat resin (T0), 1 wt% Cloisite 20A (T1C), 3 wt% Cloisite 20A (T3C), 5 wt% Cloisite 20A (T5C), 10 wt% Cloisite 20A (T10C), and pure Cloisite 20A (C) for a lower angular range are shown in [11].

Mishra et al. claim that the peak at roughly 7.1°C corresponds to a fraction of unmodified clay remaining in the composite, while the peak at roughly 3.7°C corresponds to the modified clay. The WAXD intensity spectra for the full angular range for Cloisite is shown in [11]. Compare the WAXD results for Laponite at low and full angular ranges in the neat resin (T0), 1 wt% Laponite (T1L), 3 wt% Laponite (T3L), 5 wt% Laponite (T5L), 10 wt% Laponite (T10L), and pure Laponite (L), as shown in [11]. No prominent peaks are observed because of the lack of a well-defined stacking of platelets in Laponite.

The mixed results reported by WAXD are confirmed by the TEM images. Although Mishra et al. only report images for T5C and T3L, regions of aggregation, intercalation, and exfoliation are all revealed. Figure 7.13 shows two different scales for T5C (5 wt% Cloisite) [11]. Results are even more mixed for the 3 wt% Laponite (Fig. 7.14) [11].

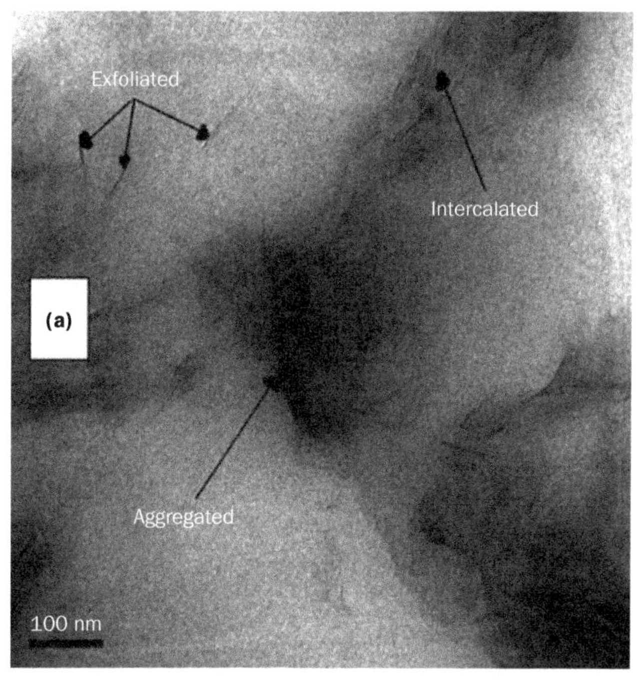

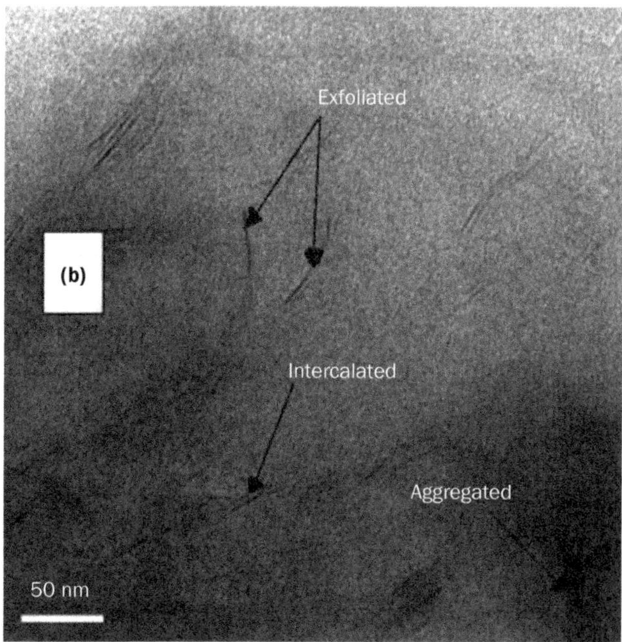

FIGURE 7.13 TEM of 5 wt% Cloisite TPU at (a) 100 nm (b) and 50 nm [11].

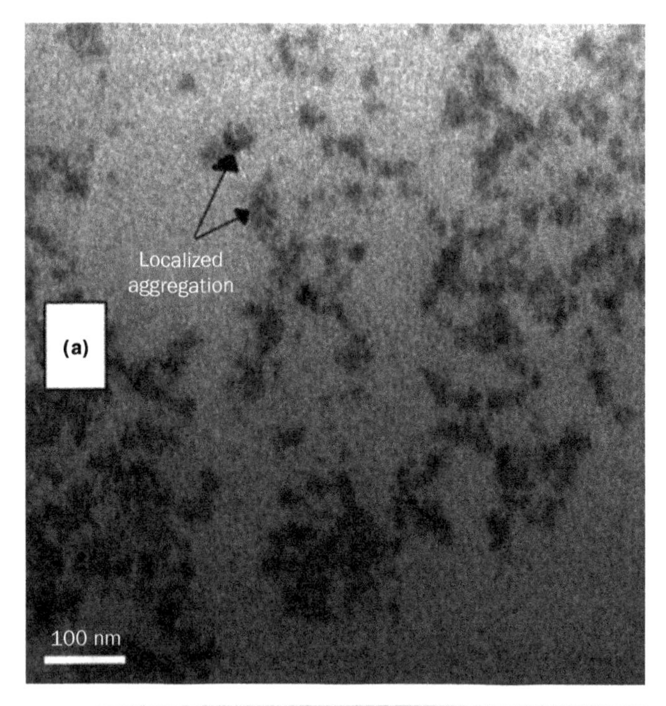

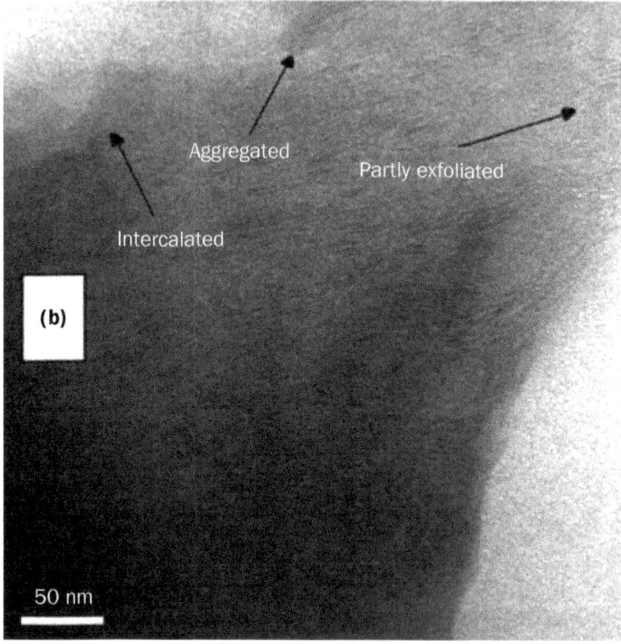

FIGURE 7.14 TEM of 3 wt% Laponite at (a) 100 nm and (b) 50 nm [11].

From the TEM images, it seems that better dispersion was achieved with the Cloisite. TGAs of the Cloisite- and Laponite-reinforced TPU show essentially delayed thermal degradation as the amount of Cloisite or Laponite was increased. This applies to the onset of degradation, which was delayed 17.5°C by the 1 wt% Cloisite loading and 8.3°C by the 1 wt% Laponite loading. As the loading was increased, further increases in the temperature of degradation onset were observed in both types of clay. TGA results are shown in Fig. 7.15 [11].

DSC thermograms reveal almost no change in T_g as Cloisite/Laponite compositions varied, remaining at roughly about 40°C (trends confirmed by DMA, though the

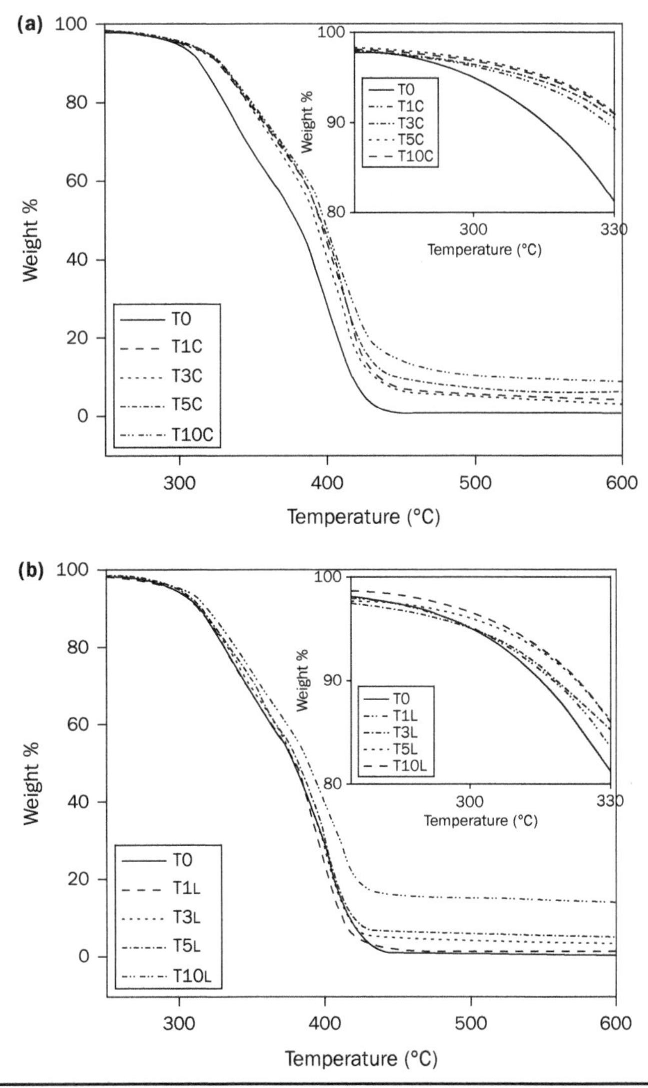

Figure 7.15 TGA results for various samples of (a) Cloisite (C, top) and (b) Laponite (L, bottom) reinforced TPU [11].

value of T_g measured there is about 22°C). Increasing Cloisite loading generally made heat flow more negative, as shown in the thermogram in [11]. This was also the case for increasing Laponite loading, except for the 10 wt% case, as observed in [11].

7.3.4 Poly(dimethyl siloxane)/Boron Nitride

The silicone rubber used with boron nitride (BN) was vinyl-terminated poly(dimethyl siloxane) (PDMS). The filler loading of different types of BN was varied at 0, 10, 30, and 50 wt%. Figure 7.16 shows the results for CTE as well as thermal conductivity for the PDMS/BN composite with varying filler loading [8]. The incorporation of boron nitride particles into silicone elastomer results in a reduction of the CTE. The reason for this is the restriction of mobility of the silicon molecules from the adsorption of the BN surfaces. As the particle size of boron nitride decreased, the CTE also decreased for any given loading level. Therefore, the lowest values for CTE occurred when using nano-sized fillers. This is likely due to the high surface-to-volume ratio of nano-sized fillers, which promotes a higher level of interaction at the interface.

For thermal conductivity, however, we see the opposite trend. For this measurement, it is seen that as particle size decreases, we obtain the less desirable result of a lower thermal conductivity as opposed to the micrometer-sized fillers. This is because the ability to form a conductive network of the filler in the matrix depends on the aspect ratio and concentration of the filler. When concentration and aspect ratio of the filler increase, after a critical concentration, particles percolate and touch one another to form a conductive network. Due to the lower aspect ratio of the nano-sized fillers, we see lower conductivities with these fillers. Regardless of BN type, the addition of filler material enhances the thermal conductivity of the composite [8].

7.3.5 Polyethylene/Single-Walled Carbon Nanotubes

For this study, multiple polymer matrices were used to host the SWNTs. These polymers were low-density and high-density polyethylene (PE) with PE crystallinities of

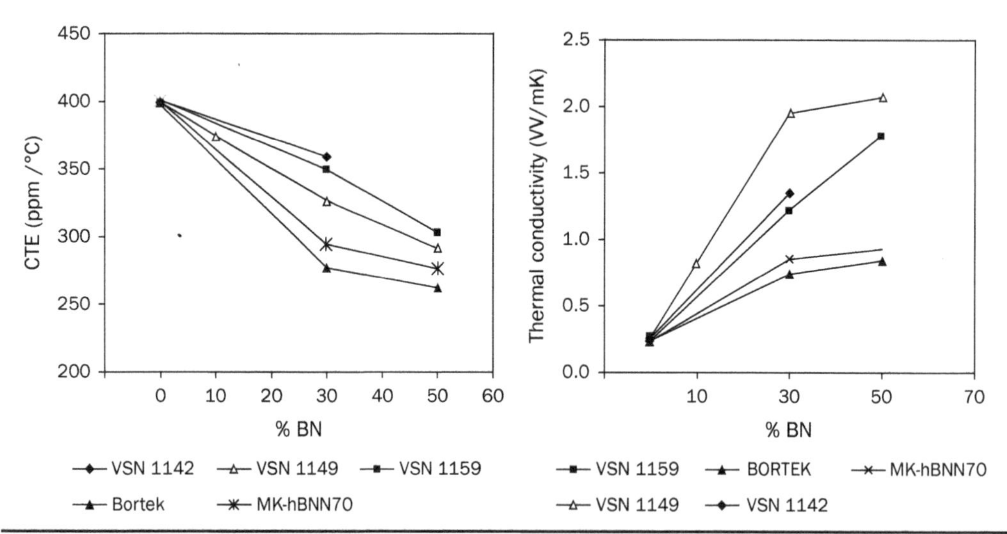

Figure 7.16 Effect of BN-type filler concentration on (a) coefficient of thermal expansion and (b) thermal conductivity [8].

33% and 78%, respectively. The composites were prepared with SWNT volume fractions of 1 and 30 vol% for low-density polyethylene (LDPE) and 1, 5, 10, and 30 vol% for high-density polyethylene (HDPE) [12].

The thermal conductivity of neat LDPE is 0.26 W/m-K and neat HDPE is 0.5 W/m-K. For both HDPE/SWNTs as well as LDPE/SWNTs we see only moderate increases in thermal conductivity at low SWNT volume fractions. This absence of a significant increase in thermal conductivity at low loadings is attributed to the modest difference of the constituent thermal conductivities and the large interfacial thermal resistance between the polymer and SWNTs, with this large interfacial thermal resistance likely due to weak thermal coupling. At the highest SWNT loading we see a significantly large increase in the HDPE/SWNT samples. This is likely due to changes at the interface between PE and SWNT, which reduces the interfacial thermal resistance [12].

7.3.6 Ethylene Propylene Diene Monomer/ZnO

The polymer used to host the nano-sized ZnO was ethylene propylene diene monomer (EPDM). The filler amount of ZnO was calculated on a volume basis with values ranging from 0 to 225 loading of filler/phr (parts per hundred polymer). The thermal conductivity results for both untreated and in situ modified composites are shown in [13].

Thermal conductivity of both untreated and in situ modified nano-ZnO–filled composites increase linearly with filler loading. Maximum thermal conductivity obtained is 0.386W/m-K at a loading of 210 phr. This is over two times greater than that of a 10 phr nano-ZnO–filled composite. It is also observed that, at the same filler content, the thermal conductivity of untreated nano-ZnO is a little higher than that of the in situ modified nano-ZnO system. One possible reason for this is that in situ modification improves the dispersion, which harms the formation of thermal conductive chains composed of filler particles and increases the total quantity of filler–rubber interface thermal resistance. Also, it is possible that parts of Si69 molecules grafted on the particle surface can enhance the original thermal resistance and decrease the thermal conductivity for the untreated nano-ZnO composite. One thing to note from these results is that excessive pursuit of higher thermal conductivity will result in a deterioration of the mechanical properties [13].

7.3.7 Summary of Thermal Properties of Thermoplastic Elastomer-Based Nanocomposites

This section provided an overview of the current technology in enhancing the thermal properties of TPEs by using nanoadditives. These studies analyzed the thermal properties of different TPEs reinforced by carbon black, synthetic, and organic clays. Consistently, nanoadditives in increasing weight percentages resulted in delayed and/or reduced thermal degradation (as measured by TGA) and reduced heat flow (as measured by DSC). Sometimes, transition temperatures (T_g) were affected by the additives, but often they were unchanged. Different levels of dispersion were achieved in the different studies, and some performance differences can be attributed to that.

Organic clays have become a mainstay as a nanoadditive because of their low cost and consistent performance. The comparison with synthetic clay (Laponite) is important to review where synthetics stand with respect to organics. Additionally, of the three-dimensional (3D) nanoadditives, carbon black is one of the least expensive, and so is a more accessible material. The economies of these reinforcements make it important to study them, as other, more expensive nanomaterials will not be practical for industrial implementation for some time. In the case of TPU, the organic modifier made the nanoclays preferentially attracted to the softer segments, and as a result, affected the properties of these softer segments much more than the harder segments. In the case of polymer foams, the porosity of the foam was also a large contributor to the thermal conductivity.

On the basis of reviews in this section the effects of nanomaterials can range from extremely useful to negligible, and are highly dependent on the type of filler used as well as its relative size. For all cases, there were enhanced properties with increasing loading of filler, leading us to believe that the more filler, the better properties your composite will display. Interestingly enough, nano-sized fillers are not always ideal when compared to micrometer-sized fillers. For all studies in which thermal conductivity data were taken for both nano- and micrometer-sized fillers, the micrometer fillers performed better. Based on these results, we see that many factors can affect the thermal properties of polymer nanocomposites.

7.4 Thermoset-Based Nanocomposites

Thermosets are class of polymers known for the curing behavior. They begin as in a malleable or liquid form, and after being cured the material becomes hard and remains in that form. This means that once the material is cured, it cannot be reheated and reshaped or recycled. The curing behavior also means that this material holds up well under elevated temperatures expanding its range of applications. Thermosets are highly cross-linked, which gives them good mechanical properties, such as a high modulus and strength, and good resistance to creep, but they have poor ductility due to their rigid behavior [14]. As a polymer, they have a low thermal stability, high electrical resistivity, and low thermal conductivity [15]. Many thermoset polymers have been known for their excellent thermal performance, which makes them prime candidates for the primary base in which to develop further advancements with nanocomposites.

7.4.1 Epoxy Nanocomposites

Epoxy is known for its excellent chemical and corrosion resistance, good adhesive properties, low shrinkage, and low price [15]. As mentioned above, curing gives them a highly cross-linked microstructure, which results in a high modulus and strength, good resistance to creep, and good performance at elevated temperature, but poor ductility [15]. To address different shortcomings, nanofillers have been investigated. With the addition of ceramic, the stiffness of the material is improved, and the cost is reduced, but the material becomes even more brittle [15]. Soft particle addition results in a tougher material but at the sacrifice of the stiffness [15]. A popular epoxy system is based on diglycidyl ether of bisphenol A (DGEBA) or bisphenol F (DGEBF) [16].

POSS in Modified Epoxy Nanocomposites

In this study performed by Nagendiran et al. [17], DGEBA was mixed with tetraglycidyl diamino diphenyl methane (TGDDM), and was then formed with 4,4′-diaminodiphenylsulfone (DDS) in presence with octa-aminophenyl silsesquioxane (OAPS) form of POSS [17]. The addition of TGDDM to the DGEBA contributes to the increased thermal stability of the composite through the former's higher epoxy functionality. The POSS and DGEBA in this case are formed through in situ polymerization. Even though POSS composites can be formed through physical blending, other studies found that this preparation method has a tendency to show immiscibility toward the epoxy resin [17].

Characterization of the thermal properties of this epoxy-POSS composite was performed with DSC and TGAs. Calorimetry was performed on 10-mg samples with a continuous flow of nitrogen, heating ranges from ambient to 400°C at 10°C/min [17]. Glass transition temperature was taken as the midpoint of this change in capacity. TGA was performed on 50-mg samples that were heated from ambient to 900°C under continuous nitrogen flow at a rate of 10°C/min. The structure of the composite was also observed using XRD and SEM micrographs [17]. The OAPS form of POSS was synthesized in a three-step reaction characterized by the scheme shown in Fig. 7.17 [17].

The SEM micrographs taken of the composite do not show the usual dispersion among thermoset nanocomposites (Fig. 7.18) [17]. The OAPS appears to have blended into a striated unidirectional form in the polymer matrix, which is more indicative of intercalated structure than it is of the more common exfoliated structure.

The TGA curves show a fairly steady improvement to the percent mass loss at final decomposition for all weight percentages of OAPS-epoxy combinations tested (Fig. 7.19) [17]. This is concurrent with—and supportive of—the other previously reviewed research on this subject. Though it shows similar trends, the overall improvement of thermal decomposition from that of neat epoxy resin does not appear to be significant, or even different, from the other thermal stability studies seen here. Furthermore, the overall improvement is someone subjective. The best improvement,

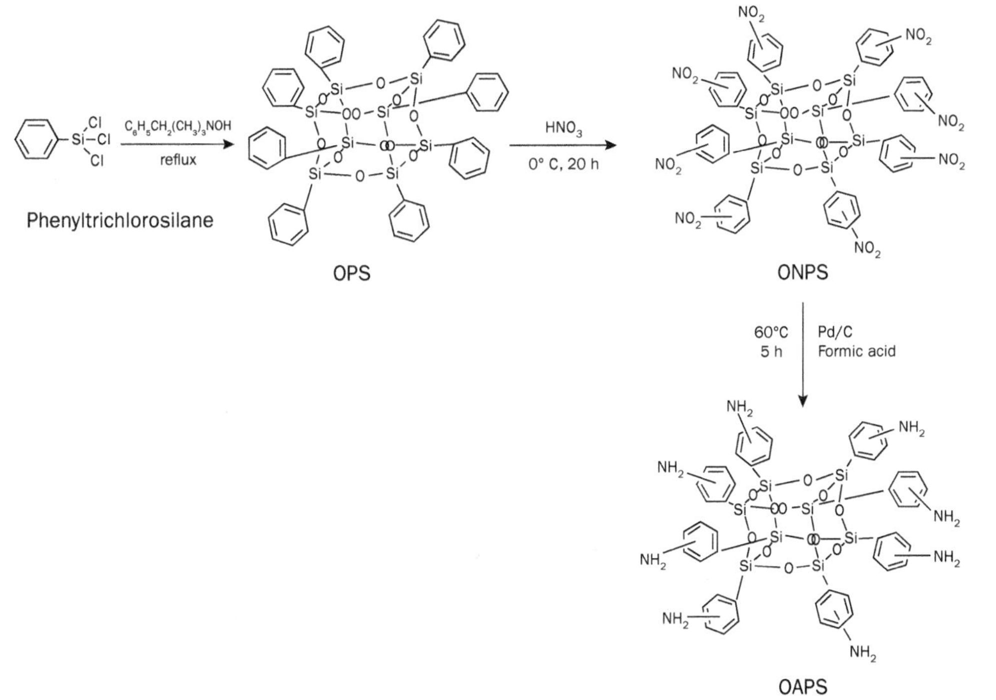

Phenyltrichlorosilane

$C_6H_5CH_2(CH_3)_3NOH$

reflux

OPS

HNO_3

0° C, 20 h

ONPS

60°C 5 h

Pd/C Formic acid

OAPS

Figure 7.17 Three-step process denoting the reactions to form OAPS [17].

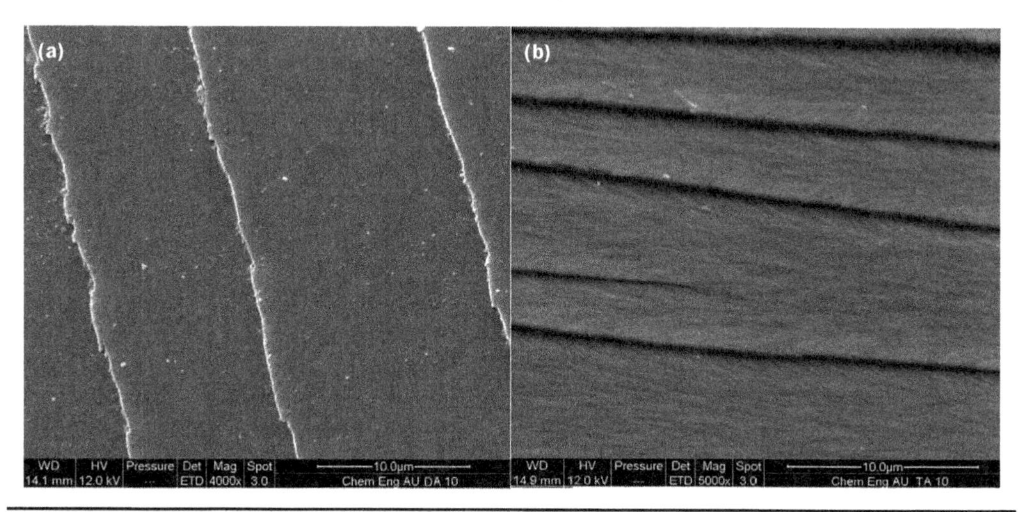

Figure 7.18 SEM micrographs of (a) 10 wt% OAPS in DGEBA and (b) 10 wt% OAPS in TGDDM [17].

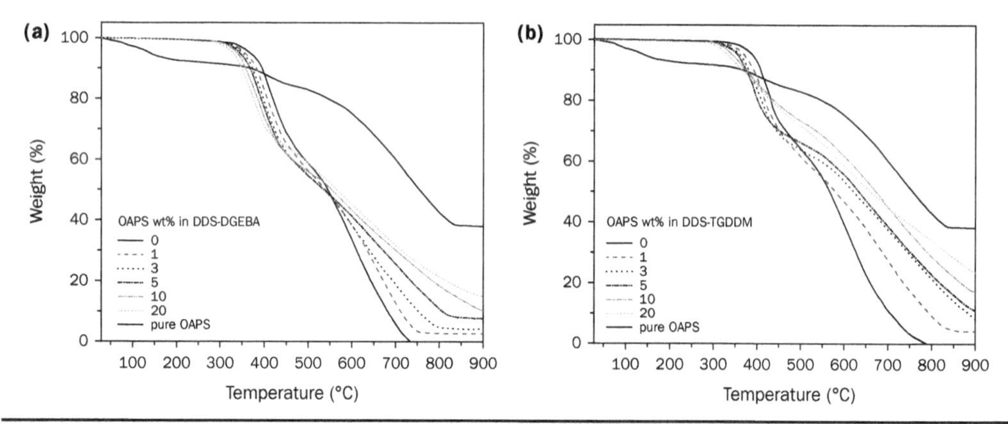

Figure 7.19 TGA curves for OAPS-epoxy nanocomposites [17].

aside for pure POSS material, is seen at 20 wt% OAPS. The issue in this case is that it is a high weight percentage for a nanocomposite, since most general nanocomposites, as a rule of thumb, are limited to a weight percentage of 5 or less. This could create issues with cost-effectiveness or take its toll on other mechanical or electrical properties.

TiO_2-SiO_2 in Epoxy (DGEBA) Nanocomposites

Omrani et al. [18] investigated the effect of TiO_2-SiO_2 in an epoxy resin (DGEBA) to form an organic-inorganic hybrid system. These composites have become of increasing interest recently because of the balanced property improvement the inorganic particles provide to an organic polymer matrix. Commonly explored inorganic particles include TiO_2, SiO_2, Al_2O_3, clay, SiC, and many others. TiO_2 has been of interest in the research of thermal properties where it has been seen to greatly improve a polymer's

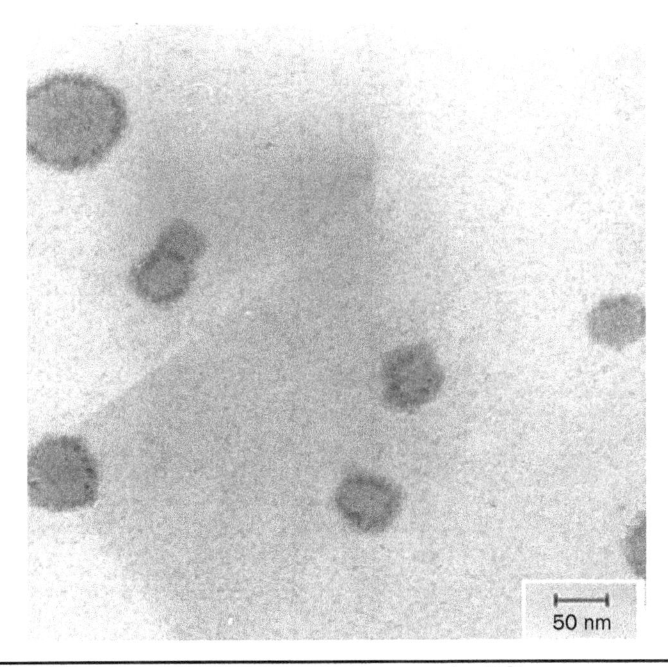

FIGURE 7.20 TEM image of the hybrid TiO_2-SiO_2 nanocomposite [18].

characteristics. Though they have an improving effect on host polymers, TiO_2 nanoparticles have demonstrated instances of poor mechanical properties and some less thermally stable structure behaviors. To smooth out some of the variant characteristics SiO_2 has been added to the TiO_2 filler particles. Note that forming a hybrid with both particles cannot create a material that embodies all the benefits of both materials; however, the mix of resulting properties that stem from each can further widen their current areas of application to broader horizons. The polymer matrix used was DGEBA that had been cured using diaminodiphenyl sulfone. Mixing this solution with diamine hardener produces a series of cross-linking reactions between the ends of epoxide groups.

TEM images for the microstructure show a very low level of particulate saturation within the polymer matrix (Fig. 7.20) [18]. Given the scale is 50 nm, the principal cause does not seem to be the result of agglomeration or otherwise poor dispersion. If each dark circle is a hybrid TiO_2-SiO_2 nanomolecule, then the dispersion, at least from the limited scope of the TEM, appears to be fairly good. Thermal stability was measured using TGA analysis up to 1,000°C at a rate of 10°C/min. SEM micrographs were taken to document the resulting structures and dispersion of the nanoparticles in the polymer matrix. Tests were performed on only one nanocomposite sample, and a sample of the neat epoxy.

The TGA data show that the TiO_2-SiO_2 hybrid composite had an increase in the onset thermal decomposition temperature of 16°C from the neat epoxy (Fig. 7.21) [18]. This can possibly be explained by its contribution to catalyze during curing reactions and its behavior as a thermal stabilizer [18]. The char yield of the nanocomposite was 25% and 0% for neat epoxy at 600°C. This high yield differences could strongly be attributed to the interactions of polymer-chain inorganic particles. Overall, the

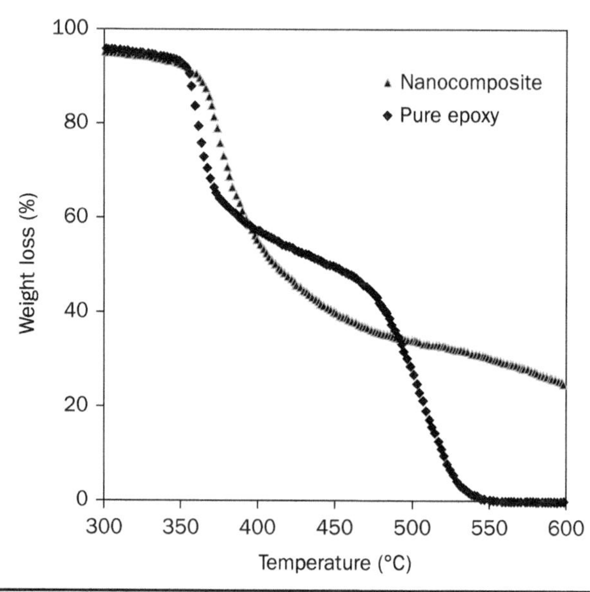

FIGURE 7.21 TGA curves for the two cured samples [18].

desired results from the experiment were to some level achieved, though no real designed magnitude was addressed or stated so we can only attribute it to a positive result of overall principle.

Polythiophene in Diglycidyl Ether Bisphenol (Epoxy) Nanocomposites

Polythiophene (PT) is a type of conducting polymer that exhibits low density, environmental stability, and good electrical conductivity [19]. Polythiophene has gained interest in the development of polymer nanocomposites because it shows signs of good thermal stability while maintaining good electrical conductivity, which makes it incredibly promising as filler for electrically conducting polymers. Prior to research on polythiophene, electrically conductive polymers were primarily using metallic particles or carbon black as conducting filler. The problem with these fillers is that metallic particles have an unfortunately common tendency to form an oxide layer on its edges. Carbon black does not suffer this problem, but it is far more expensive [19].

Zabihi et al. [19] conducted thermal testing on samples of polythiophene in a diglycidyl ether bisphenol matrix that was cured with phthalic anhydride. Thermal tests were conducted using TGA and DSC. Additionally, Kissinger plots were used to map heating rate against maximum temperature (Fig. 7.22) [19]. TGA tests were performed on samples of 1, 5, 10, and 20 wt% PT, then another TGA test was performed on samples of 1 wt% at heating rates of 2.5, 5, 10, and 20°C (Fig. 7.23) [19].

The thermal tests used a composite sample with 1 wt % PT. The Kissinger plot in Fig. 7.22 shows that the addition of 1 wt% PT was consistently able to reach a higher maximum temperature at each heating rate than the pure epoxy sample [19]. The difference in maximum temperature between the neat epoxy and epoxy nanocomposite

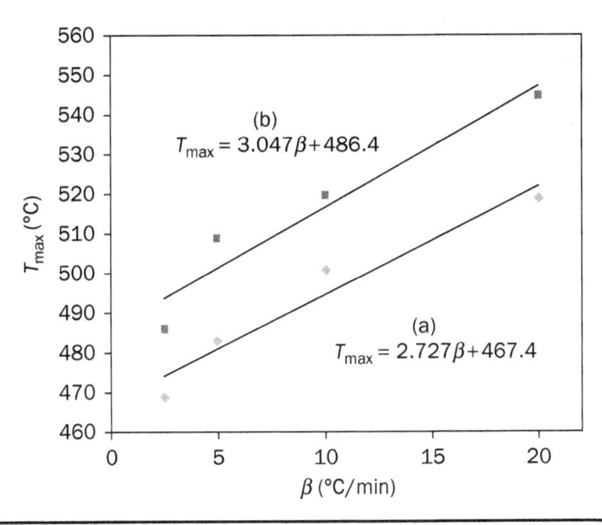

FIGURE 7.22 T_{max} versus heating rates of (a) pure epoxy and (b) epoxy-PT nanocomposite [19].

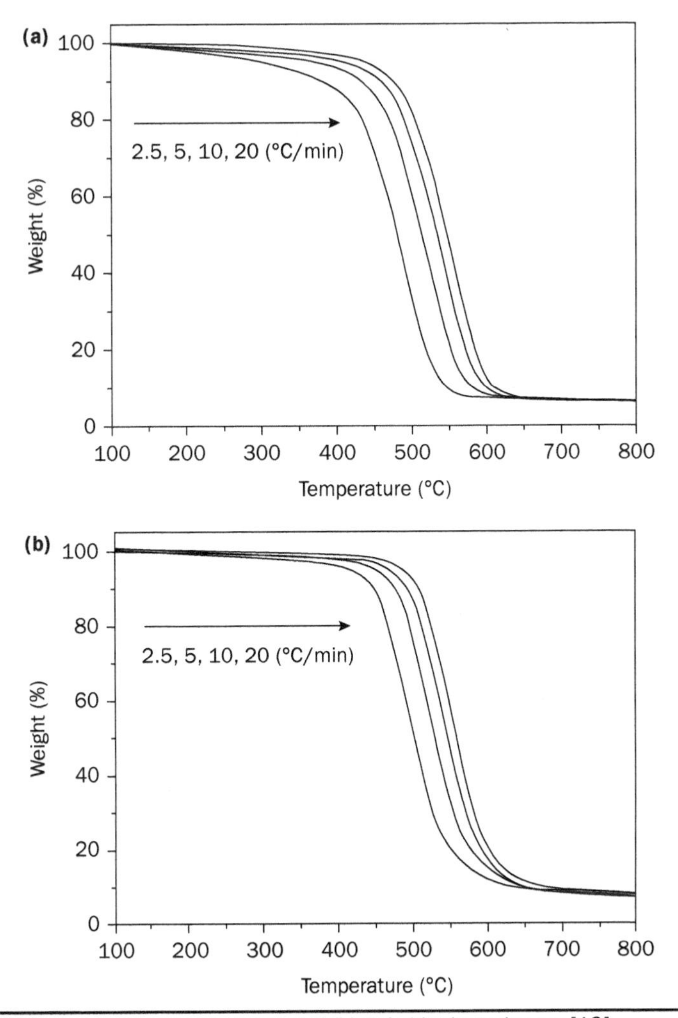

FIGURE 7.23 TGA curves for (a) pure epoxy and (b) polythiophene/epoxy [19].

Table 7.4 TGA Data for Neat Epoxy and Epoxy/PT at a Heating Rate of 10°C/min [19]

Type	$T_{initial}$ (°C)	T_{max} (°C)	IPDT (°C)
Neat epoxy	346	501	464
Epoxy/1% nano-PT	384	520	489
Epoxy/5% nano-PT	380	509	473
Epoxy/10% nano-PT	366	487	459
Epoxy/20% nano-PT	320	439	436

Table 7.5 TGA Data for Thermal Degradation at Different Heating Rates [19]

β (°C/min)	$T_{initial}$ (°C)	T_{max} (°C)
Neat epoxy		
2.5	248	469
5	296	483
10	346	501
20	386	519
Epoxy nanocomposite		
2.5	302	486
5	345	509
10	383	520
20	410	545

remained relatively stable for each heating rate tested (Tables 7.4 and 7.5) [19]. This confirms that polythiophene nanoparticle additives can enhance the conducting performance of a material; however, since no data are currently available about the range of temperatures that conductive materials need to reach in application, the true applicable value of these results is indeterminate. Further, since data are only being published for 1 wt% PT, there is only a very narrow view of the possible property enhancements made by the incorporation of PT nanocomposites. This limits the value of the research by limiting the extent to which it can be applied to real applications. The data show that the addition of PT nanoparticles to the polymer matrix increases the final decomposition temperature by 5%. The test results incorporated into the TGA on the effect of heating rate seems to suggest that thermal stability in the material is higher with a higher heating rate. From the TGA data (Fig. 7.23), one can observe that the difference between heating rates remains largely the same for both pure epoxy and the epoxy nanocomposite, and that the two effects are mutually exclusive.

7.4.2 Thermal Conductivity of Epoxy-Based Nanocomposites

Thermally conductive thermoset polymers are an important material when dealing with electronics. As new technology is created to miniaturize electronics, the material used to package the technology, thermal interface materials (TIMs), must be able to quickly and effectively dissipate heat in order to maintain the operating temperature of devices [20]. As the components become smaller, the material must also be able to be made smaller without the loss of function.

Common thermally conductive fillers are AlN, BN, Al_2O_3, and SiC, which are cost-effective but require very high loading (about 60 vol% or higher) in order to satisfy the percolation thresholds while still achieving the thermal conductivity and heat conducting network that is desired [20]. As a result of the high loading, these nanocomposites have a high bulk density and poor mechanical properties which are undesirable [20].

Carbon-based nanofillers have been very popular due to their excellent mechanical, electrical, and thermal properties [12–24]. In general, carbon nanofillers in epoxy resins improve the thermal conductivity significantly without destroying the density or mechanical properties [20]. This section discusses the effects of carbon-based nanofillers on the thermal conductivity of epoxy-based nanocomposites.

Han and Fina published an excellent review on "Thermal conductivity of carbon nanotubes and their polymer nanocomposites" [24]. They summarized the status of research of CNTs and their polymer nanocomposites through 2011. The dependence of thermal conductivity of CNTs on the atomic structure, the tube size, the morphology, the defect, and the purification are reviewed. The roles of particle–polymer and particle–particle interfaces on the thermal conductivity of polymer/CNT nanocomposites are discussed in detail, as well as the relationship between the thermal conductivity and the micro- and nano-structure of the composites.

Carbon Nanotube-Based DGEBA Epoxy

Carbon-based nanofillers have excellent mechanical, electrical, and thermal properties. Specifically, CNTs have been reported to have a stiffness of 1 TPa, strength up to 100 GPa, thermal conductivity greater than 3000 W/m-K [16], as well as a high aspect ratio [18,19]. Due to these extraordinary characteristics, the addition of CNTs has the potential to improve the mechanical, electrical, and thermal properties of certain materials. A high aspect ratio and good thermal conductivity of the nanofiller are important in improving the overall thermal conductivity of a material, because these characteristics determine the potential for an effective heat flow network, which is the essence of thermal conduction [19].

Unfortunately, due to the van der Waals interactions, CNTs have a tendency to bundle or aggregate when mixed into a polymer system. This aggregation results in poor dispersion, which hinders the property enhancements by creating defects in the material. Certain methods have been explored in order to promote dispersion including mechanical methods (high-shear mixing and shortened CNTs), surfactants, acid oxidation, in situ polymerization, and other chemical methods [19]. The advantages and disadvantages are described in more detail in the following sections, along with the resulting thermal conductivity evaluations.

Balakrishnan and Saha [16] researched MWNTs at low weight percentages in toughened epoxy resins and the resulting tensile behavior and thermal conductivity. They used unmodified MWNTs with a diameter of 5–10 nm and length of 10–20 μm. DGEBA was used as the polymer material. In order to toughen the epoxy, rubber particles with a diameter of about 250 nm were added. MWNTs were dispersed into the epoxy resin by sonication at loadings of 0.2, 0.4, 0.6, and 1 wt%. In order to measure thermal conductivity, the 3ω method was used and the results are shown in Fig. 7.24 [16]. Compared to the neat system, the addition of MWNTs improved the thermal conductivity at all temperatures. The most improved thermal conductivity for all temperatures was the addition of 0.4 wt% MWNTs, which gave about a 16% increase in thermal conductivity. It is also noted that the thermal conductivity variance between loading content for all temperatures was about the same.

The loading of 0.2, 0.6, and 1 wt% of MWNTs had about the same increase in thermal conductivity compared to the baseline. TEM images were taken in order to visualize the quality of dispersion of MWNTs achieved in their study. Figure 7.25 shows the

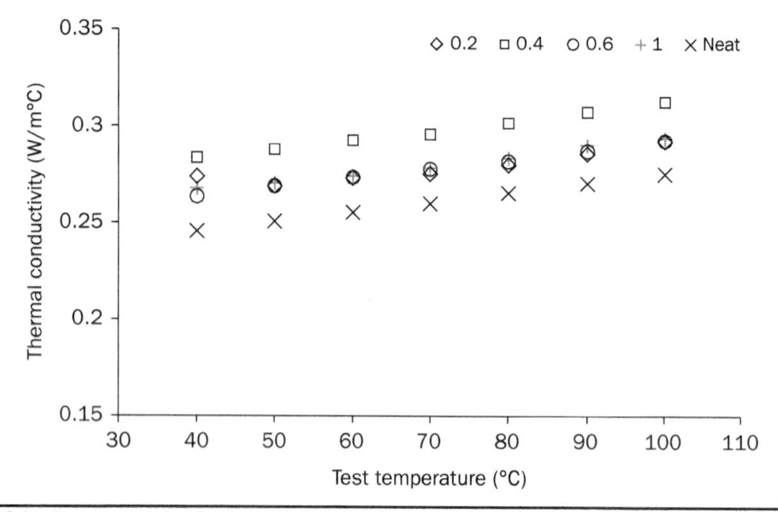

Figure 7.24 Results of the 3ω method measuring thermal conductivity of MWNT/epoxy composite system [16].

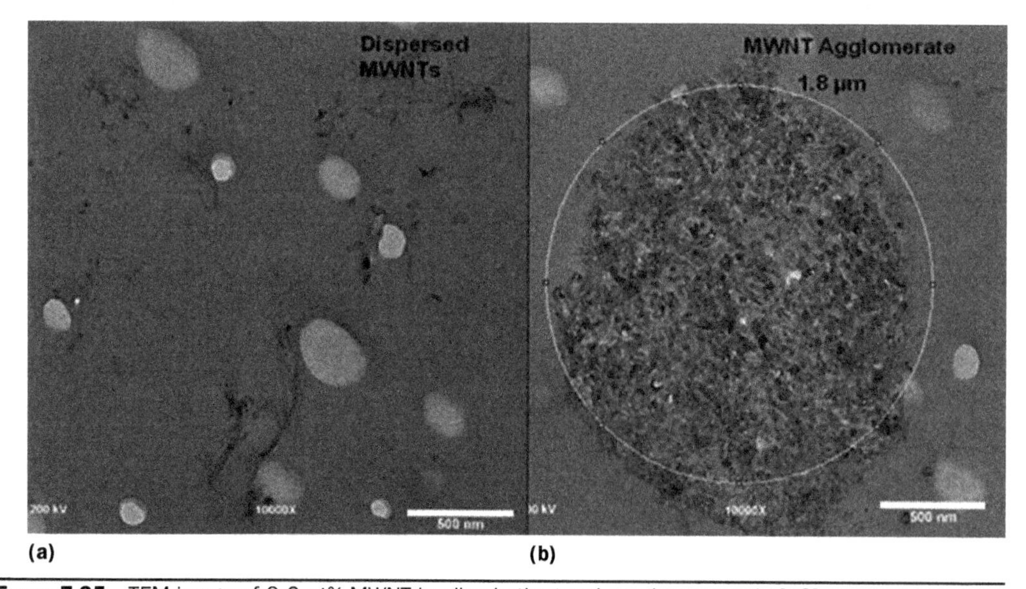

Figure 7.25 TEM image of 0.6 wt% MWNT loading in the toughened epoxy resin [16].

several states of dispersion found in the material [16]. Also, as the content of MWNTs increased in the epoxy system, larger aggregates were found. From these observations and the results of the thermal conductivity analysis it can be assumed that something is interfering with heat transport; perhaps the rubber particles get in the way or the MWNT–epoxy interface scattered the phonons. The *effective medium approach* was used to measure the thermal interfacial resistance with and without the addition of MWNTs, and Balakrishnan and Saha found that the interfacial resistance was not decreased by the addition of MWNTs as expected. It was concluded that this reason was the cause for the low thermal conductivity values.

Yang et al. [23] conducted the effect of functionalized MWNTs in an epoxy composite on the thermal conductivity. They used the Friedel–Craft method of functionalization in order to avoid damaging the MWNTs thereby enhancing the thermal conductivity more than previous functionalized MWNTs [23]. The MWNTs were functionalized by mechanically stirring a mixture of benzenetricarboxylic acid (BTC), MWNTs, polyphosphoric acid (PPA), and P_2O_5 and purging with nitrogen. The weight ratio of BTC to MWNTs was 1:1, while the overall solid content in PPA was controlled below 6 wt%. The mixture was heated and mechanically stirred for 2 hours. Then PA was added and slowly stirred for 24 hours. After being cooled in an ice bath, PPA, P_2O_5, and PA were dissolved in distilled water and the solution was filtered through a polytetrafluoroethylene (PTFE) membrane. The collected powder was mixed with acetone and filtered from the PTFE membrane two or three more times in order to filter out all the excess uncreated BTC [23]. The BTC-MWNT/epoxy composite was characterized by the hot disk method using a TPS2500 at room temperature and the results were compared with the Nan's model:

$$\frac{K_e}{K_m} = 1 + \frac{f \times P}{3} \frac{\dfrac{K_c}{K_m}}{P + \dfrac{2a_k K_c}{d \times K_m}}$$

where K_e is the effective thermal conductivity of the composites, K_m is the thermal conductivity of the epoxy resin (0.13 W/m-K), K_c is the thermal conductivity of the CNT (2,500 W/m-K), f is the content of CNTs filled, a_k is the Kapitza radius, and P is the aspect ratio of CNTs (about 350). This model considers the interface effect [25]. The results of the experiment are shown in Fig. 7.26 [23].

For all MWNT types tested, the addition of the nanotube increased the thermal conductivity of the material. While the thermal conductivity with pristine-MWNTs (P-MWNTs) increased, the values reported are lower than the Nan's model, which can be explained by the formation of agglomerates, and confirmed by SEM and TEM images shown in Figs. 7.27 and 7.28, respectively [23]. The formation of MWNT bundles in the composite restricts the transport of phonons and can cause a phenomenon called *reciprocal phonon vector*, which describes the creation of a heat reservoir that restricts the diffusion of heat. Aggregates can cause this because they reduce the aspect ratio, decrease contact area, and trap or scatter phonons. BTC-MWNT/epoxy composite resulted in the highest thermal conductivity enhancement (a 684% increase with the addition of 5 vol% of BTC-MWNT), and the experimental values were higher compared to the theoretical values predicted by the Nan's model. With only 1 vol% BTC-MWNTs, the thermal conductivity was increased by 315%. This can be attributed to the good dispersion, which can be seen in Figs. 7.27c,d and 7.28c,e [23].

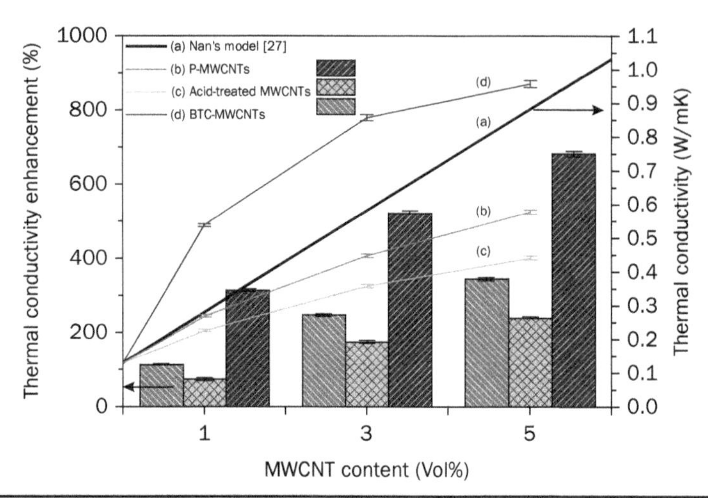

FIGURE 7.26 Thermal conductivity of MWNT/epoxy composite with varying content of pristine-MWNTs, acid-treated MWNTs, and BTC-MWNTs, as well as the Nan's model [23].

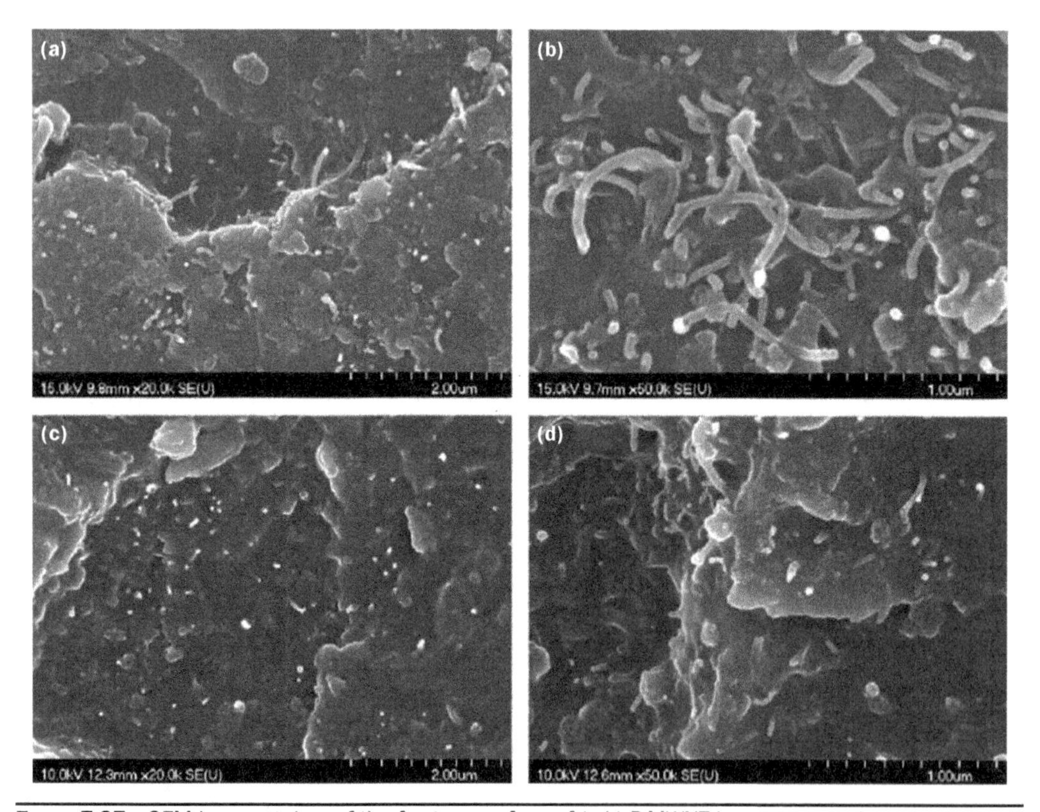

FIGURE 7.27 SEM images taken of the fracture surface of (a,b) P-MWNT/epoxy composites and (c,d) BTC-MWNT/epoxy composites [23].

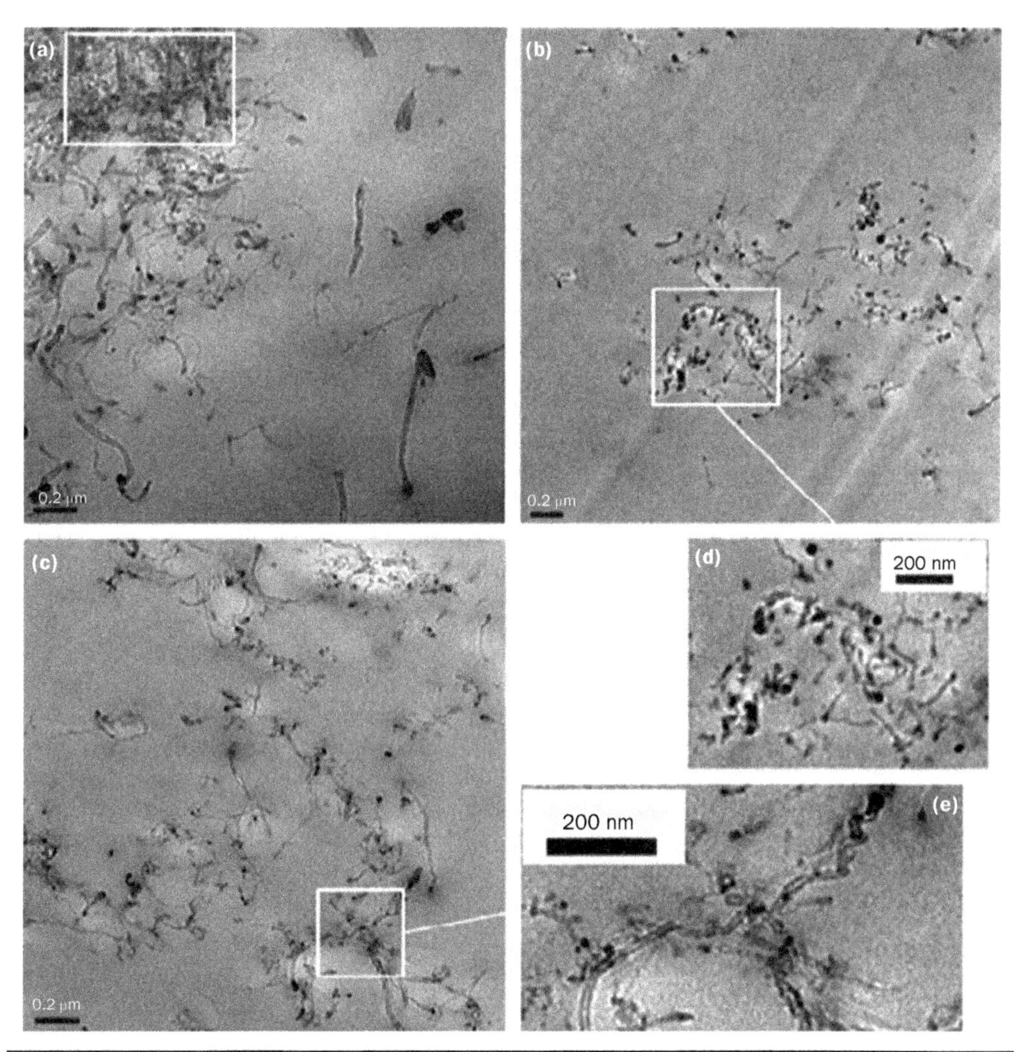

Figure 7.28 TEM images of (a) P-MWNT/epoxy composites, (b) acid-treated MWNT/epoxy composites, and (c) BTC-MWNT/epoxy composites (d) showing an enlarged region from (b) and (e) an enlarged region from (c) [23].

In the case of BTC-MWNT/epoxy composites, the interface has been improved by covalent bonding between the epoxide group of the epoxy and the carbonyl group of the BTC-MWNTs, thereby reducing the thermal interfacial resistance and increasing the thermal conductivity [24]. From these results, the Friedel–Craft functionalization method was shown to prevent damage on the surface of MWNTs and resulted in enhancement of thermal conduction in the material.

Graphite-Based Expanded Graphite (EG) Flakes

Graphite is a naturally occurring and synthetically producible form of carbon that is crystalline and highly conductive [15]. Expanded or exfoliated graphite (EG) is a form of graphite that has been treated by acid and rapidly heated in order to create an expanded material. In order to make exfoliated graphite, natural graphite flakes are mixed with nitric acid and sulfuric acid in order to form intercalations [15]. The result is an intercalated graphite compound. After being washed and dried, the compound is quickly heat treated at a high temperature so that the intercalations decompose, and expansion of the compound occurs resulting in exfoliated graphite. The idea behind exfoliating graphite is that a lower concentration of the material will be needed in order to achieve the desired thermal conductivity.

Debelak and Lafdi investigated the effect of the size of exfoliated graphite nanofillers on the physical properties of EG/epoxy composites [15]. In this experiment, exfoliated graphite flakes were separated into large, medium, and small flakes and 0.1, 0.5, 1, 2, 4, 8, 12, 16, and 20 wt% EG were prepared. The polymer used was EPON 862 resin, which is an epoxy bisphenol F resin. To prepare the composite, the EG flakes were placed in a solvent and highly sheared under a homogenizer in order to disperse the flakes. This process was followed by ultrasonication to further improve dispersion. Then the composite was molded and cured [15]. The EG/epoxy composite was characterized using an LFA 447 Nanoflash apparatus, which measured the thermal diffusivity of the specimen. Based on the Nan's model equation mentioned earlier, the thermal conductivity was determined. The results are shown in Fig. 7.29 [17]. For all particle sizes, the increase in EG flake content resulted in an increase in thermal conductivity.

At 4 wt% of large EG flakes, the thermal conductivity increased 300% from the baseline epoxy resin. The largest increase in thermal conductivity occurred from 8 to 12 wt% of the large flake content. For medium EG flakes, a higher content of EG flakes was required at lower weight percentages to achieve the same increase in thermal conductivity as the large flakes. The same is true for small flakes—to achieve the same increase in thermal conductivity, a higher loading of small EG flakes was required compared to medium and large flakes. At 20 wt% loading, there was a 2,000% increase in thermal conductivity for all particle sizes, which is about a thermal conductivity of

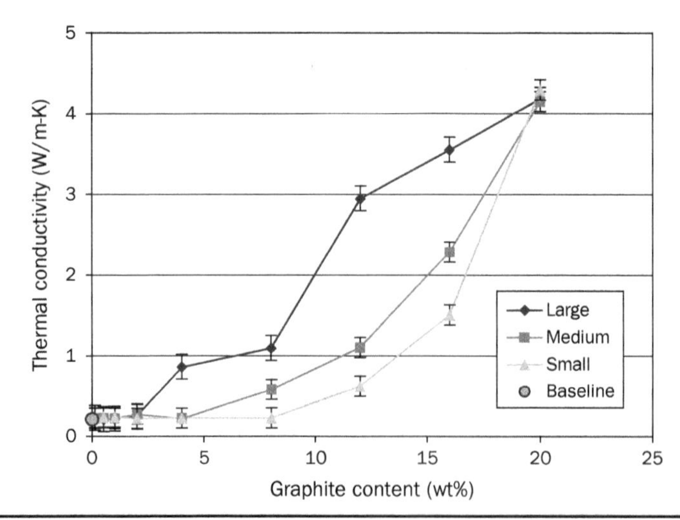

FIGURE 7.29 Thermal conductivity results for EG/epoxy composite with large, medium, and small EG flakes [15].

4.3 W/m-K. These results show that an adequate conductive network has been formed by the nanofillers for heat transport, but perhaps at 20 wt% loading there is enough filler that the particle size no longer matters for thermal conductivity. Each particle size had a different percolation threshold of 3, 6, and 10 wt% for large, medium, and small EG flakes, respectively [15]. The reason for the drastic increase in thermal conductivity can be attributed to the high aspect ratio of EG flakes [15].

Another study was conducted on the effect of EG flakes in an epoxy resin on the thermal conductivity of the material. This study focused on the difference between treated and untreated EG flakes and the relation to the thermal conductivity. Ganguli et al. [21] believed that by chemically functionalizing the EG flakes, thermal conductivity of the composite would improve due to enhanced interaction between the nanofiller and matrix. For the polymer, they also used Epon 862 as the polymer, plus Epicure W (an aromatic amine curing agent) and Epicure 537 (an organic accelerating agent used to quicken the rate of curing) in order to capture the dispersed morphology [21]. The chemical modification was carried out in a mixture of 25 vol% water and 75 vol% ethanol. First, 3 g of 3-aminopropoxyltriethoxy silane was added into 1,000 mL of the water/ethanol mixture while the temperature was maintained at 80°C. It was followed by the addition of 10 g of EG whereby the grafting reaction took place under shearing. Afterwards, the reaction product was filtered, washed, and freeze-dried before being grounded and stored. In order to create the nanocomposite, a Flacktek Speedmixer with very high shear was used to disperse the EG flakes in the epoxy resin, then the curing and accelerating agents were mixed into the nanocomposite. Samples were loaded with 2, 4, 8, 16, and 20 wt% of treated or untreated EG flakes for testing. The thermal conductivity results are shown in Fig. 7.30 [21].

From the graph it is apparent that the functionalized EG flakes increased the thermal conductivity at lower loadings, but the addition of modified or unmodified EG flakes lead to an increase in thermal conductivity. At loadings before 4 wt% for both types of

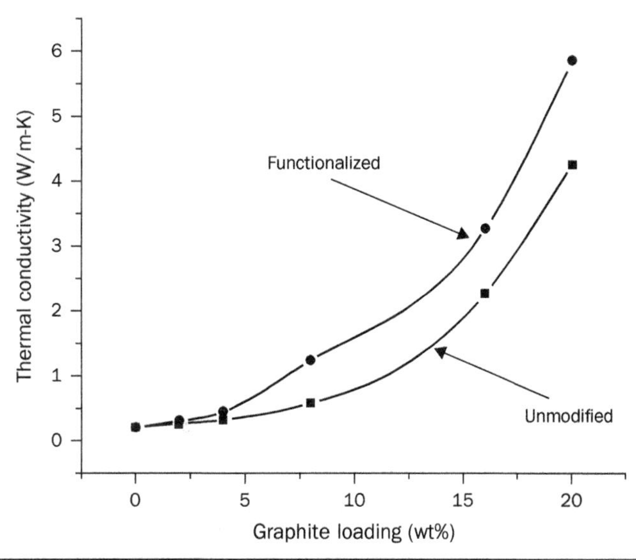

FIGURE 7.30 Thermal conductivity of unmodified and functionalized EG flakes in EG/epoxy composite [21].

EG flakes, the thermal conductivity was barely affected, and the chemically modified and unmodified flakes behaved similarly. At 8 wt% loading of untreated EG flakes, the thermal conductivity increased from 0.2 to 0.5 W/m-K, but the treated EG flakes at the same loading resulted in a much better improvement in thermal conductivity. Overall, for untreated EG flake nanofillers, there was a 19-fold increase in thermal conductivity at 20 wt% loading, which was a 4.0 W/m-K measurement of thermal conductivity. This result shows agreement with the previous study by Debelak and Lafdi [15] who reported a high thermal conductivity of 4.3 W/m-K for their untreated EG flakes. The chemically modified EG flakes saw a 28-fold enhancement of thermal conductivity at 20 wt% which measured to be 5.8 W/m-K.

SEM micrographs (Figs. 7.31 and 7.32) show the microstructure of the unmodified and chemically modified EG flakes at 4 wt% loading used in this experiment [21]. Both SEM micrographs show that agglomerations formed in both composites, but better

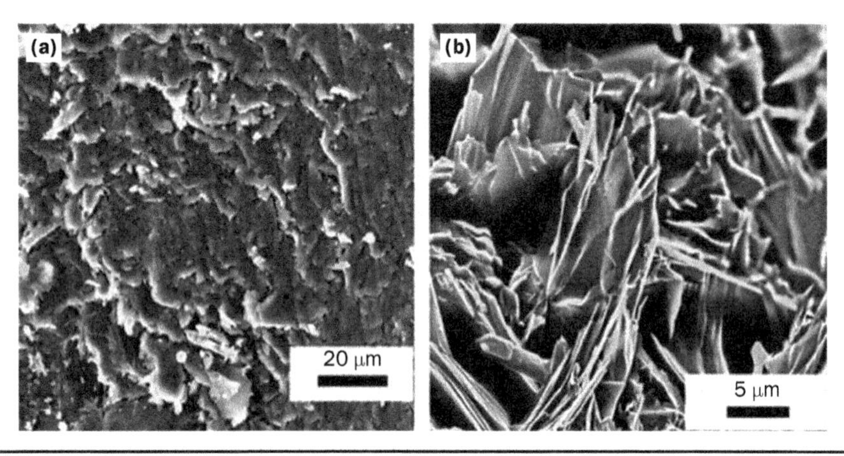

FIGURE 7.31 SEM micrograph of untreated EG flakes at 4 wt% loading in the EG/epoxy composite [21].

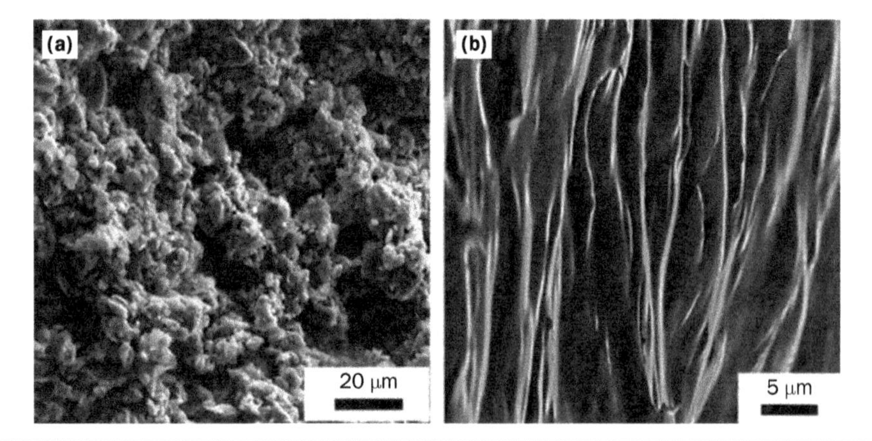

FIGURE 7.32 SEM micrograph of chemically treated EG flakes at 4 wt% loading in the EG/epoxy composite [21].

EG–epoxy interactions were made in the chemically modified sample compared with the unmodified composite. These results corroborate the findings that the composite with functionalized EG flakes resulted in higher thermal conductivity and also shows the importance of achieving good interaction between nanofiller and matrix. The fact that agglomerations were present at a loading of 4 wt% and thermal conductivity improved at higher loadings challenges the idea that a uniform dispersion is absolutely required for enhanced thermal conductivity. While it is known that the dispersion is important in determining a material's properties, perhaps efforts should be focused more on improving the interaction between nanofiller and epoxy matrix. The interface between the nanofiller and the polymer is where the heat transfer occurs; therefore, to improve thermal conduction in a material, the interface must be optimal. This will reduce the acoustic impedance mismatch between the two materials, which minimizes the interfacial phonon scattering resulting in an improved thermal conductivity.

Expanded Graphene Nanoplatelets (EGNPs)

Graphene has been determined to have a high aspect ratio, two-dimensional geometry, good stiffness, and low interface thermal resistance [14], which have resulted in the ability to improve thermal conductivity of materials with poor heat dissipation qualities. For this reason, Chatterjee et al. decided to "investigate the influence of expanded graphene nanoplatelets (EGNPs) on the tensile, nano-mechanical and thermal properties in epoxy" [14]. Natural graphite flakes were used to make the EGNPs. First, the natural graphite flakes were immersed in sulfuric acid and nitric acid to cause intercalation of the sulfuric acid into graphene layers of the flakes. Then the flakes were washed, neutralized, and dried before being thermally shocked to expand the graphene layers and create EGNPs. The EGNPs were further functionalized with dodecyl amine (DDA) to create amine-EGNPs. To create the nanocomposite, the amine-EGNPs were dispersed in acetone under ultrasonication in order to break up aggregation of the material, then the EGNPs were continued to be dispersed in a high-pressure homogenizer and placed in an ultrasonic bath. Afterwards, the dispersed EGNPs were incorporated into the epoxy resin via three-roll mill calendaring. Samples of 0.1, 0.5, 1, 1.5, and 2 wt% EGNPs were measured prior to hardening. The composite was hardened, molded, and then cured at 80°C for 12 hours then post-cured at 120°C for 4 hours. In order to characterize the thermal conductivity, the thermal diffusivity was measured using an LFA 447 Nanoflash apparatus [14].

Chatterjee et al.'s study determined that a conductive network was formed in the epoxy matrix thereby increasing the thermal conductivity of the composite [14], as shown in Fig. 7.33. As the content of EGNP increased in the composite, the thermal conductivity increased as well, showing an almost linear relationship between EGNP content and thermal conductivity. While other researchers added much more than 2 wt% EGNP, the trend shows that the addition of more than 2 wt% EGNP will continue to increase the thermal conductivity. Compared to pristine epoxy, the addition of 2 wt% of EGNP leads to a 36% increase in thermal conductivity value.

This study also investigated the dispersion of the material because it has been shown to be correlated to the evaluation of a material's properties. From TEM images, they concluded that uniform dispersion was attained at up to 0.5 wt% EGNP loading, but as the content of EGNP increased the realization of uniform dispersion became more challenging. This reason explains the decrease in thermal conductivity improvement from 0.5 to 1.0 wt% of EGNP. Without uniform dispersion, thermal interface resistance

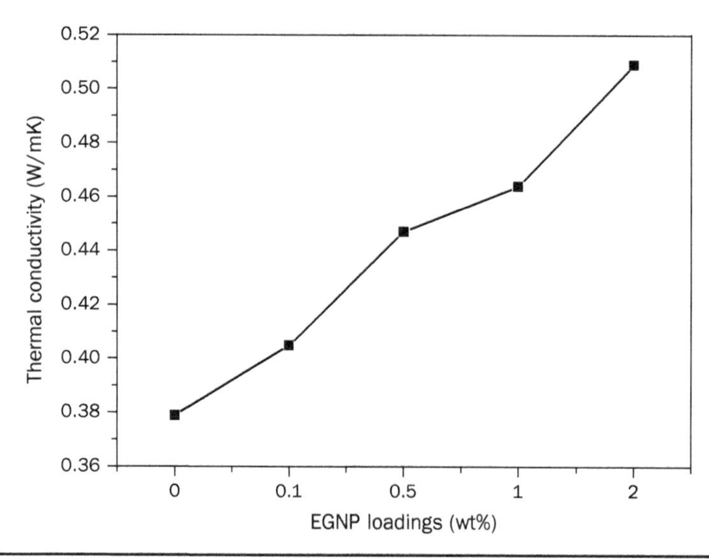

Figure 7.33 Resulting measurements for thermal conductivity of EGNP/epoxy composite [14].

may increase, and the acoustic phonons may not have a network path to travel or can be scattered. For these reasons, dispersion must be perfected.

Graphene Oxide with MWNT Hybrid

Im and Kim conducted a study on the thermal conductivity of graphene oxide (GO)/MWNT hybrid in an epoxy composite [20]. The polymer matrix for this experiment was made out of DGEBF. The MWNTs were synthesized by chemical vapor decomposition (CVD) and were then carboxyl acid-functionalized in order to improve the interface between nanotubes and the matrix. The GO was made using the Hummers method. In order to make the GO/MWNT/epoxy composite, a wetting process was used. Both GO and MWNTs were suspended separately and then mixed in order to create hybrid fillers, and the MWNT content was varied among fillers. The hybrid fillers were poured into a glass mold and filtered, then annealed to remove the solvent. Epoxy was then dropped onto the GO/MWNT mold and the "dropped epoxy penetrated into the fabricated cake easily because of sufficient wettability between GO and the MWNTs" [20]. Then the composite was cured at 120°C for 30 minutes. A schematic of the fabrication process is shown in Fig. 7.34 [20].

In order to characterize the thermal conductivity, LFA and DSC were used. Im and Kim found that at 50 wt% of filler the thermal transport properties were optimal due to the fixed pore volume [20]. At less than 50 wt%, an insulating layer of epoxy could form on the upper surface because the amount of filler used to make the cake was insufficient. Above 50 wt% filler, there is not enough epoxy to permeate the whole GO/MWNT network. With the hybrid/epoxy ratio fixed at 50 wt% each, the results of thermal conductivity were measured with varying GO/MWNT content, which are shown in Fig. 7.35 [20]. At 0.36 wt% of MWNTs, the filler achieved the maximum thermal conductivity. Below 0.4 wt% MWNT, the thermal conductivity was higher than expected; while above 0.4 wt% MWNT, the thermal conductivity was lower than predicted.

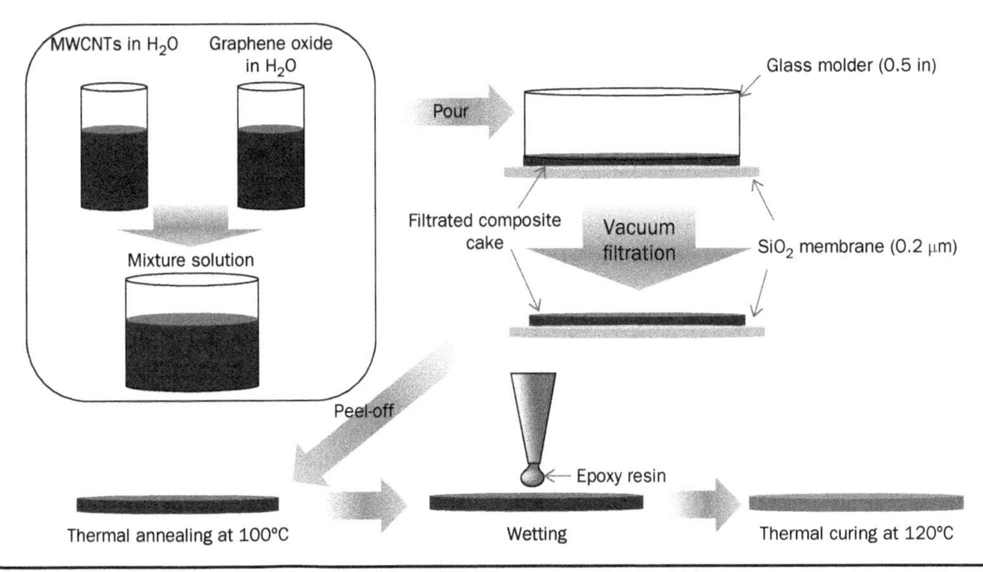

Figure 7.34 The fabrication process for GO/MWNT/epoxy composite [20].

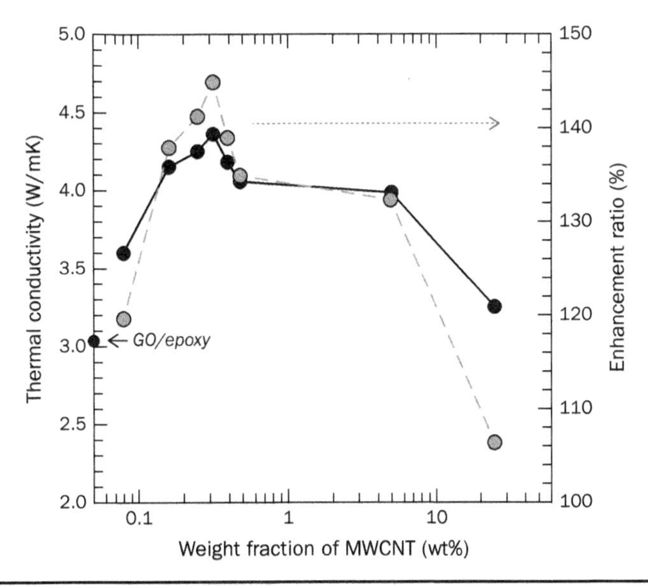

Figure 7.35 Thermal conductivity of GO/MWNT/epoxy composite with varying ratio of GO to MWNTs, shown by black dots [20].

Possible explanation of these results is that the hybrid fillers may have formed a three-dimensional heat conduction pathway, which resulted in the higher thermal conductivity. Because MWNTs are one-dimensional and GO is two-dimensional, the MWNTs could have acted like bridges between the GO particles due to the MWNT's high aspect ratio, and this created an efficient heat transport network and good thermal conduction. But increasing the MWNT content resulted in an excess of MWNT, disrupting the network and scattering the phonons in order to decrease thermal conductivity.

Al_2O_3 in HTTE/DGEBA Epoxy

Research by Chen et al. [25] into the thermal behaviors of Al_2O_3 in DGEBA was aimed at obtaining a nanocomposite that enhanced thermal conductivity, thermal stability, and mechanical toughness. Particles such as AIN, Al_2O_3, ZnO, and SiO_2 are often used as filler material in composite blends over other commonly used particles, such as organoclays, because of their greater thermally conductive performance. The negative drawback of these filler particles is their high tendency to agglomerate. The inability to disperse into the polymer matrix is a limiting factor on possible property enhancements. In order to overcome this hurdle, the DGEBA epoxy resin was mixed with epoxy-terminated low-viscosity liquid thermosetting aromatic polyester hyper-branched epoxy resin (HTTE) [25]. HTTE is a hyper-branched polymer that has often been used for its ability to increase polymer toughness. Because of its unique structures, such as lower melt and solution viscosities, small molecular dimensions, and high functional end group density, HTTE is able to increase the performance of desired properties in the DGEBA resin without sacrificing other key properties or overall processability. Additionally, the high end-group density, cavity structure, and low liquid viscosity can greatly improve the dispersion of nanoparticles in the epoxy resin matrix [25].

For experimentation, the polymer matrix was composed of 85 wt% DGEBA and 15 wt% HTTE. The matrix and subsequent composite were formed through a series of low- and high-temperature mixing and magnetic stirring, before being degassed, cured, and then post-cured to ensure complete reaction of the thermoset. Nanoparticle compositions of 5 wt% and 40 wt% were tested in the HTTE/DGEBA matrix.

Measurements of the thermal conductivity utilized a thermal conductivity analyzer. Thermal stability data were collected through TGA. SEM was used to investigate the structure and dispersion of the nanocomposite formation [25]. For the TGA, 10-mg samples were heated to 1,000°C at a rate of 10°C/min under atmospheric conditions. The initial decomposition temperature was taken as the temperature at which 5% mass loss occurred. Samples of unfilled neat epoxy, 15 wt% HTTE/85 wt% DGEBA, 5 wt% Al_2O_3, and 40 wt% Al_2O_3 were tested.

From the SEM micrographs of the DGEBA/Al_2O_3 composite, we can see large regions of well-dispersed particles and the results of the thermal conductivity experiments are presented in Table 7.6 [25]. From Table 7.6, the thermal conductivity of the

Table 7.6 Thermal Conductivity Results of the Al2O3-DGEBA Composite [25]

		Thermal conductivity (W/m-K)				
Filler	**Filler content**	**EP**	**HTTE-1/EP**	**HTTE-2/EP**	**HTTE-3/EP**	**HTTE-4/EP**
Unfilled	0	0.20	0.20	0.20	0.20	0.20
Al_2O_3-A151	5 wt%	0.3	0.31	0.26	0.27	0.29
	40 wt%	0.36	0.45	0.43	0.44	0.45
Al_2O_3-KH570	5 wt%	0.26	0.26	0.25	0.24	0.29

HTTE/DGEBA IPNs is 0.20 W/m-K. Al_2O_3/HTTE/DGEBA had an optimum conductivity value of 0.31 W/m-K at 5 wt%. At 40 wt% they achieved a maximum value of 0.45 W/m-K [25]. This corresponds to a 55% maximum increase in conductivity for the 5 wt%, and a 125% increase in conductivity for the 40 wt% composite. Since conductivity depends only on level of dispersion, and not on the dispersed structure formation, the increased conductivity could be seen as productive of the hyper-branched epoxy creating better particle dispersion among the polymer matrix.

7.4.3 Heterogeneously Structured Conductive Resin Matrix/Graphite Fiber Composite for High Thermal Conductive Structural Applications

Advanced composite materials have been widely used as structural materials for aerospace, military, and industrial applications due to their high stiffness and strength to weight ratios. Currently, low thermal conductivity of composites restricts their ability to replace metallic structures involving thermal management functions, such as the leading edges of supersonic aircraft wings, the inlet or exhaust areas of gas turbine engines, lightweight heat exchangers, electronics packaging materials, hydraulic pump enclosures, and electromagnetic interference (EMI) enclosures [26]. Carbon fiber is widely used as reinforcement in advanced polymer composites due to its superior mechanical and physical performances. Polyacrylonitrile (PAN)–based fiber and pitch-based fiber are the two of most commonly used fibers. Both types of carbon fibers possess higher thermal conductivity in its axial direction than the transverse direction due to long and continuous crystal structures existing along fiber axis direction to promote phonon transport. As a result, laminated composites show a higher in-plane thermal conductivity along fiber axial direction than through-thickness directions transverse to the fiber axis. The absence of fibers in the through-thickness direction and insulating resin rich areas between fiber tows and layers further result in low through-thickness thermal conductivity (TTTC).

Extensive studies have been investigated to improve the TTTC of PMCs. Schuster et al. [27] and Sharp et al. [28] achieved a noticeable increase in TTTC using 3D fiber reinforcements. Metal and carbon-based particles in nanoscale or microscale, such as silver [29, 30], copper [31], aluminum [32], aluminum nitride [33], carbon black [34], CNTs [35], and a combination of them [36, 37], have also been investigated.

The Florida State University and KAI, LLC team reports a new approach to enhance the TTTC of laminated carbon fabric–reinforced composites by utilizing a size synergy advantage of nanoscale and microscale silver particles [38–43]. The goal of this investigation is to develop heterogeneously structured and continuous conductive paths of the silver particles along through-thickness direction to effectively increase thermal conductivity, as shown in Fig. 7.36 [40]. The effects of silver particle size, concentration, and fiber type on TTTC values, and tensile performance were investigated [40–43].

In Phase I of this project, high conductivity of 6.62 W/(m-K) with a 5.1 vol% silver volume fraction was achieved by incorporating these nanoscale and microscale silver particles in EWC-300X/Epon862 composite [38–40]. Silver flakes were distributed within the inter-tow area, while nanoscale silver particles penetrated into the fiber tows. The combination of different sizes of silver fillers is able to effectively form continuous through-thickness conduction paths penetrating fiber tows and bridging the large inter-tow resin rich areas. Figure 7.37 illustrates the proposed microstructures of the different composite designs [42, 43]. All samples were

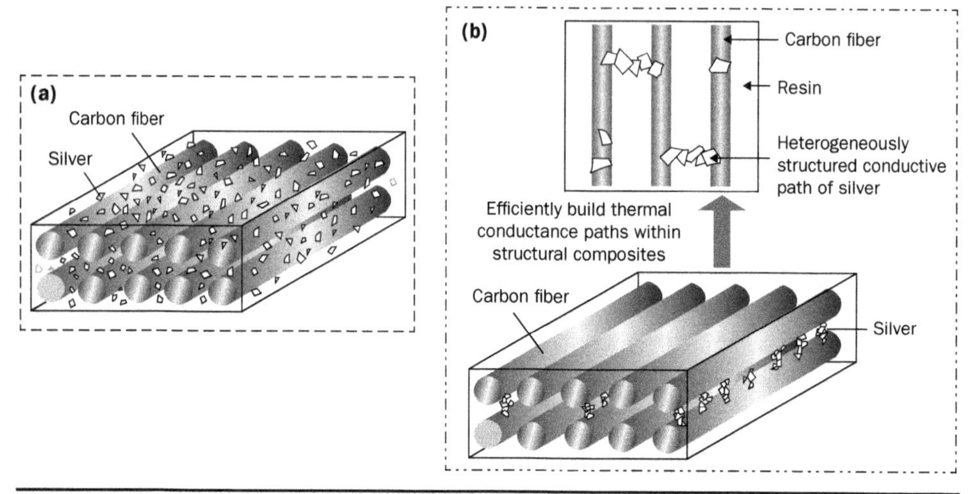

FIGURE 7.36 (a) Conventional homogeneously structured filler-enhanced CFRP and (b) heterogeneously structured continuous conductive paths to improve the through-thickness thermal conductivity of structural composites [41].

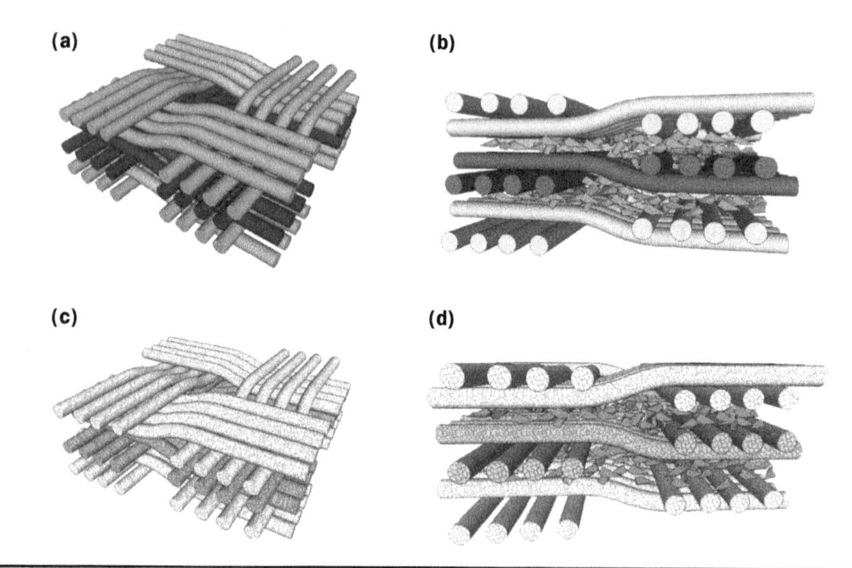

FIGURE 7.37 Composite sample designs of (a) control CFRP: carbon fabric composites without silver fillers, (b) ms-CFRP: silver flake–filled composites, (c) ns-CFRP: silver nanoparticle–coated fiber fabric composites, and (d) mns-CFRP: silver nanoparticle–coated fiber fabric and silver flake–filled composites [43].

fabricated using a hot-press process. The samples were cured at 190°C for 3 hours. Positive hybrid effects to thermal conductivity were found in IM7/EWC300X/silver particle hybrid composites. In addition, microscale fillers in resin-rich areas showed less impact on tensile performance than nanoscale particles applied directly on fiber surface [44].

In Phase I study [38–40], an attempt was made to enhance the through-thickness thermal conductivity (TTTC) of laminated graphite fiber fabric–reinforced composites by applying nanoscale and microscale silver particles to construct heterogeneous thermally conductive paths along the composite's through-thickness direction. By separately applying 5 vol% microscale silver flakes and nanoscale silver particles, the TTTC of EWC300X/Epon862 composite increased to 3.51 and 4.33 W/(m-K), respectively. The corresponding morphology indicated that silver flakes were distributed within the inter-tow area, while nanoscale silver particles penetrated into the fiber tows and formed a coating layer on the fiber surfaces through sintering. Neither filler alone can form continuous conductive paths due to their size limitations. A higher conductivity of 6.62 W/(m-K) of EWC300X/Epon862 composites under a similar total silver volume fraction of approximately 5 vol% can be achieved by the formation of continuous through-thickness conduction paths by jointly applying nanoscale and microscale silver particles and sintering them partly. The TTTC increased with increasing silver volume fraction, and a TTTC of 10.61 W/(m-K) was achieved with the silver volume fraction of 15 vol% and a density of 2.85 g/cm^3 in EWC300X/Epon862 composite.

In contrast, the IM7 fiber composites exhibited lower thermal diffusivity and conductivity than the EWC300X pitch-based fiber composites due to its low fiber conductivity. Hybrid IM7/EWC300X composites were explored to seek the synergy of IM7 composites' higher mechanical properties and EWC300X composites' attractive thermal conductivity. The hybrid composites demonstrated promising thermal conductivity and tensile performance. Further study to optimize the heterogeneous microstructure and improve interfacial bonding between the silver fillers with resin matrix in the hybrid composites could lead to attractive structural carbon fiber composites with thermal management multifunctionality.

In the Phase II study [41–43], the team focused on studying and comparing four different pitch-based graphite fibers to determine the candidate that offers the best improvements in the TTTC with adequate mechanical performance. Of the four candidates, silver-filled EWC600X (pitch-based graphite fiber)–reinforced composites showed the highest TTTC of about 8.0 W/m-K at 25°C due to the formation of unique conductive paths using a combination of both nanoscale and microscale silver fillers. The mechanical performance was relatively lower with 134 MPa in strength and 60 GPa in modulus, because many fibers were broken along the longitudinal direction due to highly aligned graphite crystallites resulting in weak transverse direction properties. Silver-filled EWC300X graphite fiber–reinforced composites had the second highest thermal conductivity of about 6.6 W/m-K and good mechanical performance of 155 MPa in strength and 82 GPa in modulus. Using EWC300X graphite fiber, we successfully delivered two 12×12 in^2 silver-filled composite large panels, achieving a through-thickness thermal conductivity of about 10.0 W/m-K at 25°C.

Materials and Manufacturing Process

Four pitch-based graphite fibers, EWC300X and EWC600X (Cytec Industries Inc., USA), YS95A (Nippon Graphite Fiber Corporation, Japan), and K13C (MITSUBISHI Chemical, Japan), were selected as candidates in this research. Epon862 epoxy resin with curing agent EPICURE-W was used to fabricate the composite samples [44]. Nanoscale silver inks were purchased from Cabot Inc. with particle sizes of less than 50 nm. A customized silver nanoparticle ink that was made in the FSU High-Performance Materials

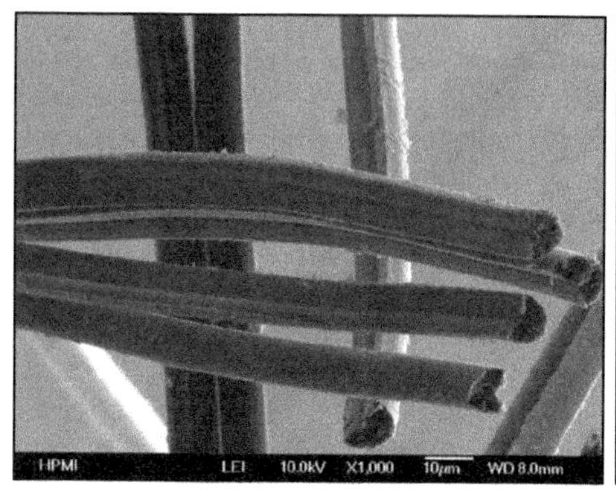

FIGURE 7.38 SEM images of EWC600 individual graphite fiber showing fractured fiber morphology [43].

Institute (HPMI) lab was used for producing two 30.48 × 30.48 cm^2 (12 × 12 in.^{2}) large composite panels. Microscale silver flakes were supplied by Alfa Aesar Company with particle size about 20 μm [43].

Nanoscale silver ink was first applied on four pieces of graphite fiber fabrics. The fabrics were cut into sample sizes of 10.16 × 10.16 cm^2 (4 × 4 in.^{2}). The amount of silver nanoparticles was 20 wt% of graphite fiber fabrics. A pressure of 0.5 MPa was applied on nano-silver–filled graphite fiber fabrics for 15 minutes to ensure good particle distribution. The nano-silver–filled graphite fiber fabrics were sintered at 200°C under vacuum for 1 hour. Microscale silver flakes were added into the resin with the curing agent. Resin mixture was degassed after stirring. The amount of silver flakes was 40 wt% in the resin system. Epoxy resin with silver flakes was applied on the graphite fiber fabrics with a 50/50 fiber/resin weight ratio. Composite panels were hot-pressed at 121°C and 0.21 MPa for 30 minutes, and followed by 177°C and 0.69 MPa for 4 hours. The final volume fraction of silver fillers was about 5 vol%.

Composite Microstructure

As shown in Fig. 7.38, the EWC600X graphite fiber is about 9 μm in diameter. However, most of the EWC600 graphite fibers were fractured into half along the longitudinal direction [43]. These SEM images were taken from as-received materials. One possible reason could be that the EWC600X fiber has high aligned graphite crystallites and much weaker radial direction properties.

Figure 7.39 shows the distribution of nanoscale silver particles and microscale silver flakes in graphite/epoxy composites [43]. As pointed out by red arrows in Fig. 7.39, most of the silver flakes were concentrated in the resin-rich areas, such as the interlayers or inter-tows. The amount of aggregated silver flakes was influenced by the fiber diameter and inter-tow porosity. At the highest magnification, some silver flakes penetrated into the space along individual fiber, forming a through-thickness conductive path of silver flakes due to the silver packing. However, some voids, as shown in Fig. 7.39, were also found. One possible reason is that the silver clusters fell off the

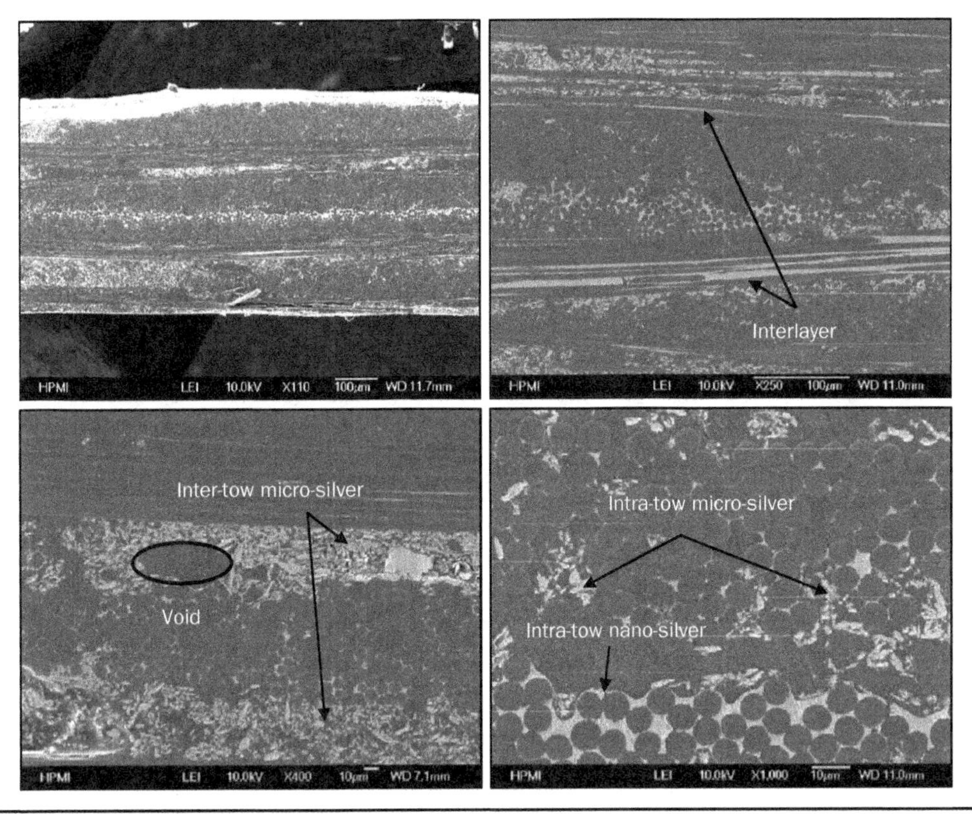

FIGURE 7.39 Microstructure of the graphite/epoxy composites with nanoscale and microscale silver fillers at different magnifications (final silver filler volume fraction was about 5.0 vol%) [43].

surface during polishing due to the poor interfacial bonding between the resin and silver fillers.

Due to their nanoscale size, the silver nanoparticles can easily penetrate into the spaces along the individual graphite fiber and coated on the fiber surface. The silver nanoparticles were mainly attributed to the enhancement of the intra-tow thermal conductivity. This procedure allowed us to utilize the excellent dispensability of silver nanoparticles in the ink media to establish thermally conductive paths within fiber tows. Unfortunately, Cobat Inc. has discontinued the silver nanoparticle ink production. Therefore, we made our own nanoscale silver ink using the silver nanoparticle powder that we purchased from Novacentrix Inc. (Austin, TX) to make the 30.48 × 30.48 cm^2 (12 × 12 $in.^2$) large composite panels. The success of a custom-made silver ink could substantially reduce the cost compared to the use of commercial inks.

Thermal Conductivities

Figure 7.40 and Table 7.7 show the through-thickness thermal conductivities of the four graphite fiber composites [43]. As shown in Fig. 7.40a, the EWC600X composite control samples demonstrated the highest thermal conductivity of 5.0 W/m-K at 25°C

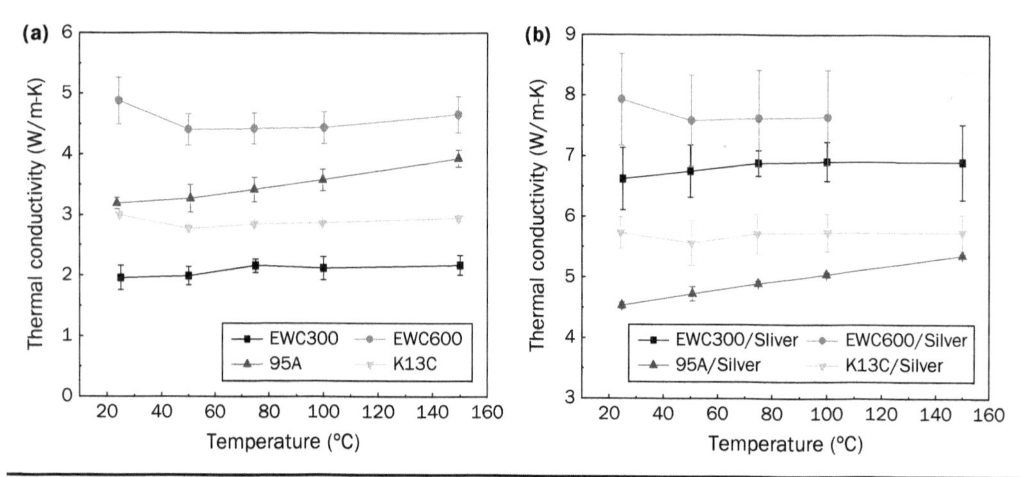

FIGURE 7.40 Comparison of thermal conductivity of EWC300X, EWC600X, YS95A, and K13C pitch-based graphite fiber composites: (a) control samples; (b) samples with silver fillers [43].

TABLE 7.7 Summary of the Thermal Conductivity and Mechanical Properties [43]

	Density	Fiber vol.%	Thermal conductivity at 25°C	Tensile strength	Young's modulus
Sample	**g/cm³**	**%**	**W/m-K**	**MPa**	**GPa**
EWC300X	1.70	58.1	1.96 ± 0.20	273.0 ± 19.6	91.2 ± 5.2
EWC600X	1.74	56.4	5.00 ± 0.22	103.8 ± 11.5	52.9
YS95A	1.74	55.2	3.19 ± 0.10	248.6 ± 31.0	101.8 ± 1.1
K13C	1.71	53.2	3.00 ± 0.05	358.5 ± 34.0	106.9 ± 10.8
EWC300X/silver	2.18	58.3	6.62 ± 0.51	154.9 ± 17.9	82.2 ± 4.7
EWC600X/silver	2.17	52.6	7.93 ± 0.76	134.0 ± 3.3	59.5 ± 5.9
YS95A/sliver	2.10	44.8	4.54 ± 0.05	168.5 ± 5.2	72.7 ± 7.7
K13C/sliver	2.11	45.8	5.73 ± 0.26	238.5 ± 33.1	50.3 ± 8.3

in the control group. This was an impressive result for graphite fiber composites without conductive fillers. YS95A composites ranked second with a thermal conductivity of 3.2 W/m-K due to its high thermal diffusivity. In contrast, EWC300X composites showed the lowest thermal conductivity of 2.0 W/m-K, due to the low thermal conductivity of EWC300X graphite fibers and low thermal diffusivity of EWC300X composites.

As seen in Fig. 7.40b, silver-filled EWC600X composites had the best thermal conductivity among the four graphite fibers. Surprisingly the thermal conductivity of silver-filled EWC300X composite achieved 6.6 W/m-K. The thermal conductivity of the silver-filled YS95A composite was expected to be low since the large inter-tow space was filled by insulation resin, which reduced the overall thermal conductivity of the YS95A composite. In addition, the large voids significantly decreased the thermal conductivity. By adding the silver fillers, the thermal conductivities of EWC300X, EWC600X, YS95A, and K13C graphite fiber composites increased to 238%, 59%, 42%, and 91% at 25°C, respectively, compared to their control samples. Similar to thermal diffusivity, silver-filled EWC300X composites still showed the highest percentage improvement on thermal conductivity, indicating that the formation of conductive path by silver fillers was most successful on EWC300X graphite fibers. The results also proved the good reproducibility of the high through-thickness conductivity we discovered in our previous research effort.

Mechanical Properties

Table 7.7 summarizes the composite mechanical properties and thermal conductivities [43]. The silver-filled EWC600X composite had the highest thermal conductivity as well as mechanical performance. The silver-filled EWC300X and K13C composites had medium thermal conductivity, and silver-filled YS95A composite demonstrated the least thermal performance. Silver-filled K13C composite had the second highest mechanical performance, followed by silver-filled EWC300X and YS95A composites. Although EWC600X composites potentially have the highest thermal conductivity, the mechanical performance was not satisfactory due to the fiber breakage. Therefore, the EWC600X was not suitable for this project. EWC300X pitch–based graphite fiber was found to be the best candidate for this research effort. Moreover, the manufacturer, Cytec Industries Inc., is a domestic supplier.

The team conducted a series of studies on the preparation of silver nanoparticle inks using commercially available silver nanoparticle dry powder to reduce the cost. The FSU HPMI-made silver ink was comparable to the silver ink from Cabot Inc., which was used in the Phase I research. The cost of FSU HPMI-made ink is lower than the Cabot fibers ink and can be further reduced. By studying and comparing five commercially available silver ink/powders, raw materials, and formulas, the team determined that producing our own silver ink would be the best option. The team also tried to synthesize the silver nanoparticles to potentially further reduce cost, achieving 65-nm average particle size and about 5% silver loading in the ink.

Thermal Conductivity of Hybrid Composites

To obtain the synergy of high mechanical performance of IM7 composites and high conductivity properties of the pitch-based carbon fiber composites, a hybrid composite was explored. Two types of pitch-based carbon were used: EWC300X and EWC600X. Figure 7.41 compares the thermal diffusivity and conductivity of IM7/EWC300X and IM7/EWC600X composites with the ply stacking sequence [IM7/EWC]s. IM7/EWC300X and IM7/EWC600X composites had densities of 2.31 and 2.38 g/cm^3, respectively [43]. IM7/EWC600X showed a thermal conductivity of 5.94 W/m-K, which was larger than 5.29 W/m-K of IM7/EWC300X. This was attributed to the higher thermal conductivity of EWC600X fabric. The thermal conductivity of hybrid composite was in between those of PAN- and pitch-based fiber-reinforced composites.

Other Research Activities

The team studied the Aldila AX resin—a CYCOM 977-3 like epoxy resin supplied by Aldila Composite Materials (now wholly owned subsidiary of Mitsubishi Chemical Holdings Corporation) for potential scale-up applications. The thermal conductivity,

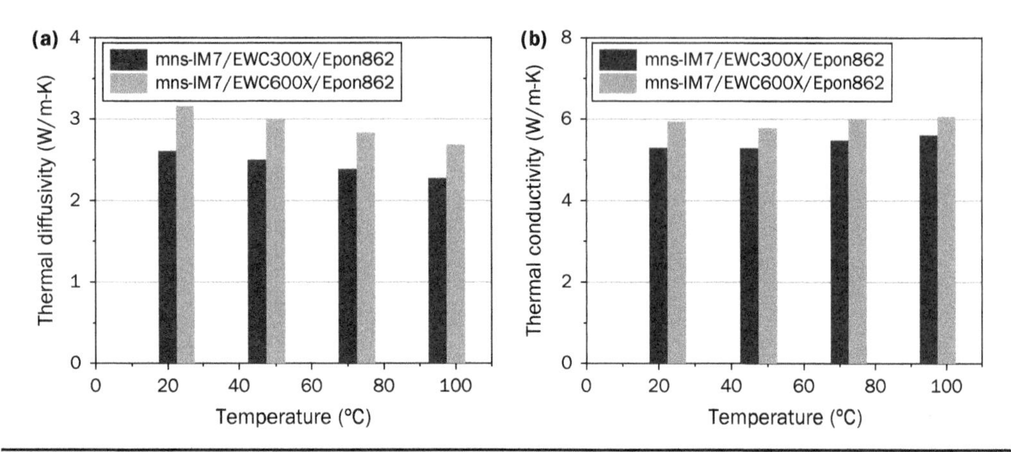

FIGURE 7.41 Comparisons of thermal diffusivity and conductivity of (a) EWC300X/Epon862 and (b) EWC600X/Epon862 composites [43].

mechanical properties, and morphologies of non-silver and silver-filled resin were investigated. The team found that the Aldila AX resin was competitive to the CYCOM 977-3 epoxy resin in terms of glass transition temperature, thermal conductivity, and fracture toughness. By adding 40 wt% of the microscale silver flakes in the resin, DSC analysis showed the glass transition temperature decreased from 183°C to 145°C. However, the thermal conductivity of silver-filled AX resin was four times higher than that of the neat AX resin.

The failure mechanisms of laminated composites with and without silver fillers were also investigated. The process quality control and reduction of material density were preliminarily studied. In addition, surface treatment on silver fillers by a silane coupling agent (SCA) was also studied to optimize interfacial bonding in the resultant composites for mechanical property improvement. By adding 4 wt% of SCA in the resin system, the strength of the EWC300X/Epon862 ms-CFRP increased 111%, compared to the ms-CFRP without SCA treatment. The rheological behavior of Epon862/curing agent resin and silver flake–filled Epon862/curing agent resin were studied and compared for the standardization and quality control of large panel fabrication. By adding 40 wt% silver flakes in the resin system, the viscosity greatly increased. The lowest viscosity of 40% silver flake–filled Epon862/curing agent resin system occurred at 65°C; therefore, the curing process was modified accordingly.

Finally, the team was able to deliver three 12 × 12 in.^{2} composite panels to the Air Force Research Laboratory (Dayton, Ohio), one of which was EWC300X graphite fiber composite control samples with through-thickness thermal conductivity of about 3.0 W/m-K, and the other two were silver-filled composite panels with the TTTC achieved about 10.0 W/m-K, using the FSU HPMI-made silver ink. The progress of large panel manufacturing, cost reduction of silver ink, and performance reproducibility gained the essential expertise for scale-up, and communalization of the technology in the next step.

The scale-up demonstration of the developed technology was also conducted at the prepreg manufacturing facility of Aldila Composite Materials. A modified

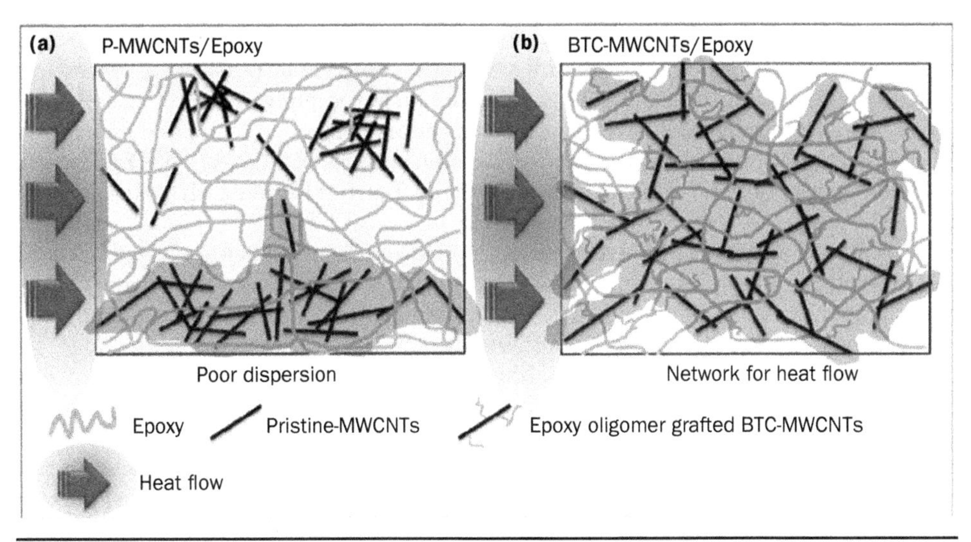

FIGURE 7.42 A model of heat flow in (a) unmodified MWNT/epoxy composites and (b) modified MWNT/epoxy composites [25].

manufacturing process was developed to handle high-viscosity challenge of high silver filler concentration, and the test panels were successfully fabricated with comparable thermal and mechanical properties as demonstrated at the Florida State University. The demonstrated technology for substantially improving through-thickness conductivity of structural composites can be potentially scaled-up for pilot production at commercial settings.

7.5 Summary of Thermal Conductivity Properties of Thermoset-Based Nanocomposites

In general, good thermal conductivity in an epoxy nanocomposite is achieved by adding highly conductive nanofiller and creating a uniformly dispersed, conductive network, and good nanofiller-epoxy interface. Epoxy is not a conductive material and as such it requires fillers if it is to be used in applications requiring heat dissipation.

In amorphous polymers like epoxy, heat flows through lattice vibrations known as *phonons* [21]. Each material has its own acoustic impedance and at the interface of two materials there is an acoustic impedance mismatch. This mismatch causes the phonon to scatter depending on the degree of mismatch. The interfacial thermal resistance is a measure of the compatibility between the two contacting surfaces and it is another way to characterize the interface and determine thermal conductivity. The lower interfacial thermal resistance, the better the thermal conductivity of the material will be.

One way to improve the interface is by functionalizing the nanomaterial. This will create a bridge between the nanofillers and the matrix. For MWNTs, the Friedel–Craft method resulted in the BTC-MWNTs that could covalently bond with the epoxy matrix [23]. A schematic of this interaction is shown in Fig. 7.42 [23]. For exfoliated graphite, functionalization improved the thermal conductivity, but the viscosity of the resulting material at higher loadings was determined to be outside the processing window for conventional composite processing [21]. Another way to enhance thermal conduction of a material is to use the wetting process described by Im and Kim [20]. Using their method, they created a heat conduction path composed of MWNT bridges to GO fillers, but phonon scattering may occur if the balance between MWNTs and GO is not optimal.

Based on their extensive review [24], Han and Fina concluded that CNTs remain one of the most promising thermally conductive filler types for polymers. However, significant advances are still needed to obtain thermally conductive composites sufficiently efficient to meet the requirements of most market applications.

The FSU/KAI team showed their silver-filled EWC600X (pitch-based graphite fiber)–reinforced composites had the highest TTTC of about 8.0 W/m-K at 25°C due to the formation of unique conductive paths using a combination of both nanoscale and microscale silver fillers [43]. The mechanical performance was relatively lower, because many fibers were broken along the longitudinal direction due to highly aligned graphite crystallites resulting in weak transverse direction properties. Silver-filled EWC300X graphite fiber–reinforced composites had the second highest thermal conductivity of about 6.6 W/m-K and good mechanical performance of 155 MPa in strength and 82 GPa in modulus. Using EWC300X graphite fiber, two 12 × 12 in^2 silver-filled composite large panels, a through-thickness thermal conductivity of about 10.0 W/m-K at 25°C was achieved. This technology has been scaled-up by Aldila Composite Materials.

7.6 Phenylethynyl Polyimide–Graphene Oxide Nanocomposites

A high volume of research has gone into novel applications of graphene, ever since it was first discovered due its potential to improve electrical, mechanical, and thermal properties [45]. The problem is that pure graphene has very poor dispersibility, effectively eliminating its practical use in the realm of composites. However, changing the processing of graphite by adding oxidative treatment helps establish large functional surface moieties in the graphene oxide structure that greatly improves dispersibility [45] and reopens an approach to use graphene oxide as composite nanofiller.

Hong et al. [45] performed a series of experiments exploring the thermal stability of fully exfoliated graphene oxide, a phenylethynyl-terminated polyimide thermosetting polymer matrix. Testing was performed using TGA from 20°C to 800°C and TEM imaging. The TEM image of the graphene oxide particles in polyimide matrix shows that this particular micro-scale test area exhibits fairly decent dispersion (Fig. 7.43) [45]. The dark thin lines in the light gray matrix signify the sides of the graphene platelets. However, there are regions, such as the top left corner, where the overall darker gray appearance suggests higher concentrations of graphene platelets stacked against each in comparison to the much lighter background color of the neat polymer. The data are further summarized in Table 7.8 [45].

The TGA showed that thermal stability increased as graphene oxide nanoparticles were increasingly added into the composite at higher wt%. This trend reached a plateau at 3 wt% graphene oxide, after which point thermal stability began to decrease. The increased thermal stability is hypothesized to be a result of the graphene oxide form trapping polyimide moieties [45], which induces chemical reactions in the graphene network that increases the overall material's stability. Once the optimal 3 wt% of

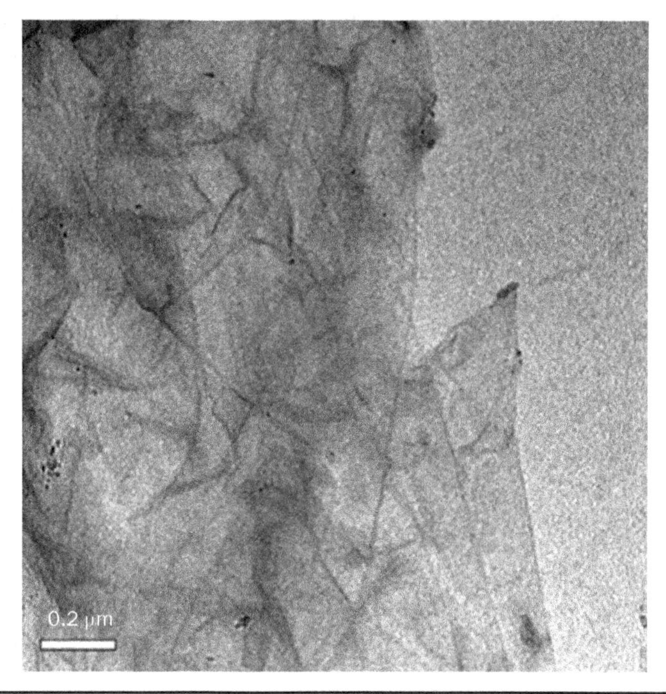

FIGURE 7.43 TEM of layered graphene oxide platelets in a polyimide matrix [45].

TABLE 7.8 TGA Data for 5% Decomposition, 10% Decomposition, Final Temperatures, and Residual Weight [45]

Sample	*T*(5%) °C	*T*(10%) °C	*T*(final) °C	Residual weight
Pure PI	480	542	750	0%
1 wt% GO	505	550	800	15%
3 wt% GO	505	551	800	29%
5 wt% GO	530	571	800	9%

graphene oxide is surpassed, the excess nanoparticles block and decrease the amount of network cross-linking because of its reduced ability to thoroughly disperse in the polymer matrix. This reduction in cross-linking makes the material more susceptible to decomposition as thermal energy is imported into the system.

7.7 Summary of Thermal Properties of Thermoset-Based Nanocomposites

Thermal properties of thermoset-based nanocomposites are very important features that hold significant implications for the future performance of systems. There exists an astounding number of different thermoset nanocomposite combinations, all capable of an array of different properties. It cannot be universally stated which polymer or nanoparticle composite is the greatest for any specific thermal property. The data seen also suggest that unlike other polymer forms, thermoset-based nanocomposites do not always abide by the golden upper limit of 5 wt% nanoparticles content being a compositional maximum [46]. Another key finding is that oftentimes the greatest difference in property enhancements is not due to the addition of nanofiller material, but to the treatment or chemical additives applied to the nanofiller material [47]. This was especially true in the case of fire retardancy, where sizable differences could be seen between the same nanocomposites that only differed in processing and chemical additives applied to the nanofiller [47, 48]. Comparatively, thermal stability can be improved greatly, but at the risk of overweighing the composite with nanofiller material and increasing deleterious performance of mechanical properties. In terms of future development, it seems DGEBA and different forms of clay, such as hectorite or montmorillonite, hold the greatest potential for widespread application, due in part to more thorough research on the materials, ease of processing, and a quickly growing repertoire of results leading to increasingly predictable, and clearly enhanced, thermal characteristics.

In general, the addition of nanoclay improves the overall thermal properties of the resulting composite. Interestingly, the organic modifier not only helps in dispersion of the nanoclays, but also induces hydrogen bonding with the polymer chains and results in greater interaction between the filler and the polymer. However, at higher loadings, the amount of the organic modifier is large enough to detrimentally affect the thermal properties due to its combustibility. The best thermal performance was observed for loading between 3 and 5 wt%. However, the addition of the nanoclay affects the processability of the composites as the viscosity increases as well as the melting temperature and heat of fusion, yet these properties are generally desired for the finished application.

The use of MWNTs for improving the thermal properties did not seem effective from the Rice University study. The group performed scanning electron microscopy and showed that the MWNTs seemed reasonably dispersed. Since the TGA was performed in an inert environment, the residual mass in an oxygen environment was not obtained. It was interesting to note that the random dispersion of the MWNTs showed a higher change in T_g, which is not surprising because of greater interfacial coverage. Further work must be conducted in order to determine what loadings are required in order to obtain the best trade-off between properties. Additionally, there may be some synergistic effects from combining certain nanoparticles which may improve overall properties that also need to be explored.

The polymer structure is a large factor in the thermal performance of these composites. Both the nature of the material and the compatibility of the filler will determine the composite effects of combining the nanomaterial and the polymer. It was seen that sometimes the addition of a nanocomposite will exhibit multifunctionality behavior of improving multiple properties, while other times the increase in one property results in the decrease of another. Interestingly, properties, such as thermal stability and fire retardancy are sometimes do not have a positive correlation despite intuition. It was also observed that at certain loadings, interactions between the filler and the polymer form certain complexes which serve as an inflection point for the properties, such as the crystallinity which in turn affects other properties. It is probably good practice to require a TEM micrograph of the composites produced, as it is difficult to compare results whenever it is uncertain whether the nanoparticle is well dispersed.

7.8 Overall Summary

An overview of the current technology in enhancing the thermal properties of thermoplastic-, TPE-, and thermoset-based polymer nanocomposite systems was analyzed. In general, their thermal performance can be enhanced by the addition of nanoparticles. The mechanical and electrical performance of the polymeric matrix can also be improved by the addition of nanoparticles. Sometimes simply maintaining good thermal performance while improving other properties is a significant feat.

Organoclays have become a mainstay as a nanoadditive because of their low-cost and consistent performance. The comparison with synthetic clay (Laponite) is important to review where synthetics stand with respect to organics. In general, the addition of nanoclay improves the overall thermal properties of the resulting composite. Interestingly, the organic modifier not only helps in dispersion of the nanoclays, but also induces hydrogen bonding with the polymer chains and results in greater interaction between the filler and the polymer. However, at higher loadings, the amount of the organic modifier is large enough to detrimentally affect the thermal properties due to its combustibility. The best thermal performance was observed for loading between 3 and 5 wt%. However, the addition of the nanoclay affects the processability of the composites as the viscosity increases as well as the melting temperature and heat of fusion, yet these properties are generally desired for the finished application.

The effects of nanomaterials can range from extremely useful to negligible and are highly dependent on the type of filler used and its relative size. For all cases, there were enhanced properties with increasing loading of filler leading us to believe that the more filler the better properties your composite will observe. Interestingly enough, nano-sized fillers are not always ideal when compared to micrometer-sized fillers. For all studies where thermal conductivity data were taken for both nano- and micron-sized fillers, the micrometer fillers performed better. The combined micron-sized silver flakes and nano-sized silver particles reinforced with graphite fiber provided a TTTC composite with good mechanical properties. Based on these results, we see that many factors can affect the thermal properties of polymer nanocomposites.

It was found that the performance of nanoparticles on the basis of improved properties is highly dependent on the dispersion and exfoliation of the nanoparticles in the polymeric matrix. Certain surface treatments to nanoparticles and specialized processing techniques can help the dispersion and exfoliation of nanofibers in the composite. Polymer nanocomposite is a promising and growing field of study. It has already led to important advancements and continues to be in high demand in the industry.

7.9 Study Questions

7.1 How does the nanomaterial dispersion affect the thermal properties of polymer nanocomposites in general? Provide examples to illustrate your claims.

7.2 How does the thermal properties of polymer-clay nanocomposites influenced by the type of clay and surface treatment of the clay, and how clay is processed in the polymer matrix? Provide examples.

7.3 Enhancement of thermal properties of thermoplastic-based nanocomposites using montmorillonite clay, carbon-based (CNF, SWNT) nanomaterials, LHDH, and clay/woven glass fibers were included in this chapter. What other nanomaterials can be used for thermoplastic-based nanocomposites? Provide examples and comment on the thermal properties of these nanocomposite systems.

7.4 Enhancement of thermal properties of thermoplastic elastomer–based nanocomposites using montmorillonite clay, synthetic clay (Laponite), carbon-based (carbon black, SWNT) nanomaterials, boron nitride, zinc oxide (ZrO), nanosilica, and aluminum oxide were included in this chapter. What other nanomaterials can be used for thermoplastic elastomer–based nanocomposites? Provide examples and comment on the thermal properties of these nanocomposite systems.

7.5 Enhancement of thermal properties (thermal stability and thermal conductivity) of epoxy-based nanocomposites using POSS, TiO_2-SiO_2, conducting polymer (polythiophene), carbon-based (MWNT, expanded graphite flakes, expanded graphene nanoplatelets, graphene oxide/MWNT hybrid) nanomaterials, Al_2O_3, and Ag were included in this chapter. What other nanomaterials can be used for epoxy-based nanocomposites? Provide examples and comment on the thermal properties of these nanocomposite systems.

7.6 What will be the effects of hybrid systems, such as montmorillonite clay-CNT, CNT-nanographene platelet, and others? What will be a good strategy to optimize the thermal properties in the polymer nanocomposites? Comment on the thermal properties of these nanocomposite systems.

7.10 References

1. Ellis, T. S., and D'Angelo, J. S. (2003) Thermal and mechanical properties of a polypropylene nanocomposite. *Journal of Applied Polymer Science,* 90:1639–1647.
2. Sandlera, J., Wernerb, P., Shaffera, M. S. P., Demchukc, V., Altstatd, V., and Windlea, A. H. (2002) Carbon-nanofibre-reinforced poly(ether ketone) composites. *Composites: Part A,* 33:1033–1039.
3. Lozano, K., and Barrera, E. V. (2000) Nanofiber-reinforced thermoplastic composites. Thermoanalytical and mechanical analyses. *Journal of Applied Polymer Science,* 79:125–133.
4. Liu, J., Chen, G., and Yang, J. (2008) Preparation and characterization of poly(vinyl chloride)/layered double hydroxide nanocomposites with enhanced thermal stability. *Polymer,* 49:3923–3927.
5. Piszczyk, Ł., Strankowski, M., Danowska, M., Haponiuk, J. T., and Gazda, M. (2012) Preparation and characterization of rigid polyurethane–polyglycerol nanocomposite foams. *European Polymer Journal,* 48:1726–1733.
6. Shen, S. Z., Bateman, S., McMahon, P., Dell'Olio, M., Gotama, J., Nguyen, T., and Yuan, Q. (2010) The effects of clay on fire performance and thermal mechanical properties of woven glass fiber reinforced polyamide 6 nanocomposites. *Composites Science and Technology,* 70:2063–2067.
7. Wu, J. H., Li, C. H., Wu, Y. T., Leu, M. T., and Tsai, Y. (2010) Thermal resistance and dynamic damping properties of poly(styrene-butadiene-styrene)/thermoplastic polyurethane composites elastomer material. *Composite Science and Technology,* 70:1258–264.
8. Kemaloglu, S., Ozkoc, G., and Aytac, A. (2010) Properties of thermally conductive micro and nano size boron nitride reinforced silicon rubber composites. *Thermochimica Acta,* 499:40–47.

9. Barick, A. K., and Tripathy, D. K. (2009) Effect of organoclay on the morphology, mechanical, thermal, and rheological properties of organophillic montmorillonite nanoclay based thermoplastic polyurethane nanocomposites prepared by melt blending. *Polymer Engineering & Science*, 50:484–498.
10. Silva, G. G., Rodrigues, M. F., Fantini, C., Borges, R. S., Pimenta, M. A., Carey, B. J., Lijie, C., et al. (2010) Thermoplastic polyurethane nanocomposites produced via impregnation of long carbon nanotube forests. *Macromolecular Materials and Engineering*, 296:53–58.
11. Mishra, A. K., Nando, G. B., and Chattopadhyay, S. (2008) Exploring preferential association of Laponite and Cloisite with soft and hard segments in TPU-Clay nanocomposite prepared by solution mixing technique. *Journal of Polymer Science: Part B: Polymer Physics*, 46:2341–2354.
12. Haggenmueller, R., Guthy, C., Lukes, J., Fischer, J., and Winey, K. (2007) Single wall carbon nanotube/polyethylene nanocomposites: Thermal and electrical conductivity. *Macromolecules*, 40:2417–2421.
13. Wang, Z., Lu, Y., Liu, J., Dang, Z., Zhang, L., and Wang, W. (2011) Preparation of nano-zinc oxide/EPDM composites with both good thermal conductivity and mechanical properties. *Journal of Applied Polymer Science*, 119:1144–1155.
14. Chatterjee, S., Wangb, J. W., Kuo, W. S., Tai, N. H., Salzmann, C., Li, W. L., Hollertz, R. et al. (2012) Mechanical reinforcement and thermal conductivity in expanded graphene nanoplatelets reinforced epoxy composites. *Chem. Phys. Lett.*, 531:6–10.
15. Debelak, B., and Lafdi, K. (2007) Use of exfoliated graphite filler to enhance polymer physical properties. *Carbon*, 45:1727–1734.
16. Balakrishnan, A., and Saha, M. C. (2011) Tensile fracture and thermal conductivity characterization of toughened epoxy/CNT nanocomposites. *Mater. Sci. Eng. A.*, 528:906–913.
17. Nagendiran, S., Alagar, M., and Hamerton, I. (2010) Octasilsesquioxane-reinforced DGEBA and TGDDM epoxy nanocomposites: Characterization of thermal, dielectric, and morphological properties. *Acta Materialia*, 58:3345–3356.
18. Omrani, A., Afsar, S., and Safarpour, A. (2010) Thermoset nanocomposites using hybrid nano TiO_2SiO_2. *Materials Chemistry and Physics*, 122:343–349.
19. Zabihi, O., Khodabandeh, A., and Mostafavi, S. M. (2012) Preparation, optimization and thermal characterization of a novel conductive thermoset nanocomposites containing polythiophene nanoparticles using dynamic thermal analysis. *Polymer Degradation and Stability*, 97(1):3–13.
20. Im, H., and Kim, J. (2012) Thermal conductivity of a graphene oxide-carbon nanotube hybrid/epoxy composite. *Carbon*, 50:5429–5440.
21. Ganguli, S., Roy, A. K., and Anderson, D. P. (2008) Improved thermal conductivity for chemically functionalized exfoliated graphite/epoxy composites. *Carbon*, 46:806–817.
22. Abdalla, M., Dean, D., Robinson, P., and Nyairo, E. (2008) Cure behavior of epoxy/MWCNT nanocomposites: The effect of nanotube surface modification. *Polymer*, 49:3310–3317.
23. Yang, S., Ma, C. M., Teng, C., Huang, Y., Liao, S., Huang, Y., Tien, H. et al. (2010) Effect of functionalized carbon nanotubes on the thermal conductivity of epoxy composites. *Carbon*, 48:592–603.
24. Han, Z., and Fina, A. (2011) Thermal conductivity of carbon nanotubes and their polymer nanocomposites: A review. *Progress in Polymer Science*, 26:914–944.

25. Chen, L.-Y., Chen, Y., Fu, J.-F., Shi, L.-Y., and Zhong, Q.-D. (2010) Thermally conductive nanocomposites based on hyperbranched epoxy and nano-Al_2O_3 particles modified epoxy resin. *Polymer Advanced Technologies*, 22:1032–1041.
26. Schuster, J., Heider, D., Sharp, K., and Glowania, M. (2008) Thermal conductivities of three dimensionally woven fabric composites. *Compos. Sci. Technol.*, 68(9):2085–2091.
27. Schuster, J., Heider, D., Sharp, K., and Glowania, M. (2009) Measuring and modeling the thermal conductivities of three-dimensionally woven fabric composites. *Mech. Compos. Mater.*, 45:165–174. doi:10.1007/s11029-009-9072-y.
28. Sharp, K., Bogdanovich, A. E., Tang, W., Heider, D., Advani, S., and Glowiana, M. (2008) High through-thickness thermal conductivity composites based on three-dimensional woven fiber architectures. *AIAA J.*, 46:2944–2954.
29. Kim, W. J., Taya, M., and Nguyen, M. N. (2009) Electrical and thermal conductivities of a silver flake/thermosetting polymer matrix composite. *Mech. Mater.*, 41:1116–1124. doi:10.1016/j.mechmat.2009.05.009.
30. Pashayi, K., Fard, H. R., Lai, F., Iruvanti, S., Plawsky, J., and Borca-Tasciuc, T. (2012) High thermal conductivity epoxy-silver composites based on self-constructed nanostructured metallic networks. *J. Appl. Phys.*, 111:104310. doi:10.1063/1.4716179.
31. Chan, K. L., Mariatti, M., Lockman, Z., and Sim, L. C. (2011) Effects of the size and filler loading on the properties of copper- and silver-nanoparticle-filled epoxy composites. *J. Appl. Polym. Sci.*, 121:3145–3152. doi:10.1002/app.33798.
32. Kumlutaş, D., Tavman, İ. H., and Turhan, Ç. M. (2003) Thermal conductivity of particle filled polyethylene composite materials. *Compos. Sci. Technol.*, 63:113–117. doi:10.1016/S0266-3538(02)00194-X.
33. Choi, S., and Kim, J. (2013) Thermal conductivity of epoxy composites with a binary-particle system of aluminum oxide and aluminum nitride fillers. *Compos. Part B Eng.*, 51:140–147. doi:10.1016/j.compositesb.2013.03.002.
34. Han, S., Lin, J. T., Yamada, Y., and Chung, D. D. L. (2008) Enhancing the thermal conductivity and compressive modulus of carbon fiber polymer–matrix composites in the through-thickness direction by nanostructuring the interlaminar interface with carbon black. *Carbon*, 46:1060–1071. doi:10.1016/j.carbon.2008.03.023.
35. Wang, S., and Qiu, J. (2010) Enhancing thermal conductivity of glass fiber/polymer composites through carbon nanotubes incorporation. *Compos. Part B Eng.*, 41:533–536. doi:10.1016/j.compositesb.2010.07.002.
36. Hwang, S. H., Bang, D.-S., Yoon, K. H., Park, Y.-B., Lee, D.-Y., and Jeong, S.-S. (2010) Fabrication and characterization of aluminum-carbon nanotube powder and polycarbonate/aluminum-carbon nanotube composites. *J. Compos. Mater.*, 44:2711–2722. doi:10.1177/0021998310369590.
37. Kang, C. H., Yoon, K. H., Park, Y.-B., Lee, D.-Y., and Jeong, S.-S. (2010) Properties of polypropylene composites containing aluminum/multi-walled carbon nanotubes. *Compos. Part. Appl. Sci. Manuf.*, 41:919–926. doi:10.1016/j.compositesa.2010.03.011.
38. Koo, J. H., Liang, R. et al. (2012) Heterogeneously Structured Conductive Resin Matrix/Graphite Fiber Composite for High Thermal Conductive Structural Applications, AFOSR STTR Phase I Final Report, KAI, LLC, Austin, TX, submitted to AFOSR, Arlington, VA, July 30, 2012.
39. Wang, S., Liang, R., Koo, J. H., et al. (2012) Through-thickness Thermal Conductivity Improvement of Carbon Fiber Laminates by the Joint Use of Nanoscale and Microscale Silver Particles, *Proc. SAMPE 2012 ISTC*, Charleston, SC, October 22–25, 2012.

40. Wang, S., Haldane, D., Gallagher, P., Liu, T., Liang, R., and Koo, J. H. (2014) Heterogeneously structured conductive carbon fiber composites using multi-scale silver particles. *Composites Part B*, 66:172–180, doi: 10.1016/j.compositesb.2014.01.049.
41. Hao, A., Wong, S., Home, J., Yang, M., Liang, R., and Koo, J. H. (2015) Microstructure and high through-thickness thermal conductivity of graphite fiber composite for structural applications, AIAA-2015-0123, *AIAA SciTech 2015*, Kissimmee, FL, January 5–9, 2015.
42. Liang, R., and Koo, J. H. (2015) Heterogeneously structured conductive resin matrix/graphite fiber composite for high thermal conductive structural applications, AFOSR STTR Phase II Final Report, KAI, LLC, Austin, TX, submitted to AFOSR, Arlington, VA, October 2015.
43. Hao, A., Liang, R., Koo, J.H., et al. (2016) Heterogeneously structured conductive resin matrix/graphite fiber composite for high thermal conductive structural applications. *National Space & Missile Materials Symposium (NSMMS)*, Westminster, CO, June 20–23, 2016.
44. Product Bulletin-EPIKOTE Resin 862/EPIKURE Curing Agent W System, Hexion/Miller-Stephenson, Morton Grove, IL.
45. Hong, L., Li, Y., Wang, T., Wang, Q. (2012) In-situ synthesis and thermal, tribological properties of thermosetting polyimide/graphene oxide nanocomposites. *Journal of Materials Science*, 47(4):1867–1874.
46. Shojaei, A., and Faghihi, M. (2010) Physico-mechanical properties and thermal stability of thermoset nanocomposites based on styrene-butadiene rubber/phenolic resin blend. *Materials Science & Engineering, A: Structural Materials: Properties, Microstructure and Processing*, A527, 4–5:917–926.
47. Zheng, X., and Wilkie, C. (2003) Flame retardancy of polystyrene nanocomposites based on an oligomeric organically-modified clay containing phosphate. *Polymer Degradation and Stability*, 81:539–550.
48. Chigwada, G., Jash, P., Jiang, D., and Wilkie, C. (2004) Synergy between nanocomposite formation and low levels of bromine on fire retardancy in polystyrenes. *Polymer Degradation and Stability*, 88:382–393.

CHAPTER 8

Flammability Properties of Polymer Nanocomposites

8.1 Introduction

This chapter provides an overview on the fundamentals, properties, and applications of flammability of polymer nanocomposites (PNCs). The structures, properties, and surface treatment of different types of commonly used flame-retardant (FR) nanoadditives were introduced in Chap. 2. These FR nanomaterials are subdivided based on their nanoscale dimensions: one-dimensional nanoscale, such as nanoclay, nanographene platelets, and layered double hydroxides (LDH); two-dimensional nanoscale, such as carbon nanofibers (CNFs), carbon nanotubes (CNTs), halloysite nanotubes (HNT®), and aluminum oxide nanofibers (Nafen™); and three-dimensional nanoscale, such as nanosilica, nano-alumina, nano–magnesium hydroxide, and polyhedral oligomeric silsesquioxanes (POSS®). The test methods that are commonly used to evaluate thermal and flammability properties of PNCs were briefly reviewed in Chap. 5. The effects of one-dimensional, two-dimensional, three-dimensional, combined nanoadditives, as well as combined nanoadditives with conventional FR additives on thermal and flammability properties of the PNCs are discussed in this chapter. The mechanisms of the effect of nanoadditives on flammability of the PNCs are proposed. The section on concluding remarks and trends for the study of PNCs completes this chapter. The readers who need to have more detailed description of the nanomaterials and their detailed influence on a specific matrix polymer may pursue some reviews, books, and papers in this emerging research area, which are listed in the references section of this chapter and other chapters in this book.

8.2 Thermal and Flame Retardancy Properties of Polymer Nanocomposites

Although PNCs show improved thermal and flame retardancy properties, it has been shown that different nanoadditives have different effects on improvement in flame retardancy. The thermal and flammability properties of PNCs depend on the types of nanofillers, types of polymers, and the nanostructures obtained. In this section, the thermal and flammability properties of these PNCs will be subdivided into (a) one-dimensional nanomaterial PNCs, which include clay-based and graphene-based nanocomposites; (b) two-dimensional nanomaterial PNCs, which include CNF-based,

CNT-based, HNT–based PNCs; and (c) three-dimensional nanomaterial PNCs which include nanosilica-based, nano-alumina–based, nano-magnesium hydroxide–based, and POSS-based PNCs. Different types of polymers, such as thermosets, thermoplastics, and elastomers, are included.

8.2.1 One Nanoscale Dimension–Based Nanocomposites

Polymer-Clay Nanocomposites

Clay-based nanomaterial is most commonly used for nanocomposites because it has been around the longest and can be obtained in high purity at low cost. In the following discussions, the clay used to create the PNCs is montmorillonite (MMT) unless otherwise identifying differently. More information can be found in several recent books on polymer-clay nanocomposites [1–3].

The polyamide 6–clay nanocomposite was the first discovered PNCs. Flame retardancy properties of polymer-clay nanocomposites have received much attention because flame retardancy was one of the first desirable properties noted by researchers. Initially, all the polymers exhibit improved flame retardancy, as evaluated by cone calorimeter, upon the incorporation of clay. This section summarizes the research progress on clay nanocomposites using different polymers.

Flammability of Polyamide-Clay Nanocomposites The polyamide 6 (PA6)–clay nanocomposite was the first discovered PNC and studied by Gilman et al. [4] at NIST (United States). Cone calorimeter was used as the characterization tool to evaluate the PA6-clay nanocomposites. The peak heat release rate (PHRR) is reduced by 32% and 63% in a PA6-clay nanocomposite containing, respectively, 2% and 5% of the clay. The PA6-clay nanocomposite showed not only improved FR properties, but also improved mechanical properties compared to neat PA6. Foster Corp. (the United States, website: www.fostercomp.com) demonstrated that high levels (13.9%) of clay can be added to PA12 polymers to achieve UL 94 V-0 ratings at 1/8-in. (3.2 mm) thickness. The clay used as a char former allows the typical 50% loading of halogen/antimony oxide FR system to be cut in half, which reduces the detrimental effects on physical properties of the polymer. Foster Corp. introduced PA12-clay nanocomposites for tubing and film in 2001.

Patel et al. [5] used organically modified montmorillonite (OMT) to disperse catalytically active platinum (Pt) nanoparticles that influence the char formation processes of polyamide 6 (PA6). The Pt nanoparticles of varying particle size are generated in organically modified montmorillonite (Pt-OMTs), which are subsequently melt-blended with PA6. The average particle size of Pt nanoparticles was found ranging from 9.2 to 21.6 nm using x-ray diffraction and field emission scanning electron microscopy (SEM) techniques. The char formation mechanism was studied by thermogravimetric analysis coupled with Fourier transform infrared spectroscopy (TG-IR). Infrared spectroscopy on the residue reveals the formation of unsaturated hydrocarbons during decomposition of PA6 nanocomposite, which are transformed into a highly conjugated structure. The Pt nanoparticle intercalated OMTs enhanced the formation of char by the formation of unsaturated compounds that are cross-linked during pyrolysis in nitrogen atmosphere forming larger amounts of char compared to nanocomposites without Pt nanoparticles. The degradation of PA6 nanocomposites in the presence of Pt nanoparticles in oxidative conditions resulted in larger amounts of CO_2 evolution in two steps. Microscale combustion calorimeter analyses demonstrate

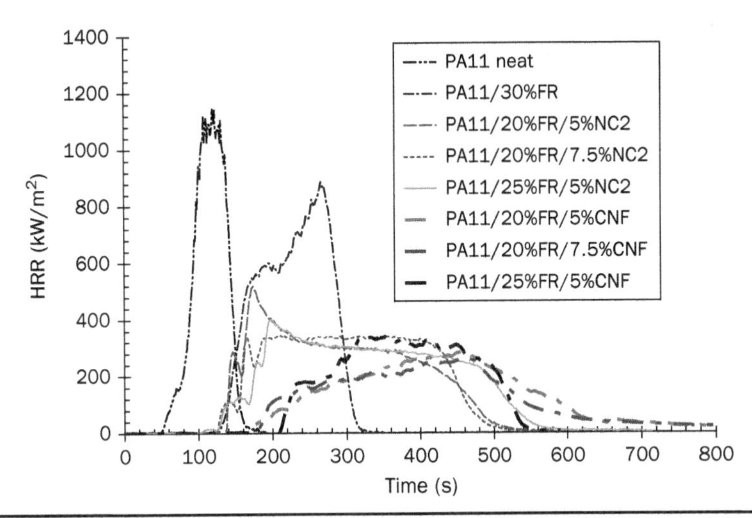

Figure 8.1 Heat release rate of neat and nanoparticle-filled PA11 with and without FR additive as a function of time (cone calorimeter, 35 kW/m^2). FR, flame-retardant additives; NC2, nanoclay; CNF, carbon nanofiber [6].

the decrease of evolved volatiles in heating conditions, similar to what would be expected in a fire scenario.

Lao and colleagues recently have carried out extensive research on FR PA11-clay and PA12-clay composites for selective laser sintering application [6–8]. The Koo Research Group demonstrated polyamide 11 (PA11) at 7.5% and 10% of nanoclay (NC), and the PHRR was reduced by 68% and 73%, respectively [6]. Figure 8.1 shows the PHRR of the neat PA11 and several PA11 polymer blends at a radiant heat flux of 35 kW/m^2 [6]. When 20% of conventional intumescent FR additives and 7.5% NC were added to PA11, PHRR was reduced to 70% and achieving a UL 94 V-0 rating [6]. The heat deflection temperatures (HDTs) of neat PA11 and PA11/20% FR/7.5% NC were 48°C and 97°C, respectively. A synergistic effect exists between conventional FR intumescent additives with nanoclay allowing Lao and colleagues to lower the FR intumescent additive from 30% to 20% by incorporating 5–7.5% NC in PA11. See Sec. 8.2.4 for additional FR PA11 and FR PA12 results.

Flammability of Epoxy-Clay Nanocomposites Epoxy resins are used in a variety of applications, such as coatings, adhesives, and electronics or in composites in the aerospace and transportation industries. Although the polyfunctional reactivity of most epoxy systems leads to a high cross-link density, brittleness of epoxy systems is always a concern. In most applications, the polymer matrix is combined with one or more phases, such as short or long fibers or a rubbery phase for toughening to form epoxy fiber–reinforced composites. Epoxy nanocomposites have attracted a lot of attention within nanocomposites research to enhance modulus, strength, fracture toughness, impact resistance, and gas and liquid barrier, and improve flame retardancy. However, like all organic materials, epoxy is flammable, and its use for replacing traditional nonflammable materials has increased the fire hazard in the past years. Evolved smoke and toxic gases from these fires create hazards for both people and the environment. At present, nanoclays are by far the most investigated nanofillers in flame retardancy. In this section, we focus on epoxy nanocomposites based on nanoclays.

The additive effect of resorcinol bis(diphenyl phosphate) (RDP) and nanoclay on epoxy was studied by Katsoulis et al. [9]. The epoxy resin, tetraglycidy1-4,4′-diaminodiphenylmethane (TGDDM) and the curing agent, 4,4′-diaminodiphenyl sulfone (DDS), RDP flame-retardant, and nanoclay of different loadings were synthesized. Transmission electron microscopy (TEM) images of epoxy-RDP-I.30E (80/15/5) at low magnification show intercalated nanoclay with gallery spacing at 5–10 nm [9]. Formulations with RDP alone at 5% and 10% gave the best flame resistance in LOI and UL 94 experiments, while going beyond 10% RDP had an adverse effect. In cone calorimeter, the addition of RDP at 10% results in remarkable reductions in PHRR and enhanced char formation. Addition of nanoclay alone did not improve the thermal stability of the epoxy resin, and neither did it improve its flame retardancy using cone calorimeter. The addition of clay with RDP did not show an advantage. It also did not produce any significant improvement in the fire properties as measured by cone calorimeter and limited oxygen index (LOI). This may be attributed to possible antagonistic interactions between the constituent components. The addition of a known flame-retardant, RDP, and nanoclay yielded mixed results with respect to thermal stability and improved fire performance.

In nitrogen, residue yields from epoxy-clay nanocomposite reveals little improvement in the carbonaceous char yield once the presence of silicate in the residue is accounted for [9–12]. In air, the char yield increase is more relevant [13]. Hussain et al. [14] studied the effect of clay and an organophosphorus epoxy modifier DOPO on DGEBA and TGDDM epoxies cured with amine curing agent. They prepared 3% P-containing epoxies by reacting DGEBA or TGDDM with DOPO. Standard and modified epoxies were used to prepare clay nanocomposites which show a mixed intercalated-exfoliated structure. The clay enhanced char formation significantly. The results are shown in Fig. 8.2 [14]. The char yields at 600°C increase from 14% in the

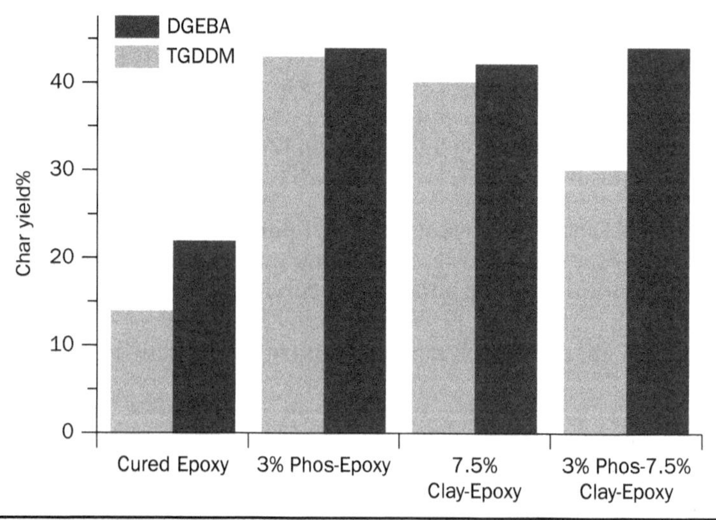

FIGURE 8.2 Char yields at 600°C for DGEBA and TGDDM control samples, 3% P-epoxies, and 7.5% clay nanocomposites [14].

near DGEBA to 38% and 42% for 5% clay-DGEBA nanocomposite and 3% P-modified epoxy, respectively. Similar results were observed for TGDDM-based formulations.

Polymer nanocomposites showed that clays can reduce the amount of traditional flame-retardants required to comply with the standards; the approach to nanocomposites itself alone is not sufficient to facilitate passing commercial tests. In terms of LOI, the improvement achieved by using clays is generally not relevant. A typical increase between 1% and 2% in LOI values is seen with 5% nanoclay [15]. Hussain et al. [14] reported an increase in LOI values from 25% for neat DGEBA to 32.7% and 34.5% for 5.0% and 7.5% nanocomposites, respectively. For a tetrafunctional TGDDM epoxy, the LOI values increase from 26.3% in neat polymer to 35.1% and 36.7% in nanocomposites containing 5% and 7.5% clay, respectively.

Cone calorimeter results of nanoclay in DGEBA cured with either methylenedianiline (MDA) or benzyldimenthylamine (BDMA) [16] were impressive. The PHRR and the average heat release rate (HRR) are significantly improved with 6% nanoclay. The HRR plots for DGEBA-MDA and the DGEBA-MDA clay nanocomposites are compared in [16]. The PHRR, HRR, and average mass loss rate (MLR) decrease by about 40%. The heat of combustion, smoke obscuration, and carbon monoxide yields are unchanged. The results suggest that the clay nanocomposite is operating primarily in the condensed phase. A shorter ignition time observed in the nanocomposite may be due to the low stability of the organic modifier.

Flammability of Elastomer-Clay Nanocomposites The combustion behavior of thermoplastic elastomer has been studied by Koo and colleagues [17, 18]. Two types of commercially available thermoplastic polyurethane (TPU) elastomer (Pellethane® and Desmopan®) were melt-blended with various loadings (2.5–10%) of MMT nanoclay (Cloisite® 30B) using twin-screw extrusion. The morphological, physical, thermal, flammability, and thermophysical properties, and kinetic parameters of these two families of polymer-clay nanocomposites were characterized. The *processing-structure-property* relationships of this class of novel thermoplastic polyurethane elastomer-clay nanocomposites (TPUNs) were established. The wide-angle x-ray diffraction (WAXD) plots of the four different nanoclay loading of Desmopan TPUNs are shown in Fig. 8.3 [17, 18]. No peaks were observed in WAXD, which indicated the all loadings of nanoclays were exfoliated in the TPU polymer. No stacks are observed in any of the three 40 kX TEM images in Fig. 8.4; scale bar is 100 nm [17]. The lack of platelet stacks indicates that the nanoclay is exfoliated and well dispersed in the Desmopan TPU material.

Cone calorimeter experiments at 50 kW/m^2 test the neat Desmopan TPU, neat Pellethane TPU, Desmopan-5% Cloisite 30B TPUN, and Pellethane-5% Cloisite 30B TPUN. Two or three experiments are performed to test for repeatability. The neat material experiments are the least repeatable due to the fact that the material melted and dripped off the test fixture; for this reason the test with the higher PHRR is used in the comparisons. The HRR graph of the materials is shown in Fig. 8.5 [17]. The flammability properties are presented in Table 8.1 [17]. The material's time to sustained ignition is not changed drastically with the addition of Cloisite 30B, it is +2 seconds for Pellethane TPUN and –1 second for the Desmopan TPUN. However, a dramatic decrease in PHRRs is observed. Pellethane TPU PHRR decreases by 73% and Desmopan TPU PHRR decreases by 50% with the addition of 5 wt% Cloisite 30B. In the first 60 seconds, the HRR is higher in the TPUN. This is due to a higher rate of decomposition at this stage observed in the mass loss graph.

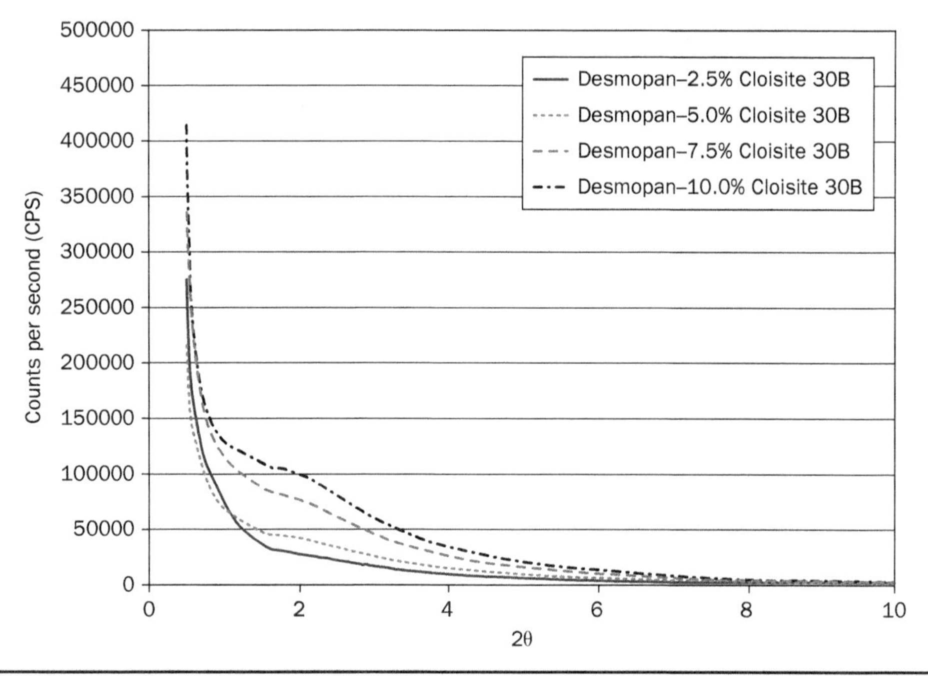

FIGURE 8.3 Wide-angle x-ray diffraction of Desmopan-clay TPUNs [17].

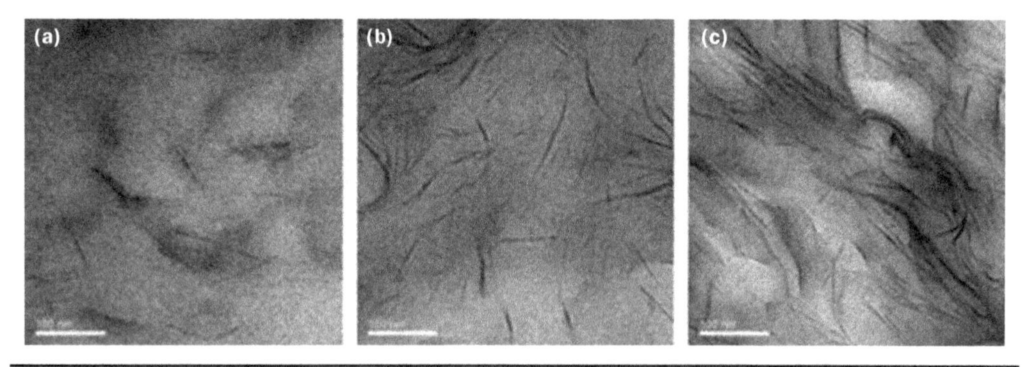

FIGURE 8.4 TEM micrographs at 40 kX of Desmopan with (a) 2.5%, (b) 5%, and (c) 10% Cloisite 30B nanoclay TPUNs with scale bar of 100 nm [17].

Polymer-Graphene Nanocomposites

Researchers started using nanographene (NG) and graphene oxide (GO) in polymer in recent years aiming to enhance the electrical and thermal conductivities of polymer resins. Both NG and GO also have the potential to work as heat barrier and char promoter when the PNCs are subjected to fire burning, thus improving the flammability properties. However, this research area is not yet well-developed. The majority of exploratory work on the use of NG and GO as flame-retardant has been done when the two were added to epoxy resin.

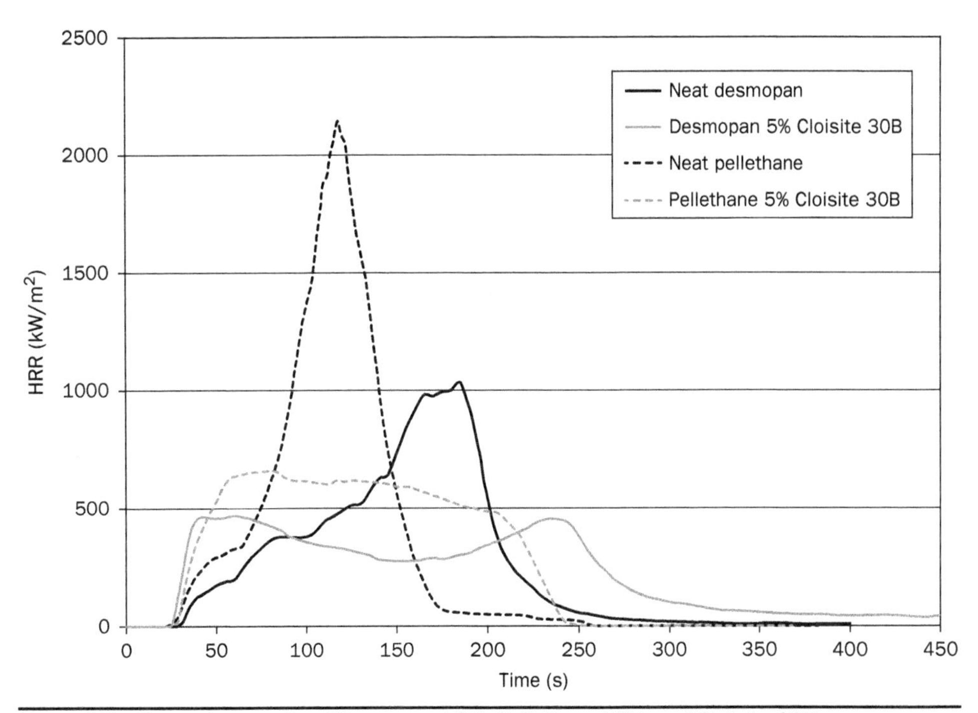

FIGURE 8.5 Heat release rate of Desmopan TPU, Pellethane TPU, and Cloisite 30B TPUNs [17].

TABLE 8.1 Summary of Cone Calorimetry Data at Irradiance Heat Flux of 50 kW/m^2

Material	t_{ig} (s)	PHRR (kW/m^2)	Average HRR, 60s (kW/m^2)	Average HRR, 180s (kW/m^2)	Average efficiency H_c (MJ/kg)	Average SEA (m^2/kg)
Neat Pellethane	32	2290	406	653	30	237
Pellethane-5% Cloisite 30B	34	664 (71% reduction)	560	562	25	303
Neat Desmopan	28	1031	228	515	27	311
Desmopan-5% Cloisite 30B	27	518 (50% reduction)	442	376	28	256

HRR, heat release rate after ignition; H_c, heat of combustion; PHHR, peak heat release rate; SEA, specific extinction area; t_{ig}, time to sustained ignition.

Flammability of Epoxy-Graphene Nanocomposites Avila et al. [19] incorporated 3 wt% of NG into epoxy resin and showed that fire resistance of the resin was enhanced. Specimens were burned under a large heat flux using an oxyacetylene torch for 2 minutes. The unburned thickness of the neat resin and nanocomposites at 60 seconds was about 1 mm and about 3.2 mm, respectively, while at 120 seconds it was 0.16 mm and 2.74 mm. Also, the peak mass loss in TGA of the neat resin and the NG nanocomposite was 1.04 mg/min and 0.63 mg/min, respectively. These results indicated that the nanoparticle could act effectively as thermal barrier which protected the resin underneath the char layer. Koo et al. [20] tested Avila's epoxy-graphene materials and showed similar results using cone calorimeter. Guo et al. [21] also found that incorporating 0.3–5 wt% of NG, GO, and FGO (functionalized graphene oxide) into epoxy resin increased the char yield under TGA of 14–16%. The PHRR and total heat release (THR) under microscale combustion calorimeter (MCC) was reduced up to 45% and 24%, respectively. Although all three nanoparticles could reduce the heat release, NG provided the greatest reduction. Guo concluded that there was a good FR effect of graphite sample in epoxy composites and the nanoparticles were good char promoters. Wang et al. [22] showed that an epoxy nanocomposite with 1 wt% NGO showed 28.5% and 11% reduction in THR and PHRR under a heat flux of 50 kW/m^2 in a cone calorimeter, respectively.

Flammability of HIPS-Graphene Oxide Nanocomposites Higginbotham et al. [23] incorporated 1, 5, and 10 wt% of GO in high-impact polystyrene (HIPS). As the weight loading of GO increased in the HIPS resin (0–10 wt%), the THR reduced from 37.5 to 34 kJ/g (10% reduction), PHRR reduced from 840 to 610 W/g (28% reduction), and the char yield increased from 1 to 6 wt% (500% improvement) as observed from the MCC results. However, since HIPS is a very flammable material, dripping was observed when the HIPS-GO nanocomposites were burned in a vertical flame test and fire did not self-extinguish. Therefore, no correlation between flame retardancy and flammability properties was observed in this nanocomposite system.

Flammability of ABS-Graphene Oxide Nanocomposites Higginbotham et al. [23] incorporated 1, 5, and 10 wt% of the same GO as in the HIPS nanocomposites into the acrylonitrile butadiene styrene (ABS) resin. The GO had similar effects on the THR, PHRR, and char yield on the ABS polymer. As the weight loading of GO increased, the THR decreased (up to 11%), PHRR decreased (up to 17%), and the char yield decreased (up to 600 wt%). ABS-GO nanocomposites also exhibit dripping under the vertical flame test. Although self-extinguishment of fire was also observed, there was still no correlation of flame retardancy and flammability properties in this system.

Flammability of PC-Graphene Oxide Nanocomposites Higginbotham et al. [23] also incorporated the same GO into polycarbonate (PC) with the same weight loadings. A general trend of decreases in THR (up to 13%) and PHRR (up to 13%) with increasing GO weight loading was observed, with a couple of exceptions that were probably some random errors. The char yield increased 21 wt% when 10 wt% of GO was added. Since the PC itself was not very flammable (fire self-extinguished in 14 seconds in a vertical flame test), the addition of 1 wt% GO decreased the time to self-extinguish to 4 seconds, and thus no dripping was observed. The incorporation of 5–10 wt% of GO even made the PS nonflammable.

8.2.2 Two Nanoscale Dimensions-Based Nanocomposites

Polymer Carbon Nanofiber Nanocomposites

Flammability of Thermoplastic-CNF Nanocomposites PA11 was melt-blended by dispersing low concentrations of CNFs by Lao and Koo [6] via twin-screw extrusion. To enhance their thermal and flame retardancy properties, an intumescent FR additive was added to the mechanically superior PA11-CNF formulations. Good dispersion of the CNFs and

FR additives in the PA11 polymer was achieved. For neat and CNF-reinforced PA11 as well as for PA11 reinforced by both intumescent FR and select CNF, decomposition temperatures by TGA, flammability properties by UL 94, and cone calorimeter values were measured. All PA11 polymer systems infused with CNFs and FR additive had higher decomposition temperatures than those infused with solely FR additive. For the PA11/FR/CNF formulations, all Exolit® OP 1311 (FR1), OP 1312 (FR2), and OP 1230 (FR3) FR additives passed the UL 94 V-0 requirement with 20%.

Based on the decomposition temperatures at 50% mass loss, thermal stability of the PA11 was enhanced by CNF significantly. All FR/CNF-reinforced PA11 blends had higher decomposition temperatures than those reinforced with CNF only. The FR3 intumescent additive provided the best synergistic effect to CNF since PA11/20%FR3/7.5%CNF sample had a significant increase of $T_{50\%}$ ($\Delta \cong 66°C$).

It was shown that 30% FR3 was needed for PA11 to achieve a UL 94 V-0 rating. We also observed that a minimum of 20% of either FR1, FR2, or FR3 is needed with either 5 or 7.5 wt% of CNF to achieve the V-0 rating. All three FR additives worked well with CNF. Since 15 wt% of FR intumescent additive will only achieve a V-1 rating, the optimal percent of intumescent FR additive that is needed with CNF lies between 15% and 20% to achieve a V-0 rating. The CNF blend had no combustion at the first flame application and a lower total combustion time than the NC blend. Therefore, the FR/CNF blend outperformed its NC counterpart. Cone calorimeter PHRRs of FR/CNF blends are lowered than that of the FR/NC blends. This may suggest CNF has a good synergistic effect with the FR2 additive.

Polymer Carbon Nanotube Nanocomposites

Carbon nanotube–based (single-walled and multiwalled) PNCs have been studied intensively by Kashiwagi and others [24, 25]. HRRs of three different samples—poly(methyl methacrylate) (PMMA), PMMA-SWNT (0.5%, good dispersion), and PMMA-SWNT (0.5%, poor dispersion)—were measured in a cone calorimeter at an external radiant flux of 50 kW/m^2, as shown in Fig. 8.6 [24]. The 0.5% SWNTs (good dispersion) showed more than 50% reduction in the PHRR compared with neat PMMA. It was proposed that the SWNTs formed a continuous network that acts as a heat shield to slow the thermal degradation of PMMA. The effects of SWNT concentration on flammability properties of PMMA-SWNT nanocomposites that have good dispersion of SWNT at levels from 0.1% to 1% prepared by the coagulation method are shown in Fig. 8.7 [24]. The addition of a 0.1% SWNTs did not significantly reduce the HRR of PMMA. The most reduction in HRR was achieved by 0.5% SWNTs. A 60% PHHR reduction was achieved by 0.5% SWNTs.

Fina et al. [26] studied the chemical activity of CNTs and polyhedral oligomeric silsesquioxane during thermal degradation and combustion of PNCs. Polymer–nanofiller systems may exhibit chemical effects capable of thermal stabilization of polymers as well as reduction of combustion rate and heat released, owing to catalytic effects induced by the nanofillers at high temperature. CNTs in the presence of oxygen are shown to promote oxidative dehydrogenation in polyethylene with production of a stable surface layer of carbon char that provides an effective oxygen barrier effect. A similar action is performed by metal-containing polysilsesquioxanes dispersed in polypropylene. With either CNTs or metal POSS, partial carbonization of the polymer matrix occurs during combustion, subtracting part of the organic polymer from combustion, targeting one of the major fire-retardancy aim. HRR plots for PP/MWNT, PB/MWNT, and PE/MWNT, compared with their reference polymers, are shown in Fig. 8.8 [26]. All

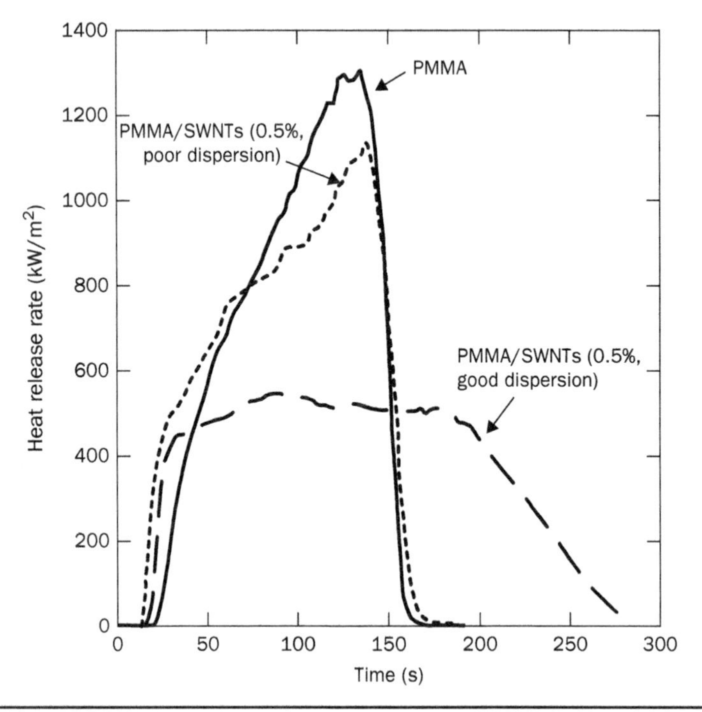

Figure 8.6 Effect of SWNT dispersion on heat release rate of PMMA-SWNTs (0.5%) nanocomposites at an external radiant flux of 30 kW/m² [24].

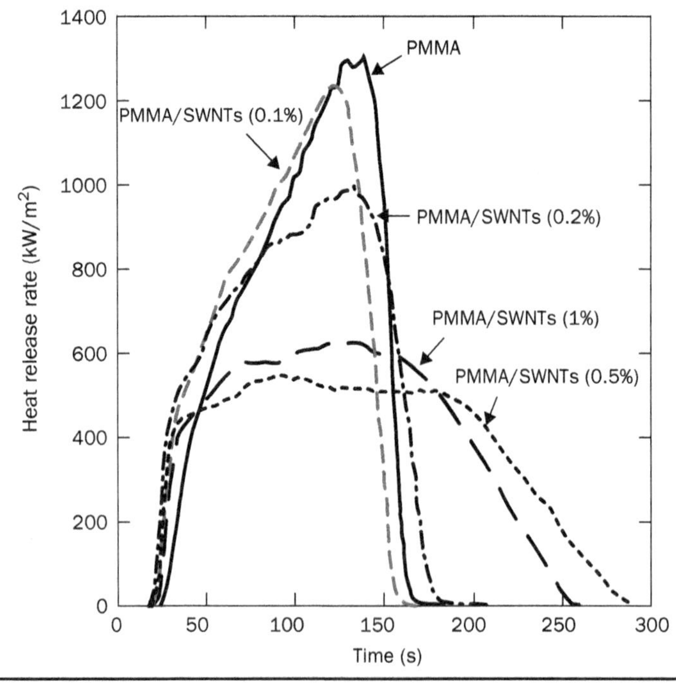

Figure 8.7 Effect of SWNT concentration on the heat release rate of PMMA-SWNTs at 50 kW/m² [24].

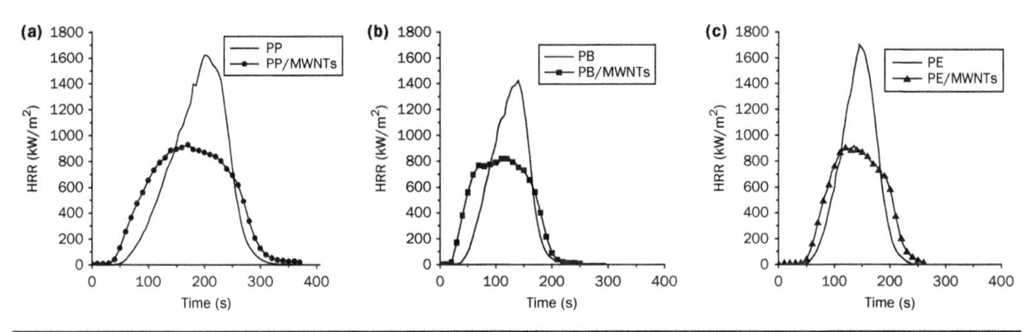

Figure 8.8 HHR plots for (a) PP/MWNTs, (b) PB/MWNTs, and PE/MWNTs compared with their reference polymers [26].

three referenced polymers showed the typical thermally thin behavior, with an almost symmetric HRR curve and a single HRR peak. The plots for the corresponding MWNT nanocomposites show significant reductions of the peak HRR compared with reference polymers, amounting to about 42% for both PP and PB and about 46% for PE. Such a behavior is typical of partially charring system producing physical effects during combustion with data reported in the literature for polyolefin/CNT nanocomposites.

Bocchini et al. [27] studied the multiwalled carbon nanotubes (MWNTs)/linear low-density polyethylene (LLDPE) nanocomposites to understand the stabilization mechanism for their thermal and oxidative degradation. Thermogravimetry coupled with infrared evolved gas analysis and pyrolysis gas chromatography–mass spectrometry demonstrate that presence of MWNTs slightly delays thermal volatilization (15–20°C) without modification of thermal degradation mechanism, whereas thermal oxidative degradation in air is delayed by about 100°C independently from MWNT concentration in the range used here (0.5–3.0 wt%). In thermal degradation, the stabilization of LLDPE is due to a thin film of MWNT net formed by migration at the surface and accumulation due to LLDPE volatilization. In oxidative conditions, the MWNT surface layer favors formation of a polyaromatic carbon char to produce a MWNT/char composite layer protecting the underlying polymer from oxygen shown by SEM and ATR FTIR of degradation residues. Thus, the thermal stabilization of MWNT/LLDPE nanocomposites can be reached with low concentration of MWNTs (0.5%).

Polymer Halloysite Nanotube Nanocomposites

Butler et al. [28] developed an enhanced PA11 composite with improved mechanical, thermal, and flammability properties through the addition of halloysite nanotubes (HNT) and a conventional intumescent FR additive. Polyamide 11 (nylon 11) is of particular interest, as it is one of the major polymers used in the selective laser sintering (SLS) process. Unfortunately, PA11 has poor flammability properties that impose limitations upon the applications for this particular material via the additive manufacturing process. HNTs are an attractive additive as they occur naturally, are relatively easy to process, and are less detrimental to the elongation and toughness properties of the polymeric material compared to other additives. HNTs and FR additives were melt-compounded via twin-screw extrusion processing in PA11. Test specimens in this study were injection-molded. The principal objective of this work is to study the combined effects of the addition of HNTs and a conventional intumescent FR additive to PA11. A matrix of nine formulations of PA11-HNT-FR was studied; properties are characterized using tensile and Izod impact testing, UL 94 flammability testing, thermogravimetric analysis (TGA), and SEM.

None of the PA11-HNT nanocomposites (2.5, 5, 7.5, and 10 wt%) were able to achieve a UL 94 V-0 rating. When a Clariant FR intumescent component (Exolit OP 1312) was added to the PA11-FR-HNT nanocomposites, they appear to be valuable high-performance materials for the SLS process. With only 2.5 wt% HNT in PA11, the tensile modulus is increased from 424 to 1867 MPa, and an elongation to break of 40% is obtained. In traditional composites, a substantial increase in modulus almost always results in a brittle material. A 40% elongation at break may be considerably lower than the neat PA11, which had an elongation at break of 227%. However, 40% elongation at break is still a considerable toughness for a material and is much higher than many other PNCs which incorporate nanoparticles, such as MMT nanoclays [6–8], CNFs [6–8], and CNTs [29]. Traditional PNCs often see elongations at break around 20% and lower for a similar increase in modulus. The combination of stiffness and toughness in the PA11 nanocomposites is therefore very intriguing. In the PA11 nanocomposite formulation comprising 20 wt% FR and 5 wt% HNT, a nondripping formulation that achieved the UL 94 V-0 rating, 940 MPa tensile modulus, and 19% elongation at break was achieved. This balance of flame retardancy, stiffness, and toughness is also very promising. A Design of Experiment approach was adopted in this study to identify additional PA11-FR-HNT formulations. Future work will involve the development of surface treatments for HNT to improve the interfacial adhesion and thus the tensile strength and elongation at break of FR-HNT nanocomposites.

Hao et al. at The University of Texas at Austin [30] studied the influence of different filler loadings on the elongation at break and flammability properties of the PA11/FR/HNT nanocomposites. PA11/FR additives/HNT nanocomposites were melt-compounded via twin-screw extrusion for all the compositions. Three FR additive loadings (15, 20, and 25 wt%) and three HNT loadings (2.5, 5, and 10 wt%) were selected. The formula with 25% FR and 2.5% HNT had the lowest additive content and the highest elongation at break of 10.22% among all UL 94 V-0-rated formulas. A homogeneous dispersion of HNTs in PA11 matrix was observed by TEM (Fig. 8.9). Figure 8.9 shows

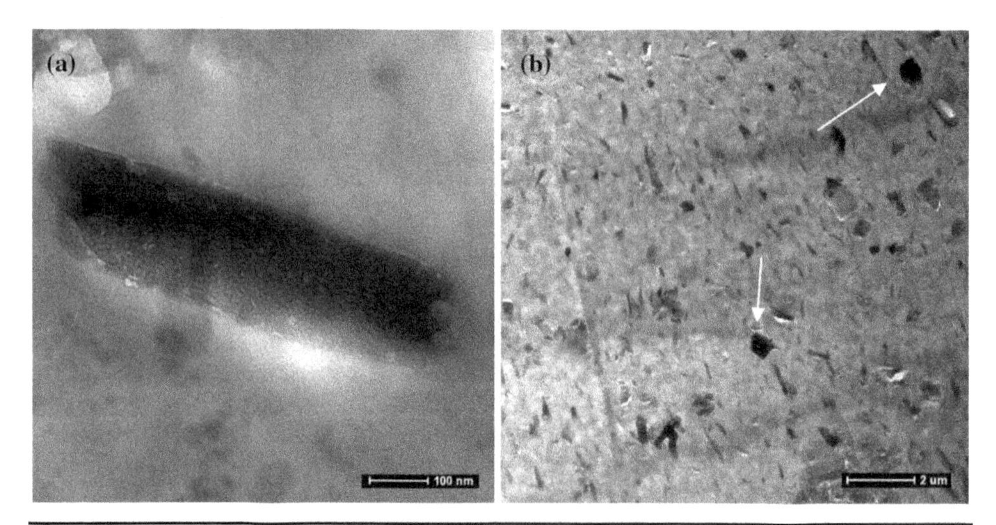

FIGURE 8.9 TEM images showing (a) individual HNT and (b) distribution of HNT and FR (pointed by arrows) in PA11 nanocomposites [30].

that individual HNTs and FR additives were both uniformly dispersed in the polymer matrix. Two arrows in Fig. 8.9b indicate the micrometer-sized FR particles. Differential scanning calorimeter measurements indicated that HNTs behaved as nucleating agents by accelerating the rate of crystallization, thus increasing crystallization temperature. The Young's modulus of the PA11 nanocomposites was enhanced with the addition of HNTs. MCC results demonstrated that the addition of HNTs also decreased the PHRR of the nanocomposites. These results indicate the effectiveness of HNTs on the mechanical, thermal, and FR performance of PA11/FR/HNT nanocomposites.

Overall, PA11/FR/HNT nanocomposites appear to be valuable high-performance compositions for the SLS AM process. The mechanical properties data suggest that an engineered combination of mechanical strength, stiffness, and toughness is plausible by HNT reinforcement. Formula FR/HNT 25/2.5 had the lowest additives content and the highest elongation at break of 10.22% among all UL 94 V-0-rated formulas. DSC measurements indicated that HNTs behaved as nucleating agents by accelerating the rate of crystallization, thus increasing crystallization temperature. MCC results proved the effectiveness of FR and HNT in reducing the thermal combustion activity. In addition, the activation energy of formula FR/HNT 25/2.5 is 2.44 times of neat PA11. Future work will involve the addition of an elastomer component in the PA11/FR/HNT nanocomposites to increase the elongation at break.

8.2.3 Three Nanoscale Dimensions-Based Nanocomposites

Polymer Nanosilica Nanocomposites

Nanosilica can have a huge interfacial area as long as the diameter of the particles is in the range of nanometers. Although they do not have the narrow gallery structure of the layered nanoclay, improvement in physical properties [31–34] and in thermal stability [35, 36] by the addition of nanosilica to polymer was reported. It was also reported that the addition of mesoscale silica to different polymers significantly reduced the HRR of polymers [37, 38]. Flammability properties of PMMA-nanosilica [39–41] and polyimide-nanosilica nanocomposites [41] have been reported. Dispersion of the particles in a polymer is critical for obtaining better FR performance. Roughly 50% reduction in PHRR was reported with the addition of 13 wt% of silica particles (Fig. 8.10) [40]. Little to no improvement was reported in LOI measurement with up to 10 wt% of nanosilica of diameter of 7 nm. Although the LOI values increased from 36 to 44, the addition of 20 wt% of nanosilica (diameter 50–300 nm) was required. The addition of nanosilica hardly reduced the HRR at the early stage of burning, and it was demonstrated that the addition of nanosilica did not significantly modify the UL 94 rating [41]. The overall FR effectiveness of nanosilica appears to be less than that of the nanoclays, as described in Sec. 8.2.1.

Polymer Nano-Alumina Nanocomposites

Three types of nano-alumina (X-0 needle, X-25SR, and X-0SR) with different organic treatments manufactured by SASAL NA were incorporated into PA11 via twin-screw extrusion by Lao et al. [42]. A total of 10 formulations of PA11/*n*-alumina nanocomposites, including the neat PA11, were melt-compounded and injection-molded to test specimens. TEM demonstrated that *n*-alumina surface treated particles are well dispersed in the PA11 matrix. TGA showed that *n*-alumina improved the thermal stability of the nylon 11 substantially. The UL 94 test showed that all formulations did not pass the ideal V-0 rating. All of them exhibited flammable dripping that ignited the cotton

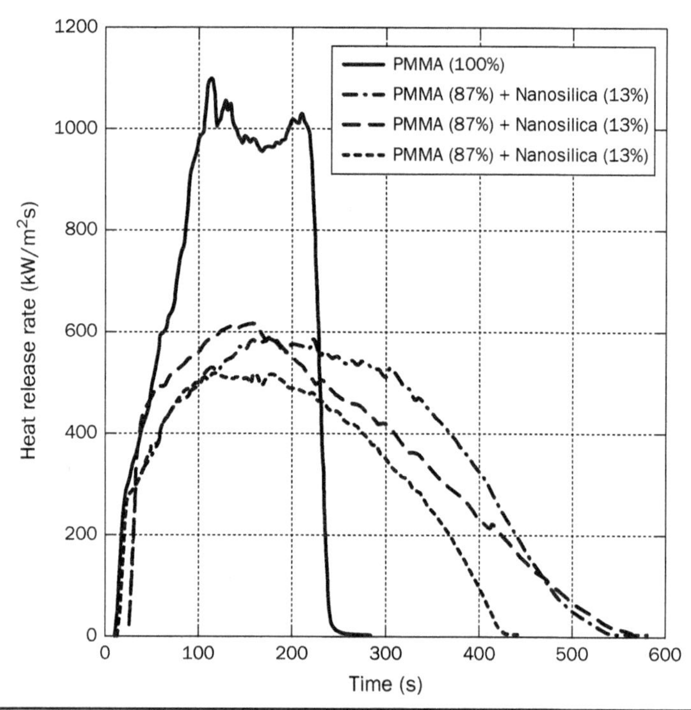

FIGURE 8.10 Effects of nanosilica addition on heat release rate of PMMA at 40 kW/m^2. The dashed lines are the results of three replica of nanocomposites made at three different times [41].

under the specimens during the test. The X-0 needle formulations performed the best, followed by X-25 SR, and the X-0 SR performed the worst. The addition of selective FR additives or antidripping agents should be incorporated into the system together with the *n*-alumina for enhanced flammability properties in the future.

Fahy et al. used alumina nanofibers (ANF; Nafen™) to compound with nonhalogenated FR additives and PA11 via twin-screw extrusion [43]. Characterizations including TGA, DSC, MCC, UL 94, and microstructural analysis using SEM were conducted and discussed. The design of experiment was used to study the *processing-structure-property* relationships of the Nafen and FR additives. Eleven formulations were selected to study this relationship. The formulations were extruded into 1.75-mm filaments for the purpose of fused filament fabrication (FFF) additive manufacturing. The filament was successfully printed into test bars for flammability and mechanical properties characterizations of the PA11 nanocomposite. The results successfully demonstrated the enhanced compatibility of PA11 with chemically functionalized Nafen (*f*-Nafen). The MCC tests show that the PHRR blends with *f*-Nafen better than their nonfunctionalized counterparts. In addition, the onset heat release for the *f*-Nafen formulations occurred at a higher temperature demonstrating improved thermal stability. This trend is verified with the TGA data in which we see that *f*-Nafen formulations reach higher temperatures before 10% and 50 % mass loss. The *f*-Nafen should be considered over the nonfunctionalized Nafen when thermal stability is desired.

The compounding of each formulation was successful in producing quality filament with a consistent diameter of 1.75 ± 0.10 mm. The formulations with highest loading, 20 wt% FR, and 5 wt% Nafen were used to print test bars for both UL 94 and tensile testing. This process could potentially be used as an alternative to traditional means of test bar production such as injection molding. Three formulations were down-selected for 3D printing based on the initial MCC and TGA analysis. Additional formulations such as 80N_15F_5A (*f*-Nafen) will be printed and tested in a future follow-up study. This work will also be cross-referenced with prior PA11 research by the Koo Research Group at The University of Texas at Austin with an approach to better optimize the FR potential of PA11.

Rallini et al. reported using nano-alumina (average diameter of 13 nm) to improve the flame resistance of carbon fiber–reinforced composites [44]. Nano-alumina was dispersed in an epoxy resin as a matrix for carbon fiber reinforced by simple mechanical stirring. The 5 wt% nano-alumina allowed the Rallini et al. to obtain a fiber-reinforced composite with improved fire resistance. After exposure to high temperature, the nano-alumina acts as a high-temperature adhesive coating and has sintered onto the carbon fibers. This coating binds together the carbon fibers promoting residual structural integrity of the burnt composite. TGA data also demonstrated that the sintered alumina coating of the fibers acts as a barrier for the oxidation of carbon fibers. The morphological investigation of the samples burnt in a muffle has shown further evidence on the role of alumina particles. The 1 wt% content is insufficient to improve the residue strength of the burnt sample, whereas the 5 wt% content is sufficient to produce a ceramic skeleton on the entire volume of the composite.

Polymer Nano-Magnesium Hydroxide Nanocomposites

Magnesium hydroxide [$Mg(OH)_2$/MH] has been used in polymer as FR additives. However, it commonly requires 60–70 wt% of the micrometer sized (2–5 μm) MH to be added into polymer to make it truly flame-retardant [45]. Since FR efficiency of MH is related to interparticle distance which decreases with decreasing particle size at a given volume fraction of filler [46, 47], it is believed that by using MH of smaller particle size—that is nano-MH—effective flame retardancy can be achieved at a lower loading of filler.

Flammability of EVA-$Mg(OH)_2$ Nanocomposites Kalfus and Jancar [45] incorporated 26 wt% of nano-MH (comparing to conventional loading of 60–70 wt% of micro-MH) into ethylene vinyl acetate (EVA) resin and found that the rate of weight loss was about the same as the micro-MH of the same weight loading. However, the thermal stability of the EVA-MH nanocomposite was slightly better. Gui et al. [48] added 33 wt% nano-MH into EVA and observed 43% reduction in PHRR, 43% reduction in mean HRR, and 182% increase in residue. Ly and Liu [49] also compared the effect of particle size on the flammability of EVA. Nano-MH and micro-MH of 80, 100, 120, and 150 phr (parts per hundred polymer) were incorporated into EVA, respectively. It was found that the LOI of the composite was increased from 18% to 39% by micro-MHs and to 46% by nano-MHs, respectively. However, all of the EVA nanocomposites, besides the 150 phr, failed the UL 94 burning test (while EVA-150 phr nano-MH achieved V-2 rating). The addition of 100 phr nano-MH increased the time to ignition (TTI) from 87 seconds to 148 seconds (70%), decomposition temperature from 438°C to 470°C (7%), and char yield from 0 to 6 wt%; it also reduced the PHRR from 1,655 to 423 kW/m^2 (74%) and average HRR from 439 to 251 kW/m^2 (43%). Ly and Liu then incorporated one more component [microcapsulated red phosphorus (MRP)] into the nanocomposites and found a synergistic effect. While EVA with 100 phr nano-MH failed the UL 94 test, EVA with 100 phr of nano-MH and MRP (e.g., 95 phr nano-MH and 5 phr MRP) achieved a V-0 rating. The synergism also increased the LOI to 60%, TTI to 176 seconds (102% increase compared to neat EVA), decomposition temperature to 490°C (12%), and char yield to 16 wt%, and reduced the PHRR to 148 kW/m^2 (91%) and HRR to 72 kW/m^2 (83.5%).

Flammability of PP-$Mg(OH)_2$ Nanocomposites Mishra et al. [50] incorporated 2, 4, 8, and 12 wt% of nano-MH into polypropylene (PP). It was found that the rate of burning per second was reduced with the increasing nanofiller composition. The reduction in the flammable property was as high as 35% compared to the neat PP. Mishra et al. claimed that the reduction in flammability was due to the endothermic nature of the evenly dispersed nano-MH in the PP matrix.

Flammability of EPDM-$Mg(OH)_2$ Nanocomposites Zhang et al. [51] mixed 20–100 phr MH into ethylene-propylene-diene monomer (EPDM) rubber with particle sizes of 800 mesh, 1250 mesh, 2,500 mesh, and nano-MH. A decrease in PHRR with decreasing particle sizes was observed. Song et al. [52] incorporated 100 MP (mass portion) of nano-MH into EPDM (100 MP)/Paraffin (100 MP) system and found an increase in LOI from 17% to 28%. Song et al. claimed that greater LOI value was obtained with higher content of nano-MH added, and it was obvious that the addition of the nano-MH promoted flame resistance.

Flammability of ABS-$Mg(OH)_2$ Nanocomposites Cao et al. [53] investigated the acrylonitrile-butadiene-styrene (ABS) system with nano-MH. The ABS/1 wt% nano-MH nanocomposite had similar flammability properties as the neat ABS polymer. However, when incorporating 5 wt% of the nano-MH, the PHRR reduced from 663 to 430 kW/m^2 (35%), and THR from 107 to 96.8 MJ/m^2 (9.5%); TTI increased from 30 to 67 seconds (123%), and time to reach PHRR increased from 169 to 245 seconds (45%). However, the combustion time was extended from 400 to 580 seconds (45%).

Flammability of SBR-$Mg(OH)_2$/PBR-$Mg(OH)_2$ Nanocomposites Patil et al. [54] incorporated 2–10 wt% of both conventional MH and nano-MH into styrene-butadiene rubber (SBR) and polybutadiene rubber (PBR), respectively. It was observed that the flame retardancy (improvement) in a burning test increased with increasing MH from 1.5 to 2.1 mm/s while conventional MH was used and from 1.6 to 2.3 mm/s when nano-MH was used. Similar trend was observed in the PBR/MH systems. The flame retardancy increased from 1.38 to 1.54 mm/s and 1.41 to 1.58 mm/s when conventional MH and nano-MH were used in the PBR resin, respectively. According to the data, flame retardancy was noticeably enhanced by the MHs. However, the difference between the conventional MH and the nano-MH in both polymer systems was insignificant, although Patil et al. claimed that the nano-MH provided significant increase in flame retardancy in the SBR resin comparing to the conventional filler.

Flammability of Epoxy-$Mg(OH)_2$ Nanocomposites Suihkonen et al. [55] compared the flammability of epoxy resins filled with micro-MH and nano-MH. Weight loadings of 1%, 5%, and 10% of both fillers were incorporated into the epoxy. TTI of all samples were about the same. However, PHRR was reduced significantly from 933 (neat) to 572 kW/m^2 (10% micro-MH) and 539 kW/m^2 (10% nano-MH). THR and TSP (total smoke produce) were also decreased, but the effects of micro-MH and nano-MH were about the same. THR was reduced from 124 to 114 MJ/m^2 while TSP reduced from 40 to 35–36 m^2. There was a significant decrease (71%) in the burning rate of the sample in the UL 94HB burning test when 10 wt% of nano-MH was added, comparing to the neat epoxy. Suihkonen et al. [54] also commented that the poor dispersion of the microparticles did not obstruct the functionality of the FR filler, since the efficiency of MH in flammability was not significantly affected by the particle size of the filler or the surface treatment in this study.

Polymer POSS Nanocomposites

Polyhedral oligomeric silsesquioxanes (POSS)–based nanocomposites have received increasing attention because of the unique three-dimensional structure of the POSS macromonomer [56], as discussed in Chap. 2. Early examples were presented with siloxane [57, 58] followed by numerous applications showing enhancement of thermal stability and improving the flammability properties of polymers. Thermal analysis study demonstrated the enhancement of thermal stability of polymer-POSS nanocomposites and suggested that there is a potential to improve the flammability properties of matrix polymers. Studies clearly demonstrating such improvement by means of the use of POSS-based nanocomposites are limited. Kashiwagi and Gilman [59] studied polytetramethylene ether-glyco-b-polyamide-12, 1% polyamide-12 (PTME-PA), polystyrene-polybutadiene-polystyrene (SBS), and polypropylene (PP) with POSS ranging from 10% to 20 % via solution blending in tetrahydrofuran (THF). For comparison purposes, composites based on other silicone compounds, such as polycarbosilane (PCS) and polysilastyrene (PSS), were also prepared by solution blending. The flammability properties of these polymer blends were characterized using a cone calorimeter. The results presented in Table 8.2 [59] reveal that both PCS and POSS are reasonably effective for reducing the HRR measured at 35 kW/m^2. However, the THR of the nanocomposites was not significantly reduced from that of the polymer matrix. Furthermore, the residual yields are about the same as the theoretical yields (Table 8.2). This means that the addition of POSS to the nanocomposites does not significantly increase the yield of carbonaceous char. The residue is mainly the inorganic component of the POSS.

Fina and colleagues focused on the preparation and characterization of POSS-based polymer hybrids by melt-blending and reactive-blending techniques [26, 60–64]. The chemical activity of POSS and CNT during thermal degradation and combustion of PNCs was studied by Fina et al. [64]. Polymer-nanofiller systems may exhibit chemical effects capable of thermal stabilization of polymers and reduction of combustion rate and heat released, owing to catalytic effects induced by the nanofillers at high temperature. POSS in the presence of oxygen is shown to promote oxidative dehydrogenation in polyethylene (PE) with the production of a stable surface layer of char that provides an effective oxygen barrier effect. The combustion of polypropylene (PP) is strongly affected by metal-containing POSS, with a mechanism that cannot be related to the simple accumulation of a ceramic phase on the surface as reported for metal-free polysilsesquioxanes [61]. In particular, Al-isobutyl POSS was reported to be an effective flame-retardant for PP, resulting in lower HRR and lower effective heat of combustion (EHC), owing to a catalytic effect of the Al moieties, promoting secondary reactions

Table 8.2 Summary of Cone Calorimeter Data of PP, PTME-PA, and SBS with Siloxanes and POSS at 35 kW/m²

Sample	Residue yield[a] (%)	Mean mass loss rate (g/m²-s)	Peak HRR [kW/m² (Δ%)]	Mean HRR [kW/m² (Δ%)]	H_c (MJ/kg)	SEA (m²/kg)	Mean CO yield (kg/kg)
PP	0	25.4	1466	741	34.7	650	0.03
PP/POSS 80/20	17 (16)	19.1	892 (40%)	432 (42%)	29.8	820	0.03
PTME-PA	0	34.2	2020	780	29.0	190	0.02
PTME-PA/PCS 80/10	15 (15)	14.8	699 (65%)	419 (46%)	28.5	260	0.02
PTME-PA/POSS 90/10	6 (8)	19.8	578 (72%)	437 (44%)	25.5	370	0.02
SBS	1	36.2	1405	976	29.3	1750	0.08
SBS/PCS 80/20	20 (15)	18.5	825 (42%)	362 (63%)	26.4	1550	0.07
SBS/POSS 90/10	6 (8)	31.2	1027 (27%)	755 (23%)	26.9	1490	0.07

H_c, mean heat of combustion; SEA, specific extinction area (smoke measurement). Uncertainties: ±5% of reported value of residue yields for heat release rate (HRR) and H_c data; ±10% for carbon monoxide and SEA data.

[a]Theoretical residue yields in parentheses.

during polymer degradation, and leading to partial charring instead of complete volatilization. To investigate the material surface evolution upon heating, PP, PP/oib-POSS, and PP/Al-POSS were heated to 350°C in air with a 15°C/min heating ramp and the residues were studied by means of SEM/EDS and ATR-FTIR analyses of the surface [61].

The PP and PP/oib-POSS undergo a large weight loss about 44% and 37% of the initial weights, respectively, and appear to be strongly degraded as discussed in [61]. PP/Al-POSS evidenced a higher thermal stability and a limited weight loss (5% of the initial weight). SEM micrographs of the surface of residue for PP/oib-POSS and PP/Al-POSS shown in [61] proved that a thin physical barrier may be obtained when treating PP/oib-POSS, this being responsible for a limited delay in the weight loss in air [61]. However, SEM/EDS results evidence that thermal stabilization of PP/Al-POSS cannot be ascribed to a pure physical effect by inorganic accumulation on the surface, confirming the occurrence of chemical effects induced by the metal-POSS embedded in PP matrix [61].

8.2.4 Multicomponent FR Systems: Polymer Nanocomposites Combined with Additional Materials

Polymer-Clay with Conventional FR Additive Nanocomposites

A combination of 65 wt% alumina trihydrate (ATH) and 35 wt% of a high-level accepting polymer matrix, such as poly(ethylene-co-vinyl acetate) (EVA), must often be used for cable outer sheaths [65]. The performances of two polymer blends were compared

Table 8.3 Fire Performance of FRNH Coaxial Cables with EVA–ATH and EVA–ATH–Organoclay Outer Sheaths

UL-1666 requirement	EVA–ATH compound	EVA–ATH–Organoclay compound
Maximal temperature at 12 ft: <850°F	1930°F	620°F
Maximal flame height <12 ft	>12 ft	6 ft

by Beyer [66, 67]. One blend was made from 65 wt% ATH and 35 wt% EVA (28 wt% vinyl acetate content), and a second blend was made from 60 wt% ATH, 5 wt% organoclay, and 35 wt% EVA. Both blends were investigated with TGA in air and by cone calorimeter at 50 kW/m^2. TGA in air clearly showed a delay in degradation by a small amount of organoclay [66, 67]. The char of the EVA–ATH–organoclay compound generated by the cone calorimeter was very rigid and showed only few small cracks. However, the EVA–ATH compound was much less rigid and with many big cracks. This could be the reason that the PHRR of nanocomposite was reduced to 100 kW/m^2 as compared to 200 kW/m^2 for the EVA–ATH compound. To obtain the same decrease for PHRR by using just ATH in EVA, the content of ATH has to be increased to 78 wt% within the EVA–ATH system. To maintain 200 kW/m^2 as a sufficient PHRR level, ATH could be decreased from 65 to 45 wt% by the presence of 5 wt% organoclay within the EVA–ATH–clay system. The reduction of ATH resulted in improved mechanical and rheological properties of the EVA-based nanocomposite.

This research resulted in the commercial product FRNH cables passing UL 1666 [67]. The outer sheath was based on a nanocomposite with an industrial EVA–ATH–organoclay composition. The analogous coaxial cable was tested with an outer sheath based on EVA–ATH. Table 8.3 shows the results of the fire performance of these two coaxial cables [67]. The insulating and no burning char reduced the emission of volatile produce from polymer degradation into the flame area, and thus minimized the maximal temperature and height of the flames.

Polymer-Carbon Nanotubes with Conventional FR Additive Nanocomposites

EVA with purified MWNTs, crude MWNTs, and organoclay were investigated by Beyer and colleagues [66–71] by cone calorimeter at 35 kW/m^2. All compounds were melt-blended in a Brabender mixing chamber. It is evident from the results presented in Table 8.4 that all the filled polymers had improved flammability properties. For EVA and EVA-based nanocomposites containing 2.5 phr of filler, the PHRR decreased as follows: EVA > organoclays ~ purified MWNTs. For EVA and EVA-based composites containing 5.0 phr of filler, the PHRR decreased as follows: EVA > organoclays > purified MWNTs = crude MWNTs. Crude MWNTs were as effective as purified MWNTs in the reduction of PHRR as purified MWNTs! Increasing the filler content from 2.5 to 5.0 phr caused an additional FR effect that was most significant when purified or crude MWNTs were used.

A synergistic effect for flame retardancy between MWNTs and organoclays was observed for a nanocomposite containing 2.5 phr of purified MWNTs and 2.5 phr of organoclays [71]. The latter sample was found to be the best FR compound. The variation of screw velocity from 45 rpm (sample A) to 120 rpm (sample B) did not change the flame retardancy properties for composites containing 5.0 phr of crude MWNTs. There was also no reduction in time to ignition for the EVA MWNT–based composite in contrast to the EVA organoclay–based composite (Fig. 8.11).

TABLE 8.4 Peak Heat Release Rates at 35 kW/m² for Various Compounds with Organoclays and Multiwalled Carbon Nanotubes

Sample	EVA[a] (parts resin)	MWNTs (Purified) (phr)	MWNTs (Crude) (phr)	Organoclay[b] (phr)	PHRR (kW/m²)
EVA[a]	100.0	–	–	–	580
1[c]	100.0	2.5	–	–	520
2	100.0	5.0	–	–	405
3	100.0	–	–	2.5	530
4	100.0	–	–	5.0	470
5[c,d]	100.0	2.5	–	2.5	370
6a[c]	100.0	–	5.0	–	403
6b[e]	100.0	–	5.0	–	405

[a] Eaxorene UL-00328 with 28 wt% vinyl acetate content.
[b] Nanofil 15.
[c] The screw velocity was 45 rpm and the mass temperature was 136°C.
[d] The nanotubes and the organoclay were premixed before addition.
[e] The screw velocity was 120 rpm and the mass temperature was 142°C.

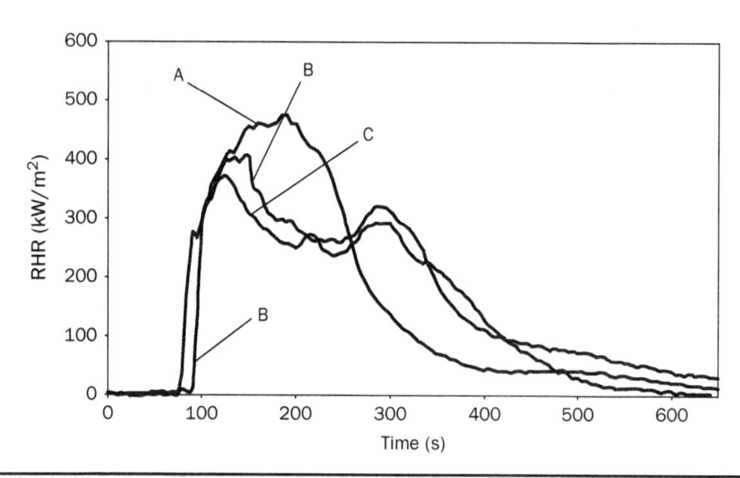

FIGURE 8.11 Heat release rates at 35 kW/m² for various EVA-based materials: (a) EVA + 5.0 phr organoclay; (b) EVA + 5.0 phr pure MWNT; (c) EVA + 2.5 phr organoclay + 2.5 phr pure MWNTs. EVA is Escorene UL 00328 with 28 wt% vinyl acetate content; organoclay is Nanofil 15 [71].

Polymer-Clay and Polymer-Carbon Nanotubes with Conventional FR Additive Nanocomposites

Johnson et al. [72] studied the combined effects of MMT clay, CNF, and FR additives on the mechanical and flammability properties of these PA11 nanocomposites for application in selective laser sintering (SLS) for additive manufacturing. Test specimens of PA11 containing various percentages of intumescent FR additive, MMT clay, and CNF

Table 8.5 UL 94 Vertical Burn Ratings for PA11 Nanocomposites

Formulation	MMT	CNF	FR	UL 94 rating
Neat PA11	0%	0%	0%	V-2
#1	2.5%	2.5%	0%	Fail
#3	2.5%	2.5%	15%	V-0
#4	2.5%	2.5%	20%	V-0
#2	3.5%	3.5%	0%	Fail
#5	3.5%	3.5%	15%	V-0
#6	3.5%	3.5%	20%	V-0

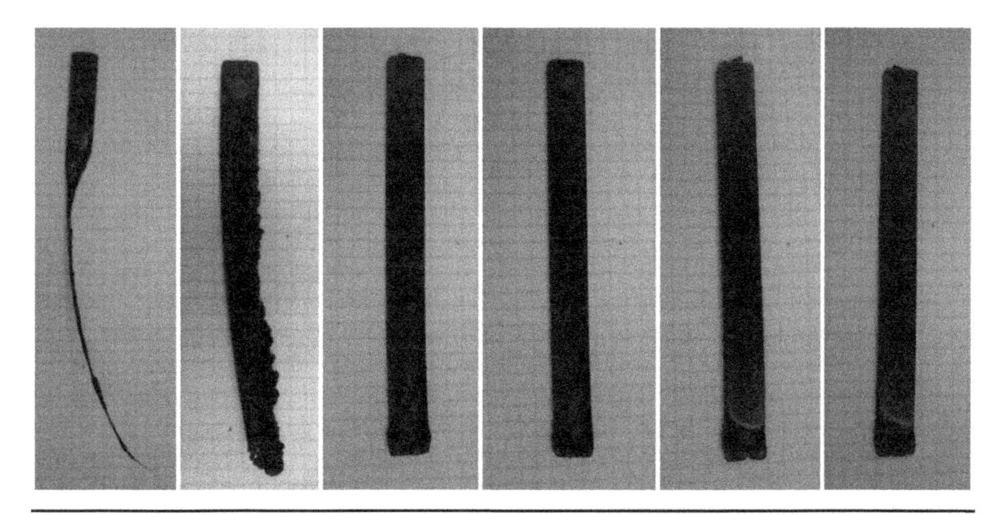

Figure 8.12 Posttest photos of UL 94 test specimens. Formulations 1 and 2 (left-most two photos) can be seen to have burned significantly, while formulations 3–6 (in ascending order from left to right) show no significant burn damage [72].

were prepared via the twin-screw extrusion technique. Izod impact testing, tensile testing, and SEM analysis are used to characterize mechanical properties. UL 94 and SEM analysis of char surfaces are used to characterize the flammability properties of these materials. Results are analyzed to determine any synergistic effects among the additives to the material properties of PA11.

The results for the UL 94 vertical burn test of their materials are shown in Table 8.5. The baseline PA11 material received a V-2 rating because it produced burning drips, but was still self-extinguishing in less than 30 seconds. The two formulations with MMT clay and CNF, but no FR additive failed the test. In the PA11/2.5% MMT/2.5% CNF formulation, the material burned quickly once ignited and produced numerous viscous drips. The PA11/3.5% MMT/3.5% CNF also burned readily once it was ignited, but it only produced drips on 6 out of 10 tests and the drips were very viscous. All of the formulations with FR additive passed the test with V-0 ratings. They all immediately self-extinguished and produced no drips during testing. Photos of the posttest specimens are shown in Fig. 8.12 [72].

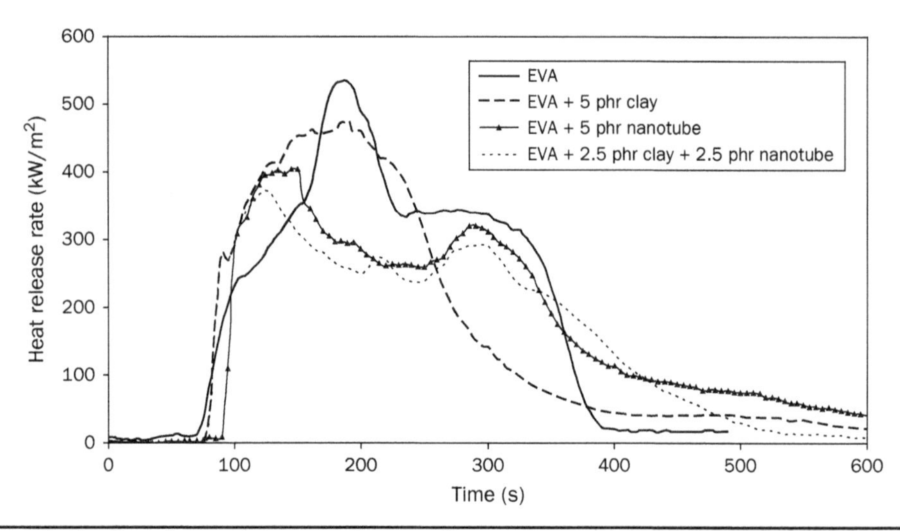

Figure 8.13 The relationship between the rate of heat release rate and burning time for EVA and EVA composites containing 5% MWNTs, 5% organoclay, and the combination of 2.5% nanotubes and 2.5% clay, respectively. The data are obtained using cone calorimetry under 35 kW/m^2 heat flux [73].

The results for the formulations without FR additive are as expected. It is not unusual for nanomaterial additives alone to fail to improve a material's flammability rating and they are typically most beneficial when used to improve the effectiveness of conventional FR additives [72]. The results for the remaining formulations are quite promising. In previous tests, where MMT clay and CNF were combined with intumescent FR additives separately, such significant improvements in flammability properties were not observed with such low percentages of FR. In those experiments there were no formulations that were able to reach V-0 rating with only 15% FR, and there were several formulations that failed to reach it even at 20% FR [72]. These results suggest that there is a synergistic effect between MMT clay and CNF in improving the flammability of PA11 composites when combined with conventional FR additives.

Gao et al. [73] reported the clay- and CNT-enhanced EVA composites. Ternary and binary PNCs were investigated through the cone calorimeter test [73]. Figure 8.13 depicts the HRR of three types of EVA composites containing CNT, organoclay, and mixtures of clay and MWNTs [73]. The results indicate that CNT/clay nanofillers were more effective in reducing HRR than the binary PNCs. Figure 8.14 shows the morphology of the chars of the PNCs from the cone calorimeter test, the natural burning experiment, and the burning test at 600°C for 20 minutes, respectively [73]. The ternary composites show the well-formed char and the smooth surface (Fig. 8.14a). In the case of natural burning (Fig. 8.14b), it presents reduced crack extent compared to the binary composite. Figure 8.14c exhibits the similar morphology. Hence, the adding of clay into the CNT-EVA composites has an effect on enhancing the flammability properties. This could be explained by the fact that the clay promotes the formation of graphitic structure and increases the resistance of char oxidation.

Peeterbroeck et al. [74] studied and announced the effect of the CNT/organo-modified clay on the thermal and mechanical properties of EVA nanocomposites. Table 8.6 presents the result of the cone calorimeter test. The largest reduction of the PHRR (36%) was obtained for the ternary PNCs. Figure 8.15 shows the thermal

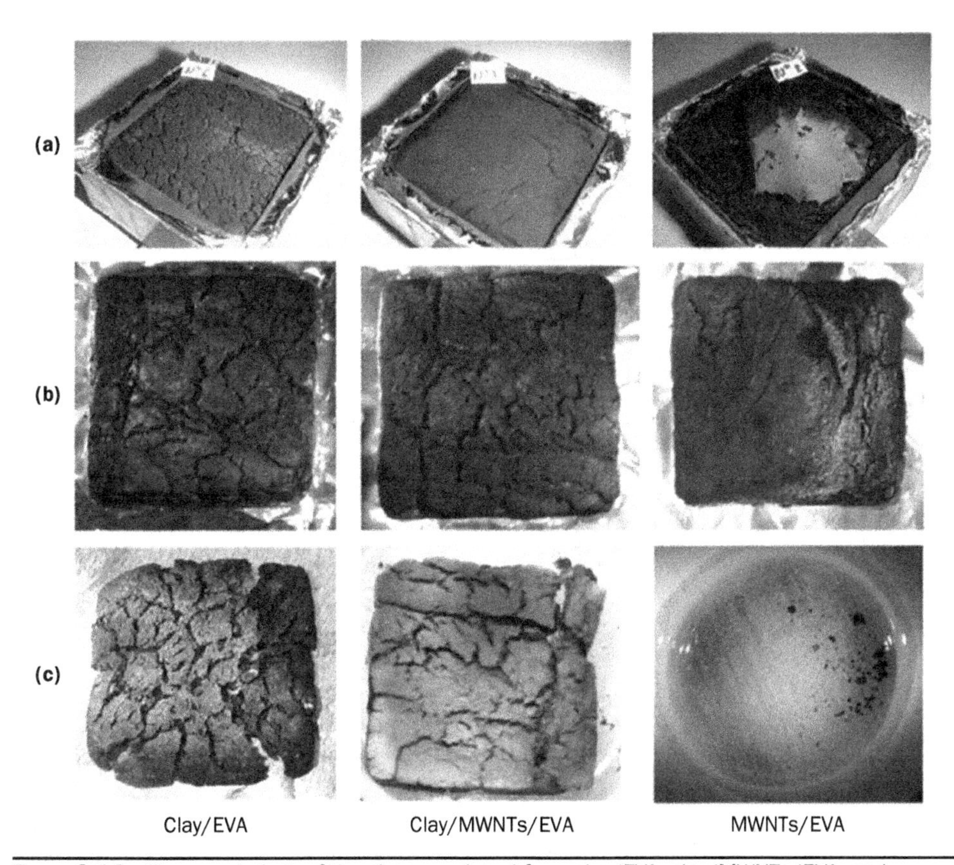

FIGURE 8.14 The morphology of the chars produced from clay/EVA, clay/MWNTs/EVA, and MWNTs/EVA nanocomposites following (a) the cone calorimeter test, (b) the natural burning, and (c) the burning at 600°C for 20 minutes [73].

TABLE 8.6 Comparison between Binary and Ternary PNC

Sample	Ignition time (s)	PHRR (kW/m²)	Crack density
EVA	84	580	+++
EVA/OM clay (4.8%)	67	470	+
EVA/CNT (4.8%)	83	405	++
EVA/OM clay (2.4%)/ CNT (2.4%)	71	370	0

+++, almost complete elimination/volatilization of EVA matrix; ++, formation of holes; +, presence of cracks; 0, almost no visible cracks on the surface.

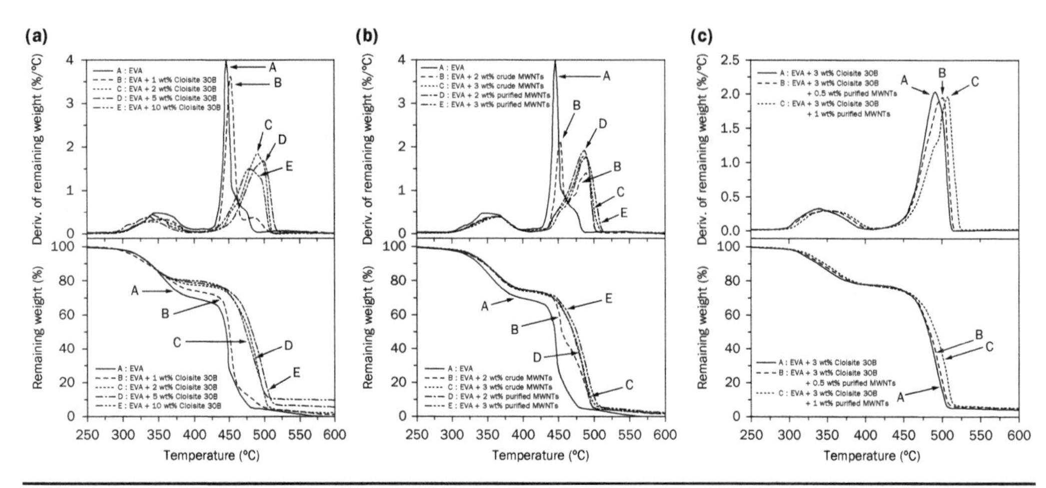

FIGURE 8.15 Thermogravimetric analysis of (a) binary EVA nanocomposites based on various amounts of Cloisite 30B, (b) binary EVA nanocomposites based on various amounts of crude and purified MWNTs, and (c) EVA nanocomposites filled with 3 wt% Cloisite 30B and either 0, 0.5, or 1.0 wt% purified MWNTs under air flow at 20 K/min [74].

degradation property using the remaining weight and its derivative [74]. The positive effect on EVA thermal stabilization resulting from the organo-modified clay/CNT is illustrated in Fig. 8.15c. Thus, the synergetic influence of the ternary system was observed for the flame retardancy properties.

Mechanical properties of the binary and ternary system were evaluated by tensile tests. Binary PNCs present the better strength compared to the ternary ones [74]. However, the ternary PNCs still exhibits the enhanced mechanical properties than the pure EVA does.

Wu et al. developed a multifunctional PA11 with balanced thermal, flammability, and mechanical properties for SLS additive manufacturing [75, 76]. In the first study [75], two sets of formulations were prepared by twin-screw extrusion: the first set examined the effect of maleic anhydride modified elastomers on flammability and mechanical properties, whereas the second set added various amount of nanoclay and discussed thermal stability, flammability, and mechanical properties. The addition of 20 wt% elastomer brought the elongation at break up to 40%. Reduction in heat release capacity as high as 49% was achieved; all nanocomposite samples passed UL 94 V-1 rating. The addition of nanoclay improved the tensile modulus by up to 78%, the elongation at break for all the formulations was negatively affected by the addition of the FR additive and nanoclay [75, 77]. In the second study [76], Wu et al. explored the synergism between two nanoparticles, nanoclay and MWNTs, to see whether better FR properties can be achieved. TEM micrographs indicate that both nanoclay and MWNTs achieved high level of dispersion. Flammability results showed that all formulations achieved UL 94 V-0 rating, which is a significant improvement from the previous formulations without MWNTs. Char morphology characterization indicated that a solid carbonaceous char layer was reinforced by nanoclay and MWNTs.

A total of seven formulations were twin-screw extruded with different loadings (wt%) of FR additive, elastomer, nanoclay, and MWNTs (Table 8.7) [76]. A Thermo

Table 8.7 Formulation Matrix for the FR PA11 Nanocomposites

Sample ID	Formulation	PA11 (wt%)	Flame-retardant (wt%)	Kraton (wt%)	Nanoclay (wt%)	MWNTs (wt%)
0	PA11	100	–	–	–	–
1	15FR_10K_2.5NC_2.5MWNT	70	15	10	2.5	2.5
2	15FR_15K_2.5NC_2.5MWNT	65	15	15	2.5	2.5
3	15FR_20K_2.5NC_2.5MWNT	60	15	20	2.5	2.5
4	15FR_10K_3.5NC_3.5MWNT	68	15	10	3.5	3.5
5	15FR_15K_3.5NC_3.5MWNT	63	15	15	3.5	3.5
6	15FR_20K_3.5NC_3.5MWNT	58	15	20	3.5	3.5

Table 8.8 UL 94 Results for PA11/FR/K/NC/MWNT Blends

Formulation	Average first burn flaming combustion duration (s)	Averaged second burn flaming combustion duration (s)	Flaming drip	UL 94 rating
PA11	4	–	Yes	V-2
15FR_10K_2.5NC_2.5MWNT	2	3	No	V-0
15FR_15K_2.5NC_2.5MWNT	2	2	No	V-0
15FR_20K_2.5NC_2.5MWNT	3	1	No	V-0
15FR_10K_3.5NC_3.5MWNT	3	1	No	V-0
15FR_15K_3.5NC_3.5MWNT	2	1	No	V-0
15FR_20K_3.5NC_3.5MWNT	2	2	No	V-0

Scientific Process 11 Parallel Twin Screw Extruder was used for compounding. PA11 was dried at 80°C for 24 hours prior to processing. The FR additive, elastomer, nanoclay, and MWNTs were used as received. To ensure a homogenous dispersion, each formulation was pre-mixed by physical stir mixing prior to melt-compounding. The extruded formulations were made into small pellets, air cooled, then dried at 80°C for 24 hours before injection-molding.

Table 8.8 summarizes the UL 94 test results for all the FR nanocomposite PA11 formulations. All formulations passed the V-0 rating, which means they all exhibit nondrip and self-extinguishing properties. Figure 8.16 shows the posttest samples. All samples show slight swelling at the tip due to the intumescent FR additives. Similar to the MCC results, the difference in elastomer, nanoclay, and MWNT concentrations seems to have negligible effect in the UL 94 tests. The posttest specimens shown in Fig. 8.16 have no difference and the combustion duration for all the nanocomposite formulations are also very close. Compared to the FR PA11 samples in the previous study where no MWNTs were used [75], it is possible that there is a synergistic effect between nanoclay and MWNTs in improving the flame retardancy of PA11. Because replacing nanoclays

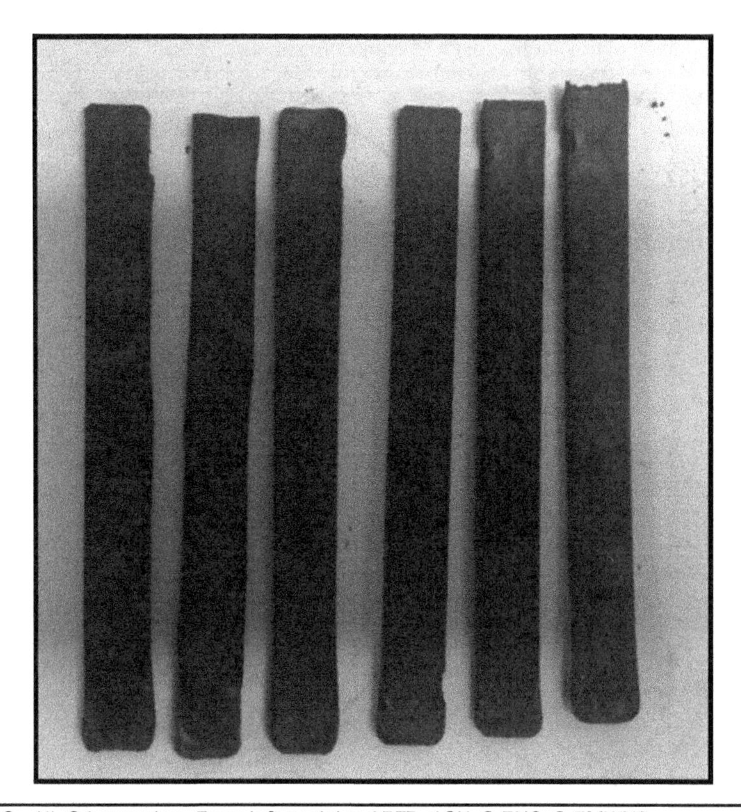

FIGURE 8.16 UL 94 samples. From left to right: 15FR_10K_2.5NC_2.5MWNT, 5FR_15K_2.5NC_2.5MWNT, 15FR_20K_2.5NC_2.5MWNT, 15FR_10K_3.5NC_3.5MWNT, 15FR_15K_3.5NC_3.5MWNT, and 15FR_20K_3.5NC_3.5MWNT [75].

TABLE 8.9 Summary of Tension Test Results for PA11/FR/K/NC/MWNT Blends

Formulation	Tensile strength (MPa)	SD	Modulus (MPa)	SD	Elongation at break (%)	SD
PA11	49	3	1,380	41	164	74
15FR_20K_2.5NC_2.5MWNT	33	1.2	1,460	33.5	30	1.7
15FR_15K_3.5NC_3.5MWNT	33	0.9	1,716	18.2	17	3.3
ALM	46	-	1,345	–	38	–

with MWNTs does not change decomposition and heat release behaviors, it is believed that this synergistic effect comes from the formation of a better barrier effect from the charred surface.

Based on the MCC results, two formulations were tensile tested: 15FR_20K_2.5NC_2.5MWNT and 15FR_15K_3.5NC_3.5MWNT. These two formulations have the lowest heat release capacity out of all the formulations in this study [76]. Table 8.9 summarizes the room temperature mechanical properties. From our previous studies [75], it is known that the main impact of the FR and nanoclay on mechanical

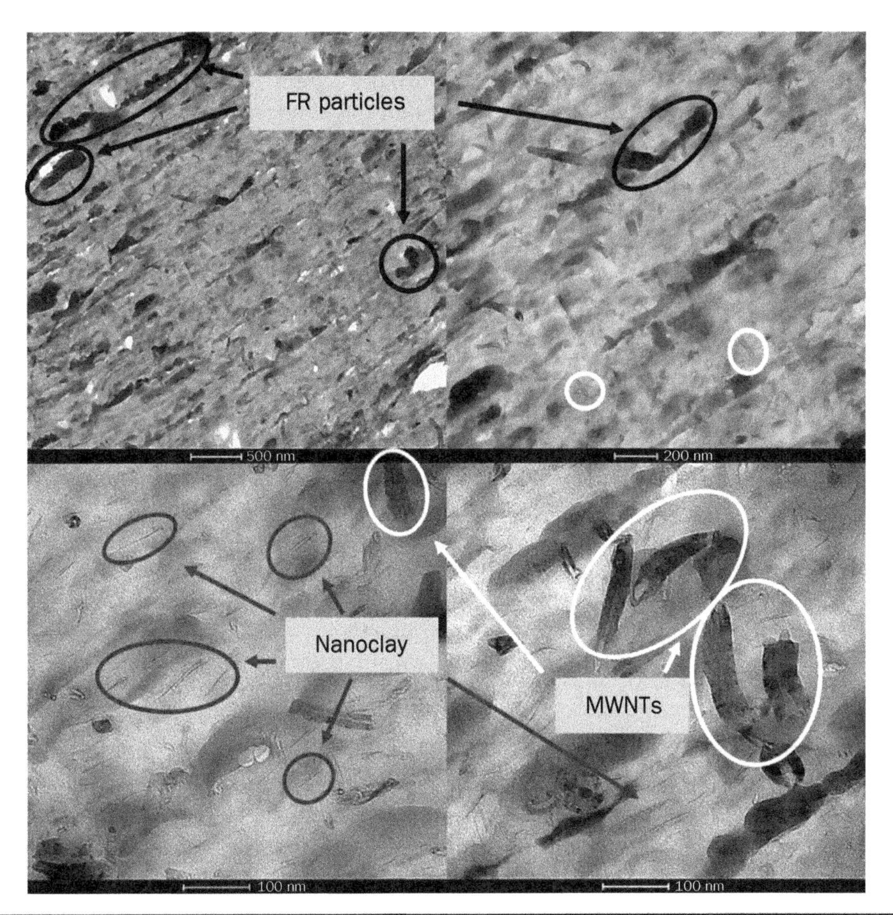

FIGURE 8.17 TEM of formulation #1, containing 15 wt% flame-retardant additive, 10 wt% elastomer, 2.5 wt% nanoclay, and 2.5 wt% carbon nanotube [75].

properties lies in the elongation at break, which is typically decreased by more than 90%. Compared to neat PA11, there is a decrease in the tensile strength for the two formulations tested. PA11 has a tensile strength of 49 MPa and both formulations have a tensile strength of 33 MPa. This is a similar trend observed in our previous studies where the tensile strength decreased with the addition of the FR additive, elastomer, and a nanoparticle. The addition of both nanoclay and MWNTs, however, increased the modulus for both formulations. Although the elongation at break decreased from 164% to 17% and 30%, these values are significantly higher than the 4% that Johnson et al. reported in [72]. According to the company's datasheet, ALM's material has higher tensile strength and elongation at break than the FR PA11 nanocomposite formulations, but lower modulus. A more appropriate comparison between ALMs and our formulations should be made after SLS specimens are made. The summary of tension test results for PA11/FR/K/NC/MWNT blends is presented in Table 8.9.

To examine the dispersion of the nanoparticles, the TEM micrograph of a representative formulation containing both CNT and nanoclay is used (Fig. 8.17) [75]. At lower magnifications, large FR particles can be identified and nanoclays together with

MWNTs are evenly distributed throughout the sample. Higher magnification images clearly show that the MWNTs are individually separated, and the majority of the nanoclay has been exfoliated. This indicates that the nanoparticles have good compatibility with the PA11 matrix and the twin-screw extrusion process achieved a good dispersion of the additives.

In summary, elastomer-toughened FR PA11 nanocomposite formulations with varying elastomer and nanoclay/MWNT concentrations were prepared by twin-screw extrusion. Thermal, flammability, and mechanical properties, and morphological microstructural analysis were performed. Good dispersion of both nanoclay and MWNTs was achieved. The addition of FR additive, nanoclay, and MWNTs improved the thermal stability and flame retardancy properties of the materials. Taking advantage of the synergism between nanoclay and MWNTs, all FR nanocomposite formulations achieved UL 94 V-0 rating. Solid char layers made of FR additive, nanoclay, and MWNTs formed on the surface after combustion. Due to poor interfacial bonding between flame-retardant additive and the PA11 matrix, the tensile strength of the nanocomposite FR PA11 is lower than the neat polymer. However, the FR and nanoclay/MWNTs significantly improved the modulus of the materials. Elongation at break as high as 30% was reported. In this study, no significant difference was found among the FR PA11 formulations throughout all characterizations. Further investigations into the relationship among the added components as well as SLS additive manufacturing process for the selected formulations are needed.

8.3 Flame-Retardant Mechanisms of Polymer Nanocomposites

The mechanism of nanomaterials on the thermal and flammability properties of PNCs has been widely investigated and different mechanisms have been proposed [78], but a comprehensive understanding of this phenomenon is not yet available. The enhancement of the thermal and flammability properties for the PNCs is very dependent on the matrix polymer as well as how the different types of nanomaterials are used to fabricate the PNCs. For polymer-clay nanocomposites, it is hypothesized that migration of the nanoclays to the surface of the matrix polymer occurs during combustion [78, 79]. One mechanism is based on the force of the volatile products produced in the bulk of the polymeric material. The numerous rising bubbles during combustion push the nanoclays upward from the burning area. Another mechanism is based on the recession of polymer from the surface during pyrolysis, leaving the de-wetted nanoclay platelets behind. The decomposition of the organic modified in the clay will lead to aggregation of the clay. The migration of clay onto the surface was also confirmed by XRD and attenuated total reflectance Fourier transform infrared (ATR-FTIR) measurements on the isothermally heated samples [80]. As observed, a minimum amount of clay platelets are needed to be present in order to have an effective char formation on the surface of the PNC.

It is proposed that the accumulation of clay at the surface of the degraded polymer acts as a barrier (ceramic shield) to both mass transport of degradation productions and thermal transfer of energy from the heat source to the polymer underneath the clay platelets. Figure 8.18 shows a schematic of the mechanism on how the clay barrier forms [78]. This clay platelet barrier slows the rate of mass loss and will in turn lower the HRR in a cone calorimeter experiment. The barrier function of the clay platelets can provide thermal insulation for the condensed phase and thus increased

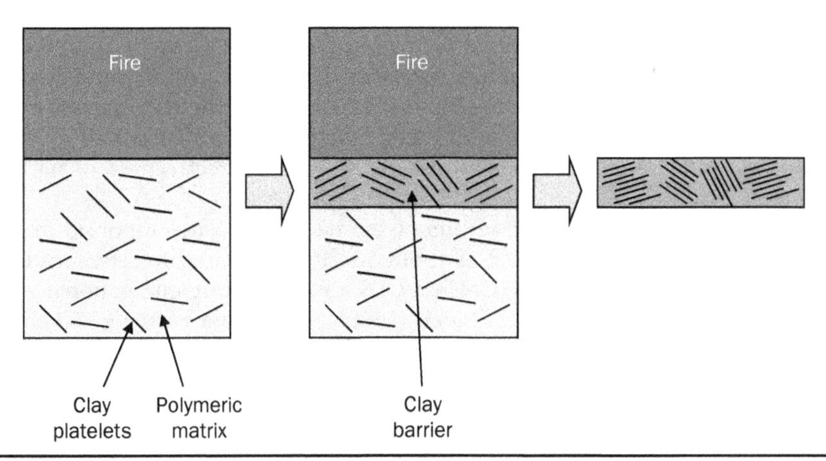

Figure 8.18 Schematic showing the mechanism of how the clay barrier forms.

the thermal stability of the matrix polymer. No char formation was observed by burning the polyamide 6, but a significant amount of char was formed for polyamide 6 with 2 and 5 wt% clay. If the char is continuous, it will be more effective in enhancing flame retardancy [81].

The protective mechanism for CNFs and CNTs is similar, since they are nanoscale-diameter carbon fibers, but they are different from MMT nanoclay platelets. The carbonaceous residues from CNFs and CNTs form a continuous carbon network, as a result the CNFs and CNTs need to be uniformly distributed in the PNC. As the polymer degrades, a carbon network is formed. As observed, a minimum amount of carbon network needs to be present in order to have an effective char formation on the surface of the PNC. Formation of tougher char layer during the combustion process can also be observed [81] for PP/MWNTs and is essential to enhance the flammability properties of the PNCs. As the surface of the matrix polymer is the primary pathway by which the nanomaterials can escape from the polymer, crack-free and carbonaceous network barriers can significantly shield the heat and oxygen from the matrix polymer. This mechanism will reduce the HRR by slowing down the combustion process [81].

Fina et al. prepared an excellent article describing the thermal behavior of nanocomposites and their fire test performance [82]. They addressed the mechanisms of thermal evolution of nanocomposites occurring during combustion, reviewed the present state of the art, and provided a critical discussion of experimental results from different widely used fire tests. The mechanisms involved in PNC's flame retardancy is based on (a) *physical effects*—formation of a ceramic protective floccules on the sample surface acting as a thermal insulation layer, and (b) *chemical effects*—formation of a carbonaceous char catalyzed by the nanofiller during combustion.

Combustion of PNCs in different testing scenarios was examined by Fina and colleagues [82, 83]. At present, the three different fire tests used extensively are vertical UL 94, LOI, and cone calorimeter. UL 94 and LOI are generally referred to as flammability test, in terms of the capability of the materials to ignite and to self-sustain a flame, thus representing a scenario in which the material is at the fire's origin. The cone calorimeter test is representing a forced combustion, in which the material is forced to burn under controlled heat flux. This test addresses the ignition time, the rate of combustion, and total heat released. These flammability and combustion tests also differ for the specimen positioning. With PNCs, the differences in performance obtained in the flammability and forced combustion test are usually very significant, having caused ongoing discussions on the actual effectiveness of nanoparticles or flame-retardants.

Typical cone calorimeter results of the nanocomposites reported in the literature show a decrease in peak HRR of about 50–70% compared to reference unfilled polymers, with either nanoclays, CNTs, POSS, or other inorganic nanoparticles [83]. The general observation is that the presence of dispersed nanoparticles switches the typical non-charring behavior of most thermoplastic polymers to that of charring materials to develop a surface protective layer when exposed to heat. This often leads to the formation of a solid residue at the end of the test, with variable degree of compactness. The compactness of the residue during burning is generally related to the efficiency in HRR reduction: the higher the compactness, the lower HRR. The simple comparison of the peak HRR values to quantitatively assess relative fire risk of virgin polymer and corresponding PNC may be misleading. In most cases, the reduction of combustion rate of polymer matrix in PNCs is explained with the barrier effect on nanoparticle and with polymer char accumulation on the burning surface by mechanisms described previously in this section.

Among flammability tests, the most widely used are the UL 94 and LOI tests. UL 94 was originally developed for devices and appliances. From the scientific point of view, there is little knowledge on the phenomenological behavior of material in the simple pass/fail result. To better describe the flaming behavior during the test, such as the height of the flames, the burning rate, and the end-of-test residue, UL 94 and LOI tests should be used together. Complementary information can be obtained via LOI test, which represents the opposite configuration compared to vertical UL 94—the specimen being ignited from the top. Relatively few papers reported LOI test for PNCs, generally showing either a decrease or no significant increase in the LOI values compared to their reference polymers. The LOI setup appears to be critical for the production of an effective ceramic protective layer on the material surface, as the specimen below the flame front is not licked by the flame and there is no external irradiation. This prevents the formation of the protective char-ceramic surface layer on the material prior to being reached by the flame front.

Nanocomposites thus take a step forward toward reduction of fire risk and hazard for polymers because they avoid flame spreading by flaming dripping and reduce the rate of combustion. For the first time in polymer FR records, we have polymeric materials available in which fire risk and hazard are reduced with simultaneous improvement of other polymer material properties and using an environmentally friendly technology. Most of the flame-retardants used now have to be loaded into polymers in relatively large amount (10–70 wt%), with negative effects on polymer physical and mechanical properties and on environmental issues.

8.4 Concluding Remarks and Trends of Polymer Nanocomposites

Nanotechnology has a very bright future to improve the flame retardancy of polymeric materials. Although the nanomaterial by itself cannot achieve the fire properties required in industrial applications, the presence of the nanomaterial will enhance other desirable properties, such as mechanical, barrier, or electrical properties, and in combination with other conventional flame-retardants can provide better FR solutions. Finding new ways of applying nanomaterials cost-effectively in polymers with desired properties, such as thermal and flammability properties, has been of increasing interest and is very challenging. Adaption of nanomaterials in polymers to promote the thermal and flame retardancy as well as other desired properties is warranted [84].

To promote the thermal and flame retardancy, and multifunctional properties of the polymer, one trend is to combine two or more nanomaterials into the polymer matrix to achieve a PNC that will consist of multiple components. If synergism exists between these nanomaterials, the resulting PNC will exhibit better performance. Currently, MMT clay is considered as one component of the combined nanomaterials due to its low cost, wider availability, and its comprehensive performance. Of increasing interest is the use of HNTs and synthetic clays, such as fluorinated synthetic mica, magadiite, and layered double hydroxides (LDHs), to fabricate PNCs. However, the high cost and limited sources of the synthetic clays have been an obstacle for the wide usage of these nanomaterials.

Another trend is the use of other conventional FR additives, which are micrometer-sized, to enhance the flame retardancy of the polymer matrix, such as using an additive that can form a glassy barrier to seal the crack in the char, resulting in a self-extinguishing PNCs. One common and practical way is to use nanomaterials in the presence of conventional flame-retardants, such as halogen-, or phosphorus-based, or a nano-sized component in a conventional FR system to achieve a FR polymeric composite. This includes nanoparticles, such as nano-$Mg(OH)_2$ and nano-$Al(OH)_3$, obtained from conventional inorganic FR additives that have original particle size in micrometers. Again, clay will be of great interest because of its excellent performance, such as the antidripping and charring properties, and/or the synergism with many conventional FR additives.

Other nanomaterials, such as CNFs, CNTs, POSS, NGPs, and alumina nanofibers, although expensive when comparing with MMT clay when used alone in the polymer matrix, can also be combined with conventional FR system to reduce the raw material cost and to improve the flame retardancy of the resulting PNCs.

Due to the thermal instability of the organic modifier of the nanoclays, it is most likely that advances in the future will occur with advanced polymers that can be processed below about 200°C. As new surfactants are developed and commercialized, FR PNCs that can be used for polymers that must be processed at higher temperatures may be developed. As the price of CNFs, CNTs, POSS, and NGPs are coming down, they may be used as the next generation of FR nanomaterials. The PNC does offer many advantages—enhanced barrier, flame, mechanical, and electrical properties—so they will be used in multifunctional materials and the enhanced flame retardancy will be an added "bonus."

Alongi et al. [85] conducted a comprehensive review on current emerging techniques to impart flame retardancy to fabrics. They reviewed the state of the art on the novel and emerging techniques recently developed in the textile field for conferring flame retardancy properties to natural and synthetic fibers. The current emerging techniques reviewed by Alongi et al. are showing very significant potentialities for effectively improving the FR properties of different natural and synthetic fibers/fabrics. They may represent a possible efficient alternative to the conventional FR systems, some of which (e.g., halogenated products) are encountering limitation in use due to their suspected/assessed toxicity and poor eco-sustainability. The in situ assembly of nanocoating on the fabric surface through nanoparticle adsorption, layer by layer deposition, sol-gel, and dual-cure processes, or plasma treatments has clearly demonstrated that nanotechnology represents a valued tool for the textile field. The unusual use of proteins and nucleic acids may overcome the current limitations of flame-retardants toward novel, sustainable, and green products.

Industrial development of PNCs seems now underway because since the appearance of the first clay-based nanocomposites two decades ago, a wide selection of nanofillers has become available on a commercial basis. A parallel, extensive worldwide research effort has strongly improved our knowledge on preparation and characterization of PNCs, now available for industrial exploitation.

8.5 Study Questions

8.1 How does the nanomaterial dispersion affect the flammability properties of polymer nanocomposites in general? Provide examples to illustrate your claims.

8.2 How does the flammability properties of polymer-clay nanocomposites influenced by the type of clay and surface treatment of the clay, and how clay is processed in the polymer matrix? Provide examples.

8.3 Enhancement of flammability properties of thermoplastic-based nanocomposites using one-, two-, and three-dimensional-scale nanomaterials were included in this chapter. Which types of nanomaterials are more effective for thermoplastic-based nanocomposites? Provide examples and summarize the flammability properties of these thermoplastic-based nanocomposites.

8.4 Enhancement of flammability properties of thermoplastic elastomer–based nanocomposites using one-, two-, and three-dimensional-scale nanomaterials were included in this chapter. Which type of nanomaterials is more effective for thermoplastic elastomer–based nanocomposites? Provide examples and summarize the flammability properties of these thermoplastic elastomer–based nanocomposites.

8.5 Enhancement of flammability properties of thermoset-based nanocomposites using one-, two-, and three-dimensional-scale nanomaterials were included in this chapter. Which type of nanomaterials is more effective for thermoset-based nanocomposites? Provide examples and summarize the flammability properties of these thermoset-based nanocomposites.

8.6 What will be the effects of these multicomponent FR systems with two or more nanomaterials and conventional FR additive systems? What will be a good strategy to optimize the flammability properties in these polymer nanocomposites? Summarize the flammability properties of these multicomponent FR nanocomposite systems.

8.7 What are the flame-retardant protective mechanism for these one-, two-, and three-dimensional-scale nanomaterials, as well as the hybrid FR systems? Summarize and compare them.

8.8 Are there any correlations or observed trends when these polymer nanocomposites were evaluated with different flammability test methods?

8.6 References

1. Pinnavaia, T. J., and Beall, G. W., eds. (2000) *Polymer-Clay Nanocomposites*. West Sussex, England: John Wiley & Sons.
2. Utracki, L. A. (2004) *Clay-Containing Polymeric Nanocomposites*. Shropshire, England: Rapra Technology Limited.
3. Bhattacharya, S.N., Gupta, R. K., and Kamal, M. R. (2007) *Polymeric Nanocomposites—Theory and Practice*. Düsseldorf: Carl Hanser Verlag.
4. Gilman, J.W., Kashiwagi, T., and Lichtenhan, J. D. (1997) Nanocomposites: a revolutionary new flame retardant approach. *SAMPE J.*, 33(4):40.
5. Patel, H. A., Bocchini, S., Frache, A., and Camino, G. (2010) Platinum nanoparticle intercalated montmorillonite to enhance the char formation of polyamide 6 nanocomposites. *J. Mater. Chem.*, 20:9550–9558.
6. Lao, S.C., Wu, C., Moon, T. J., Koo, J. H., Morgan, A., Pilato, L. A., and Wissler, G. (2009) Flame-retardant polyamide 11 and 12 nanocomposites: thermal and flammability properties." *Journal of Composite Materials*, 43(17):1803.
7. Lao, S.C., Yong, W., Nguyen, K., Moon, T. J., Koo, J. H., Pilato, L. A., and Wissler, G. (2010) Flame-retardant polyamide 11 and 12 nanocomposites: processing, morphology and mechanical properties." *Journal of Composite Materials*, 44(25):2933–2951.
8. Lao, S.C., Koo, J. H., Moon, T. J., Londa, M., Ibeh, C. C., Wissler, G. E., and Pilato, L. (2011) Flame-retardant polyamide 11 nanocomposites: further thermal and flammability studies. *Journal of Fire Sciences*, 29. doi: 10.1177/0734904111404658.
9. Katsoulis, C. K., Kandare, E., and Kandola, B. K. (2009) Thermal and fire performance of flame-retarded epoxy resin: investigating interaction between resorcinol bis (diphenyl phosphate) and epoxy nanocomposites. In *Fire Retardancy of Polymers-New Strategies and Mechanisms*. Hull, T. R., and Kandola, B. K., eds. Cambridge, England: The Royal Society of Chemistry, pp. 184–205.
10. Zammarano, M. (2007) Thermoset fire retardant nanocomposites. In *Flame Retardant Polymer Nanocomposites*. Hoboken, NJ: John Wiley & Sons, pp. 235–285.
11. Wang, Z., and Pinnavaia, T. J. (1998) Hybrid organic-inorganic nanocomposites: exfoliation of magadiite nanoclays in an elastomeric epoxy. *Polymer*, 10:1820–1826.
12. Triantafillidis, C.S., LeBaron, P. C., and Pinnavaia, T. J. (2002) Thermoset epoxy-clay nanocomposites: the dual role of alpha, omega-diamines as clay surface modifiers and polymer curing agents." *J. Solid State Chem.*, 167:354–362.
13. Camino, G., Tartaglion, G., Frache, A., Manferti, C., and Costa, G. (2005) Thermal and combustion behavior of layered silicate-epoxy nanocomposites. *Polym. Degrad. Stab.*, 90:354–362.
14. Hussain, M., Varley, R. J., Mathys, Z., Cheng, Y. B., and Simon, G. P. (2004) "Effect of organophosphorus and nano-clay materials on the thermal and fire performance of epoxy resins." *J. Appl. Polym. Sci.*, 91:1233–1253.
15. Hartwig, A., Purtz, D., Schartel, B., Bartholmai, M., and Wenschuh-Josties, M. (2003) Combustion behavior of epoxide based nanocomposites with ammonium and phosphonium bentonites. *Maromol. Chem. Phys.*, 204:2247–2257.
16. Gilman, J. W., Kashiwagi, T., Nyden, M., Brown, J. E. T., Jackson, C. L., Lomakin, S., Giannelis, E. P., et al. (1990) Flammability studies of polymer-layered silicate nanocomposites: polyolefin, epoxy, and vinyl ester resins. In *Chemistry and Technology of Polymer Additives*. Ak-Malaika, S., Colovoy, A., and Wilkie, C. A., eds. Malden, MA: Blackwell Science, pp. 249–265.

17. Koo, J.H., Nguyen, K., Lee, J. C., Ho, W. K., Bruns, M. C., and Ezekoye, O. A. (2010) Flammability studies of a novel class of thermoplastic elastomer nanocomposites. *Journal of Fire Sciences*, 28(1):49–85.
18. Koo, J. H., Ezekoye, O. A., Lee, J. C., Ho, W. K., and Bruns, M. C. (2011) Rubber-clay nanocomposites based on thermoplastic elastomers. In *Rubber Clay Nanocomposites*. Galimberti, M., ed. New York, NY: John Wiley & Sons, pp. 489–521.
19. Avila, A.F., Dias, E. C., Lopes da Cruz, D. T., Yoshida, M. I., Bracarense, A. Q., Carvalho, M. G. R., and Avila, J. (2010) An investigation on graphene and nanoclay effects on hybrid nanocomposites post fire dynamic behavior. *Mater. Res*, 13:143–150.
20. Koo, J. H., Leo, H., Clay, W., and Conaway, J. (2011) Methodology to evaluate epoxy nanocomposites for fire protection application. *AIAA Paper No. 2011-1799*, Reston, VA, 2011.
21. Guo, Y., Bao, C., Song, L., Yuan, B., and Hu, Y. (2011) In situ polymerization of graphene, graphite oxide, and functionalized graphite oxide into epoxy resin and comparison study of on-the-flame behavior. *Ind. Eng. Chem. Res.*, 50: 7772–7783.
22. Wang, Z., Tang, X. Z., Yu, Z. Z., Guo, P., Song, H. H., and Du, X. S. (2011) Dispersion of graphene oxide and its flame retardancy effect on epoxy nanocomposites. *Chinese J. Polym. Sci.*, 3:368–376.
23. Higginbotham, A. L., Lomdea, J. R., Morgan, A. B., and Tour, J. M. (2009) Graphite oxide flame-retardant polymer nanocomposites. *Appl. Mater Interfaces*, 1:2256–2261.
24. Kashiwagi, T., Du, F., Winey, K. I., Groth, K. M., Shield, J. R., Bellayer, S. P., Kim, H., and Douglas, J. F. (2005) Flammability properties of polymer nanocomposites with single-walled carbon nanotubes: effects of nanotube dispersion and concentration. *Polymer*, 46:471–481.
25. Kashiwagi, T. (2007) Progress in flammability studies of nanocomposites with new types of nanoparticles. In *Flame Retardant Polymer Nanocomposites*. Morgan, A. B., and Wilkie, C. A., eds. Hoboken, NJ: John Wiley & Sons, pp. 285–324.
26. Fina, A., Tabuani, D., Frache, A., and Camino, G. (2005) Polypropylene-polyhedral oligomeric silsesquioxanes (POSS) nanocomposites. *Polymer*, 46:7855–7866.
27. Bocchini, S., Frache, A., Camino, G., and Claes, M. (2007) Polyethylene thermal oxidative stabilization in carbon nanotubes based nanocomposites. *European Polymer Journal*, 43:3222–3235.
28. Butler, S., Kim, G., Koo, J. H., et al. (2011) Polyamide 11-halloysite nanotube nanocomposites: mechanical, thermal, and flammability characterization. Paper presented at *2011 SAMPE ISTC*, SAMPE, Covina, CA.
29. Lao, S. C., Kan, M. F., Lam, C. K., Koo, J. H., Moon, T., Londa M., Takatsuka T. E., et al. (2010) Polyamide 11-carbon nanotubes nanocomposites: processing, morphological, and property characterization. Paper presented at *2010 Solid Freeform Fabrication Symposium*, the University of Texas at Austin, Austin, TX.
30. Hao, A., Wong, I., Wu, H., Lisco, B., Ong, B., Sallean, A., Butler, S., et al. (2014) Mechanical, thermal, and flame-retardant performance of polyamide 11-halloysite nanotube nanocomposites. *Journal of Material Science*, doi: 10.1007/s10853-014-8575-7.
31. Landry, C.J.T., Coltrain, B. K., Landry, M. R., Fitzgerald, J. J., and Long, V. K. (1993) Poly(vinyl acetate) silica filled materials: material properties of in-situ vs. fumed silica particles. *Macromolecules*, 26: 3702–3712.
32. Hajji, P., David, L., Gerard, J. F., Pascault, J. P., and Vigier, G. (1999) Synthesis, structure, and morphology of polymer-silica hybrid nanocomposites based on hydroxyethyl methacrylate. *J. Polym. Sci. B.*, 37:3172–3187.

33. Ou, Y., Yang, F., and Yu, Z. Z. (1998) New conception on the toughness of nylon 6/silica nanocomposite prepared via in situ polymerization. *J. Polym. Sci. B.*, 36: 789–795.
34. Reynaud, E., Jouen, T., Gauthier, C., Vigier, G., and Varlet, J. (2001) Nanofiller in polymeric matrix: a study on silica reinforced PA6. *Polymer*, 42: 8759–8768.
35. Hsiue, G. H., Kuo, W. J., Huang, Y. P., and Jeng, R. J. (2000) Microstructural and morphological characteristics of PS-SiO_2 nanocomposites. *Polymer*, 41:2813–2825.
36. Liu, Y. L., Hsu, C. Y., Wei, W. L., and Jeng, R. J. (2003) Preparation and thermal properties of epoxy-silica nanocomposites from nanoscale colloidal silica. *Polymer*, 44: 5159–5167.
37. Kashiwagi, T., Gilman, J. W., Butler, K. M., Harris, R. H., and Shields, J. R. (2000) Flame retardant mechanism of silica gel/silica. *Fire Mater.*, 24(6): 277–289.
38. Kashiwagi, T., Shields, J. R., Harris, R. H., and Davis, R. D. (2003) Flame-retardant mechanism of silica: effect of resin molecular weight. *J. Appl. Polym. Sci.*, 87:1541–1553.
39. Yang, F., and Nelson, G. L. (2004) PMMA/silica nanocomposite studies: synthesis and properties. *J. Appl. Polym. Sci.*, 91:3844–3850.
40. Kashiwagi, T., Morgan, A. B., Antonucci, J. M., Van Landingham, M. R., Harris, R. H., Awad, W. H., and Shields, J. R. (2003) Thermal and flammability properties of a silica-poly(methyl methacrylate) nanocomposite. *J. Appl. Polym. Sci.*, 89:2072–2078.
41. Yang, F., Yngard, R., and Nelson, G. L. (2005) Flammability of polymer-clay and polymer-silica nanocomposites. *J. Fire Sci.*, 23: 209–226.
42. Lao, S. C., Koo, J. H., Moon, T. J., Hadisujoto, B., Yong, W., Pilato, L., and Wissler, G. (2009) Flammability and thermal properties of polyamide 11-alumia nanocomposites. Paper presented at *2009 Solid Freeform Fabrication Symposium*, University of Texas at Austin, Austin, TX.
43. Fahy, W. P., Wu, H., Koo, J. H., Kim, S., and Kim, H. (2018) Flame retardant polyamide 11 and alumina nanocomposites for additive manufacturing. *Proc. 2018 SAMPE ISTC*, Long Beach, CA, May 21–24.
44. Rallini, M., Monti, M., Natali, M., Kenny, J. M., and Torre, L. (2011) Alumina nanoparticles as a filler of carbon fibre/epoxy composites for improved fire resistance. *Proc. 2011 SAMPE ISTC*, SAMPE. Covina, CA.
45. Kalfus, J., and Jancar, J. (2010) Effect of particle size on the thermal stability and flammability of $Mg(OH)_2$/EVA nanocomposites. *Comp. Interfaces*, 17:689–703.
46. Vesely, K., Rychly, J., Kummer, M., and Jancar, J. (1990) Flammability of highly filled polyolefins. *Polym. Deg. Stab.*, 30: 101–105.
47. Rychly, J., Vesely, K., Gal, E., Kummer, M., Jancar, J., and Rychla, L. (1990) Use of thermal methods in the characterization of the high-temperature decomposition and ignition of polyolefins and EVA copolymers filled with Mg(OH)2, Al(OH)3, and CaCO3. *Polym. Deg. Stab.*, 30: 57–62.
48. Gui, H., Zhang, X. H., Gao, J. M., Dong, W. F., Song, Z. H., Lai, J. M., and Qiao, J. L. (2007) An EVA/unmodified nano-magnesium hydroxide/silicone rubber nanocomposite with synergistic flame retardancy. *Chinese J. Polym. Sci.*, 25: 437–440.
49. Ly, J. P., and Liu, W. H. (2007) Flame retardancy and mechanical properties of EVA nanocomposites based on magnesium hydroxide nanoparticles/microcapsulated red phosphorus. *J. Appl. Polym Sci.*, 105:333–340.
50. Mishra, S., Sonawane, S. H., Singh, R. P., Bendale, A., and Patil, K. (2004) Effect of nano-Mg(OH)2 on the mechanical and flame-retarding properties of polypropylene composites. *J. Appl. Polym. Sci.*, 94: 116–122.

51. Zhang, Q., Tian, M., Wu, Y., Lin, G., and Zhang, L. (2004) Effect of particle size on the properties of Mg(OH)2-filled rubber composites. *J. Appl. Polym. Sci.*, 94: 2341–2346.
52. Song, G., Ma, S., Tang, G., Yin, Z., and Wang, X. (2010) Preparation and characterization of flame retardant form-stable phase change materials composed by EPDM, paraffin and nano magnesium hydroxide. *Energy*, 35: 2179–2183.
53. Cao, H., Zheng, H., Yin, J., Lu, Y., Wu, S., Wu, X., and Li, B. (2010) Mg(OH)2 complex nanostructures with superhydrophobicity and flame retardant effects. *J. Phys. Chem. C*, 114: 17362–17368.
54. Patil, C.B., Kapadi, U. R., Hundiwale, D. G., and Mahulikar, P. P. (2008) Effect of nano-magnesium hydroxide on mechanical and flame-retarding properties of SBR and PBR: a comparative study. *Polym. Plast. Tech. Eng.*, 47: 1174–1178.
55. Suihkonen, R., Nevalainen, K., Orell, O., Honkanen, M., Tang, L., Zhang, H., Zhang, Z., et al. (2012) Performance of epoxy filled with nano- and micro-sized magnesium hydroxide. *J. Mater. Sci.*, 47(3):1480–1488. doi: 10.1007/s10853-011-5933-6.
56. Hybrid Plastics, Inc. "Home page." Last modified 2010. http://www.hybridplastics.com.
57. Mantz, R. A., Jones, P. F., Chaffee, K. P., Lichtenhan, J. D., Gilman, J. W., Ismail, I. M. K., and Burneister, M. J. (1996) Thermolysis of polyhedral oligomeric silsesquioxane (POSS) macromers and POSS-siloxane copolymers. *Chem. Mater.*, 8:1250–1259.
58. Schwab, J. J., and Lichtenhan, J. D. (1998) Polyhedral oligomeric silsesquioxane (POSS)-based polymers. *Appl. Orgamet. Chem.*, 12:707–713.
59. Kashiwagi, T., and Gilman, J. W. (2000) Silicon-based flame retardants. In *Fire Retardancy of Polymeric Materials*. Grand, A. F., and Wilkie, C. A., eds. New York: Marcel Dekker, pp. 353–389.
60. Baldi, F., Bignotti, F., Fina, A., Tabuani, D., and Ricco, T. (2007) Mechanical characterization of polyhedral oligomeric silsesquioxane/polypropylene blends. *Journal of Applied Polymer Science*, 105:935–943.
61. Fina, A., Bocchini, S., and Camino, G. (2008) Catalytic fire retardant nanocomposites. *Polymer Degradation and Stability*, 93:1647–1655.
62. Monticelli, O., Fina, A., Ullah, A., and Waghmare, P. (2009) Preparation, characterization, and properties of novel PSMA-POSS systems by reactive blending. *Macromolecules*, 42:6614–6623.
63. Fina, A., Tabuani, D., Peijs, T., and Camino, G. (2009) POSS grafting on PPgMA by on-step reactive blending. *Polymer*, 50: 218–226.
64. Fina, A., Monticelli, O., and Camino, G. (2010) POSS-based hybrids by melt/reactive blending. *Journal of Materials Chemistry*, 20:9297–9305.
65. Herbert, M. J., and Brown, S. C. (1992) New developments in ATH technology and applications. Paper presented at *Flame Retardants '92 Conference*, London, January 12–13, pp. 100–119.
66. Beyer, G. (2007) Flame retardant properties of organoclays and carbon nanotubes and their combinations with alumina tri hydrate. In *Flame Retardant Polymer Nanocomposites*. Morgan, A. B., and Wilkie, C. A., eds. Hoboken, NJ: John Wiley & Sons, pp. 163–190.
67. Beyer, G. (2005) Flame retardancy of nanocomposites: from research to technical products. *J. Fire Sci.*, 23:75–87.
68. Beyer, G. (2002) Carbon nanotubes as flame retardants for polymers. *Fire Mater.*, 26:291–293.

69. Beyer, G. (2002) Improvements of the fire performance of nanocomposites. Paper presented at *13th Annual BCC Conference on Flame Retardancy for Polymers*, Stamford, CT, June 3–6.
70. Beyer, G. (2005) Filler blend of carbon nanotubes and organoclays with improved char as a few flame retardant system for polymers and cable application. *Fire Mater.*, 29: 61–69.
71. Beyer, G., Gao, F., and Yuan, Q. (2005) A mechanistic study of fire retardancy of carbon nanotube/ethylene vinyl acetate copolymers and their clay composites. *Polym. Degrad. Stab.*, 89: 559–564.
72. Johnson, B., Allcorn, E., Baek, M. G., and Koo, J. H. (2011) Combined effects of montmorillonite clay, carbon nanofiber, and fire retardant on mechanical and flammability properties of polyamide 11 nanocomposites. Paper presented at *2011 SAMPE ISTC*, SAMPE, Covina, CA.
73. Gao, F., Beyer, G., and Yuan, Q. (2005) A mechanistic study of fire retardancy of carbon nanotube/ethylene vinyl acetate copolymers and their clay composites. *Polymer Degradation and Stability*, 89(3): 559–564.
74. Peeterbroeck, S., Alexandre, M., Nagy, J., Pirlot, C., Fonseca, A., Moreau, N., Philippin, G., et al. (2004) Polymer-layered silicate–carbon nanotube nanocomposites: unique nanofiller synergistic effect. *Composites Science and Technology*, 64(15):2317–2323.
75. Wu, H., Ortiz, R., and Koo, J. H. (2018) Rubber toughened flame retardant (FR) polyamide 11 nanocomposites. Part 1: the effect of SEBS-g-MA elastomer and nanoclay. *Flame Retardancy and Thermal Stability of Materials*, 1(1):25–38.
76. Wu, H., Ortiz, R., and Koo, J. H. (2019) Rubber toughened flame retardant (FR) polyamide 11 nanocomposites. Part 2: synergy between multi-walled carbon nanotube (MWNT) and MMT nanoclay. *Flame Retardancy and Thermal Stability of Materials*, accepted.
77. Wu, H., Ortiz, R., and Koo, J. H. (2018) self-extinguishing and non-drip flame retardant polyamide 6 nanocomposite: mechanical, thermal, and combustion behavior. *Flame Retardancy and Thermal Stability of Materials*, 1(1):1–13.
78. Gilman, J. W. (2007) Flame retardant mechanism of polymer-clay nanocomposites. In *Flame Retardant Polymer Nanocomposites*. Morgan, A. B., and Wilkie, C. A., eds. Hoboken, NJ: John Wiley & Sons, pp. 67–87.
79. Kashiwagi, T., Harris, R. H., Zhang, X., Briber, R. M., Cipriano, B. H., Raghavan, S. R., Awad, W. H., et al. (2004) Flame retardant mechanism of polyamide 6-clay nanocomposites. *Polymer*, 45:881–891.
80. Lewin, M., et al. (2005) Nanocomposites at elevated temperatures: migration and structural changes. *Polym. Adv. Tech.*, 4:928.
81. Kashiwagi, T., Du, F., Winey, K. I., Groth, K. M., Shields, J. R., Bellayer, S. P., Kim, H., et al. (2005) Flammability properties of polymer nanocomposites with single-walled carbon nanotubes: effects of nanotube dispersion and concentration. *Polymer*, 46:471–481.
82. Fina, A., Bocchini, S., and Camino, G. (2009) Thermal behavior of nanocomposites and fire testing performance. In *Fire and Polymers*. Wilkie, C. A., Morgan, A. B., and Nelson, G. L., eds. Washington, DC: American Chemical Society, pp. 10–24.
83. Fina, A., Canta, F., Castrovinci, A., and Camino, G. (2009) Significant assessment of nanocomposites' combustion behaviour by the appropriate use of the cone calorimeter. In *Fire Retardancy of Polymer—New Strategies and Mechanism*. Hull, T. R., and Kandola, B. K., eds. Cambridge, UK: RSC Publishing, pp. 147–159.

84. Morgan, A. B., and Wilkie, C. A. (2007) Practical issues and future trends in polymer nanocomposite flammability research. In *Flame Retardant Polymer Nanocomposites*. Morgan, A. B., and Wilkie, C. A., eds. Hoboken, NJ: John Wiley & Sons, pp. 355–399.
85. Alongi, J., Carosio, F., and Malucelli, G. (2013) Current emerging techniques to impart flame retardancy to fabrics: an overview. *Polymer Degradation and Stability*, doi: dx.doi.org/10.1016/j.polymdegradstab.2013.07.012.

CHAPTER 9

Ablation Properties of Polymer Nanocomposites

9.1 Introduction

A comprehensive review was conducted on the research and development (R&D) of polymer nanocomposites used as thermal protection system (TPS) for reentry vehicles as well as nozzle and internal insulation materials for propulsion systems. This chapter summarizes the significant R&D efforts on the studies of nanostructured ablative materials by scientists from China, India, Italy, Iran, and the United States for the past decade. Thermoset, thermoplastics, elastomers, and thermoplastic elastomers were used as polymer matrices. Nanomaterials, such as montmorillonite (MMT) organoclays, carbon nanofibers (CNFs), polyhedral oligomeric silsesquioxanes (POSS®), nanosilicas (*n*-silicas), nano-alumina (*n*-alumina), multiwalled carbon nanotubes (MWNTs), and nano-ceramics (*n*-ceramics) as well as classical fillers, such as carbon black (CB) and ceramics, were incorporated into the different resin matrices as nano- or micro-sized fillers. Conventional fibers, such as asbestos, carbon, glass, silica, and quartz were used as reinforcements. Ablation mechanisms were studied and proposed by the researchers on their nanostructured ablatives.

Thermal protection materials are required to shield structural components of space vehicles during the reentry stage, solid rocket motors (SRMs), liquid rocket engines, and missile launching systems. A thorough literature survey was conducted by Koo and colleagues [1–4] to review the development of these thermal protection materials for different military and aerospace applications. The series of literature surveys were grouped into (a) numerical modeling [1], (b) materials thermophysical properties characterization [2], (c) experimental testing [3], and (d) polymer nanocomposites [4].

The introduction of inorganic nanomaterials as additives into polymer matrix systems has resulted in polymer nanocomposites (PNCs) exhibiting multifunctional, high-performance polymer characteristics beyond what traditional polymer composites possess [5–13]. Multifunctional features attributable to PNCs consist of improved thermal resistance, flame retardancy, ablation resistance, moisture resistance, decreased permeability, chemical resistance, charge dissipation, and electrical/thermal conductivity [5–13]. Through control/adjustment of the additive at the nanoscale level, one is able to maximize property enhancement of selected polymer systems to meet or exceed the requirements of current military, aerospace, and commercial applications.

One of the most important and pioneering research studies on the use of the MMT for poly(caprolactam) (nylon 6) was performed by Vaia et al. [14] in 1999 to 2 wt% and 5 wt% loaded layered silicate/poly(caprolactam) nanocomposite. These materials were synthesized by the in situ polymerization of ε-caprolactam in the presence of dispersed organically modified montmorillonite (OMMT) nanoclay. This concept was further developed by many researchers worldwide using different types of polymers, such as nylon, phenolic, elastomer, and thermoplastic elastomer, with different types of nanomaterials, such as nanoclay, POSS, CNF, multiwalled carbon nanotube (MWNT), nano-alumina, nanosilica, nano-graphene, nano-kaolinite, nano-ceramic, and others. Traditional reinforcements, such as asbestos, carbon, glass, Kevlar, phenolic, silica, quartz were also used. This chapter reviews the open literature on how researchers processed, characterized, and tested their PNCs as thermal protection materials for low- and high-energy hyperthermal environments.

9.2 Behavior of Thermal Protection Materials

Polymeric composites have been used as ablative TPS for a variety of military and aerospace applications. Thermal protection materials, such as carbon-phenolic (C-Ph) and carbon-carbon composites are widely used for spacecraft heat shields during the reentry stage, thermal protection of missile launching system, and in SRMs as nozzle materials. Thermosetting polymer matrix composites are constructed of layers of woven carbon, glass, silica, or quartz fibers impregnated with a resin matrix. Phenolic resin is the most commonly used polymer matrix for ablatives. Phenolic resins undergo facile carbonization into carbon-carbon composites at elevated temperatures. These ablative materials are exposed to a thermochemical flow and subjected to high temperatures in excess of 3,000°C with very high heating rates.

Thermochemical ablation refers to the phenomenon of surface recession of an ablative due to severe thermal attack by an external heat flux. The initial heat transfer into the ablative occurs by pure conduction, and the resulting temperature rise causes material expansion (swelling) which may be attributed to pure thermal expansion as well as vaporization of any traces of moisture and other volatiles generated by resin decomposition. When the material reaches a sufficiently high-temperature, thermochemical degradation or pyrolysis of the polymer matrix begins. The pyrolysis reactions result in the production of decomposition gases and solid carbonaceous char residue. The thermal expansion and the disappearance of solid material due to the decomposition result in an increase in porosity and permeability of the polymeric material. Thus, pyrolysis gases begin to escape through the polymeric material. For the thermal protection to be effective all these reactions must be strongly endothermic. The gases which flow through the char structure remove energy by convention, thus attenuating the conduction of heat to the reaction zone. This phenomenon is depicted in Fig. 9.1 by Laub and Venkatathy [15].

9.3 Polymer Nanocomposite Review

There are only a small number of research groups globally that are investigating PNCs as advanced thermal protection materials:

- China: Beijing University of Chemical Technology, Harbin Institute of Technology, National University of Defense Technology, and Xi'an Jiaotong University
- India: Defence Research and Development Organisation (DRDO) and Indian Institute of Technology (IIT)

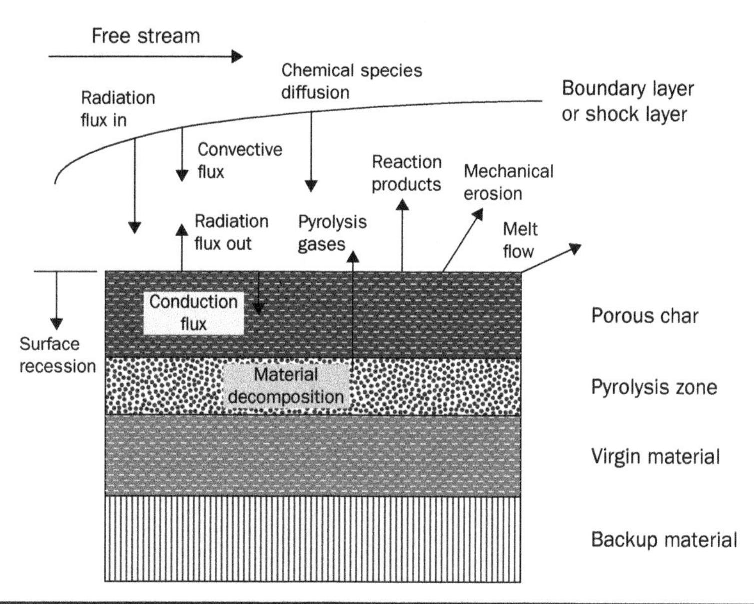

FIGURE 9.1 Energy accommodation mechanisms of ablative TPS materials [15].

- Iran: Islamic Azad University and Tarbiat Modares University
- Italy: University of Perugia, Politecnico di Torino, and University of Rome
- South Korea: Gyeongsan National University
- Pakistan: National University of Sciences and Technology
- United States: Air Force Research Laboratory (AFRL), Mississippi State University, Texas State University-San Marcos, and The University of Texas at Austin

It is more meaningful to subdivide these studies in terms of the types of polymer matrix used in each group's studies. Table 9.1 shows a summary of the different polymer systems, types of nanomaterials, and conventional fiber reinforcements used in their particular applications [4].

9.3.1 Thermoplastic Nanocomposite Studies

Nylon-Clay Nanocomposite Studies by AFRL, United States [Vaia et al. (1999)]

The ablative performance of poly(caprolactam) (nylon 6) nanocomposites was examined by Vaia et al. of Air Force Research Laboratory (ARFL) [14]. A relatively tough, inorganic char forms during the ablation of these nylon-clay nanocomposites resulting in at least an order-of-magnitude decease in mass loss (erosion) rate relative to the neat polymer. This occurs for a small amount of 2 wt% (about 0.8 vol%) exfoliated mica-type layered silicate. The nanoscopic distribution of silicate layers leads to a uniform char layer that enhances the ablative performance. The effect of concentration and size of the silicate layers as well as specific bonds between the polymer and the silicate on degradation kinetics and on the microstructure of the char layer has been well characterized.

Table 9.1 Summary of the Different Polymer Matrix Used by the Various Research Groups Studying Ablation Using Polymer Nanocomposites

Polymer matrix/institution			Nanoadditive/reinforcement	Remarks
A.1	Nylon 6	AFRL, USA	MMT organoclay	Improvement in ablative performance of nanocomposites is associated with the char-forming characteristics of the polymer nanocomposites
A.2	Poly(phenylene oxide)/polystyrene (PPO/PS)	Politecnico di Torino, Italy	Modified clays (MMT, hydrotalcite, sepiolite) and equiaxial, oxide and non-oxide powders	Sepiolite-added and alumina-based PPO composites showed the most promising behaviors. Only TGA, DTGA, cone calorimeter, and microstructural characterization were conducted in this study
B.1	Different Polymers	AFRL, USA	MMT organoclay	Focused on polymer/layered silicate materials for use in space applications
C.1	EPDM	AFRL, USA	POSS	Detailed the strategy employment of using sacrificial nanocomposites along with applied research in the areas of SRM insulation and space-survival materials
C.2	EPDM	The University of Texas at Austin, USA	Carbon nanofibers (CNFs)	Detailed TGA data and SEM micrographs were reported on different heating rates
C.3	EPDM	Harbin Institute, China	Organo-MMT (OMMT) with Kevlar fibers	Up to 11 phr (parts per hundred parts of EPDM) OMMT in EPDM erosion rate decreased, beyond its ablation resistance decreased.
C.4	EPDM	Indian Institute of Technology, India	Polyimide micro-particles with nanosilica in maleic anhydride grafted EPDM	Erosion rate at 3 MW/m^2 is 0.14–0.17 mm/s and heat of ablation for the corresponding sample was 4700 cal/g. Not much details on ablation data were reported
C.5	EPDM	University of Perugia, Italy	Nanosilica, aramid pulp, phenolic fibers, and silica fibers	Silica fibers yielded the greatest residual mass during TGA and the greatest mass loss during ablation testing. The aramid pulp had the lowest in-depth temperatures. The amount of fibers used should be increased for better char and more conclusive results
C.6	EPDM	National University of Sciences and Technology, Pakistan	MWCNTs and carbon black	Back-face temperature decreased significantly, ablation rate decreased sharply at low loadings, thermal conductivity increased, ultimate tensile strength and elastic modulus increased, and hardness increased with increased loadings of MWCNTs

D.1	NR	Tarbiat Modares University, Iran	MMT organoclay	Improvement in aging hardness, curing time, mechanical, flammability properties by the addition of 3 wt% of MMT clay
D.2	Hydrogenated nitrile butadiene rubber (HNBR)	Beijing University of Chemical Technology, China	Fumed silica, organically modified montmorillonite (OMMT), and expanded graphite (EG)	OMMT was found to be a better additive for ablation resistance than fumed silica and EG
E.1	Thermoplastic polyurethane (TPU)	The University of Texas at Austin, USA	MMT organoclay, CNF, and MWNTs	MMT organoclay-based TPU nanocomposite (TPUN) has the best ablation performance, followed by MWNT-based TPUN
E.2	Silicone rubber	National University of Defense Technology, China	Zirconium carbide (ZrC) and zirconia (ZrO_2) were incorporated into silicone rubber with SiO_2 and carbon fibers	40 phr (parts per hundred parts of rubber) of ZrO_2, ablation rate was reduced by 72% and the mass loss dropped by 38%. 40 phr of ZrC reduced the ablation rate by 40% and the mass loss by 32%
F.1	Phenolic	Mississippi State University, USA	CNF with carbon fibers	The nanoscale dimensions of the VGCF causes major changes in the heat transfer rates and affects the resulting combustion/decomposition chemistry
F.2	Phenolic	The University of Texas at Austin, USA	MMT organoclay, POSS, and CNF with and without carbon fibers	High loading of MMT organoclay in phenolic, low loading of POSS in phenolic impregnated with carbon fibers, and high loading of CNF in phenolic without carbon fibers all gave better ablation performance than carbon phenolic baseline
F.3	Phenolic	Harbin Institute and National University of Defense Technology, China	POSS with carbon fibers; flake graphite reinforced barium-phenolic resin composite	Phenolic-POSS-CF produced the best charred surface, hence has a best ablation performance. Composite with flake graphite diameter of about 0.5 mm has the lowest ablation rate
F.4	Phenolic	Xi'an Jiaotong University, China	Graphene oxide (GO) dispersed in phenolic with no fiber reinforcement	The GO particles were found to disperse well into the phenolic and produced a moderate increase in char yield and thermal stability even at low loadings, but no ablation testing was conducted
F.5	Phenolic	DRDO, India	Nanosilica with carbon fibers	Ablation resistance increased with the nanosilica content up to 2.0 wt%, beyond its ablation resistance decreased

Table 9.1 Summary of the different polymer matrix used by the various research groups studying ablation using polymer nanocomposites (*Continued*)

Polymer matrix/institution			Nanoadditive/reinforcement	Remarks
F.6	Phenolic	DRDO, India	Zirconia (ZrO_2) and MWCNTs with carbon fabric	No ablation resistant decreased was resulted using ZrO_2 and MWCNT coated on carbon fabric to produce composites
F.7	Phenolic	Tarbiat Modares University, Iran	MMT organoclay with asbestos fibers	The 6 wt% MMT with asbestos fibers has the best ablation performance
F.8	Phenolic	Tarbiat Modares University, Iran	Kaolinite and graphite nanopowder without carbon fibers. Micrometer-sized kaolinite, nano-sized kaolinite, and nano-sized graphite powders with chopped carbon fibers	The phenolic nanocomposites without carbon fiber reinforcements have ablation rates of 0.0184 mm/s for neat phenolic resin, 0.0164 mm/s for 3 wt% graphite, and 0.0153 mm/s for 5 wt% graphite. The 9 wt% graphite with carbon fibers decreased ablation rate (measured in g/s and not in mm/s) and thermal diffusivity coefficient by 10%, and 62%, respectively
F.9	Phenolic	Islamic Azad University, Iran	Nanosilica with chopped carbon fibers	Mass loss and ablation rates showed greater improvement at higher loadings of nanosilica
F.10	Phenolic	University of Perugia, Italy	Nanosilica, carbon black, and MWNTs with E-glass fibers	Possibilities to exploit the synergy between traditional and nano-sized fillers were investigated
F.11	Phenolic	Gyeongsang National University, Republic of Korea	MWNTs with chopped carbon fibers and woven carbon fibers	Ablation resistance was observed with the Ph/carbon nanotube chopped carbon fibers and woven carbon fiber composites.
F.12	Phenolic	Texas State University-San Marcos, USA	MWNT with carbon fibers	Mass loss, recession rate, and in-depth temperature all reduced with the increase of MWNT wt% (up to 2 wt%).
F.13	Phenolic	University of Rome	Graphitic felt and graphitic foam	All felt-based samples passed undamped the longest oxy-acetylene exposure tests

Poly(caprolactam) (nylon 6, N, Ube) was compared to 2 wt% (NCH2) and 5 wt% (NCH5) layered silicate/poly(caprolactam) nanocomposites. These materials were synthesized by in situ polymerization of ε-caprolactam in the presence of dispersed OMMT. Additionally, a nominally 5 wt% layered silicate/poly(caprolactam) nanocomposite (NLS5) formed during melt processing of poly(caprolactam) (Capron B125WP, Allied Signal) and Cloisite 30A (OMMT, Southern Clay Products) was examined. This exfoliated nanocomposite was fabricated using a co-rotating twin-screw extruder (MPC/V-30, Haake). In contrast to in situ polymerization of NCH2 and NCH5, melt processing of NLS5 is not expected to result in a large fraction of poly(caprolactam) chains tethered through covalent bonds to the organic modifiers on the silicate surface.

The ablation rate (erosion rate) of the materials was determined from exposure to combustion gases in a mock SRM firing apparatus. Fifty grams of pelletized resin were compression-molded into 7 in. (180 mm) by 3 in. (75 mm) triangular sections approximately 0.2 in. (5 mm) thick in an autoclave. To enhance retention of material during the test, a second series of plaques were produced by consolidating the resins into a 1/8 in. (0.32 mm) polyimide mat. In addition to the poly(caprolactam) nanocomposites, ethylene-propylene diene monomer (EPDM) with 30% silica filler, EPDM with 15% chopped Kevlar fiber [state-of-the-art (SOTA) internal insulation for SRMs], and phenolic impregnated canvas were also examined. Table 9.2 summarizes the char motor ablation rate results [14]. Ablation tests were performed in a 4 in. (10.2 cm) diameter char motor. Two triangular plaques were arranged along the inside of a tapered, rectangular chamber. The propellant consisted of 88% ammonium perchlorate with hydroxylated polybutadiene binder. Temperature of the combustion gases was approximately

Table 9.2 Summary of Char Motor Ablation Rate Results

	Erosion rate (mil/s)[a], ±15%	
	Low mass flux (Mach 0.013, 12.8 m/s)	High mass flux (Mach 0.041, 39.6 m/s)
	(0.10 lb/in.^{2}s) (0.69 kPa/s)	(0.40 lb/in.^{2}s) (2.75 kPa/s)
N(poly(caprolactam))	nvmr	nvmr
NCH2	5	nvmr
NCH2, mat	4	9
NCH5	4	9[b]
NCH5, mat	3	10
NLS5	5	14
NLS5, mat	5	11
Phenolic canvas	7	8
EPDM/Kevlar	6	10
EPDM/30% silica	6	11

nvmr, no origin material remaining.
[a]mil/s = 0.0254 mm/s.
[b]At 0.35 lb/in.^{2}s.

3,000°C and the test duration was 8 seconds. The thickness of the remaining virgin material (after removal of the degraded char layer) was determined from the average of at least four measurements.

Since the presence of the layered silicate does not alter primary polymer decomposition in the reaction zone, the initial nanostructure of the dispersed silicate must influence the microstructure and composition of the char layer, thus enhancing the ablation performance. A series of scanning electron micrographs of the surface char region of several ablative samples were shown. Figure 9.2 shows the scanning electron micrographs of the surface char region of NCH5 ablative sample [14]. Figure 9.3 shows

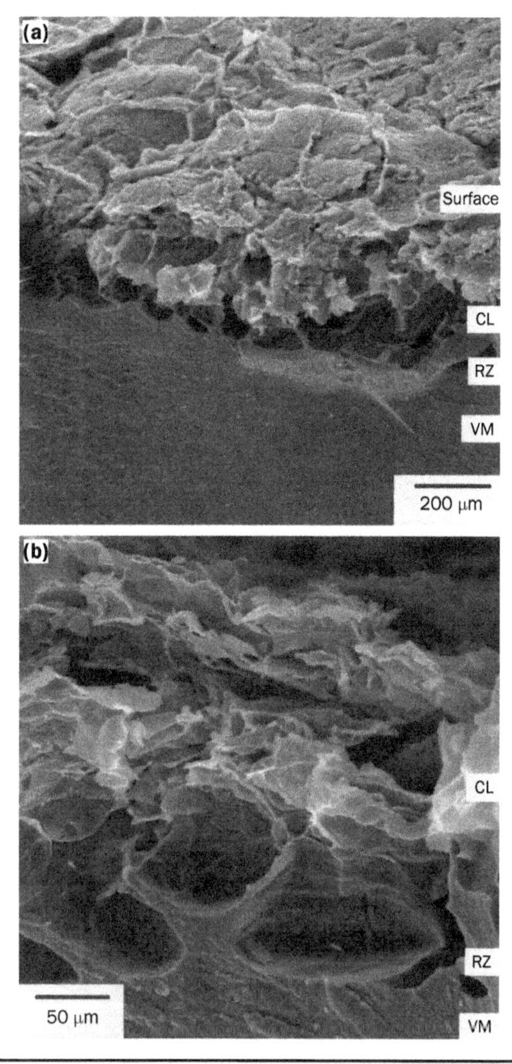

FIGURE 9.2 Cross-section SEM image of NCH5 ablative material, the charred layered silicate/ poly (caprolactam) nanocomposite. In the (a) image, the char layer (CL), the reaction zone (RZ), and the virgin material (VM) are clearly visible. In the (b) image, the nanostructured nature of the char is displayed [14].

the tiled angle to observe the cross-section and surface features of the NLS5 ablative sample [14]. In these chars, a nanoscopic (10–20 nm thick) densely packed platelet morphology is observed (Fig. 9.4) [14]. The characteristic ablation regions (virgin material, porous reaction zone, and dense char layer) are apparent in all samples. The microstructural features of these chars imply that increasing silicate content results in a tougher char. In these chars, nanoscopic (10–20 nm thick) densely packed plate-like morphology is observed. Increased silicate content results qualitatively in spatially more uniform

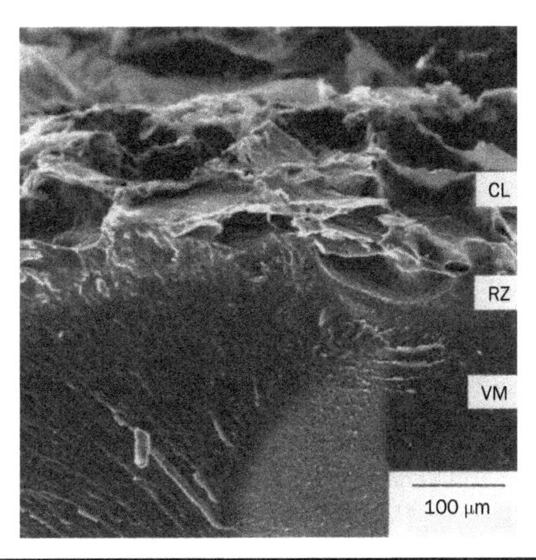

Figure 9.3 Scanning electron micrographs of surface region (titled to observe cross-section and surface features) of NLS5 ablative sample (CL, char layer; RX, reaction zone; VM, virgin material) [14].

Figure 9.4 Bright field transmission electron micrograph of NCH5, NCH5 char [14].

and thicker char. Specific characteristics of the silicate, such as layer size, uniformity of layer distribution, and strength of the interphase, result in a finer surface texture for NCH5 char than NLS5 char. In general, the spatially uniform arrangement of the silicate layers on an ultra-fine (nanometer) level facilitates the formation of a spatially uniform inorganic char. At 5 wt% the layers are on average only 20–25 nm apart. Thus, the efficient distribution of the inorganic precursors necessary to form a uniform char layer is the key structural characteristic of the nanocomposites leading to the enhanced ablative performance at remarkable low additions to the reinforcing phase.

Poly(phenylene oxide)/Polystyrene (PPO/PS) Studies by Politecnico de Torino, Italy [Lombardi et al. (2011–2012)]

Lombardi and colleagues [16–18] investigated PPO-based composites for potential ablative TPS application. The first part of their studies focused on PPO composites containing layered modified clay, such as MMT, hydrotalcite, and sepiolite [16]. A commercial PPO/PS blend, Noryl® 914 resin (GE Plastics) based on PPO blended with polystyrene PS, was used as the polymer matrix. It has a density of 1.09 g/cm^3 and a melting range of 280–310°C. Three-layered silicates were used: Cloisite® 15A (C, Southern Clay Products), a MMT clay; Perkalite® A100 (H, AkzoNobel Polymer Chemicals), an organically modified hydrotalcite; and Sepiolite (S, Tolsa), a hydrated magnesium silicate.

Brabender internal mixer was used to melt-compound these nanomaterials into the PPO/PS matrix blends. Degradation of these polymer composites were studied using thermal analyses (TGA) and cone calorimeter tests. Thermal analyses were conducted in Ar heating up to 1,500°C with heating rate of 20°C/min, and differential thermal analysis (DTA) performed up to 550°C with a heating rate of 10°C/min. Cone calorimeter tests were conducted with a heat flux of 35 kW/m^2. SEM analyses revealed all the composites still presented micrometer-sized aggregate platelets or needles. On the basis of the DTA results, PPO+S and PPO+C presented the best behavior for ablative materials, since they showed a more endothermic decomposition in comparison in the same temperature range of the neat polymer. Based on cone calorimeter results, PPO added with C and S presented a longer combustion time and a higher charring ability. Overall investigation showed the Sepiolite-added PPO composites can be selected as promising materials for innovative ablative shields.

In a second study [17, 18], equiaxial, oxide and non-oxide powders, such as magnesium aluminate spinel (SP), aluminum hydroxide (boehmite) (B), σ- and α-alumina, silicon carbide (SiC), silicon nitride (N), and molybdenum disilicide (M) were used as fillers incorporated into the PPO/PS matrix polymer. Nanometric and micrometric α-alumina and silicon carbide powders were utilized to investigate the effect of particle size. SEM analyses revealed all the composites still presented micrometer-sized aggregate platelets or needles. From DTA data, it is evident that PPO decomposition was unaffected by oxide fillers, except for PPO+SP material. Among the non-oxide fillers, SiC decreased the material ability to absorb heat during decomposition. The alumina-based composites presented the most promising behavior. DTA results indicated that PPO+S and PPO+C presented the best behavior for ablative materials. They exhibited a more endothermic decomposition in comparison to the same temperature range of the neat polymer. Cone calorimeter results showed PPO combined with C and S presented a longer combustion time and a higher charring ability. Thermal behaviors of all fillers were summarized in this study [17, 18]. The overall investigation showed the

Sepiolite-added PPO composites can be selected as promising materials for innovative ablative shields. No experimental data to simulate typical high heat flux for ablative heat shields were reported.

9.3.2 Polymer-Clay Nanocomposite Studies

Polymer-Clay Nanocomposite Studies by AFRL, United States [Lincoln et al. (2000)]

Lincoln et al.'s study [19] focuses on polymer/layered silicate materials to highlight the potential of nanocomposites for use in space applications. They also focus on fabrication methods for nanocomposites by controlled dispersion of preformed constituents, such as passivated nanoparticles or organically modified mica-type silicates. Filler particles on the nanoscale can alter the fundamental structure of the matrix material, especially in the case of semi-crystalline polymers and have a profound effect on the physical properties. The filler can also affect the interaction of materials with electromagnetic radiation. The improvement of nanocomposites over the neat polymer has the potential to offer space system designers a choice of a new, relatively less expensive material for use in their systems. These nanocomposites are polymer-containing materials. Lincoln et al. [19] discussed the benefits of using PNCs for space applications, such as environmental stability, launch systems (SRMs and liquid fuel tanks), and satellite bus (platform, deployable membranes, photovoltaic arrays, and solar concentrators). Nanocomposites are novel materials that potentially offer drastic cost savings for the space industry. They are lightweight, which is most desirable in the age where it costs thousands of dollars to place spacecraft into orbit. They are also easy to process and manufacture into useable parts, making them inherently less expensive and/or more reliable. The most attractive nanocomposite property is their potential to be space survivable. Nanocomposites can offer protection to vital systems and maintain spacecraft operational possibly for the duration of their lifetime.

9.3.3 EPDM Nanocomposite Studies

EPDM-POSS Studies by AFRL, United States [Philip et al. (2000)]

Air Force Research Laboratory researchers have studied how the incorporation of POSS can be used for protective coatings in oxygen-rich environments. Philip et al. [20] detailed the strategic employment of using sacrificial nanocomposites along with applied research in the areas of SRM insulation and space-survival materials. Proprietary POSS monomer (5–50%) was blended into the EPDM prepolymer using a standard brabender, followed by extended curing in a standard oven. The flexible POSS-EPDM (ethylene-propylene-diene monomer rubber) polymer was then cut and fitted into conical phenolic sections for testing. POSS-dianiline and POSS-polyimide were synthesized and characterized. A film was cast from the resulting POSS-poly(amic acid) on a clean glass plate and placed in a clean oven under nitrogen to cure under specific conditions. The resulting films were dried at room temperature for 24 hours.

The testing of POSS-EPDM samples for ablation performance was achieved using standard tactical and boost SRM propellants. Both the insulation standard and the POSS-EPDM sample were present in the same SRM firing. Post-firing analyses of the samples involved the use of industrial standard removal of the formed char layer followed by measurements of the remaining polymer at various stages in the sample, which represent different mass fluxes that the sample was exposed. At higher mass flux, the POSS-EPDM sample performs significantly better than the standard, with ablation rates of up to 50% less than the industry standard. These results are very significant, since a 50% reduction in the ablation rate of current SOTA materials results up to a 7.4% increase in payload to orbit. For SRM tests, analysis of post-fired samples showed the formation of a glass SiO_2 layer. Significant reduction in ablation rates was reproducibly demonstrated to the SOTA SRM insulation.

EPDM-CNF Studies by The University of Texas at Austin, United States [Koo et al. (2002)]

Polymer nanostructured ablatives were prepared by blending an EPDM elastomer polymer system with CNFs using standard brabender mixer. Thermogravimetric and SEM analyses were used to characterize these materials in a laboratory and a scaled SRM heating environment. Kevlar®-filled EPDM is used as motor insulation in solid rocket propulsion systems [21]. EPDM blends consisting of Kevlar fibers and CNFs were also included in this study.

Materials Ethylene-propylene-diene monomer rubber (EPDM) filled with silica or Kevlar has been used in propulsion systems for both thermal and ablation protection. CNFs were blended with Kevlar-filled EPDM.

Thermogravimetric Analysis (TGA) The TGA data compiled in these studies were collected using the TA Instruments Model TGA 2050. The TGA tests were conducted in an air environment, as compared to nitrogen, in order to best simulate the conditions inside SRMs. The TGA has a heating rate range of 0–200°C/min. The maximum temperature that the TGA furnace can achieve is 1,000°C. The TGA was equilibrated at 30°C before ramping to 1,000°C at the specified heating rate.

TGA Tests on EPDM/Nanoparticle Systems Mathias and Johnson [22] reported an extraordinary high heating rate (100,000°C/min) material property data of carbon cloth phenolic used in space shuttle reusable SRM nozzles. This study prompted the examination of SRM insulation samples using TGA and SEM [23–25]. Two types of materials were included in this study, Kevlar-filled EPDM (11% Kevlar) and Kevlar-filled EPDM blended with 15 wt% CNF. A low heating rate of 5°C/min and a high heating rate of 50°C/min from TGA tests, and a calculated heating rate of 150,000°C/min from AFRL's 4-in. char motor were included in the study. Figure 9.5 shows the TGA of the two samples at 5°C/min and at 50°C/min [23]. The EPDM-Kevlar-15 wt% CNF samples had less mass loss than the EPDM-Kevlar sample at both heating rates (5 and 50°C/min). The behavior of EPDM-Kevlar and EPDM-Kevlar-CNF exhibited a strong dependence on the heating rate.

SEM Analyses Scanning electron microscopy (SEM) images were produced to study the surface as well as the cross-sectional behavior of the nanostructured materials under thermal stress. Figures 9.6–9.8 show the samples at heating rates of 5°C/min (TGA), 50°C/min (TGA), and 150,000°C/min (4-in. char motor), respectively [23]. At 5°C/min there was no EPDM charring observed in Fig. 9.6, and the Kevlar fibers were

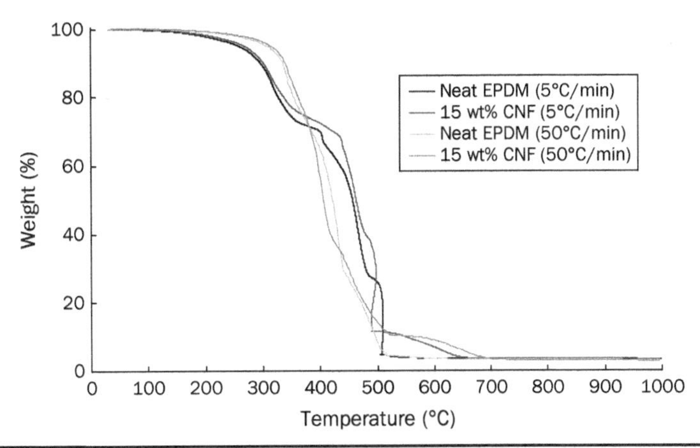

Figure 9.5 Mass loss of EPDM and EPDM/CNF composites at 5°C/min and 50°C/min.

Figure 9.6 Scanning electron microscopy image of EPDM-Kevlar exposed to heating rate of 5°C/min where the unit bar is 40 μm.

visible. Slight charring of the EPDM and Kevlar fibers was again quite visible, as shown in Fig. 9.7 [23]. In Fig. 9.8, severe EPDM charring was quite visible in the rocket motor–fired samples [23]. The SEMs show substantial difference in material behavior under these three heating conditions. Figures 9.9–9.11 show the material behavior of the three samples at different heating rates of the EPDM-Kevlar-CNF samples [23].

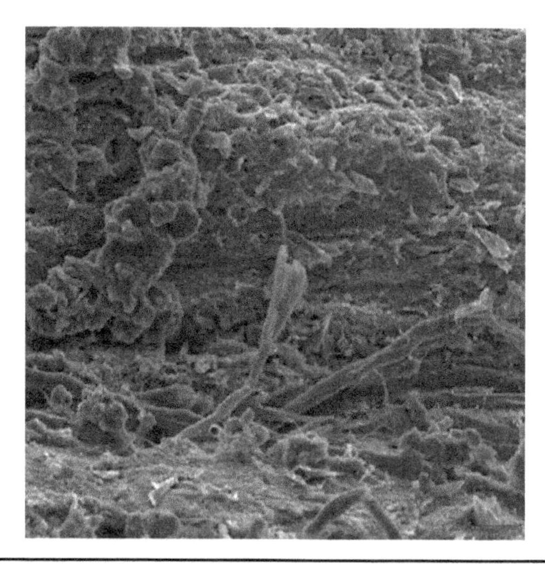

Figure 9.7 Scanning electron microscopy image of EPDM-Kevlar exposed to heating rate of 50°C/min where the unit bar is 40 μm.

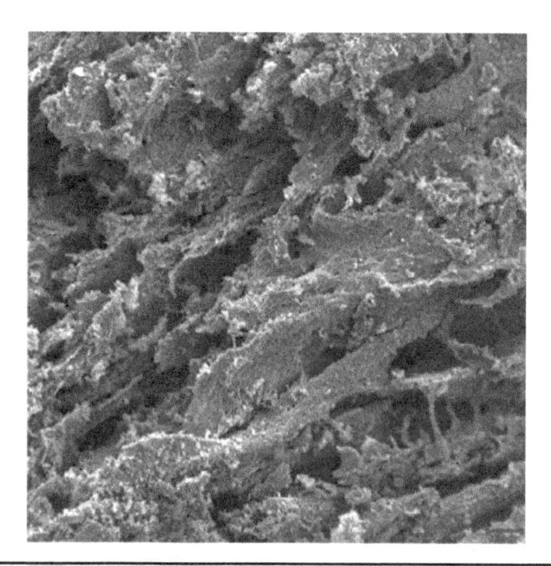

Figure 9.8 Scanning electron microscopy image of EPDM-Kevlar exposed to heating rate of 150,000°C/min where the unit bar is 40 μm.

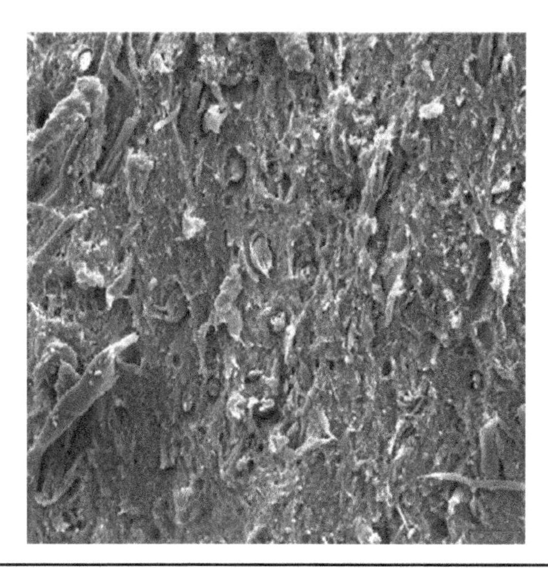

Figure 9.9 Scanning electron microscopy image of EPDM-Kevlar with carbon nanofiber exposed to heating rate of 5°C/min where the unit bar is 40 μm.

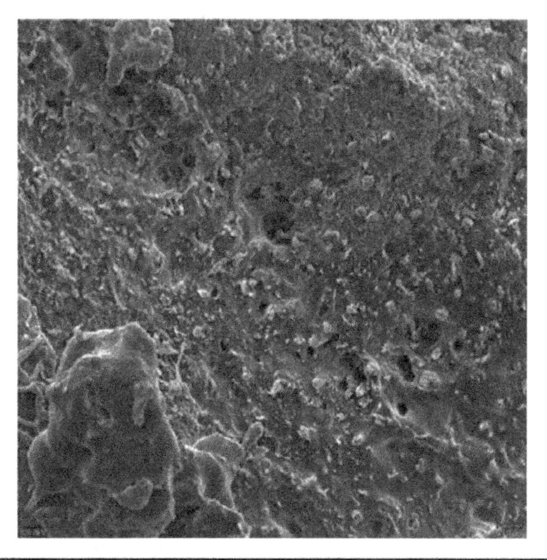

Figure 9.10 Scanning electron microscopy image of EPDM-Kevlar with carbon nanofiber exposed to heating rate of 50°C/min where the unit bar is 40 μm.

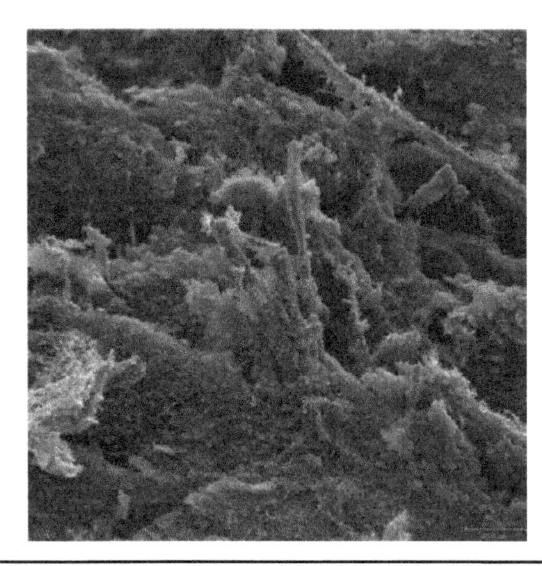

Figure 9.11 Scanning electron microscopy image of EPDM-Kevlar with carbon nanofiber exposed to heating rate of 150,000°C/min where the unit bar is 40 μm.

Conclusions Thermogravimetric analysis may not be a suitable method to study polymer nanostructured ablatives because of the extremely high heating rate that occurs inside the SRM combustion chamber. The SEM micrographs also confirmed that the polymer nanostructured ablatives behave very differently for TGA and scaled rocket motor samples.

EPDM-Clay Studies by Harbin Institute of Technology, China [Gao et al. (2010)]

Gao et al. [26] of China focused their attention to nanomodify the SOTA SRM insulation material EPDM-Kevlar by adding an organo-montmorillonite (OMMT) to the EPDM matrix and examined the synergy of these nanofillers with the traditional components of the formulation. The content of OMMT was increased from 1 to 15 phr. After all the additives—namely the nanosilica, the short fibers, the flame-retardants, the high char retention resin, and the peroxide—were well dispersed in the matrix with a two-roll mill for about 20 minutes, the materials were vulcanized (160°C and 15 MPa for 40 minutes). X-ray powder diffraction (XRD) analysis of the EPDM/OMMT was performed on the cured materials and TEM was used to further evaluate the morphology of nanocomposites. The mechanical properties of the produced materials, mainly the elongation to break, were measured and discussed. This property is very important, since when the ablative is placed in the SRM, it must absorb the mechanical stresses induced to the SRM during propellant casting, storage, transportation, and flight. It also assists in anchoring the propellant grain to the inner wall of the SRM case. As a result, the insulator should have high elongation to break (in excess of 200%). Moreover, the ablative properties were evaluated using an oxyacetylene torch according to ASTM E285-80 [27].

The results can be summarized as follows. The XRD patterns revealed that the interlayer spacing between the nanoclay layers of the OMMT was increased due to the intercalation of EPDM molecules in the filler tactoids. The TEM images also showed that the platelets were homogeneously dispersed in the matrix by the strong shear forces of rollers. In terms of mechanical properties, the elongation to break of the EPDM-Kevlar composites was analyzed as a function of the fiber content and these results were compared with the corresponding data of the EPDM-Kevlar-OMMT system with respect to the OMMT percentage. For the baseline EPDM-Kevlar, it appeared that above a fiber content equal to 6 phr, the elongation to break strongly reduced with the increasing content of the fibers: over 8 phr of fibers, the elongation to break of the EPDM-Kevlar (about 150%) became less effective to meet the requirements of SRM liners. The 12 phr EPDM-Kevlar formulation was chosen for the next step of the research in which the OMMT was introduced in the heat-shielding material. Even at this high fiber content, due to the presence of the nanoclay, the material exhibited an impressive increase of the maximum strain. This result was confirmed also at a high nanoclay content: at 13 phr OMMT the elongation to break displayed by the EPDM-Kevlar-OMMT system showed a value equal to approximately 560%.

However, with OMMT content of 7 phr, the elongation to break appeared to decrease. From 1 to 7 phr OMMT, the enhancement in the elongation at break can be directly attributed to the excellent dispersion of OMMT in EPDM matrix. The SEM images of the EPDM-Kevlar-OMMT fracture surfaces indicated that the nanoclay platelets reduced the interfacial adhesion of the aramid fibers within the matrix, and thus the OMMT acted as a lubricating agent. At the higher filler content, the reduction of the elongation to break can be attributed to the aggregation of the silicate layers. In terms of ablative properties with the increase of the nanoclay content up to 11 phr OMMT, the loss of mass as well as the erosion rate decreased. Quantitatively, the erosion rate decreased from 0.18 mm/s at 0% OMMT phr to about 0.07 mm/s at an OMMT content of 11 phr. Above this threshold, the erosion rate and the loss of mass slightly increased. These results are particularly interesting given that the OMMT-based formulations are potentially attractive candidates to replace the traditional EPDM-Kevlar heat-shielding elastomers.

EPDM-Polyimide-Nanosilica Nanocomposite Studies by Indian Institute of Technology, India [Singh et al. (2013)]

Singh et al. [28] examined the potential synergistic effect of polyimide micro-sized particles with nanosilica in maleic anhydride grafted EPDM. Several formulations were considered, in which the loadings of each filler varied from 0 to 10 wt%, as presented in Table 9.3 [28]. They were mixed in a Brabender Plasti-Corder at 70 rpm and 120°C and vulcanized with vulcanizing agents at 150°C for 30 minutes at 6 MPa.

Singh et al. characterized mechanical properties (ASTM D412), hardness (ASTM2240), density (ASTM D792), morphology via SEM, dispersion via TEM, thermal stability via TGA in N_2 at 10°C/min to 600°C, thermal conductivity (ASTM D5334-08), erosion rate via arc-jet ablation testing (ASTM E285-65T) on a selected formulation, coefficient of thermal expansion (CTE) on a selected formulation via TMA (ASTM E831-12), and specific heat via DSC.

Singh et al. observed that the density of all formulations remained nearly constant, varying within a range of only 0.066 g/cc, and that their mixing techniques had resulted in near-uniform dispersion. They found that all formulations had improved mechanical properties, and the formulation with the most improvement was the one with 10 wt% each of both fillers. For this reason, they chose that this be the one used in the arc-jet facility, producing an erosion rate of 0.155 mm/s under 3 MW/m^2 of heat flux. This formulation showed a small improvement in CTE over pure EPDM.

Table 9.3 Designation of EPDM–Based Samples

Sl. No.	Sample designation	Polyimide powder (phr)	Nanosilica (phr)
1	EP_0S_0	0.0	0.0
2	EP_5S_0	5.0	0.0
3	EP_1S_5	1.0	5.0
4	EP_3S_5	3.0	5.0
5	EP_5S_5	5.0	5.0
6	EP_7S_5	7.0	5.0
7	$EP_{10}S_5$	10.0	5.0
8	$EP_{10}S_{10}$	10.0	10.0
9	EP_5S_1	5.0	1.0
10	EP_5S_3	5.0	3.0
11	EP_5S_7	5.0	7.0
12	EP_5S_{10}	5.0	10.0

E, EPDM; P, polyimide; S, nanosilica.

TGA showed small improvements in thermal stability for all formulations. Singh et al. concluded that the material showed good promise for multiple thermal protection and insulation applications.

EPDM-Nanosilica Nanocomposite with Aramid Pulp, Chopped Phenolic Fiber, and Chopped Silica Fiber Studies by University of Perugia, Italy [Natali et al. (2013)]

Natali et al. [29] compared three microfiber reinforcement candidates for EPDM-silica nanocomposites. Three candidates were considered: aramid pulp, chopped phenolic fibers, and chopped silica fibers. These were added to EPDM at 10 wt% in addition to 20 wt% nanosilica, 15 wt% paraffin oil, and 4 wt% dicumyl peroxide, a vulcanizing agent. The formulations were mixed in a Bausano SD3025 single-screw extruder and vulcanized at 180°C for 2 hours in a heated press at a pressure of 50 bar.

The ablatives were studied for thermomechanical and thermal stability as well as ablation characteristics. They were examined via SEM before and after ablation testing. Their mechanical properties were evaluated using a dynamometer and a thermomechanical analyzer, and ablation tests were carried out using the group's oxyacetylene test bed (OTB) according to ASTM E-285-80 at 500 W/cm^2. TGA and differential scanning calorimetry (DSC) were also conducted on each.

First, the formulations were examined with SEM for dispersion of fibers. This was found to be satisfactory in each, but the dispersion of nanosilica was not verified. There appear to be some micron-sized particles visible in their EPDM matrix [29]. The researchers observed that the material was stiffer with the phenolic fibers and the aramid pulp, but tougher with silica fibers. The unreinforced material was not investigated. They found that while the silica fibers yielded the greatest residual mass during TGA, they showed the greatest mass loss during ablation testing. Posttest SEM inspection showed that the silica fiber–reinforced material produced a char that had high porosity while the phenolic fiber material had the least. It was the phenolic fiber material that showed the highest char/virgin material adhesion and the highest in-depth temperatures during ablation testing. The aramid pulp–reinforced material had the lowest in-depth temperatures. The researchers concluded that the amount of fibers included in the material should be increased for better char and more conclusive results. They also suggested mixing the fiber reinforcements into a single formulation.

EPDM-MWNT-Carbon Black Nanocomposite Studies by University of Sciences and Technology, Pakistan [Iqbal et al. (2014)]

Iqbal et al. [30] investigated nanocomposites of EPDM with MWNTs and CB. The CB was held at 40 wt% loading, while the MWNT loading varied between 0.1 and 1 wt%. In addition, a silane coupling agent was added to the mixtures at a 4 wt% loading. The formulations were mixed using dispersion kneader at 110°C for 30 minutes and then using a two-roll mill at 70°C and 40 rpm speed for 20 minutes. They were cured using a hot press at 130°C and 1,500 psi for 40 minutes.

The formulations were characterized for their ablation resistance according to ASTM E285-08 using a 1:1 oxygen/acetylene ratio with a flow rate of 0.35 m^3/h and torch distance of 10 mm for 200 seconds, dispersion via SEM and EDS, thermal stability via TGA to 830°C at 10°C/min, thermal conductivity according to ASTM E1225-99, mechanical properties according to ASTM D412-98A, and Shore A hardness.

The researchers observed a steady decrease in elongation at break, and an improvement in all other properties with increasing loadings of MWNTs. Backside temperature dropped significantly, ablation rate dropped sharply at low loadings, thermal conductivity rose, ultimate tensile strength and elastic modulus rose, and hardness rose with increased loadings of MWNTs. Iqbal et al. found that this composite is a good candidate for use as internal insulation SRM material.

9.3.4 Natural Rubber (NR) and Hydrogenated Nitrile Butadiene Rubber (HNBR) Nanocomposite Studies

NR-Clay Studies by Tarbiat Modares University, Iran [Khanlari et al. (2010)]

The effect of OMMT clay in natural rubber on the thermal, flammability, hardness, and mechanical properties using melt-compounding were investigated by Khanlart and Kokabi [31]. They have shown that the addition of 3 wt% of OMMT (Cloisite 15A) to NR results in decreased curing time by 33%, its aging hardness rise was decreased more than 55%, and elastic modulus increased by 63%. The ignition time was delayed by 150% and a reduction in peak heat release rate (PHRR) of 54% when compared to pristine NR. The thermal stability for the NR-clay nanocomposite was also enhanced. No ablation testing was conducted in this study.

HNBR–Fumed Silica, Nanoclay, and Expanded Graphite Nanocomposite Studies by Beijing University of Chemical Technology, China [Guan et al. (2011)]

Guan et al. [32] conducted an investigation into the effects of OMMT, expanded graphite (EG), and nanosilica on the thermal and ablation properties of hydrogenated nitrile butadiene rubber (HNBR)–based composites. The composites were reinforced with 40 wt% aramid pulps and mixed with some other agents for vulcanization, and so on. The nanofillers were added at a loading of 35 wt%. The formulations were mixed in a 6-in. two-roll mill at a speed ratio of 1:1.1 and cured in a molding compound at 170°C and 10 MPa.

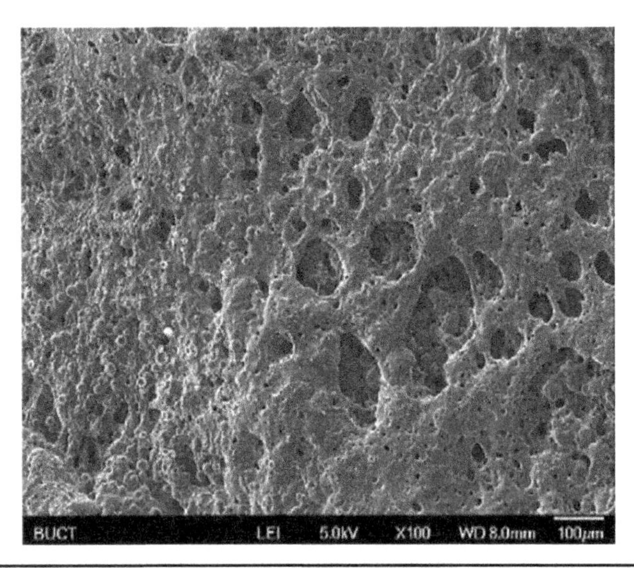

Figure 9.12 Scanning electron microscopy image of the char layer for the nanosilica formulation [32].

For characterization, Guan et al. used an oxyacetylene torch with an oxygen/acetylene ratio of 1:1 and total flow rate of 0.7 m^3/h, a torch distance of 10 mm, and an exposure time of 20 seconds to observe ablation resistance. They also used SEM and WAXD to discover morphology and dispersion of pre- and posttest samples. TGA in N_2 to 800°C was conducted at heating rates of 5, 10, 15, and 20°C/min to determine thermal stability properties.

Guan et al. found that the processing technique they had used to prepare the EG from its as-received state was insufficient to expand the layers. This prevented the filler from producing an optimal performance. However, the other nanofillers dispersed satisfactorily and Guan et al. were able to observe that the nanosilica did not form a gas barrier during ablation, resulting in a tighter, more solid char that conducted heat in-depth much more (Fig. 9.12) [32]. The OMMT was found to be a better additive for ablation resistance, since it forces the pyrolysis gases to escape through the char as it forms, making its pores much larger and its structure looser, as well as impeding the incoming torch gases (Fig. 9.13) [32]. As per Guan et al.'s conclusions, the char is compact. The fillers were not found to have significant effect on the TGA results.

9.3.5 Thermoplastic Polyurethane Nanocomposite (TPUN) Studies

TPUN Studies by The University of Texas at Austin, United States [Koo et al. (2003–2013)]

Polymer chopped-fiber–filled systems are used as insulation materials in rocket propulsion systems. This research program was aimed at developing new classes of polymer nanostructured materials that are lighter and exhibit better erosion and insulation characteristics than current insulation materials (e.g., Kevlar-filled EPDM rubber).

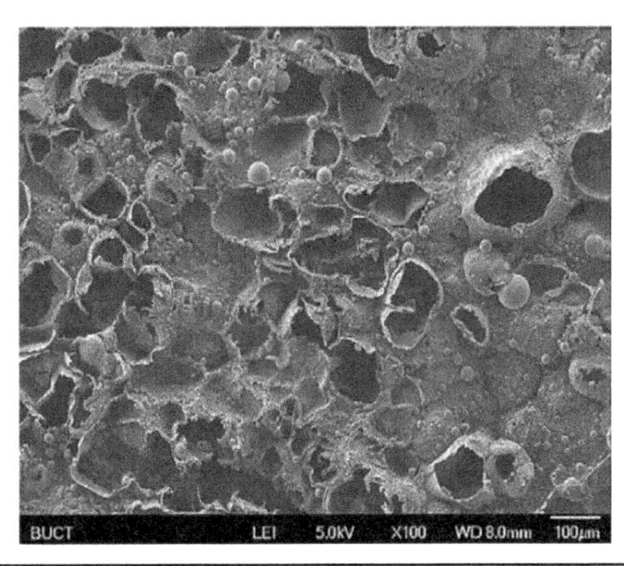

FIGURE 9.13 Scanning electron microscopy image of the char layer for the OMMT formulation [32].

TABLE 9.4 Thermoplastic elastomer/nanoparticle blends

Material	TPE (wt%)	Nanoparticle (wt%)
1	TPSiV™ X1180 (90%)	$Ph_{12}T_{12}$-POSS (10%)
2	TPSiV™ X1180 (92.5%)	Cloisite 30B (7.5%)
3	TPSiV™ X1180 (85%)	PR-24-PS CNF (15%)

Materials The TPSiV™ X1180 thermoplastic elastomer (TPE) is a polyamide-based, vulcanized silicone thermoplastic resin manufactured by Dow Corning that was selected as a potential replacement material. Its typical uses include profiles for automotive, fuel and vapor line covers, brake hose covers, and industrial applications involving extruded profiles exposed to harsh environments. Table 9.4 shows the chemical compositions of three thermoplastic elastomer/nanoparticle blends that were produced by twin-screw extrusion [33].

Characterization Transmission electron microscopy analyses were conducted on the above three blends. Figure 9.14 shows the TEM images of the Dow Corning polyamide silicone TPSiV™ X1180 in two separate phases; the dark color is the silicone phase, and the light color is the polyamide phase [33]. Figure 9.15 shows the TEM images of the 7.5 wt% Cloisite 30B/92.5 wt% TPSiV™ X1180 blend [33]. We speculate that the Cloisite 30B nanoclays are dispersed only in the polyamide phase, because the silicone phase is already cross-linked. TEM images of the 15 wt% PR-24-PS CNF/85 wt% TPSiV™ X1180 blend and 10% $Ph_{12}T_{12}$-POSS/90 wt% TPSiV™ X1180 blend showed similar results [33]. It was determined that the TPSiV™ X1180 TPE is difficult to process with any of the three selected nanoparticles. None was compatible with the cross-linked silicone.

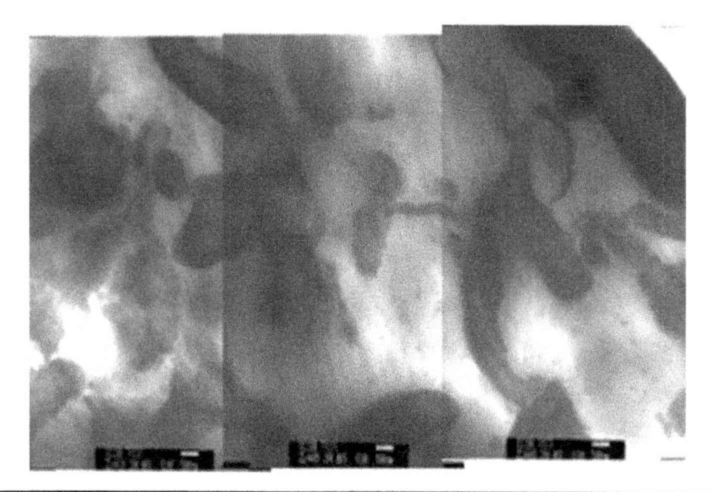

Figure 9.14 Transmission electron micrographs of polyamide silicone TPSiV™ X1180 (unit bar is 500 nm).

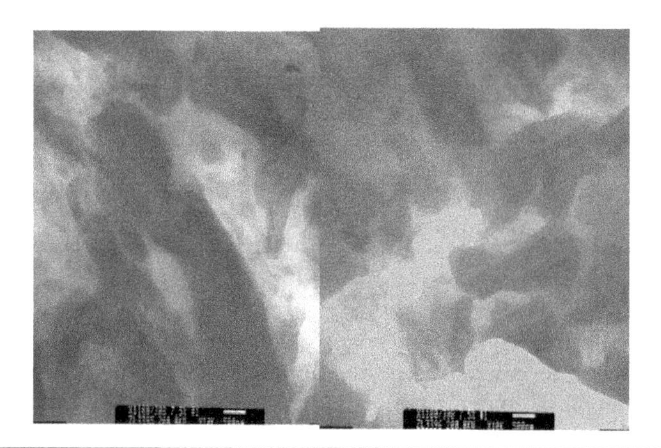

Figure 9.15 Transmission electron micrographs of 92.5% polyamide silicone TPSiV™ X1180 with 7.5% Cloisite 30B (unit bar is 500 nm).

Performance The above three blends with the baseline Kevlar-filled EPDM rubber were tested for ablation resistance using a subscale SRM. Figure 9.16 shows the ablation rate of the materials at low, medium, and high Mach number regions inside the rocket motor [33]. The baseline material outperformed all three TPEs. These observations clearly demonstrate that if the nanoparticles are poorly dispersed in the polymer matrix, no ablation performance improvement can be expected. The limited dispersion of nanoparticles into the polyamide phase was insufficient for improved ablation performance.

For advanced SRM internal insulation, Koo and colleagues [33–47] (which include Koo et al. [33-37]; Ho et al. [38, 39]; Bruns et al. [40]; Lee et al. [41–45]; and Allcorn et al. [46, 47]) used two thermoplastic polyurethane elastomers (TPUs)

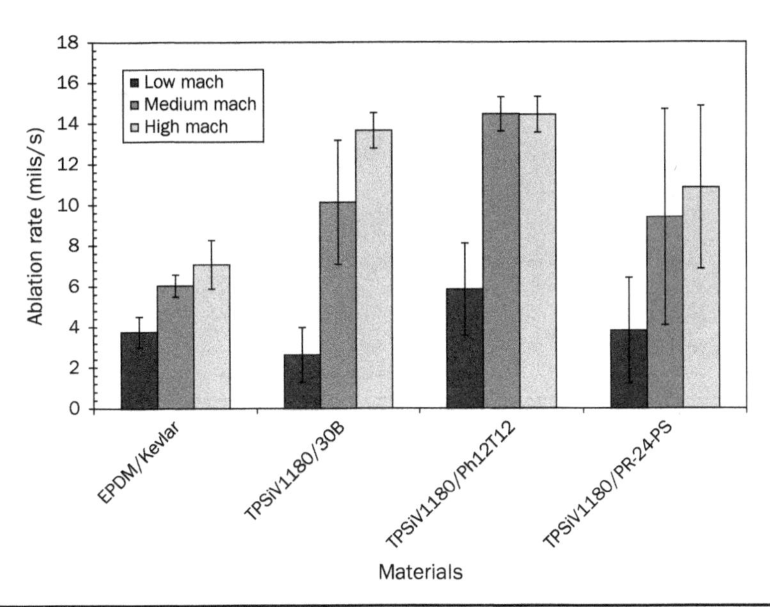

FIGURE 9.16 Ablation rate of polyamide silicone TPSiV™ X1180 with Cloisite 30B, Ph_{12}-T_{12}-POSS, and PR-24-PS CNF as compared to EPDM-Kevlar in low, medium, and high Mach number regions inside the solid rocket motor.

as the base polymer and incorporated them with different amounts of MMT clay, CNFs, POSS, and MWNTs. Selective properties improvement was observed with these TPU nanocomposites (TPUNs) as compared to industry standard Kevlar-filled EPDM insulation material. The effects of weight loadings of layered clay, CNFs, and MWNTs on the thermal and flammability performance of this novel class of materials have been explored using a variety of test protocols and methods, such as TGA, radiant panel experimental apparatus, UL 94, and cone calorimeter. Selective TPUNs were then tested using scaled SRM, oxyacetylene torch, as well as scaled hybrid rocket motors with the intent to develop them for SRM insulation application. Two commercially available TPUs were melt-blended with various loadings of these nanofillers using twin-screw extrusion. The morphological, physical, thermal, flammability, and thermophysical properties, and kinetic parameters of these three families of TPUNs were characterized. The *processing-structure-property* relationships of this class of novel TPUNs were established and are reported elsewhere.

Morphological Results TEM images of both the MWNTs and nanoclay TPUs were carried out to determine the visual degree of dispersion (Figs. 9.17 and 9.18) [33]. In the 2.5 wt% nanoclay sample, the individual platelets are easily identifiable and good dispersion is observed. The 5 and 10 wt% images show some remaining stacks; however, the majority of the images show individual nanoclay platelets. It should also be noted that it does not appear to have any directional orientation from the injection-molding process. The MWNT agglomerates are clearly debundled as shown in the two lower weight percentages. In the 10 wt% MWNT TEM image, there is a high density of MWNTs. These

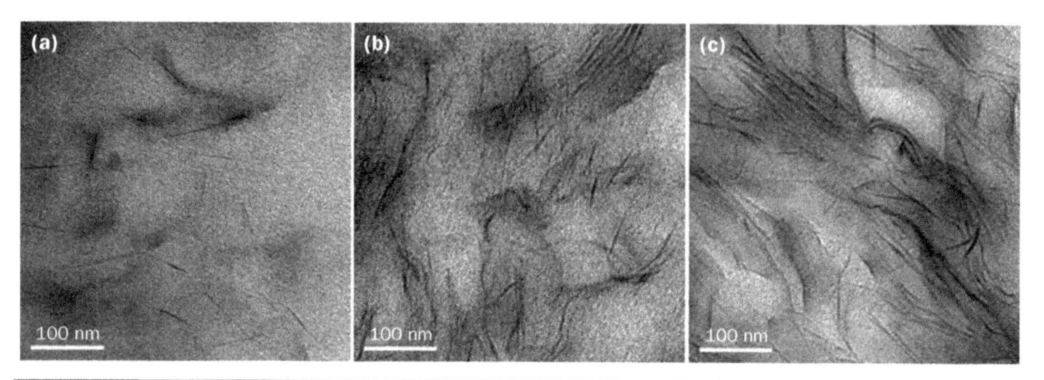

Figure 9.17 TEM images of (a) 2.5%, (b) 5%, and (c) 10% Cloisite 30B in TPUN (unit bars are 100 nm).

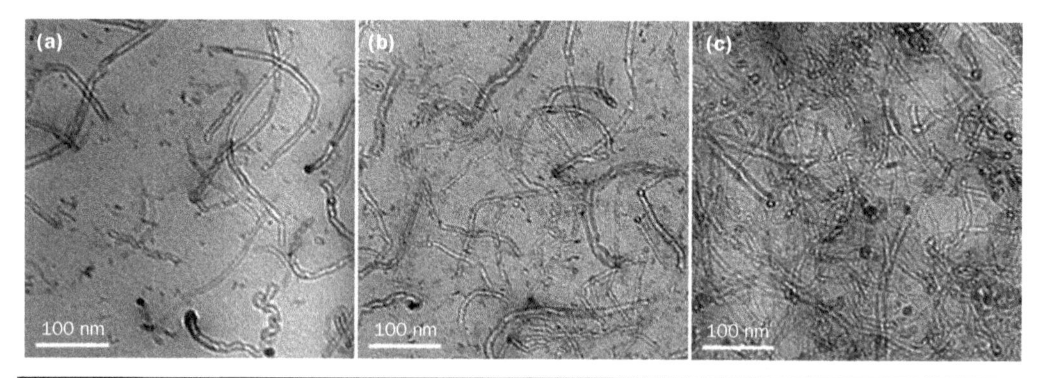

Figure 9.18 TEM images of (a) 2.5%, (b) 5%, and (c) 10% MWNTs in TPUN (unit bars are 100 nm).

are not in bundle form. The bulk material dispersion is assessed by observing TEM images at different locations and view similar images. Again, there does not appear to be any preferred orientation.

Ablation Test Results High heat flux experiments have been conducted using the OTB (see Sec. 5.13.3). The material is exposed to the hottest portion of the flame, with a 7-mm flame diameter, at a heat flux of 1,000 W/cm^2 for 25 seconds. Samples of 12.7 mm × 12.7 mm × 50 mm are cut out of a 12.7 mm × 10.16 cm × 10.16 cm (½ in. × 4 in. × 4 in.) compression-molded sheet. The torch is directed toward the square face (12.7 mm × 12.7 mm). In this orientation, the sample may be tested twice. At 5-mm increments three thermocouples (TCs) are embedded into the center of the specimen to measure the in-depth heat-soaked temperature during testing. TCs are drilled into the specimen at both ends for the possibility of testing a single specimen twice depending on the ablation rates.

Thermal Results A representative temperature profile of each specimen is shown in Fig. 9.19 [44, 45]. The specific materials are color coded and the TCs are distinguished as TC 1 solid line, TC 2 dotted line, and TC 3 dashed line. After the 25-second flame

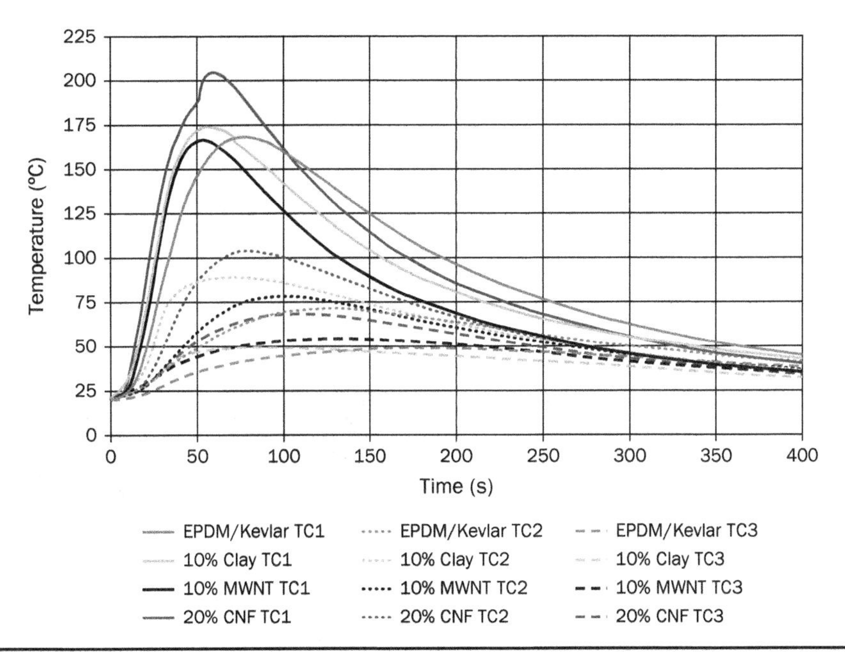

Figure 9.19 Representative temperature profile of each specimen of oxyacetylene ablation tests.

exposure is removed, the heat is transferred conductively through the specimen. In this set of tests, the peak temperature of TC 1 in the CNF sample is 205°C and occurs at 60 seconds. The CNF has the highest heat-soaked temperature. The MWNT and nanoclay TC 1 peak at 166°C and 174°C and occur at 50 seconds and 64 seconds, respectively. The EPDM-Kevlar TC 1 peaks at 168°C and shows the most delay at 75 seconds. The nanoclay, MWNT, and CNF PNCs lose 13.5%, 33.7%, and 46.5% more mass, respectively, compared to the mass loss of Kevlar-EPDM.

Post-Firing Images The samples after firing are shown in Fig. 9.20 [44, 45]. A 1-mm compact char layer is left at the end of the firing of Kevlar-EPDM. Although compact, the virgin and char material does not have good adhesion. The nanoclay PNC posttest material has about twice the char thickness of the Kevlar-EPDM. Although the char has a crack, the char was maintained well within the virgin material. Similar to the results found in the vertical UL 94 test, many cracks are observed in the MWNT char surface and result in a weak 2-mm thick char. The CNF PNC char is very thick, greater than 5 mm, as the first TC was in the char layer. The CNF char surface is flat without cracks; however, it is still very weak. The results from this set of experiments are consistent with the flammability and hybrid rocket tests [44, 45].

SEM images of the posttest samples also were taken. The Kevlar-EPDM char material is compact and shows that the individual Kevlar fibers maintain the material together (Fig. 9.21) [44, 45]. The nanoclay char is also a compact composition. A granular morphology is also observed [47]. Bubble formations observed in SEM images

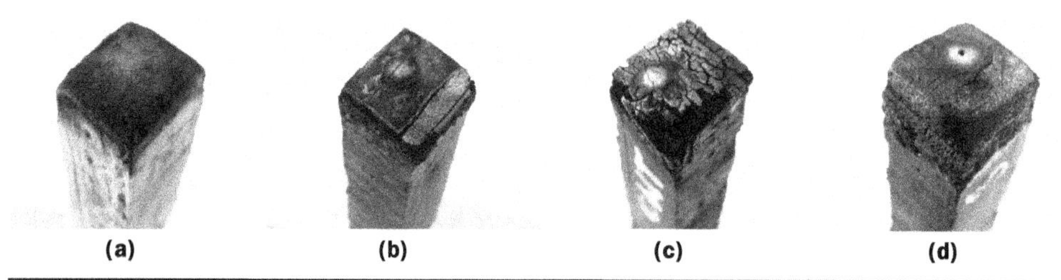

FIGURE 9.20 (a) Kevlar-EPDM, (b) 10% clay, (b) 10% MWNTs, and (d) 20% CNF PNCs post-oxyacetylene torch burn.

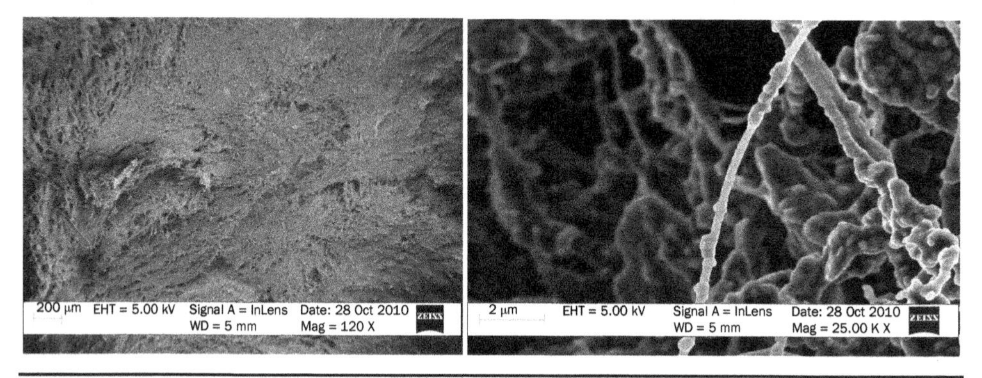

FIGURE 9.21 SEM of Kevlar/EPDM char post-oxyacetylene torch burn in progressive magnification (unit bar of the left image is 200 µm and of the right image is 2 µm).

found in the UL 94 burn tests are observed on the MWNT char surface as well as crack formations [47]. An in-depth SEM image shows the bundle form of MWNT that is also observed in the UL 94 burn test. The weak CNF char is shown to have a very porous structure [47]. Individual CNFs are also observed. More detailed SEM images of MMT clay TPUNs, MWNT TPUNs, and CNF TUNs are presented later in this section.

The initial results from the oxyacetylene torch test show that both the nanoclay PNC and MWNT PNCs char layers are about twice the thickness of the Kevlar-EPDM char. However, the MWNT char layer is weak compared to the nanoclay char. The CNF char is also very weak and was over five times the thickness of the Kevlar-EPDM char. However, the mass loss from the CNF sample is almost twice that lost from the Kevlar-EPDM sample. In-depth thermal measurements also show that the thermal protection of the PNCs is comparable to that provided by the Kevlar-EPDM, except for the CNF sample which had a much higher peak temperature due to the fact that the first embedded TC ended up in the char region. A conventional advanced fiber is therefore needed to boost up the PNC ablation performance.

Allcorn et al. [46, 47] continued to investigate the potential of TPUNs as alternative elastomeric heat shielding materials (EHSMs) in ablative applications. Three different nanofillers were tested—MMT clay, MWNTs, and CNFs—and formulations were made for each material with a range of weight loadings, as shown in Table 9.5 [47]. Small-scale ablation testing was performed using an OTB with Kevlar-filled EPDM as the baseline

Table 9.5 Composition of Nanocomposite Formulations

Nanocomposite formulation	Weight percent TPU	Weight percent MMT clay	Weight percent MWNTs	Weight percent CNFs
MMT-2.5%	97.5%	2.5%	0.0%	0.0%
MMT-5%	95.0%	5.0%	0.0%	0.0%
MMT-7.5%	92.5%	7.5%	0.0%	0.0%
MMT-10%	90.0%	10.0%	0.0%	0.0%
MWNTs-2.5%	97.5%	0.0%	2.5%	0.0%
MWNTs-5%	95.0%	0.0%	5.0%	0.0%
MWNTs-7.5%	92.5%	0.0%	7.5%	0.0%
MWNTs-10%	90.0%	0.0%	10.0%	0.0%
CNFs-5%	95.0%	0.0%	0.0%	5.0%
CNFs-10%	90.0%	0.0%	0.0%	10.0%
CNFs-15%	85.0%	0.0%	0.0%	15.0%
CNFs-20%	80.0%	0.0%	0.0%	20.0%

EHSM. Mass loss, recession, and char thickness values were measured in addition to in situ internal temperatures. The TPUN formulations with 5 and 7.5 wt% MMT clay showed the best results, outperforming the baseline in recession and peak temperature values. SEM images of the char from MMT clay containing TPUNs showed ablative reassembly, which likely served to enhance the material performance. Overall, results showed that the TPUN ablatives performed well relative to the baseline EHSM.

The testing of ablative materials involves simulating the extreme conditions they are exposed to in actual applications, while measuring various metrics. Some common metrics for ablative performance measurement include mass loss, recession and/or recession rate, front-face temperature, back-face temperature, in-depth temperature, char thickness, and off-gas composition. The metrics identify different criteria depending on the specific application of the ablative material. The material's resistance to heat is determined via measurements of the back-face, front-face, and heat-soaked (internal) temperatures during and after the testing procedure. These measurements show how well the material inhibits heat flow and can determine whether it fails to meet specific requirements regarding back-face temperature. For example, if the back-face temperature is too high, it may result in failure of the adhesive bonding the ablative to the supporting structure, and ultimately in the failure of the overall material. In addition, posttest measurements of the char thickness and material erosion can provide additional insight into the resistant capabilities of the material relative to exposure time. If the reaction progresses too quickly through the material or the protective char layer erodes too quickly due to surface shear, then the material may not provide thermal protection for a long enough time period.

For the materials tested in this study, specific settings within the OTB's range of operation were selected and all the material formulations were tested under these conditions. The nanocomposite formulations were cut into 1.27 cm × 1.27 cm × 5 cm (0.5 in. × 0.5 in. × 2 in.) samples for testing, with three samples tested for each formulation to increase the statistical confidence in the results. The square face of the samples was exposed to the oxyacetylene torch at a calibrated heat flux of approximately 1,000 W/cm^2 for duration of 25 seconds. The OTB facilitates three K-type TC measurements during testing, and for our study these TCs were placed within the samples at distances of 5, 10, and 15 mm from the exposed surface with a heat shield in place to protect the protruding wires. In addition to these in situ temperature curves, the mass loss, recession, and char thickness of each sample were determined based on pre- and posttest measurements.

Mass Loss The results of the mass loss measurements for the TPUN formulations are shown in Fig. 9.22 [47]. Mass loss values can provide an indication of the time duration that ablatives can offer effective thermal protection; low values are desired. For all three nanomaterials a similar trend is observed with a decrease in mass loss up to a certain weight percent and then a slight increase beyond that value. This trend is typical among nanocomposites as greater amounts of the additive material provide benefits to the base matrix until a certain threshold amount where excessive additive content compromises the overall matrix structure of the base material [6, 47]. For both MMT clay and MWNTs, this ideal value of minimum mass loss is somewhere between 5% and 10%, while for CNFs the ideal value is somewhere between 10% and 20%. When comparing the different nanomaterials, they all appear to have minimum mass loss values that are approximately the same. The minimum mass loss for both MMT clay and CNFs is near 400 mg while the minimum for MWNT is slightly worse at approximately 500 mg.

Overall, none of the TPUN formulations outperforms the EPDM baseline, which shows an average mass loss of slightly over 300 mg. However, it must be pointed out that for SOTA EPDM-Kevlar materials the amount of the fillers—the fibers, the micrometer-sized and nano-sized additives, such as the mineral oxides and flame retardants—is higher than in the case of the produced nanocomposites [49–51]. Consequently, since

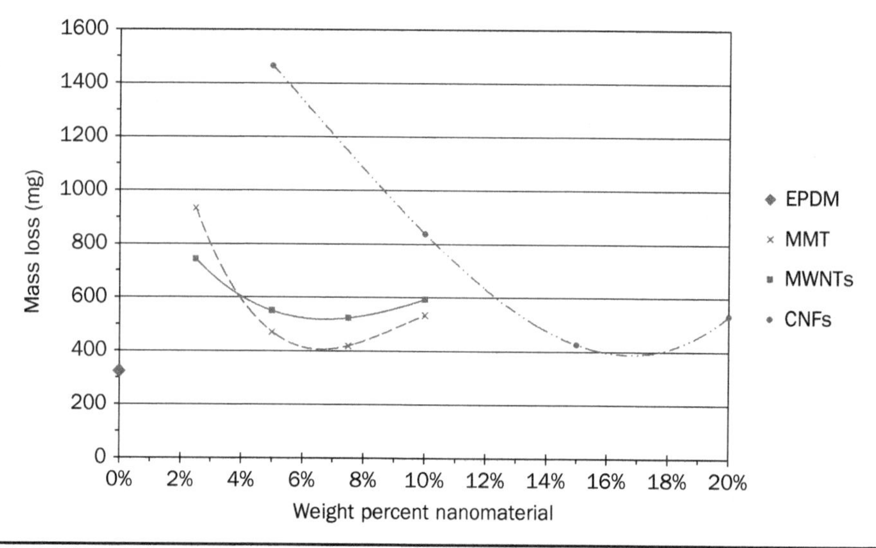

Figure 9.22 Mass loss results for OTB ablation testing of TPUN formulations.

the filler content of the EPDM-Kevlar is typically higher than the nanocomposites produced in this research, these results can be considered quite promising. The low weight percentage formulation for each nanomaterial has high mass loss because they are losing mass not only through the pyrolysis of the material, but also through physical erosion at the char surface. The char material for the MMT clay 2.5%, MWNT 2.5%, and CNF 5% TPUNs was too weak to withstand the pressure from the OTB flame and was observed to physically erode off of the sample. The CNF 10% formulation also experienced a small amount of erosion, though more limited.

Recession and Char Thickness The thickness of each sample was measured before and after testing and was used to determine the overall recession of the sample. For this measurement, the thickness after testing was considered to be the thickness of virgin material only, so the recession value is a measure of the progression of the reaction layer into the virgin material and does not factor in the thickness of the char on the surface. Recession is also an indicator of the time duration of thermal protection; so low values are once again desirable. The results are shown in Fig. 9.23 and follow roughly the same trends as the results for mass loss and the best performing weight percent values match up to those for minimal mass loss: 5–10% for MMT clay and MWNTs, and 10–20% for CNFs [47]. When the nanomaterials are compared to one another, both MWNT and CNF TPUNs have similar minimum recession values of roughly 3.5 mm. However, MMT clay performs better than the other materials with minimum recession value of around 2.5 mm. This is superior even to the baseline EPDM-Kevlar material which showed recession values near 3.5 mm.

The char thickness of each sample was also measured after testing [47]. For our measurement, we considered the char thickness to be the distance from the front surface of the posttest sample to the reaction layer [47]. The trends in this instance are quite different than those seen in other measurements. Both MMT clay and CNF TPUNs have char thickness values that generally increase with weight percent. This is likely due to the intumescent nature of the char for these materials, which causes them to swell

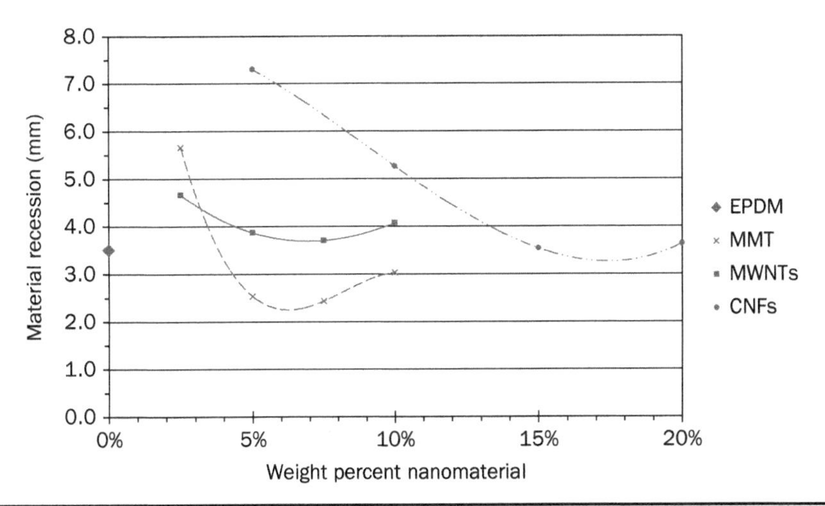

Figure 9.23 Material recession results for OTB ablation testing of TPUN formulations.

under heat exposure. For these materials, the total sample length after testing was actually several millimeters longer than before testing. This trend suggests that the amount of swelling from the char increases with weight percent up to at least 10% for MMT clay and roughly 15% for CNFs. The CNF 20% TPUN shows less char thickness than the CNF 15%, suggesting that beyond that weight percent the inclusion of more CNF inhibits the intumescent effect, possibly due to a decrease in the amount of polymer matrix. For MMT clay, no decrease is observed in the studied range, suggesting the intumescent effect is likely not inhibited until the weight percent reaches some unspecified value above 10%. The char of MWNT TPUNs behaves differently and shows a trend of decreasing thickness with greater weight percent nanomaterial. This is because MWNT char does not swell so the sample thickness before and after testing is roughly equivalent. This trend suggests that MWNT char becomes more effective at protecting the material beneath it at greater weight percentages, which slows the growth of the char layer. This assertion agrees with the observed trends in recession [47]. EPDM showed smaller char thickness than any of the TPUNs tested, though this is not necessarily an indicator of superior or inferior performance, as there appears to be little correlation between char thickness and the other performance parameters measured in this study. It should be noted that for both MMT clay and CNFs, no value was recorded for the lowest weight percent loading because the char was either completely eroded or eroded to a degree that no measurement could be taken. Finally, it should be noted that the char of the MMT-based systems was compact and qualitatively it seemed to have the highest structural integrity while the chars of both the MWNT- and CNF-based formulations were quite porous and fragile. In addition, the MMT char did not easily detach from the remaining virgin material while both the MWNT and CNF chars would detach relatively easy. Thorough analysis of the relative quantitative strengths of these chars may be useful, but was not considered in this study. However, char strength sensors are being developed by the Koo Research Group [52, 53] (see Sec. 5.13.4 for details of these char strength sensors).

Peak Temperature and Time The three embedded TCs in each sample produced a temperature versus time curve for each sample from which data could be gathered concerning the peak temperature at each depth within the sample as well as the time element to reach peak value. Peak temperature and time to peak are both indicators of the actual effectiveness of the ablative materials at providing thermal protection and are therefore some of the most important performance parameters. A material that provides effective thermal protection should have both low peak temperatures as well as long time to peak values.

The peak temperature values for each TPUN specimen in Fig. 9.24 are compared to the performance of the baseline EPDM-Kevlar composite at the left-most position in the figure [47]. The MMT clay formulations show the lowest peak temperature values, while the MWNTs and CNFs have mixed results. Some notable formulations are the 5% and 7.5% weight loadings of MMT clay and appear to outperform the baseline EPDM-Kevlar material. The 2.5% and 5% weight loadings of MWNTs also perform well, almost duplicating the performance of the baseline. The 15% weight CNF-based formulation exhibited reasonably similar results to the baseline. For both the MMT clay- and CNF-based TPUNs, the peak temperature trends seem similar to the mass loss and recession trends with mid-range weight loadings performing the best. For MWNT-based TPUNs, however, the trends appear to vary, as lower weight loadings seem to perform better in terms of peak temperature values. This may be due to the enhanced thermal protection from

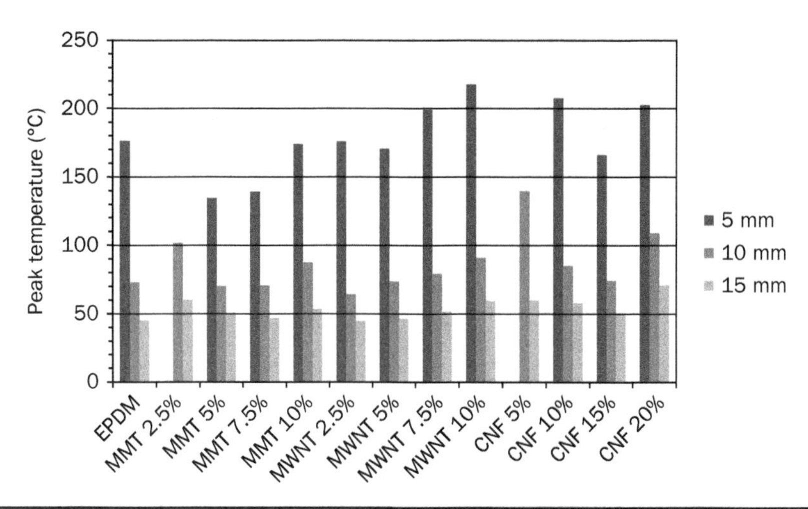

Figure 9.24 Peak temperature values during ablation testing for TPUN formulations.

the more porous char of lower weight loading formulations or the heat energy that is removed as these materials physically erode. However, while these factors may enhance the peak temperature performance, they sacrifice the length of protection that is desired.

The time to peak values for each TPUN specimen are compared to the performance of the baseline material [47]. The MMT clay formulations show the highest overall time to peak values, the MWNTs have a slightly lower overall value, and CNF formulations overall have very low values. Some notable formulations are 5% and 7.5% weight loadings of MMT clay as they appear to have the highest time to peak values among the TPUNs and roughly match the baseline material. The 10% weight loading of MMT clay as well as all the MWNT formulations also perform relatively well, though still not as well as the baseline. All the CNF formulations have very poor time to peak values. The MWNT and CNF formulations likely performed slightly worse than MMT clay because CNFs increased thermal conductivity of polymeric materials while MMT clay decreased it [54–56]. The effect is probably more prominent in the CNF materials due to their larger size that allowed for a more connected network through which heat could flow.

Microstructural Analyses SEM images taken after testing show the microstructure of the char that forms at the surface of these TPUN materials during ablation. Figure 9.25 shows the char structures for the 5, 7.5, and 10 wt% MMT clay formulations [47]. The 2.5 wt% formulation did not produce stable enough char for imaging. The char of the MMT clay TPUNs is composed of a microstructure of layered platelets, as is seen in the low-magnification images of the MMT clay formulations (Fig. 9.25d–f) [47]. Although the MMT clay is not aligned during dispersion or synthesis, the layers formed in the char structure are strongly aligned such that they lie parallel to the sample surface (ablative reassembly) [57]. This enhances the thermal protection effectiveness of the material because the layers are also aligned perpendicular to the direction of heat flow, providing no direct path for heat flow across the spacing between layers. There also appears to be evidence of pockets of gas forming bubbles within the char during ablation, such as at the top of Fig. 9.25 [47]. This may be the cause of the intumescent behavior observed in this char. Under higher magnification (Figs. 9.25a–c) the platelets appear to be composed

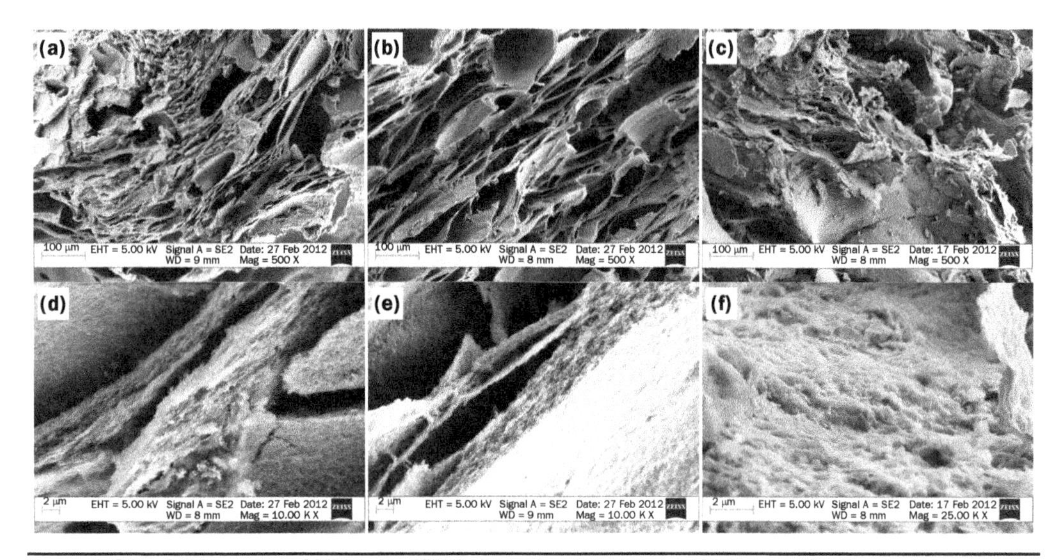

Figure 9.25 SEM images of MMT clay TPUN char microstructures at high magnification (unit bars are 100 µm) for (a) 5 wt%, (b) 7.5 wt%, and (c) and 10 wt% formulations and at low magnification (unit bars are 2 µm) for (d) 5 wt%, (e) 7.5 wt%, and (f) 10 wt% formulations.

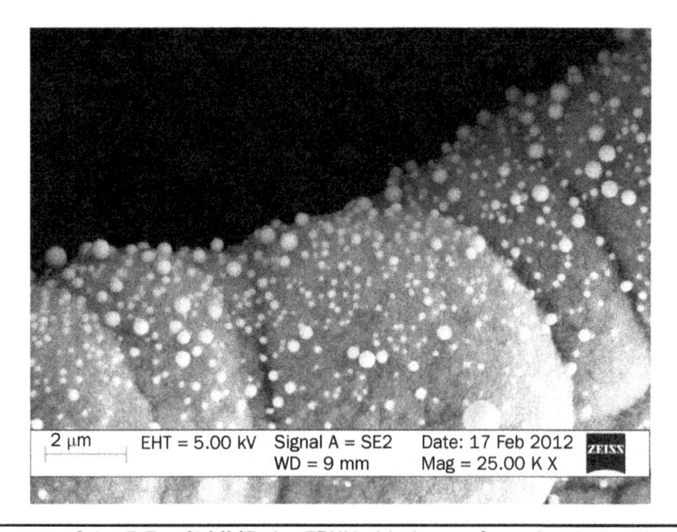

Figure 9.26 SEM images of the 7.5 wt% MMT clay TPUN ablation surface showing evidence of nanoclay sintering.

of nanoclay that has either been compacted as the polymer matrix is eroded away or sintered together to form the larger platelets [47]. It is likely that there is at least some degree of sintering within these platelets because SEM images of the ablative surface show spherical particles evident of sintering on all three formulations of MMT clay TPUN (Fig. 9.26) [47].

Figure 9.27 shows the char structures of the MWNT TPUNs [47]. The 2.5 wt% formulation has again been excluded because it did not produce structurally stable char.

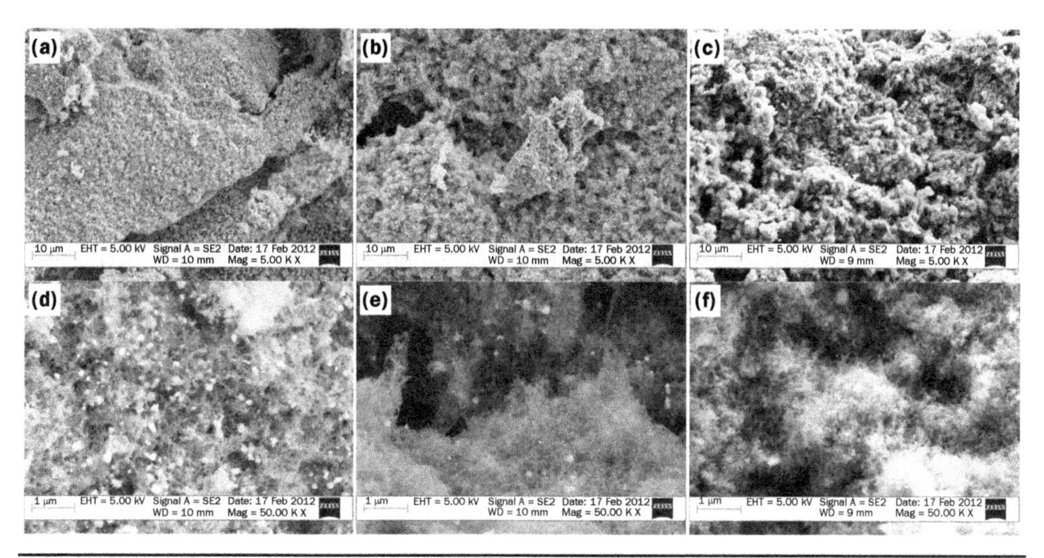

FIGURE 9.27 SEM images of MWNT TPUN char microstructures at low magnification (unit bars are 10 µm) for (a) 5 wt%, (b) 7.5 wt%, and (c) 10 wt% formulations and at high magnification (unit bars are 1 µm) for (d) 5 wt%, (e) 7.5 wt%, and (f) 10 wt% formulations.

From the low-magnification images (Fig. 9.27a–c), the MWNT char appears to have a relatively isotropic structure, contrasting the obvious directionality of the MMT clay char [47]. Under higher magnification (Fig. 9.27d–f), the MWNT char is clearly composed of a matrix of entangled MWNTs [47]. This structure of entangled conductive nanotubes likely results in a much more thermally conductive char, which could contribute to the poorer performance of the MWNT formulations overall. In addition, the high-magnification images show some bright spots that are evident of eroded MWNTs and related to the metallic catalytic precursors used during their preparation [47].

Figure 9.28 shows the char structures of the CNF TPUNs. Only the 15% and 20% weight loading formulations are shown, as both the 5 wt% and 10 wt% formulations produced unstable char [47]. From the low-magnification images (Fig. 9.28a,b) the CNF char shows a relatively isotropic structure, similar to that of the MWNT char [47]. In Fig. 9.28b there is also evidence of gas pockets within the char, which could account for the highly intumescent behavior of the CNF ablatives [47]. Higher magnification images of the char (Fig. 9.28c,d) show the char to be composed of a matrix of entangled CNFs [47]. This structure produces a thermally conductive char, resulting in poorer overall performance of the CNF TPUNs.

Summary and Conclusions Based on the results of this study, TPUNs appear to offer a potentially attractive alternative to conventional composites as a choice for SRM internal insulation materials. All three of the different nanomaterial additives enabled the polymer system to perform at a level comparable to that of SOTA insulation composite EPDM-Kevlar. Of particular importance were the TPUN formulations of 5% and 7.5% weight loading of MMT clay. These materials performed exceptionally well in every metric measured and actually outperformed the baseline EPMM-Kevlar material in recession and peak temperatures. Some other noteworthy formulations

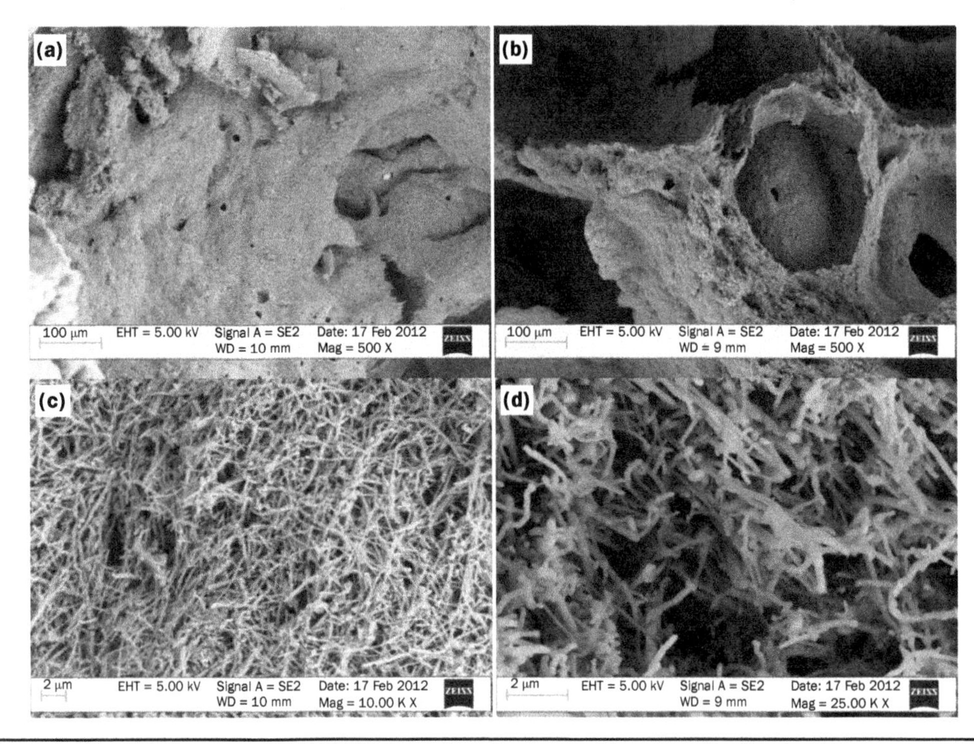

Figure 9.28 SEM images of CNF TPUN char microstructures at low magnification (unit bars are 100 µm) for (a) 15 wt% and (b) 20 wt% formulations and at high magnification (unit bars are 2 µm) for (c) 15 wt% and (d) 20 wt% formulations.

are the 5 wt% MWNT and 15 wt% CNF samples. The 5 wt% MWNT material performed well by roughly matching the baseline in all metrics, but was outperformed by the MMT clay formulations. The 15 wt% CNF material also performed well in all metrics, but was again outperformed by the MMT clay formulations and produced very poor time to peak results which reflect negatively on the overall performance of the material.

Through this study, PNCs are shown to be viable candidates for SRM insulation materials, and the relative effectiveness of various nanomaterial additives has been well documented. The next step in advancing these materials and enhancing their performance is to either focus upon the fine-tuning of nanomaterial loading, incorporation of multiple nanomaterial additives, or incorporation of a combination of conventional and nanomaterial additives. In addition, alternative processing techniques or nanomaterial types and preparations can be considered.

Ambuken et al. [59] investigated the high-temperature flammability and mechanical properties of TPU nanocomposites. From the UL 94 studies it was found that high loading of nanoparticles is needed for all the formulations to have a stable char structure. From the cone calorimeter studies on TPU nanocomposites, the lowest PHRR was observed for TPU-nanoclay formulation, while the highest PHRR was shown by TPU-CNF nanocomposites. The char obtained after firing TPU-nanoclay nanocomposites was swollen and it retained its structural integrity while TPU-MWNT and TPU-CNF exhibited fractures.

DMA testing was performed on the same materials to study complex modulus as a function of temperature up to 300°C when the char is expected to form. TPU-MWNT displayed better char modulus in comparison to TPU-nanoclay. For TPU-CNF, tests could not be completed above 220°C as the sample was transformed into a viscous melt. All samples lost modulus (soften) in the range of 120–160°C. For TPU-MWNT and TPU-nanoclay nanocomposites recovery in modulus (reinforced char formation) was observed at about 230°C with loadings above 7.5 wt%. The temperature range at which modulus had decreased but char formation was not yet significant enough to exemplify modulus recovery was termed the reinforcement gap. A correlation can be developed between crossover of dissipative versus elastic behavior (dominance of loss modulus versus storage modulus) and dripping in UL 94 tests.

Silicone Rubber–Ceramic Nanocomposite Studies by National University of Defense Technology, China [Yang et al. (2013)]

For their investigation, Yang et al. [60] created nanocomposites of silicone rubber with 30 wt% nanosilica reinforced with 8 wt% chopped carbon fibers. Then they investigated the effects of various loadings, from 0 to 40 phr (parts per hundred parts of rubber), of ZrC and ZrO_2 microparticles. The formulations were mixed for 30 minutes in a two-roll mill, using toluene to reduce the viscosity during mixing so that the carbon fibers would not be broken. Vulcanization lasted for 15 minutes at 160°C and 10 MPa in a compression-mold, and for 2 hours at 180°C at 1 atmosphere.

The researchers characterized the formulations for their mechanical properties using a universal testing machine (UTM) according to GB/T 528-1998 at a strain rate of 200 mm/min, and their ablation properties using an oxyacetylene torch with a heat flux of 4,152.9 kW/m^2. The calorimetry method used to find this heat flux was not given. They performed SEM, TGA at 20°C/min to 800°C in argon and 1,100°C in air, Fourier transform infrared spectroscopy (FTIR), and XRD on the pre- and posttest samples.

The carbon fibers were found to disperse well without breaking, but to adhere poorly to the rubber matrix [60]. For this reason, they actually had a negative impact on the mechanical properties of composites. Yang et al. speculated that it was the toluene that prevented strong bonding before and during its evaporation during processing. When the ZrO_2 content increased to 40 phr loading, the ablation rate was reduced by 72% and the mass loss dropped by 38%. The loading of 40 phr ZrC was found to reduce the ablation rate by 40% and the mass loss by 32%. The FTIR analysis revealed a few chemical reactions affecting the ceramics present in the char of each formulation. The TGA results showed a slight improvement in thermal stability with increased loadings of microparticles. Yang et al. found that the ZrO_2 particles were actually agglomerated nanoparticles, whereas the ZrC microparticles were shown as each single crystal. The authors concluded that was the reason for the better ablation performance of the ZrO_2 formulations [60].

9.3.6 Phenolic Nanocomposite Studies

Phenolic-CNF Nanocomposite Studies by Mississippi State University, United States [Patton et al. (1999)]

The ablation, mechanical, and thermal properties of vapor grown carbon fiber, VGCF (Pyrograf III™) manufactured by Applied Sciences, Inc. (Cedarville, Ohio), and phenolic resin composites were evaluated by Patton et al. [61, 62]. The objective of Patton et al.'s study was to determine the potential of using this material in SRM nozzles. Composites with varying VGCF loadings (30–50 wt%) including one sample with ex-rayon carbon fiber plies (the carbon fiber used in MX-4926, an industry standard ablative material) were prepared and exposed to a plasma torch with a heat flux of 16.5 MW/m^2 at approximately 1,650°C for 20 seconds. Low erosion rates and little char formation were observed, confirming that these materials were promising for SRM nozzle materials. When fiber loadings increased, mechanical properties and ablative properties improved. The VGCF composites had low thermal conductivities (approximately 0.56 W/m-K), indicating they were good insulating materials.

VGCF and phenolic resin SC-1008 manufactured by Borden Chemical, Inc. (now owned by Hexion) were used in this study. VGCF-phenolic resin composite specimens were prepared by high-shear mixing followed by thermal curing. Nineteen composite specimens were made by different mixing methods. Four types of samples were prepared and identified in wt/wt as follows: 30/70 VGCF-phenolic, 40/60 VGCF-phenolic, and 40/60 VGCF-phenolic (where ball milled as-received VGCF was used). Two types of fibers were used: as-received and compacted fibers. All specimens used in ablation testing were made as-received VGCF.

The flexural strengths and moduli were lower for these VGCF-phenolic samples compared to Patton et al.'s previous data of VGCF-epoxy using identical mixing techniques [62]. The flexural strength of 95.9 MPa and a modulus of 6.24 GPa for VGCF-epoxy composite [61, 62] are reported earlier. When the epoxy resin was replaced by phenolic resin, a flexural strength of 25.6 MPa and a modulus of 1.07 GPa was obtained [61]. This reduction in mechanical properties may be due to poor fiber-to-matrix adhesion for the VGCF-phenolic composites as well as lower mechanical properties of phenolic resin as compared to epoxy resin. Samples made with compacted fiber had high flexural moduli and strengths. The compacted fiber was a different batch of fiber and this may have affected also the properties of the composite samples made with it.

All the composite samples exhibited similar thermal conductivities (0.54–0.62 W/m-K) as compared to 0.28 W/m-K of cured phenolic resin without fiber. The data did not indicate that increased carbon loading or different manufacturing techniques significantly affect the thermal conductivity. The thermal conductivities in three dimensions were obtained on a 40/60 VGCF-phenolic ablation test composite sample. These values indicated that the material is slightly anisotropic with a difference of less than ±9% from the average in all three directions. The specimen has a thermal conductivity that is nearly the average of all three specimens.

Ablation was characterized by erosion rates, specimen weight loss, and the increase in load. NASA's standard nozzle material MX-4926 was used as the baseline. The MX-4926 composite is composed of 50 wt% woven ex-rayon carbon fiber, 15 wt% CB filler, and 35 wt% SC-1008 phenolic resin. The total carbon loading is 65 wt% in MX-4926. The VGCF-phenolic tested in this study contained only 30–45 wt% VGCF (only 46–69% as much carbon as MX-4926 composite). As compared to the MX-4926 baseline, the VGCF composites experienced (1) higher (58–114%) erosion rates, (2) lower weight loss (10–26%), and (3) lower load changes (60.5–68.5%). Therefore, more weight was being lost on or near the surface of the specimen due to higher erosion rates versus weight loss from subsurface thermal decomposition. The load changes for the VGCF composites are significantly lower than that of the MX-4926 composite. Load change is an indirect measure of neat penetration and char depth; increases in loading change indicate an increase in heat penetration and char depth. Thus, the VGCF composite appears to be a far better insulator than the MX-4926. The higher erosion rates of the VGCF samples probably reflect their lower carbon content. The erosion rate/wt%

of carbon present in the composites was very similar between MX-4926 and the VGCF composites.

The series of VGCF-phenolic composites exhibited extremely good erosion resistance (considering their low carbon loadings) while exhibiting less weight loss and load change than current SOTA MX-4926. Higher carbon contents are needed to match the erosion resistance of MX-4926. The carbon loadings used in this study have been much lower (30–45 wt%) than the MX-4926 composite (50 wt%). The mechanical properties of the VGCF composites need improvement. Lower void volume, better surface treatments for fiber/matrix adhesion, and mixed VGCF/carbon fiber weave/carbon filler combinations should be investigated. VGCF-phenolic composites should be prepared with higher carbon loadings (50–65 wt%) and subjected to ablation testing. The nanoscale dimensions of the VGCF cause major changes in the heat transfer rates and affect the resulting combustion/decomposition chemistry. The nano-sized fiber reduces the rate of heat transfer perpendicular to the surface into the composite compared with the rate of heat transfer in the MX-4926 material which uses continuous carbon fibers.

Phenolic Nanocomposite Studies by The University of Texas at Austin, United States [Koo et al. (2002–2005)]

The University of Texas at Austin Koo Research Group has been involved in the traditional R&D of ablative materials since the early 1990s by Koo and colleagues [64–70] (Koo et al. [63–65]; Cheung et al. [66]; Wilson et al. [67]; Shih et al. [68]; Koo et al. [69, 70]) and began the development of PNC ablatives in 2002. Koo and colleagues reported two new classes of PNCs (SRM insulation and nozzle materials) that are lighter and exhibit better ablation performance and insulation characteristics than current SOTA ablative materials. The SRM insulation PNCs have been reported in sections "EPDM-CNF Studies by The University of Texas at Austin, United States" and "Thermoplastic Polyurethane Nanocomposite (TPUN) Studies." This section relates solely to SRM nozzle PNCs.

Ablative materials are required to protect aerospace launching systems against rocket exhaust plumes. The flow environment of solid rocket exhaust is particularly hostile. System components must be protected from extreme heat flow temperatures of 1,000–4,000°C and from highly abrasive particles ejected at velocities greater than 1,000 m/s. Phenolic matrix composites have been used extensively for ablative materials. Current rocket nozzle assemblies are made of C-Ph composites, such as MX-4926 from Cytec Engineered Materials (CEM now known as Cytec Solvay Group). Three versions of C-Ph ablative MX-4926 laminates were fabricated by Cytec Solvay as controls for this study. The three versions of MX-4926 laminates were fabricated based on fiber orientation: (a) 60° shingle (SH), (b) molding compound (MC), and (c) two-dimensional (2D) fabric.

The objective of this nanocomposite rocket ablative materials (NRAMs) program is to develop a new class of nanostructured material that is lighter and has better erosion and insulation characteristics than current ablatives [71–79]. Our previous studies [71–73] have allowed us to explore other nanoparticles besides MMT nanoclays with phenolic for ablative application. The NRAMs exploit the ablation resistance of both phenolic and nanoparticles. Hexion SC-1008, a resole phenolic, was selected as the resin for this investigation. Several nanoparticles, such as Southern Clay Products (SCP) nanoclay, Applied Sciences CNFs, and Hybrid Plastics POSS have been evaluated with the SC-1008 phenolic resin. Based on WAXD and TEM, several nanoparticles were selected for further study. Cytec Solvay fabricated several MX-4926 alternates by replacing the CB filler with selected nanoparticles for ablation testing.

A small-scale, supersonic, liquid-fueled rocket motor burning kerosene and oxygen was used to study the ablation and insulation characteristics of the ablatives. Testing ablative materials in full-scale firings is not only expensive, but also requires lengthy planning and limited exposure time, thus reducing the number of samples and the variety of test conditions available. The simulated solid rocket motor (SSRM) (Fig. 5.15; see Sec. 5.13.1) has been demonstrated as a cost-effective laboratory device to evaluate different ablatives under identical conditions for initial material screening and development [63].

Experimental Approach: Simulated Solid Rocket Motor (SSRM) The SSRM [63] was used to evaluate the candidate materials (see Sec. 5.13.1). The SSRM is an established testing apparatus developed earlier that has been used extensively in our previous studies [64–70, 80–84]. The standard sample size is 4 by 4 by 1/4- or 1/2-in. thick (10.2 by 10.2 by 0.64- or 1.28-cm thick). The composite materials were bonded to 4 by 4 by 1/8-in. (10.2 by 10.2 by 0.32 cm) steel substrate. A narrow slot was machined into the bondline side of the steel substrate. The TC was embedded into this slot. The bead of the TC was placed at the center point of the plate. The temperature history of the bondline was recorded with this TC. C-clamps were used to clamp the ablatives to the steel substrate during testing. The test samples were placed in a fixture downstream from the SSRM nozzle exit. The fixture had adjustments in axial distance and impingement angle with reference to the plume centerline. A normal 90° impingement was used for this study.

Three axial standoff distances were used for this study to correspond to heat fluxes of 250, 625, and 1,000 Btu/ft^2-s (2,838, 7,094, and 11,345 kW/m^2). These distances were selected to simulate low, medium, and high heating conditions. Prior to the motor burn, the bondline temperature was recorded. Peak erosion was determined by pre- and posttest measurements using a pencil-point dial indicator. Mass loss was determined by pre- and posttest weight loss measurements. Consistent data sets were averaged for this study.

Selection of Materials: Polymer Matrix Composites MX-4926 is a rayon-based, eight-harness weave carbon fabric impregnated with a Mil R-9299 phenolic resin that contains CB fillers [85]. The approximate composition of materials for MX-4926 is 50 wt% carbon reinforcements, 35 wt% phenolic resin, and 15 wt% CB. MX-4926 is currently inserted in a typical SRM nozzle assembly [86]. The phenolic char formed at high-temperature is usually not very strong. The CB filler in the phenolic resin was replaced by different nanoparticles, so that the nanoparticles would reinforce the phenolic in the nanometer scale and strengthen the char.

Phenolic Resin Hexion SC-1008 [87] is the phenolic resin used in the MX-4926. This was our phenolic resin of choice for this investigation. SC-1008 is a solvent (isopropanol)-based resole. Table 9.6 shows some properties of the phenolic resin [87].

Table 9.6 Phenolic Resin Properties

Resin	Supplier	Formulation	Final cure temperature (°C)	T_g (°C) by DMTA	Specific gravity
SC-1008[12]	Momentive	60–64% solid isopropanol solvent	140	110	1.28

MMT Nanoclays Cloisite 30B, with loadings of 5, 10, and 15 wt%, was dispersed in the phenolic resole (SC-1008). The TEM analyses indicated the degree of dispersion and exfoliation of the nanoclay before committing to a 20 lb (9.1 kg) run at Cytec Solvay of these nanoparticle-resin mixtures to make prepregs. This proved to be a very cost-effective and efficient technique for screening different formulations. This nanoclay has been used with other resin systems, such as cyanate ester [88–90], thermoplastic elastomer [91–93], epoxy [94–96], and polyamide [97–105], by the Koo Research Group. For more details of MMT nanoclay, refer to Chap. 2.

Carbon Nanofibers (CNFs) CNFs are a form of vapor-grown carbon fiber, which is a discontinuous graphitic filament produced in the gas phase from the pyrolysis of hydrocarbons. The CNFs PR-19-PS and PR-24-PS were used in this study. More details of CNFs can be found in Chap. 2.

Polyhedral Oligomeric Silsesquioxanes (POSS) Representing a merger between chemical and filler technologies, POSS nanostructured materials can be used as multifunctional polymer additives, acting simultaneously as molecular-level reinforcements, processing aids, and flame retardants. Trisilanolphenol POSS was selected for this study. More details of POSS can be found in Chap. 2.

Discussion of Results The candidate materials were evaluated for degree of dispersion using WAXD and TEM prior to full ablation testing using the SSRM. Our first attempts were to modify the phenolic resin by the incorporation of MMT organoclays. Blends of phenolic and nanoparticles were dispersed using high-shear, nonsparking paint-mixing equipment. Neat resin castings (no fiber reinforcement) were prepared for WAXD and TEM analyses.

Transmission Electron Microscopy Analyses and Blending Experiments Each type of nanoparticles was dispersed in 0.5% IPA solvent overnight to create specimens for TEM analyses. The TEM image in Fig. 9.29a shows the CB particles agglomerated in the IPA [33]. The TEM image in Fig. 9.29b shows that the Cloisite 30B nanoclay layers are in an intercalated state [33]. Figure 9.29c shows the PR-24-PS CNFs are entangled [33]. Figure 9.29d,e shows the TEM images of trisilanolphenyl-POSS-SO1458 ($C_{48}H_{38}O_{12}Si_7$) and polyvinyl silsesquioxane uncured (PVSQ)-PM1285 ($(C_2H_3O_{1.5}Si)_n$) dissolved in the IPA solvent [33].

Based on TEM analyses, it was concluded that CB particles caused interference in the dispersion of all nanoparticles in the SC-1008 resin. As a result, CB was eliminated in subsequent blending. Cloisite 30B, in loadings of 5, 10, and 15 wt%, was dispersed in the SC-1008. Figure 9.30 shows the TEM images of the 5 wt% Cloisite 30B: 95 wt% SC-1008 [33]. The TEM images identified intercalation, and not exfoliation, of the nanoclay in the resin system. Similar results were observed for the 5 and 10 wt% Cloisite 30B in SC-1008 resin. The TEM analyses provided guidance to determine the degree of dispersion and exfoliation of the nanoclay before committing to a 20 lb (9.1 kg) run at Cytec Solvay of these nanoparticle–resin mixtures for the preparation of prepregs. This proved to be a very cost-effective and efficient technique for screening different formulations on a small scale.

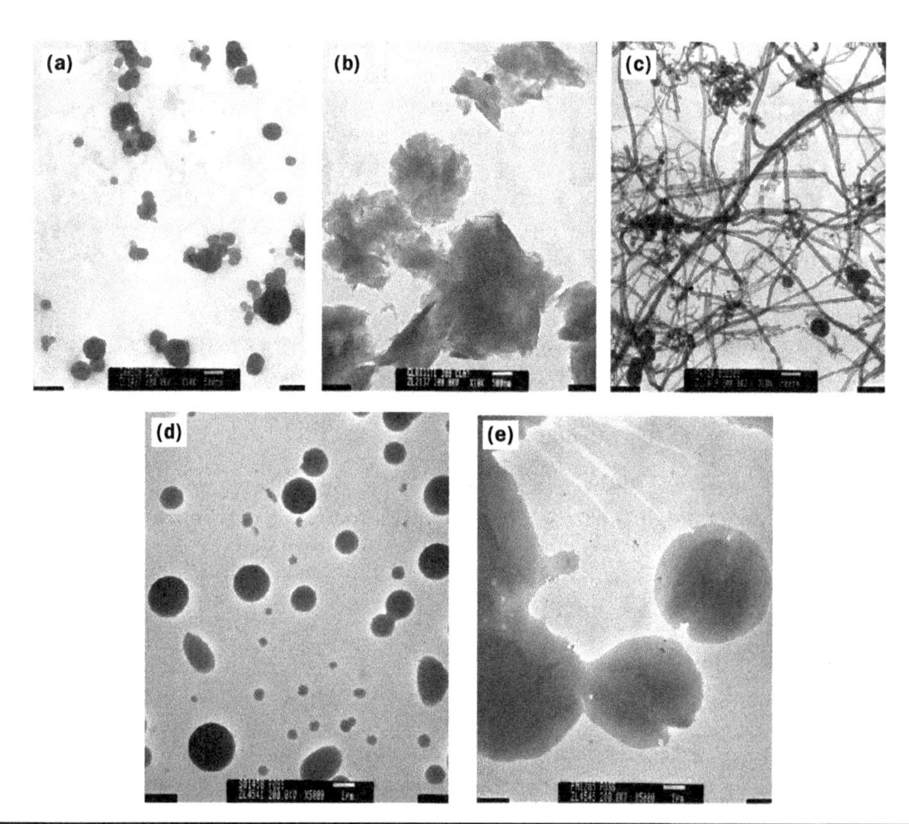

Figure 9.29 Transmission electron micrographs of (a) carbon black, (b) Cloisite 30B, and (c) PR-24-PS CNF, (d) trisilanolphenyl-POSS-SO1458 ($C_{48}H_{38}O_{12}Si_7$), and (e) (PVSQ)-PM1285 $(C_2H_3O_{1.5}Si)_n$ dissolved in the IPA solvent. Unit bars are 500 nm for (a)–(c) and unit bars are 1 μm for (d)–(e).

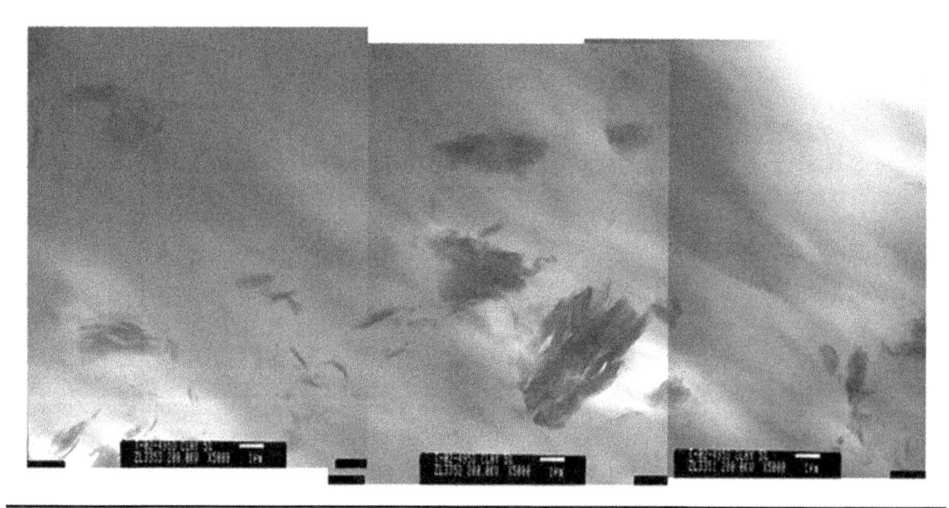

Figure 9.30 Transmission electron microscopy images of 5 wt% Cloisite 30B in 95 wt% SC-1008 (unit bars are 1 μm).

SSRM Test Conditions for Ablation Test Three test conditions to simulate thermal effect of the exhaust plumes using low, medium, and high heat flux were selected. These different conditions represent different regions in a rocket nozzle assembly. Alumina (Al_2O_3) particles were added to the exhaust stream to simulate the particle impingement effect of the solid rocket exhaust plumes from the aluminized solid propellants. The following six test conditions used are listed in the order of their severity:

- 1,000 Btu/ft^2-s (11,345 kW/m^2) with Al_2O_3 particles
- 625 Btu/ft^2-s (7,094 kW/m^2) with Al_2O_3 particles
- 250 Btu/ft^2-s (2,838 kW/m^2) with Al_2O_3 particles
- 1,000 Btu/ft^2-s (11,345 kW/m^2) without Al_2O_3 particles
- 625 Btu/ft^2-s (7,094 kW/m^2) without Al_2O_3 particles
- 250 Btu/ft^2-s (2,838 kW/m^2) without Al_2O_3 particles

MX-4926 as Controls Cytec Solvay fabricated three sets of C-Ph laminates (MX-4926) based on the SC-1008 resin system. Three versions of MX-4926 were used as controls for this investigation. They were made based on fiber orientation:

- Molding compound (1/2 in. squares of chopped fabric, designated by MX-4926 MC)
- Two-dimensional fabric (designated by MX-4926 2D)
- 60° shingle (the most effective orientation for rocket nozzle application designated by MX-4926 SH)

Samples of MX-4926 MC were tested at six conditions, as listed above, with an Al_2O_3 particle flow rate of 15 g/min for a test duration of 15 seconds. Table 9.7 shows a

Table 9.7 Summary of Ablation Data for MX-4926 MC

Ablation data/heat flux	1,000 Btu/ ft^2-s with particles	625 Btu/ ft^2-s with particles	250 Btu/ ft^2-s with particles	1,000 Btu/ ft^2-s without particles	625 Btu/ ft^2-s without particles	250 Btu/ ft^2-s without particles
Peak erosion (in.)	**0.235***	0.234	0.073	0.03	0.029	0
Residual mass (%)	85.0	86.3	90.2	88.2	88.9	91.6
MBHS temperature rise (°C)	315	159	133	175	164	133
Time to MBHS temperature rise (min)	0.3	0.5	1	1	0.9	0.9

*Bold entry stands for total erosion (burned through).

summary of the ablation data for MX-4926 MC specimens when the MX-4926 MC was exposed to the six test conditions [33]. As shown in Table 9.7, the peak erosion with Al_2O_3 was significantly higher than those specimens without particles. The residual mass increased with the decrease of heat flux for test cases with and without particles. The maximum backside heat-soaked (MBHS) temperature rise also decreased with the decrease of heat flux.

Figure 9.31 shows that MX-4926 SH has the lowest peak erosion, followed by MX-4926 2D, and MX-4926 MC has the highest peak erosion [33]. The residual mass of the three versions of MX-4926 specimens is similar for the three different heat fluxes. This is an insensitive criterion for comparing material candidates. The data indicated that MX-4926 MC has the best insulation property, followed by MX-4926 2D, and MX-4926 SH the worst insulation property.

In summary, MX-4926 SH has the most erosion-resistant (lowest peak erosion) characteristic and the worst insulation property (highest MBHS temperature rise); MX-4926 MC has the worst erosion-resistant characteristic and the best insulation property; and MX-4926 2D has the moderate peak erosion and insulation properties. For ease of manufacturing, MX-4926 SH is the most difficult and MX-4926 2D is the easiest. MX-4926 MC is more representative than MX-4926 2D. The molding compound (MC) specimens were selected for this study.

Scale Ablation Test SC-1008 with 5, 10, and 15 wt% Cloisite 30B were selected to replace 15 wt% of carbon black in the original MX-4926 formulation. Cytec Solvay prepared three versions of MX-4926 alternates and were designated as MX-4926 ALT 5%, MX-4926 ALT 10%, and MX-4926 ALT 15%. Three loadings of PR-24-PS in 20, 24, and 28 wt% were dispersed in phenolic without the rayon-carbon fiber reinforcement. Three loadings of trisilanolphenyl-POSS in 2, 6, and 10 wt% were also dispersed in SC-1008 phenolic resin. The POSS/SC-1008 mixture was used with the rayon-carbon fabric to produce prepregs in a manner similar to the nanoclay mixture. Table 9.8 shows the chemical compositions of the laminates used for ablation testing. Figure 9.32 compares the densities of three types of NRAMs with nanoclay, CNFs, and POSS at various loading levels

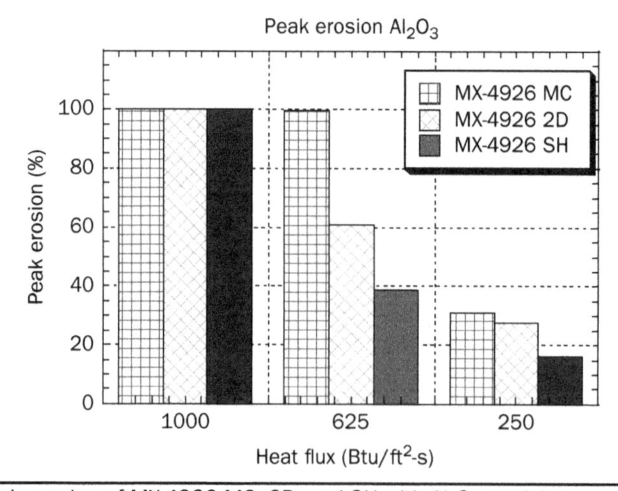

FIGURE 9.31 Peak erosion of MX-4926 MC, 2D, and SH with Al_2O_3 particles.

Table 9.8 Specimen Configuration for SSRM Laminate fabrication

Material ID	Density (g/cc)	Rayon-carbon fiber reinforcement (wt%)	Resin SC-1008 phenolic (wt%)	Filler (wt%)
MX-4926 (Control)	1.44	50	35	15 carbon black (CB)
MX-4926 ALT Clay 5%	1.42	50	47.5	2.5 Cloisite® 30B [$(HE)_2$MT]
MX-4926 ALT Clay 10%	1.43	50	45	5 Cloisite® 30B [$(HE)_2$MT]
MX-4926 ALT Clay 15%	1.43	50	42.5	7.5 Cloisite® 30B [$(HE)_2$MT]
PR-24-PS 20%/ SC-1008	1.35	None	80	20 PR-24-PS CNF
PR-24-PS 24%/ SC-1008	1.38	None	76	24 PR-24-PS CNF
PR-24-PS 28%/ SC-1008	1.41	None	72	28 PR-24-PS CNF
MX4926 ALT SO-1458 2%	1.41	50	49	1 Trisilanolphenyl-POSS® [SO-1458]
MX4926 ALT SO-1458 6%	1.38	50	47	3 Trisilanolphenyl-POSS® [SO-1458]
MX4926 ALT SO-1458 10%	1.40	50	45	5 Trisilanolphenyl-POSS® [SO-1458]

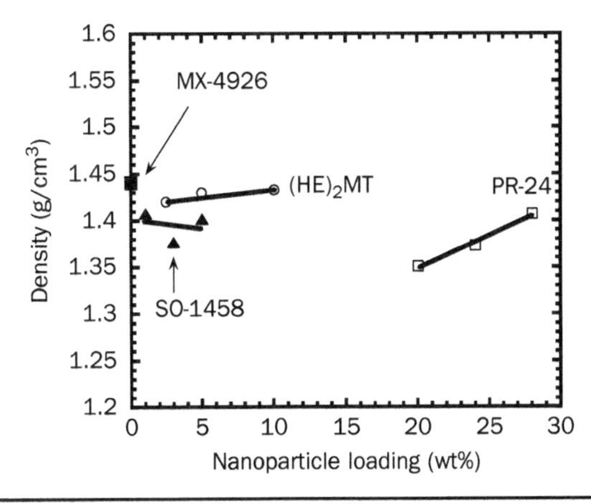

Figure 9.32 Density of MX-4926 and NRAMs with different nanoparticle loadings.

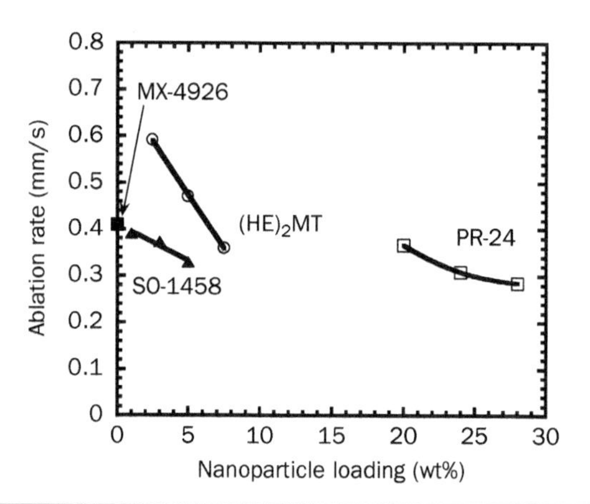

FIGURE 9.33 Ablation rate of MX-4926 and NRAMs with different types of nanoparticles at various loading levels.

of nanoparticles [33]. All clay-NRAMs, CNF-NRAMs, and POSS-NRAMs have densities lower than MX-4926.

At the suggestion of W. Luehmann [86] of Pratt & Whitney Space Propulsion/Chemical System Division (P&WSP/CSD), a lower flow rate of Al_2O_3 particles in the SSRM to simulate a total erosion of 200 mils (0.20 in.) of MX-4926 SH at 1000 Btu/ft^2-s was adopted. Several MX-4926 SH and MX-4926 2D were tested under these new test conditions, using a flow rate of 4 g/min Al_2O_3 particles [39]. For this study, test conditions of 1,000 Btu/ft^2-s heat flux with a flow rate of 4 g/min Al_2O_3 particles for a test duration of 15 seconds were adopted.

Figure 9.33 shows the ablation rates of MX-4926 and all three groups of NRAMs: clay-NRAM [$(HE)_2MT$ nanoclay], CNF-NRAM (PR-24), and POSS-NRAM (SO-1458) [34]. The ablation rate of MX-4926 was about 0.4 mm/s. For the clay-NRAM group, only the 7.5 wt% clay-NRAM had a lower ablation rate than MX-4926. The 15 wt% clay and 85 wt% phenolic becomes 7.5 wt% clay-NRAM when it is transformed into laminate, since all clay-NRAM contains 50 wt% carbon fiber reinforcements (Table 9.8) [34]. For the CNF-NRAM group, all three loadings had lower ablation rates than MX-4926, with 28 wt% CNF-NRAM being the lowest. For the POSS-NRAM group, all three loadings had lower ablation rates than MX-4926, with 5 wt% POSS-NRAM being the lowest. The loadings of POSS were 1, 3, and 5 wt%, with the 5 wt% exhibiting the lowest value of the three NRAM groups. The loadings of clay were 2.5, 5, and 7.5 wt%, the medium values of the three NRAM groups. The loadings of CNFs were 20, 24, and 28 wt% without the rayon-carbon reinforcements, the highest of the three NRAM groups.

Figure 9.34 shows the residual masses of MX-4926 and all the NRAMs [34]. The residual mass of MX-4926 was about 92 wt%. The POSS-NRAM group had the most residual mass, about 93 wt% for all three loadings. The clay-NRAM group had about the same residual mass as the MX-4926. The CNF-NRAM group had about 86–88 wt% residual mass, the lowest of all materials, control, and NRAMs.

Figure 9.35 shows the maximum backside heat-soaked temperature rise of MX-4926 and the NRAMs [34]. All NRAMs had lower maximum backside heat-soaked

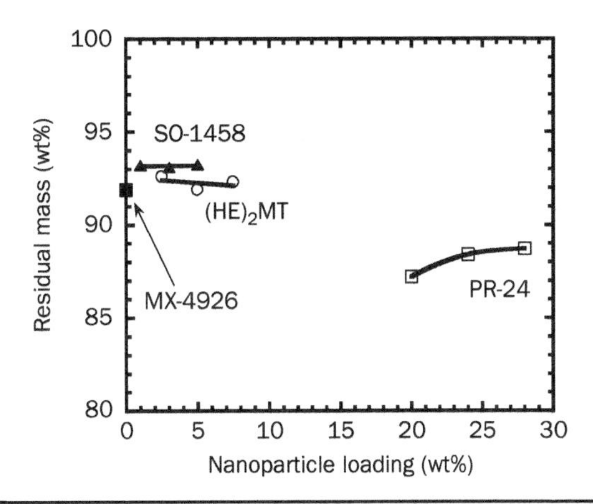

FIGURE 9.34 Residual mass of MX-4926 and NRAMs with different types of nanoparticles at various loading levels.

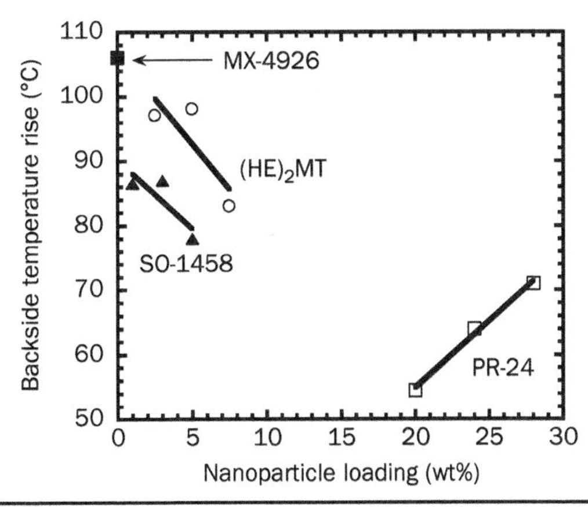

FIGURE 9.35 Backside heat-soaked temperature rise of MX-4926 and NRAMs with different types of nanoparticles at various loading levels.

temperature rises than MX-4926. The backside heat-soaked temperature rise of MX-4926 was about 106°C. It is obvious that the CNF NRAM group had substantial lower maximum backside heat-soaked temperature rise than MX-4926. It was from 54°C to 72°C. The result for the POSS-NRAM group was from 75°C to 86°C, the second lowest. The clay-NRAM group had the third lowest, from 82° C to 98°C.

An IR pyrometer was used to measure the surface temperatures of all materials during SSRM firings. Figure 9.36 shows the surface temperatures of MX-4926 and the NRAMs [34]. The surface temperature of MX-4926 was about 1,700°C. Surface temperatures of the CNF-NRAM samples were higher than those of MX-4926, the clay-NRAMs, and the POSS-NRAMs. This finding suggests that we may have better radial heat transfer than axial heat transfer, supported by the glowing heat of the surface observed during material testing. This phenomenon was observed by other researchers [61] and NRAMs had a significant effect on the surface temperature of the POSS-NRAMs.

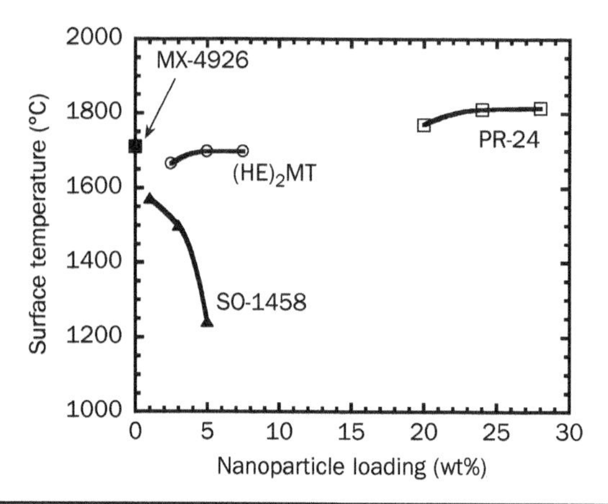

FIGURE 9.36 Surface temperature of MX-4926 and NRAMs with different types of nanoparticles at various loading levels.

FIGURE 9.37 Prefired Pi-K rocket motor nozzle assemblies fabricated from MX-4926, clay-NRAM, CNF-NRAM, and SM-8029 materials. This nozzle assembly shown as the nozzle part in Fig. 5.16 slides into the back end of the heavy-walled Pi-K rocket motor.

Air Force Research Laboratory Pi-K Solid Rocket Motor Ablation Test Based on the SSRM test data, selective NRAMs were scaled to 20 lb (9.1 kg) quantities at Cytec Solvay. These nanoparticle–resin mixtures were transformed into prepregs using T300 carbon fabric in a semi-production R&D facility in Winona, Minnesota. These prepregs were fabricated into molding compound and compression-molded into cylindrical billets [about 4.25 in. (10.8 cm) in diameter and 4 in. (10.16 cm) in length]. The billet was machined into a nozzle assembly similar to the ones shown in Fig. 9.37 [33]. Figure 9.37 shows

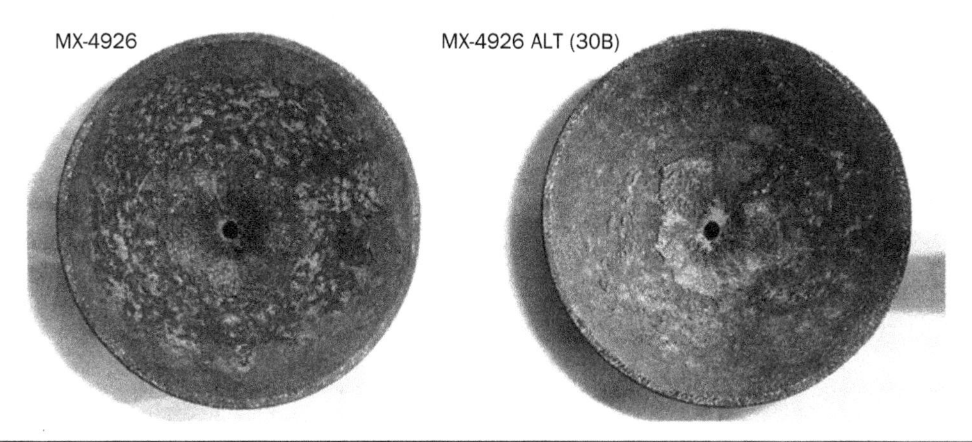

Figure 9.38 Pictures of the MX-4926 (left) and clay-NRAM (MX-4926 ALT/7.5 wt% Cloisite 30) (right) Pi-K rocket nozzles after firing at AFRL/Edwards AFB.

a MX-4926 with SC-1008 phenolic resin, a clay-NRAM with SC-1008 phenolic resin, a CNF-NRAM with SC-1008 phenolic resin, and a SM-8029 (quartz fabric–silicone resin composite) machined nozzle assembly of the AFRL/Edwards Pi-K SRM.

At Edwards AFB, California, AFRL/PRSM has developed a very cost-effective char motor to screen SRM insulation materials [91–93]. A schematic of the Pi-K SRM experimental setup is illustrated in Fig. 5.16 [33]. The motor is operated and fired in the upward direction. It usually holds up to 2 lb (0.91 kg) of solid propellant in an end-grain burning configuration (see Sec. 5.13.2).

For nozzle material testing, the cylindrical cone is removed, and the whole nozzle assembly at the backend of Pi-K SRM is replaced by NRAM candidate materials. The whole assembly is a convergent/divergent nozzle, and the nozzle throat is a machined graphite nozzle insert (see Fig. 5.17). The nozzle throat area is replaced by a graphite insert to avoid excess throat erosion during the motor firing. Using this experimental setup, reasonably accurate testing of materials designed to be used for the entrance region and the exit cone of the solid rocket nozzle assembly was anticipated.

Ablation Results Four successful Pi-K motor firings were conducted. The clay-NRAM specimen with 7.5 wt% Cloisite 30B survived the Pi-K motor firing with no obvious damage and exhibited a better appearance than MX-4926 in visual examination (Fig. 9.38) [106]. A deposit of aluminum oxide was formed near the center of the nozzle that resulted from the motor firing (Fig. 9.39) [106]. This deposit was removed for later analyses, resulting in some apparent material loss in the shape of the nozzle. A series of dimensional measurements were recorded on the clay-NRAM nozzles (Fig. 9.40) and are expressed relative to the MX-4926 control [106]. In all regions but right at the transition from the graphite insert to the NRAM materials (where the aluminum oxide deposit was removed, as noted earlier), there was an advantage to the NRAM in eroded shape.

Figure 9.40 is a photo of a posttest CNF-NRAM (without T300 carbon fiber reinforcement) after a Pi-K motor firing [106]. This sample fractured during test or during motor cooldown, as can be seen on the edges of the nozzle in Fig. 9.41. This CNF-NRAM specimen consists of 24 wt% PR-24-PS CNFs and 76 wt% SC-1008 phenolic resin and has no carbon fiber reinforcement. It is obvious that the conventional carbon fibers are needed as reinforcement for nozzle ablation.

Figure 9.39 Deposit of aluminum oxide in CNF-NRAM (SC-1008/PR-24 CNF).

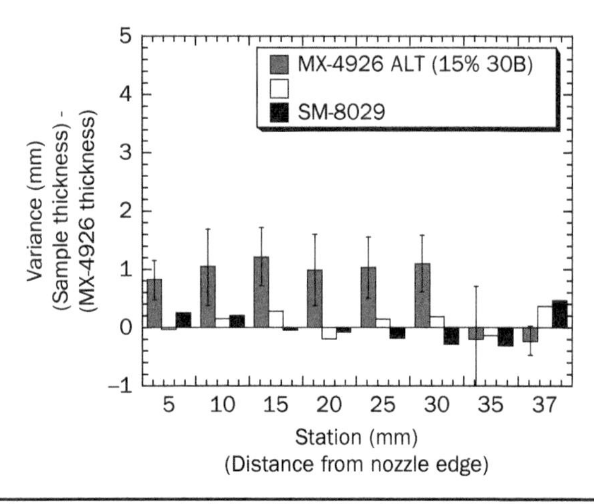

Figure 9.40 Dimension measurement of clay-NRAM [MX-4926 ALT (15% 30B)].

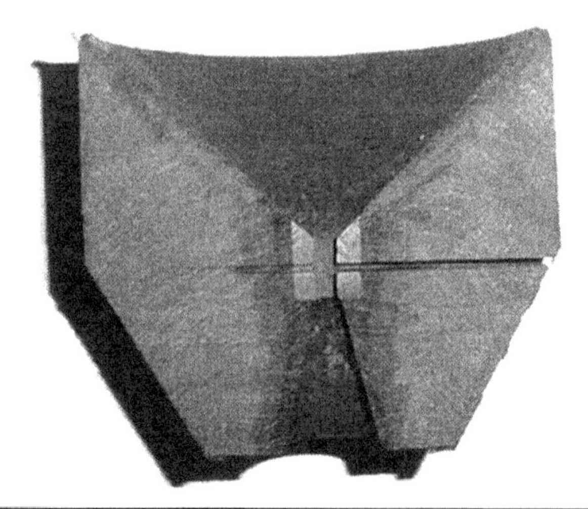

Figure 9.41 Photo of posttest CNF-NRAM (without T300 carbon fiber reinforcement) after Pi-K firing.

Mechanical Properties Results Peak stress and tensile modulus of MX-4926 control, 7.5 wt% Cloisite-NRAM, 24 wt% PR-19-PS CNF-NRAM, and 24 wt% PR-24-PS CNF-NRAM specimens were measured [106]. Figure 9.42 shows that only the tensile modulus of 7.5 wt% Cloisite 30B–NRAM exceeded the MX-4926 control [106]. Table 9.9 shows the mechanical properties of MX-4926 and MX-4916 ALT 7.5 wt% 30B [106]. The increased elongation and strain of the 7.5 wt% Cloisite 30B–NRAM resulted in a much high energy level needed to reach the peak load than MX-4926 control, but this was accompanied by a small decrease in modulus as compared to the control. Certainly nothing in the tensile data for the clay-NRAM would preclude it from service based on a comparison to MX-4926.

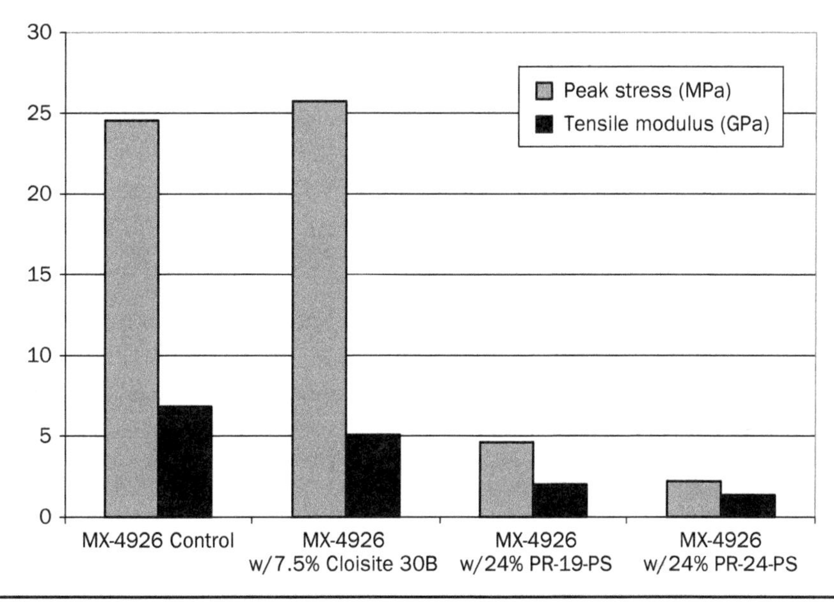

Figure 9.42 Peak stress and tensile modulus of MX-4926 control, MX-4926/7.5% Cloisite 30B, MX-4926/24% PR-19 CNFs, and MX-4926/24% PR-24 CNFs.

Table 9.9 Clay-, CNF-, and POSS-NRAM and MX-4926 Mechanical Properties

Nanoparticles	Peak stress (MPa)	Elongation at peak load (mm)	% Strain at peak load (%)	Energy to peak load (N-mm)	Elongation at break (mm)	Modulus (GPa)
MX-4926	24.55	0.25	0.41	82.02	0.27	6.85
MX-4926 ALT 7.5% 30B	25.74	0.52	0.87	427.65	0.53	5.05
24% PR-19-PS/SC-1008	4.66	0.15	0.25	13.88	0.15	2.00
24% PR-24-PS/SC-1008	2.19	0.15	0.24	4.60	0.15	1.35
POSS-NRAM	–	–	–	–	–	–

As expected, the CNF-NRAMs have very inferior mechanical properties than the MX-4926 since there was no carbon fiber–reinforcement. Table 9.9 shows the tensile test results for CNF-NRAMs with PR-19-PS or PR-24-PS [106]. The lack of reinforcement in the CNF-NRAMs made them significantly weaker than MX-4926. No POSS-NRAM mechanical properties were measured.

Summary and Conclusions For this study, Koo et al. [71–79] concluded the following:

- The feasibility of using NRAMs in rocket nozzle assemblies was clearly demonstrated using SSRM subscale and SRM Pi-K char motor ablation testing.
- MMT organoclays, CNFs, and POSS can be implemented into the existing semi-production line at Cytec Solvay to manufacture fiber-reinforced prepregs.
- A higher loading of MMT, POSS, and CNFs improves erosion resistance, and 28 wt% CNF has the lowest erosion rate in the absence of carbon fiber–reinforcing agent using SSRM.
- Backside heat-soaked temperatures of all NRAMs were lower than that of baseline MX-4926, and CNF-NRAM as a group had lower backside heat-soaked temperatures than the MMT-NRAM and POSS-NRAM using SSRM.
- Backside heat-soaked temperature trends were reversed for MMT and CNFs, whereas higher loadings of CNFs appeared to conduct heat better using SSRM.
- Peak erosion of POSS-NRAM is 20% lower than MX-4926 at very low loadings using SSRM.
- Peak erosion of CNF-NRAM (28 wt% CNF) without rayon fabric is 42% lower than MX-4926 using SSRM.
- Backside heat-soaked temperatures of CNF-NRAM are 68% lower than MX-4926 using SSRM.
- Backside heat-soaked temperatures of clay-NRAM are 28% lower than MX-4926 using SSRM.
- Backside heat-soaked temperatures of POSS-NRAM are 26% lower than MX-4926 using SSRM.
- Four solid rocket Pi-K char motor firings were tested with MX-4926, clay-NRAM (7.5 wt% clay), and CNF-NRAM (24 wt% CNF) at the Air Force Research Laboratory/Edwards AFB (AFRL/PRSM).
- Clay-NRAM specimen with 7.5 wt% Cloisite 30B survived the Pi-K motor firing with no obvious damage and exhibited a better appearance than MX-4926; mechanical properties of clay-NRAM are comparable to MX-4926, which make this clay-NRAM a suitable replacement of MX-4926.
- Physical, mechanical, and thermophysical properties of selective NRAMs need to be measured.
- Optimal nanoclay dispersion in SC-1008 phenolic resin needs further study.
- Synergism between CNF and other nanoparticles in SC-1008 phenolic resin needs further study.

- Selective high-performance thermosetting resin system, such as cyanate ester resin, can be a potential replacement of phenolic resin due to its high T_g and relative ease of manufacturing than phenolic resin used for advanced NRAM development.

Phenolic Nanocomposite Studies in China [Liu et al. (2009) and Yu et al. (2012)]

Based on the current developments, it clearly appears that the Chinese scientific community is diligently conducting the study of nanostructured ablative materials. Liu et al. [107] have focused their attention on the study of C-Ph composites (PR-CF) in the presence of POSS. Two types of functionalized POSS were employed as nanomodifiers of the phenolic resin: a commercial octamethyl-POSS (Me-POSS) and a phenyl-POSS (Ph-POSS). The Ph-POSS was produced with commercial OctaTrimethylsiloxy-POSS (TMS-POSS) as starting material and using styrene according to a procedure developed by Laine [108]. For comparison purposes, a silica precursor tetraethoxysilicate (TEOS) was also selected to produce another nanomodified resin [109] used to impregnate the C-Ph composite (T-SiO_2/PR/CF). The exact compositions of the different formulations in terms of filler percentages were not reported. The ablation performances of the produced composites are T-SiO_2/PR/CF, Me-POSS/PR/CF, and Ph-POSS/PR/CF. These materials were studied using an OTB. The following conditions were selected: the distance from torch tip to sample was 20 mm and the exposure time was 20 seconds.

In terms of linear ablation rate as well as mass loss, the hierarchy displayed by the tested materials was the following: T-SiO_2/PR/CF > Me-POSS/PR/CF > Ph-POSS/PR/CF. With respect to the type of functionalization attached to the POSS, the Ph-POSS displayed a significantly lower erosion and mass loss than the Me-POSS. This result can be directly related to the higher carbon content of the phenyl groups over methyl structures. The better thermal stability of Ph-POSS over Me-POSS embedded in a polymeric matrix was also confirmed by other studies [110, 111]. The SEM analysis of the eroded samples also confirmed that the Ph-POSS/PR/CFs produced the best charred surface: a homogeneous hybrid carbonaceous/silica skin on the top region of the Ph-POSS/PR/CF samples protected the inner layers of the composites, whereas the other two formulations displayed a porous and irregularly eroded char, leaving the carbon fibers exposed to the flame.

A novel flake graphite–reinforced barium-phenolic resin composite was made by roller coating technology and its thermal ablation was studied by Yu and Wan [112] under long pulse laser irradiation. Results show that ablation rate of the flake graphite–reinforced barium-phenolic composite is 32.8 μg/J at 1,700 W/cm^2 irradiation power density, which is lower than that of the barium resin and the barium-phenolic carbon fiber-reinforced composites. When the composites are exposed to the laser radiation, the flake graphite acts as a mirror and reflects part of the laser energy. As a result, the laser radiation deposited on the composites is reduced. The size of the flake graphite also affects the ablation rate. The composite with flake graphite diameter of about 0.5 mm has the lowest ablation rate.

Phenolic-Graphene Oxide Studies by Xi'an Jiaoton University, China [Si et al. (2009)]

For their experiment, Si et al. [113] prepared nanocomposites of graphene oxide (GO) in phenolic resin for thermal characterization. They prepared GO from graphite at three different temperatures and mixed it into the phenolic resin via ultrasonic bath at loadings ranging from 0.1 to 2 wt%. The mixtures were degassed at room temperature and pressure overnight and then at 50°C and 0.095 MPa until no more weight loss was observable. Finally, the composites were cured at 120°C for 2 hours, 165°C for 4 hours, and 220°C for 2 hours.

Si et al. conducted FTIR to verify the success of the transformation to GO, and particle size analyses on the unmixed GO as well. For thermal characterization, TGA in N_2 to 800°C at 10°C/min and DSC in N_2 to 300°C at 10°C/min were conducted on each formulation. Finally, SEM and XRD were conducted on the composites for dispersion, composition, and morphology verification.

Si et al. found that the transformation technique was most effective at higher temperatures, and that the particles were comparable to nanoclay in size. The GO particles were found to disperse well into the resin and produced a moderate increase in char yield and thermal stability even at low loadings. The researchers recommended that the GO-phenolic nanocomposite be investigated for ablation resistance in the future.

Phenolic–Nanosilica Nanocomposite Studies by Defense Research and Development Organisation, India [Srikanth et al. (2010)]

The most significant research of the Indian scientific community on nanostructured polymeric ablatives was carried out at the Defence Research and Development Organisation (DRDO). Srikanth et al. [114] studied the effect of nanosilica on the ablation resistance and on the interlaminar shear strength (ILSS) of carbon–phenolic (C-Ph) composites. The through-thickness thermal conductivity was also measured at 300°C under steady-state conditions (ASTM E–1225-09) [115]. Moreover, the microstructure of the ablated materials was investigated by SEM. Rayon-based carbon fabric was used as a reinforcing material and a resole was used as the matrix resin. The nanosilica had an average diameter of 40–50 nm. Four different formulations were prepared by adding 0.0 (control), 0.5, 2.0, and 4.0 wt% of nanosilica to the phenolic resin. The pristine resin and the produced nanocomposites were used to impregnate the carbon fabric. Three samples from each of the laminate were machined to a size of 10 mm diameter by 20 mm (the length of the sample was perpendicular to the direction of the fabric). The specimens were exposed to a plasma arc jet at a flame velocity of about Mach 1 and at a stagnant flux of 2.5 MW/m^2 for 20 seconds. The ablation performance was determined by dividing the weight loss during the test by the exposure time.

Due to the presence of nanosilica, the ILSS increased with the nanosilica content up to a filler percentage equal to 2.0 wt%: above this threshold the ILSS decreased. This result indicated that nanosilica particles embedded in the matrix acted as an anti-slippage agent. It was also speculated that above 2.0 wt% the nanosilica decreased the adhesion between the fabric layers. An increased ILSS allows the composites to be more resistant to the high aerodynamic shear forces and is expected to result in an enhanced ablation resistance. Plasma arc-jet test results were in good agreement with the ILSS trend: in fact, the ablation resistance increased with the nanosilica content up to 2.0 wt%, beyond which it was lower than the control sample. The addition of nanosilica also reduced the thermal conductivity of the C-Ph composites: the thermal conductivity depends on the lattice vibrations and the microstructure (micro-cracks, grain boundaries, porosity, etc.). The introduction of nanosilica increased the number of grain boundaries, resulting in lower thermal conductivity: similar observations for the C–SiC composites have been reported by Kumar et al. [116]. As a result of a reduced thermal conductivity, the ablation resistance was improved.

Since the formation of the char is an endothermic process, the amount of heat dissipated during the ablation depends on the efficiency of this process. Moreover, the char dissipates a large fraction of incident heat through surface reradiation. Accordingly, a complete charring of the resin combined with high char retention provides better heat dissipation during ablation. The SEM microstructure studies showed that the damage to the carbon fibers in the blank–Ph is considerably higher compared to the damage occurring in the presence of nanosilica. The ablated surfaces of nanostructured C-Ph composites showed that SiC was formed on carbon fibers during ablation. This result was also confirmed by XRD analysis. Kumar et al. proposed two main mechanisms for the ablation results and the related microstructural and compositional observations. First, under plasma arc-jet testing, the temperature can reach 4,000°C. In the case of C-Ph composites, the matrix cannot effectively maintain the carbon layers together, leading to severe erosion losses. In the case of nanosilica loaded C-Ph composites, two unique processes occur when the temperature increases beyond 1,600°C: the silica becomes molten [117] and forms a viscous layer that acts like a high-temperature binder able to hold the underlying phenolic matrix and the char. Thus, it ensures a higher radiant emission and leads to a complete charring of the resin for absorbing more heat (first process). Moreover, the molten silica reacts with the carbon, forming a hard SiC phase [118] (second process). The resulting carbon–silicon carbide (C–SiC) can withstand the aerodynamic shear forces under high-temperature erosive environments, thereby enhancing the ablation resistance [119].

Phenolic-Zirconia and MWNT Nanocomposite–Modified Carbon Fabric Composite Studies by Defense Research and Development Organisation (DRDO), India [Srikanth et al. (2013)]

Srikanth et al. [120] conducted a novel investigation into the effects of ZrO_2 as a coating on carbon fabric for use in C-Ph composites. MWNTs were also used in nanocomposites of C-Ph without ZrO_2. Then, the best performer of the ZrO_2-coated composites was combined with MWNTs for a nanocomposite with ZrO_2. A solution of $ZrOCl_2$, ethanol, and water was sprayed on rayon carbon fabric and allowed to stay for 96 hours, while the $ZrOCl_2$ and H_2O reacted to produce ZrO_2 and HCl, in a coating on each fiber. By modulating the amount sprayed, the researchers produced formulations of up to 9.5 wt% of ZrO_2. Resole-type phenolic resin was applied to the fabric to produce prepregs, and these were cut into 60 squares of 15 cm side length for hand layup and autoclaving at 120°C and 5 bars for 2 hours, and then at 180°C and 5 bars for 4 hours. MWNTs were mixed into the resin at 0.5 wt% and ball milled for 3 hours at 250 rpm. This resin was applied to uncoated fabric and processed the same way. When the best performer of the ZrO_2 formulations was found, 40 layers of ZrO_2-coated carbon fabric prepregs were laid on 9 layers of uncoated carbon fabric prepregs with MWNTs.

Srikanth et al. [120] characterized the formulations for fiber volume fraction (ASTM D 3171), flexural strength (ASTM D 790), ILSS (ASTM D2344), thermal conductivity (ASTM E1225-99), and ablation resistance via exposure to a heat flux of 4.0 MW/m^2 produced by plasma arc jet for 30 seconds at a flow velocity of Mach 1. Backside temperature was recorded during and for 40 seconds after the test, which was not enough time to reach peak temperature. The posttest ablation samples were examined via SEM and XRD.

Srikanth et al. found that the ZrO_2 coating significantly reduced the density, fiber volume fraction, flexural strength, and ILSS of the composite, so much so that the highest loading made the composite so easily delaminated that it could not be machined into usable samples. SEM showed that the ZrO_2 coating often cracked, flaked, or grew in nodules. Higher loadings of ZrO_2 were found to decrease thermal conductivity and increase the ablation rate of the composites (Table 9.10) [120]. While the MWNTs were dispersed well and did improve mechanical properties and increase thermal conductivity, they did not have a significant effect on the ablation rate. Table 9.10 shows the ablation rate and back-face temperature of composites tested in plasma arc-jet test. The blank C-Ph composite (control sample) has the best ablation performance.

Phenolic-Clay Nanocomposite Studies by Tarbiat Modares University, Iran [Bahramian et al. (2006–2012)]

The main contribution of the Iranian scientific community on the topic of nanostructured ablatives has been the efforts of researchers at Tarbiat Modares University. They focused mostly on the improvement of asbestos-phenolic ablatives using numerical modeling and experimental approach. Particularly, Bahramian and colleagues [121–126] studied the numerical and experimental aspects of the ablative performance, thermal decomposition, and temperature distribution through the thickness of phenolic/layered silicate nanocomposites impregnated with asbestos cloth and compared it with traditional asbestos-phenolic ablators.

Bahramian et al. [121] modeled the ablating, charring, and thermal degradation behavior of a resole-type phenolic resin/asbestos cloth composite using an oxyacetylene flame test. Their model requires solving heat transfer equations with moving boundaries. Explicit forward finite difference method (FDM) is used for the heat transfer calculation. Moving boundaries are fixed by the Landau transformation. The ablation equation is solved numerically. This model demonstrates the variation of thermophysical properties, such as thermal conductivity, density, and specific heat at different temperatures. Bahramian et al. showed that their model is adequately confirmed by the experimental data of the thermophysical and ablation properties of an asbestos-phenolic composite and can be used as a simple tool in the design of the thickness of a heat shield. In their test case study using conditions of 800 W/cm^2 heat flux and 3,000 K temperature of hot gas, and a test duration of 20 seconds; the back-face temperature of char surface of a 4-mm-thick asbestos-phenolic composite is 468 K, while three oxyacetylene flame show temperatures of 431, 438, and 423 K.

In Bahramian et al.'s work [123], the combination of solution and in situ intercalation methods was used to fabricate the nanocomposite. Ethyl alcohol was used to disperse the layered silicates (Cloisite 15A) and dissolve the phenolic resin. The phenolic resin in

Table 9.10 Ablation Rate and Back-Face Temperature of Composites in Plasma Arc-Jet Test

Composite description	Ablation rate (mm/s)	Back-face temperature (°C)*
Blank C-Ph	0.054	169
3.5 wt% Zr-C-Ph (zirconia-coated C-fabric)	0.078	114
6.5 wt% Zr-C-Ph	0.105	104
FG-C-Ph (functionally-graded C-Ph)	0.103	135

* Back-face temperature at the end of the test (30 seconds).

an excess of ethyl alcohol could be adsorbed into the nanoclay tactoids. However, once the ethyl alcohol was removed, the phenolic chains remained trapped in the layered silicate sandwiched structure. The asbestos cloth was impregnated with the nanocomposite system. Three composites having different nanoclay percentages were studied: 3 wt% (NMA1), 4 wt% (NMA2), and 6 wt% (NMA3). Flat panel samples were prepared for the oxyacetylene flame test (heat flux 900 W/cm^2). K-type TCs were placed through the thickness of the sample at 2, 4, and 6 mm from the surface: the surface erosion measurement was carried out on cylindrical shaped samples (10 mm diameter and 25 mm height). XRD analysis was used to evaluate the dispersion of the nanoclay in the composites. The specific heat capacity measurements were performed according to ASTM E-1269 [127]. Thermogravimetric and DSC analyses were also performed on the resulting materials.

The XRD patterns of nanocomposites exhibited a behavior typical of an exfoliated morphology, confirming the penetration of the resol molecules between the clay sheets. According to the TGA data, the presence of clay improved the thermal stability of the composites. Many factors are responsible for this improvement. Among them, the dispersed large aspect ratio platelets hindered the diffusion of the decomposed volatile products, as a direct result of the decreased permeability. For the thermal oxidation process, the mass transfer of oxygen into the gallery may also be reduced, decreasing the degradation of the matrix. The ablative reassembly of nanoclay platelets and the sintering of them at elevated heat fluxes led to the formation of a homogeneous silicate-carbonaceous char layer that functions as a heat shield during erosion. This is widely accepted as the main flame retardancy mechanism common in this class of nanocomposites [128–131]. It follows that in the range of the quantity of nanoclay content, the higher the nanoclay content, the better the thermal stability. TGA results were also used to acquire the parameters used to produce a model of the ablation process. The specific heat capacity measurements were carried out to a temperature of 500 K: the values of the specific heat of the asbestos-phenolic composite and of the NMA3 nanocomposite are in the range of 1,000–1,500 and 3,300–5,800 J/kg-K, respectively. The higher value of the NMA3 nanocomposite is one of the key factors to provide the better insulation performance of the NMA3 over the asbestos-phenolic composite. The dynamic DSC scans also indicated that the heat of ablation of the NMA3 composite was considerably higher (about 950 kJ/kg) than the control material (45 kJ/kg).

The experimental temperatures of the top surface and through the thickness of asbestos-phenolic and NMA3 samples were also determined. During the same experimental conditions, the steady-state surface temperature of the asbestos-phenolic composite and of the NMA3 were about 1,500°C and 2,100°C, respectively. This difference clearly shows that the nature of top surface of the traditional composite and of the nanocomposite-impregnated asbestos fabric is completely different. The higher value of surface temperature of the NMA3 is another key factor to anticipate a higher insulation performance than the traditional asbestos-phenolic composite. Since the reradiated energy follows a T^4 law, a decreased heat flux is transferred into the material thus increasing the lifetime of the NMA3 material. Bahramian et al. also evaluated the thermal diffusivity of the material as a function of the depth. The effective thermal diffusivity of the asbestos-phenolic composite is twice the NMA3 composite. This phenomenon suggests a clear difference in the char layer formation and in the ablation mechanism of the composite and of the nanocomposite-based system.

The above study can be summarized as follows: (a) at 900 W/cm^2 and 3,400 K of hot gas, the effective thermal diffusivity of the asbestos-phenolic and asbestos-phenolic-6 wt% MMT clay composites was estimated. The ablation performance of NMA3 nanocomposite heat shield is 100% more than the composite counterpart, and (b) in cone calorimeter test at 80 kW/m^2 radiation heat flux, the NMA3 nanocomposite shows 51% heat release rate (HRR) and 40% mass loss lower than the asbestos-phenolic composite.

In Paydeyesh et al.'s work [126], a comparative study of ablation performance of highly filled phenolic-asbestos-MMT nanocomposite was conducted and evaluated with the comparison material. The heat diffusion through the thickness and erosion rate were measured for both systems. Ablation mechanism, thermal degradation kinetics, and thermophysical properties of highly filled ablative nanocomposites were also investigated. The mathematical model was evaluated by the oxyacetylene test data (heat flux 800 W/cm^2). MMT clay (Cloisite 15A) content of 0, 40, 50, 60, and 70 phr (0, 14, 16, 17, and 20 wt% of the total composite weight) was studied with varying amounts of phenolic resin and asbestos cloth. To prepare the highly filled nanocomposites, a combination of solution and in situ interaction methods was used in their study.

Figure 9.43 shows the SEM micrographs of the ablative nanocomposite [containing 50 phr (16 wt%) MMT clay] specimens after oxyacetylene flame test, respectively [126].

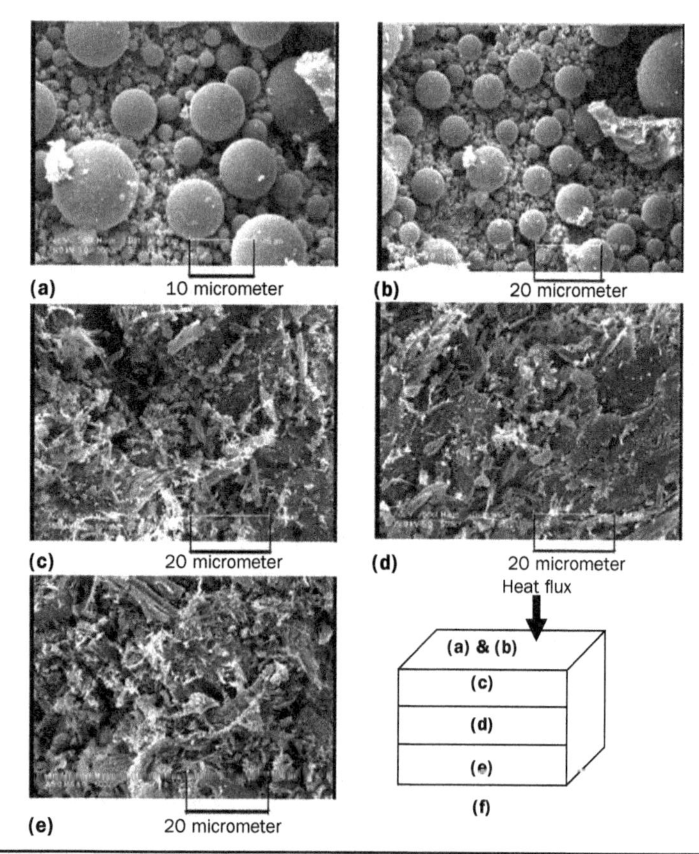

FIGURE 9.43 Scanning electron micrographs of ablative composite containing 50 phr MMT clay after ablation test: top surface (a,b); lateral surfaces (c–e); and heat flux direction (f) [126].

The figure shows the top surface (surface of char zone), lateral surfaces, and heat flux direction. The figure is intended to show the characteristics of various regions: virgin zone, porous pyrolysis zone, and char zone. In the char zone, nanoscopic porous morphologies are observed. The molten combustion products formed particles with special shapes. It was noted that the char formed on the surface of composite sample has higher porosity than those formed on the surface of nanocomposite sample [126].

Several conclusions can be considered from the above study [126]:

1. Ablation performance of highly filled nanocomposites is better than their composite counterpart. Nanocomposite sample containing 50 phr of MMT clay (asbestos-phenolic-clay ratio is 52/32/16 wt%) has the best erosion rate (mm/s) of 135% lower than that of the asbestos-phenolic composite sample.
2. Better performance of highly filled nanocomposites is due to the formation of ceramic char layer on the ablator surface. The formation of char layer is easier in the highly filled than the low filled nanocomposites.
3. No significant difference is observed between thermal stability of highly filled nanocomposite and its composite comparison sample.
4. Mathematical model is adequately confirmed by the experimental data of the thermophysical and ablation properties of nanocomposites. It can be used as a design tool to estimate optimal thickness of a nanocomposite heat shield.
5. Asbestos fibers are banned in many countries throughout the world because they are carcinogenic.

Phenolic-Kaolinite and Graphite Nanocomposite Studies by Tarbiat Modares University, Iran [Bahramian et al. (2013–2014)]

For this investigation, Bahramian [132] developed a mathematical model for the HRR and thermal conductivity curves of an ablative and prepared nanocomposites of novolac-type phenolic resin with graphite to compare results. The graphite was dispersed by solution and in situ intercalation in methanol with phenolic resin and dried, then held at 80°C for 180 minutes, 120°C for 15 minutes, 150°C for 60 minutes, and 160°C for 30 minutes. Formulations with 0, 3, and 5 wt% graphite were prepared.

Bahramian inspected the samples for morphology via SEM, composition via XRD, dispersion via TEM, specific heat via DSC at 10°C/min according to ASTMD-1269, thermal stability via TGA, thermal conductivity according to ASTM E1225-87, density and porosity according to ASTMD-4018, flammability via cone calorimetry according to ASTM E1354 at a heat flux of about 8×10^4 W/m^2, and ablation resistance via oxyacetylene flame test according to ASTM-E-285-80 at 2.5×10^6 W/m^2.

Bahramian found that the graphite was dispersed well in the polymer matrix. He plotted HRR and thermal conductivity curves from the models that agreed well with the experimental results for all three formulations [132]. The peak HRRs were reduced by 41% and 58% for 3 and 5 wt% graphite, respectively. The compositions were found to have ablation rates of 0.0184 mm/s for neat phenolic resin, 0.0164 mm/s for 3 wt% graphite, and 0.0153 mm/s for 5 wt% graphite. The graphite was found to have a positive effect on the thermal stability.

For their most recent investigation, Bahramian and Astaneh [133] prepared composites of 20 wt% kaolinite microparticles, 6–9 wt% kaolinite nanoparticles, and 6–9 wt%

graphite nanoparticles in novolac-type phenolic resin for ablation, mechanical, and tribological testing. In addition, chopped carbon fibers were used for reinforcement. The formulations were prepared by dissolving phenolic powder in isopropyl alcohol and stirring in the kaolinite or graphite particles at 1,500 rpm for an hour and sonicating for 20 minutes. Then, 40 wt% carbon fibers were added and mixed by hand. Finally, the mixtures were degassed at 350 K for 30 minutes and hot press molded at 370 K for 1 hour and then 410 K for 2 hours.

The formulations were inspected for their morphology via XRD, ablation resistance according to ASTM-E-285-80, thermal stability via TGA at 10°C/min to 950°C, flexural strength according to ASTM D 790, dispersion via TEM, and thermal insulation/diffusivity and abrasion resistance via the team's own methods. The ablation specimens were 10 mm in diameter and 20 mm tall and were exposed to 8.5×10^6 W/m^2 for 20 seconds. The abrasion testing was conducted by measuring mass loss after a certain amount of exposure to an abrasive surface, and the thermal testing was performed by measuring backside temperature response to a heating jacket placed on the sample.

Bahramian and Astaneh found that for all loadings, the graphite nanoparticles produced a greater ablation resistance than the kaolinite nanoparticles. No nanoparticles were found to have a severely negative effect on the mechanical properties, but the kaolinite microparticles were found to reduce the flexural strength by over 75%. At the same time, the kaolinite nanoparticles produced the same ablation resistance at their highest loading as the microparticles. The thermal insulation and diffusivity of the 9 wt% graphite formulation were anomalously low, lower even than the kaolinite microparticles. This formulation also showed the least decrease in abrasion resistance, so the researchers concluded that it was the formulation with the best dispersion. One TEM picture was taken of this formulation [133]. The dispersion shown is not excellent. Finally, the TGA results showed no significant change in thermal stability properties.

Phenolic-Nanosilica with Carbon Fiber Studies by Islamic Azad University, Iran [Mirzapour et al. (2014)]

Mirzapour et al. [134] investigated nanocomposites of resole-type phenolic resin with nanosilica and chopped carbon fibers for ablative properties. A loading of 50 wt% carbon fiber was used for all formulations, while the nanosilica loading varied from 0 to 5 wt%. The nanosilica was added to the resin with methanol and simultaneously sonicated and stirred at 25°C for 1 hour. The chopped carbon fiber was then added and mixed by hand. The formulations were degassed in a compression press for 30 minutes at 120°C and 1 bar, before the press was opened to allow the gases to escape. Then it was replaced and the formulations were cured at 160°C and 10 bars for 5 hours. After staying for 1 week in atmosphere, they were post-cured at 200°C for 2 hours.

Mirzapour et al. inspected the mechanical properties of the formulations via the three-point bending testing at a strain rate of 1 mm/min, thermal stabilities via TGA in air at 10°C/min to 1,000°C, and ablation performance via oxyacetylene flame testing according to ASTM E285-80 at a heat flux of 8 MW/m^2. Pre- and posttest specimens were examined via SEM, XRD, and EDS.

Mirzapour et al. found that the nanosilica was dispersed well in the matrix, with particles averaging 50 nm across. They found that the thermal stability and mechanical properties improved somewhat at the lowest loading of nanosilica but increases in loading beyond that produced marginal improvements at best. However, the mass loss and ablation rates showed greater and greater improvement at higher loadings of nanosilica. XRD patterns revealed the existence of a small amount of SiC in the char, and SEM revealed protective coatings of resolidified silica on the ablated carbon fibers [134].

Phenolic Nanocomposite Studies by University of Perugia, Italy [Torre, Kenny, and Natali (2011–2012)]

For more than 15 years, the University of Perugia—STM group (Terni)—has extensively worked in the field of ablative materials. At the end of 1990, in collaboration with ALENIA Spazio S.p.A, the group developed and tested a medium-density silicone-based ablative (ALS-051) for the heat shield of the Italian Space Agency recoverable capsule (CARINA). An accurate ablation model of the studied material was also developed by Torre and colleagues [135–137]. Since the beginning of 2000, the researchers also focused on the study of nanostructured ablatives. Nanosilica [138,139] and MWNTs [140] were the main filler candidates. The possibility to exploit the synergy between traditional and nano-sized fillers was also investigated.

Glass and silica fiber phenolic composite ablators are a very important class of ablatives. As an example, they can be effectively used in the production of passively cooled rocket combustion chambers. To reduce the erosion phenomena of these ablators, the studies were focused on the selection of the best types of fiber reinforcement, in terms of short or long fibers. Research has shown that in many hyperthermal conditions, liners produced with chopped reinforcements—fibers or fabrics cut in small squares [e.g., 1.27 cm by 1.27 cm (1/2 in. by 1/2 in.)]—could match the performance of configurations based on the use of more expensive long fibers [139]: accordingly, small parts as well as significant portions of a rocket combustion chamber can be effectively produced using ablative bulk molding compounds (BMCs) [141, 142]. A wide range of commercial ablators are sold as molding compounds [143, 144]. Research on glass or silica-phenolic liners also showed that silica powder performs a crucial role in the improvement of the ablative performance of composite liners [145–147]. In fact, when exposed to high-temperature, micrometer-sized silica particles form a highly viscous melt layer which covers the charred surface, working as an anti-oxidizing protective barrier and without inhibiting radiative cooling. The melt layer also continues to adsorb a significant amount of heat by virtue of endothermic processes and due to the phase transitions of silica.

Natali et al. [138] investigated the producing of glass-phenolic composites with nano-sized silica as an alternative to micrometer-sized silicon dioxide. Typically, in glass-phenolic or silica-phenolic liners composed with micrometer-sized silica, the amount of SiO_2 particles is of about 8–10 wt% and the fiber content is of about 60 wt% [138, 139]. For example, Cytec Solvay MX-2600 (an ablative used in SRM nozzle entrance region) consists of 31 wt% resin, 61 wt% chopped strands of silica, and 8 wt% of micrometer-sized silica particles. Natali et al. first studied two nanosilica-composed nanocomposites [138]: the resulting filler percentages were 5 and 20 wt%. Then the pristine resin and the nanocomposites were used as matrices for the impregnation of E-glass chopped fibers. In all the cases, the fiber percentage remained constant at a value equal to 60 wt%. The amount of nanosilica on the total weight was 8 wt% (BMC3) when the 20 wt% loaded nanocomposites were used to impregnate the short fibers: this composition strongly resembles the formulation of the MX-2600 with the difference that the micrometer-sized silica was replaced by the nano-sized material. To further investigate the effects of nanosilica, another BMC composition containing a nanofiller percentage equal to 2 wt% was studied (BMC2): this viscous mixture was produced using the

5 wt% loaded nanocomposite. A BMC composition prepared with no nanosilica was used as a control material (BMC1). Oxyacetylene torch testing (ASTM-E-285-80) [27] was used to study the resulting materials (heat flux equal to 850 W/cm^2 and an exposure time of 60 seconds). The flame was applied to one of the bases of the cylindrical shaped samples. In-depth temperatures were acquired at 5 mm (T1) and 10 mm (T2) from the surface. The morphological structure of charred materials was investigated by SEM analysis.

The charred surface of the control composition (BMC1) summarizes the ablation mechanism for this material [138]. In absence of nanosilica, the glass fibers become molten into low-viscosity material under exposure to the flame, leading to a high erosion rate. Unbounded spheres of glass easily drifted away from the zone impacted by the torch plume leaving the charred surface unprotected: the glass droplets moved to a well-defined corona on the external edge of the flame. In terms of the contact angle within the charred surface, the marked spherical shape of the glass droplets clearly identified the poor compatibility of the melt with the substrate.

The results of the BMC2 composition are presented in [138]. Under exposure to the flame, the glass fibers and nanosilica become molten together to form a blend. Even this low nanosilica percentage was able to increase the viscosity of the blend. The droplets of molten glass fibers exhibited better adhesion on the charred surface. However, this percentage of nanofiller did not completely immobilize the glass droplets under the zone impacted by the flame as summarized in [138]: the shear forces produced by the flame gases were sufficiently vibrant to displace the molten material from the zone impacted by the flame [138].

However, in comparison to the BMC1 composition, glass droplets moved to a corona with a reduced diameter, confirming the enhanced viscosity of the charred surface. The contact angle of the molten droplets within the charred surface indicated that the charred substrate increased its adhesive capability and compatibility within the molten system. Accordingly, the erosion rate experienced by the BMC2 material was decreased. As a matter of fact, the most interesting result was obtained from the BMC3 composition (Fig. 9.44a) [139]. In this case, the higher nanosilica content was able to effectively maintain a high viscosity, silica rich char, along with droplets of molten glass precisely in the zone impinged by the flame (Fig. 9.44b) [139]. The protection of the charred substrate and the preservation of the surface noticeably increased, thus leading to significantly lower erosion. Observing the contact angle of the molten droplets, it was possible to maintain the high compatibility between the charred surface and the droplets of melt. The adhesion capability of the charred surface was very high, confirming the effectiveness of the use of silica nanoparticles dispersed in an organic matrix. Since the charred substrate was held above the virgin material for a long period, the re-radiation efficiency was increased. Figure 9.44 shows the results of the SEM characterization carried out on the charred samples confirming the observations with traditional microscopy [139].

The sintering mechanism of silica nanoparticles was more intently investigated by Natali et al. in a previous work on silica-phenolic nanocomposites [139]. The in-depth temperatures showed that as the nanosilica content increased, the BMC viscous mixture exhibited improved insulation. The higher the amount of silica nanoparticles, the higher the amount of heat absorbed by the filler. Moreover, increasing the nanosilica content increased the thickness of charred region above the TCs, reducing the transferred heat.

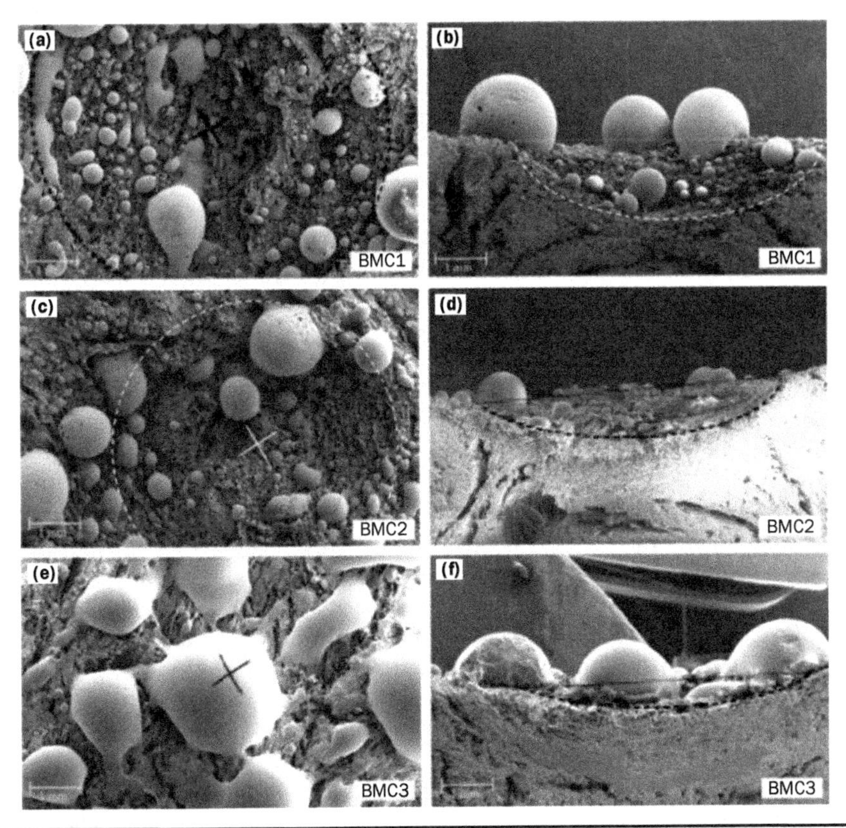

FIGURE 9.44 SEM micrographs of the charred samples: (a) and (b) refer to BMC1; (c) and (d) refer to BMC2; and (e) and (f) refer to BMC3. The crosses on the images (a), (c), and (e) mark the point where the core of the torch impinged the sample. In all the pictures, the dashed lines represent the eroded areas [139].

For silica-rich chars exposed to a non-oxidizing hyperthermal environment, a reaction producing silicon carbide (SiC) is proposed: the presence of SiC can enhance the ablation performance of the char [149–151]. The production of SiC is possible only when a carbothermal reduction between silica and carbon takes place [138]. During oxyacetylene tests, at the core of the plume, in the presence of a neutral flame, the charred surface is relatively protected against the oxidation and SiC can be produced. However, since the temperature of the core exceeds about 2,700°C, the SiC phase undergoes decomposition.

Natali et al. [140] also investigated the ablation properties of two carbon nanofillers. In particular, CB and MWNTs were examined to produce highly loaded (50 wt%) phenolic composites. This study is important because it allowed a distinction of the main differences in terms of ablative performance among traditional carbonaceous filler such as CB and relatively new filler, such as MWNT. In fact, prior to the concept of "nanocomposites" and its uses, CB was extensively used in the production of polymeric ablators. For example, Cytec Solvay MX-4926, a standard C-Ph nozzle material, is composed of 50 wt% of carbon fibers, 15 wt% of CB, and 35 wt% of phenolic resin. Recently, in the class of nano-sized carbon fillers, nanofilaments, such as CNFs have been studied as filler for polymeric TPS (see the Section "Phenolic Nanocomposite Studies by The University of Texas at Austin, United States"). The aim and novelty of the Italian research was to evaluate carbon nanotubes (MWNTs) as potential fillers for ablatives. In fact, MWNTs have been shown to be very effective flame-retardant additives for polymers [152]: the improved flame resistance was attributed to the formation of a continuous protective nanotube felt layer functioning as a heat shield [143–145]. In analogy with the mechanism of layered silicates, such a protective barrier reduces the mass loss rate and the flammability of the material. In agreement with the above-mentioned mechanism, flame retardancy was improved at higher MWNT loading. Moreover, the MWNTs embedded in the charred surface reemit a large amount of the incident radiation into the gas phase, decreasing the heat transferred into the inner layers and thus reducing the polymer pyrolysis rate [156, 157]. As a drawback, carbon-based reinforcements are easily decomposed in an oxidized hyperthermal environment. Nonetheless, to date, after considering the flame-retardant properties of MWNTs, there appeared to be a lack of research on MWNTs used as nanofillers for ablatives. The research of Natali et al. [140] is important not only because the ablative response of MWNTs embedded in a polymeric matrix was investigated, but also because a direct comparison with CB certainly represents a valuable contribution for the scientific community and the industry. In this study, two different mixtures were produced, both consisting of 50 wt% of the phenolic matrix and 50 wt% of the nanofiller: one with CB (Vulcan 7H) referred to as PR-CB, constituted by 50%-PR and 50%-CB, the other with MWNTs, constituted by 50% PR and 50% MWNTs (Arkema Graphistrength C100) referred to as PR-MWNT. The results concerning the TGA, the evaluation of the heat capacity, the oxyacetylene test, and the postburning morphology are summarized below.

In terms of thermal stability, at a heating rate of 20°C/min the TGA results in air and nitrogen showed that PR-CB and PR-MWNT exhibited a parallel behavior. The study of the heat capacity as a function of the temperature can be very important for the understanding of the ablation mechanism and it can also be used for the modeling of the process. It appeared that in the whole range of temperatures, both composites exhibited a higher heat capacity than the pristine matrix, with the PR-MWNT system having the highest values. The post-charred images of the tested materials showed that in terms of the erosion rate, the PR-MWNT composite exhibited higher erosion than the PR-CB system. The PR-MWNT samples suffered a marked removal of material and the flame produced a crater exactly in a zone confined under the plume of the torch. During the tests, small flakes of the material eroded off the surface of the PR-MWNT samples: this behavior was also found by other researchers of CNF-based NRAMs. The PR-MWNT samples also showed a very thin charred layer above the virgin material. The PR-MWNT samples showed a very limited number of in-depth cracks propagating from the mantel surface to the center of the sample. For the PR-CB composites, the flame produced a broad and deep charred region having relatively good integrity: however, the flame did not produce any crater. Also, during the test, this material did not show any macroscopic release of the burning material. The PR-CB charred samples exhibited a remarkable number of macro-cracks emanating from the mantel of the samples up to the lower limit of the charred region. It can be concluded that the PR-CB system suffered char penetration extended on a higher and deeper region of the sample.

The morphology of the charred surfaces was also investigated by SEM. At low magnification, the PR-CB revealed the cracking phenomena also identified with the optical analysis. The top surface of the PR-CB showed a morphology characterized by micrometer- and submicrometer-sized open and closed bubble-like structures: the CB particles were not distinguishable from the charred layer. The morphology of the zones beneath the surface was very close to that of the surface, but in some cases a uniform and compact medium with no voids or bubble was found. When compared to the PR-CB system, the surface of the PR-MWNT samples showed a homogeneous morphology free of any macro-cracks: the ablation appeared to be a surface phenomenon with a very thin transition zone—the char—between the virgin and charred material. The char was constituted by many irregular aggregates mainly composed of entangled MWNTs. Beyond the undamaged and uncharred MWNT filaments, SEM also demonstrated many bright spots clearly related to the metal catalyst precursors used to produce the MWNTs. The temperature profiles showed that the PR-MWNT system exhibited higher thermal diffusivity than the PR-CB composition. Natali et al. [140] also concluded that when compared to the PR-CB system, the PR-MWNT composite possessed a higher thermal conductivity in the direction of the flame plume. Loss of mass data indicated that both materials experienced a similar removal of mass. As a consequence, since the surfaces of PR-MWNT samples experienced more severe erosion than the PR-CB specimens and considering that the amount of loss of mass is of the same order-of-magnitude for both materials, it can be concluded that the ablation mechanisms of the two materials are considerably different. For the PR-MWNT, the loss of mass was due to the physical erosion of material and the loss of mass can be attributed to the higher-degree char of the organic matrix since the PR-CB did not erode appreciably. In conclusion, CB and MWNTs exhibit a very different charring behavior.

Phenolic-MWNT Nanocomposite Studies by Gyeongsang National University, Korea [Park et al. (2014)]

In this investigation by Park et al. [158], carbon nanotubes and carbon fibers were used in conjunction with phenolic resole-type resin to produce articles for thermal characterization. Three composites were produced: one with 30 wt% chopped carbon fibers, one with 30 wt% woven carbon fiber fabric, and one with 30 wt% woven carbon fiber fabric as well as 0.5 wt% carbon nanotubes. The chopped carbon fibers were stirred into the phenolic resin with acetone as a diluting agent for 3 hours at 100 rpm. The carbon nanotubes were dispersed into the resin according to a method developed previously by the team, and the mixture was applied to the carbon fabric by a hand lay-up method. All neat and composite samples were cured by autoclave at 690 kPa at 70°C for 1.5 hours, 100°C for 1 hour, and then 140°C for 4 hours.

To characterize the formulations, the flame retardancy testing, ablation testing, and TGA were conducted on each. For the flame retardancy investigation, a butane flame was applied to the specimen according to ASTM E84-11, while a thermal conductivity analyzer using a modified transient plane source technique recorded conductivity data. Thermal distortion and surface damage were both inspected after testing. For the ablation testing, an oxygen-kerosene flame with a 1:1 fuel ratio was applied to the sample. Distance and heat flux values were not reported. For the TGA, samples were taken from the front and back sides of the posttest specimens for data on the amount of carbonization of the material during testing.

It was observed that the composites each had increased thermal conductivity as compared to the neat resin, with the chopped CF composite having the least increase, and the CF fabric-carbon nanotube composite with the greatest. This trend persisted in the ablation testing results, where the same ranking occurred for increase in ablation resistance. The researchers conclude that the two properties are related. In addition, it was found that the carbonization of the material during testing may also be indicative of ablation resistance, since it does not occur without sustained elevated temperatures in the material. The neat resin had no carbonization on the back side, but the chopped CF composite had some and the two CF fabric composites had nearly identical carbonization on their front and back sides.

Phenolic-MWNT Nanocomposite Studies by Texas State University-San Marcos, United States [Tate and colleagues (2012)]

This research by Tate and colleagues was a collaborative effort with Cytec Solvay as an attempt to enhance the properties of the SOTA ablative, MX-4926, by the introduction of small amounts of MWNT [159, 160]. Test panels were manufactured using processes and materials similar to those used by Cytec Solvay. MX-4926 MC (Molding Compound) is rayon precursor–based carbon fabric with a MIL-R-9299 phenolic resin. Composite panels were manufactured using rayon-based carbon fabric supplied by Cytec Solvay, which is 8-harness satin weave. The fabric has a weight of 261 g/m^2, a specific gravity of 1.84, and a thickness of 0.48 mm. Break strength in the warp direction is 0.496 and 0.599 MPa in the fill direction, with 1.96 picks/mm warp and 2 picks/mm fill. SC-1008 phenolic resin, supplied by Hexion Specialty Chemicals, meets the MIL-R-9299 specification. The viscosity is about 180–300 cps depending on storage conditions. SC-1008 contains about 20–25% IPA as a solvent. Graphistrength™ C100 MWNTs were supplied by Arkema, Inc. Typical diameter of these MWNTs is 10–15 nm and a length between 1 and 10 μm. Graphistrength™ C100 MWNTs appear to be agglomerated with dimensions of approximately 50–900 μm. High-shear mixing and sonication have been used in previous studies to provide good dispersion and exfoliation of nanoparticles [161–163]. Tate and colleagues investigated the dispersion characteristics using sonication, high-shear mixer, and combination of sonication and high-shear mixer in previous studies [159]. TGA results and SEM analysis suggested that dispersion of MWNTs within IPA using sonication followed by high-shear mixing produces optimal dispersion characteristics for use in ablative panels. This method was adapted to produce composite panels with MWNT loading of 0.5, 1, and 2 wt%.

Oxyacetylene test bed was developed at The University of Texas at Austin to perform a small-scale testing of newly developed ablatives. Samples of size 12.7 mm × 12.7 mm ×10 mm were prepared and two holes of approximately 1.59 mm (1/16 in.) diameter drilled at depth of 2.7 mm and 7.7 mm for TC placement. The samples were exposed to a heat flux of approximately 1,000 W/cm^2 (6 mm distance from the OTB) for 45 seconds [160]. Six samples were tested in each category. Ablation performance of the material was analyzed based on the percentage mass loss, recession, peak temperatures at different depth in ablative samples, and SEM to analyze dispersion of MWNTs in phenolic resin. There was gradual decrease in percentage mass loss as wt% of MWNTs increased. For control samples, average mass loss was 26%; and for 2 wt% MWNT average mass loss was 23%; for 0.5 and 1 wt% MWNT average mass losses were 25.5% and 25%, respectively. There is considerable reduction in mass loss as weight percentage of MWNTs increased. Typically, mass loss is a combined pyrolysis of the material and char erosion. Similarly, recession was also high for low wt% formulations and decreased gradually as wt% of MWNTs increased. Average recession was 0.83 mm for control samples, while it was 0.38 mm for 2 wt% samples. Average peak temperatures were reduced considerably with the addition of MWNT to the composite [159, 160].

In summary, the combination of sonication and high-shear mixing was used for uniform dispersion and exfoliation of MWNT. Ablative samples were tested on OTB for 45 seconds and at a heat flux of 1,000 W/cm^2. The samples were compared on the basis of percentage mass loss, recession, and peak in situ temperatures at depths of 10 and 5 mm from flame front. The percentage mass loss for control sample was 26% and 23% for 2 wt% MCNT sample. Average recession was 0.83 mm for control samples, while it was reduced to 0.39 mm for 2 wt% MWNT samples. The peak temperatures at depths of 10 and 5 mm from flame front showed decreasing trends as the MWNT weight percentage increased. Based on these results, it was concluded that the increase in wt% of MWNT improved ablation and insulation performance of phenolic-MWNT nanocomposite.

Phenolic-Graphitic Felt and Phenolic-Graphitic Foam Studies by University of Rome, Italy [Pulci et al. (2010), Marra et al. (2010), and Pulci et al. (2012)]

Sapienza University of Rome group developed, manufactured, and tested a family of C-Ph ablative TPS materials with the aim of fulfilling the thermal and mechanical requirements corresponding to the actual loads experienced by reentry vehicles [164–166]. Experimental investigations were conducted on two different composite systems: a resole resin coupled with a graphitic felt and graphitic foam. Their aim was optimization of the manufacturing procedure and to characterize the mechanical behavior and the insulation performance of the fabricated composites.

A commercially available resole phenolic resin was employed and two carbon-based reinforcements were selected for impregnation with the phenolic resin. Graphitic foam (Grafoam®, manufactured by Graftech International, Parma, Ohio, website: www.graftech.com) and a rigid graphitic felt (Sigratherm®, manufactured by SGL Carbon SE, Wiesbaden, Germany) were used. Grafoam carbon foam is an open-cell, rigid, carbon-based material that can be produced in densities ranging from 0.03 to 0.56 g/cm^3 and in a variety of sizes. Sigratherm felt has high thermal stability in vacuum up to 2,500°C and under inert atmospheres up to 3,000°C. It is well known that above 2,000°C physical properties of the ablative can change. Figure 9.45 shows the microstructures of carbon-based reinforcements.

Infiltration process was performed by total immersion procedure: both reinforcements were totally immersed for 30 minutes in a phenolic solution with an isopropanol/resin ratio tailored in order to obtain the specific density (0.5 g/cm^3) after the

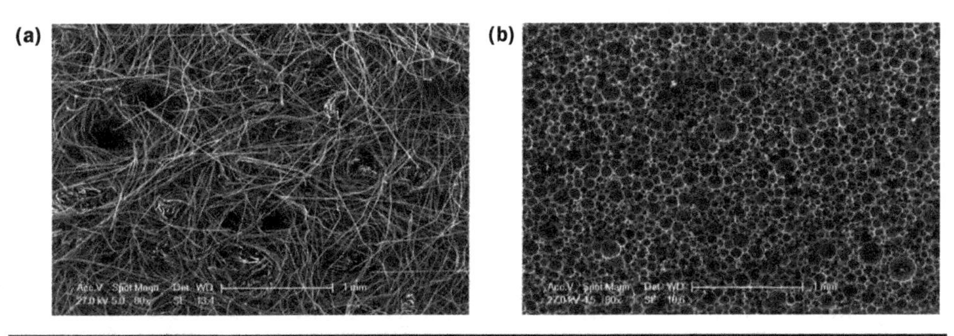

FIGURE 9.45 Microstructures of carbon-based reinforcements: (a) Sigratherm felt and (b) Grafoam. (*Courtesy of Prof. Valente.*)

cure process. After resin cure, a resin/composite volume ratio of 0.61 and 0.59 was calculated for the Sigratherm and Grafoam composites, respectively. Resin preheating (50°C) and vacuum infiltration (3 minutes at 10^{-3} bar) after soaking were mandatory for removing excess solvent and efficient impregnation. All tests were carried out under these conditions.

Preliminary ablation tests were performed on the two fabricated systems according to the ASTM E285-80 standard [27]. The size of samples tested with oxyacetylene torch was 100 mm × 100 mm × 20 mm or 50 mm × 50 mm × 20 mm; the distance between nozzle and samples was 20 mm and the temperature of the front and the back surface of the tiles was monitored by a two-color digital fiber-optics noncontact pyrometer (Impac Infrared, Frankfurt, Germany) and by a type-K TC, respectively. Heat flux, as measured by a calorimeter (HFM 1000-Vatell Corp., Christiansburg, VA) during experimental tests, was 1.5 MW/m^2. Four exposure times were investigated: 1, 5, 10, and 20 minutes. Exposure tests were performed on five impregnated and two unimpregnated samples. Maximum temperature of about 4,000 K was registered on the front face of the exposed samples.

Higher hardness values were noted on the surface rather than in the cross-sectional areas. Impregnated samples showed a homogeneous resin distribution throughout the thickness where the cross section of a 3-cm-thick Sigratherm composite is observed [164]. Magnified views of cross sections of the impregnated samples, showing suitable degree of resin impregnation, were achieved for both reinforcements [164]. All specimens exhibited a linear elastic behavior; Grafoam and Sigratherm composites showed similar values of elastic modulus (about 2 GPa), but the felt composite showed a much higher (+144%) flexural strength than the foam composite. Pure resin has an elastic modulus of 4 GPa, whereas unimpregnated materials have a modulus of 0.25 and 0.5 GPa for the felt and the foam, respectively. It is evident that the resin provides most of the composite stiffness. Results of four-point bending tests indicated a higher flexural strength of impregnated Sigratherm with respect to Grafoam.

Thermophysical properties (thermal diffusivity, thermal conductivity, and specific heat) of the composite samples have been reported by Pulci et al. [164]. Similar behaviors were observed for the two types of materials. Foam composites exhibited a slightly lower thermal conductivity at high temperature (1,500–1,900°C), while impregnated felts exhibited lower thermal conductivity at temperatures lower than 800°C. Unimpregnated Grafoam and Sigratherm reinforcements showed a room temperature thermal conductivity of 0.12 and 0.22 W/(m-K), respectively. The results of a typical oxyacetylene exposure test on Sigratherm and Grafoam composite samples, in terms of front and back surface temperature, are reported by Pulci et al. [164]. The two materials show the same temperature characteristics with time, indicating comparable insulation properties. However, the analysis of samples after tests showed that marcroporous cells (Grafoam) were fractured during the flame exposure, whereas all felt-based samples (Sigratherm) passed undamaged during the longest oxyacetylene exposure tests.

There appears to be a clear indication that a C-Ph composite material made of rigid nonwoven graphitic felt impregnated with a resole phenolic resin can act as a very effective insulator in a very wide range of temperatures (from room temperature to about 2,000°C), making it suitable for forming/machining operations. A very accurate optimization of the main steps of the manufacturing procedures is required to achieve the above. The research multistep curing cycle of 12 days at different temperatures is required to achieve optimal microstructure and highest mechanical properties of the composite. The outstanding performance of the manufactured composite system exposed to the intense heat flux of the oxyacetylene flame was achieved. Sigratherm-based ablators were selected for arc jet testing in a Plasma Wind Tunnel (Scirocco, CIRA ScoA, Italy) to simulate reentry conditions.

In order to support the material/component design and to offer new and differential tools for the screening of TPS materials, this research group has developed a computational fluid dynamics (CFD) model of the oxyacetylene test apparatus as per the ASTM E285-80 standard. The CFD model simulates the combustion of the oxygen–acetylene mixture occurring in a conventional welding torch and calculates the heat flux incident on the material sample and the related temperature fields. The validation of the CFD model was obtained by performing suitable instrumented tests with direct measurements of both incident heat flux and temperatures of the front and back surfaces of an ablative sample. Two different models using Fluent® software have been developed for the simulation: the first model is intended for the evaluation of the heat flux actually generated by oxyacetylene torch, and the second model simulates a real test on an ablative material. Two different experimental setups were prepared in order to validate the results obtained by the numerical simulations. Good agreement was found between numerical results and experimental measurements [164, 165]. For the configuration corresponding to the real test of an ablative material, the comparison between simulation and test applies to the temperatures of the exposed face, as measured by the pyrometer, and of the sample back-face. The range of measured front-face temperature is 1,970–2,070 K [165]. The results of CFD simulation of this configuration are shown in [165]. The temperature value calculated at the center of the irradiated sample is about 1,900 K and is in good agreement with the measured value.

Recently, additional lightweight ablative materials development data were obtained from Pulci et al. [166]. Figure 9.46 shows the (a) rigid graphite felt (long fibers) and (b) rigid graphite felt (short fibers) [166]. Innovative lightweight ablative materials using carbon-based reinforcement (felts) infiltrated with different volumetric fractions of phenolic resin to obtain different densities were investigated (Table 9.11) [166]. Mechanical properties (elastic modulus and flexural strength) were characterized using four-point bend test. Microstructural characterization was investigated using SEM analysis (Fig. 9.46) [166]. Thermal performance by oxyacetylene exposure tests were conducted with 2-color pyrometer to measure front face temperature and TCs for back-face temperature, while heat flux sensor HFM 1000 was used to calibrate heat flux [166].

Table 9.11 Carbon-Based Reinforcements (Felts) Infiltrated with Different Volumetric Fractions of Phenolic Resin to Obtain Different Densities

Felt	Matrix	Density (g/cc)	Additives
Carbon felt—long fibers	Phenolic	0.5	CNT mixture
Carbon felt—long fibers	Phenolic	0.25	No
Carbon felt—long fibers	Phenolic	0.5	Nano-Al_2O_3
Carbon felt—long fibers	Phenolic	0.5	Multilayer with graphitic adhesive
Carbon felt—short fibers	Phenolic	0.5	No
Carbon felt—short fibers	Phenolic	0.25	Nano-Al_2O_3

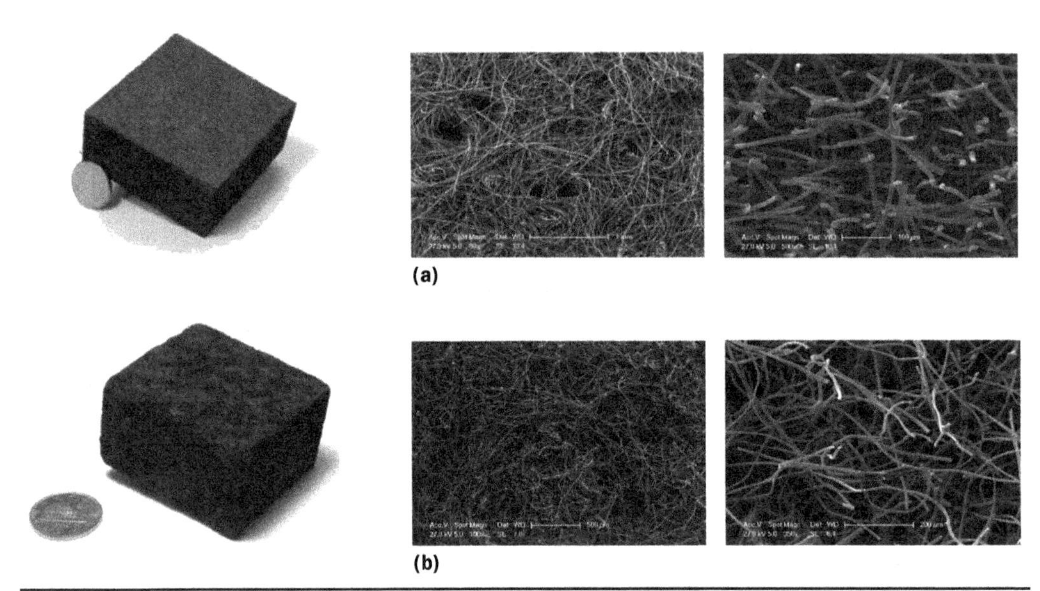

Figure 9.46 Microstructural characterization of a (a) long-fiber felt and (b) short-fiber felt. (*Courtesy of Prof. Valente.*)

Test conditions of heat flux 2 MW/m^2 and exposure time of 10 minutes were used. All samples exhibited good insulation performance with a surface temperature of about 1,700°C and back-face temperature always below 500°C after 10 minutes exposure time using a sample thickness of 30 mm. Samples exhibited somewhat little recession of about 1.8–3.4 mm. Multilayer long-fiber felt shows damage at the interface between joined layers, probably due to oxidation. The density of 0.25 g/cc samples appears to exhibit the worst thermal behavior in terms of insulation and recession. MWNT dispersion in the phenolic resin is a critical point, but their addition seems to improve the thermal performance. The posttest observation of surfaces reveals that nano-alumina forms alumina-like layer on the sample's surface. This phenomenon could be effective to improve ablation resistance of charred layer.

9.4 In Situ Ablation Sensing Technology

In this research by Koo et al. [167], a break-wire–like ablation recession and thermal sensor based on commercial, affordable TCs were designed, fabricated, and tested. The objective was to demonstrate the capability of this novel technique to provide real-time measurement of in-depth temperature and recession rates. The in situ ablation recession and thermal sensors were assembled on ablatives ranging from low to high densities (approximately 0.30–1.45 g/cc). The different types of ablatives adapted for this study were phenolic impregnated carbon ablator (PICA), AVCOAT, C-Ph, and carbon-carbon composites. These ablation recession and thermal sensors were tested under an oxidative hyperthermal environment at a heat flux in the range of about 900–1,000 W/cm^2 created by an OTB (see Sec. 5.13.3). The results were very promising, demonstrating that the proposed approach can provide accurate recession rate data of a variety of ablatives. The sensing technology developed is very versatile and can be applied to the TPS of spacecraft, and can also be exploited on rocket nozzles. The key features of this sensor system are the use of inexpensive, commercially available TCs and industry-standard drilling techniques demonstrated on different ablatives.

During the past 40 years, many different ablation recession sensors have been implemented for entry vehicle applications: most of them were primarily with high-density ablative TPS materials. Some intrusive sensors have been based on the use of break-wire arrangements. In a break-wire sensor arrangement, some pairs of wires are embedded in a TPS plug with their junctions located at different prescribed depths. As the TPS surface erodes, the increasing heat melts each wire, opening an electric circuit. However, since the charred material is electrically conductive, it can promote false signals, leading to an improper reading of the TPS status.

The Koo Research Group through AFOSR and KAI, LLC funding envisioned a low-intrusiveness sensor based on the break-wire–like method with the aim to be cost-effective, easy to manufacture, and scalable. In the break-wire–like approach, the metal wires typically used in this arrangement were replaced with ultrafine commercial TCs. Even if TCs were successfully used to instrument TPS materials, the approach introduced in this research presents some distinctiveness, which deserves to be described in detail. In fact, only commercially available processing techniques and raw materials were considered; this is a more affordable and reliable approach as compared to other equivalent solutions, in which each single constituent of the measurement chain was produced ad hoc as for the ARAD sensor. Low-intrusiveness K-type TCs with a stainless-steel outer sheath and a diameter of only 250 μm (0.25 mm) were chosen for the initial phase of this research. The TCs were embedded in the TPS, perpendicularly at the TPS surface. Each sensing head of the TC was positioned at a well-defined depth from the surface (see Fig. 5.31).

During the heating of the ablator, first the TCs would work as a temperature sensor acquiring data about the state of the TPS. When the temperature of the plug would rise above the melting point of the metal sheath and of the Seebeck junction, the TC would experience a break. In principle, the rupture of the Seebeck junction marks the spatial region, in which the melting temperature of the gage is reached. However, it correlates with the surface recession since in order to reach temperature in excess of about 1500°C, the surface has to be removed, that is, the head of the TC has to be exposed to the direct plume of the torch which, moreover, is thermo-oxidizing, thus promoting the removal of the steel outer case and degradation of the inner junction. Due to this twofold nature of the sensing heads—as a Seebeck junction and as a position marker—it would be possible to obtain a wide range of data on the recession state of the TPS. This technology was tested on several SOTA ablative materials, such as the following:

1. *Low-density carbon-carbon (LDCC):* A low-density, two-dimensional (2D) carbon-carbon composite manufactured by SMJ Carbon was used with a density of 1.34 g/cc.
2. *High-density carbon-carbon (HDCC):* A high-density, 2D carbon-carbon composite manufactured by DACC, Changwon, South Korea was used with a density of 1.70 g/cc.

3. *Carbon-phenolic (C-Ph):* The phenolic resin reinforced with rayon carbon fibers, MX-4926, is an industry-standard ablative for solid rocket nozzle with a density of 1.45 g/cc. It is manufactured by Cytec Solvay Group.
4. *AVCOAT 5026-39:* AVCOAT is a low-density epoxy-phenolic matrix filled with silica fibers and phenolic microspheres. The compound was placed in the cavities of a fiberglass-reinforced phenolic honeycomb. It is a material that NASA used on the Apollo command module, and was used on the Orion command module as an ablative heat shield. It is manufactured by Textron Specialty Materials with a density of 0.53 g/cc.
5. *Phenolic impregnated carbon ablator (PICA):* Fiber Materials, Inc. (FMI) produced PICA material from a low-density carbon fiber preform impregnated with a phenolic polymer. The carbon fiber preform, Fiberform®, is also produced by FMI. PICA has a density of about 0.30 g/cc.

The technology was preliminary tested using 0.25-mm-diameter K-type TCs on LDCC/HDCC, and then to speed up the manufacturing procedures and reduce the testing cost, 0.50-mm-diameter K-type TCs were used to extend this study on the other TPS materials (C-Ph, PICA, and AVCOAT). The difference between the four- and eight-level sensors is strictly related to cost constraints. The greatest efforts were spent on the carbon-carbon composites instrumented with the 0.25-mm-diameter TCs. In that case, the technology showed its maximum level of TRL due to the low-intrusiveness of the transducers and due to the high level of the obtained results. All the following tests with the 0.50-mm-diameter TCs were aimed at showing the possibility to extend the concept to other materials. However, the technology was not optimized with the use of the smallest TCs. The authors simply wanted to show that, by using TCs placed perpendicular to the exposed surface, as an alternative to the classical layout in which they are placed on isothermal surfaces, it is possible to get valuable data at a fraction of the cost. Material and sensor fabrication can be obtained from [166].

9.4.1 A Comparison Among the Temperature Profiles of High-, Mid-, and Low-Density Materials

A comparison was undertaken that was aimed to highlight the similarities and differences on the in-depth temperature of high-, mid-, and low-density TPS materials used by NASA, which was the aim of this research [167]. In particular, the in-depth temperature profiles of LDCC (about 1.34 g/cm^3), of AVCOAT 5026-39 (about 0.53 g/cm^3), and of PICA (about 0.30 g/cm^3) were confronted. Because these materials were tested at a comparable heat flux, this comparison was scientifically feasible. In light of Fig. 9.47, the slope and shape of the profiles are influenced by thermal diffusivity (α) of the different materials (with α defined as $\alpha = \lambda/\rho C_p$ where λ is the thermal conductivity, ρ is the density, and C_p is the heat capacity of the considered material) [168]. However, the temperature profiles were also directly influenced by the intrinsic erosion rate of each material: in general, the higher the erosion rate (which typically tends to increase at low densities), the lower the amount of material left to shield the embedded TCs. As shown in Fig. 9.47, at the given experimental conditions, the AVCOAT material resulted had the lowest thermal insulation capability since it displayed the temperature curves with the highest slope. The LDCC was the best insulator as compared to PICA and AVCOAT. This evidence supports the conclusion that the thermal conductivity, the density, the

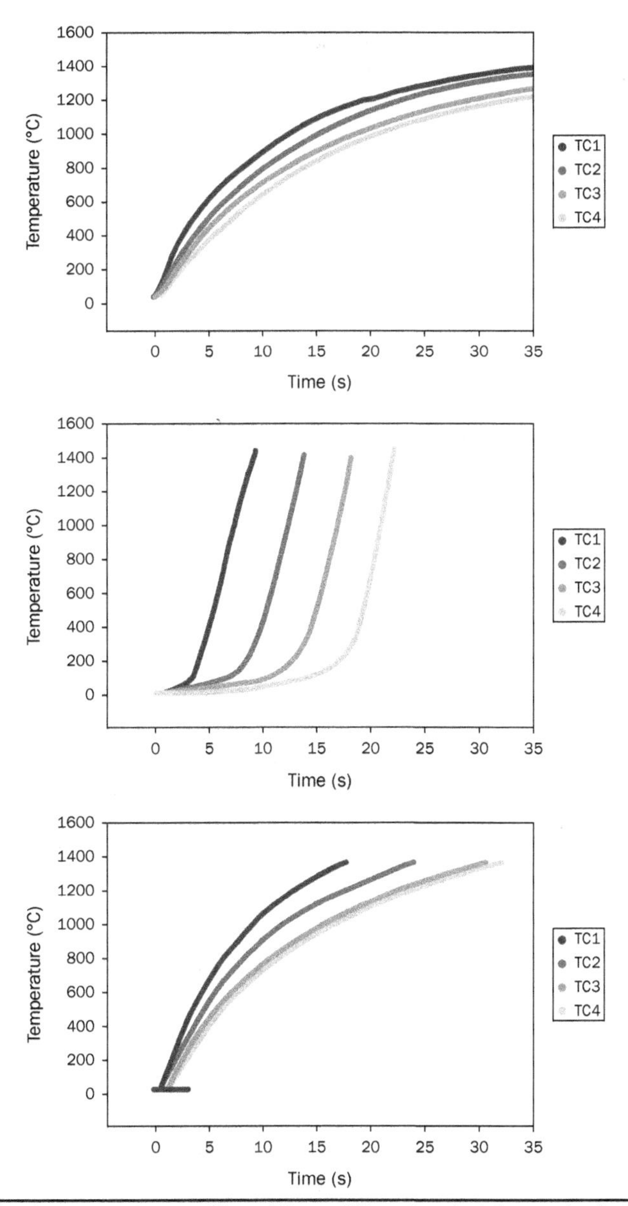

FIGURE 9.47 In-depth temperature profiles of three different materials acquired with the system proposed in [167]: (a) LDCC composite, (b) AVCOAT 5026-39, and (c) PICA. TC1 was closest to the surface; TC4 was placed deepest in the samples.

heat capacity, and the erosion resistance of the material play a relevant role on its insulation capability; these profiles would suggest that the higher erosion resistance of the LDCC combined with its intrinsic λ, ρ, and C_p seems to reduce the thermal diffusivity of the material. These results also encouraged us to undertake a dedicated test campaign aimed at evaluating the λ, ρ, and C_p of the different materials as a function of the temperature. In such a way, the possibility to explain these data will be improved.

It must also be pointed out that the effective heat-rejection capability of each material strongly depends on the testing conditions, such as heat flux, pressure of the gas, presence of oxidizing species, etc. As an example, the thermal performance and ablation characteristics of PICA were evaluated in an oxidizing environment at NASA [168]. The materials were studied in a heat flux range from about 350 to 3,350 W/cm^2 and surface pressures from about 0.1 to 0.43 atm. PICA shown that, from 350 to 450 W/cm^2, the ablation mechanism was oxidation-rate controlled; accordingly, the higher the stagnation pressure, the higher the concentration of oxygen diffusing into the boundary layer, thus increasing the oxidation rate and consequently the surface recession. Above 450 W/cm^2 and up to about 1,600 W/cm^2, the main mechanism of heat rejection was reradiation. Unfortunately, a direct comparison between NASA and our testing conditions is not feasible. However, since the PICA samples tested with OTB (heat flux of about 1,000 W/cm^2 and 3.4:1 oxygen/acetylene ratio) evinced high erosion, it is reasonable to infer that the ablation mechanism was mainly oxidation-rate controlled.

9.4.2 Summary and Conclusions of Ablation Recession Rate of Different Types of Ablatives and Future Outlook

Table 9.12 shows a summary of the ablation recession rate (RR) and the density of the different types of ablatives evaluated in this research. The developed in situ ablation recession and thermal sensing technology has demonstrated its adaptability to a variety of ablatives under a severe hyperthermal environment for ablative TPS application. Table 9.12 also shows how the low-density materials tended to exhibit a higher error on the estimation of the RR. This evidence also supports the general conclusion that lower-density materials tend to display lower char strength [138] and, as a result, they are more prone to produce a higher uncertainty in the RR.

In the future, the compressive and shear strength of the char layers of these ablative materials will be evaluated [169, 170]. These investigations will contribute to the goal of quantitatively modeling the performance of each class of ablator with a consistent evaluation technique, so that suitable materials can be selected for every heat shielding application. This ablation recession and thermal sensing technology will be also used for the Koo Research Group's polymer nanocomposite ablatives research. The surface temperatures, ablation recession rates, and in-depth thermal data will be used to validate the "material response (MR)" models using NASA's *FIAT* [171] and *CHAR* [172, 173], Penn State University's *MESA* [66, 81,174], and NGC's *ITRAC* [175] and *HERO* [176] ablation codes that are currently in progress.

In conclusion, the proposed ablation recession and thermal sensing technique were applied to low-density ablators (PICA has a density of 0.30 g/cc and AVCOAT has a density of 0.53 g/cc), medium-density ablators (LDCC density: 1.34 g/cc; C/Ph, MX-4926 density: 1.43 g/cc), and high-density ablators (HDCC density: 1.70 g/cc). The 0.25-mm TC-based ablation recession sensor paradigm was demonstrated successfully using two types of carbon-carbon composites (four and eight levels) and PICA (four level). Four-level carbon-carbon, C-Ph, PICA, and AVCOAT sensors were also tested with 0.50-mm TCs. Different methods to evaluate RR of ablatives were demonstrated: the linear regression of the combined ablation test data approach was reported in this research.

Table 9.12 Summary of Ablation RR of the Different TPS Materials [166]

Material	Type of TC (mm)	Layout	Density (g/cc)	RR* (mm/s)	$\overline{SD}$ (mm/s)	RR* (mm/s)	$\overline{SD}$ (mm/s)	$\overline{ARR}$ (mm/s)	$\overline{SD}$ (mm/s)
LDCC	0.25	4 levels	1.34	0.058	*0.009*	0.057	*0.003*	0.056	*0.007*
HDCC	0.25	4 levels	1.70	0.018	*0.006*	0.025	*0.003*	0.023	*0.005*
HDCC	0.25	8 levels	1.70	0.045	*0.004*	0.030	*0.001*	0.043	*0.003*
HDCC	0.5	9 levels	1.70	0.026	*0.003*	0.028	*0.002*	0.033	*0.007*
C/Ph	0.5	4 levels	1.45	0.069	*0.008*	0.059	*0.004*	0.060	*0.003*
AVCOAT	0.5	4 levels	0.53	0.319	*0.025*	0.279	*0.015*	0.365	*0.020*
PICA	0.5	4 levels	0.30	0.292	*0.047*	0.274	*0.025*	0.283	0.082

$\overline{RR}$, Recession Rate Mean Value; $\overline{ARR}$, average recession rate measured as a function of the final depth of the crater and of the exposure time; $\overline{SD}$, Standard Deviation Mean Value; * Evaluation of the recession rate considering the SUM method, i.e. all the data of all the experiments

A feasible and reliable microdrilling technique has been demonstrated for low- and mid-density ablators. The electrical discharge machining (EDM) technique was very useful to drill holes for carbon-carbon composites. Other techniques, such as laser drilling, should be investigated for the 0.25-mm TC-based sensor using PICA and AVCOAT materials.

Additionally, it is worth to remark that the erosion rate is strongly influenced by the hyperthermal environment in which the materials are tested: most of the tests performed on PICA and AVCOAT by NASA have been gathered using arc-jet torches. It makes no sense to compare data obtained with the arc-jet torch and the oxyacetylene torch. Consequently, the value of this study is that all the materials were tested using a unique source of hyperthermal conditions, and thus within this framework all data can be compared consistently. This research is the only published resource in the form of open literature that provides such a valuable comparison.

9.5 Overall Summary and Conclusions

A comprehensive review of polymer nanocomposite ablatives designed for advanced thermal protection materials has been conducted. It has been demonstrated that the improvement in ablation performance of these polymer nanocomposites relative to traditional filled system is associated with the char-forming characteristics. The spatially uniform distribution of nanoscaled inorganic filler results in the formation of a uniform inorganic char at a relatively low fraction of inorganic char (about a few percent by weight). The nanostructured ablative materials will provide multifunctional properties, such as ablation resistance, thermal stability, insulation, and weight saving. Ablation mechanisms have been studied and proposed by the researchers on their nanostructured ablatives.

A breakwire-like ablation recession and thermal sensor based on commercial, affordable TCs were introduced. The objective was to demonstrate the capability of this novel sensing technique to provide real-time measurement of in-depth temperature and recession rates.

Recently, Duffa published an excellent book *Ablative Thermal Protection Systems Modeling* [177]. This book discusses the broad physical knowledge of modeling of phenomena, including (from microscopic to macroscopic) atomic physics, thermodynamics, gas kinetics, radiative transfer, physical and chemical reactions (both homogenous and heterogeneous), fluid mechanics, and turbulence applied to previously studied areas, such as rough walls. Duffa commented that many books were devoted on each of these research areas, but very little literature was devoted on the synthesis of ablation problem. Ablation research is undoubtedly strongly coupled between various physical domains. For an overview of this particular research, the reader is referred to the book by Duffa entitled *Ablative Thermal Protection System Modeling*, in the Further Reading section of Chap. 1. The objective of this book by Duffa is to develop physical skills in the research applied to the ablation modeling of TPS. It consists of 13 chapters and 7 appendices that provide comprehensive technical discussions on ablation modeling phenomena, testing and specific test facilities, and selective thermal protection materials.

In order to develop material response (MR) models to predict ablative performance with any confidence, detailed material thermophysical properties must be accurately characterized as stated by Koo et al. [178, 179]. Thermophysical properties of ablatives are required to understand how they affect the ablative performance and to describe thermal behaviors of the ablatives under hyperthermal environments. These properties are also the input parameters required for the MR models. ASTM test standards are used to characterize each thermophysical property at virgin and charred states of the ablative materials. Koo et al. [178, 179] provide a summary of material thermophysical properties traditionally used in the creation and operation of MR models for the ablation codes. Ablative MR modeling requires a multitude of material properties with respect to temperature in order to provide accurate and reliable results. Ewing et al. [175], Koo et al. [178, 179], and NASA report [180] summarized the specific material properties that have to be characterized in each category: (1) Virgin Properties: Virgin density, virgin specific heat, virgin thermal conductivity; (2) Pyrolysis Properties: TGA analysis, char fraction, pyrolysis kinetics, elementary compositions, pyrolysis gas enthalpy, heat-of-formation, heat-of pyrolysis; (3) Char Thermal Properties: Char density, char-specified specific heat, char thermal conductivity; and (4) Surface Radiation Properties: Virgin material emissivity, virgin material absorptivity, char material emissivity, and char material absorptivity. This research area of to characterize the thermophysical properties of these newly developed ablative polymer nanocomposites is urgently needed in order to advance the material response modeling.

9.6 Study Questions

9.1 Discuss the thermochemical behavior of thermal protection materials under severe thermal attack in high heat-flux and high-velocity extreme environments.

9.2. In the thermoplastic polyurethane nanocomposite studies (Sec. 9.3.5), compare the ablation performance of (a) MMT clay, (b) carbon nanofiber (CNF), and (c) multiwalled carbon nanotubes (MWNTs) and explain their respective ablation mechanisms. Postulate the ablation performance when a combination of (a) MMT clay–CNF, (b) MMT clay–MWNT, and (c) CNF-MWNT were used in the TPUNs. What percentage of MMT clay, CNF, and MWNT should be used in (a) to (c), and what is the sequence of polymer blends compounding?

9.3 In an extension to the thermoplastic polyurethane nanocomposite studies (Sec. 9.3.5), what would be the mechanical properties of the TPUN-clay, TPUN-CNF, TPNU-MWNT, TPUN-clay-CNF, TPUN-clay-MWNT, and TPUN-CNF-MWNT nanocomposite systems? What are the important mechanical properties you should characterize?

9.4 What other nanofillers you would propose in the thermoplastic polyurethane nanocomposite studies to enhance the ablation performance, insulation, and mechanical properties?

9.5 In the phenolic nanocomposite studies (Sec. 9.3.6), (a) CNF, (b) MMT clay, (c) POSS, (d) nanosilica, (e) carbon black, and (f) MWNT were used as fillers in the phenolic resin and reinforced with carbon or glass fiber reinforcements. Discuss the protective mechanisms for each of the fillers in their composites. Discuss the important mechanical properties of these phenolic nanocomposites and which mechanical properties you should characterize.

9.6 What other nanofillers you would propose in the phenolic nanocomposite studies to enhance its ablation performance, insulation, and mechanical properties?

9.7 In the poly(phenylene oxide)/polystyrene (PPO/PS) studies [see the section "Poly(phenylene oxide)/Polystyrene (PPO/PS) Studies by Politecnico de Torino,

Italy"], what polymer-processing technique and sequence you would propose to better incorporate these ceramic particles into the polymer matrix? What dispersion technique you would propose to examine the degree of particle dispersions in the polymer matrix? What ablation testing you would propose to compare the ablation performance and heat transfer characteristics of these nanocomposites?

9.8 In an extension to the poly(phenylene oxide)/polystyrene (PPO/PS) studies [see the section "Poly(phenylene oxide)/Polystyrene (PPO/PS) Studies by Politecnico de Torino, Italy"], characterize the important mechanical properties of these PPO/PS nanocomposites.

9.9 Discuss what other ablation test facilities, such as arc-jet or inductively coupled plasma (ICP), are available in the industry to study thermal protection materials as well as the locations, and the majority of test parameters used in these ablation testing facilities.

9.7 References

1. Koo, J. H., Ho, W. K., and Ezekoye, O. A. (2006) A review of numerical and experimental characterization of thermal protection materials—Part I. Numerical modeling, AIAA-2006-4936, *42nd AIAA/ASME/SAE/ASEE Joint Propulsion Conference*, July 9–12. Sacramento, CA.
2. Koo, J. H., Ho, W. K., Bruns, M., and Ezekoye, O. A. (2007) A review of numerical and experimental characterization of thermal protection materials—Part II. Material properties characterization, AIAA-2007-2131, *48th AIAA/ASME/ASCE/AHS Structures, Structural Dynamics, and Materials Conference*, April 23–26. Honolulu, HI.
3. Koo, J. H., Ho, W. K., Bruns, M., and Ezekoye, O. A. (2007) A review of numerical and experimental characterization of thermal protection materials—Part III. Experimental testing, AIAA-2007-5773, *43rd AIAA/ASME/SAE/ASEE Joint Propulsion Conference*, July 8–11. Cincinnati, OH.
4. Koo, J. H., Natali, M., Tate, J., and Allcorn, E. (2013) Polymer nanocomposites as advanced ablatives—a comprehensive review. *International Journal of Energetic Materials and Chemical Propulsion*, 12(2):119–162.
5. Pinnavaia, T. J., and Beall, G. W., eds. (2000) *Polymer-Clay Nanocomposites*. New York: Wiley & Sons.
6. Koo, J. H. (2006) *Polymer Nanocomposites: Processing, Characterization, and Applications*. New York: McGraw-Hill.
7. Morgan, A. B., and Wilkie, C. A. eds. (2007) *Flame Retardant Polymer Nanocomposites*. Hoboken, NJ: Wiley & Sons.
8. Gupta, R. A., Kennel, E., and Kim, K. J., eds. (2010) *Polymer Nanocomposites Handbook*. Boca Raton, FL: CRC Press.
9. Mittal, V., ed. (2010) *Polymer Nanotube Nanocomposites: Synthesis, Properties, and Applications*. Wiley, Hoboken, NJ: Wiley & Sons.
10. Mittal, V., ed. (2010) *Optimization of Polymer Nanocomposites Properties*. Weinheim, Germany: Wiley-VCH.
11. Mittal, V., ed. (2011) *Thermally Stable and Flame Retardant Polymer Nanocomposites*. Cambridge, UK: Cambridge University Press.

12. Mittal, V., ed. (2012) *Characterization Techniques for Polymer Nanocomposites.* Weinhein, Germany: Wiley-VCH.
13. Beall, G. W., and Powell, C. B. (2011) *Polymer-Clay Nanocomposites.* Cambridge, UK: Cambridge University Press.
14. Vaia, R. A., Price, G., Ruth, P. N., Nguyen, H. T., and Lichtenhan, J. L. (1999) Polymer/layer silicate nanocomposites as high performance ablative materials, *Applied Clay Sciences,* 15:67–92.
15. Laub, B., and Venkatathy, E. (2003) Thermal protection system technology and facility needs for demanding future planetary missions. *Proc. of the International Workshop on Planetary Probe Atmospheric Entry and Descent Trajectory Analysis and Science.* Lisbon, Portugal.
16. Lombardi, M., Fino, P., Malicelli, G., and Montanaro, L. (2012) Exploring composites based on PPO blend as ablative thermal protection systems – part I: the role of layered fillers. *Composite Structures,* 94:1067–1074.
17. Lombardi, M., Fino, P., and Montanaro, L. (2014) Influence of ceramic particles features on the thermal behavior of PPO-matrix composites. *Science and Engineering of Composite Materials,* 23–28, ISNN: 2191-0359.
18. Fino, P., Lombardi, M., Antonini, A., Malucelli, G., and Montanaro, L. (2012) Exploring composites based on PPO blend as ablative thermal protection systems – part II: the role of equiaxial fillers. *Composite Structures,* 94:1060–1066.
19. Lincoln, D. M., Vaia, R. A., Brown, J. M., and Benison Tolle, T. H. (2000) Revolutionary nanocomposite materials to enable space systems. *Proc. of 21st Century Aerospace Conference IEEE.* Big Sky, MT, vol. 4, pp. 183–192.
20. Philip, S. H., Gonzales, R. I., Blanski, R. L., and Viers, B. D. (2002) Hybrid inorganic/organic reactive polymers for sever environment protection. *Proc. of 47th SAMPE ISSE,* May 12–16. SAMPE, Covina, CA.
21. Young, M. H., and Glaittli, S. R. (2001) *50th JANNAF Propulsion Meeting.* CPIA Pub. 705, vol. II, pp. 1–14.
22. Mathias E. C., and Johnson, T. N. (1999). *20th JANNAF Rocket Nozzle Subcommittee Meeting.* CPIA Pub. 694, pp. 237–266.
23. Koo, J. H., Polidan, J., et al. (2002) An investigation of polymer nanocomposite ablatives characterization. Invited lecture at the *30th Annual Conference of the North American Thermal Analysis Society,* September 23–25. Pittsburgh, PA.
24. Koo, J. H., Blanski, R., et al. (2003) Nanostructured ablatives for rocket propulsion system–recent progress, AIAA-2003-1769, *44th AIAA/ASME/ASCE/AHS Structures, Structural Dynamics, and Materials Conference,* April 7–10. Norfolk, VA.
25. Koo, J. H., and Pilato, L. (2005) Polymer nanostructured materials for propulsion systems, AIAA-2005-3606, *41st AIAA/ASME/SAE/ASEE Joint Propulsion Conference,* July 10–12. Tucson, AZ.
26. Gao, G., Zhang, Z., Li, X., Meng, X. Q., Zheng, Y., and Jin, Z. (2010) Study on mechanical and ablative properties of EPDM/OMMT thermal insulating nanocomposites. *Journal of Nanoscience and Nanotechnology,* 10:7031–7035.
27. ASTM E-285-80. (2008) *Standard Test Method for Oxyacetylene Ablation Testing of Thermal Insulation Materials.* Annual Book of ASTM Standards.
28. Singh, S, Guchhait, P. K., Bandyopadhyay, G. G., and Chaki, T. K. (2013) Development of polyimide–nano-silica filled EPDM based light rocket motor insulator compound: Influence of polyimide–nano-silica loading on thermal, ablation, and mechanical properties. *Composites: Part A,* 44:8–15.

29. Natali, M., Rallini, M., Puglia, D., Kenny, J., and Torre, L. (2013) EPDM based heat shielding materials for solid rocket motors: a comparative study of different fibrous reinforcements. *Polymer Degradation and Stability,* 98:2131–2139.
30. Iqbal, N., Sagar, S., Khan, M. B., and Rafique, H. M. (2014) Elastomeric ablative nanocomposites used in hyperthermal environments. *Polymer Engineering and Science,* 54:255–263.
31. Khanlart, S., and Kokabi, M. (2010) Thermal stability, aging properties, and flame resistance of NR-based nanocomposites. *Journal of Applied Polymer Science,* 119:855–862.
32. Guan, Y., Zhang, L. X., Zhang, L. Q., and Lu, Y. L. (2011) Study on ablative properties and mechanisms of hydrogenated nitrile butadiene rubber (HNBR) composites containing different fillers. *Polymer Degradation and Stability,* 96:808–817.
33. Koo, J. H. (2006) *Polymer Nanocomposites: Processing, Characterization, and Applications.* New York: McGraw-Hill, pp. 214–228.
34. Koo, J. H., Pilato, L., and Wissler, G. (2007) Polymer nanostructured materials for propulsion systems. *Journal of Spacecraft and Rockets,* 44(6):1250–1262.
35. Koo, J. H., Ezekoye, O. A., et al. (2009) Characterization of polymer nanocomposites for solid rocket motor—recent progress. *Proc. of SAMPE 2009 ISSE.* SAMPE, Covina, CA.
36. Koo, J. H., et al. (2010) Flammability studies of a novel class of thermoplastic elastomer nanocomposites. *Journal of Fire Sciences,* 28(1):49–85.
37. Koo, J. H., Ezekoye, O. A., Lee, J. C., Ho, W. K., and Bruns, M. C. (2011) Rubber-clay nanocomposites based on thermoplastic elastomers. In *Rubber-Clay Nanocomposites,* Galimberti, M., ed. Hoboken, NJ: Wiley and Sons, pp. 489–521.
38. Ho, W. K., Koo, J. H., and Ezekoye, O. A. (2009) Kinetics and thermophysical properties of polymer nanocomposites for solid rocket motor insulation. *Journal of Spacecraft and Rockets,* 46(3):526–544.
39. Ho, W. K., Koo, J. H., and Ezekoye, O. A. (2010) Thermoplastic polyurethane elastomer nanocomposites: Morphology, thermophysical, and flammability properties. *Journal of Nanomaterials,* doi: 10.1155/2010/583234.
40. Bruns, M. C., Koo, J. H., and Ezekoye, O. A. (2009) Population-based models of thermoplastic degradation: Using optimization to determine model parameters. *Polym Deg Stab,* 94:1013–1022.
41. Lee, J. C., Koo, J. H., and Ezekoye, O. A. (2009) Flammability studies of thermoplastic polyurethane elastomer nanocomposites, AIAA-2009-2544, *50th AIAA/ASME/ASCE/AHS/ASC Structures, Structural Dynamics, and Materials Conference,* May 4–7. Palm Spring, CA.
42. Lee, J. C., Koo, J. H., Ezekoye, O. A, et al. (2009) Heating rate and nanoparticle loading effects on thermoplastic polyurethane elastomer nanocomposite kinetics, AIAA-2009-4096, *AIAA Thermophysics Conference,* June 22–25. San Antonio, TX.
43. Lee, J. C., Koo, J. H., and Ezekoye, O. A. (2009) Thermoplastic polyurethane elastomer nanocomposites: Density, hardness, and flammability properties correlations, AIAA-2009-5273, *AIAA Joint Propulsion Conference,* August 2–5. Denver, CO.
44. Lee, J. C. (2010) *Characterization of Ablative Properties of Thermoplastic Polyurethane Elastomer Nanocomposites.* Ph.D. dissertation, The University of Texas at Austin, Department of Mechanical Engineering, Austin, TX.
45. Lee, J. C., Koo, J. H., and Ezekoye, O. A. (2011) Thermoplastic polyurethane elastomer nanocomposite ablatives: characterization and performance, AIAA-2011-6051, *47th AIAA/ASME/SAE Joint Propulsion Conference,* August 1–4. San Diego, CA.

46. Allcorn, E., Natali, M., and Koo, J. H. (2011) Ablation performance and characterization of thermoplastic elastomer nanocomposites. *Proc. of SAMPE 2011 ISTC*, October 17–20. Fort Worth, TX.
47. Allcorn, E., Natali, M., and Koo, J. H. (2013) Ablation performance and characterization of thermoplastic elastomer nanocomposites. *Composites: Part A*, 45:109–118.
48. De Heer, W. A. (2004) Nanotubes and the pursuit of applications. *Materials Research Society Bulletin*, 29(4):281–285.
49. Bell, M. S., and Tam, W. (1992) ASRM case insulation design and development. NASA-CR-191947. Presented at the Society for the Advancement of Materials and Process Engineering, May 10–13. Anaheim, CA.
50. Bhuvaneswari, C. M., Kakade, S. D., Deuskar, V. D., Dange, A. B., and Gupta, M. (2008) Filled ethylene-propylene diene terpolymer elastomer as thermal insulator for case-bonded solid rocket motors. *Defence Science Journal*, 58(1):94–102.
51. Bhuvaneswari, C. M., Sureshkumar, M. S., Kakade, S. D., and Gupta, M. (2006) Ethylene-propylene diene rubber as a futuristic elastomer for insulation of solid rocket motors. *Defence Science Journal*, 56(3):309–320.
52. Jaramillo, M., Koo, J. H., and Natali, M. (2014) Compressive char strength of polyurethane elastomer nanocomposites. *Polymers for Advanced Technology*, 25(77):742–751. doi: 10.1002/pat.3287.
53. Lewis, J., Koo, J. H., et al. (2017) Development of a shear char strength sensing technique to study thermoplastic polyurethane elastomer nanocomposites. *Polymer for Advanced Technologies*, 28(12):1707–1725, doi: 10.4044.1002/pat.
54. Moniruzzaman, M., and Winey, K. I. (2006) Polymer nanocomposites containing carbon nanotubes. *Macromolecules*, 39(16):5194–5205.
55. George, J. J., and Bhowmick, A. K. (2008) Fabrication and properties of ethylene vinyl acetate-carbon nanofiber nanocomposites. *Nanoscale Research Letters*, 3:508–515.
56. Lee, S. H., et al. (2004) Thermal properties of maleated polyethylene/layered silicates nanocomposites. *International Journal of Thermophysics*, 25:1585–1595.
57. Schartel, B., Weib, A., Sturm, H., Kleemeier, M., Hartwig, A., Vogt, C., and Fischer, R. X. (2010) *Polym Adv Technol*, doi: 10.1002/pat.1644.
58. Natali, M., Monti, M., Puglia, D., Kenny, J. M., and Torre, L. (2011) Ablative properties of carbon black and MWNT/phenolic composites: a comparative study. *Composites Part*, 43:174–182.
59. Ambuken, P., Stretz, H., Koo, J. H., Lee, J., and Trejo, R. (2012) High temperature flammability and mechanical properties of thermoplastic polyurethane nanocomposites. In *Fire and Polymers VI: New Advances in Flame Retardant Chemistry and Science*, Morgan, A. B., Wilkie, C. A., Nelson, G. L., eds. ACS Books Series, Washington, DC: American Chemical Society.
60. Yang, D., Zhang, W., Jiang, B. Z., and Guo, Y. (2013) Silicone rubber ablative composites improved with zirconium carbide or zirconia. *Composites: Part A*, 44:70–77.
61. Patton, R. D., Pittman, C. U., Jr., Wang, L., and Hill, J. R. (2002) Ablation, mechanical and thermal conductivity properties of vapor grown carbon fiber/phenolic matrix composites. *Composites Part A*, 33(2):243–251.
62. Patton, R. D., Pittman, C. U., Jr., Wang, L., and Hill, J. R. (1999) Vapor grown carbon fiber composites with epoxy and poly(phenylene sulfide) matrices. *Composites: Part A*, 30:1081–1091.
63. Koo, J. H., Kneer, M., et al. (1992) A cost-effective approach to evaluate high-temperature ablatives for military applications. *Naval Engineers Journal*, 104(3):166–177.

64. Koo, J. H., Lin, S., et al. (1992) Performance of high-temperature polymer composite ablatives under a hostile environment. *Science of Advanced Materials and Process Engineering Series,* vol. 37. SAMPE, Covina, CA, pp. 506–520.
65. Koo, J. H., Miller, M., et al. (1993) Evaluation of fiber-reinforced composites ablatives for thermal protection. *Science of Advanced Materials and Process Engineering Series,* vol. 38. SAMPE, Covina, CA, pp. 1085–1098.
66. Cheung, F. B., Koo, J. H., et al. (1993) Modeling of one-dimensional thermo-mechanical erosion of high-temperature ablatives. *Journal of Applied Mechanics,* 60:1027–1032.
67. Wilson, D., Beckley, D., and Koo, J. H. (1994) Development of silicone matrix-based advanced composites for thermal protection. *High Performance Polymer,* 6(2):165–181.
68. Shih, Y. C., Cheung, F. B., and Koo, J. H. (2003) Numerical study of transient thermal ablation of high-temperature insulation materials. *Journal of Thermophysics and Heat Transfer,* 17(1):53–61.
69. Koo, J. H., Miller, M. J., Weispfenning, J., and Blackmon, C. (2011) Silicone polymer composites for thermal protection system: fiber reinforcements and microstructures. *Journal of Composite Materials,* 45(13):1363–1380.
70. Koo, J. H., Miller, M. J., Weispfenning, J., and Blackmon, C. (2011) Silicone polymer composite for thermal protection of naval launching system. *Journal Spacecraft and Rockets,* 48(6):904–919.
71. Koo, J. H., Stretz, H., Bray, A., and Wootan, W. (2001, September) Next generation nanostructured ablatives for rocket propulsion system. AFOSR Contract No. F49620-00-C-0045, STTR Phase I Final Report, Submitted to AFOSR, Arlington, VA.
72. Koo, J. H., Stretz, H., Bray, A., Wootan, W., Mulich, S., Powell, B., Grupa, T., and Weispfenning, J. (2002) Phenolic-clay nanocomposite for rocket propulsion systems. *Int'l SAMPE Symposium and Exhibition,* vol. 47. SAMPE, Covina, CA, pp. 1085–1099.
73. Koo, J. H., Stretz, H., and Bray, A. (2002, September) Nanocomposite rocket ablative materials. AFOSR Contract No. F49620-00-C-0045, STTR Phase II Annual Report, Submitted AFOSR, Arlington, VA.
74. Koo, J. H., Stretz, H., Bray, A., Weispfenning, J., Luo, Z. P., and Wootan, W. (2003) Nanocomposites rocket ablative materials: Processing, characterization, and performance. *Proc. of 2003 SAMPE ISSE,* vol. 48. SAMPE, Covina, CA, pp. 1156–1170.
75. Koo, J. H., Stretz, H., Bray, A., Weispfenning, J., Luo, Z. P., and Wootan, W. (2004) Nanocomposite rocket ablative materials: processing, microstructure, and performance, AIAA-2004-1996 paper, *44th AIAA/ASME/ASCE/AHS Structures, Structural Dynamics, and Materials Conference,* April 19–22. Palms Springs, CA.
76. Koo, J. H., Stretz, H., Weispfenning, J., Luo, Z. P., and Wootan, W. (2004) Nanocomposite rocket ablative materials: subscale ablation test. *Int'l SAMPE Symposium and Exhibition,* vol. 49. SAMPE, Covina, CA, pp. 1000–1014.
77. Koo, J. H., Chow, W. K., Stretz, H., Cheng, A. C.-K., Bray, A., and Weispfenning, J. (2003) Flammability properties of polymer nanostructured materials. *Int'l SAMPE Symposium and Exhibition,* vol. 48. SAMPE, Covina, CA, pp. 954–963.
78. Koo, J. H., Pilato, L., and Wissler, G. E. (2005) Polymer nanostructured materials for high-temperature applications. *SAMPE J.,* 41(2):7.
79. Koo, J. H., and Pilato, L. A. (2006) Thermal properties and microstructures of polymer nanostructured materials. In *Nanoengineering of Structural, Functional, and Smart Materials,* Schulz, M. J., Kelkar, A., and Sundaresan, M. J., eds. Boca Raton, FL: CCR Press, pp. 409–441.

80. Miller, M. J., Koo, J. H, et al. (1993, January) Evaluation of different categories of composite ablative for thermal protection, AIAA-93-0839, *31st AIAA Aerospace Sciences Meeting*. Reno, NV.
81. Yang, B.C., Cheung, F. B., and Koo, J. H. (1995, January) Prediction of thermo-mechanical erosion of high-temperature ablatives in the SSRM facility, AIAA-95-0254, *33rd Aerospace Sciences Meeting*. Reno, NV.
82. VanMeter, M., Koo, J. H., et al. (1995) Mechanical properties and material behavior of a glass silicone polymer composite. *Proc. 40th International SAMPE Symposium*. SAMPE, Covina, CA, pp. 1425–1434.
83. Koo, J. H., et al. (1998) Effect of major constituents on the performance of silicone polymer composites. *Proc. 30th International SAMPE Technical Conference*. SAMPE, Covina, CA.
84. Koo, J. H., et al. (1999) Thermal protection of a class of polymer composites. *Proc. 44th International SAMPE Symposium*. SAMPE, Covina, CA, pp. 1431–1441.
85. MX-4926 Technical Data Sheet, Cytec Engineered Materials (now known as Cytec Solvay Group), Winona, MN.
86. Luehmann, W. Pratt & Whitney Space Propulsion/Chemical Systems Division, San Jose, CA, personal communication.
87. SC-1008 Technical Data Sheet, Borden Chemical, Louisville, KY (now owned by Hexion).
88. Koo, J. H., Pittman, C. U., Jr., Liang, K., Cho, H., Pilato, L. A., Luo, Z. P., Pruett, G., and Winzek, P. (2003) Nanomodified carbon/carbon composites for intermediate temperature: processing and characterization, 2003 *Int'l SAMPE Technical Conference*, vol. 35, SAMPE, Covina, CA, pp. 521–534.
89. Koo, J. H., Pilato, L. A., Pittman, C. U., and Winzek, P. (2004, January) Nanomodified carbon/carbon composites for intermediate temperature, AFOSR Contract No. F49620-02-C-0086, STTR Phase I Final Report, submitted to AFOSR. Arlington, VA.
90. Koo, J. H., Pilato, L. A., Winzek, P., Shivakumar, K. Pittman, C. U., Jr., and Luo, Z. P. (2004) Thermo-oxidative studies of nanomodified carbon/carbon composites. *Int'l SAMPE Symposium and Exhibition*, vol. 49. SAMPE, Covina, CA, pp. 1214–1228.
91. Blanski, R., Koo, J. H., et al. (2004, May) Polymer nanostructured materials for solid rocket motor insulation–ablation performance. *Proc. 52nd JANNAF Propulsion Meeting*. CPIAC, Columbia, MD.
92. Koo, J. H., Marchant, D., et al. (2004, May) Polymer nanostructured materials for solid rocket motor insulation–processing, microstructure, and mechanical properties. *Proc. 52nd JANNAF Propulsion Meeting*. CPIAC, Columbia, MD.
93. Ruth, P. Blanski, R., and Koo, J. H. (2004, May) Preparation of polymer nanostructured materials for solid rocket motor insulation. *Proc. 52nd JANNAF Propulsion Meeting*. CPIAC, Columbia, MD.
94. Koo, J. H. Pilato, L., et al. (2005) Epoxy nanocomposites for carbon fiber-reinforced composites. *Proc. SAMPE 2005 Int'l Symposium*, May 1–5. SAMPE, Covina, CA.
95. Koo, J. H., Pilato, L., et al. (2005) Nanocomposites for carbon fiber-reinforced polymer matrix composites, AIAA-2005-1928, *46th AIAA/ASME/ASCE/AHS Structures, Structural Dynamics, and Materials Conference*, April 18–21. Austin, TX.
96. Koo, J. H., Pilato, L., et al. (2004, October) Nanocomposite for carbon fiber reinforced polymer matrix composites, AFOSR STTR Phase I Final Report, submitted to AFOSR. Arlington, VA.

97. Koo, J. H., Pilato, L., Wissler, G. E., and Luo, Z. P. (2005, May) Flammability and mechanical properties of nylon 11 nanocomposites. *Proc. Int'l SAMPE 2005 Symposium and Exhibition (ISSE).* SAMPE, Covina, CA.
98. Koo, J. H., Pilato, L. A., and Wissler, G. E. (2005) Fire retardant polymer nanocomposites for selective laser sintering processing, submitted USPO patent application on July 27, 2005.
99. Cheng, J., Lao, S., Nguyen, K., Ho, W., Cummings, A., and Koo, J. H. (2005, August) SLS processing of nylon 11 nanocomposites. *Proc. 17th Solid Freeform Fabrication Symposium.* The University of Texas at Austin, Austin, TX.
100. Koo, J. H., Pilato, L., et al. (2005, August) Innovative selective laser sintering rapid manufacturing using nanotechnology. *Proc. 2005 Solid Freeform Fabrication Symposium.* The University of Texas at Austin, Austin, TX.
101. Lao, S., Ho, W., Nguyen, K., Cheng, J., and Koo, J. H. (2005) Microstructural analyses of nylon 11 nanocomposites. *Proc. 37th Int'l SAMPE Technical Conference (ISTC),* October 31–November 3. Seattle, WA.
102. Cummings, A., Shi, L., and Koo, J. H. (2005) Thermal conductivity measurements of nylon 11-carbon nanofiber nanocomposites. *Proc. of IMECE2005 (2005 ASME International Mechanical Engineering Congress and Exposition),* November 5–11. Orlando, FL.
103. Lao, S. C., Moon, T., Koo, J.H., et al. (2009) Flame-retardant polyamide 11 and 12 nanocomposites: thermal and flammability properties. *Journal of Composite Materials,* 43(17):1803–1816.
104. Lao, S. C., Koo, J. H., et al. (2010) Flame-retardant polyamide 11 and 12 nanocomposites: processing, morphology, and mechanical properties. *Journal of Composite Materials,* 44(25):2933–2951.
105. Lao, S. C., Koo, J. H., et al. (2011) Flame-retardant polyamide 11 nanocomposites: further thermal and flammability studies. *Journal of Fire Sciences,* 29(6):479–498.
106. Bray, A., Beal, G., and Stretz, H. (2004) Nanocomposite rocket ablative materials. AFOSR STTR Phase II Contract F49620-02-0013, Final Report, submitted to AFOSR, Arlington, VA.
107. Liu, Y., Lu, Z., Chen, X., Wang, D., Liu, J., and Hu, L. (2009) Study on phenolic-resin/carbon-fiber ablation composites modified with polyhedral oligomeric silsesquioxanes. *Proc. of 2009 4th IEEE International Conference on Nano/Micro Engineered and Molecular Systems,* January 5–8. Shenzhen, China.
108. Laine, R. M. (2005) Nanobuilding blocks based on the [OSiO] (x=6, 8, 10) octasilsesquioxanes. *J. Mater. Chem.,* 15:3725–3744.
109. Hu, L., Zhang, X., and Sun, Y. (2005) Hardness and elastic modulus profiles of hybrid coating. *J. Sol-Gel Sci. Tech.,* 34:41–46.
110. Tanaka, K., Adachi, S., and Chujo, Y. (2009) Structure–property relationship of octa-substituted POSS in thermal and mechanical reinforcements of conventional polymers. *Journal of Polymer Science: Part A: Polymer Chemistry,* 47:5690–5697.
111. Franchini, E., Galy, J., Gèrard, J. F., Tabuani, D., and Medici, A. (2009) Influence of POSS structure on the fire-retardant properties of epoxy hybrid networks. *Polymer Degradation and Stability,* 94:1728–1736.
112. Yu, Q-C., and Wan, H. (2012) Ablation capability of flake graphite reinforced barium-phenolic resin composite under long pulse laser irradiation. *Journal of Inorganic Materials,* 27(2):157–161 (in Chinese).

113. Si, J., Li, J., Wang, S., Li, Y., and Jing, X. (2013) Enhanced thermal resistance of phenolic resin composites at low loading of graphene oxide. *Composites: Part A*, 54:166–172.
114. Srikanth, I., Daniel, A., Kumar, S., Padmavathi, N., Singh, V., Ghosal, P., Kumar, A., and Devi, G. R. (2010) Nano silica modified carbon–phenolic composites for enhanced ablation resistance. *Scripta Materialia*, 63:200–203.
115. ASTM E1225-09. (2008) *Standard Test Method for Thermal Conductivity of Solids By Means of the Guarded-Comparative-Longitudinal Heat Flow Technique*. Annual Book of ASTM Standards.
116. Kumar, S., Kumar, A., Shukla, A., Devi, G. R., and Gupta, A. K. (2005) Thermal-diffusivity measurement of 3D-stitched C–SiC composites. *J. Eur. Ceram. Soc.*, 29(3):489–495.
117. Knacke, O., Kubaschewski, O., and Hesselman, K. (1991) *Thermo-Chemical Properties of Inorganic Substances*, 2nd ed. Berlin: Springer-Verlag, Berlin, pp. 1836–1837.
118. Xiao, J., Chen, J-M., Zhou, H-D., and Zhang, Q. (2007) Study of several organic resin coatings as anti-ablation coatings for supersonic craft control actuator. *Mater. Sci. Eng. A.*, 452–453:23–30.
119. Lee, Y. J., and Joo, H. (2004) Ablation characteristics of carbon fiber reinforced carbon (CFRC) composites in the presence of silicon carbide. *Surf. Coat. Tech.*, 180–181:286–289.
120. Srikanth, I., Padmavathi, N., Kumar, S., Ghosal, P., Kumar, A., and Subrahmanyam, C. (2013) Mechanical, thermal and ablative properties of zirconia, CNT modified carbon/phenolic composites. *Composites Science and Technology*, 80:1–7.
121. Bahramian, A. R., Kokabi, M., Navid Famili, M. H., and Beheshty, M. H. (2006) Ablation and thermal degradation behavior of a composite based on resol type phenolic resin: process modelling and experiment. *Polymer*, 47:3661–3673.
122. Bahramian, A. R., Kokabi, M., Navid Famili, M. H., and Beheshty, M. H. (2007) Thermal degradation process of resol type phenolic resin matrix/kaolinite layered silicate nanocomposite. *Iranian Polymer Journal*, 16(6):375–387.
123. Bahramian, A. R., Kokabi, M., Navid Famili, M. H., and Beheshty, M. H. (2008) High temperature ablation of kaolinite layered silicate/phenolic resin/asbestos cloth nanocomposite. *Journal of Hazardous Materials*, 150:136–145.
124. Bahramian, A. R., and Kokabi, M. (2009) Ablation mechanism of polymer layered silicate nanocomposite heat shield. *Journal of Hazardous Materials*, 166:445–454.
125. Bahramian, A. R., and Kokabi, M. (2011) Numerical and experimental evaluations of the flammability and pyrolysis of a resole-based nanocomposite by cone calorimeter. *Iranian Polymer Journal*, 20(5):399–411.
126. Paydeyesh, A., Kokabi, M., and Bahramian, A. R. (2012) High temperature ablation of highly filled polymer-layered silicate nanocomposites. *J. Applied Polymer Science*, doi: 10.1002/APP.377588.
127. ASTM E1269-11. (2008) *Standard Test Method for Determining Specific Heat Capacity by Differential Scanning Calorimetry*. Annual Book of ASTM Standards.
128. Bartholmai, M., and Schartel, B. (2004) Layered silicate polymer nanocomposites: new approach or illusion for fire retardancy? Investigations of the potentials and the tasks using a model system. *Polym. Adv. Technol.*, 15:355–364.
129. Zanetti, M., Camino, G., and Mülhaupt, R. (2001) Combustion behavior of EVA/flourohectorite nanocomposites. *Polym. Degrad. Stab.*, 74:413–417.

130. Duquesne, S., Jama, C., Bras, M. L., Delobel, R., Recourt, P., and Gloaguen, J. M. (2003) Elaboration of EVA-nanoclay systems-characterization, thermal behavior and fire performance. *Compos. Sci. Technol.*, 63:1141–1148.
131. Gilman, J. W., Jackson, C. L., Morgan, A. B., Harris, R., Manias, E., Giannelis, E. P., Wuthenow, M., et al. (2000) Flammability properties of polymer layered silicate nanocomposites polypropylene and polystyrene nanocomposites. *Chem. Mater.*, 12:1866–1873.
132. Bahramian, A. R. (2013) Pyrolysis and flammability properties of novolac/graphite nanocomposites. *Fire Safety Journal*, 61:265–273.
133. Bahramian, A. R., and Astaneh, R. A. (2014) Improvement of ablation and heat shielding performance of carbon fiber reinforced composite using graphite and kaolinite nanopowders. *Iranian Polymer Journal*, 23:979–985.
134. Mirzapour, A., Asadollahi, M. H., Baghshaei, S., and Akbari, M. (2014) Effect of nano-silica on the microstructure, thermal properties and bending strength of nano-silica modified carbon fiber/phenolic nanocomposite. *Composites: Part A*, 63:159–167.
135. Torre, L., Kenny, J. M., and Maffezzoli, A. M. (1998) Degradation behavior of a composite material for thermal protection systems Part I–Experimental characterization. *Journal of Materials Science*, 33:3137–3143.
136. Torre, L., Kenny, J. M., and Maffezzoli, A. M. (1998) Degradation behavior of a composite material for thermal protection systems part II process simulation. *Journal of Materials Science*, 33(12):3145–3149.
137. Torre, L., Kenny, J. M., Boghetich, G., and Maffezzoli, A. M. (2000) Degradation behavior of a composite material for thermal protection systems. Part III: Char characterization. *Journal of Materials Science*, 35(18):4563–4566.
138. Natali, M., Monti, M., Kenny, J., and Torre, L. (2011) Synthesis and thermal characterization of phenolic resin/silica nanocomposites prepared with high shear rate-mixing technique. *Journal of Applied Polymer Science*, 120:2632–2640.
139. Natali, M., Monti, M., Kenny, J., and Torre, L. (2011) A nanostructured ablative bulk moulding compound: development and characterization. *Composite Part A: Applied Science and Manufacturing*, doi:10.1016/j.compositesa.2011.04.022.
140. Natali, M., Monti, M., Puglia, D., Kenny, J., and Torre, L. (2012) Ablative properties of carbon black and MWNT/phenolic composites: a comparative study. *Composites Part A: Applied Science and Manufacturing*, 43(1):174–182.
141. Pavli, A. J. (1968) Experimental evaluation of several advanced ablative materials as nozzle sections of a storable propellant rocket engine. NASA TM X-1559.
142. Warga, J. J. (1979) Low cost fabrication techniques for solid rocket nozzles. *Proc. of national aeronautics and space engineering and manufacturing meeting*, October 5–9. Los Angeles, CA, pp. 700–796.
143. LR1504 Prepreg datasheet. Website: http://www.lewcott.com/ablatives.html.
144. Cytec Engineered Materials. Website: http://www.cytec.com/engineered-materials/selectorguide.htm.
145. D'Aelio, G. F., and Parker, J. A. (1971) *Ablative Plastics*. New York: Marcel Dekker.
146. Sutton, P., and Biblarz, O. (2000) *Rocket Propulsion Elements*. New York: Wiley-IEEE.
147. Peterson, D. A., Winter, J. M., and Shinn, A. M., Jr. (1969) Rocket engine evaluation of erosion and char as functions of fabric orientation for silica-reinforced nozzle materials. NASA TM X-1721.
148. ASTM E457-08. (2008) *Standard Test Method for Measuring Heat-Transfer Rate Using a Thermal Capacitance (Slug) Calorimeter*. Annual Book of ASTM Standards.

149. Wu, C. S., Liu, Y. L., and Chiu, Y. S. (2002) Epoxy resins possessing flame retardant elements from silicon incorporated epoxy compounds cured with phosphorus or nitrogen containing curing agents. *Polymer,* 43(15):4277–4284.
150. Wang, W. J., Perng, L. H., Hsiue, G. H., and Chang, F. C. (2000) Characterization and properties of new silicone-containing epoxy resin. *Polymer,* 41(16):6113–6122.
151. Zheng, S., Wang, H., Dai, Q., Kuo, X., Ma, D., and Wang, K. (1995) Morphology and structure of organosilicon polymer-modified epoxy resins. *Makromol. Chem. Phys.,* 196(1):269–278.
152. Monti, M., Natali, M., Petrucci, R., Puglia, D., Terenzi, A., Valentini, L., and Kenny, J. M. (2011) Advanced fiber reinforced composites based on nanocomposite matrices. In *Wiley Encyclopedia of Composites,* 2nd ed., Nicolais, L., and Borzacchiello, A., eds. doi: 10.1002/9781118097298.weoc025.
153. Kashiwagi, T., Du, F., Winey, K. I., Groth, K. M., Shields, J. R., Bellayer, S. P., et al. (2005) Flammability properties of polymer nanocomposites with single-walled carbon nanotubes: effects of nanotube dispersion and concentration. *Polymer,* 46:471–481.
154. Kashiwagi, T., Du, F., Douglas, J. F., Winey, K. I., Harris, R. H., and Shields, J. R. (2005) Nanoparticle networks reduced the flammability of polymer nanocomposites. *Nat. Mater.,* 4:928–933.
155. Cipiriano, B. H., Kashiwagi, T., Raghavan, S. R., Yang, Y., Grulke, E. A., Yamamoto, K., et al. (2007) Effects of aspect ratio of MWNT on the flammability properties of polymer nanocomposites. *Polymer,* 48:6086–6096.
156. Zhao, Z., and Gou, J. (2009) Improved fire retardancy of thermoset composites modified with carbon nanofibers. *Sci. Technol. Adv. Mater.,* 10:015005.
157. Rahatekara, S. S., Zammarano, M., Matko, S., Koziol, K. K., Windle, A. H., Nyden, M., et al. (2010) Effect of carbon nanotubes and montmorillonite on the flammability of epoxy nanocomposites. *Polym. Degrad. Stabil.,* 95:870–879.
158. Park, J. M., Kwon, D. J., Wang, Z. J., Roh, J. U., Lee, W. I., Park, J. K., and DeVries, K. L. (2014) Effects of carbon nanotubes and carbon fiber reinforcements on thermal conductivity and ablation properties of carbon/phenolic composites. *Composites: Part B,* 67:22–29.
159. Tate, J. S., Jacobs, C. J., and Koo, J. H. (2011) Dispersion of MWCNT in phenolic resin using different dispersion techniques and evaluation of thermal properties. *Proc. of 2011 SAMPE ISSE,* , May 23–26. Long Beach, CA.
160. Tate, J. S., Gaikwad, S., Theodoropoulou, N., Trevino, E., and Koo, J. H. (2013) Carbon/Phenolic nanocomposites as advanced thermal protection material in aerospace applications. *Journal of Composites,* http://dx.doi.org/10.1155/2013/403656.
161. Thostenson, E. T., Li, C., and Chou, T. W. (2005) Nanocomposites in context. *Composites Science and Technology,* (65):491–516.
162. Cheng, J. (2006) *Polysyanate Ester/Small Diameter Carbon Nanotubes Nanocomposite.* Master's thesis, Department of Mechanical Engineering, The University of Texas at Austin, Austin, TX.
163. Safadi, R. A. (2002) Multiwalled carbon nanotube polymer composites: synthesis and characterization of thin films. *Journal of Applied Polymer Science,* 84:2660–2669.
164. Pulci, G., Tirillo, J., Marra, F., Fossati, F., Bartuli, C., and Valente, T. (2010) Carbon/phenolic ablative materials for reentry space vehicles: manufacturing and properties. *Composites: Part A,* 41:1483–1490.

165. Marra, F., Pulci, G., Tirillo, J., Bartuli, C., and Valente, T. (2011) Numerical simulation of oxyacetylene testing procedure of ablative materials for reentry space vehicles. *Proc. IMechE Vol. 225 Part L: J. Materials: Design and Applications*, 225:32–40.
166. Pulci, G., Tirillo, J., and Valente, T. (2012) Unpublished data of "Ablative Materials for Thermal Protection Systems," University of Rome, Rome, Italy.
167. Koo, J. H., Natali, M., Lisco, B., Yao, E., and Schellhase, K. (2018) In situ ablation recession and thermal sensor for thermal protection systems. *Journal of Spacecraft and Rockets*, 55(4):783–796. doi: 10.2514/1.A33925.
168. Tran, H. K., Johnson, C. E., Rasky, D. J., Hui, F. C. L., Hsu, M. T., Chen, T., Chen, Y. K., Paragas, D., and Kobayash, L. (1997, April) Phenolic impregnated carbon ablators (PICA) as thermal protection systems for discovery missions. NASA TM 110440. doi: 10.2514/6.1996-1911.
169. Jaramillo, M., Koo, J. H., and Natali, M. (2014) Compressive char strength of thermoplastic polyurethane elastomer nanocomposites. *Polymers for Advanced Technology*, 25:742–751. doi: 10.1002/pat.3287.
170. Lewis, J., Jaramillo, M., and Koo, J. H. (2017) Development of a shear char strength sensing technique to study thermoplastic polyurethane elastomer nanocomposites. *Polymer for Advanced Technologies*, 28(12):1707–1725. doi: 10.4044.1002/pat.
171. Milos, F. S., and Chen, Y. K. (2010) Ablation and thermal response property model validation for phenolic impregnated carbon ablator. *Journal of Spacecraft and Rockets*, 47(5):786–804. doi.org/10.2514/1.42949.
172. Amar, A. J., Oliver, A. B., Kirk, B. S., Salazar, G., and Droba, J. (2016) Overview of the CHarring Ablator Response (*CHAR*) Code. AIAA-2016-3385. *46th AIAA Thermophysics Conference*, June 13–16, Washington, DC.
173. Salazar, G., Droba, J., Oliver, B, and Amar, A. J. (2016) Development and verification of enclosure radiation capability in the charring ablator response (*CHAR*) code. AIAA-2016-3388. *46th AIAA Thermophysics Conference*, June 13–16, Washington, DC.
174. Yang, B. C. (1992, December) *Theoretical Study of Thermochemical Erosion of High-Temperature Ablatives*. Ph.D. dissertation, Pennsylvania State University, State College, PA.
175. Ewing, M. E., Laker, T. S., and Walker, D. T. (2013) Numerical modeling of ablation heat transfer. *Journal of Thermophysics and Heat Transfer*, 27(4):615–632.
176. Ewing, M. E., Walker, D. T., Isaac, D. A., Phipps, B. E., and Dewey, H. H. (2013, April) Heat Transfer and Erosion Analysis Program, HERO 2010, Theory Manual, ATK, Aerospace System, Brigham City, UT.
177. Duffa, G. (2013) *Ablative Thermal Protection Systems Modeling*. AIAA, Reston, VA.
178. Koo, J. H., and Mensah, T. (2018) Novel Polymer Nanocomposite Ablative Technologies for Thermal Protection of Propulsion and Reentry System for Space Application. In T. Mensah, B. Wang, J, Winter, and V. Davis, eds., *Nanotechnology Commercialization: Manufacturing Processes and Products*, Wiley, New York, pp. 177–244, doi: 10.1002/9781119371762.ch6.
179. Koo, J. H. and Langston, J. (2019) Polymer Nanocomposite Ablative Technologies for Solid Rocket Motors. In G-Q. He, Q-L. Yan, P-J. Liu, and M. Gozin, eds., *Nanomaterials in Rocket Propulsion Systems*, Elsevier, Oxford, UK, pp. 423-494, doi: 10.1016/B978-0-12-813908-0.00001-0.
180. PICA Material Property Report C-TPSA-A-DOC-158 (2009) for Crew Exploration Vehicle Block II Heatshield Advanced Development Project, NASA Report, Rev. 1, June 8, 2009.

CHAPTER 10

Electrical Properties of Polymer Nanocomposites

10.1 Introduction

This chapter provides an overview for the electrical properties of polymer nanocomposites. The nano-sized fillers for the conductive polymer nanocomposites are introduced in two groups: inorganic nanomaterials which are mainly used with intrinsically conductive polymers, which are not included in this chapter; organic nanomaterials including carbon nanotubes (CNTs), carbon black (CB), carbon nanofibers (CNFs), and graphite. The commonly used techniques for the characterization of electrical conductivity are presented in Chap. 5, and also the mechanism of conductive network formation and some theories that explain the phenomenon.

The different processing techniques and characterization of electrical properties of polymer nanocomposites are discussed in Chaps. 4 and 5, respectively. Based on the different kinds of polymers used, the electrical properties of nanocomposites are discussed within three categories: thermoplastic-based nanocomposites, thermoset-based nanocomposites, and thermoplastic elastomer–based nanocomposites.

10.2 Electrical Properties of Thermoplastic-Based Nanocomposites

In this section, electrical properties of thermoplastic-based nanocomposites are discussed in groups of CNT-, CNF-, graphene-based reinforced thermoplastic nanocomposites.

10.2.1 Carbon Nanotube–Reinforced Thermoplastic-Based Nanocomposites

One of the key challenges to fabricate CNT thermoplastic nanocomposites that possess good electrical and mechanical properties lies in the good processing to obtain excellent interconnecting network of CNT [1]. Lew and Luizi [1] investigated the effect of the parameters of twin-screw extruders, such as mixing residence time, molding temperature, and screwing speed on the volume and surface resistivity of MWNT/PP nanocomposites. They explained the possible reason behind the decrease in resistivity with higher value of some of these parameters from the perspective of exfoliation of multiwalled carbon nanotubes (MWNTs) [1]. The delamination of agglomerate into primary aggregates caused the decrease in electrical resistivity. With high shear, the primary aggregates further go into secondary aggregates that are loosely entangled and finally exfoliated CNTs.

With the understanding of the aggregation of CNTs that facilitate the formation of conductive network, there are more examples about the application of CNT-reinforced nanocomposites that use thermoplastic, such as polyamide (PA). The influence of mixing speed, melt temperature, or residence time on the electrical conductivity has been investigated for MWNT/PA nanocomposites [2–4]. Since most literature was dealing with the PA6 [2, 5, 6] or PA6,6 [2] as matrix for MWNTs, it was found that percolation threshold for PA6,6 was 1 wt%, and 2.5 wt% for PA6, which was slightly higher. However, Socher et al. [7] studied the electrical properties of various PA12 nanocomposites with different types of MWNTs. The PA12 can be categorized into low- and high-viscosity with two kinds of amine groups. Four different types of MWNTs and one type of CB (Pritx® XE2) were used for comparison (Table 10.1) [7].

The electrical volume conductivity of three different MWNTs and CB is shown in Fig. 10.1 [7]. Obviously, the percolation thresholds of MWNTs are all lower than that of CB because of their higher aspect ratio. Among different types of MWNTs, the Nanocyl™ has the lowest percolation of 0.7 wt%, followed by Future Carbon MW-K

Table 10.1 Properties of the CNTs Used in Socher Et al.'s Study [7]

Material	Diameter (nm)	Length (μm)	Carbon purity (%)	Bulk density (kg/m³)
Baytube® C150P	13–16	1–>10	>95	120–170
Nanocyl™ NC7000	9.5 (average)	15 (average)	>90	66
Future Carbon CNT-MW-K	15	–	90	28
Printex® XE2	30–35	n.a.	>99	100–400

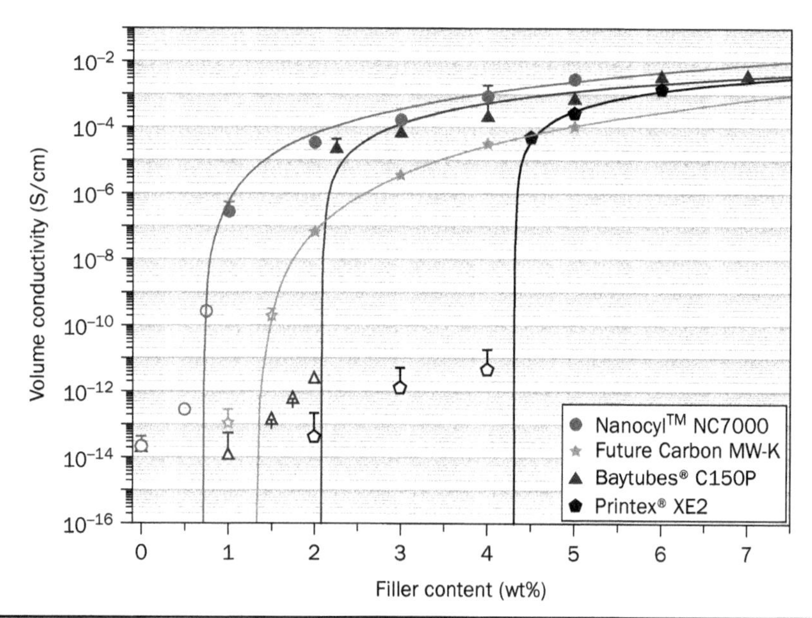

Figure 10.1 Electrical volume conductivity of polyamide 12 composites containing different MWNTs and CB [7].

of 1.3 wt%, and Baytubes® of 2.1 wt%. The electrical percolation threshold for CB (Pritx XE2) is determined at 4.3 wt%. Therefore, the influence of different types of carbon-based nanofillers on the electrical property of PA12 nanocomposites can be decided.

The electrical percolation behavior of different PA12 composites containing Baytubes C150P is presented in Fig. 10.2 [7]. The composites with low viscous PA12 and acid termination have the lowest electrical percolation threshold of 2.1 wt%, and the one with low viscous and amine excess has significantly higher percolation of 3.7 wt%, which indicates the influence of end group termination on the electrical percolation behavior, also reflected in high-viscosity system. However, with the same end group, the ones with low-viscosity present lower electrical percolation, suggesting the influence of viscosity on the electrical percolation behavior.

10.2.2 Carbon Nanofiber-Reinforced Thermoplastic-Based Nanocomposites

One of the issues for CNF polymer nanocomposites is the poor interaction of fiber and polymer interface. Although shear mixing is an efficient way to achieve better dispersion, it causes severe damage to nanofibers and leads to low aspect ratios [8, 9]. Another way to disperse CNFs in a polymer matrix is to functionalize CNFs, which will improve the dispersion of CNFs in matrix [10]. However, the usage of oxidized CNFs with PMMA of chaotic mixing leads to decrease in electrical conductivity with the improvement in dispersion partially attributed to the low intrinsic conductivity of the functionalized CNFs.

Again, with the same mixing method, Jimenez and Jana [11] showed that the electrical conductivity of polymethylmethacrylate (PMMA)-CNF nanocomposites is dependent on dispersion of CNF and mixing time at low shear, especially at nanofiber loading (2 wt%)

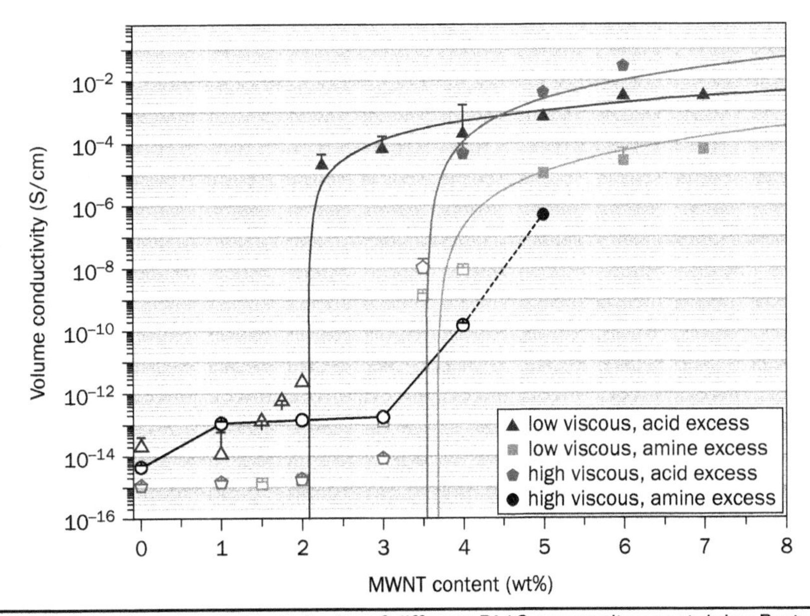

FIGURE 10.2 Electrical percolation behavior of different PA12 composites containing Baytubes C150P [7].

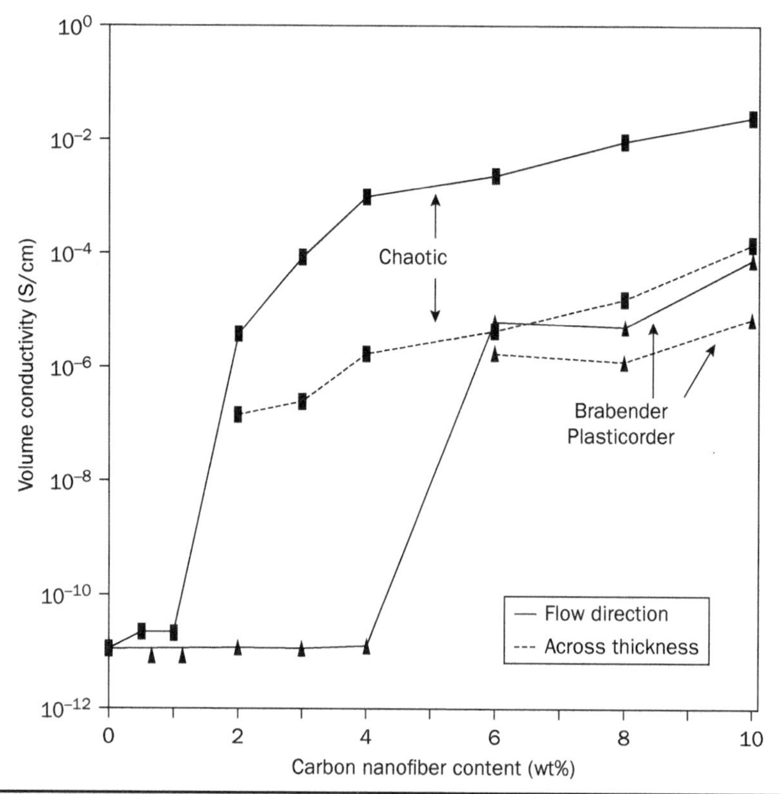

FIGURE 10.3 Volume electrical conductivity of PMMA-CNF composites. Results from materials mixed in a chaotic mixer and Brabender® Plasticorder® are identified. Measurements reflect a maximum standard deviation of 10% [11].

at percolation threshold. Figures 10.3 and 10.4 show that with better dispersion of nanofibers, the electrical conductivity of nanocomposites, using both methods, increases significantly after reaching the percolation threshold and nanocomposites mixed using chaotic method have superior conductivity [11]. Figure 10.5 illustrates that longer mixing time lowers the electrical conductivity despite better dispersion; for example, for 2 wt% formulation, the material is conductive up to 4 minutes but becomes insulative for 10 minutes [11]. One possible explanation is that the nanostructures of CNFs are broken after long-time mixing, which leads to decrease in conductivity.

10.2.3 Graphite-Reinforced Thermoplastic-Based Nanocomposites

The electrical properties of graphite-reinforced thermoplastic-based polymers, such as polypropylene nanocomposites, have been reported [12]. Li and Shimizu [13] added different loadings of expanded graphite (EG) into polyvinylidene fluoride (PVDF) with direct melt mixing and interestingly introduced the alternating current (AC) electrical measurement at various frequencies and the biased random walk approach to explain the formation of conductive network at such a condition [14]. As a result, the electrical conductivity of EG/PVDF system was found to be frequency and temperature dependent, which was known as tunnel effect in quantum mechanics [15]. Besides the above factors, other factors, such as filler aspect ratio, content, and morphology can affect the

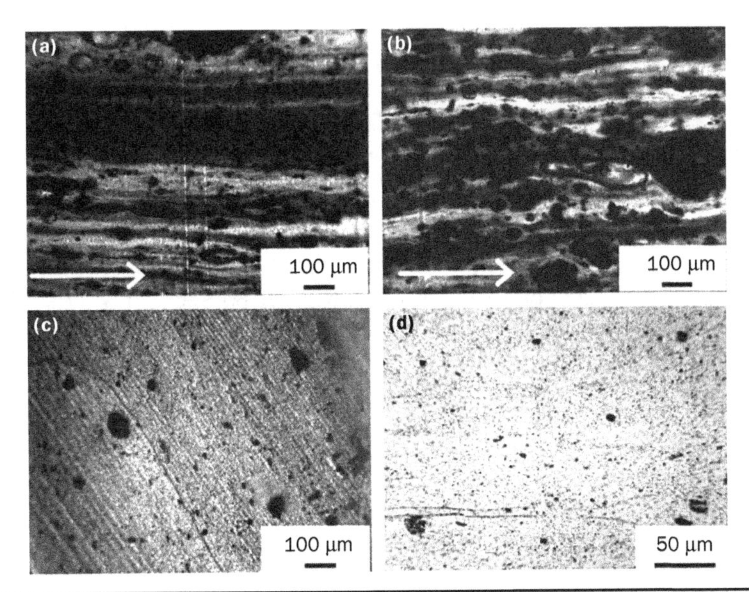

FIGURE 10.4 Optical micrographs of PMMA-CNF composites. Film thickness for (a), (b), and (c) was about 50 lm, and 1 lm for (d). The arrow indicates flow direction. Traces of lines in (c) are due to microtoming. (a) Chaotic 1 wt%, (b) chaotic 2 wt%, (c) Brabender Plasticorder 4 wt%, and (d) Brabender Plasticorder 6 wt% [11].

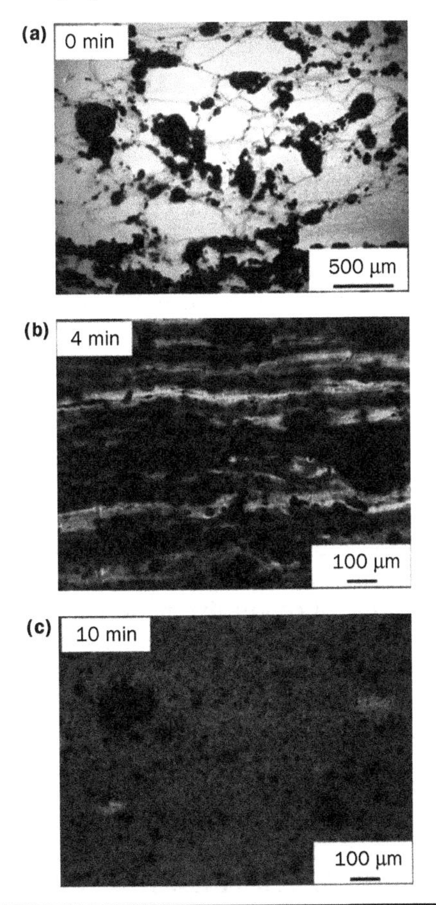

FIGURE 10.5 Effect of mixing time on nanofiber dispersion in 2 wt% CNF composite prepared in a chaotic mixer. Mixing times are indicated in (a) to (c) [11].

electrical conductivity of nanocomposites. Goyal et al. [16] claimed that even for the same conductive filler, such as EG, the percolation threshold also depends on the types of polymer used in the system. It is because the good interaction between matrix and fillers is necessary to form excellent conductive network. For instance, the percolation for PMMA/EG was claimed to be 3.5 wt% [17], and the electrical conductivity at 25°C and applied voltage of 10 V can be as high as 60 S/cm at 8 wt% [15]. The electrical conductivity of nylon 6/EG reached 10^{-4} S/cm when the graphite content was 2.0 vol% [18]—a conductivity worse than poly(phenylene sulfide) (PPS)/EG, which had an electrical conductivity percolation threshold of 1 wt% (0.6 vol%) [16].

Via et al. [19] compared the electrical properties of different carbon-based fillers of CB, MWNT, and exfoliated graphite nanoplatelets (xGnP) in polycarbonate (PC). The above nanocomposites are used in producing electrostatic dissipative materials and semiconductors, which require relatively low electrical conductivity; for example, fuel gauges require electrical conductivity in the range of 10^{-2} to 10^{-1} S/cm. All the fillers exhibited low percolation threshold: CNT/PC was the lowest at 1.2 vol%, 2.4 vol% for CB/PC, and highest for xGnP/PC at 4.6 vol%. Additionally, for the first time, Via et al. applied the general effective media (GEM) models for these composite systems and found that the modeling results agreed with the experimental results—that with higher concentration of fillers the conductivity increases.

The synergistic effect of conductive networks formed by various carbon nanofillers in thermoplastic-based nanocomposites has been investigated. For instance, Zhang et al. [20] produced conductive polypropylene-based nanocomposites with MWNT/CB mixtures in melt-blending. The nanocomposites with only CB had higher percolation threshold than that of nanocomposites with only MWNT because of the higher aspect ratio of MWNTs [20]. With 0.25 wt% of MWNTs and 0.25 wt% of CB hybrid nanofillers, the nanocomposites became conductive, but it took 2.5 wt% of MWNTs for the nanocomposites with only MWNTs to become conductive. $MWNT_1$-CB_1 nanocomposites were more conductive than $MWNT_1$-CB_4 at the same total filler content. The resistivity of the composites containing hybrid nanofillers is higher than MWNT composites at the same filler content. However, the resistivity of $MWNT_4$-CB_1 and $MWNT_1$-CB_1 was similar to that of MWNT nanocomposites. As a result, the synergistic effect between CB and MWNTs makes possible the partial replacement of high-aspect-ratio and high-priced nanofillers (MWNTs) with low-aspect-ratio and low-priced CB. The result is in correspondence with the theoretical modeling of electrical percolation of mixed MWNT/CB- and MWNT/graphene-reinforced polypropylene or polyoxymethylene nanocomposites [21].

As for the mechanism of formation of conductive network, it is easy to understand that for pure MWNT- or CB-reinforced nanocomposites, nanofillers were responsible for the formation of conductive network. In the hybrid system, there are three mechanisms for the formation of conductive networks: Mechanism 1: shortcut of network branches caused by new active branches of networks co-formed by CB; Mechanism 2: CBs are introduced into MWCT networks by bridging MWNT local networks with CB; Mechanism 3: bridging local contacts between CB clusters through MWNTs. For nanocomposites containing low filler content [20], conductive networks are formed in $MWNT_1$-CB_1 because there are bridges formed by CB to MWNT local networks. For $MWNT_4$-CB_1, the MWNTs are more entangled, but are not efficiently used to build up conductive network. For $MWNT_1$-CB_4, the amount of MWNTs, which have high-aspect-ratio filler, is too low to build conductive networks. As a result, $MWNT_1$-CB_1 has much lower percolation threshold than the rest of the composites.

10.3 Electrical Properties of Thermoset-Based Nanocomposites

In the category of thermoset polymers, epoxy is widely used for advanced composites, because they display characteristics, such as good stiffness, specific strength, dimensional stability, chemical resistance, and strong adhesion with reinforcement [22]. Numerous research results on the electrical conductivity of CNF-reinforced epoxy nanocomposites are available. Cyanate ester (CE), a class of high-performance thermosetting resins, has received considerable attention due to its good mechanical properties, thermal stability, flammability properties, ease of process, and volatile-free curing process; these properties of CE are discussed in this section [23].

10.3.1 Carbon Nanotube-Reinforced Thermoset-Based Nanocomposites

Battisti et al. [24] used the high-shear mixing method to process MWNT/unsaturated polyesters with CNT loading ranging from 0.5 to 3 wt%. The electrical measurement was performed with AC impedance spectroscopy and DC conductivity measurement. Figure 10.6a shows the microstructure of the cured nanocomposites of 0.15 wt%, at which concentration the reaggregation at microscale and nanoscale can be compared [24]. The low bright regions, which show high content of CNTs, cover a large area of the sample, and can indicate that a different kind of aggregations has been formed that is different from those originally mixed in the matrix because these aggregations are not tightly packed as the original clusters. Figure 10.6b shows the charge contrast imaging used to demonstrate the morphology of the secondary agglomerate in matrix [24]. It is obvious that nanotubes were entangled and loosely packed. So the reaggregation of CNTs during the cure process can explain well the low percolation threshold of 0.026 wt%, which was lower than the theoretical value of about 0.96 wt%. Compared with Vera-Agullo et al.'s study [25] about the electrical properties of untangled MWNTs or unsaturated polyester nanocomposites processed without solvent, which have a percolation threshold of 0.1 wt%, the electrical properties of nanocomposites in Battisti el al.'s [24] study improve about four times.

Other similar reports about the CNT-reinforced thermoset polymer nanocomposites are available. For instance, Martin et al. [26] reported the effect of aspect ratio of fillers on the percolation threshold and analyzed the cluster formation that was responsible for the conductive network formation. Thostenson et al. [27] manufactured the MWNT/vinyl ester nanocomposites using three-roll milling technique and discovered that the conductive network of MWNTs formed in the matrix at the loading of more than 1 wt%. Sandler et al. [28] obtained similar percolation threshold as Battisti et al.

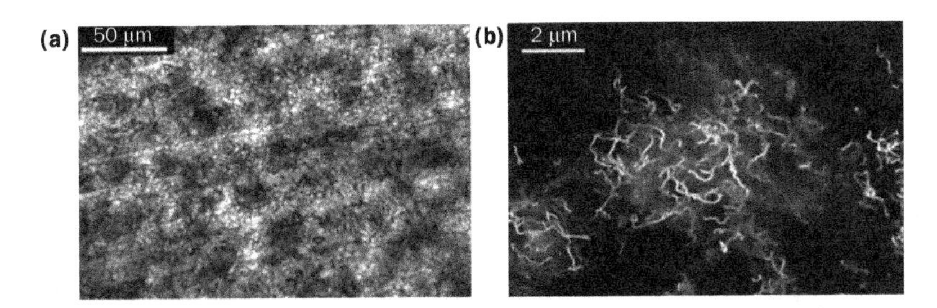

Figure 10.6 Nanocomposite containing 0.15 wt% of CNTs: (a) transmission light micrograph and (b) charge contrast imaging [24].

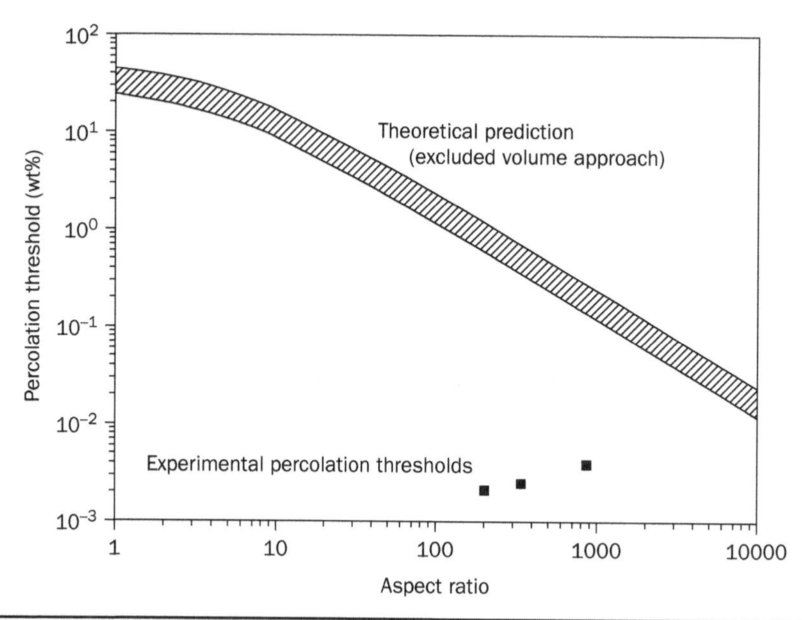

FIGURE 10.7 The electrical percolation threshold with the function of aspect ratios [31].

[24] for the MWNT/epoxy nanocomposites produced by direct high-shear mixing with aligned MWNTs at 0.025 wt%. Other researchers [29, 30] investigated the effect of treatment on the electrical property of CNT/epoxy nanocomposites, and concluded that SWNTs needed more concentration than MWNTs to attain percolation threshold, and the treatment or functionalization did improve the mechanical properties of nanocomposites because of better interconnection between fillers and matrix [31, 32], but it decreased the electrical conductivity by making the aspect ratio smaller (Fig. 10.7) [31].

Cyanate ester, a class of high-performance thermoset, has good mechanical properties, thermal stability, flammability properties, ease of process, and volatile-free curing process [23]. The relatively high glass transition temperature (more than 250°C) of CE falls between that of epoxy and polyimide. With proper nano-modification and cure, CE has great potential to replace the hard-to-process polyimide resin. MWNTs are often selected to create polymer nanocomposites, due to their unique combination of excellent mechanical, electrical, and thermal properties [33]. It is anticipated that the incorporation of MWNTs will enhance the mechanical properties and thermal properties as well as reduce the electrical resistivity of the resulting CE-MWNT nanocomposites. Ultimately, the high-performance carbon fiber–reinforced CE-MWNT nanocomposites will be fabricated for structural applications in the aerospace, aviation, and defense industries, which require excellent mechanical, thermal, and flammability properties as well as good electrical conductivity.

Recently, Lao conducted CE nanocomposite research to determine the extent to which several standard processing techniques will affect/improve MWNT dispersion and attendant nanocomposite properties, such as the thermal, mechanical, and electrical properties, as well as flammability [34]. A *processing-structure-property* relationship, as well as performance of this class of carbon-based CE nanocomposite, is established by Lao. Effects of different standard functionalization of MWNTs (i.e., –OH, –COOH, and $-NH_2$) on MWNT dispersion in the CE resin are studied. In addition to aiding dispersion, this functionalization has the potential to differentially affect/improve nanocomposite properties (such as mechanical properties, flammability properties, electrical conductivity, and thermal conductivity) due to the resulting improved interfacial bonding or adhesion between the CE resin and MWNTs. Preliminary results of carbon fiber–reinforced CE-MWNT composites are also reported [34].

Most processing techniques for nano-modifications aim to disperse nanomaterials uniformly in the polymer resin. Dispersion techniques vary for different polymer resins and different nanomaterials. Common methods include emulsion polymerization, in situ polymerization, melt compounding, solvent blending, ultrasonication, mechanical mixing, and high-shear mixing. For example, polyhedral oligomeric silsesquioxane (POSS®) is usually functionalized, specifically tailored to the chosen polymer resin, and processed by in situ polymerization. However, nanoclay usually requires shearing force, so that layers of clay platelets can be sheared apart. In the case of CNTs, vigorous stirring or high-shear mixing is believed to debundle the nanotube aggregates. More details regarding these processing techniques and property enhancement of the resulting CE-MWNT nanocomposites are discussed by Lao.

In Lao's study, several processing instruments—including ultrasonicator, planetary centrifugal mixer, high-shear mixer, three-roll mill, and stand mixer—were used to disperse the CNTs into the CE resin. The PNC morphology and MWNT dispersion are characterized by transmission electron microscopy (TEM) and scanning electron microscopy (SEM). Further, the thermal stability is assessed by thermogravimetric analysis (TGA), while combustion characteristics are examined by microscale combustion calorimetry (MCC). The glass transition temperature is determined by dynamic mechanical analysis (DMA), while the electrical conductivity is measured by megohmmeter and plate electrodes. This section presents only the electrical results of the Lao's study. Electrical conductivity/resistivity was calculated from the voltage drop measured by the Hioki Super Megohmmeter and plate sample electrode. Plate specimens of 1 mm thickness were used in the electrical volume resistivity measurement.

Neat MWNT-5 was mixed with CE and cured with 200 ppm Fe^{3+} catalyst or coupling agents, CA-1 and CA-2, followed by a post-cure at 250°C for 2 hours [34–37]. Specimens of 1 mm thickness were made for the electrical resistivity measurement. Figure 10.8 shows that the addition via stand mixer of less than 0.5 wt% or 0.3 vol% MWNTs lowers the neat CE's electrical resistivity. These measurements reveal that the percolation threshold is about 0.4 wt% or 0.24 vol% and the resistivity reaches a plateau beyond 1 wt% or 0.6 vol%. Furthermore, the electrostatic dissipation (ESD) requirement (less than 10^{11} ohm-cm) [38] is fulfilled by incorporating as little as 0.3 wt% or 0.18 vol% MWNTs into the CE matrix. The resulting percolation threshold matches Hu et al.'s review [39], which reported percolation thresholds of 1 wt% MWNTs or SWNTs or less in other resin systems. Similar percolation threshold of 1 wt% was reported by Ma et al.'s review [40]. In fact, the viscosity of the polymer becomes too high to fabricate void-free polymer nanocomposites when the content of nanotubes is higher than 1 wt%. Therefore, it is crucial to use processing techniques that can produce polymer-MWNT nanocomposites with a percolation threshold of less than 1 wt%. Among the processing techniques used in Ma et al.'s study, the stand mixer was able to achieve that threshold.

Interestingly, similar electrical resistivity drops were not seen in samples processed by other techniques, even at the same MWNT weight loadings. For instance, the electrical resistivity of the CE-MWNT nanocomposites is still higher than the ESD requirement at a loading of 1 wt% or 0.6 vol% (Fig. 10.9), although the addition of 1.5 wt% or 0.91 vol% of MWNTs lowers the electrical resistivity by about four orders of magnitudes relatively to the 1 wt% or 0.6 vol% sample. While it is likely that the percolation

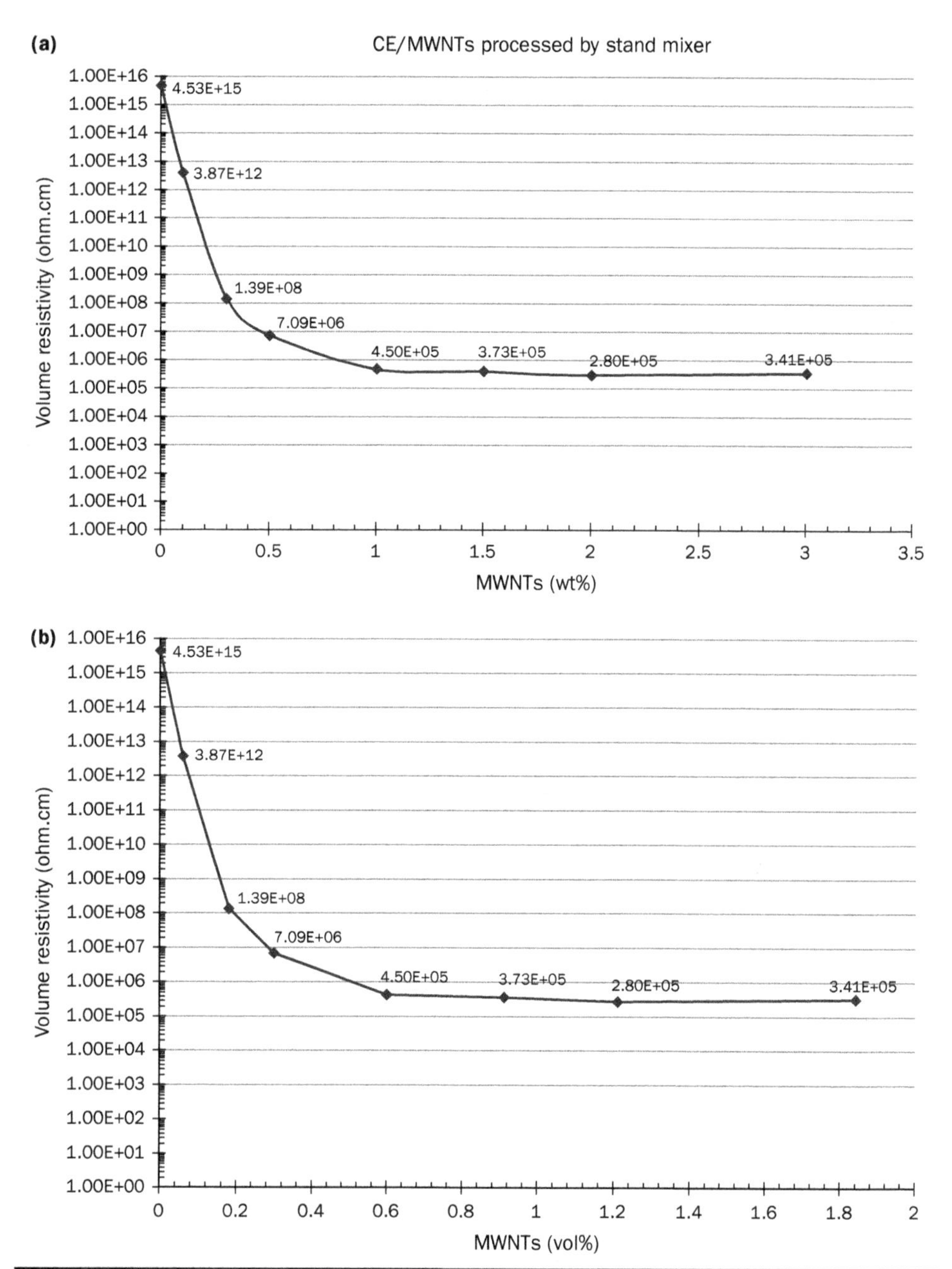

FIGURE 10.8 Electrical resistivity of CE-MWNT-5 (with 200 ppm Fe^{3+} catalyst) polymer nanocomposites processed by the stand mixer with various MWNT loadings in (a) wt% and (b) vol% (number of samples per MWNT wt%, $N = 1$) [34].

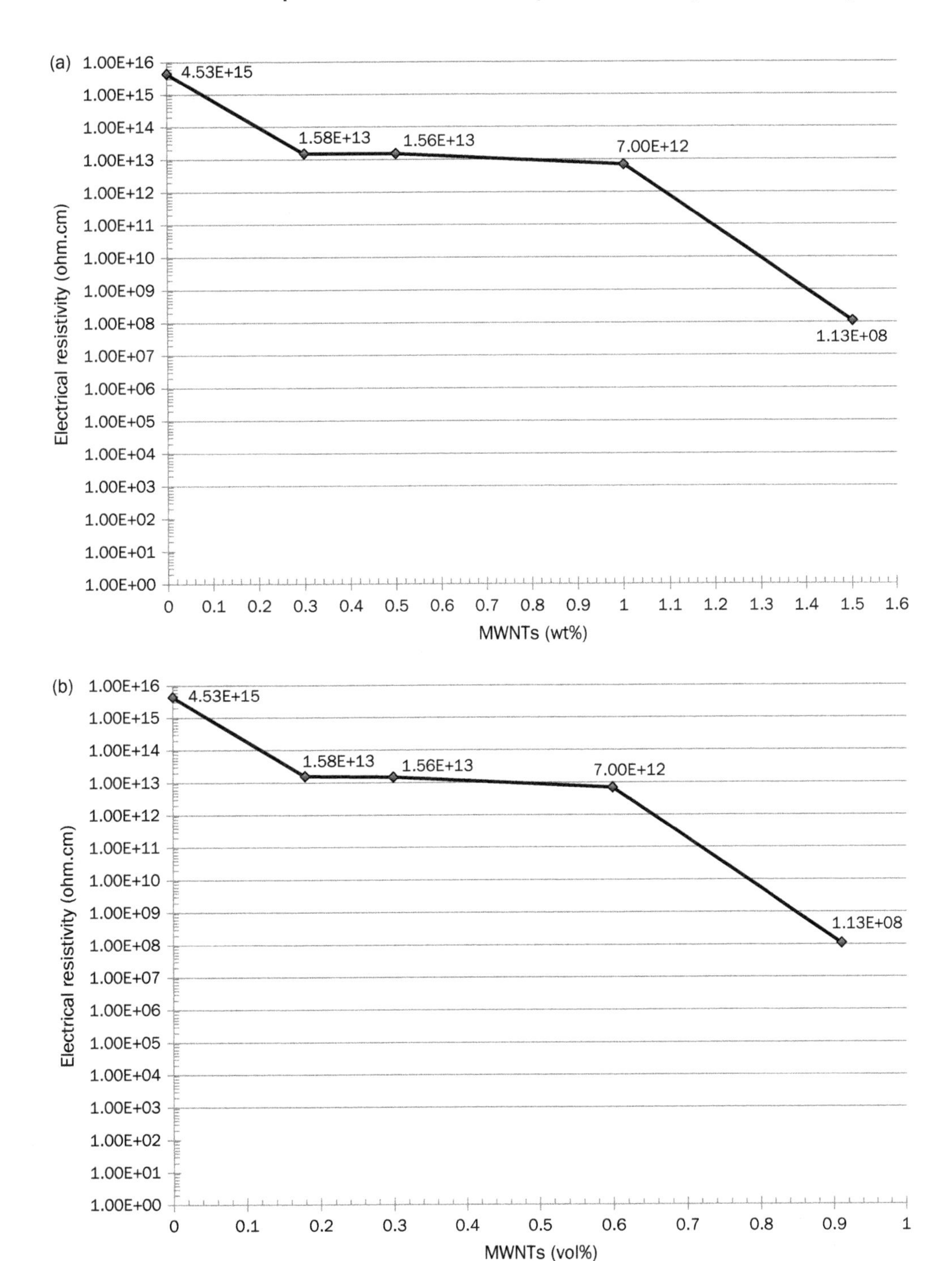

FIGURE 10.9 Electrical resistivity of CE-MWNT-5 (with 200 ppm Fe^{3+} catalyst) polymer nanocomposites processed by three-roll mill with various MWNT loadings in (a) wt% and (b) vol% (number of samples per MWNT wt%, $N = 1$) [34].

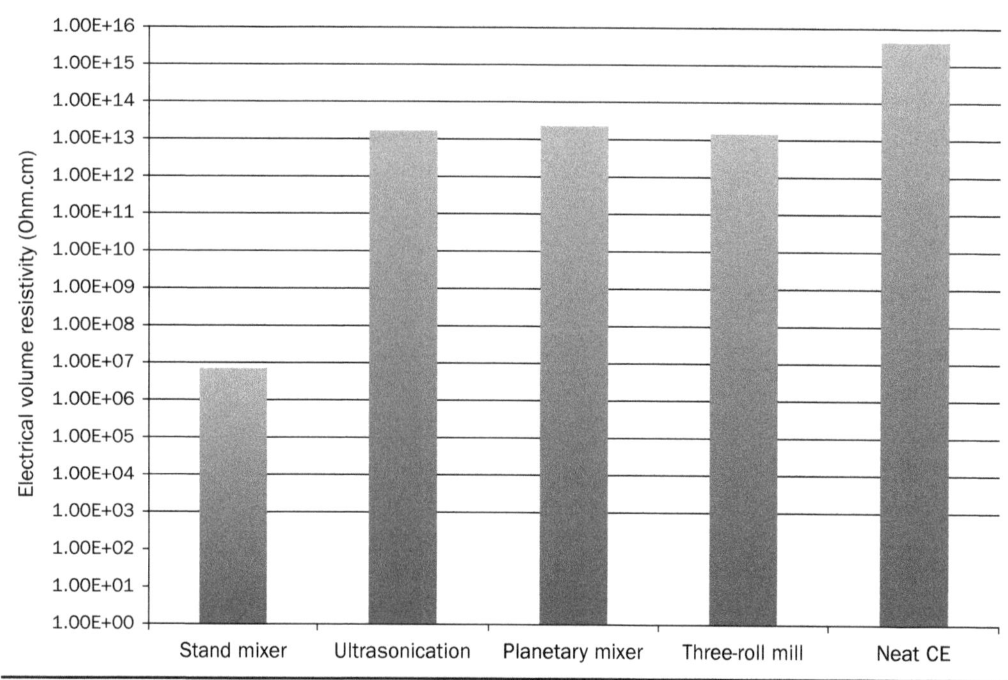

FIGURE 10.10 Electrical resistivity of CE-0.5 wt% MWNT-5 (with 200 ppm Fe^{3+} catalyst) polymer nanocomposites processed by five different processing techniques (number of samples per processing technique, $N = 1$) [34].

threshold is between 1 and 1.5 wt% of MWNTs, additional data are needed to determine this value. Although the percolation threshold is above 1 wt%, void-free polymer nanocomposites can still be fabricated, since air bubbles are removed during the three-roll milling process because of the small gap distance between rollers.

Figure 10.10 compares the electrical resistivity of a CE-0.5 wt% MWNT-5 polymer nanocomposite processed by five different techniques. The electrical resistivity values of samples prepared by three-roll mill, ultrasonication, and planetary centrifugal mixer are two orders of magnitude lower than that of neat CE, but still higher than the ESD requirement. Perplexingly, their readings are about six orders of magnitudes higher than the one processed by the stand mixer. This indicates that the MWNTs in the sample processed by the stand mixer were dispersed and debundled enough to form a much better conductive path for the electron to pass through the sample than the other samples, just as Ma et al. commented on the conducting behavior of composites consisting of conducting fillers and insulating matrices using percolation theory [40].

Figure 10.11 shows an illustration of the relationship between nanotube dispersion and electrical conductivity. In the sample with better nanotube dispersion, as shown in Fig. 10.11a, individual debundled nanotubes may appear in the polymer matrix uniformly. However, they may be separated as electrons may not have a continuous conductive path to travel through the polymer matrix. On the contrary, as shown in Fig. 10.11b, connected paths may be formed in the sample with lower degree of nanotube dispersion, resulting in higher electrical conductivity.

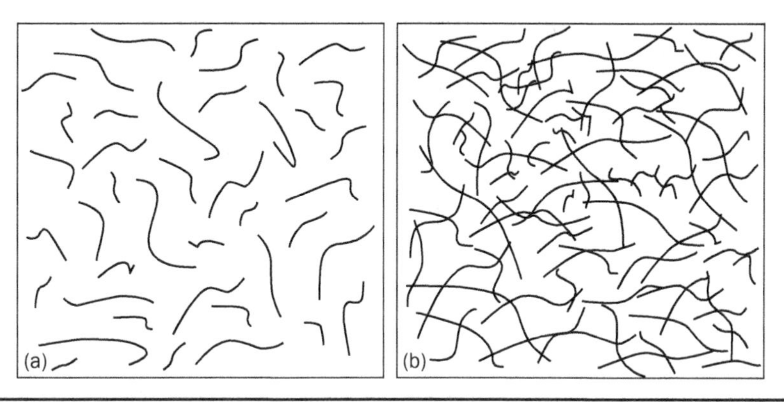

FIGURE 10.11 Illustration showing correlation of degree of dispersion and electrical conductivity: (a) more debundled and better dispersed nanotubes may not form continuous conductive path, and (b) conductive path may form in the polymer matrix with lower degree of nanotube dispersion [34].

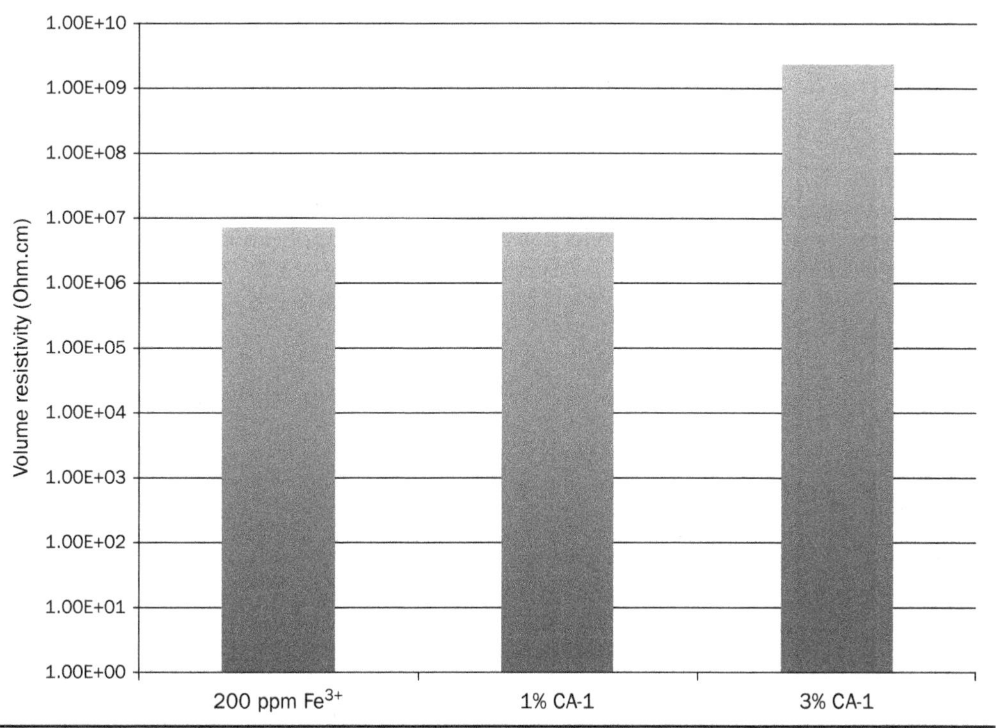

FIGURE 10.12 Electrical resistivity of CE-0.5 wt% MWNT-5 polymer nanocomposites, cured with Fe^{3+} catalyst (200 ppm) or coupling agent CA-1 (1% or 3%), and processed by the stand mixer (number of samples per resin modifier, $N = 1$) [34].

Figure 10.12 shows that when 1 wt% of CA-1 is used instead of the Fe^{3+} catalyst (200 ppm), the electrical resistivity is the same. However, when higher weight loading (3 wt%) of CA-1 is used, the electrical resistivity is two orders of magnitudes higher.

In summary [34], the morphology of the CE-MWNT-5 nanocomposites was characterized by the SEM and TEM. The quality of MWNT dispersion was also examined using both visual observation and a method of quantification. Samples processed using different processing techniques show different aggregate sizes and various degrees of debundling with aggregates. Inconsistency exists between the visual observation and dispersion quantification methods and further work is needed. DMA shows that both the catalyst and coupling agents had detrimental effects on the mechanical properties and glass transition temperatures of the CE-MWNT nanocomposites. However, the glass transition temperature is not affected by the mixing methods or weight loadings of MWNTs. The thermal stability of the nanocomposites was not significantly affected by the processing techniques or the weight loadings of MWNTs. MCC shows that the onset temperature of heat release rate remains unaffected by mixing techniques, modifiers, or weight loadings of MWNTs. Electrical resistivity of CE was reduced by the incorporation of MWNTs. Sample processed using the stand mixer has the lowest electrical resistivity among those processed using other techniques.

10.3.2 Carbon Nanofiber-Reinforced Thermoset-Based Nanocomposites

Smrutisikha [41] focused on the study of the critical content (0.5–1 wt%) of CNFs in epoxy that makes the composites conductive. The results of this study showed that the nanocomposites with different contents of CNTs have the conductivity in the range of 2×10^{-6} to 4×10^{-3}S/cm (Fig. 10.13)—that is, three to six orders of magnitude greater than that of neat epoxy (2.8×10^{-9} S/cm) [41]. The samples were tested in three directions, and Fig. 10.13 shows that most fibers aligned in the x-direction; since the conductivity of this nanocomposite is higher in the x-direction than the other two directions, this nanocomposite can be used in electrostatic dissipation application. Ardanuy et al. [42] obtained similar result of surface electrical conductivity of CNF/epoxy nanocomposites around 1×10^{-6} S/cm at 2 wt% loading, which indicates eight orders of magnitude increase over neat epoxy.

So far, as presented earlier in this chapter, most processing methods involved adding conductive fillers into polymer melts, which were convenient, but the conductivity

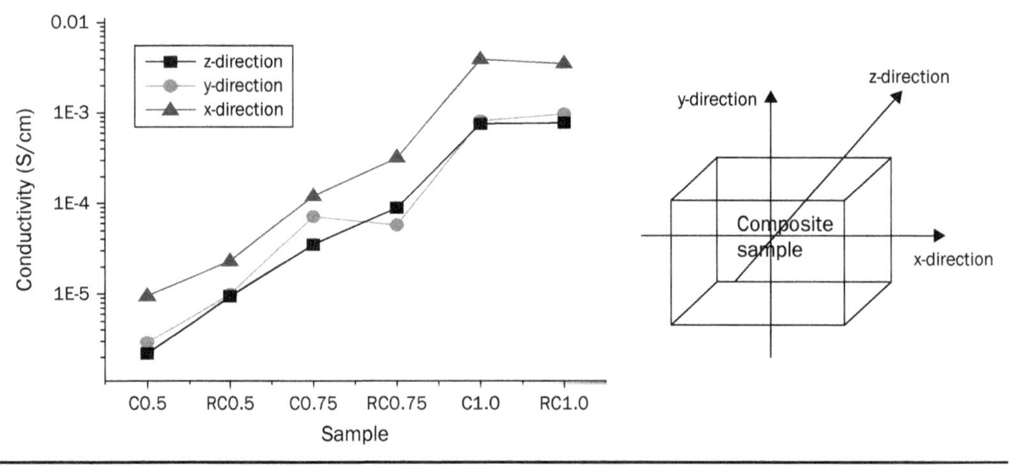

Figure 10.13 Electrical conductivity of CNF/epoxy nanocomposites in three different directions along with schematic of the sample indicating directions [41].

of these materials would be lost during the melting process [43]. Cipriano et al. [43] proposed a new way, melt-annealing, to achieve higher conductivity of MWNT/polystyrene and CNF/polystyrene nanocomposites. Figure 10.14 shows the effect of annealing temperature on the conductivity of two kinds of nanocomposites [43]. At 170°C, the annealing effect of MWNT at 4 wt% is moderate but not suitable for other loadings (Fig. 10.14a). At annealing temperatures of 230°C, the conductivity increased significantly for most loadings, especially at 4 wt%, which approached 1 S/cm, with about two orders-of-magnitude increase over the conductivity of composites at annealing temperature of 170°C. Similar trend was observed in CNT-reinforced nanocomposites

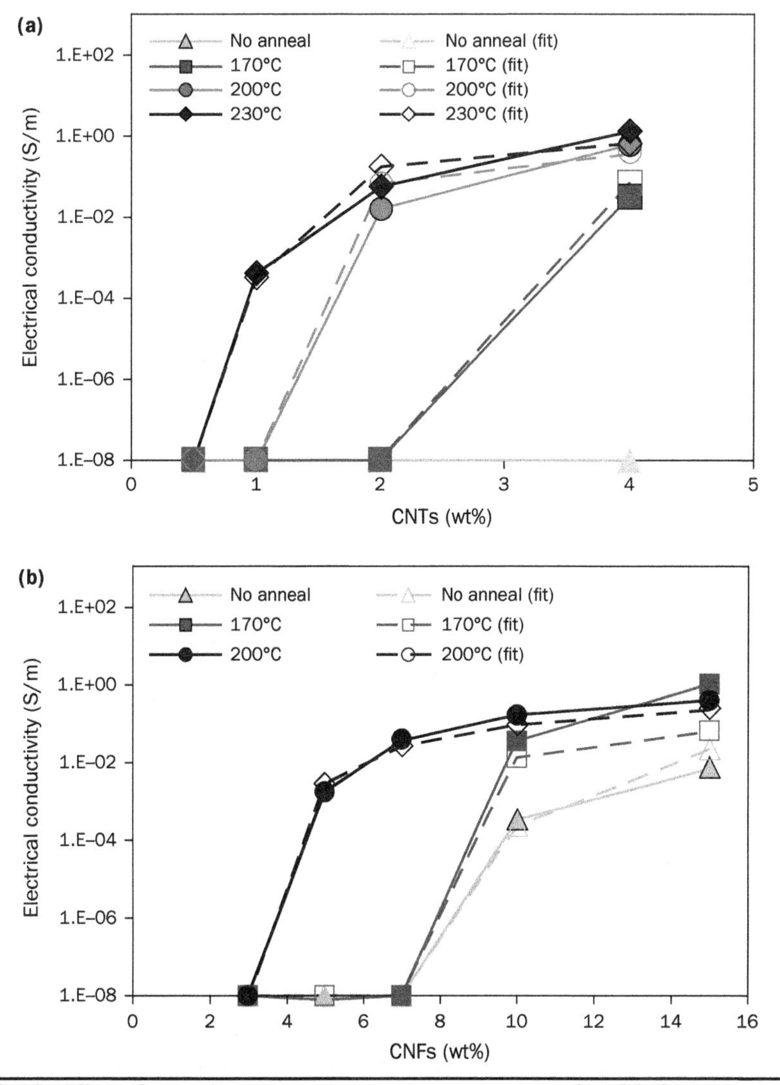

Figure 10.14 Effect of annealing temperature on the conductivity of (a) PS/CNT and (b) PS/CNF nanocomposites. Samples were annealed for 30 minutes [43].

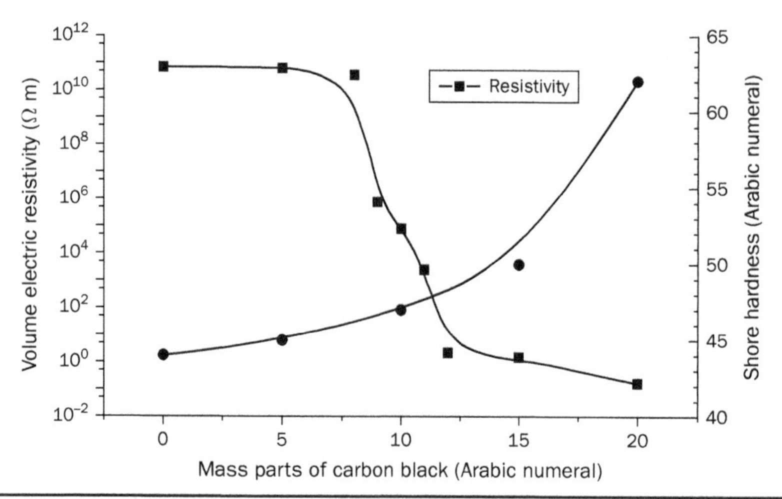

FIGURE 10.15 Bulk resistivity of the composite and Shore A hardness as functions of mass parts of the extraconductive carbon black added to 100 mp of polyisoprene [44].

(Fig. 10.4b). The reason for this result is that during the melt processing, such as twin-screw extruder and compressing molding, the particles became aligned in the flow direction, which resulted in the poorer formation of the conductive network and decreased conductivity. However, the annealing process was able to reestablish the conductive network, leading to recovered electrical conductivity.

10.3.3 Carbon Black-Reinforced Thermoset-Based Nanocomposites

The concept of tensoresistive effect was introduced by Knite et al. [44]: when the composites are stretched, the resistivity would increase because the electroconductive network would be disrupted and the whole process would be irreversible. To solve this problem, Knite et al. made polyisopropene (PI)/CB nanocomposites on micro- and nanoscales to observe the relationship between electrical and elastic properties. To estimate the percolation threshold, the researchers prepared composites of 5, 8, 9, 10, 11, 12, 15, 20, and 30 mass parts (mp) of CB with 100 mp of polyisoprene. Figure 10.15 shows that at micro level, 10 mp, the volume electric resistivity decreased intensely, which indicated the percolation around this value [44]. Upon a 40% stretch, the resistivity changed more than fourfold. When the samples were released, the resistivity went back to its previous level. This proved that the tensoresistive effect was reversible. However, at 5, 8, and 20 mp, there was no evident variation in electrical resistance, which is in accordance with the prediction that at low and high concentration of filler there would be weak or no tensoresistive effect.

At the nano level, Knite et al. examined the tensoresistive effect by looking at the AFM images (Fig. 10.16) [44]. The black spots of CB formed protrusions on the surface of the polymer matrix, which can be related to the electroconductive channels that formed in the composites. The 10 mp of CB formulation had three electroconductive channels in the nanoscale map, and with higher concentration of CB, there were more and more channels, which agrees with the result that higher concentration of CB leads to low resistivity.

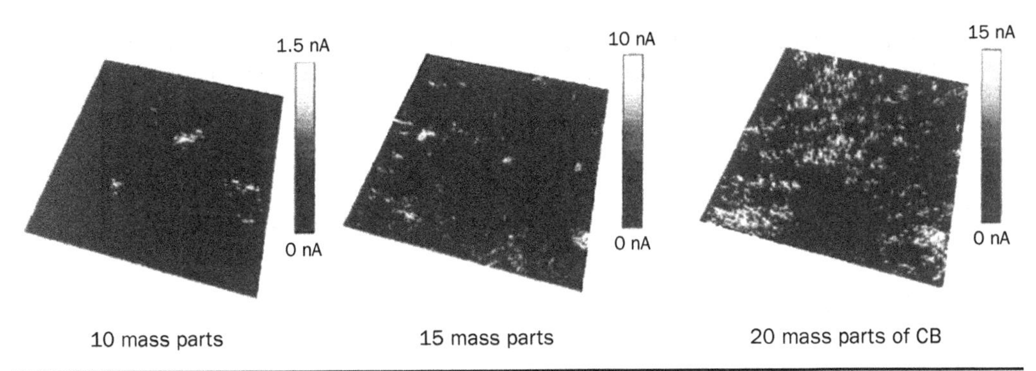

FIGURE 10.16 Nanoscale maps of electroconductive channels of carbon black in the polyisoprene matrix of composites with 10, 15, and 20 mp of carbon black obtained by an atomic force microscope with a conductive tip [44].

10.3.4 Graphite-Reinforced Thermoset-Based Nanocomposites

Tkalya et al. [45] used the latex technology to prepare graphene/polystyrene nanocomposites. The percolation threshold was very low, which was less than 1 wt% of graphene, and a maximum conductivity of about 15 S/m was achieved for 1.6–2 wt% of the nanofiller. As has been discussed before, with higher loading of nanofillers, the conductivity goes up [45].

The expandable graphite (EG) used to make nanocomposites with polystyrene has the most advantage of improving the electrical property of thermoset matrix because of the nanostructure of the conductive filler [46]. By referring to the Chen et al. research [46], it is not hard to find out that the percolation threshold was 1.8 wt%, which was higher than graphite powder with micrometer size (50 wt%) and the highest value achieved was 10^{-2} S/cm at 5 wt%, about 14 orders of magnitude improvement over the conductivity of the pure polystyrene, which was 10^{-16} S/cm. Similar results can be observed in EG-reinforced poly(phenylene sulfide) (PPS) nanocomposites by melt-blending [47]. Compared with other fillers, graphite has the worm forms or filaments that are made up of a large number of nanosheets, which is advantageous to form a good conductive network in the thermoset matrix.

Furthermore, Kim and Drzal [48] tried to compare the different properties of exfoliated graphite nanoplatelets (xGnP)/ethylene vinyl acetate (EVA) and CNT/EVA nanocomposites. Despite the fact that CNT/EVA was electrically conductive with only 5 wt% CNT loading, and the percolation threshold for xGnP/EVA was about 14–16 wt%, Kim and Drzal still recommended to use xGnP to make polymer nanocomposites because of improvement in other properties [49]. With the same thermoset matrix of EVA, EG-reinforced nanocomposites had lower percolation threshold of 6 vol% than that of unexpanded graphite of 15 vol% [50].

Similar to the thermoplastic system discussed earlier, the synergetic effects of hybrid fillers have been also investigated by researchers. Sumfleth et al. [51] obtained the synergistic effect by looking at the conductivity values of CB/MWNT-reinforced epoxy system compared with MWNT and CB systems and explained the phenomenon based on the conductive network formation mechanism that was similar to what had been discussed in [52]. Ma et al. [53] developed epoxy-based nanocomposites containing CNTs and CB in order to gain good electrical and mechanical properties at lower cost.

It turned out that in addition to the improvement in electrical conductivity at low total loading, CB effectively contributed to the enhancement of fracture toughness and ductility of nanocomposites and maintenance of high flexural modulus and strength, which indicate the synergistic effect of CB.

10.4 Electrical Properties of Thermoplastic Elastomer-Based Nanocomposites

10.4.1 Inorganic Filler in Thermoplastic Elastomer-Based Nanocomposites

Most electrically conductive polymer nanocomposites are made using thermoplastic or thermosetting polymers, whereas the less commonly used polymer is thermoplastic elastomers. According to Karttunen et al. [54], it was the very first time that the electrical conductivity of nanocomposites consisting of silver particles and thermoplastic elastomers was reported. As per their article, the agglomerated (average size from 1 to 10 μm) and non-agglomerated silver particles (average size ≤100 nm) were added into thermoplastic elastomer styrene-ethylene-propylene-styrene copolymer (SEPS) to form nanocomposites at different filling rates ranging from 5 to 23.6 vol%. The percolation result is plotted in Fig. 10.17 [54]. For the well-distributed silver particles, the minimum resistivity was 0.18 ohm-cm at 16 vol% and it was stable at higher filling rates; for the agglomerated silver particles, the minimum resistivity was higher, 2.9×10^3 ohm-cm. As the size of the agglomerates goes up, the resistivity increases and the percolation threshold shifts to higher filling rates.

10.4.2 Organic Fillers in Thermoplastic Elastomer-Based Nanocomposites

Since an extremely low loading of CNT can efficiently achieve electrical conductivity that is similar to conventional filled system while maintaining the other properties of matrix polymers, CNTs are considered as the best filler for an elastomer matrix to manufacture

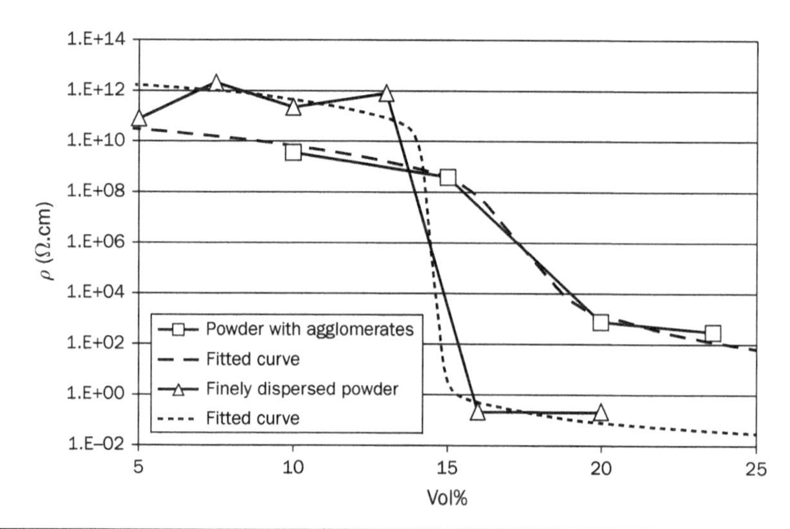

Figure 10.17 Resistivity versus degree of filling rate for nanocomposites containing uniformly distributed and agglomerated silver particles [54].

conductive nanocomposites, where stretchability should be well maintained at low filler loading [55]. Koerner and colleagues [56, 57] successfully fabricated MWNT/thermoplastic polyurethane (PU) elastomer nanocomposites, and discovered that the elastomer nanocomposites possessed high electrical conductivity of about 1–10 S/cm with good mechanical properties (Fig. 10.18) and only a small loading of CNTs (0.5–10 vol%) [56]. Figure 10.18 shows that the initial modulus E, and engineering stress at the apparent yield point σ_Y, increases as the volume fraction of CNTs goes up in the system. The strain hardening, λ_H, decreases with CNT loading after an initial increase with respect to the neat resin. The stress, σ_H, also increases. The elongation at rupture

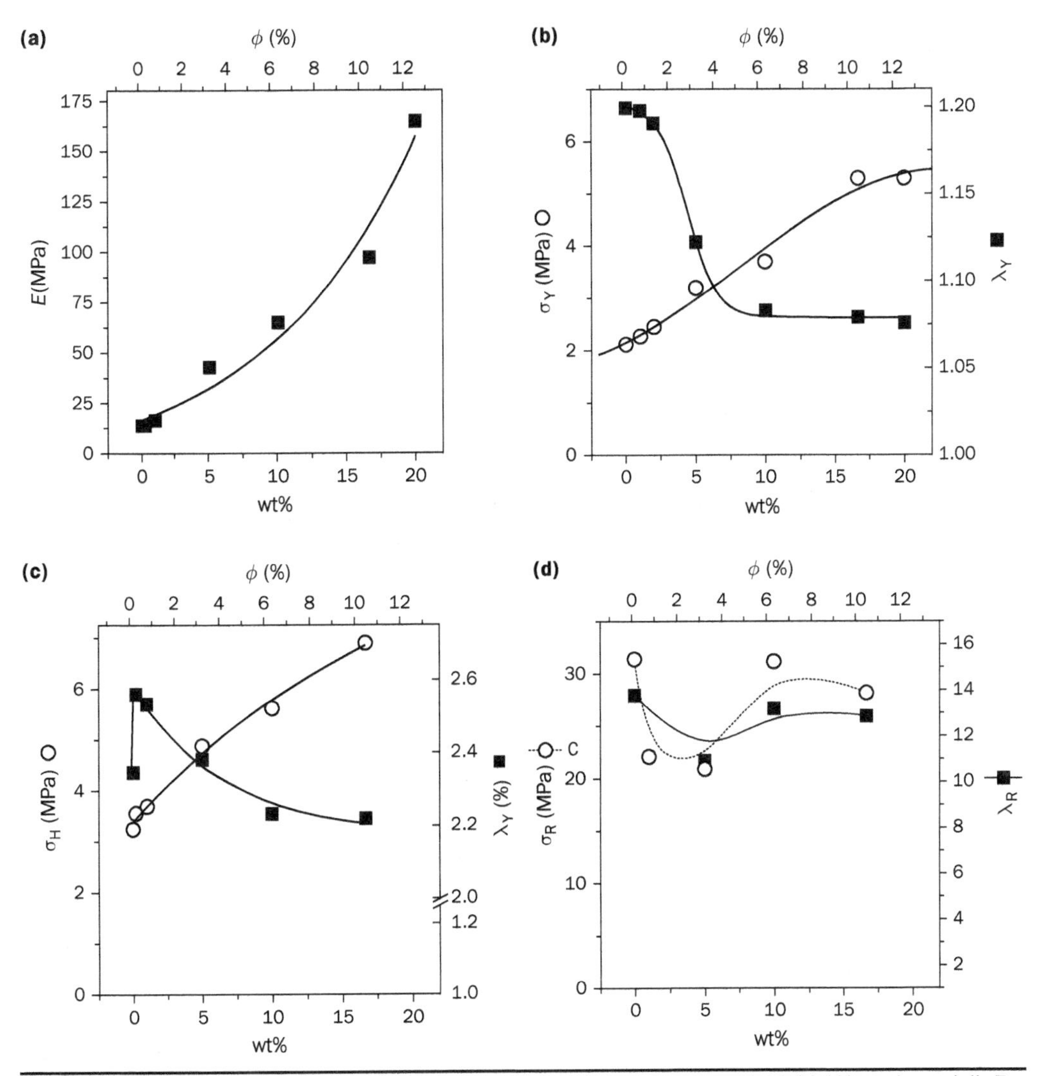

Figure 10.18 Mechanical characteristics of the CNT/PU nanocomposites: (a) initial Young's moduli, E; (b) stress and elongation at yield (σ_Y, λ_Y); (c) stress and elongation at hardening (σ_H, λ_H); and (d) stress and elongation at rupture (σ_R, λ_R). Curves express qualitative data trends. Open circles indicate stress(s) and filled squares depict elongation (l) [56].

decreases slightly with CNT loading; however, even at 17 wt% (10.2 vol%) CNT, it still results in a thermoplastic elastomer with 1,200% elongation before failure. The reason for this is that the mechanical properties of the thermoplastic elastomer rely on the soft-segment crystalline content, and the addition of small amount of CNTs will greatly increase the soft-segment crystalline content, which leads to better mechanical properties.

To enhance electrical conductivity, Sekitani et al. [58] fabricated SWNT/fluorine rubber nanocomposites with ionic liquid by solution blending. At the same time, the resulted nanocomposites showed very good stretchability of more than 100%. Specifically, it was found that the conductivity of elastomer conductor made of fluorinated copolymer G801 and SWNTs was as high as 30 S/cm, but the elasticity and flexibility were not good, so it was suggested to add some ionic liquid to the nanocomposites. It is shown that when the content of SWNTs is equivalent to that of the ionic liquid, the film has the highest conductivity about 50 S/cm at the bucky gel content of 30 wt% [58]. But with further increase in bucky gel content, the electrical conductivity goes down because the films become fragile and porous membranes are resulted.

In terms of processing technique, most of the thermoplastic elastomer–based nanocomposites were prepared by solution blending or other methods that may have negative influence on environment.

10.5 Summary

This chapter briefly reviews the electrical property of different nonconductive polymer-based nanocomposites reinforced by different types of conductive nanofillers. The characterization methods of electrical properties of nanocomposites have been demonstrated in detail in Chap. 5. The electrical properties of polymer nanocomposites based on three categories of polymers have been discussed and some factors, such as polymer, types of nanofillers, and processing conditions that can influence the electrical properties of nanocomposites have been introduced. The basic mechanism of formation of conductive networks which contributes to the conductivity of nanocomposites has been examined and should be thoroughly understood to explain this phenomenon.

10.6 Study Questions

10.1 How does the nanomaterial dispersion affect the electrical properties of polymer nanocomposites in general? Provide examples to illustrate your claims?

10.2 This chapter looked at the enhancement of electrical properties of thermoplastic-based nanocomposites using carbon-based nanomaterials, such as MWNTs, carbon black (CB), CNFs, graphite, and mixed MWNTs-CB. Briefly discuss the mechanism of the formation of conductive network of each type of nanofiller. Would surface treatment of these carbon-based nanomaterials enhance electrical conductivity?

10.3 Which types of nanomaterials are more effective for thermoplastic-based nanocomposites? Provide examples. Besides the aforementioned nanomaterials, what other nanomaterials would you propose for thermoplastic-based nanocomposites? Discuss the mechanical properties of these electrically conductive thermoplastic-based nanocomposites.

10.4 This chapter explored the enhancement of electrical properties of thermoset-based nanocomposites using MWNT, CNF, CB, graphite, and CB/MWNT nanomaterials. Which type of nanomaterials are more effective for epoxy-based nanocomposites? Provide examples. Discuss the mechanical properties of these electrically conductive thermoset-based nanocomposites.

10.5 This chapter also explored the enhancement of electrical properties of thermoplastic elastomer–based nanocomposites using silver particles, MWNTs, and SWNTs. Which type of nanomaterials are more effective for thermoplastic elastomer–based nanocomposites? Provide examples. Besides these nanomaterials, what other nanomaterials can be used in thermoplastic elastomer to enhance its electrical conductivity? Discuss the mechanical properties of these electrically conductive thermoplastic elastomer–based nanocomposites.

10.6 What will be the effects of the multicomponent systems with two or more nanomaterials and conventional conductive polymer systems? What will be a good strategy to optimize the electrical properties in these polymer nanocomposites?

10.7 What are the electrical property mechanisms for these different types of nanomaterials in the polymer nanocomposites and the hybrid material systems?

10.7 References

1. Lew, C. Y., and Luizi, C. M. The influence of processing conditions on the electrical properties of polypropylene nanocomposites incorporating multiwall carbon nanotube. Published online at http://www.nanocyl.com [cited November 18, 2013].
2. Krause, B., Pötschke, P., and Häußler, L. (2009) Influence of small scale melt mixing conditions on electrical resistivity of carbon nanotube-polyamide composites. *Composites Science and Technology,* 69:1505–1515.
3. Pujari, S., Ramanathan, T., Kasimatis, K., Masuda, J. I., Andrews, R., Torkelson, J. M., Brinson, L. C. and Burghardt, W. R. (2009) Preparation and characterization of multiwalled carbon nanotube dispersions in polypropylene: melt mixing versus solid-state shear pulverization. *Journal of Polymer Science Part B: Polymer Physics,* 47:1426–1436.
4. Villmow, T., Pötschke, P., Pegel, S., Häussler, L., and Kretzschmar, B. (2008) Influence of twin-screw extrusion conditions on the dispersion of multi-walled carbon nanotubes in a poly(lactic acid) matrix. *Polymer,* 49:3500–3509.
5. Logakis, E., Pandis, C., Peoglos, V., Pissis, P., Pionteck, J., Pötschke, P., Mičušík, M., and Omastová, M. (2009) Electrical/dielectric properties and conduction mechanism in melt processed polyamide/multi-walled carbon nanotubes composites. *Polymer,* 50:5103–5111.
6. Meincke, O., Kaempfer, D., Weickmann, H., Friedrich, C., Vathauer, M., and Warth, H. (2004) Mechanical properties and electrical conductivity of carbon-nanotube filled polyamide-6 and its blends with acrylonitrile/butadiene/styrene. *Polymer,* 45:739–748.
7. Socher, R., Krause, B., Boldt, R., Hermasch, S., Wursche, R., and Pötschke, P. (2011) Melt mixed nano composites of PA12 with MWNTs: Influence of MWNT and matrix properties on macrodispersion and electrical properties. *Composites Science and Technology,* 71:306–314.

8. Carneiro, O. S., Covas, J. A., Bernardo, C. A., Caldeira, G., Van Hattum, F. W. J., Ting, J. M., Alig, R. L., and Lake, M. L. (1998) Production and assessment of polycarbonate composites reinforced with vapour-grown carbon fibres. *Composites Science and Technology*, 58:401–407.
9. Tibbetts, G. G., Lake, M. L., Strong, K. L., and Rice, B. P. (2007) A review of the fabrication and properties of vapor-grown carbon nanofiber/polymer composites. *Composites Science and Technology*, 67:1709–1718.
10. Jimenez, G. A., and Jana, S. C. (2007) Oxidized carbon nanofiber/polymer composites prepared by chaotic mixing. *Carbon*, 45:2079–2091.
11. Jimenez, G. A., and Jana, S. C. (2007) Electrically conductive polymer nanocomposites of polymethylmethacrylate and carbon nanofibers prepared by chaotic mixing. *Composites Part A: Applied Science and Manufacturing*, 38:983–993.
12. Kalaitzidou, K., Fukushima, H., and Drzal, L. T. (2007) A new compounding method for exfoliated graphite–polypropylene nanocomposites with enhanced flexural properties and lower percolation threshold. *Composites Science and Technology*, 67:2045–2051.
13. Li, Y., and Shimizu, H. (2008) Conductive PVDF/PA6/CNTs nanocomposites fabricated by dual formation of cocontinuous and nanodispersion structures. *Macromolecules*, 41:5339–5344.
14. Kilbride, B. E., Coleman, J. N., Fraysse, J., Fournet, P., Cadek, M., Drury, A., Hutzler, S., Roth, S., and Blau, W. J. (2002) Experimental observation of scaling laws for alternating current and direct current conductivity in polymer-carbon nanotube composite thin films. *Journal of Applied Physics*, 92:4024–4030.
15. Wang, W.-P., Liu, Y., Li, X.-X., and You, Y.-Z. (2006) Synthesis and characteristics of poly(methyl methacrylate)/expanded graphite nanocomposites. *Journal of Applied Polymer Science*, 100:1427–1431.
16. Goyal, R. K., Samant, S. D., Thakar, A. K., and Kadam, A. (2010) Electrical properties of polymer/expanded graphite nanocomposites with low percolation. *Journal of Physics D-Applied Physics*, 43.
17. Zheng, W., and Wong, S.-C. (2003) Electrical conductivity and dielectric properties of PMMA/expanded graphite composites. *Composites Science and Technology*, 63:225–235.
18. Pan, Y.-X., Yu, Z.-Z., Ou, Y.-C., and Hu, G.-H. (2000) A new process of fabricating electrically conducting nylon 6/graphite nanocomposites via intercalation polymerization. *Journal of Polymer Science Part B: Polymer Physics*, 38:1626–1633.
19. Via, M. D., King, J. A., Keith, J. M., and Bogucki, G. R. (2012) Electrical conductivity modeling of carbon black/polycarbonate, carbon nanotube/polycarbonate, and exfoliated graphite nanoplatelet/polycarbonate composites. *Journal of Applied Polymer Science*, 124: 182–189.
20. Zhang, S. M., Lin, L., Deng, H., Gao, X., Bilotti, E., Peijs, T., Zhang, Q., and Fu, Q. (2012) Synergistic effect in conductive networks constructed with carbon nanofillers in different dimensions. *eXPRESS Polymer Letters*, 6:159–168.
21. Sun, Y., Bao, H.-D., Guo, Z.-X., and Yu, J. (2008) Modeling of the electrical percolation of mixed carbon fillers in polymer-based composites. *Macromolecules*, 42:459–463.
22. Puglia, D., Valentini, L., and Kenny, J. M. (2003) Analysis of the cure reaction of carbon nanotubes/epoxy resin composites through thermal analysis and Raman spectroscopy. *Journal of Applied Polymer Science*, 88:452–458.

23. Christodoulou, L., and Venables, J. D. (2006) Multifunctional material systems: the first generation. *Journal of Materials*, 55: 39–45.
24. Battisti, A., Skordos, A. A., and Partridge, I. K. (2010) Percolation threshold of carbon nanotubes filled unsaturated polyesters. *Composites Science and Technology*, 70:633–637.
25. Vera-Agullo, J., Glória-Pereira, A., Varela-Rizo, H., Gonzalez, J. L., and Martin-Gullon, I. (2009) Comparative study of the dispersion and functional properties of multiwall carbon nanotubes and helical-ribbon carbon nanofibers in polyester nanocomposites. *Composites Science and Technology*, 69:1521–1532.
26. Martin, C. A., Sandler, J. K. W., Shaffer, M. S. P., Schwarz, M. K., Bauhofer, W., Schulte, K., and Windle, A. H. (2004) Formation of percolating networks in multi-wall carbon-nanotube–epoxy composites. *Composites Science and Technology*, 64: 2309–2316.
27. Thostenson, E. T., Ziaee, S., and Chou, T.-W. (2009) Processing and electrical properties of carbon nanotube/vinyl ester nanocomposites. *Composites Science and Technology*, 69:801–804.
28. Sandler, J. K. W., Kirk, J. E., Kinloch, I. A., Shaffer, M. S. P., and Windle, A. H. (2003) Ultra-low electrical percolation threshold in carbon-nanotube-epoxy composites. *Polymer*, 44:5893–5899.
29. Gojny, F. H., Wichmann, M. H. G., Fiedler, B., Kinloch, I. A., Bauhofer, W., Windle, A. H., and Schulte, K. (2006) Evaluation and identification of electrical and thermal conduction mechanisms in carbon nanotube/epoxy composites. *Polymer*, 47:2036–2045.
30. Moisala, A., Li, Q., Kinloch, I. A., and Windle, A. H. (2006) Thermal and electrical conductivity of single- and multi-walled carbon nanotube-epoxy composites. *Composites Science and Technology*, 66:1285–1288.
31. Gojny, F. H., Wichmann, M. H. G., Köpke, U., Fiedler, B., and Schulte, K. (2004) Carbon nanotube-reinforced epoxy-composites: enhanced stiffness and fracture toughness at low nanotube content. *Composites Science and Technology*, 64:2363–2371.
32. Gojny, F. H., Wichmann, M. H. G., Fiedler, B., and Schulte, K. (2005) Influence of different carbon nanotubes on the mechanical properties of epoxy matrix composites—a comparative study. *Composites Science and Technology*, 65:2300–2313.
33. Moniruzzaman, M., and Winey, K. I. (2006) Polymer nanocomposites containing carbon nanotubes. *Macromolecules*, 39:5194–5205.
34. Lao, S. C. (2013, December) *Multifunctional Cyanate Ester/MWNT Nanocomposites: Processing and Characterization*. Ph.D. dissertation, The University of Texas at Austin, Austin, TX.
35. Liang, K., Li, G., Toghiani, H., Koo, J. H., Pittman Jr., C. U., and Dave C. (2006) Cyanate ester/polyhedral oligomeric silsesquioxane (POSS) nanocomposites: synthesis and characterization. *Chemistry of Materials*, 18(2):301–312.
36. Cho, H. S., Liang, K., Chatterjee, S., and Pittman Jr., C. U. (2005) Synthesis, morphology, and viscoelastic properties of polyhedral oligomeric silsesquioxane nanocomposites with epoxy and cyanate ester matrices. *Journal of Inorganic and Organometallic Polymers and Materials*, 15(4):541–553.
37. Liang, K., Toghiani, H., Li, G., Pittman Jr., C. U., and Dave, C. (2005) Synthesis, morphology, and viscoelastic properties of cyanate ester/polyhedral oligomeric silsesquioxane nanocomposites. *Journal of Polymer Science, Part A: Polymer Chemistry*, 43(17):3887–3898.
38. ESD Association. (2010) *Fundamentals of Electrostatic Discharge, Part One—An Introduction to ESD*. Rome, NY: ESD Association.

39. Hu, N., Masuda, Z., and Fukunaga, H. (2009) *Carbon Nanotubes: New Research*. New York, NY: Nova Science Publishers, Inc., pp. 175–222.
40. Ma, P.-C., Siddiqui, N. A., Marom, G., and Kim, J.-K. (2010) Dispersion and functionalization of carbon nanotubes for polymer-based nanocomposites: A review. *Composites Part A*, 41:1345–1367.
41. Smrutisikha, B. (2010) Experimental study of mechanical and electrical properties of carbon nanofiber/epoxy composites. *Materials Design*, 31:2406–2413.
42. Ardanuy, M., Rodríguez-Perez, M. A., and Algaba, I. (2011) Electrical conductivity and mechanical properties of vapor-grown carbon nanofibers/trifunctional epoxy composites prepared by direct mixing. *Composites Part B: Engineering*, 42:675–681.
43. Cipriano, B. H., Kota, A. K., Gershon, A. L., Laskowski, C. J., Kashiwagi, T., Bruck, H. A., and Raghavan, S. R. (2008) Conductivity enhancement of carbon nanotube and nanofiber-based polymer nanocomposites by melt annealing. *Polymer*, 49:4846–4851.
44. Knite, M., Teteris, V., Polyakov, B., and Erts, D. (2002) Electric and elastic properties of conductive polymeric nanocomposites on macro- and nanoscales. *Materials Science & Engineering C-Biomimetic and Supramolecular Systems*, 19:15–19.
45. Tkalya, E., Ghislandi, M., Alekseev, A., Koning, C., and Loos, J. (2010) Latex-based concept for the preparation of graphene-based polymer nanocomposites. *Journal of Materials Chemistry*, 20:3035–3039.
46. Chen, G.-H., Wu, D.-J., Weng, W.-G., He, B., and Yan, W.-L. (2001) Preparation of polystyrene–graphite conducting nanocomposites via intercalation polymerization. *Polymer International*, 50:980–985.
47. Zhao, Y. F., Xiao, M., Wang, S. J., Ge, X. C., and Meng, Y. Z. (2007) Preparation and properties of electrically conductive PPS/expanded graphite nanocomposites. *Composites Science and Technology*, 67:2528–2534.
48. Kim, S., and Drzal, L. T. (2009) Comparison of exfoliated graphite nanoplatelets (xGnP) and CNTs for reinforcement of EVA nanocomposites fabricated by solution compounding method and three screw rotating systems. *Journal of Adhesion Science and Technology*, 23:1623–1638.
49. Kalaitzidou, K., Fukushima, H., and Drzal, L. T. (2007) Mechanical properties and morphological characterization of exfoliated graphite–polypropylene nanocomposites. *Composites Part A: Applied Science and Manufacturing*, 38:1675–1682.
50. Tlili, R., Boudenne, A., Cecen, V., Ibos, L., Krupa, I., and Candau, Y. (2010) Thermophysical and electrical properties of nanocomposites based on ethylene–vinylacetate copolymer (EVA) filled with expanded and unexpanded graphite. *International Journal of Thermophysics*, 31:936–948.
51. Sumfleth, J., Adroher, X., and Schulte, K. (2009) Synergistic effects in network formation and electrical properties of hybrid epoxy nanocomposites containing multi-wall carbon nanotubes and carbon black. *Journal of Materials Science*, 44:3241–3247.
52. Zhang, S. M., Lin, L., Deng, H., Gao, X., Bilotti, E., Peijs, T., Zhang, Q., and Fu, Q. (2012) Synergistic effect in conductive networks constructed with carbon nanofillers in different dimensions. *Express Polymer Letters*, 6:159–168.
53. Ma, P.-C., Liu, M.-Y., Zhang, H., Wang, S.-Q., Wang, R., Wang, K., Wong, Y.-K., Tang, B.-Z., Hong, S.-H., Paik, K.-W., and Kim, J.-K. (2009) Enhanced electrical conductivity of nanocomposites containing hybrid fillers of carbon nanotubes and carbon black. *ACS Applied Materials & Interfaces*, 1090–1096.

54. Karttunen, M., Ruuskanen, P., Pitkanen, V., and Albers, W. M. (2008) Electrically conductive metal polymer nanocomposites for electronics applications. *Journal of Electronic Materials*, 37:951–954.
55. Li, Y. J., and Shimizu, H. (2009) Toward a stretchable, elastic, and electrically conductive nanocomposite: Morphology and properties of poly[styrene-b-(ethylene-co-butylene)-b-styrene]/multiwalled carbon nanotube composites fabricated by high-shear processing. *Macromolecules*, 42:2587–2593.
56. Koerner, H., Price, G., Pearce, N. A., Alexander, M., and Vaia, R. A. (2004) Remotely actuated polymer nanocomposites—stress-recovery of carbon-nanotube-filled thermoplastic elastomers. *Nature Materials*, 3:115–120.
57. Koerner, H., Liu, W. D., Alexander, M., Mirau, P., Dowty, H., and Vaia, R. A. (2005) Deformation-morphology correlations in electrically conductive carbon nanotube thermoplastic polyurethane nanocomposites. *Polymer*, 46:4405–4420.
58. Sekitani, T., Noguchi, Y., Hata, K., Fukushima, T., Aida, T., and Someya, T. (2008) A rubberlike stretchable active matrix using elastic conductors. *Science*, 321:1468–1472.

CHAPTER 11

Widespread Properties of Polymer Nanocomposites

11.1 Introduction

This chapter discusses several properties of the polymer nanocomposites that were not mentioned in previous chapters. This chapter covers the (a) tribological properties of thermoplastic polymer materials containing several different nanoparticles including nanoclay, single- and multiwalled carbon nanotubes (SNWTs and MWNTs), and graphene platelets; (b) permeability properties of polymer nanocomposites; and (c) applications of nanotechnology and nanomaterials in the oil field. Tribological property characterization techniques are included in this chapter since these are more specialized techniques.

11.2 Tribological Properties of Polymer Nanocomposites

11.2.1 Abrasion, Wear, and Scratch Resistance Characterization Techniques

Taber Abraser

The abrasion resistance test is detailed in the ASTM D4060 standard [1]. This test is performed using a Taber abraser [1, 2]. Two abrading wheels and the specimen rotate simultaneously to produce abrasion paths at a variety of angles. Figure 11.1 shows a typical Taber abraser [2]. Loads of 1,000 g are set for each abrasion wheel which is composed of rubber and aluminum oxide abrasive particles. The specimen is then subjected to a specified number of cycles. For tests covered in this section, the abrasion resistance is quantified by the weight loss during the process [1–4]. A lower weight loss indicates more wear/abrasion resistance.

Pin-on-Disc Tribometer

Several variations on the pin-on-disc tribometer were used in some of the following experiments to measure abrasion resistance. The tribometer consists of a stationary pin under a load in contact with a rotating disc similar to the diagram in Fig. 11.2. The pin is generally fabricated from the polymer to be studied, and an abrasive paper is mounted to the disc. The frictional force and the change in dimensions of the pin can be measured to quantify the abrasion resistance of the material [5, 6].

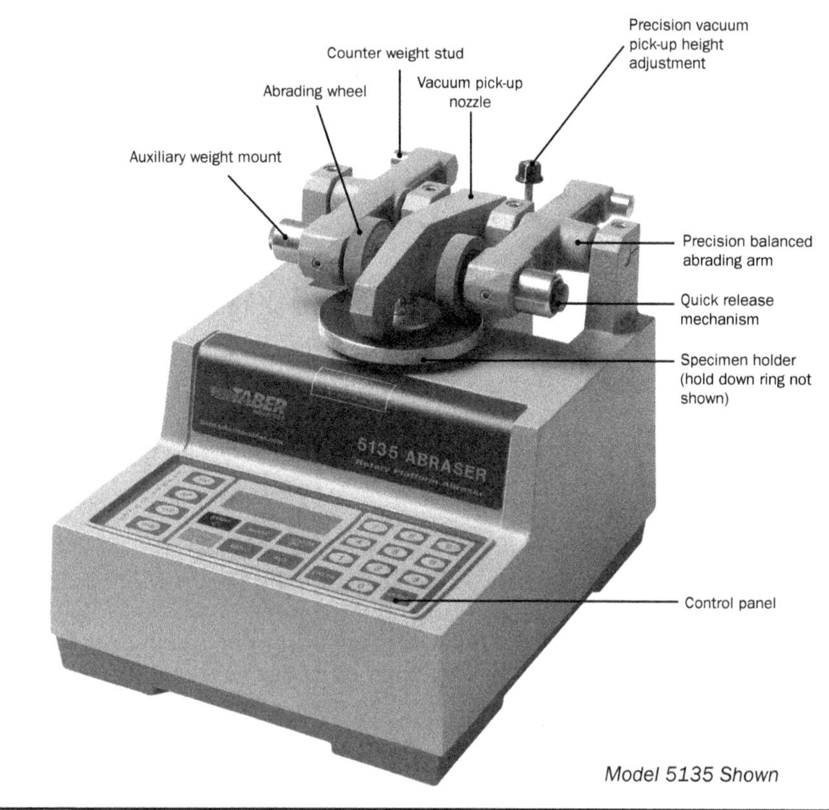

FIGURE 11.1 Taber abraser [2].

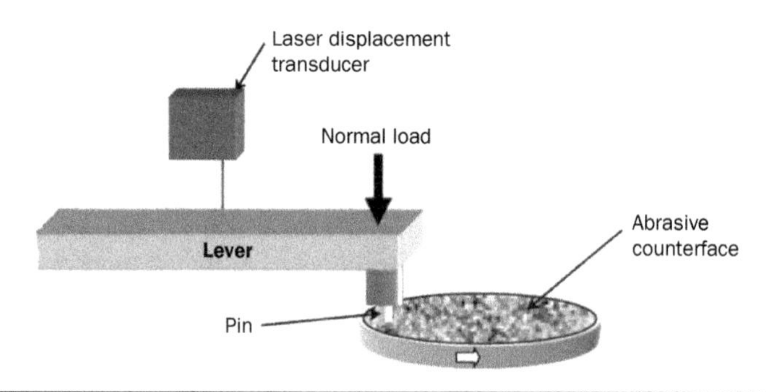

FIGURE 11.2 Pin-on-disc tribometer [5].

Scratch Test

"The Standard Test Method for Evaluation of Scratch Resistance of Polymeric Coatings and Plastics Using an Instrumented Scratch Machine" as described by the ASTM D7027-05 standard is used to quantify the scratch resistance of polymers [7]. Variations in this test method are common, but in general, the method involves using a conical indenter (usually diamond) under a load to scratch a specified length into the polymer. One scratch, or pass, of the indenter is used to quantify scratch resistance, but multiple

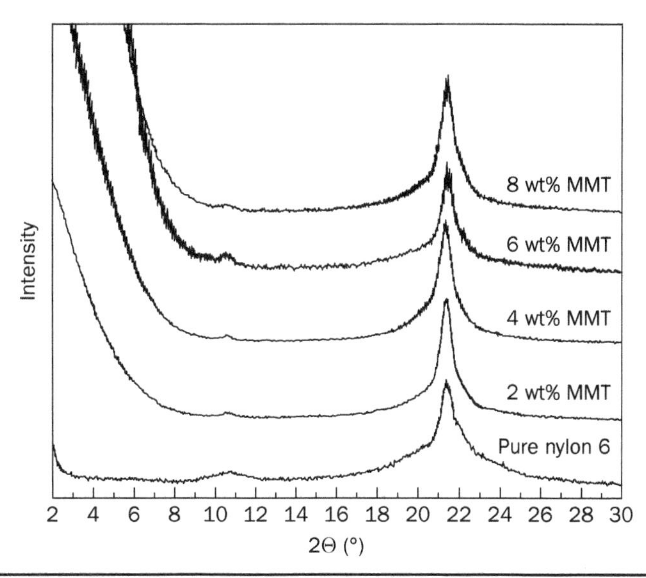

Figure 11.3 WAXS patterns of pure nylon 6 and nylon 6/nanoclay nanocomposites [3].

passes can be used to quantify wear resistance. The scratch width and depth, frictional force, and normal force during the test are all parameters that can be analyzed and used to find the scratch hardness [7].

11.2.2 Wear and Abrasion Resistance of Polymer-Clay Nanocomposites

Nylon 6 is a polymer utilized in a wide array of industries. Nylon 6/nanoclay nanocomposites have been used, since the Toyota Research Group first utilized them in the early 1990s. The thermal, mechanical, and permeability properties were much improved by the addition of nanoclay, but the abrasion resistance suffered [3]. Since abrasion resistance is an important property of nylon 6, understanding the abrasion resistance properties of nylon 6/nanoclay nanocomposites is important.

The abrasion resistance of nylon 6/nanoclay nanocomposites was studied by Zhou et al. [3] using the ASTM D4060 standard abrasion test. Before the test, wide-angle x-ray scattering (WAXS) was performed on the nanocomposite to determine the structure. Nylon 6 is polymorphic, and two crystalline modifications can exist at room temperature—the α phase and γ phase. The WAXS data are shown in Fig. 11.3 [3]. Diffraction peaks are shown at $2\Theta = 11°$ and $21.5°$, which correspond to the diffraction peaks characteristic of the γ crystal. No diffraction peaks corresponding to the α phase exist, indicating that the dominant crystalline structure for the neat nylon 6 as well as the nylon 6/nanoclay nanocomposites is the γ phase. Furthermore, diffraction peaks indicating the presence of nanoclay are generally seen at $2\Theta < 10°$, so the absence of peaks in this region indicates that the nanoclay is fully exfoliated. The abrasion results are shown in Fig. 11.4 [3]. The weight loss increases almost linearly with the percentage of nanoclay ranging from 8.03 mg for the neat nylon 6 to 19.4 mg for the 8 wt% nanoclay nanocomposite [3]. The abrasion resistance decreases with increasing clay percentage.

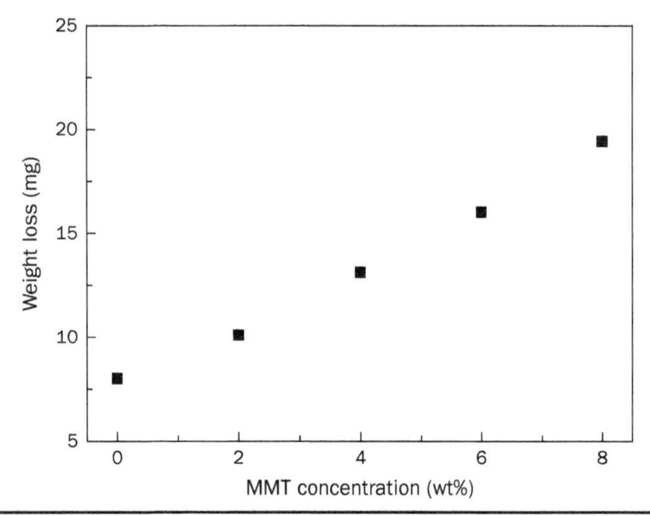

Figure 11.4 Abrasion weight loss of nylon 6/nanoclay nanocomposites versus nanoclay concentrations up to 8% [3].

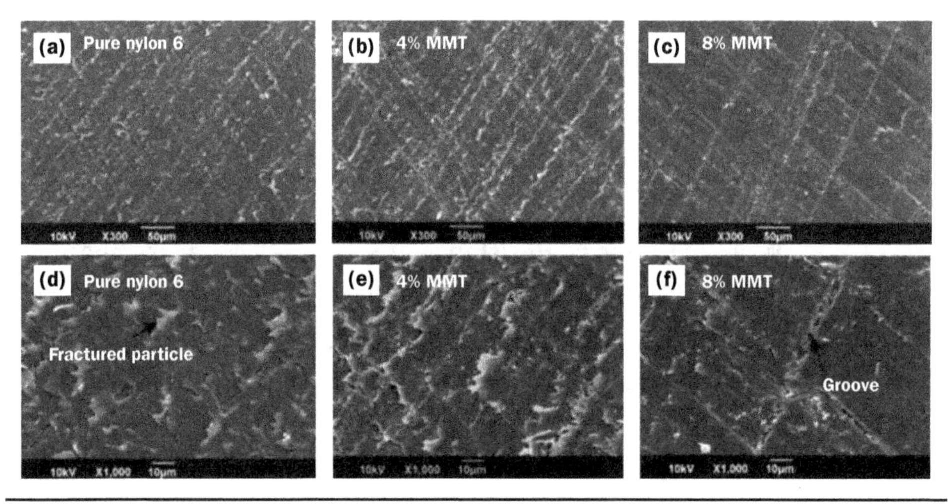

Figure 11.5 SEM micrographs of the abrasion surfaces of nylon 6 and the nylon 6/nanoclay nanocomposites [3].

The scanning electron microscopy (SEM) results shown in Fig. 11.5 reveal the abrasion mechanisms for the neat nylon 6 and the 4 wt% and 8 wt% nanoclay after the abrasion test [3]. Grooves seen specifically in the 8 wt% nanoclay, but also in the 4 wt% nanoclay, suggest that the cutting function is the main abrasion mechanism at work. The fractured particles evident in the neat nylon 6 and some in the 4 wt% nanoclay are due to the abrasive tractive stress. This pattern shows that as the nanoclay content increases, less fractured particles and more grooves will be seen on the abraded surface [3].

After the abrasion test, FTIR-ATR measurements were performed to understand the crystalline transformation of the nylon 6 during abrasion. The results showed that after abrasion, on the abraded surface of the neat nylon 6 as well as the nylon 6/nanoclay

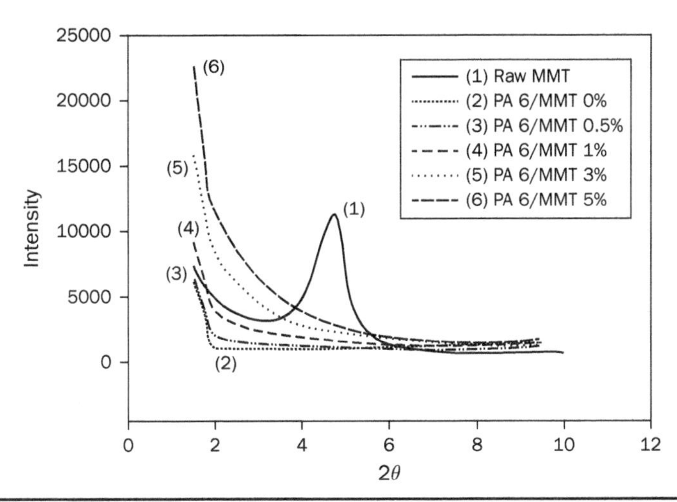

FIGURE 11.6 WAXD data for raw nanoclay, neat nylon 6, and nylon 6/nanoclay nanocomposites [4].

nanocomposites, there was an increase in α crystals and a decrease in γ crystals. The magnitude of the increase in α crystals was also highest for the neat nylon 6 and lowest for the 8 wt% nanoclay. Due to the tensile forces during abrasion causing stretching of the crystal chains, the γ crystals were transformed into α crystals. From this data, it can be concluded that nanoclay additives decrease the ductility of nylon 6 and cause more cutting than tensile tearing [3].

X-ray photoelectron spectroscopy (XPS) was also used to characterize the composition of the sample surface before and after abrasion [3]. Before abrasion, the ratio of silicon to nitrogen atoms is relatively the same for every nanoclay concentration. After abrasion, the ratio increases overall and also increases with increasing nanoclay concentration. This leads to the conclusion that the presence of nanoclay particles makes it easier for the polymer to be removed from the surface. Nanoclay would increase the density of surface defects, making the nylon 6/nanoclay nanocomposite less abrasion resistant [3].

Liu et al. [4] also performed abrasion tests on nylon 6/nanoclay nanocomposites using the ASTM D4060 standard, but their results were different. The WAXD method of characterization (Fig. 11.6) and the TEM images (Fig. 11.7) show the dispersion to be exfoliated for nanoclay up to 3 wt%, but the 5 wt% nanoclay is somewhat intercalated or unmixed [4].

In Liu et al.'s wear resistance test, the wear mass-loss decreased with increasing clay content up to 3 wt% nanoclay, and the change was more significant for the higher wearing speed rate (Fig. 11.8) [4]. However, the wear mass-loss increased for 5 wt% nanoclay for both wearing speed rates. This increase was probably due to the aggregated (intercalated or unmixed) dispersion at the higher clay concentration. The SEM images show a rough abrasion surface for the neat nylon 6 as compared to the nylon 6/nanoclay nanocomposites (except for the 5 wt% nanoclay which is relatively rough). This would indicate that the inclusion of exfoliated nanoclay in nylon 6 enhances the wear resistance of nylon 6, but once the nanoclay concentration gets above a certain point, the wear resistance is lessened. The exact cause of this optimum clay content

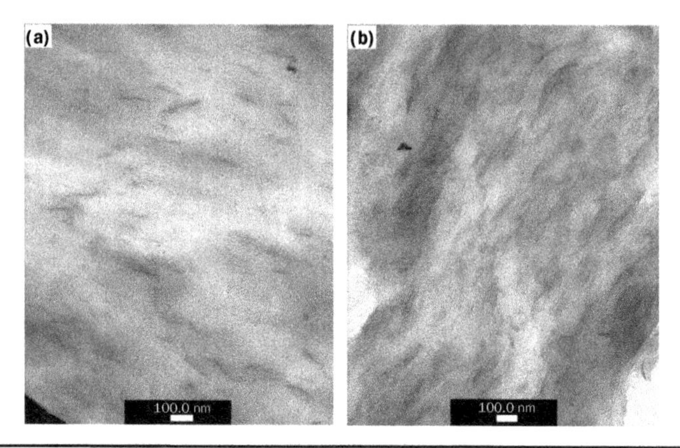

FIGURE 11.7 TEM images of (a) 3 wt% nanoclay and (b) 5 wt% nanoclay [4].

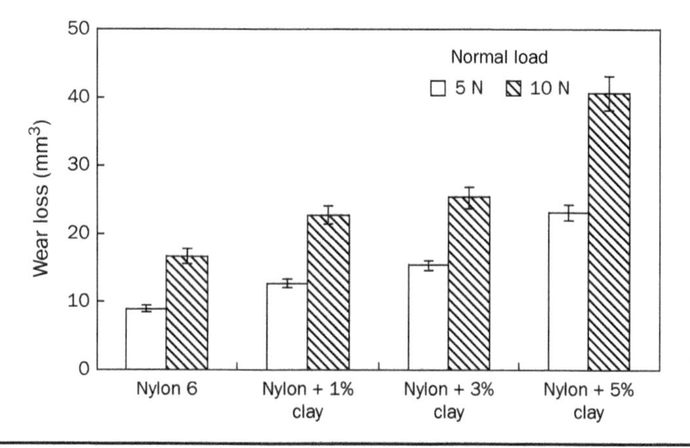

FIGURE 11.8 The effect of nanoclay concentration on wear loss for loads of 5 and 10 N (grit #180, velocity 0.2 m/s) [4].

is not definitively known, but it is hypothesized that the phenomenon results from the competitive effects of transfer film development and the formulation of abrasive aggregates. Adding nanoclay to neat nylon 6 leads to increased hardness, so the corresponding increase in surface wear resistance could also be attributed to this increased hardness [4].

Srinath and Gnanamorthy [5] conducted abrasive wear tests on nylon 6/nanoclay composites using the pin-on-disc test method. The specimen pins were injection-molded, and various grades of abrasive paper were mounted to the steel disc. The friction force and dimensional change of the pin were continuously measured during the test. The effect of grit size, load, and sliding velocity on the wear resistance of the nylon 6/nanoclay composites was documented and can be reviewed in [5]. For this section, only the effect of nanoclay concentration on wear resistance is reviewed. WAXD was carried out on the specimens prior to testing and showed good exfoliation of the

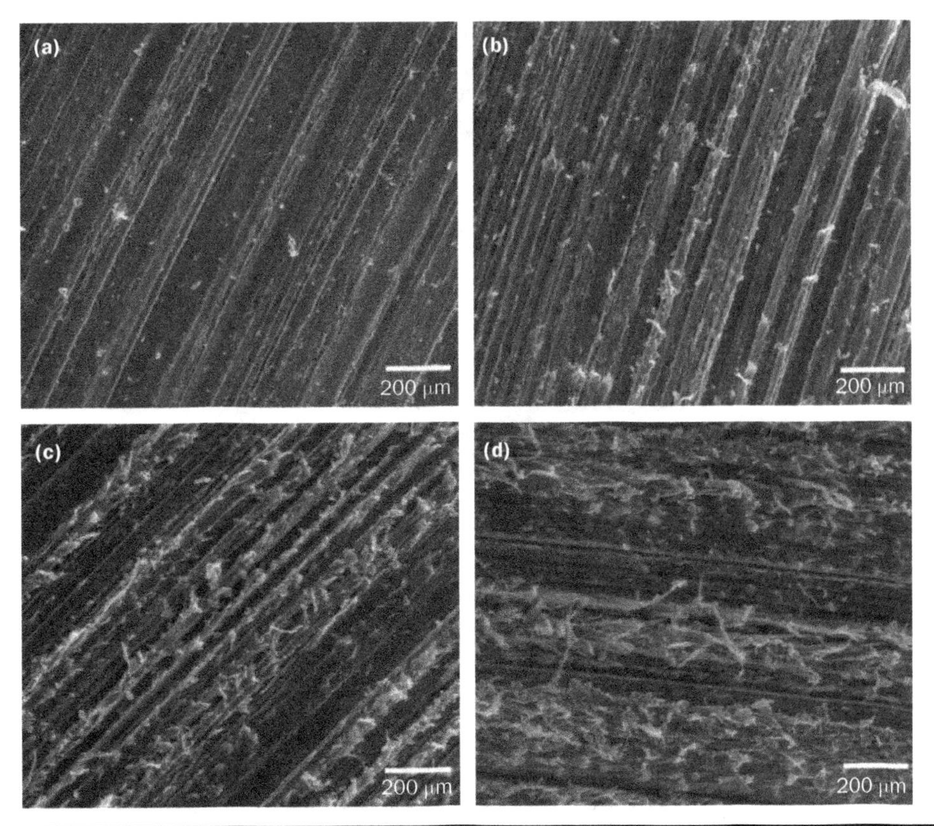

Figure 11.9 SEM of the worn pins (grit #180, 5 N, 0.2 m/s): (a) neat nylon 6, (b) 1% nanoclay, (c) 3% nanoclay, and (d) 5% nanoclay [5].

nanoclay in the nylon 6 polymer matrix. The wear loss versus nanoclay concentration ranging from neat nylon 6 to 5 wt% nanoclay is discussed in [5]. The neat nylon 6 shows less wear loss than any of the nylon 6/nanoclay nanocomposites, and the wear loss increases as more nanoclay is added for both the 5 N and 10 N normal loads [5].

The SEM images (Fig. 11.9) show the abrasion surface of neat nylon 6 and 1, 3, and 5 wt% nanoclay [5]. Neat nylon 6 is ductile, and grooves are seen resulting from the dominant ploughing mechanism. As nanoclay is added to the neat polymer, the material becomes harder and more brittle, and the cutting function dominates, leading to the grooves and fractured particles (Fig. 11.9b–d) [5].

In modified pin-on-disc experiment, Mu et al. [6] showed that the abrasion resistance of nylon 6 is heavily affected by the addition of nanoclay particles. In their test, the slider was a steel spherical ball, and the polymer specimens were injection-molded into small blocks in place of the disc. Before the test, WAXD was performed and indicated that the samples were a mix between exfoliated and intercalated nanoclay platelets. The TEM micrograph shown in Fig. 11.10 confirms the dispersion [6].

The wear rate initially decreases only slightly from that of the neat polymer with the inclusion of nanoclay particles. Above 5 wt% nanoclay, the wear rate increases drastically and is much higher than the wear rate of the neat polymer (Fig. 11.11) [6]. A similar

FIGURE 11.10 TEM micrograph of nylon 6/5 wt% nanoclay nanocomposite [6].

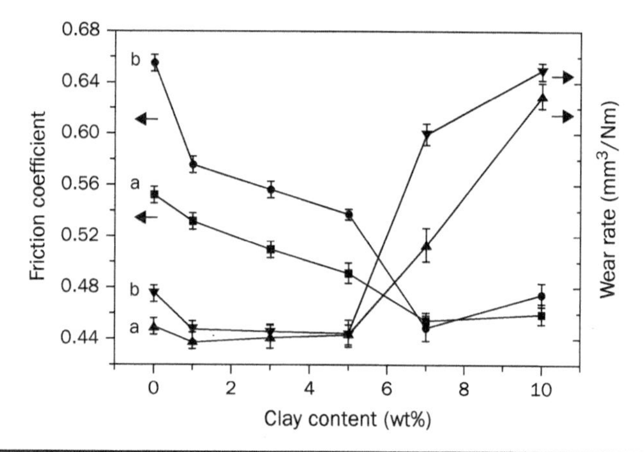

FIGURE 11.11 Friction coefficient and wear rate for nylon 6/nanoclay nanocomposites at different temperatures. Solid circles indicate friction coefficient, while solid triangles indicate wear rate (5 N, 0.6 m/s) at (a) 25°C and (b) 60°C [6].

trend is seen for the experiment run with a 10 N normal load, but the wear rate begins increasing at only 3 wt% nanoclay. The interfacial strengthening of the nanoclay in the nanocomposites acts to carry load and resist wear at lower nanoclay concentrations, but at larger concentrations the abrasion action of the nanoclay aggregates damages the transfer film and causes additional wear on the abrasion surface [6]. Since the dispersion of the nanoclay platelets was a mix of intercalated and exfoliated particles, the trend in wear rate versus nanoclay content is likely related to the type of dispersion in each sample. Additional experiments in which the nanoclay is fully exfoliated could prove useful.

SEM images of the worn surfaces of nanocomposites and the steel ball pins showed the mechanisms of wear. The wear mechanism for neat nylon 6 is severe scuff and adhesion. With low concentrations of nanoclay, the adhesion of nylon 6 appears lower, and at high nanoclay concentrations, fatigue crack is the main wear mechanism [6].

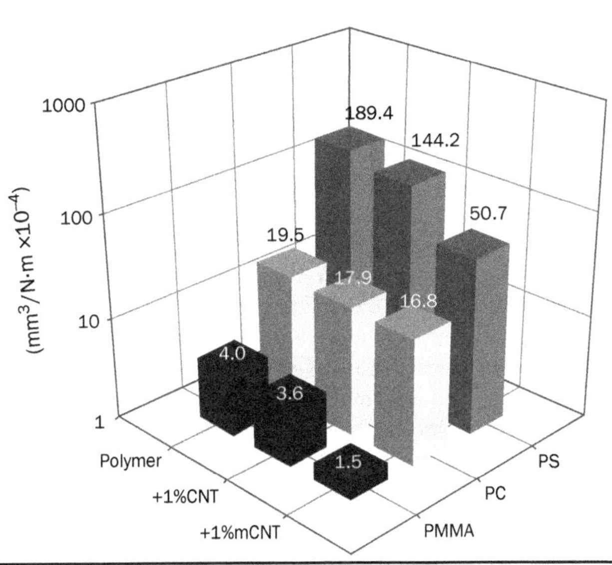

Figure 11.12 Specific wear rates for polymers and nanocomposites [8].

11.2.3 Wear and Scratch Resistance of Polymer-Carbon Nanotube Nanocomposites

Carbon nanotube–polymer nanocomposites have been studied quite extensively since 1994 when Ajayan first published his carbon nanotube (CNT) research [8]. As previously mentioned, CNTs are known for their thermal, mechanical, and electrical properties, but studies focusing on their wear and scratch resistance are harder to find. Some experiments also yield contradicting evidence on whether they improve tribological performance of polymers or not [9]. Regardless, understanding the tribological effects of CNTs in polymers is important, since CNTs are widely used to improve other properties in polymers.

Carrión et al. [8] performed tribology experiments on multiple types of polymers to study the antiwear properties of thermoplastics with ionic liquid–modified CNT additives. The study tested the wear resistance of polystyrene (PS), polymethylmethacrylate (PMMA), and polycarbonate (PC). The properties of each of the neat polymers were compared to those of the polymer with SWNT additives. Some of the SWNTs were added as-received from NanoAmor Inc. (referred to as CNTs in this section), and some were modified with the ionic liquid, [OMIM]BF_4 (mCNT). The processing and preparation methods are detailed in the article by Carrión et al. [8]. Pin-on-disc tests were performed using polymer discs and stainless-steel spherical pins. Raman spectroscopy was carried out to assess the degree of nanoparticle dispersion in the polymers, and the PS and PMMA show a more uniform dispersion than that of PC [8].

Figure 11.12 shows that the specific wear rates for all polymers decrease with the addition of SWNTs [8]. Furthermore, the nanocomposites containing the SWNTs treated with the ionic liquid have the lowest specific wear rates, thus confirming that the ionic liquid has an effective antiwear effect. PS has the highest specific wear rates; the addition of CNTs to the neat PS reduces the specific wear rate by 24%, while the addition of mCNTs reduces the specific wear rate by 74% with respect to the neat polymer. The addition of CNTs to neat PC reduces the specific wear rate by 8%, while the addition of mCNTs reduces the specific wear rate by 14% with respect to the neat polymer. The addition of CNTs to neat PMMS reduces the specific wear rate by 35%, while the addition of mCNTs reduces the specific wear rate by 63% with respect to the neat polymer. PMMA has significantly lower specific wear rates overall than PC or PS. Overall, PS + mCNTs and PMMA + mCNTs show the maximum reduction in specific wear rates with respect to the neat polymer. The PC likely does not show as high of a reduction because of the less uniform dispersion of nanoparticles. The wear reduction is caused by the uniform dispersion of SWNTs in the polymer, which increases resistance to crack propagation and fracture, and the ionic liquid surface modification improves the lubricating ability of the nanoparticle additives [8].

Carrión et al. performed another set of experiments focusing only on PS to determine the abrasive wear under multi-scratching of PS. The effects of sliding friction and the same ionic liquid modification ([OMIM]BF_4) were studied. The previous study analyzed the effect of friction and wear under sliding conditions, while this study uses the ASTM D7027-05 scratch test method to determine the abrasive wear behavior of the nanocomposites. The specifics of sample preparation and test parameters are detailed in the article by Bermúdez et al. [9].

Figure 11.13 shows the multiple scratch behaviors of neat PS, PS + NT, and PS + NTm [9]. After nine scratches, the neat PS starts showing more severe surface damage, indicating that strain hardening does not occur with increasing scratches. Conversely, the nanocomposites display an asymptotic response to increasing number of scratches, indicating that the CNTs induce strain hardening. Neat PS is quite brittle, and the addition of CNTs acts to reduce this brittleness and increase instantaneous penetration. SEM micrographs, surface topography, and cross-section profiles show a rough surface with

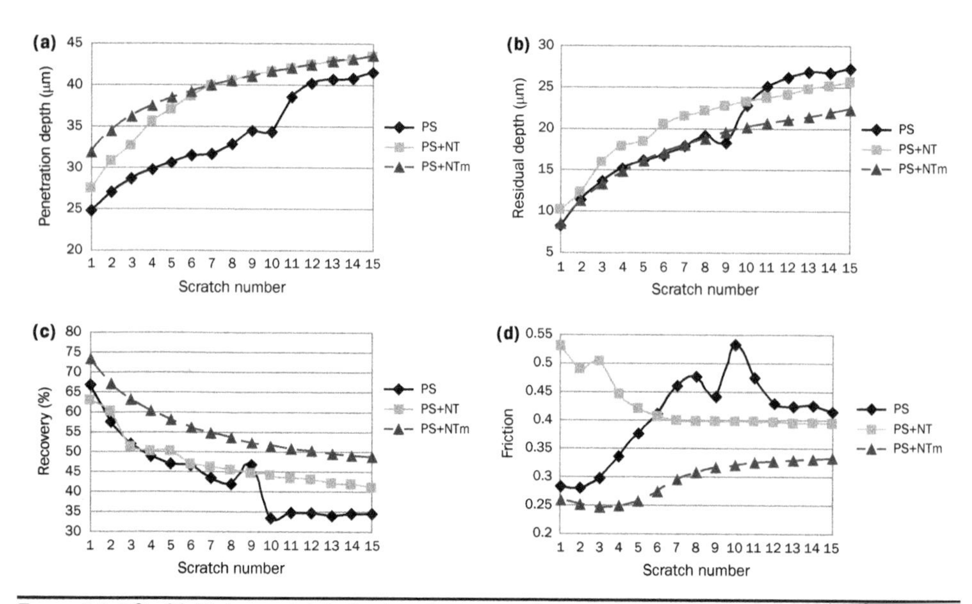

FIGURE 11.13 Multiple scratch behavior of compression-molded discs [9].

cracks and debris formation for the neat PS, whereas both nanocomposites showed a lower surface roughness [9].

In Fig. 11.14, the nanocomposite compression-molded scratch response is compared with the longitudinal and transverse (with respect to injection flow) scratch response of injection-molded coupons [9]. The result shows that the injection-molded coupons exhibit an anisotropic response, since the nanocomposite scratch behavior is dependent on the scratch direction. A scratch direction parallel to injection flow yields the highest scratch resistance and lowest friction coefficients, while the lowest scratch resistance and highest friction coefficient values are found for a scratch direction perpendicular to injection flow. The scratch resistance of the compression-molded discs is always intermediate to the longitudinal and transverse data. The polymer chains and nanoparticles align themselves in the longitudinal direction when they are injection-molded; thus, they have a higher mobility in this direction and a hindered mobility in the transverse direction leading to the absence of strain hardening [9].

Giraldo et al. [10] also studied scratch and wear resistance using the scratch test, but they used MWNTs dispersed in a nylon 6 polymer matrix. Figure 11.14 shows the instantaneous penetration depth (R_p) and residual healing depth (R_h) as a function of scratch number for the neat nylon 6 and nanoclay nanocomposites [10]. The instantaneous depth is deepest for the neat nylon 6 and decreases with increasing MWNT concentration. Conversely, the residual depth is deepest for the 1 wt% nanocomposite and decreases with decreasing MWNT concentration. Both the instantaneous depth and the residual depth increase asymptotically with increasing scratches due to strain hardening. This phenomenon may be attributed to the high aspect ratio of MWNTs, since they act as reinforcing particles but may also induce more nucleation sites, thus hampering viscoelastic recovery [10].

SEM images confirm that the neat nylon 6 deforms in a relatively soft manner, and adding MWNTs increases the stiffness to better resist the deformation from the indenter. Overall, the addition of MWNTs increases abrasion resistance [10].

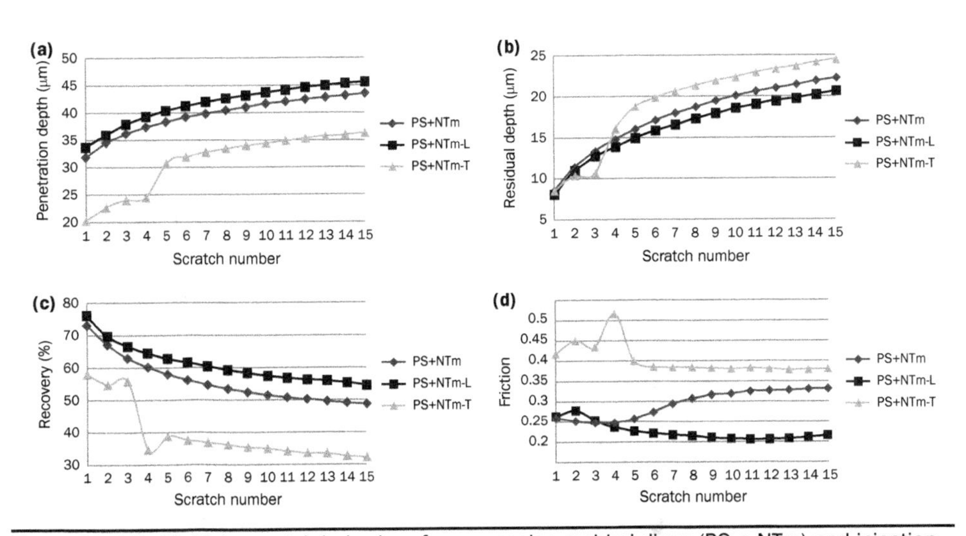

FIGURE 11.14 Multiple scratch behavior of compression-molded discs (PS + NTm) and injection-molded coupons in the longitudinal (L) and transverse (T) directions with respect to injection flow [10].

11.2.4 Wear Resistance of PTFE-Graphene Nanocomposites

Polytetrafluoroethylene (PTFE) is a commonly used thermoplastic solid lubricant with a wide range of industrial applications where using fluid lubricants is not feasible. PTFE provides low friction in dry sliding conditions, but is limited in its use by a very high wear rate. Micro-sized particles have been shown to reduce the wear rates of PTFE by several orders of magnitude, but the nano-sized graphene, with such large interfacial areas, is expected to reduce wear rates even more dramatically [11].

Kandanur et al. [11] reported on the wear resistance properties of PTFE-graphene nanocomposites. They used a tribometer to perform pin-on-disc experiments, and the test and preparation methods are detailed in their article. SEM was used to look at the dispersion of the graphene platelets in the PTFE matrix [11]. While TEM would give a better idea of dispersion, there is no indication of agglomeration of graphene platelets in the SEM image (Fig. 11.15) [11].

Figure 11.16 shows that the wear volume versus sliding distance decreases as the concentration of graphene increases [11]. The neat PTFE and nanocomposites with graphene concentrations up to 0.12 wt% show relatively the same wear rate as their data crowd near the vertical axis. However, once the graphene concentration reaches 0.32 wt%, the wear resistance increases dramatically with the 10 wt% showing only about 1 mm^3 of wear volume even after sliding 80 km. The second graph also shows that the wear volume for 2, 5, and 10 wt% graphene seems to increase mainly during the first 7–15 km of sliding. Here, once the wear rate starts to decrease after the concentration reaches 0.12 wt%, the wear rate is found to decrease with approximately the square of the graphene platelet content. Furthermore, the wear rate does not appear to reach a minimum, so it is feasible that lower wear rates could be achieved with higher graphene platelet concentration [11].

Figure 11.17 shows the SEM images of the surface and counter surface of neat PTFE and 10 wt% graphene-PTFE nanocomposite [11]. The wear debris on the counter surface of the neat PTFE is hundreds of micrometers in in-plane dimensions, while the

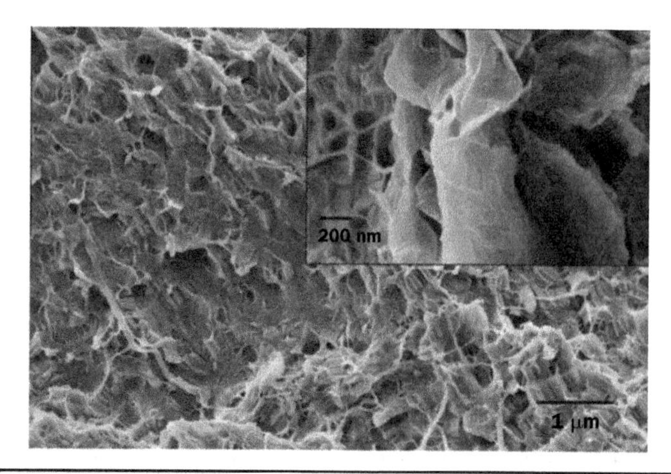

FIGURE 11.15 SEM image of 2 wt% graphene-PTFE composite (inset shows the wavy edges of graphene platelets) [11].

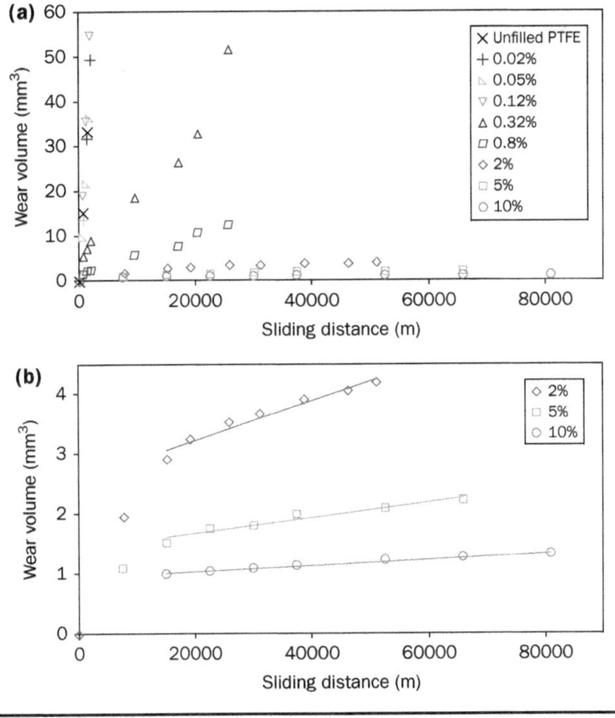

FIGURE 11.16 Wear volume of neat PTFE and graphene platelet–PTFE nanocomposites versus sliding distance [11].

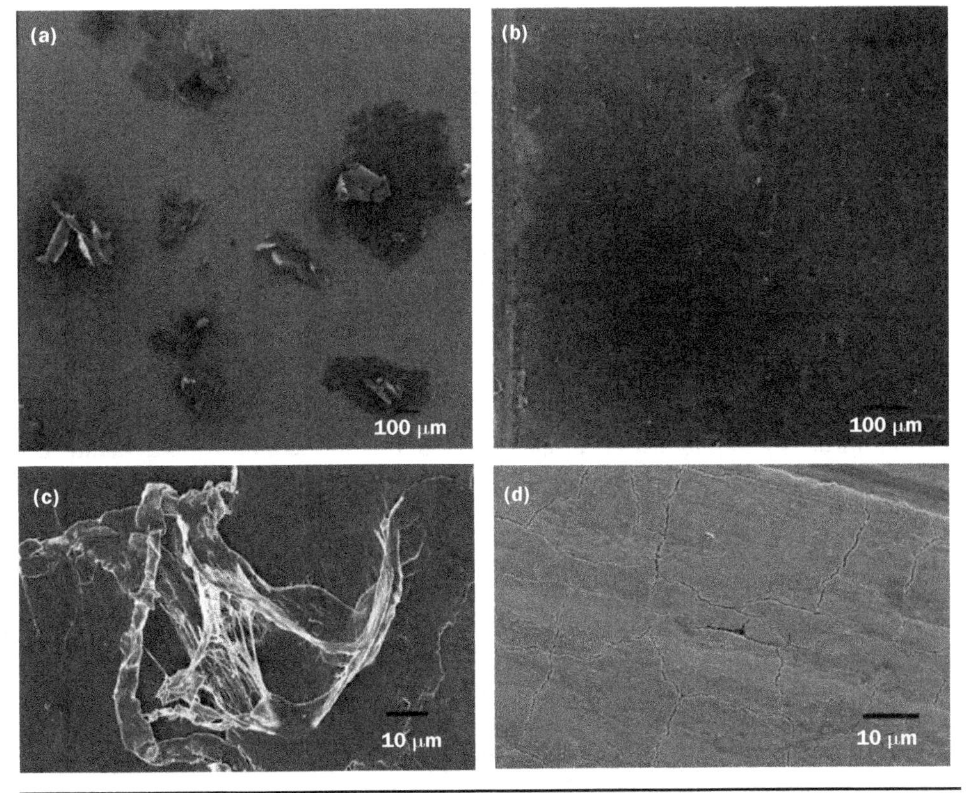

FIGURE 11.17 SEM images of the (a) counter surface of neat PTFE, (b) counter surface of 10 wt% graphene-platelet PTFE, (c) wear surface of neat PTFE, and (d) wear surface of 10 wt% graphene platelet–PTFE nanocomposites [11].

wear debris on the counter surface of the 10 wt% graphene-PTFE nanocomposite is much finer. Similarly, the wear surface of neat PTFE shows large, plate-like debris, whereas the 10 wt% graphene-PTFE nanocomposite shows only *mudflat* features. Neat PTFE is known to wear by ejecting large, plate-like debris likely due to propagation of subsurface cracks parallel to the sliding surface. The addition of graphene platelets could inhibit these subsurface cracks by deflecting them through a path of high-aspect-ratio graphene platelets. Graphene platelets show a definite increase in wear resistance when added to PTFE [11].

11.2.5 Summary of Tribological Properties of Polymer Nanocomposites

There are generally fewer studies on tribological properties of polymer nanocomposites than other, more sought-after properties such as thermal, mechanical, or electric properties. However, thermoplastic polymer nanocomposites generally exhibit better abrasion, wear, and scratch resistance than their unfilled polymer counterparts. These properties are often highly dependent on the degree of nanoparticle dispersion achieved in a polymer, since enhanced properties are often attributed to the extremely high aspect ratio of nanoparticles. In the case of nanoclay, more research is needed to determine exactly how to achieve better abrasion resistance, since different studies seem to produce contradictory results. The formation of nanoclay aggregates almost certainly decreases abrasion resistance, so an exfoliated dispersion within the polymer is required. Both CNTs and graphene platelets show a positive effect on the tribological properties. The mechanisms of wear in polymers is often crack propagation or debris formation on the abraded surface of the polymer, and well-dispersed nanoparticles often act to inhibit these mechanisms through better polymer-nanoparticle adhesion. Abrasion, wear, and scratch resistance are important properties for most polymers, and the inclusion of nanoparticles can create robust thermoplastic and metallic nanocomposites with enhanced tribological properties presented in Paulo Davim's book [12]. Chapter 1 of Paulo Davim's book [12] provides the tribology of bulk PNCs and nanocomposite coatings. Chapter 2 is dedicated to nano- and micro-polyethersulphone (PTFE) for surface lubrication of carbon fiber–reinforced PTFE. Chapter 3 describes tribology of molybdenum disulfide (MoS_2)–based nanocomposites. Chapter 4 describes the friction and wear of Al_2O_3-based composites. Chapter 5 is dedicated on wear of multi-scaling phase–reinforced composites.

11.3 Permeability Properties of Polymer Nanocomposites and Applications of Nanotechnology and Nanomaterials in the Oil Field

Permeability in polymers is of great interest due to the increasing demands of better membranes with decreased fouling and degradation. Separations, barriers, and even combustion prevention are dependent on good polymer films. A way to improve on the transport properties of polymer is with the addition of nano-sized particles for the production of polymer nanocomposites. Although organic nanofillers have been always added in polymers, such as rubber, studies of gas transport in polymer nanocomposites are fairly recent [13]. It is the purpose of this section to highlight the use of nanomaterials in polymers for the enhancement of permeability. Interested readers should refer to the list of articles listed in the Further Reading section.

Nanotechnology is one of the fastest growing, most promising research fields. Specifically, nanotechnology has seen significant advances in the electronics, biomedical, materials and manufacturing, pharmaceutical, aerospace, photography, and energy industries [14]. Relatively few applications of nanotechnology are currently seen in the oil field. The oil and gas industry could greatly benefit from nanotechnology innovations in the oil field since strong, stable materials are needed in the high-temperature, high-pressure environment of an oil well. Additionally, with fossil fuels growing scarcer, there is a greater need for effective equipment to extract oil from unconventional oil wells [15]. Further Reading section is provided at the end of this chapter to explore the benefits of nanocomposite materials, nanofluids, and nanosensors with emphasis on their potential applications in the oil field.

11.4 Overall Summary

This chapter briefly reviews the tribology properties of different types of polymer nanocomposites. For tribology properties, nanoclay, carbon nanotubes (MWNTs and SWNTs), functionalized CNTs, and nanographene platelets were used to enhance the abrasion, wear, and scratch resistance of the unfilled polymers. These properties are often highly dependent on the degree of nanoparticle dispersion achieved in a polymer, since enhanced properties are often attributed to the extremely high aspect ratio of the nanoparticles. Until this demand is fulfilled, polymer nanocomposites will continue to be explored to find an optimal material for this application. Permeability properties of polymer nanocomposites and nanotechnology in oil field applications are briefly introduced in this chapter.

11.5 Study Questions

11.1 How does the nanomaterial dispersion affect the tribological properties of polymer nanocomposites in general? Provide examples to illustrate your claims.

11.2 How does the tribological properties of polymer-clay nanocomposites influenced by the type of clay and surface treatment of the clay, and how clay is processed in the polymer matrix? Provide examples.

11.3 This chapter reviewed the enhancement of tribological properties of thermoplastic-based nanocomposites using *montmorillonite* (MMT) clay, MWNTs, SWNTs, and graphene nanomaterials. Which types of nanomaterials are more effective for thermoplastic-based nanocomposites? Provide examples. Discuss the protective mechanism of these nanomaterials to provide better tribological properties. Besides the above nanoparticles, what other types of nanomaterials are used to improve tribological properties of polymer nanocomposites?

11.4 Discuss other commonly used characterization techniques, such as (a) Taber abraser, (b) pin-on-disc tribometer, and (c) scratch test detailed in Sec. 11.2.1 to evaluate tribological properties of polymer nanocomposites.

11.5 Conduct a literature search on how the tribological properties of thermoset-based and thermoplastic elastomer–based nanocomposites are improved.

11.6 In addition to the nanocomposites mentioned in Question 11.5, what other types of nanoparticles can be used to enhance tribological properties of polymer nanocomposites?

11.6 References

1. International, ASTM. (2010) Standard test method for abrasion resistance of organic coatings by the taber abraser.
2. Taber Rotary Abraser 5135/5155. edited by Taber Industries, 2008.
3. Zhou, Q., Wang, K., and Loo, L. S. (2009) Abrasion studies of nylon 6/montmorillonite nanocomposites using scanning electron microscopy, Fourier transform infrared spectroscopy, and x-ray photoelectron spectroscopy. *Journal of Applied Polymer Science*, 113(5):3286–3293.
4. Liu, S.-P., Hwang, S.-S., Yeh, J.-M., and Hung, C.-C. (2011) Mechanical properties of polyamide-6/montmorillonite nanocomposites—prepared by the twin-screw extruder mixed technique. *International Communications in Heat and Mass Transfer*, 38(1):37–43.
5. Srinath, G., and Gnanamoorthy, R. (2006) Two-body abrasive wear characteristics of nylon clay nanocomposites—effect of grit size, load, and sliding velocity. *Materials Science and Engineering: A*, 435–436:181–186.
6. Mu, B., Wang, Q., Wang, T., Wang, H., and Jian, L. (2008) The friction and wear properties of clay filled Pa66. *Polymer Engineering & Science*, 48(1):203–209.
7. International, ASTM. (2009) The standard test method for evaluation of scratch resistance of polymeric coatings and plastics using an instrumented scratch machine.
8. Carrión, F. J., Espejo, C., Sanes, J., and Bermúdez, M. D. (2010) Single-walled carbon nanotubes modified by ionic liquid as antiwear additives of thermoplastics. *Composites Science and Technology*, 70(15):2160–2167.
9. Bermúdez, M. D., Carrión F. J., Espejo C., Martínez-López, E., and Sanes J. (2011) Abrasive wear under multiscratching of polystyrene + Single-walled carbon nanotube nanocomposites. Effect of sliding direction and modification by ionic liquid. *Applied Surface Science*, 257(21):9073–9081.
10. Giraldo, L. F., Brostow, W., Devaux, E., Lopez, B. L., and Perez, L. D. (2008) Scratch and wear resistance of polyamide 6 reinforced with multiwall carbon nanotubes. *Journal of Nanoscience and Nanotechnology*, 8(6):3176–83.
11. Kandanur, S. S., Rafiee, M. A., Yavari, F., Schrameyer, M., Yu, Z.-Z., Blanchet, T. A., and Koratkar N. (2012) Suppression of wear in graphene polymer composites. *Carbon*, 50(9):3178–3183.
12. Paulo Davim, J., ed. (2013) *Tribology of Nanocomposites.* Berlin Heidelberg: Springer-Verlag.
13. Bhattacharya, M., Biswas, S., and Bhowmick, A. K. (2011) Permeation characteristics and modeling of barrier properties of multifunctional rubber nanocomposites. *Polymer*, 52:1562–1576.
14. Kong, X., and Ohadi, M. (2010) Applications of micro and nano technologies in the oil and gas industry—overview of the recent progress. Paper presented at *Abu Dhabi International Petroleum Exhibition and Conference*, Abu Dhabi, UAE, November 1–4. Society of Petroleum Engineers, SPE-138214-MS. doi: 10.2118/138241-MS.
15. Singh, S. K., Ahmed, R. M., and Growcock, F. (2010) Vital role of nanopolymers in drilling and stimulations fluid applications. Paper presented at *SPE Annual Technical Conference and Exhibition*, Florence, Italy. September 20–22. ISBN: 978-1-55563-300-4.

11.7 Further Reading

Bhattacharya, M., Biswas, S., and Bhowmick, A. K. (2011) Permeation characteristics and modeling of barrier properties of multifunctional rubber nanocomposites. *Polymer,* 52:1562–1576.

Cai, J., Chenevert, M. E., Sharma, M. M., and Friedheim, J. E. (2012) Decreasing water invasion into atoka shale using nonmodified silica nanoparticles. *Society of Petroleum Engineers,* 27(1):103–112.

Castarlenas, S., Gorgojo, P., Casado-Coterillo, C., Masheshwari, S., Tsapatsis, M., Tellez, C., and Coronas, J. (2013) Melt compounding of swollen titanosilicate JDF-L1 with polysulfone to obtain mixed matrix membranes for H_2/CH_4 separation. *Industrial & Engineering Chemistry Research,* 52(5):1901–1907.

Chang, J. H., An, Y. K., and Sur, G. S. (2003) Poly(lactic acid) nanocomposites with various organoclays. I. Thermomechanical properties, morphology, and gas permeability. *Journal of Polymer Science Part B,* 41(1):94–103.

Clark, M. (2009) Minimizing environmental footprint by utilizing prevention technology. Paper presented at SPE Annual Technical Conference and Exhibition.

Czichos, H., Saito, T., and Smith, L. E., eds. (2006) Mechanical properties. In *Springer Handbook of Materials Measurement Methods.* Springer, pp. 283–397.

Dasari, A., Lim, S., Yu, Z., and Mai, Y. (2007) Toughening, thermal stability, flame retardancy, and scratch–wear resistance of polymer–clay nanocomposites. *Australian Journal of Chemistry,* 60(7):496–518.

Dasari, A., Yu, Z-Z., and Mai, Y-W. (2009) Fundamental aspects and recent progress on wear/scratch damage in polymer nanocomposites. *Materials Science and Engineering: R: Reports,* 639(2):31–80.

Endo, M., Noguchi, T., Ito, M., Takeuchi, K., Hayashi, T., Kim, Y. A., Wanibuchi, T., Jinnai, H., Terrones, M., and Dresselhaus, M. S. (2008) Extreme-performance rubber nanocomposites for probing and excavating deep oil resources using multi-walled carbon nanotubes. *Advanced Functional Materials,* 18(21):3403–3409.

Ito, M., Noguchi, T., Ueki, H., Takeuchi, K., and Endo, M. (2011) Carbon nanotube enables quantum leap in oil recovery. *Materials Research Bulletin,* 46(9):1480–1484.

Karkhanechi, H., Kazemian, H., Nazockdast, H., Mozdianfard, M. R., and Bidoki, S. M. (2012) Fabrication of homogenous polymer-zeolite nanocomposites as mixed-matrix membranes for gas separation. *Chemical Engineering & Technology,* 35(5):885–892.

Koo, J. H. (2006) *Polymer Nanocomposites: Processing, Characterization, and Applications.* McGraw-Hill, 2006.

Koo, J. H. (2016) Fundamentals, Properties, and Applications of Polymer Nanocomposites. Cambridge University Press, Cambridge: UK, 2016.

Nabhani, N., Emami, M., and Taghavi Moghadam, A. B. (2011) Application of nanotechnology and nanomaterials in oil and gas industry. *AIP Conference Proceedings,* 1415(1):128–131.

One colossal discovery: Materials science breakthrough creates nanostructured material of immense proportions. *Connexus,* 2011;2(2):8–13.

Pinto, A. M., Cabral, J., Pacheco Tanaka, D. A., Mendes, A. M., and Magalhaes, F. D. (2013) Effect of incorporation of graphene oxide and graphene nanoplatelets on mechanical and gas permeability properties of poly(lactic acid) films. *Polymer International,* 62:33–40.

Poreba, R., Spirkova, M., Brozova, L., Lazic, N., Pavlicevic, J., and Strachota, A. (2013) Aliphatic polycarbonate-based polyurethane elastomers and nanocomposites. II.

Mechanical, thermal, and gas transport properties. *Journal of Applied Polymer Science*, 329–341.

Pourafshary, P., Azimipour, S. S., Motamedi, P., Samet, M., Taheri, S. A., Bargozin, H., and Hendi, S. S. (2009) Priority assessment of investment in development of nanotechnology in upstream petroleum industry. Paper presented at SPE Saudi Arabia Section Technical Symposium.

Prusty, G., and Swain, S. K. (2013) Dispersion of multiwalled carbon nanotubes in polyacrylonitrile-co-starch copolymer matrix for enhancement of electrical, thermal, and gas barrier properties. *Polymer Composites*, 34(3):330–334.

Savino, V., Fallatah, G. M., and Mehdi, M. S. (2010). Applications of nanocomposite materials in the oil and gas industry. *Advanced Materials Research*, 83–86:771,772–776.

Sinha, S. K., Song, T., Wan, X., and Tong, Y. (2009) Scratch and normal hardness characteristics of polyamide 6/nanoclay composite. *Wear*, 266(7–8):814–821.

Valix, M., Mineyama, H., Chen, C., Cheung, W. H., Shi, J., and Bustamante, H. (2011) Effect of film thickness and filler properties on sulphuric acid permeation in various commercially available epoxy mortar coatings. *Water Science & Technology*, 64(9):1864–1869.

van Rooyen, L. J., Karger-Kocsis, J., Vorster, O. C., and Kock, L. D. (2013) Helium gas permeability reduction of epoxy composite coatings by incorporation of glass flakes. *Journal of Membrane Science*, 430:203–210.

Verdejo, R., Bernal, M., Romasanta, L. J., and Lopez-Manchado, M. A. (2011) Graphene filled polymer nanocomposites. *Journal of Materials Chemistry*, 21(10):3301–3310.

Wu, Q., Zhu, W., Zhang, C., Liang, Z., and Wang, B. (2010) Study of fire retardant behavior of carbon nanotube membranes and carbon nanofiber paper in carbon fiber reinforced epoxy composites. *Carbon*, 48:1799–1806.

Xu, Z., Agrawal, G., and Salinas, B. J. (2011) Smart nanostructured materials deliver high reliability completion tools for gas shale fracturing. Paper presented at SPE Annual Technical Conference and Exhibition.

PART 3

Opportunities and Trends for Polymer Nanocomposites

CHAPTER 12

Opportunities, Trends, and Challenges for Nanomaterials and Polymer Nanocomposites

12.1 Introduction

Just because new technology works does not necessarily mean that it will be successful. There are a lot of other aspects that many people do not generally consider. This situation can be observed in many contemporary technologies that are being advocated for mass production and use, such as solar and wind energy. Currently, the startup cost is too high for many companies to consider switching over to these forms of energy. The pathway is similar for polymer nanocomposites in that although they do yield some exciting properties, they are difficult and expensive to produce on a mass scale. This chapter discusses the research funding and commercial market opportunities, developments currently being made to bring polymer nanocomposites to a larger consumer market, and the challenges of manufacturing nanomaterials and polymer nanocomposites.

12.2 Government and Commercial Research Opportunities

12.2.1 U.S. Government Research Opportunities, Program Plans, and Progress

Since its inception in 2001, the National Nanotechnology Initiative (NNI; website: www.nano.gov) has received cumulatively more than $27 billion in government grants [1, 2]. Twenty federal departments, independent agencies, and commissions work together toward the shared vision of a future in which the ability to understand and control matter at the nanoscale leads to a revolution in technology and industry that benefits society. The President's 2019 budget provides $1.4 billion for the NNI, a continued investment in basic research, early-stage applied research, and technology transfer efforts that will lead to the breakthroughs of the future [2]. The NNI investments in 2017 and 2018 and those for 2019 reflect a sustained emphasis on broad, fundamental research in nanoscience to provide continuity to new discoveries that will enable future transformative commercial products and services.

This support reflects the continued importance of investments that advance our fundamental understanding and control material at the nanoscale, as well as the translation of the knowledge in nanotechnology's potential to significantly improve our fundamental understanding and control of matter at the nanoscale and to translate that knowledge into solutions for critical national needs. Through sustained support of basic and early-stage applied research, the President's budget for nanotechnology will further the progress of the NNI to strengthen the national security innovation base, transform health care, modernize American's infrastructure, advance manufacturing, educate a future-focused workforce, and lead to job growth and economic prosperity [2].

Federal agencies are committed to promoting the transition of nanotechnology-based discoveries from the laboratory to market. Although there is a renewed emphasis on activities aimed at promoting accelerated translation of nanotechnologies into commercial products, NNI agencies remain committed to strong and sustained support for fundamental research in nanoscience. The President's 2019 budget supports nanoscale science, engineering, and technology research and development (R&D) at 12 agencies. The five federal organizations with the largest investments (representing 95% of the total) are as follows [2]:

- National Institutes of Health (NIH): nanotechnology-based biomedical research at the intersection of life and physical sciences
- National Science Foundation (NSF): fundamental research and education across all disciplines of science and engineering
- Department of Energy (DOE): fundamental and applied research provided a basis of new and improved energy technologies
- Department of Defense (DOD): science and engineering research advancing defense and dual-use capabilities
- National Institute of Standards and Technology (NIST): fundamental R&D of measurement and fabrication tools, analytical methodologies, metrology, and standards of nanotechnology

Table 12.1 shows the budget allocation to the NNI for nanotechnology R&D from 2017 to 2019 by federal agencies [2]. Figure 12.1 shows the NNI annual funding by agencies over time since the inception of the NNI in 2001 [2].

The program component area (PCA)-wise breakdown of the 2019 NNI funding is as follows [2]:

- *PCA 1:* Nanotechnology Signature Initiatives and Grand Challenges (13%)
- *PCA 2:* Foundational Research (39%)
- *PCA 3:* Nanotechnology-Enabled Applications, Devices, and Systems (28%)
- *PCA 4:* Research Infrastructure and Instrumentation (16%)
- *PCA 5:* Environmental, Health, and Safety (5%).

Table 12.2 lists the proposed PCA-wise 2019 NNI funding in millions of dollars.

The discussion on progress in each of the four goals of the NNI is presented as follows (for more information and additional details, please refer to Nano.gov).

Table 12.1 NNI Budget, by Agency, 2017–2019 (in Millions of Dollars) [2]

Agency	2017 Actual	2018 Estimated*	2019 Proposed
CPSC	1.9	1.0	1.0
DHS	0.4	0.4	0.3
DOC/NIST	80.7	70.1	57.9
DOD	143.3	143.7	125.9
DOE**	341.2	327.2	324.1
DOI/USGS	0.1	0.1	0.0
DOJ/NIJ	2.0	1.7	1.7
DOT/FHWA	0.3	1.5	1.5
EPA	10.7	10.7	4.5
HHS (total)	472.3	469.1	464.3
FDA	11.6	12.5	12.5
NIH	449.6	445.5	440.7
NIOSH	11.1	11.1	11.1
NASA	9.5	7.8	6.0
NSF	465.7	420.8	387.7
USDA (total)	24.2	23.3	20.7
ARS	3.0	3.0	2.0
FS	6.2	5.3	3.7
NIFA	15.0	15.0	15.0
TOTAL	1552.3	1477.4	1395.6

CPSC, Consumer Product Safety Commission; DHS, Department of Homeland Security; DOC, Department of Commerce; DOD, Department of Defense; DOE, Department of Energy; DOI, Department of the Interior; DOJ, Department of Justice; EPA, Environmental Protection Agency; HHS, Department of Health and Human Services; NASA, National Aeronautics and Space Administration; NSF, National Science Foundation; USDA, Department of Agriculture.

* 2018 numbers are based on annualized 2018 continuing resolution levels and are subject to change based on final appropriations and operating plans.

** Funding levels for DOE include the combined budgets of the Office of Science, the Office of Energy Efficiency and Renewable Energy, the Office of Fossil Energy, the Office of Nuclear Energy, and the Advanced Research Projects Agency–Energy.

Source: National Nanotechnology Initiative (NNI) [2].

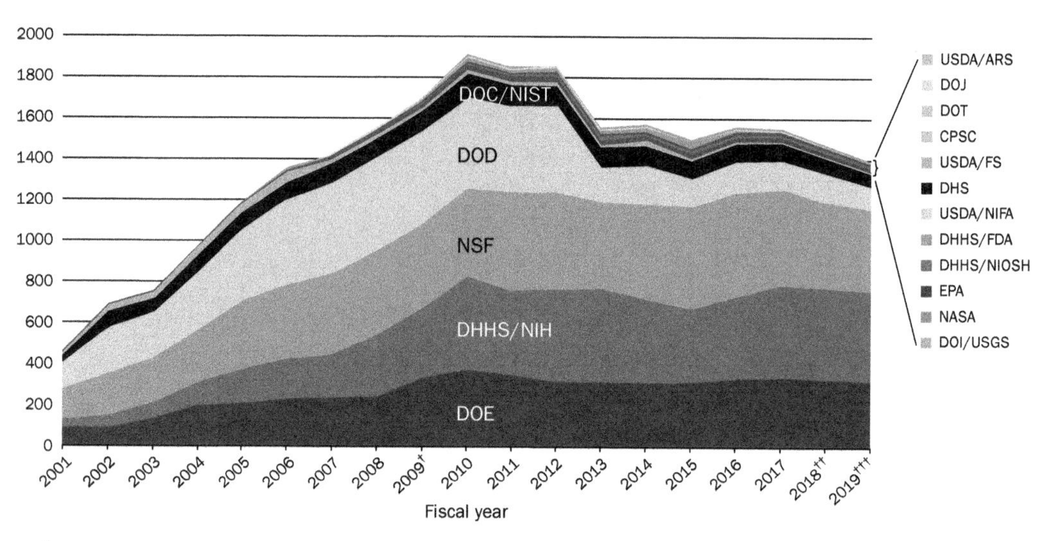

† 2009 figures do not include American Recovery and Reinvestment Act funds for DOE ($293 million), NSF ($101 million), NIH ($73 million), and NIST ($43 million).

†† 2018 estimated funding is based on annualized 2018 continuing resolution levels and subject to change based on final appropriations and operating plans.

††† 2019 Budget.

FIGURE 12.1 NNI funding by agencies, 2001–2019 (millions of dollars) [2]. (*Source: NNI.*)

Goal 1. Advance a World-Class Nanotechnology Research and Development Program: The following points exemplify the NNI research pushing the boundaries of scientific understanding in nanoscience and nanotechnology:

- Understanding the electrical, optical, and magnetic properties of nanomaterials
- Expanding the boundaries of material science
- Advancing sensing and metrology
- Developing new tools and devices
- Other advances

Goal 2. Foster the Transfer of New Technologies into Products for Commercial and Public Benefit: The focus of Goal 2 is to facilitate the transfer of nanotechnology R&D breakthroughs into applications that the private sector can bring to market. The following points illustrate the transfer of new nanotechnologies into application for public benefit:

- Fostering entrepreneurship
- Transiting information and communication technologies
- Bringing nanotechnology to aerospace and automotive markets
- Translating nanotechnology from bench to bedside

Table 12.2 Proposed 2019 PCA-Wise Agency Funding (in Millions of Dollars)

Agency	1.Nanotechnology Signature Initiatives (NSIs) and Grand Challenges (GCs)	1a. Nanomanufacturing NSI	1b. Nanoelectronics NSI	1c. NKI NSI	1d. Sensors NSI	1e. Water NSI	1f. Future Computing GC	2. Foundational Research	3. Nanotechnology-Enabled Applications, Devices, and Systems	4. Research Infrastructure and Instrumentation	5. Environment, Health, and Safety	NNI Total
CPSC	0.0	0.0	0.0	0.0	0.0	0.0	0.0	0.0	0.0	0.0	1.0	1.0
DHS	0.3	0.0	0.0	0.0	0.3	0.0	0.0	0.0	0.0	0.0	0.0	0.3
DOC/NIST	19.0	3.7	10.0	0.9	0.8	0.3	3.3	11.4	4.6	20.7	2.3	57.9
DOD	22.5	0.5	15.7	1.1	0.9	2.2	2.1	72.7	26.5	0.7	3.5	125.9
DOE	0.0	0.0	0.0	0.0	0.0	0.0	0.0	182.1	9.2	132.9	0.0	324.1
DOI/USGS	0.0	0.0	0.0	0.0	0.0	0.0	0.0	0.0	0.0	0.0	0.0	0.0
DOJ/NIJ	0.0	0.0	0.0	0.0	0.0	0.0	0.0	0.0	1.2	0.5	0.0	1.7
DOT/FHWA	0.0	0.0	0.0	0.0	0.0	0.0	0.0	0.0	1.5	0.0	0.0	1.5
EPA	0.0	0.0	0.0	0.0	0.0	0.0	0.0	0.0	0.0	0.0	4.5	4.5
HHS (total)	14.4	0.0	0.0	1.7	12.7	0.0	0.0	91.8	294.7	22.8	40.5	464.3
FDA	0.0	0.0	0.0	0.0	0.0	0.0	0.0	0.0	0.0	0.0	12.5	12.5
NIH	14.4	0.0	0.0	1.7	12.7	0.0	0.0	91.8	294.7	22.8	16.9	440.7
NIOSH	0.0	0.0	0.0	0.0	0.0	0.0	0.0	0.0	0.0	0.0	11.1	11.1
NASA	1.3	0.0	0.8	0.0	0.5	0.0	0.0	3.4	1.2	0.0	0.0	6.0
NSF	111.2	28.0	32.0	19.5	7.5	11.0	13.2	185.4	38.0	42.6	10.6	387.7
USDA (total)	8.3	4.3	0.0	0.0	3.0	1.0	0.0	2.3	7.0	1.0	2.1	20.7
ARS	0.0	0.0	0.0	0.0	0.0	0.0	0.0	0.0	2.0	0.0	0.0	2.0
FS	3.3	3.3	0.0	0.0	0.0	0.0	0.0	0.3	0.0	0.0	0.1	3.7
NIFA	5.0	1.0	0.0	0.0	3.0	1.0	0.0	2.0	5.0	1.0	2.0	15.0
TOTAL	**177.0**	**36.5**	**58.5**	**23.2**	**25.8**	**14.5**	**18.6**	**549.0**	**383.9**	**221.3**	**64.4**	**1395.6**

CPSC, Consumer Product Safety Commission; DHS, Department of Homeland Security; DOC, Department of Commerce; DOD, Department of Defense; DOE, Department of Energy; DOI, Department of the Interior; DOJ, Department of Justice; EPA, Environmental Protection Agency; HHS, Department of Health and Human Services; NASA, National Aeronautics and Space Administration; NSF, National Science Foundation; USDA, Department of Agriculture.
Source: NNI [2].

- Working with standards developing organizations
- Advancing technologies for energy and infrastructure

Goal 3. Develop and Sustain Education Resources, a Skilled Workforce, and a Dynamic Infrastructure and Toolset to Advance Nanotechnology: The following examples illustrate activities that support the NNI physical, cyber, and human infrastructure:

- Developing tools and instrumentation for nanotechnology
- Providing research infrastructure, including centers, user facilities, and education resources

Goal 4. Support Responsible Development of Nanotechnology: The following points showcase the NNI agency activities that support the responsible development of nanotechnology:

- Detecting and understanding the effects of nanomaterials in the environment and the human body
- Understanding and mitigating potential impact in the workforce
- Developing and disseminating information

12.2.2 Commercial Market Opportunities

The data on commercial market opportunities and trends of nanomaterials and polymer nanocomposites are provided to the author by Lucintel [3] (consulting firm from Irving, Texas; website: www.lucintel.com); Pyrograf Products, Inc., a division of Applied Sciences, Inc. [4] (carbon nanofiber manufacturer based at Cedarville, Ohio; website: www.pyrografproducts.com); an excellent article published by Vance et al. entitled, "Nanotechnology in the real world: Redeveloping the nanomaterial consumer products inventory" [5]; and Allied Market Research's website www.alliedresearch.com on "Global Polymer Nanocomposites Market—Opportunities and Forecasts, 2014–2022 Report" [6]. The executive summary of recent nanomaterials/polymer nanocomposites market research reports from Lucintel [3] and Applied Market Research state the following [6]:

- The global industrial applications of the nanomaterials had reached at $1.7 billion in 2010, with an average annual growth rate of 10.4% over the last 5 years
  - Top 10 players cover about 80% of the global nanomaterial market
  - Nanoclays and carbon nanotubes (CNTs) are the dominant nanomaterials representing about 90% of the market value
- The industrial use of the nanomaterials in North America accelerated by about 25% in 2010 and in Europe by about 22%, while in Asia and other regions the use increased by 32% in 2010. This growth is because of the following factors:
  - Active participation of governments in nanotechnology R&D funding
  - Advancement in production processes
  - Heavy investment in R&D by major players

- Global polymer nanocomposite/nanomaterial market was valued at $5.276 billion in 2015, and is expected to reach a $11.549 billion business by 2022
  - Expected to grow at a CAGR of 10.9% during the forecast period of 2016–2022
  - Driven by health and personal care, energy, and electrical and electronics (E&E) applications
  - Emerging applications include field emission display (FED), drug delivery, and biofuels.
  - Polymer nanocomposites are multiphase materials that contain dispersion of nanomaterials, such as nanoclay, carbon nanotubes, metal oxides, and ceramics among others.
- Health-care and energy industries are expected to surpass the E&E industry in size over the coming 2–3 years, driven by nanotechnology advancements in biomedical and solar energy applications

The term *commercial product* is defined as the product that is both *offered for sale* and *used* in the regular production of a device or component in general commence. This section includes numerous examples of *commercial* and *potential commercial* nanomaterials. Table 12.3 shows an overview of current available nanotechnology products, the developer/manufacturer, and the benefits of these nanomaterials applied to their prospective applications [7]. Table 12.4 illustrates some of the nanotechnology potential applications under development by various composite manufacturers combined with anticipated benefits [7].

An excellent article by Sanchez et al. summarizes the current and potential commercial applications of hybrid organic-inorganic nanocomposites [8]. They pointed out that organic-inorganic hybrid materials not only represent a creative way to design new materials and compounds for academic research, but the improved and unusual features of these methods will result in the development of innovative industrial applications. Most of the hybrid materials that are currently commercially available are synthesized and processed by using conventional soft-chemistry-based routes developed in the 1980s. Sanchez et al. grouped all hybrid material processes into three routes:

- Copolymerization of functional organosilanes, macromonomers, and metal alkoxides
- Encapsulation of organic components within sol-gel–derived silica or metallic oxides
- Organic functionalization of nanofillers with lamellar structures, and so forth

The chemical strategies [self-assembly, nanobuilding block approaches, hybrid MOF (Metal Organic Frameworks), integrative synthesis, coupled processes, bioinspired strategies, etc.], allow the development of new chemistry to direct the assembling of a large variety of structurally well-defined nano-objects into complex hybrid architectures hierarchically organized in terms of structure and functions. These new generations of hybrid materials will open very promising applications in optics, electronics, ionics, mechanics, energy, environment, biology, medicine, functional smart coatings, fuel and solar cells, catalysts, sensors, and so on. The Sanchez et al.'s study [8] identifies selective examples of hybrid materials with an implied emerging potential commercialization as well as functional hybrids with real commercial opportunities.

Table 12.3 An Overview of Current Commercially Available Nanotechnology Products

Product	Developer/Manufacturer	Benefit/Function
Auto sideboards	GM/Basell/Southern Clay Products/Blackhawk	Nano TPO-clay provides stronger, lighter sideboards
Nylon film	Honeywell/Nancor/Ube/Unitika	Low oxygen barrier film for wrapping
Performance plastics masterbatch	Hyperion	EMI components for office and home appliances
Trousers	Nanotex/Burrlington Mills/Eddie Bauer	Stainless trousers
Mattress covers	Nanotex/Seally	Shield for stains
Floor coverings	Nanophase	Vinyl flooring with nanocoating abrasion resistant
Sunscreen	Nanophase	Nanocrystals allow sunscreen to be transparent
Tennis ball	InMat™/Wilson	Coating for butyl rubber core improves pressure seal for longer life
Red stain and grease removal	Eastman/P&G	Peroxide and nanoparticle to remove stain and grease
Fire-retardant power cable	Kabelwerk Eupen	Combining ATH-based compound with organoclay improves on fire performance and smoke density
Basketball shoe pouch	Converse/Triton System	Nanocomposite pouch filled with helium inserts that fits into basketball shoe
Medical tubing	Foster-Miller	Nylon nanocomposite tubing with improved modulus for medical application
Chemical, mechanical planarization	DuPont/Air Products/Others	Nanosilica, nanocerium oxide, smooth chips
Nano-wax particle	BASF	Self-cleaning surfaces
Polymer-clay masterbatch	Nanocor	Molded parts
Nanoclay additives	Sud Chemie	Flame-retardant additives
Polycarbonate/ABS	Bayer	Flame-retardant polymer blends
Optical coating	Nanofilm	Antistreaking windshield coating
Polycarbonate panels	GE/Cabot	Nanogel translucent aerogel
Rocket fuel	Mach I	Nanoferric oxide additive to rocket fuel

Recently, selective applications of nanoclays and CNTs are also reported by Lucintel (Fig. 12.2) [3]. Polymer-clay nanocomposites (PNCs) are used in (a) bottling of O_2 and CO_2 sensitive products, such as beer and soft drink bottles; (b) food and packaging industries; (c) motor compartment of vehicle for casting and connectors; and (d) wire and cable applications. Polymer-CNT nanocomposites are used in (a) timing belt covers in automotive, (b) scratch-resistant coatings in automotive and

TABLE 12.4 An Overview of Potential Commercial Nanotechnology Products

Product	Developer	Benefit/Function
Tires	InMat	Keep tires from losing air pressure, up to 2× as current seals
Power cells	Ballard	Fullerenes for H cells, notebook, auto computer
Drug delivery	Microchip/ChipRx	Medical devices, biosensors for drug delivery
Robots	NYU/USC	Tiny machine to clean ocean, attack gems, space research

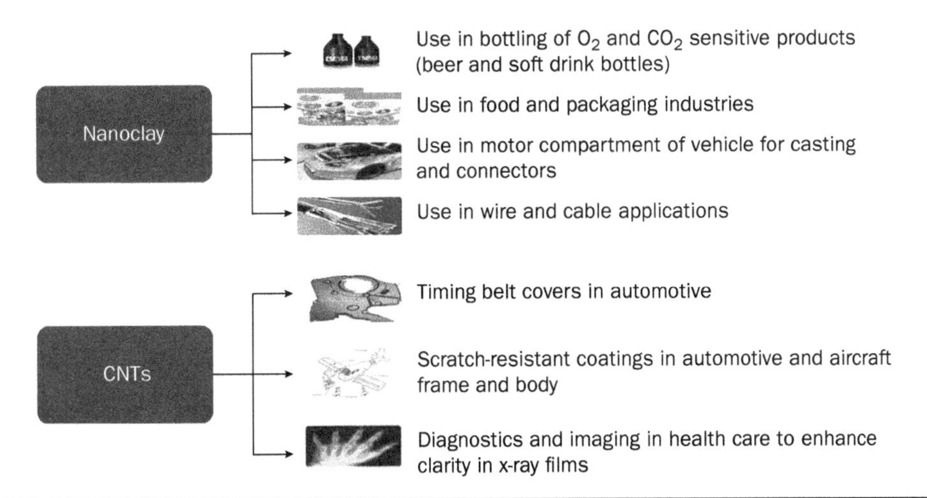

FIGURE 12.2 Selective applications for polymer-nanoclays and polymer-CNT nanocomposites [3]. *(Courtesy of Lucintel.)*

aircraft frame and body components, (c) EMI/RFI shielding and coatings in electronic devices to control radiation, (d) CMP slurries in electronic chips and wafers to enhance electrical conductivity, and (e) diagnostics and imaging in health care to enhance clarity in x-ray films.

Several current and potential applications shown in Fig. 12.3 can be subdivided into market segments of (a) transportation, such as engine and fuel systems, scratch resistant exterior parts and coatings, car interior, aircraft structure and framing, and wear-resistant paints and coatings for defense vehicles; (b) construction, such as conductive flooring, pipes, insulating material for roofs and thatches, house and building siding, and self-cleaning windows; (c) packaging, such as meat and food packaging, computers and electronics, medicines and pharmaceuticals, and soft drink and beer bottles; (d) consumer goods, such as home appliances, sporting goods and toys, furniture, and others; (e) energy, such as battery electrodes, fuel cell membranes, and super capacitors; (f) electrical and electronics, such as sensors, semiconductors, and hard-disk storage in computers; (g) health care, such as body implants, medical devices, and dental filling materials; and (h) others, such as anti-foul coatings for marine ships, industrial equipment to increase strength, and fire-resistant clothes [3].

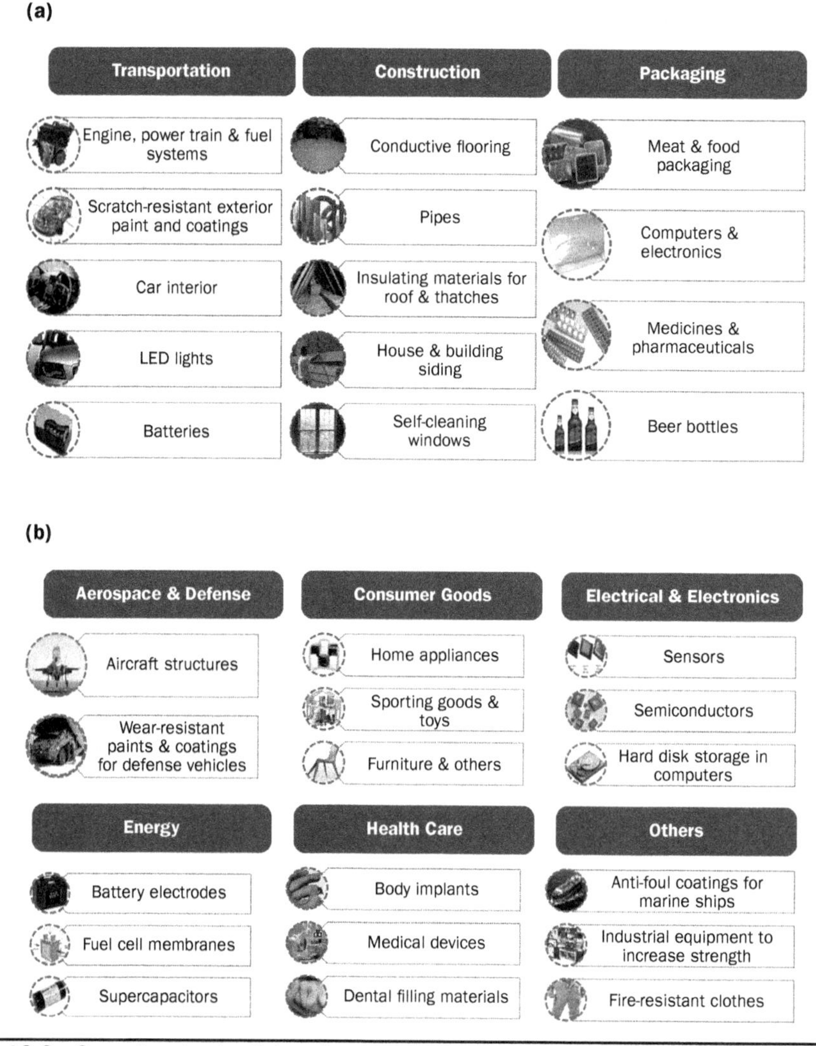

Figure 12.3 Current and potential applications for polymer nanocomposites in different market segments, such as transportation, construction, packaging, consumer goods, energy, electrical and electronics, health care, and others [3]. (*Courtesy of Lucintel.*)

Polymer–carbon nanofiber (CNF) nanocomposite applications in terms of technology readiness level (TRL) are shown in Fig. 12.4 [4]. At TRLs 9 and 8, applications, such as aerospace structural components, sporting goods, anti-corrosion/erosion coatings, audio equipment, conductive inks, thermal conductive grease, and electrostatic dissipation (ESD) components are demonstrated. At TRLs 7 and 6, applications, such as ESD carpets/fabrics, electronic device foam, fire-retardant foam, and automotive panels are used. At TRL 5, multifunctional/structural composites are used. At TRL 4, biosensors/smart composites are considered. At TRL 3,

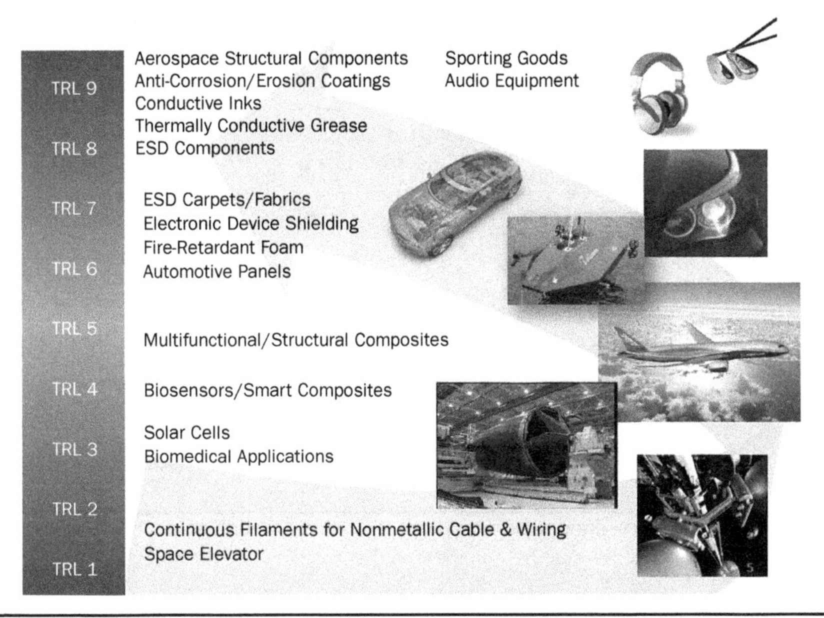

Figure 12.4 Polymer–carbon nanofiber nanocomposites from TRL 9 to 1 [4]. (*Courtesy of Applied Sciences, Inc.*)

solar cells and biomedical applications are investigated. At the lowest TRLs 1 and 2 continuous filaments for nonmetallic cable and wiring, and space elevator are considered.

Hybrid Material for Sporting Goods

Due to its strength and light weight, several polymer nanocomposites are currently used in several sporting goods products (Fig. 12.5) [3]. Silicon dioxide (SiO_2) is used in badminton racket. CNTs are used in tennis racquet, baseball bat, bicycle frame, and hockey sticks. Titanium dioxide (TiO_2) is used in fishing rods.

Hybrid Compounds for Cosmetics

New hybrid compounds are targeted for cosmetics and controlled release of *active ingredients* for applications concerning skin care and protection. They are based on the encapsulation of active organic components within porous silica microcapsules. Undesirable skin alternations and melanomas created by sun-induced premature skin aging require the use of efficient but biocompatible sunscreens. Conventional commercial sunscreens are directly applied to the skin and usually contain an extremely high amount of active ingredients. These can be detrimental to health especially when they are absorbed by skin and are not photostable, thus generating free radicals that my cause damage to the DNA. Sol-Gel Technologies Ltd. based in Bet Shemesh, Israel (website: www.sol-gel.com) [9] provides a safe solution by encapsulating the active organic ultraviolet (UV) filters in silica microcapsules. As a result, contact of these potent chemicals with skin is reduced and damage from free radicals within the porous host is prevented. These *UV-pearls* are incorporated into a suitable cosmetic vehicle to achieve high sun

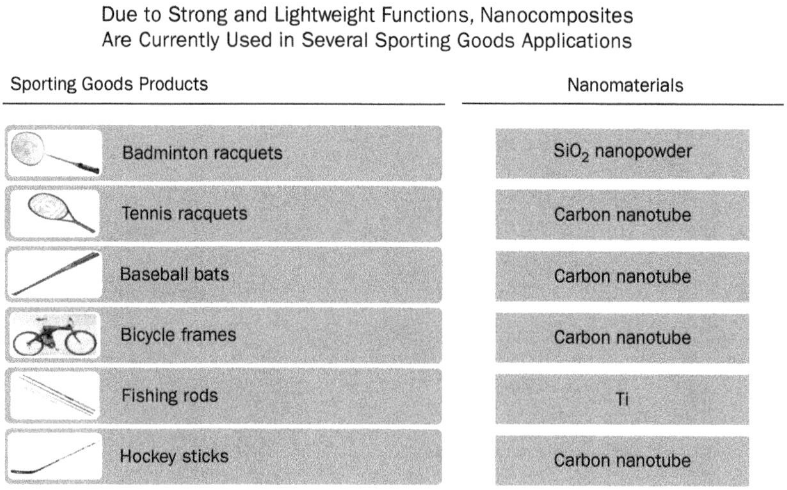

Due to Strong and Lightweight Functions, Nanocomposites Are Currently Used in Several Sporting Goods Applications

Sporting Goods Products	Nanomaterials
Badminton racquets	SiO_2 nanopowder
Tennis racquets	Carbon nanotube
Baseball bats	Carbon nanotube
Bicycle frames	Carbon nanotube
Fishing rods	Ti
Hockey sticks	Carbon nanotube

Figure 12.5 Several nanomaterials currently used in sporting goods products [3]. (*Courtesy of Lucintel.*)

protection factors (SPF) while providing an improved safety profile as the penetration of the absorbed UV is conveniently reduced. These newly developed products have already been marketed by companies for sunscreens and daily-wear cosmetics. Figure 12.6a shows the UV pearls and Fig. 12.6b shows the core-shell capsule for carrying functional contents [9].

Sol-Gel Technologies Ltd. also developed silica rainbow pearls containing organic dyes for cosmetics applications. These silica active–pearls contain an effective acne medication, such as benzoyl peroxide (BP), which is as effective as antibiotics but does not cause bacterial resistance or stomach upset. BP in direct contact with skin causes skin irritation, dryness, and hyperpigmentation in many patients. Encapsulating the BP-active ingredient in a silica shell prevents it from contacting the epidermis while gradually delivering the BP into the follicular region where acne bacteria are found. Sol-gel active-pearls are presently a successful commercial development.

Hybrid Materials for Protective and Decorative Coatings

Many interesting and highly productive applications of inorganic-organic hybrid materials have occurred in automotive coatings. Hybrid coatings provide coloration, scratch resistance, and protection from UV and chemical attacks for the substrate. DuPont with Generation 4® has developed a complex mixture of two hybrid polymers cross-linked simultaneously during curing to form a polymer network. This unique network is partially grafted and partially interpenetrated as a decorative coating for the automotive industry [10]. This hybrid material consists of a high-density acrylate polymer core with organically modified alkoxysilane groups and residual unsaturations. It confers a high modulus and scratch-resistant function, dispersed in a low cross-linked density polymer, which provides film-forming properties. The superior scratch and environmental etch resistance of these hybrid coatings led to their acceptance as topcoats for

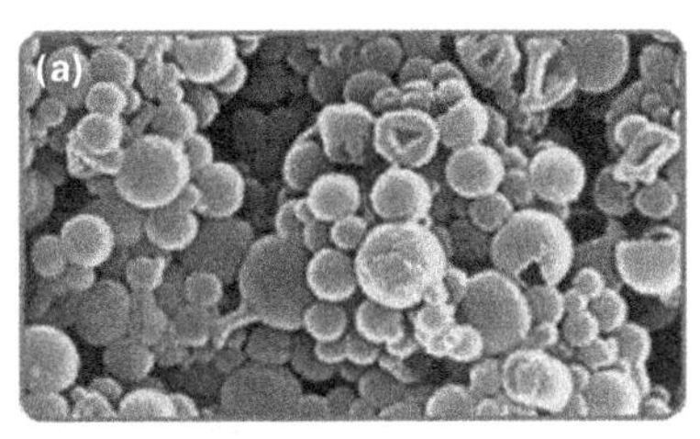

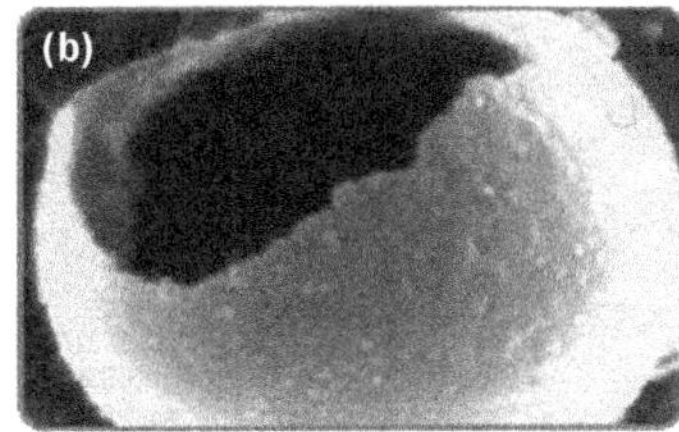

FIGURE 12.6 (a) Microcapsules of UV pearls and (b) micrograph of a core-shell capsule for carrying functional contents [9].

eight of the ten top-selling automobiles for 1997, including the Ford Taurus, Toyota Camry, and Honda Civic/Del Sol.

Decorative hybrid coatings also encompass an industrial area of interest, such as packaging glass, glass sheets for furniture and sanitary appliances, and glass in architecture and the building industry. For example, a dishwasher-safe ORMOCER® color coating is used on glass (DEKOSIL®). The main advantages in terms of processing these coatings are the easy reproducibility of the colors and the finishing by conventional wet painting procedures (low curing temperatures less than 200°C). From an aesthetic point of view, some requirements, such as unlimited color palette and additional decorative effects, such as color gradients and partial coloration are also considered as key features. Another example is the titania siloxane–based hybrid organic-inorganic materials developed by several Japanese and European glass packaging industries [11]. These coatings provide a variety of colors, enhance the consumer appeal, and improve the mechanical properties of these glass bottles. These dyed colored bottles are recyclable as uncolored glass, because in contrast to conventional glass bottles these glass bottles are color coated. The coloration does not originate from transition metals that are normally dissolved in glass and difficult to remove upon remelting glass for recycling. Conventional colored glass remains colored glass on remelting.

Hybrid Materials for Dental Applications

Inorganic-organic hybrid materials can be used as filling composites in dental applications [12–15]. As shown in Fig. 12.7a, these composites feature tooth-like properties (appropriate hardness, elasticity, and thermal expansion behavior) and are easy to use by dentists as these composites can easily penetrate into the cavity and harden quickly under the effect of blue light [12]. These materials feature minimum shrinkage, are nontoxic, and sufficiently nontransparent to x rays.

Traditional plastic filling composites had long-term adhesion problem and a high degree of polymerization shrinkage resulting in marginal fissures. The dual character of the ORMOCER®s as inorganic-organic copolymers is the key for improving the properties of dental filling composites. The organic, reactive monomers are bounded in the sol-gel process by the formation of an inorganic network. Thus, in the subsequent curing process, polymerization takes place with less shrinkage. Abrasion resistance is significantly enhanced by the existing inorganic Si–O–Si network.

Commercial filling composites based on dental ORMOCER®s from Fraunhofer ISC, such as Definite® and Admira®, are as shown in Fig. 12.7b. In the case of the Admira product, a specifically designed dentine-enamel bonding, an adhesive ORMOCER® developed in cooperation with VOCO GmbH, is used to make this product especially attractive and

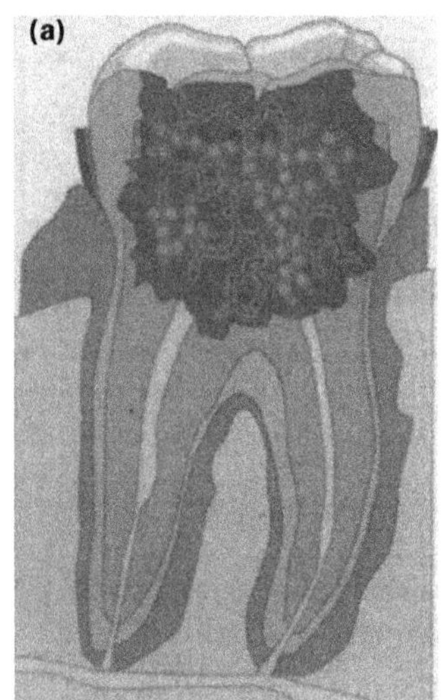

Requirements:
-Easy to handle/model
-Biocompatibility
-Reduced polymerization shrinkage
-Perfect marginal adaptation
-Strong adhesion to dentine and enamel
-Tooth-like aesthetics
-Tight wear resistance
-Elasticity
-High x-ray absorption

Applications:
-Matrix materials
-Composites
-Glass ionomer cements
-Bondings
-Filler/particle (monodisperse)

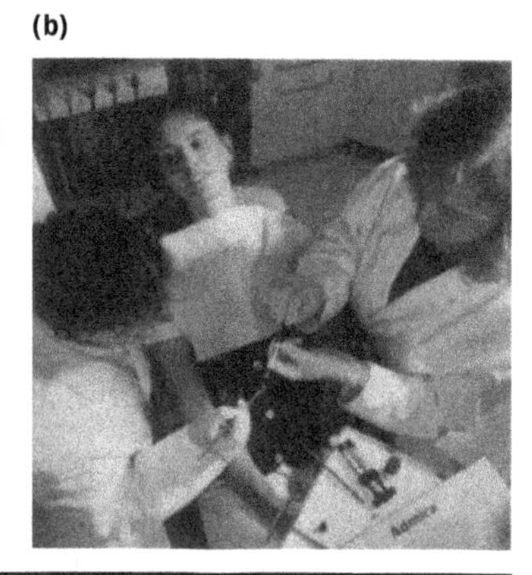

Figure 12.7 (a) Requirements and possibilities of dental applications of ORMOCER®s and (b) application of an ORMOCER as dental filling material [12].

Figure 12.8 Partners in development: (a) GM/Basell/Southern Clay Products/Blackhawk for the first commercial nano TPO–clay nanocomposite step-assist on Astro and Safari vans in the automotive industry and (b) GM R&D manager stepping into a van with a step-assist [21]. (*Courtesy of SCP.*)

brilliant in appearance. In the glass ionomer cement–based dental composites, blue light–polymerized carboxyl-functionalized ORMOCER®s have been developed.

Hybrid Materials (Polymer-Clay Nanocomposites) with Structural Properties

The first successful development of polymer-clay nanocomposites (PCNs) was pioneered by Toyota's researchers [16–20] for high-performance reinforced plastic applications in automobiles. The earliest successes in the automotive industry are credited to the sonication process developed by Ford Corporation to better disperse the clay within the polymer. Related to these early disclosures, General Motors Corporation (GM) developed a step-assist component [21] (Fig. 12.8) for 2002 GM Safari and Chevrolet Astro vans, which are made of thermoplastic polyolefin–clay nanocomposites. This material is lighter and stiffer, more robust at low temperatures and easily recyclable.

GM also used these PCNs in the lateral protection wire of the 2004 Chevrolet Impala since they are 7% lighter in weight and present a better surface appearance.

New applications of hybrid PCNs are envisaged by the automotive industry for commercialization: PP-clay for bodywork with anti-scratch properties (Dow Plastics/Magma); acetal-clay for ceiling lights (Showa Denko); PP-clay for panes of doors, consoles, and interior decoration (Ford, Volvo) due to aesthetics, recyclability, and weight saving properties; nylon-clay for bumpers with enhanced mechanical and weight saving properties (Toyota); and nylon-clay for fuel reservoir with airtight properties (Ube America).

Besides the above properties, the weight advantage provides a significant impact on the environment. It has been reported [22] that widespread use of PCNs by U.S. automotive industry could save 1.5 billion liters of gasoline in 1 year of automotive production and reduced related carbon dioxide emissions by more than 10 billion pounds.

For environmental applications, nanoclays are used as effective reinforcing agents in *green nanocomposites* [23]. These renewable-resource–based biodegradable nanocomposites consist of cellulose plastics (plastics made from wood); polylactic acid (corn-derived plastic), or polyhydroxyalkanate (bacterial polyester). They are attractive substitute materials for petroleum feedstock in manufacturing the biodegradable plastic for the commercial market [24].

Hybrid Materials with Gas Barrier Properties

Through the incorporation of nanoclays into polymer matrices, it is possible to create a labyrinth within the structure, which physically delays the passage of molecules of gas. The excellent barrier properties of nanoclays against gas and vapor transmission have resulted in their applications in food and beverage packaging, and also as barrier liners in storage tanks and fuel lines for cryogenic fuels in aerospace systems [25]. Researchers at Bayer Polymers have developed a hybrid material combining silicate nanoparticles with polyamide 6 (PA6) and ethylene vinyl alcohol (EVOH), which is significantly better than neat PA6 [26].

Nanocor and Mitsubishi Gas Chemical Company developed the MXD6 high-barrier semi-aromatic nylon nanocomposites [26, 27]. Mitsubishi marketed this PCN under the trade name Imperm®. Imperm features oxygen and carbon dioxide barrier properties that were previously unachievable. Imperm's barrier properties are superior to EVOH and its processing characteristics are ideal for multilayer films, bottles, and thermoformed containers (Fig. 12.9) [26, 27].

Honeywell has developed a polyamide-clay nanocomposite with specific active and passive barrier properties against oxygen [26]. The passive barriers are the nanoclays, which render it more difficult for oxygen to transmit inside the composite, but also conduct the oxygen molecules to specific oxygen-captors. The clay incorporation limits the oxygen transmission up to 15–20% of the value for the pure polymer.

The beverage industry is not the only one that can benefit from implementing the aforementioned technologies. The proper storage capacity of food and beverage can also be enhanced using the same improved barrier properties. These materials are always in high demand as they reduce the amount of food spoilage. Triton Systems and the U.S. Army collaborated using PCNs as nonrefrigerated package materials, which can retain food freshness for 3 years [27, 28]. Currently, the use of polymer nanocomposites to enhance the preservation of military meal-ready-to-eat (MRE) package is being considered, as till now metal interfaces are used to preserve the food items (Fig. 12.10). If the right filler and polymer combination would be used, the cost of the MRE packages could decrease by as much as 30% [29].

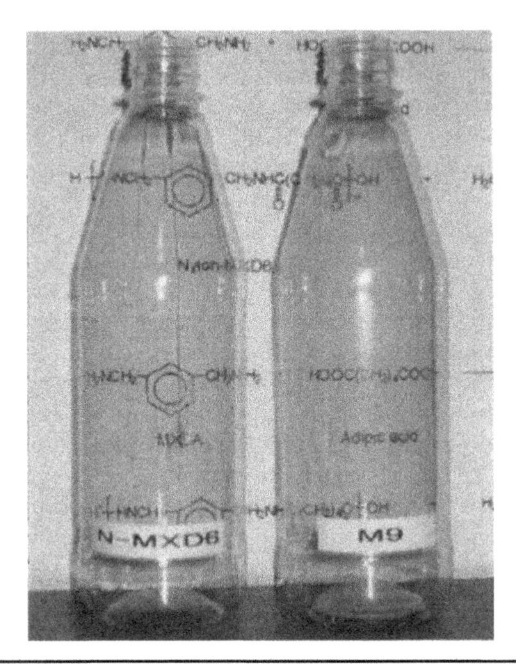

Figures 12.9 Nano N-MXD6 and N-MXD6 M9 are high gas barrier resins commercially available from Mitsubishi Gas Chemical Company [26].

Figure 12.10 An MRE food pouch sealed in a polymer bag [29].

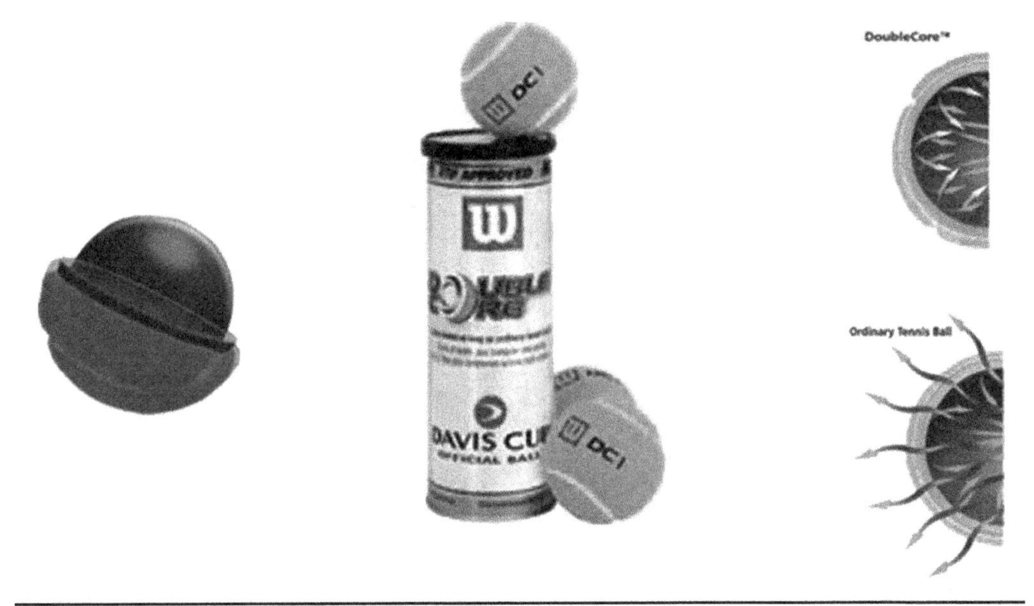

Figure 12.11 The core of this Wilson tennis ball is covered by a polymer-clay nanocomposite coating that acts as gas barrier, doubling its shelf-life [30]. (*Courtesy of SCP.*)

InMat Inc. has developed several applications for their nanocomposite coatings [30]. The nanocomposite coating of the Wilson tennis balls maintains the internal pressure for an extended period of time. Figure 12.11 shows the Wilson tennis balls containing a double core. This core is coated by a polymer-clay (vermiculite) nanocomposite coating that acts as a gas barrier, doubling its shelf life [30]. It is anticipated that this technology could be extended to the rubber industry and it could be incorporated into soccer balls or in automobile or bicycle tires.

The barrier properties of PCNs have also been extended to liquids (solvents) or molecules, such as water, since these liquids are up to some extent responsible for polymer deterioration. De Bievre and Nakamura of Ube Industries reveal significant reduction in fuel transmission through polyamide 6/66 polymers by incorporation of nanoclay, resulting in reduced fuel emission for fuel tank and fuel line components [26, 31]. InMat Inc. collaborates with the U.S. Army with regard to the concept of liquid–hydrocarbon barrier [30]. Some envisaged applications are as protective gloves and as coatings for combustible and fuel tanks. Triton System Inc. and Converse created a PCN pouch–filled helium insert that fits into the baseball shoe.

Hybrid Materials with Flame-Retardant Properties

Gilman et al. have reported that thermal stability and flame retardancy properties of nanoclays have motivated researchers to use them as flame-retardant additives for commodity polymers [26, 32]. Gitto (polymer manufacturer) and Nanocor (nanoclay manufacturer) are developing flame-retardant polyolefin-clay nanocomposites, since this polymer has a wide range of industrial applications [26, 33]. The heat distortion temperature (HDT) of PCNs can be increased up to 100°C, extending their usage to higher temperature applications, such as automotive under-the-hood parts. Figure 12.12 shows a nanocomposite power cable manufactured by Kabelwerk

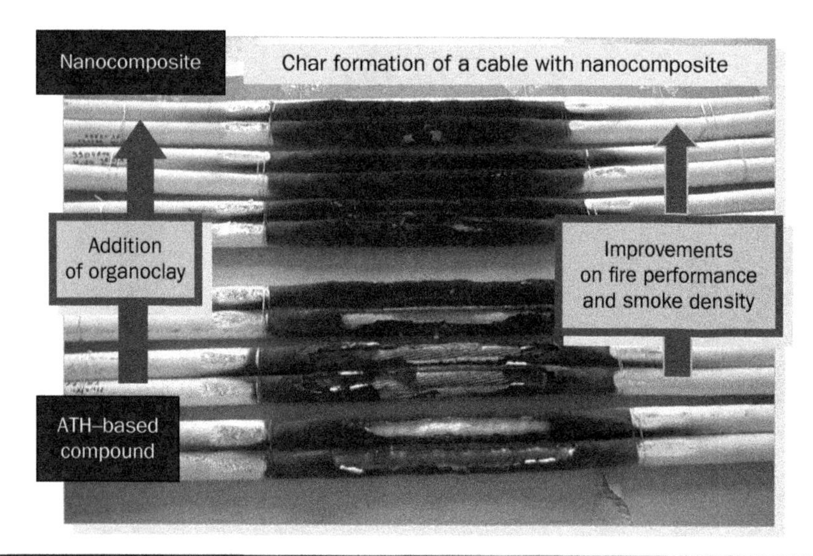

FIGURE 12.12 Nanocomposite power cable by Kabelwerk Eupen showing char forming is more efficient, with improvement in fire performance and smoke density. (*Courtesy of SCP.*)

Eupen with enhanced fire performance and smoke density. For additional flammability studies of polymer nanocomposites refer to Chap. 8.

Hybrid Materials with Other Properties

The significance of PCN technology relates to its value-added performance in terms of its mechanical, barrier, or flame-retardant properties which are not present in the neat polymer; these properties occur when the organically modified layered silicate (OLS) nanomaterials are introduced in low volume. For example, in contrast to conventional filled polymers that are opaque and exhibit compromised mechanical properties, the introduction of a low-volume fraction of OLS that is of nanoscopic dimension and uniformly dispersed within the polymer results in PCNs with optimum transparency and clarity similar to the neat polymer. Further, these PCNs exhibit multifunctionality behavior that is unavailable in the neat polymer. Nylon 12 nanocomposite for medical applications with improved modulus has been demonstrated.

In the area of fuel cells, polymer nanocomposites have been proposed as a viable material for producing electrical energy and reducing the solvent transmission to the fuel. This is an important area for any engine operating on fossil fuel. The implementation of polymer nanocomposites into the fuel lines will greatly lower the material costs. It may, in some cases, increase the fuel efficiency of the engine. This is because the polymer nanocomposite will act against the fuel mixing with the solvent in the walls of the tube. Also, similar nanomaterials have been researched to inhibit the ability of certain polymers to absorb and transmit water. The same idea applies as in the case with the fuel cells. The nanofillers will act as a block and inhibit the water particles from permeating the matrix of the polymer. This technology is ideal in the area of environmental protection [34].

Protective coatings are a huge area of opportunity for polymer nanocomposites. First, UV light protection has recently become an important application. In unreinforced polymers, the high-energy UV beams can and will break the weak bonds holding monomers in a chain, resulting in the rapid degradation of polymers exposed to sunlight (i.e., photo-degradation). This results in loss of mechanical properties. Nanomaterials solve this problem by functioning as internal absorbers of UV light, preventing the damage that would normally take place [35, 36].

The use of nano-reinforcements in radar-absorbing materials (RAMs) has become another, new application. In practice, polymer nanocomposites can be tuned to absorb electromagnetic signals of specific frequency bands. As such, they can be used for shielding of sensitive electronic equipment or in stealth technology. RAMs were used in the development of the USAF F-117 and B-2 to reduce the radar signature of the aircraft. Other uses include antennas, car radios, cellular telephones, and other communications technologies [37].

Research into *hard yet tough* coatings is being driven by industry and will produce nanomaterial coatings that have a resistance to plastic deformation without being vulnerable to cracking. Focus for this application is on a dual-phase coating: a nanocrystalline phase along with a CNT-embedded matrix phase. The first provides the hardness and the second provides the toughness. A myriad of uses for this type of coating exist, including everything from automotive engine components to tools for machining [38].

Additionally, nanomaterials have been used for some time in the reinforcement of tires. The majority of the carbon black produced is used in tires. The nanomaterials enhance the mechanical properties and also provide for heat dissipation. Likewise, carbon black is used in other rubber products, such as belts, hoses, and pads. Although the industry has expanded into other applications, the focus for carbon black production will likely remain on tires [36, 39].

Because of their enhanced mechanical properties, polymer nanocomposites have also been researched for use in commercial vehicles. There are many areas in an automobile where these materials can be used [35, 40].

The use of carbon black in tire technology has been discussed earlier, but the automotive industry also is constantly searching for ways to make the frame lighter without sacrificing durability and safety. This research rose partially from the need to increase the gas mileage of cars. Most car manufacturers decrease the weight of their cars by adding aluminum pieces to replace heavier steel pieces. However, this can also be accomplished using polymer nanocomposites with similar mechanical strength but significantly lower weight. Polymer nanocomposites were introduced into the car industry when the Toyota Research Group decided to use nylon 6–clay nanocomposites in their engine belt covers. Since then, GM has made a one-piece, compression-molded rear floor assembly. Similar advancements can be made to systems such as the suspension, using polymer nanocomposites, to stabilize the frame. CNTs are also starting to make their way into the market due to their ability to enhance mechanical strength in polymer matrices. The next step in the incorporation of nanocomposites into the automotive industry is to create biodegradable materials. The development of such materials could replace petroleum-based materials currently used in the assembly and therefore reduce the overall carbon footprint of the manufacturer [40].

Corrosion plays a large part in the design of many products which deal with memory storage. For example, hard drives used to store mass amounts of data in a computer contain metal disks that are magnetically encoded with the data. The purity

Materials	Strength	Elasticity	Price	Weight	Surface area	Wear resistance	Barrier property	Electrical conductivity
Nanomaterials	●	●	○	○	○	●	●	●
Ceramic	◔	◕	●	◕	◕	◑	◑	◔
Kevlar (Aramid)	◕	◑	◕	◔	◔	◕	◑	◔
Glass Fibers	◑	◑	●	◑	◔	◕	◕	◔
Carbon Fibers	◕	◕	◑	◔	◔	◕	◕	●

● *High* ◕ *Medium* ◑ *Low-Medium* ◔ *Low* ○ *Least*

Nanomaterials have highest strength, modulus, wear resistance, barrier properties, and electrical conductivity relative to other materials

FIGURE 12.13 Cost and property analysis of nanomaterials with respect to other conventional materials [3]. (*Courtesy of Lucintel*)

of these data and the ability to keep them depend highly on the condition of the disks over time. Since the disks are currently made of metals, they are prone to corrode, which affect the ability to encode the disks. Currently, there is research being conducted in order to find chemically stable polymer nanocomposites that may replace the metal disks in the hard drives. The same chemical instability occurs in many sensor equipment, such as magnetic field sensors and explosive chemical sensors. A lightweight and highly electrical conductive polymer nanocomposite is needed to reduce the amount of oxidation that the metal sensors experience. Furthermore, similar magnetic and electrically conductive properties can be used to make electrochromism-based smart windows [41].

12.2.3 Cost and Property and Geographical Breakdown Analyses

The cost and property analysis of nanomaterials with respect to the other conventional materials, such as ceramic, Kevlar®, glass fibers, and carbon fibers are shown in Fig. 12.13 [3]. As one can see, the key features of nanomaterials are highest strength, modulus, wear resistance, barrier properties, and electrical conductivity relative to other materials. In the areas of price competitiveness, weight, and surface area, nanomaterials score very low when compared with other conventional materials.

The geographical breakdown of the market of nanomaterials in 2010 is shown in Fig. 12.14 [3]; North America shared 38% of the market, Europe 37%, and Asian and other regions 25%. Some key insights are as follows: (a) North America and Europe were the two largest markets for nanomaterials in 2010; (b) Europe has significant usage of nanomaterials in pharmaceuticals, since many pharmaceutical companies are based there, and (c) good growth of nanomaterials was witnessed in Asian region during the last 5 years mainly due to (i) government support, (ii) continued environmental consciousness, and (iii) expected increase in demand for specialty materials.

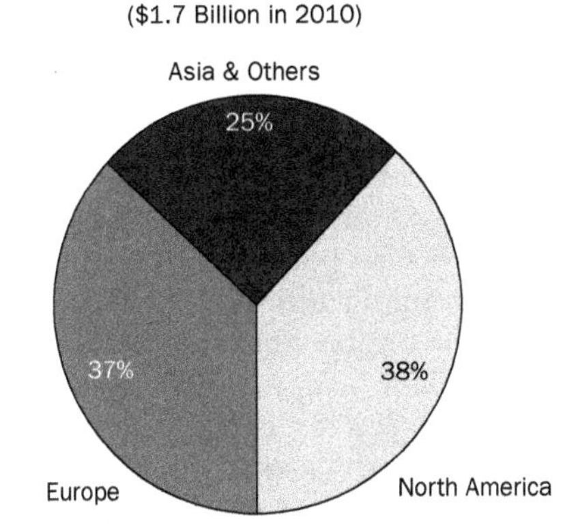

Key Insights

- North America and Europe were the two largest markets for nanomaterials in 2010.
- European region has significant use of nanomaterials in pharmaceuticals, as many pharmaceutical companies are based in Europe.
- Strong growth in nanomaterials was witnessed in Asia region during last 5 years due to:
  - Government support
  - Expected increase in demand for specialty materials
  - Presence of huge E&E industries for which the nanomaterials are best suited

FIGURE 12.14 Geographical breakdown of nanomaterial market [3]. (*Courtesy of Lucintel.*)

12.2.4 Technical and Funding Developments

The technical developments in CNT promise further capacity expansion and price reduction [3]. The price of multiwalled carbon nanotubes (MWNTs) has fallen substantially from more than $150 per gram in 2000 to under $50 per gram in 2010, and has been about $2–25 per gram in 2019 (depending on the MWNT diameter and its quality). Improved manufacturing and large-scale production by CVD process enabled the drop in price. Current technical developments in CNT fabrication may encourage further increase in production capacity and further reduction in prices.

The growing nanotechnology market will boost the growth of polymer nanocomposites. In the past 5 years a huge amount has been invested in a nanotechnology R&D by U.S. and European countries. Japan has shown strong interest in nanotechnology development, spending about $1 billion in nanotechnology R&D in 2010. Global expenditures for nanotechnology R&D expected to grow at about 23% until 2016 [3].

12.3 Nanotechnology Research Output

Typically, the research and production of polymer nanocomposites is focused on the enhanced barrier properties that they can provide. The properties that polymer nanocomposites have are significant for the food packaging industry and other air seal technologies. The mechanical properties of the resulting composite are examined closely but are often not the main goal. Although addition of nanofillers makes the polymer matrix stronger, it often can degrade the elongation property of the polymer, thus more R&D must be carried out to find ways to rectify this issue. As more research is done in the area of enhancing barrier properties, other ideas are generated from similar concepts. Recently, there has been interest in finding ways to use a sealing polymer composite in tissue bonding for materials, such as skin grafts. Since skin acts as a protective layer for your body, the same barrier properties that help keep food fresh can help to bond ends of tissue together with less risk of separation [39].

There are many physical constants in the incorporation of nanofillers into polymer matrices. These phenomena occur in almost every combination of matrix and filler used. First, the density of the polymer always increases with the incorporation of nanoparticles. This is an intuitive result considering you are packing stronger materials into a weaker matrix. Second, the volume of polymer nanocomposites under mechanical strain increases faster than the neat polymer. As the particles used in the matrix get smaller the volume increases under the same loading conditions. Also, as nanofillers are added, the polymer tends to become stiffer, and the elongation properties degrade [42]. This may be a problem, as more brittle materials are seldom useful. Other mechanical properties, such as tensile strength and modulus, tend to increase as well in most polymer matrices. A study also found that in crystalline materials the T_g actually decreased while it increased in most amorphous structures. Finally, a property that is detrimental to the processing stage is the viscosity which tends to increase as you add more nanomaterials. The increase of polymer viscosity could cause the batch to mix improperly, or in some cases the mixer may not be powerful enough to overcome this significant increase of polymer viscosity during manufacturing [42].

12.4 Trend and Forecast

The trend and forecast of nanomaterials market are described in Fig. 12.15 [3]. From 2005 to 2010, nanomaterial shipment grew from $1 billion to $1.7 billion, a healthy 11% CAGR growth. The forecast of nanomaterial shipment from 2010 to 2016 is between

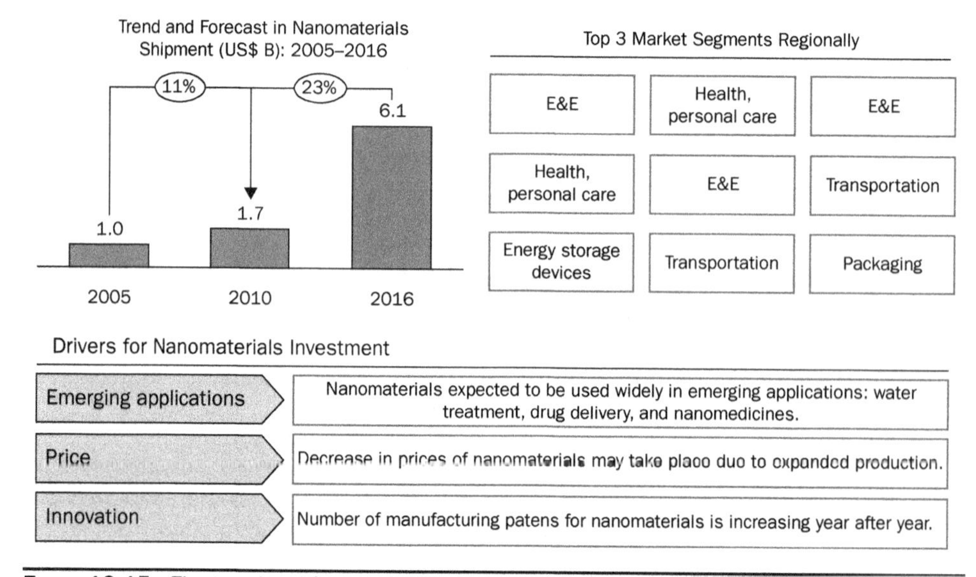

FIGURE 12.15 The trend and forecast of nanomaterial market [3]. (*Courtesy of Lucintel.*)

$1.7 billion and $6.1 billion, an extraordinary 23% CAGR growth! According to Allied Market Research's latest forecast report [6], polymer nanocomposites market was valued at $5.276 billion in 2015, and is expected to reach $11.549 billion by 2022, supported by a CAGR of 10.9%, during the forecast period 2016–2022. The electrical and electronics (E&E) market witnessed a robust growth rate due to large price decrease of CNTs and increase in mass production of nanomaterials. The drivers for nanomaterials investment can be summarized by three factors: (a) emerging application—expect nanomaterials to be extensively used in emerging applications, such as water treatment, drug delivery, and nanomedicines; (b) price—decrease in price of nanomaterials may take place due to expanded production capacity; (c) innovation—the number of patents regarding manufacturing of nanomaterials is increasing extraordinarily up to 2010 [43]. The top three market segments can be subdivided regionally and are quite different: (a) North America—electrical and electronics, health and personal care, and energy storage devices; (b) Europe—health and personal care, electrical and electronics, and transportation; and (c) rest of the world (RoW)—electrical and electronics, transportation, and packaging.

The growth opportunities for nanomaterials in various applications are shown in Fig. 12.16 [3]. It is forecasted that the health-care industry would see the most growth, followed by energy storage, electrical and electronics, transportation, and construction. Health-care industry is expected to gain market share and overtake electrical and electronics segment in size over the next 5 years. Several key forecasts are as follows: (a) nanomaterials have great potential in electrical and electronics applications because of their extraordinary electrical conductivity; (b) global recession induced cost/price

Health-care Segment Is Expected to Gain Market Share, Overtake E&E in Size within Next 5 Years

Growth Opportunities for Nanomaterials in Various Applications

Health care
Energy storage
E&E
Transportation
Construction
CAGR (2011–2016)
Market Size (US $/M)

Key Insights

- Nanomaterials have great potential in E&E applications because of their extraordinary electrical conductivity.
- Global recession induced cost/price sensitivity in E&E segment, given its potential to reduce manufacturing costs and increase product competitiveness.
- Packaging is another important segment, mainly flourishing in North America and Western Europe. Mainly used for packaging of
  - Carbonated soft drinks
  - Beer bottles
  - Meat
- Energy segment is expected to steadily grow at a double-digit growth rate.

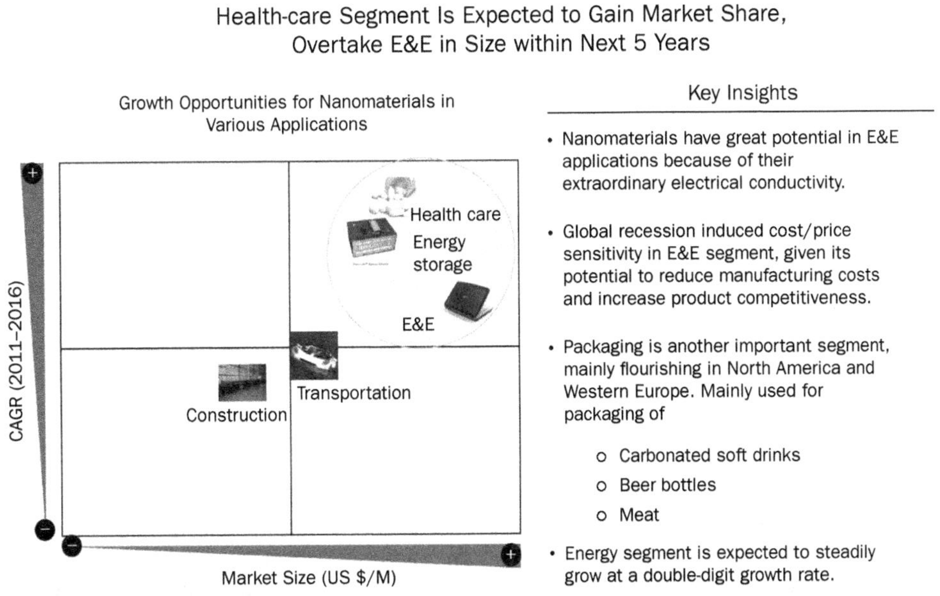

FIGURE 12.16 Growth opportunities for nanomaterials in various applications [3]. *(Courtesy of Lucintel.)*

Markets	Nano-materials	Applications	Benefits
 Aerospace & defense	Carbon nanotubes	Fuselage and wings	• High strength-to-weight ratio • High impact resistance • Radiation protection
 Wind energy	Carbon nanotubes	Wind turbine blades	• Increase mechanical efficiency • Lightning protection
 Automotive	Carbon nanotubes	External body	• Light weight and conductivity
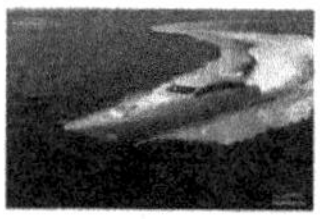 Marine	Carbon nanotubes	Boat hull	• Light weight and increasing payload

FIGURE 12.17 Several potential applications of nanomaterials in composites for different market segments [3]. (*Courtesy of Lucintel.*)

sensitivity in electrical and electronics segment, given its potential to reduce manufacturing costs and increase product competitiveness; (c) packaging is another important market segment, mainly flourishing in North American and Western Europe; and (d) energy market segment is expected to pick up a nice pace in coming year at double-digit growth rate.

Figure 12.17 shows the various potential applications for nanomaterials in composites [3]. The different market segments using CNTs include (a) aerospace and defense, for fuselage and wings, (b) wind energy, for wing blades, (c) automotive, for external body, and (d) marine, for boat hulls. Wind energy, automotive, and marine industries use CNTs for various applications, mostly due to the light weight and improved mechanical, thermal, and electrical properties they provide to various composites.

In modern aircrafts, nanomaterials are expected to show healthy growth in the next 10 years due to high strength, lightweight, and electrical conductivity properties, as shown in Fig. 12.18 [3]. From 1990 to 2010, aircraft industry was shifting from aluminum to composites. Both composites and polymer nanocomposites are currently used in Boeing B787 and Airbus A380. It is expected that from 2020 to 2030, the aircraft industry would use nanomaterials instead of titanium and increase the use of CNTs in prepreg to fabricate composites.

Figure 12.19 shows the different emerging trends in global nanomaterial and nanocomposite industry [3]. The trends are as follows: (a) growing demand for high-strength, durable structural materials; (b) new material development and product design; (c) new and emerging market applications; (d) falling prices of nanocomposites; (e) huge expansion by existing suppliers and new entrants; and (f) government support and R&D funding.

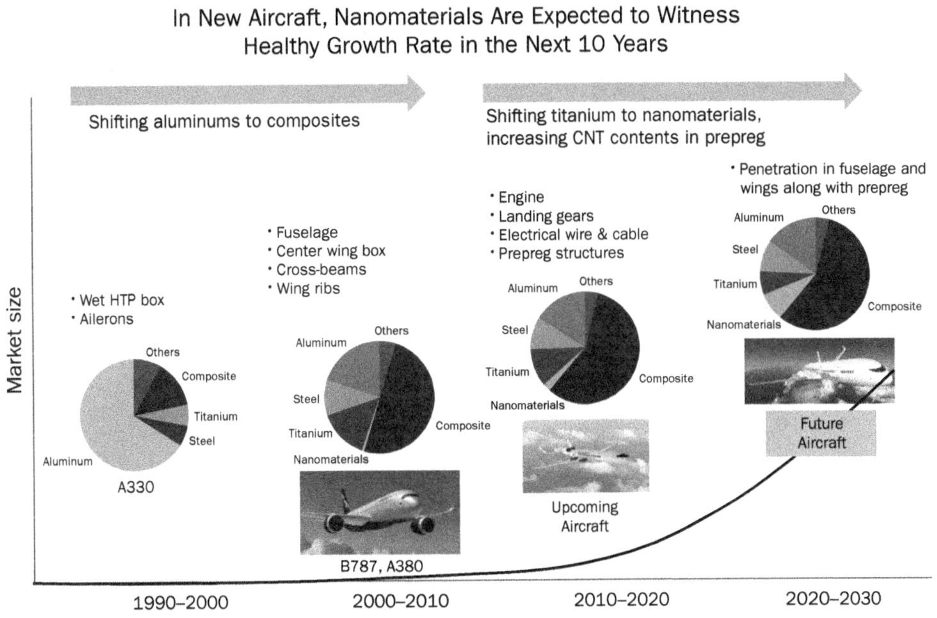

In New Aircraft, Nanomaterials Are Expected to Witness Healthy Growth Rate in the Next 10 Years

Shifting aluminums to composites

Shifting titanium to nanomaterials, increasing CNT contents in prepreg

Market size

- Wet HTP box
- Ailerons

Others
Composite
Titanium
Steel
Aluminum
A330

- Fuselage
- Center wing box
- Cross-beams
- Wing ribs

Aluminum
Others
Steel
Titanium
Nanomaterials
Composite
B787, A380

- Engine
- Landing gears
- Electrical wire & cable
- Prepreg structures

Aluminum
Others
Steel
Titanium
Nanomaterials
Composite
Upcoming Aircraft

- Penetration in fuselage and wings along with prepreg

Aluminum
Others
Steel
Titanium
Nanomaterials
Composite
Future Aircraft

1990–2000
2000–2010
2010–2020
2020–2030

Figure 12.18 Usages of materials in the aircraft industry from 1990 to 2030 [3]. (*Courtesy of Lucintel.*)

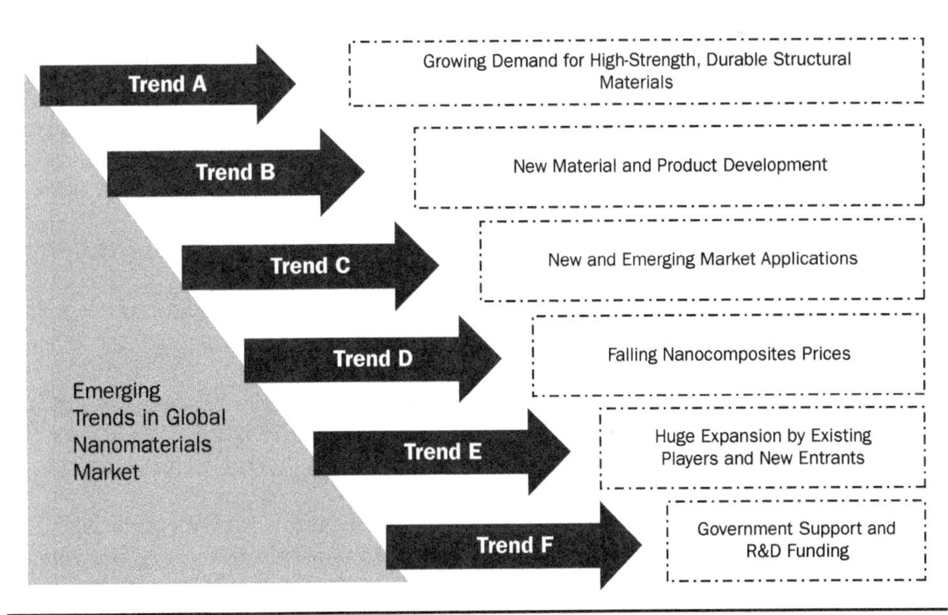

Figure 12.19 Emerging trends in global nanomaterial and nanocomposite industry [3]. (*Courtesy of Lucintel.*)

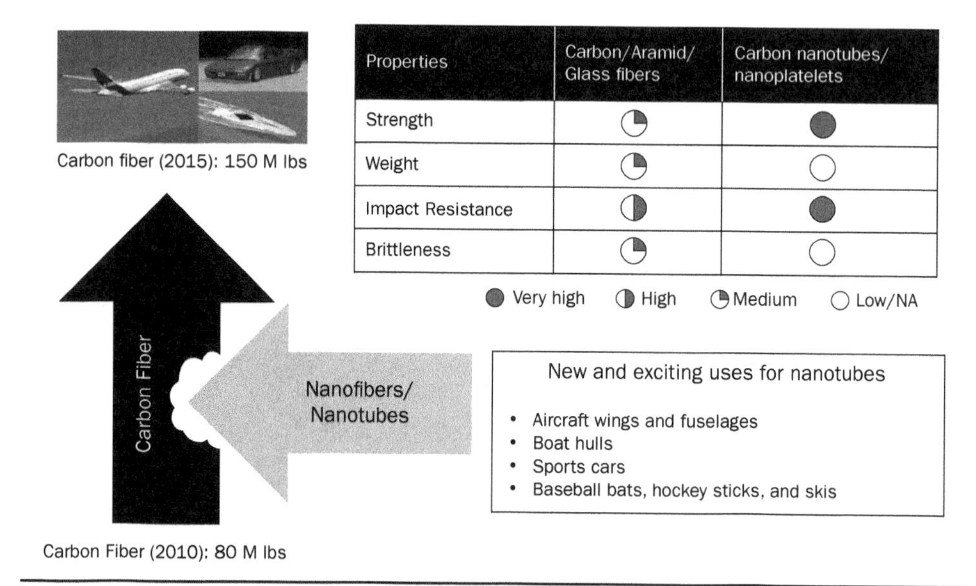

FIGURE 12.20 Growing demand for high-strength, durable materials will create good future opportunities for nanocomposites [3]. (*Courtesy of Lucintel.*)

Figure 12.20 illustrates the growing demand for high-strength and durable materials that will create good future opportunities for polymer nanocomposites [3]. In 2010, demand for carbon fiber reached 80 million pounds, with a small amount replaced by CNFs and CNTs; and the forecast for carbon fibers in 2015 was 150 million pounds. The strength and impact properties of CNTs/graphene nanoplatelets are superior for carbon/aramid/glass fibers. Other new and exciting uses of CNTs are in aircraft wings and fuselages, boat hulls, sports cars, baseball bats, hockey sticks, and skis.

Figure 12.21 shows new material development and product design [3]. These new development include (a) Nanoledge Inc.'s prepreg containing CNTs; (b) New Piranha using Zyvex nanomaterials for marine application; (c) Harbor Composites designing nanotube components for aircraft; and (d) Velozzi's Supercar using nanotube composites for structural parts.

Figure 12.22 shows growing uses of nanocomposites in various areas in transportation industry will create huge demand in future [3]. Figure 12.23 shows potential applications of nanocomposites in aircraft, though market remains nascent due to certification and overall slow adoption [3].

Figure 12.24 shows additional emerging applications for nanomaterials, including field emission display (FED), body armor, water purification, drug delivery, biofuel cell, and wind turbine blades [3]. The size of opportunities and the key drivers of these applications are described in Fig. 12.24 [3].

In addition, several polymer-CNF applications are provided by Applied Sciences [4]. ESD/wear improves pulleys of polymer-CNF nanocomposites and the benefits of using CNFs for this application [4]. Electrostatic painting using CNFs and its benefits are

Nanoledge Inc. developed prepreg containing nanotubes
- Nanotubes used to boost mechanical properties
- Developed for experimental use in boat hull

New Piranha USV using Zyvex nanomaterials
- Reduces weight while greatly increasing payload and cruising range of unmanned surface vehicle

Harbor composites–designed nanotube aircraft
- Impart additional strength and durability to ordinary composites and make them even more appealing

Velozzi's Supercar will use nanotube composites–structural parts
- Objective to improve properties and reduce weight

FIGURE 12.21 New material development and product design [3]. (*Courtesy of Lucintel.*)

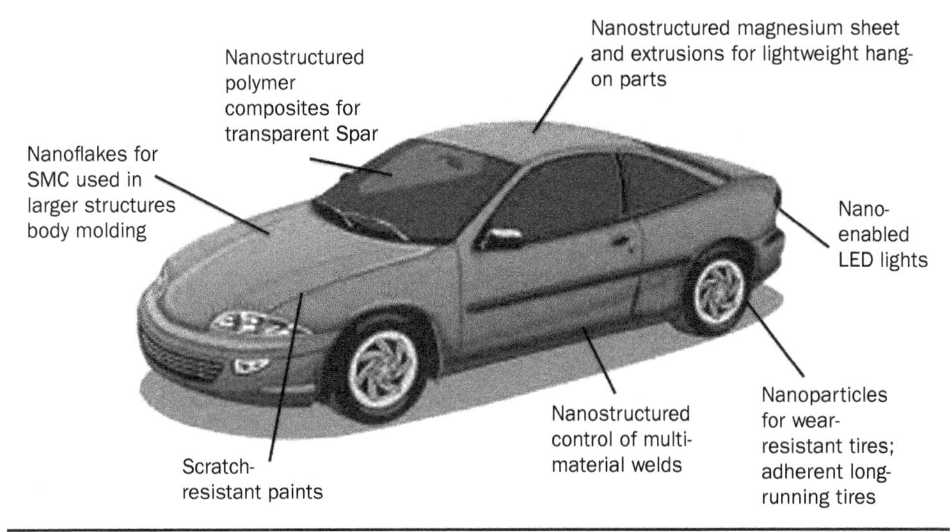

FIGURE 12.22 Growing uses of nanocomposites in transportation industry will create huge demand in future [3]. (*Courtesy of Lucintel.*)

demonstrated in [4]. CNFs are used in nanofluid applications to provide unique transport properties, such as not settling under gravity, no clogging, and large surface area. Figure 12.25 shows nanotechnology can be used for top technical thermal management challenges facing transportation, defense, microelectronics, and energy industries [4].

Another emerging field of nanotechnology is *human health and medicine*. BASF Chemical Company is researching various ways that nanotechnology can be used to

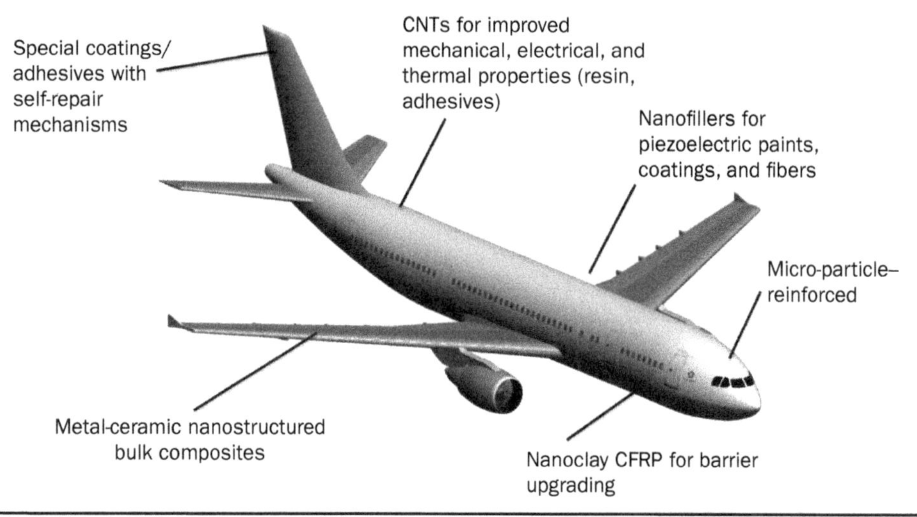

FIGURE 12.23 Potential applications of nanocomposites in aircraft, though market remains nascent due to certification and overall slow adoption [3]. CFRP, carbon fiber–reinforced plastics. (*Courtesy of Lucintel.*)

improve peoples' standard of living [44]. BASF is working on developing nanomembranes to filter bacteria, germs, and viruses from water. They have also been developing vitamins with nanomaterials that allow for better adsorption into the body. Using zinc oxide nanoparticles, BASF has been a global leader in UV absorbing products as well.

NASA has been developing a shape-shifting aircraft that changes shape according to different maneuvers and pressure variations along the wings [45]. The technology uses pressure sensors along the length of the wing and actuators to change shape, similar to bird flight. The actuators are said to use CNT-polymer composites and electrostriction to change shape by varying the stresses in the material from longitudinal to bending.

In summary, several market trend and forecast conclusions can be deduced from the market research reports of Lucintel [3] and Allied Market Research [6]:

- Nanomaterials/polymer nanocomposites market is expected to grow at about CAGR of 10.9% during 2016–2022.
- Top six players have secured two-thirds of the global nanomaterials market shipments; there are greater opportunities for new players to enter this growth market.
- Decline in demand in 2009 for nano-enabled products in the automotive industry, such as automotive lubricants, catalytic converters, sensors and filters, among others, drove the heavy downturn in market opportunities for nanomaterials, such as MWNTs and ceramic nanoparticles.
- Health-care and energy markets were the two main application areas which helped global nanomaterial industry recover from downturn experience in 2009.

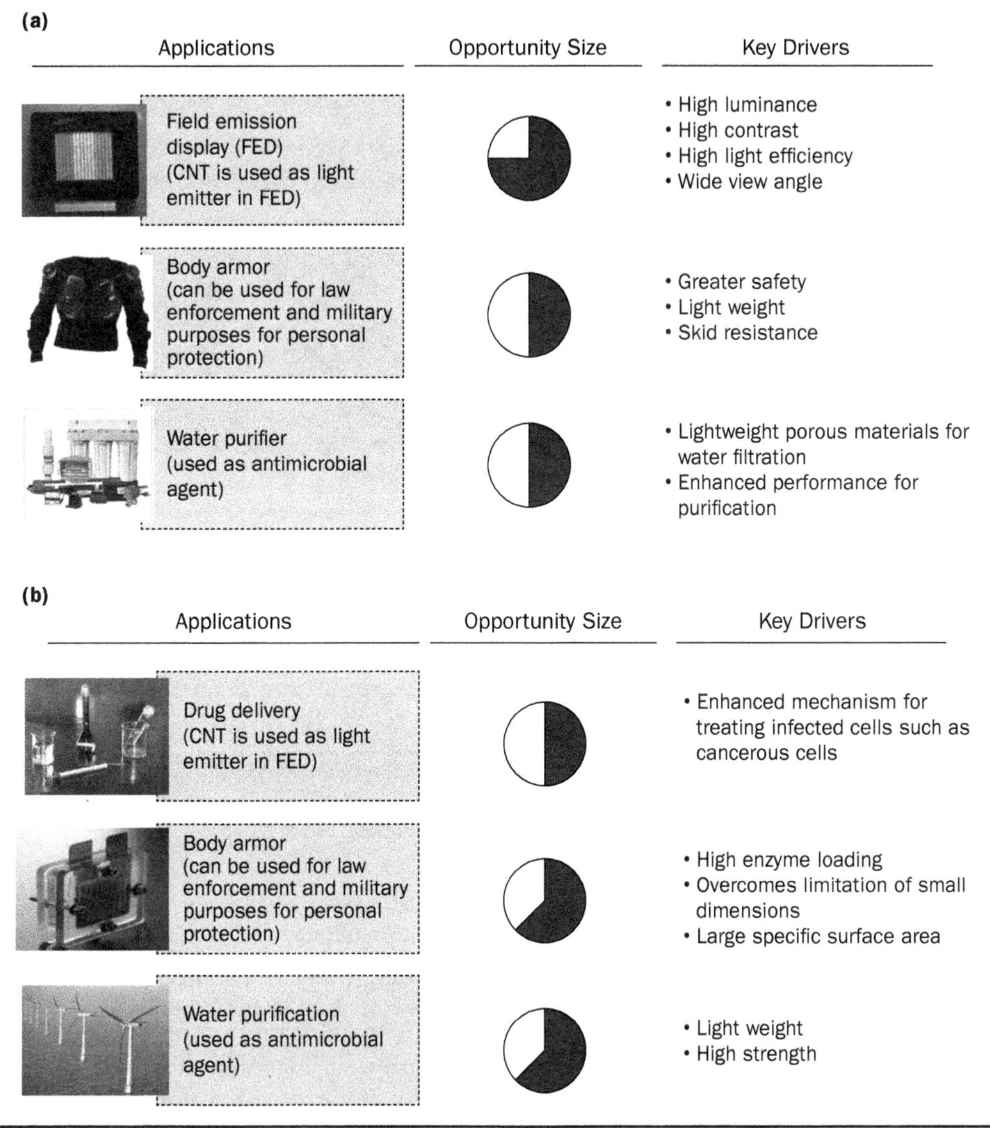

Figure 12.24 Emerging applications of nanomaterials [3]. (*Courtesy of Lucintel.*)

- Health-care industry is expected to surpass the electrical and electronic industry in size over the next 2–3 years, driven by nanotechnology advancements in the biomedical field.
- The growing automotive and packaging industries, superior mechanical and physical properties, and infrastructural growth in emerging economies are the key factors driving the growth of polymer nanocomposites market.
- Elastomeric nanocomposites are gaining momentum in the automotive industry due to their applications in the production of rolling resistance tires, which are lightweight and save fuel.

Figure 12.25 Nanotechnology can be used for thermal management of different high-tech industries [4]. (*Courtesy of Applied Sciences, Inc.*)

- North America will remain the market-leading region for several years, as it has significant ongoing R&D activities in nanomaterials.
- South Korea and Japan are the two leading polymer nanocomposite manufacturing countries in Asia-Pacific. Japan accounted for around 50% of the polymer nanocomposites alone in the Asia-Pacific in 2010. Furthermore, China and Singapore are the two fastest growing manufacturers of polymer nanocomposites owing to their investment in the respective domestic markets and the support from the national governments.
- The leading companies in this market have been proactively working toward their expansion and the launch of new polymer nanocomposite products. The major companies include Nanophase Technologies Corporation (the United States), The Arkema Group (France), DuPont (the United States), RTP Company (the United States), Showa Denko Carbon, Inc. (the United States), Inframat Corporation (the United States), Powdermet, Inc. (the United States), Nanocor, Inc. (the United States), Nanocyl S.A. (Belgium), and Evonik Industries AG (Germany).
- It is expected that nanomaterials/polymer nanocomposites will broaden their markets as they enter into more applications, such as drug delivery, armor, defense equipment, weapons, nanomedicines, and others.

12.5 Challenges

12.5.1 Manufacturability of Nanoparticles

The starting point of any polymer nanocomposite is choosing the appropriate nanomaterial. Many times, the cost of certain nanomaterial exceeds the gain from the property improvement, causing many combinations and weight percentages to be out of reach for many researchers. Finding inexpensive ways to mass-produce useful nanomaterials is an area that is highly important for the further development of all polymer nanocomposites. Current methods are successful and provide the industry with a steady supply of various nanomaterials, but other methods have potential to decrease cost and increase volume of nanomaterial production.

Currently, there are three commonly used ways to create carbon-based nanomaterials: arc discharge, laser ablation, and chemical vapor deposition. U.S. patent number 20130089738 discusses a way to create activated nanocarbon from food waste using a new method that has the potential to reduce cost and reduce food wastes [46]. This method requires crushing food waste and sieving it to create a powder. Then, the powder is impregnated with an agent and activated using an activation gas at high temperatures, between 500°C and 800°C. Finally, the activated food waste is cryo-grinded to produce nanocarbon ranging from 10 to 100 nm.

For making other nanoparticles, grinding methods are sometimes used. According to Kim and Park, grinding methods still need refinement to obtain particles smaller than the micro-scale when using more traditional dry grinding methods [47]. Typically, wet grinding processes present problems, such as water growth and size up during water removal. To overcome problems associated with traditional grinding methods, the patent describes a method that first involves dissolving sucrose in water over several hours of stirring. Then, a fatty acid and surfactant are added and stirred at slightly elevated temperatures (about 60°C). Finally, an active ingredient is added followed by drying, grinding many times, and the resulting powder stirred. Using this technique, mean particle sizes are expected to be smaller and more consistent than with traditional methods.

There is a wide variety of CNTs and CNFs provided by different manufacturers. The purity and consistency of these products are a great concern for the end-users. These parameters will affect the final properties of the manufactured polymer nanocomposites.

12.5.2 Manufacturability of Polymer Nanocomposites

The main challenges in implementing nanocomposites lie in the processing stage. Polymer nanocomposites require very precise measurements when it comes to the size of the filler materials. If the particles are too large, they may not disperse well in the matrix. Also, the orientation of the particles must be parallel to the surface of the object being constructed. It is very difficult to control the direction of the particles during the mixing process, but ignoring the orientation could have an adverse effect on the properties of the polymer nanocomposite material. It is also important to have nanofiller compatible with the polymer matrix. If they are not compatible, the nanofiller may not mix evenly within the matrix and the synthesis of a nanocomposite will have failed [48]. The interphase properties between the nanomaterials and the host polymer have two aspects: weak interfacial interaction favors network formation and strong interfacial affinity favors mechanical reinforcement.

The amount of capital required to be invested to distribute the amount of nanocomposite that is needed to even make a difference, as well as investing in new equipment and technologies, had been one of the main reasons why this technology has not already boomed. The process of making nanomaterials is expensive, which consequently makes the resulting nanocomposite expensive, and in the long run much more expensive than the standard polymer. If a successful market can be established, someone with capital needs to take a risk on the polymer nanocomposite market. If a successful market can be established, then the prices to produce these products will decrease and the market will have a much more viable alternative financially than the one that exists now [49].

Along with individual product challenges, using polymer nanocomposites on a large scale requires more knowledge in the field than currently available. Research in this field is growing, but the awareness of the technology among the public is low. Because of the lack of interest in the market currently, companies have been reluctant to provide more funding, fearing that it may take more time to produce results than they feel comfortable to invest in [38].

It is established that these materials are either already being used or could be used in the making of automobiles. There are certain challenges, however, that prevent these materials from becoming commonplace within the industry and setting new standards. First of all, with the number of automobiles produced by larger companies, there is no guarantee that the conditions of the composites will stay consistent in mass production [50]. Also, the use of these new materials would require a large amount of government oversight and permission before the company may proceed with their implementation. The start date for these projects could be pushed back if there are delays in the application process. While nanoclays provide some great additive properties, they are also thermally unstable and can degrade at temperatures as low as 170°C. This is undesirable considering that most parts are fabricated using some sort of melt-processing technique. Figure 12.26 is useful in showing how these nanostructured materials will affect the industry over time and relate them to market interests [51].

Another major obstacle to the continued use and growth of nanomaterials is safety. No all-encompassing safety regulations or procedures have been established yet (either from industry, government, or academia). Further complicating this is the plethora of competing safety strategies emerging from a variety of sources [52].

The effects of exposure to these materials are serious, and exposure can occur through many different means, such as inhalation, contact with the skin, and ingestion. Animal studies have shown that smaller particles are more toxic than larger ones because they can penetrate farther into the lungs and may be able to move to other organs through barriers considered impermeable to larger particles [53].

In 2009, the EPA awarded $4 million in grants to study the risks associated with the proliferation of nanomaterials—an amount many considered insufficient. Funded studies included absorption and toxicity of nanoparticles on skin, nanoparticles, and drinking water, effect of nanoparticles on lung tissue and airways, and effect of nanoparticles on marine life. To date, however, little effort has gone into establishing manufacturing and handling codes [54].

This lack of guidance has led to the emergence of many different suggestions, and navigating through them and choosing an appropriate procedure for one's application can be difficult. Also, some companies may hesitate to invest in technologies that do not

	Short-term impact (2011–2013)	Medium-term impact (2013–2016)	Long-term impact (2016–2021)
Growth Drivers Growth in high-performance and super-size vehicles	Demand for high-performance nanocomposites will follow the growth trends of high-performance vehicles	As market for high-end vehicles mature, demand for nanocomposites will stabilize	Innovations in new high-performance nanocomposites with low cost will drive faster growth
Challenges Automotive markets facing global contraction	Global recessionary trends 2009–2012: delay in demand for new vehicles manufacturing	Slow growth, as vehicles manufacturing picks pace	Market for nanocomposites in automotive applications will mature, decline in market share

Figure 12.26 Incorporating polymer nanocomposites into automobiles [51].

have guidelines in place out of concern over how the government will eventually react. In other words, fears that the regulations eventually created will result in unfavorable market conditions may stifle current and future investment in nanomaterials research, development, production, and implementation. Strong leadership in this area is necessary if the projected growth in the industry is to be maintained.

These challenges can be solved by committing further research into this field to develop better processing techniques. When the proper methods of fabrication are determined, what type of safety precautions can be made to overcome the concerns? Research is growing in this area, but the demand and awareness in polymer nanocomposites are still low. Educating the public and promoting awareness toward this technology are key to overcoming the challenges discussed above.

12.6 Concluding Remarks

Nanocomposites continue to play an ever-increasing role in improving products in terms of lightweight, strong, flame-retardant, conductive, and many other physical and chemical properties. This chapter has discussed just a few of the ways nanomaterials are already changing the world. In recent years, research has been increasing heavily both in the United States and the rest of the world. As technologies develop and costs continue to decrease, the world will see more areas of applications of nanomaterials.

Polymer nanocomposites are highly promising for use in a vast number of applications because of their light weight and high strength. Other emerging properties, such as electrical and thermal conductivity, and increased thermal stability, make them highly sought after in extreme environments where currently only metals and ceramics are used.

Another view is that the nanomaterial industry is at a crossroad. With opportunities and applications abound, it could become a large and profitable sector with healthy R&D. However, safety, lack of guidance, and underfunded/underpursued R&D could cause the industry to flounder for many years. It has already been discussed that the nanomaterial industry is capable of providing incredible growth. Hence, it will fall to everyone involved to help direct the fledgling sector—those in academia, government, and industry.

12.7 References

1. U.S. National Nanotechnology Initiative, website: http://www.nano.gov. Accessed November 3, 2018.
2. National Science and Technology Council (NSTC). (2018, August) *The National Nanotechnology Initiative Supplement to the President's 2019 Budget*. Washington, DC: National Science and Technology Council (NSTC), Committee on Technology (COT), Subcommittee on Nanoscale Science, Engineering, and Technology (NSET).
3. Lucintel, Irving, TX, website: www.lucintel.com.
4. Applied Sciences, Inc., Cedarville, OH, website: www.apsci.com.
5. Vance, M. E., Kuiken, T., Vejerano, E. P., McGinnis, S. P., Hochella, Jr., M. F., Rejeski, R., and Hull, M. S. (2015) Nanotechnology in the real world: redeveloping the nanomaterial consumer products inventory. *Beilstein Journal of Nanotechnology*, 6:1769–1780. doi: 10.3762/bjnano.6.181.
6. Allied Market Research (2014) Global polymer nanocomposites market-opportunities and forecasts, 2014–2022. Allied Market Research, Pune, India, website: www.alliedmarketresearch.com.
7. Koo, J. H. (2006) *Polymer Nanocomposites—Characterization, Properties, and Applications*, New York, NY: McGraw-Hill , pp. 243–253.
8. Sanchez, C., Julian, B., Belleville, P., and Popall, M. (2005) Applications of hybrid organic-inorganic nanocomposites. *J. Mater. Chem.* 15:3559–3592.
9. Sol-Gel Technologies Ltd., Bet Shemesh, Israel. website: www.sol-gel.com.
10. Hazan, I., and Rummel, M. (1992) U.S. Patent No. 5,162, 426.
11. Minami, T. (2000) *J. Sol-Gel Sci. Technol.*, 18:290.
12. Wolter, H., and Storch, W. (1994) *J. Sol-Gel Sci. Technol.*, 2:93.
13. Wolter, H., Storch, W., and Ott, H. (1994) *Mater. Res. Soc. Symp. Proc.*, 346:143.
14. Wolter, H., Glaubitt, W., and Rose, K. (1992) *Mater. Res. Soc. Symp. Proc.*, 271:719.
15. Wolter, H., Storch, W., and Gellerman, C. (1996) *Mater. Res. Soc. Symp. Proc.*, 435:67.
16. Fujiwara, S., and Sakamoto, T. (1976) Japanese Kokai Patent Application No. 109998 (assigned to Unichika, K. K., Japan).
17. Fukushima, Y., and Inagaki, S. (1987) Synthesis of an intercalated compound of montmorillonite and 6-polyamide. *J. Inclusion Phenomena*, 5:473–482.
18. Okada, A., Fukushima, Y., Kawasumi, M., Inagaki, S., Usuki, A., Sugiyama, S., Kuraunch, T., Kamigaito, O. (1988) United States Patent No. 4739007 (assigned to Toyota Motor Co., Japan).
19. Kawasumi, M., Kohzaki, M., Kojima, Y., Okada, A., Kamigaito, O. (1989) United States Patent No. 4810734 (assigned to Toyota Motor Co., Japan).
20. Usuki, A., Kojima, Y., Kawasumi, M., Okada, A., Fukusima, Y., Kurauch, T., Kamigaito, O. (1993) Synthesis of nylon 6-clay hybrid. *J. Mater. Res.*, 8:1179–1184.
21. Southern Clay Products, Gonzales, TX. website: http://www.scprod.com/gm.html.
22. National Institute of Standards and Technology. (1997) *Nanocomposites New Low-Cost, High-Strength Materials for Automotive Parts*, ATP Project 97-02-0047. Gaithersburg, MD: National Institute of Standards and Technology.
23. Mohanty, A. K., Drzal, L. T., and Misra, M. (2003) *Polym. Mater. Sci. Eng.*, 88:60.
24. Wilkinson, S. L. (2001) *Chem. Eng. News*, January 22, p. 61.
25. Beall, G. W. (2001) New conceptual model for interpreting nanocomposite behavior. In *Polymer-Clay Nanocomposites*, Pinnavaia, T. J., and Beall, G. W., eds. New York, NY: Wiley, pp.267–279.

26. Hay, J. N., and Shaw, S. J. A review of nanocomposites 2000. Abstracted version from http://www.azom.com/details.asp?ArticleID -921.
27. Multilayer containers featuring nano-nylon MXD6 barrier layers with superior performance and clarity, Mitsubishi Gas Co., Inc. and Nanocor, presented at Nava-Pac 2003. Website: http://www.nanocor.com/techpapers.asp.
28. Triton Systems, Inc., Chelmsford, MA. Website: http://www.tritonsys.com.
29. Downing-Perrault, A. *Polymer Nanocomposites Are the Future.* University of Wisconsin-Stout. Retrieved from http://www.iopp.org/files/public/DowningPerraultAlyssaUWStroutNanoStructures.pdf. Accessed December 6, 2013.
30. InMat Inc., Hillsborough, NJ. Website: http://www.InMat.com.
31. De Bievre and Nakamura, K. (2004) Laminate structure excelling in fuel permeation preventing performance. *PCT Int. Appl.*, 26. Patent No. WO20004054802.
32. Gilman, J. W. (1999) *Appl. Clay Sci.*, 15:31.
33. Nanocor, Inc., Arlington Heights, IL. Website: http://www.nanocor.com.
34. Baksi, S., Basak, P. R., and Biswas, S. Nanocomposites—Technology Trends & Application Potential. Technology Information, Forecasting and Assessment Council. Retrieved from http://www.tifac.org.in/index.php?option=com_content&view=article&id=523:nanocomposites--technology-trends-a-application-potential&catid=85:publications&Itemid=952. Accessed December 6, 2013.
35. Carbon black uses. International Carbon Black Association. Retrieved from http://www.carbon-black.org/uses.html. Accessed December 6, 2013.
36. Team Innovation. (2007) Why some rotomolded tanks are black in color—carbon black and plastics. PolyProcessing Company. Retrieved from http://www.polyprocessing.com/pdf/technical/CarbonBlackandPlastics.pdf. Accessed December 6, 2013.
37. Radar absorbing materials. (2012) *GlobalSpec.* Retrieved from http://beta.globalspec.com/learnmore/materials_chemicals_adhesives/electrical_optical_specialty_materials/radar_absorbing_materials_structures_ram_ras. Accessed December 6, 2013.
38. Simonsen, J. (2009) *Bio-based Nanocomposites: Challenges and Opportunities.* Corvallis, OR: Department of Wood Science & Engineering, Oregon State University. Retrieved from http://www.cof.orst.edu/cof/wse/faculty/simonsen/Nanocomposites.pdf. Accessed December 6, 2013.
39. Zhang, S., Wang, H. L., Ong, S. E., Sun, D., and Bui, X. L. (2007) Hard yet tough nanocomposite coatings—present status and future trends. *Plasma Processes and Polymers*, 4:219–228. Retrieved from http://onlinelibrary.wiley.com/doi/10.1002/ppap.200600179/full. Accessed December 6, 2013.
40. Polymer nanocomposites drive opportunities in the automotive sector. (2012) *nanowerk.* Retrieved from http://www.nanowerk.com/spotlight/spotid=23934.php. Accessed December 6, 2013.
41. Bonsor, K. (2012) How smart windows work. *howstuffworks.* Retrieved from http://home.howstuffworks.com/home-improvement/construction/green/smart-window4.htm. Accessed December 6, 2013.
42. Jordan, J., Jacob, K. I., Tannenbaum, R., Sharaf, M. A., and Jasiuk, I. (2005) Experimental trends in polymer nanocomposites—a review. *Materials Science and Engineering: A,* 393:1-11. Retrieved from http://www.sciencedirect.com/science/article/pii/S0921509304012328. Accessed December 6, 2013.
43. Karn, B. (2012) *NNI Research in NanoEHS.* NSF Grantees Meeting Presentation, December 4.

44. BASF Chemical Company. Website: http://www.basf.com. Accessed December 6, 2013.
45. Berger, M. (2008) NASA nanotechnology research into shape-shifting airplanes. June 16. Retrieved from http://www.nanowerk.com/spotlight/spotid=6067.php.
46. Al-Zahrani, S., Rahman, A., Ali, I., and Elliethy, R. U.S. Patent No. 2013 0089738.
47. Kim, K., and Park, J., U.S. Patent No. 2013 005643.
48. Downing-Perrault, A. *Polymer Nanocomposites Are the Future.* University of Wisconsin-Stout. Retrieved from http://www.iopp.org/files/public/DowningPerraultAlyssaUWStroutNanoStructures.pdf. Accessed December 6, 2013.
49. Baksi, S., Basak, P. R., and Biswas, S. Nanocomposites—technology trends & application potential. Technology Information, Forecasting and Assessment Council. Retrieved from http://www.tifac.org.in/index.php?option=com_content&view=article&id=523:nanocomposites--technology-trends-a-application-potential&catid=85:publications&Itemid=952. Accessed December 6, 2013.
50. Presting, H., and König, U. (2003) Future nanotechnology developments for automotive applications. *Materials Science and Engineering: C*, 23:737–741. Retrieved from http://www.sciencedirect.com/science/article/pii/S0928493103002169. Accessed December 6, 2013.
51. Polymer nanocomposites drive opportunities in the automotive sector. (2012) *nanowerk*. Retrieved from http://www.nanowerk.com/spotlight/spotid=23934.php. Accessed December 6, 2013.
52. Nanomaterial Safety Plan. (2010) SLAC National Accelerator Laboratory, Stanford University. Retrieved from http://www-group.slac.stanford.edu/esh/eshmanual/references/hazmatplannano.pdf. Accessed December 6, 2013.
53. Weiss, R. (2004) EPA back nanomaterial safety research. *The Washington Post*. Retrieved from http://www.washingtonpost.com/wp-dyn/articles/A43763-2004Nov11.html. Accessed December 6, 2013.
54. Conductive polymers and electronics: Tackling corrosion with nanotechnology. (2011) *Enginuity 2011*. Lamar University. Retrieved from http://www.lamar.edu/research/research-magazines/enginuity-2011/tackling-corrosion.html. Accessed December 6, 2013.

Index

O

X

Y